An Introduction to Advanced Fluid Dynamics and Fluvial Processes

This book covers fluid dynamics and fluvial processes, including basics applicable to open channel flow followed by turbulence characteristics related to sediment-laden flows. It presents well-balanced exposure of physical concepts, mathematical treatments, validation of the models/theories, and experimentations using modern electronic gadgets within the scope. In addition, it explores fluid motions, sediment-fluid interactions, erosion and scouring, sediment suspension and bed load transportation, image processing for particle dynamics, and various problems of applied fluid mechanics in natural sciences.

Features:

- Gives comprehensive treatment on fluid dynamics and fluvial process from fundamentals to advanced level applications in one volume.
- Presents knowledge on sediment transport and its interaction with turbulence.
- Covers recent methodologies in the study of turbulent flow theories with verification of laboratory data collected by ADV, PIV, URS, LDA, and imaging techniques, and field data collected by MMB and S4 current meters.
- Explores the latest empirical formulae for the estimations of bed load, saltation, suspension, and bedform migration.
- Contains theory to experimentations with field practices with comprehensive explanations and illustrations.

This book is aimed at senior undergraduates, engineering and applied science postgraduate and research students working in mechanical, civil, geo-sciences, and chemical engineering departments pertaining to fluid mechanics, hydraulics, sediment transportation, and turbulent flows.

An Introduction to Advanced Fluid Dynamics and Fluvial Processes

B. S. Mazumder and T. I. Eldho

CRC Press
Taylor & Francis Group
Boca Raton London New York

CRC Press is an imprint of the
Taylor & Francis Group, an **informa** business

Designed cover image: B. S. Mazumder and T. I. Eldho

First edition published 2024
by CRC Press
6000 Broken Sound Parkway NW, Suite 300, Boca Raton, FL 33487-2742

and by CRC Press
4 Park Square, Milton Park, Abingdon, Oxon, OX14 4RN

CRC Press is an imprint of Taylor & Francis Group, LLC

ISBN: 9780367428983 (hbk)
ISBN: 9781032485294 (pbk)
ISBN: 9781003000020 (ebk)

DOI: 10.1201/9781003000020

Typeset in Times
by codeMantra

Dedication

To my Late Parents, In-laws & Family Madhusri, Rahul & Ayan

Lisa & Akira

Sisters

– B. S. Mazumder

To my Late Parents (Iype & Marium); Family (Dr. Manjush, Iype & Basil); Brothers & Sisters

– T. I. Eldho

Contents

Authors

Professor B. S. Mazumder is a former Visiting Professor at the Department of Civil Engineering, Indian Institute of Technology (IIT), Bombay. After his retirement from Indian Statistical Institute (ISI), Kolkata, as a Professor and Head of the Physics and Applied Mathematics Unit, and In-Charge of Fluvial Mechanics Laboratory, Professor Mazumder joined as an Emeritus Fellow at the Department of Aerospace Engineering and Applied Mechanics, IIEST, Shibpur, and as a Visiting Professor at the Civil Engineering Department of at IIT Bombay. Soon after his Ph.D. Degree (1976) from the Indian Institute of Technology (IIT), Kharagpur in Applied Mathematics (Fluid Mechanics), B. S. Mazumder joined as a Post-doctoral Fellow at the Indian Statistical Institute (ISI), Kolkata, to work on the problems of Applied Fluid Mechanics with Professor Supriya Sengupta, Geologist and Professor J. K. Ghosh, Statistician. Since then, he has been working on various problems of inter-disciplinary nature, in the areas of *Turbulence, Fluvial sediment transport, Particle-size distributions, Combined wave-current flows over bedforms, Convection-diffusion process, Flow visualization and image processing, Navigation hydraulics, Flow and heat transfer.* He took initiatives and was instrumental to upgrade the existing laboratory, and named as Fluvial Mechanics Laboratory (FML) at ISI Kolkata with sophisticated equipment and excellent research facilities. A hi-tech laboratory is developed to elucidate various problems of statistical fluid mechanics related to turbulence and sediment transport. He received the financial supports from ISI, INSA, DST, DOD, NBHM and CSIR for several interdisciplinary projects. During the period 1989–1991, he worked as a Guest Scientist at the Illinois State Water Survey, UIUC, USA to work on turbulence generated due to navigation traffic in Mississippi and Illinois Rivers. During his tenure at ISI, Kolkata, he supervised nine Ph. D. Theses, seven Post-doctoral Fellows and ten Project Assistants.

As an Emeritus at IIEST Shibpur Prof. Mazumder was closely associated with the Hydraulics and Fluid Mechanics Laboratory. He was also involved as a Visiting Professor (2015–2021) in the Civil Engineering Department of IIT Bombay. During this period, he shared his innovative ideas in the development of the Hydraulics Laboratory in the department. He mentored 12 Ph. D. Scholars of different premier universities and institutes. He has more than 170 research papers to his credit in peer-reviewed journals, like JFM, Proc. Roy. Soc., Physics of Fluids-AIP, JGR, WRR, AWR, FDR, JHR (IAHR), JHE (ASCE), JFE (ASME), IJHMT, Sedimentology, Sedimentary Geology, SERRA, ESPL, etc. and many national and international Conference Proceedings. He visited several places in India and abroad, and delivered invited talks. He is a Fellow of Indian Society for Hydraulics (FISH). Prof. Mazumder was a Guest Scientist/Professor at different Universities, like the University of Illinois at Urbana-Champaign (UIUC), California Institute of Technology (Caltech), Stanford University, University of California-Davis (UC-Davis), UC-Santa Barbara (UCSB), Rensselaer Polytechnic Institute (RPI), New York, MIT-Cambridge, University of Minnesota, Minneapolis, SUNY-Binghamton, USA, University of Western Ontario, Simon Fraser University (SFU)-Canada, Nottingham University, Sheffield University, UCL, King's College, Cambridge University, UK, University of Stuttgart, University of Munich, University of Karlsruhe, Germany, ICTP-Italy, Technical University of Denmark, Tsinghua University, Beijing and many others. Prof. Mazumder is still deeply engrossed in his research domain.

Dr T. I. Eldho is a Professor & Former Head and Chair of the Department of Civil Engineering, IIT Bombay. He has more than 30 years of experience in the area of water resources and environmental engineering as a Scientist, Professor and Consultant. He works in the areas of Fluid Mechanics, Turbulence, Hydraulics Coastal Hydrodynamics, and Climate Change Impact on Water Resources, CFD, Groundwater Flow, and Contaminant transport. He got his Ph.D. from IIT Bombay in 1995 and worked as Postdoctoral Fellow at the Institute for Hydromechanics, University of Karlsruhe (currently Karlsruhe Institute of Technology), Germany during 1996–1998. During 1998–1999, he worked as Water Resources Modeler and Consultant at Mott MacDonald Company in Cambridge UK. Later Dr. Eldho worked as a Senior Research Fellow at Hydrotech Research Institute, National Taiwan University, Taipei during 1999–2000. In 2000, Dr. Eldho returned to India and joined at Department of Civil Engineering, IIT Kharagpur as a Faculty. Further Dr. Eldho joined IIT Bombay as a Faculty in 2001 and continued his teaching, research, and professional career. He has developed Hydraulics Laboratory of IIT Bombay with many sophisticated equipment and facilities. During the past 21 years at IIT Bombay, Dr. Eldho has guided 35 Ph.Ds., nine Postdoctoral Fellows and 60 Masters theses. He has published more than 220 research papers in reputed international and national journals and more than 350 papers in various international and national conferences. He has developed and offered 25 courses and a number of short-term courses for college teachers and working professionals. Dr. Eldho has delivered more than 200 invited lectures at various institutions in India and abroad. He has also developed two popular video courses on "Fluid Mechanics" and "Watershed Management." Prof. Eldho serves as an Editor/Associate Editor and Editorial Board Member of a number of Indian and international journals and has worked as a reviewer for more than 50 national and international journals in the recent past. He is also a research proposal reviewer for many research organizations and delivered several invited/keynote lectures at various national and international organizations and conferences. Dr. Eldho has co-authored two popular textbooks and contributed 20 book chapters in various edited books. Dr. Eldho has worked as a Consultant for many State and Central Government organizations, municipal corporations and companies in the areas of hydraulics, fluid mechanics, flood studies, sump model studies, hydraulic designs, etc. He completed more than 100 consultancy projects and more than 10 sponsored projects with agencies such as ISRO, DST, and various Ministries of State and Central Governments. Dr. Eldho is a recipient of many prestigious awards including Eminent Water Resources Scientist, constituted by the Indian Water Resources Society in 2018.

Preface

This book is written for engineering and applied science students, faculties, researchers, and professionals, from introductory to advanced level courses in fluid dynamics and fluvial processes. The state-of-the art of this book can be visualized only through a careful investigation of the development, starting from the theoretical to modern experimentation in the laboratories and fields using advanced technology. The important features of this book are that from the beginning to most up-to-date research findings in turbulence and physics of sediment transport are focused. This book is started with fundamental of fluid dynamics applicable to open channel flow followed by turbulence characteristics related to sediment-laden flows. It may also serve the expert engineers, scientists, and researchers who need the basic understanding of fluid mechanics, hydraulics, image processing, and sediment transportation in both bed load and suspension, for their research work, as well as for more advanced level work. The merit of this book lies in its well-balanced exposure of physical concepts, mathematical treatments, validation of the models/theories, and experimentations using modern electronic gadgets within the scope. Thus, it is not surprising that this book is of interest to a wide circle of scientists, including engineers, fluid dynamists, hydrologists, applied statisticians, applied mathematicians, sedimentologists, geographers and environmentalists for their development of judicious theoretical modelling that could be used for sediment transportation processes in real river systems and in geological records. This book also describes the local scour and protection measure problems around the hydraulic structures.

During the research work on interdisciplinary in nature related to fluid mechanics, hydraulics, statistics and sedimentology at the Fluvial Mechanics Laboratory of Physics and Applied Mathematics Unit (PAMU) of Indian Statistical Institute (ISI), Kolkata, and Hydraulics Engineering Laboratory at IIT Bombay, Mumbai, authors are encouraged by the responses of the interdisciplinary researchers. So, this book is an outcome of our teaching, theoretical and experimental research experiences on turbulence and sediment transport. Mostly, this book is designed for the final year Bachelor, Masters and Research students and teachers as well as for field investigators. Therefore, the primary attempt is to address fundamental aspects of fluid dynamics and basic concepts of sediment transport phenomena with experiments. In that respect, we tried to keep a level of detailed treatment for proper understanding of turbulence and fluvial processes. In addition, descriptions in the book include fluid motions, sediment-fluid interactions, erosion and scouring, sediment suspension, bed form dynamics and bed load transportation, image processing for particle dynamics, and various problems of applied fluid mechanics in natural sciences. Throughout this book, we have tried categorically to explain lucidly the whole text, mathematical calculations and data analyses for easy understanding to the readers.

An extensive referencing of available textbooks on the subject showed that there is no concise book deliberating the fluid dynamics principles on turbulence with the fluvial processes so that the researchers, teachers, and students can refer to their daily needs on the subject. Accordingly, we feel that this book will be very useful and acceptable to all researchers, teachers, students, and practicing engineers in the areas of fluid dynamics and fluvial processes.

We encourage readers to convey feedback and suggestions for the improvement of the content of the book and point out necessary corrections in it.

(Prof. B. S. Mazumder and Prof. T. I. Eldho)

Acknowledgments

The authors have gained immensely through the works of several researchers in the field of fluid mechanics, fluid dynamics, turbulence, image processing, hydraulics and fluvial processes. The authors were always fascinated by the immense works and textbooks on these topics by researchers such as H. Schlichting (*Boundary Layer Theory*); Schlichting H. and Gersten K. (*Boundary Layer theory*); S. W. Yuan (*Foundations of Fluid Mechanics*), J. O. Hinze (*Turbulence*); T. Cebeci (*Analysis of Turbulent Flows*); I. Nezu & H. Nakagawa (*Turbulence in Open Channel Flows*); P. Y. Julien (*Erosion and Sedimentation*); B. M. Sumer (*Turbulence*); W.H. Graf (*Hydraulics of Sediment Transport*); R. J. Garde & K. G. Rangaraju (*Mechanics of Sediment Transportation and Alluvial Stream Problems*); M. S. Yalin (*Mechanics of Sediment Transport*); S. M. Sengupta (*Introduction to Sedimentology*); C. S. P. Ojha, R. Berndtsson, & P. N. Chandramoulli (*Fluid Mechanics and Fluid Machinery*); S. K. Som & G. Biswas (*Introduction to Fluid Mechanics and Fluid Machines*); B. M. Sumer & J. Fredsoe (*The Mechanics of Scour in the Marine Environment*); B. R. Munson, D. F. Young, T. H. Okiishi, & W. W. Huebscs (*Fundamentals of Fluid Mechanics*); S. Dey (*Fluvial Hydrodynamics, Hydrodynamic and Sediment Transport Phenomena)*; R.A. Granger (*Fluid Mechanics*); R. C. Hibbeler (*Fluid Mechanics*), etc. We highly acknowledge their contributions and this book is based on many of the topics discussed in these books.

The authors have lectured various courses on Fluid Mechanics, Fluid Dynamics, Fluid Turbulence, Fluvial Process and Sediment Transport for over a period of 30 years or more to undergraduate and postgraduate students at ISI Kolkata and IIT Bombay and given many invited lectures at various Institutes in India and abroad. Further, the authors have supervised a number of Ph.D. and Masters students in these areas at ISI Kolkata and IIT Bombay. These activities have helped the authors to refine many ideas in the areas of fluid dynamics, turbulence, fluvial processes, and sediment transport. The authors are thankful to all their students, teachers, and colleagues. Special thanks are extended to all Masters, Ph.D. Scholars and Postdoctoral Fellows whose project-related works are extensively referred in the book by the authors. Specifically, the authors wish to express their thanks to former and present students: Dr. M. R. Saha, Dr. S. K. Das, Dr. D. C. Dalal, Dr. Rajat Maumder, Dr. Koeli Ghoshal, Dr. K. Sarkar, Dr. S. P. Ojha, Dr. H. Maity, Dr. D. Pal, Dr. R. N. Ray, Dr. Anindita Bhattacharyya, Dr. K. K. Mandal, Dr. Debasmita Chatterjee, Dr. B. A. Vijayashree, Dr. Priyanka Gautam, Dr. Gaurav Misuriya, and Sanjukta Das.

The first author (BSM) expresses his deep sense of gratitude to *Professor T. I. Eldho*, Former Head, Department of Civil Engineering of Indian Institute of Technology (IIT), Bombay, who honoured him to join as a Visiting Professor, and proposed and inspired him to write this book jointly on *Introduction to Advanced Fluid Dynamics and Fluvial Processes*. BSM is really fortunate to get such endless help and encouragement from him. BSM heartily expresses his earnest recognition to Professor Jayanta K. Ghosh, Statistician and Professor Supriya Mohan Sengupta, Geologist, who initially taught him the subjects-statistics and fluvial sediment transport in a laboratory flume at ISI Kolkata. Further, BSM is also thankful to Professor N. Datta of IIT Kharagpur, Professor R. J. Garde, Professor Roddam Narasimha, Professor S. K. Tandon, Sedimentologist, Professor Nani G. Bhowmik of UIUC, Prof. Zhao-Yin Wang of Tsinghua University, Beijing, Prof. Stephen Monismith Stanford University, Prof. Gary Parker UIUC, Prof. Marcelo H. Garcia UIUC, Prof. John Bridge Binghamton University, Professor S. R. McLean of UCSB, Professor N. J. Clifford of King's College-London, Professor Sankar Pal, Former Director of ISI, Professor D. Datta Mazumder of ISI Kolkata, Professor Chandan Chakraborty of ISI Kolkata, Professor B. S. Dandapat of ISI Kolkata, Professor Subir Ghosh of ISI Kolkata, Professor Kaustuv Debnath of IIEST, Shibpur and Dr. Barendra Purkait of GSI-Kolkata for inspiring to write project proposals, brief reports and finally to write a book on his research work conducted at ISI Laboratory, and all his friends and technicians Mr. Narayan Ray, Mr. Apurba Biswas, Mr. Samanta Paramanik and Mr. Shyam S. Samanta who worked from their heart at the laboratory. BSM received the great help from his Chotakaka (*Phanindra Ch. Mazumder*) and Chotama (*Kankan Mazumder*), and his elder sister (*Bela Roy*) and brother-in-law (*Harisadhan Roy*), while pursuing his education M.Sc., M.Tech, Ph.D. at IIT-Kharagpur. BSM cannot forget the stimulation and encouragement from his late parents (*Dr. Khagendra Jiban Mazumder, Father* and *Bidyut Prova Mazumder, Mother*). This book would not have been possible without the constant encouragement and inspiration from BSM's wife *Madhusri,* two sons *Rahul* and *Ayan*, his *father and mother-in-laws, Sujit Kaku, daughter-in-law* (*Lisa*) and *grand-daughter (Akira).* BSM is very much thankful to all his friends and colleagues of ISI, Kolkata, IIEST, Shibpur, and IIT-Bombay for their heartiest support, cooperation, and sincere help in several ways. He also expresses his sincere thanks to ISI, INSA, DST, DOD, NBHM, and CSIR members for their financial support to achieve the experimental work in the ISI laboratory.

The second author (TIE) expresses his thanks to all his former and present students, colleagues, and friends especially Late Prof. B.V. Rao, Late Prof. Gerhard H. Jirka KIT Germany, Prof. Der Liang Young NTU Taiwan and the staff of Hydraulics Laboratory, IIT Bombay. TIE expresses his deep sense of gratitude to his late parents, family (Dr. Manjush, Iype, and Basil), brothers (Late Kuriakose, Mathewkutty & Paulose), and sisters (Late Annamma, Kunjamma, and Sherly) for their moral support.

The authors wish to express sincere gratitude to Mr. Pintu Paul, Dr. Gaurav Misuriya, and Mahendra Palsania who helped us with figure drawings, to the editorial team and to all those who helped directly or indirectly and made this book possible.

1 Introduction

1.1 BACKGROUND

Fluid mechanics has a wide range of applications in almost all branches of science and engineering fields. In engineering, fluid mechanics has significant applications in almost all areas including civil, mechanical, chemical, aerospace, metallurgy, agriculture, ocean, and others. Fluid mechanics and its principles are significantly used in all branches of sciences such as geophysics, astrophysics, physical chemistry, biomechanics, biomedical, meteorology, atmospheric, applied mathematics, etc. For example, in civil engineering, the principles of fluid mechanics and its advancements are used to design water-supply networks, drainage systems, water-resisting structures like dams and reservoirs, construction and foundation of bridge piers, river management, protection of coastal and harbor areas, etc. In fluid mechanics, the topics with significant impacts on practical engineering applications include fluid dynamics, fluid turbulence, hydraulics, fluvial processes, and sediment transport.

Although a number of textbooks are available on fluid mechanics, there is no concise book deliberating the fluid dynamics principles with the fluvial processes, so that the professionals and students can refer to their daily usage on the subject. We have compiled this book in such a way that it gives a comprehensive treatment in the areas of fluid dynamics, turbulence, and fluvial processes.

1.2 FLUID DYNAMICS

In fluid mechanics, there are three important branches, such as statics, kinematics, and dynamics. In fluid statics, the fluid elements are acted upon by the forces at rest, whereas in fluid kinematics, we deal with the geometry of fluid motion in reference to translation, rotation, and deformation of a fluid particle. In fluid dynamics, we consider the forces acting on the fluid particles in motion with respect to one another that causes the acceleration of the fluid. In most of the practical problems, the dynamic analysis of fluid in motion is necessary to identify the fluid flow pattern and its surroundings in the motion. In comparison to fluid statics and kinematics, fluid dynamic analysis is more complex and challenging as we deal with fluids in motion. The dynamics of fluids are the foundation of the understanding of fluid flows in streams, pipes, and in the subsurface. The fundamental principles of motion are applied to the flow of liquids or gases, namely, conservation of mass, conservation of momentum (known as Newton's second law of motion), and conservation of energy (from the first law of thermodynamics). Fluid dynamics offers a systematic analysis of fluid flow problems that embraces empirical and semi-empirical laws derived from fundamental principles, and flow measurement is used to solve practical problems. In fluid dynamics, typical problems involve the calculation of various properties of the fluid, such as velocity, pressure, density, and temperature, as functions of space and time.

1.3 FLUID FLOW ANALYSIS

In many engineering problems, determining the flow state of fluids is required; also, ascertaining how fluids will flow through a system can be critical to its operation. Figure 1.1 shows a schematic of how we solve a fluid flow problem (Granger, 1985). A fluid flow problem can be solved experimentally or theoretically. The experimental solution can be done through a scaled model or field-level investigations. In theoretical studies, the investigation is performed through a physical analysis and then through a mathematical analysis. Depending on the problem, the physical analysis can be based on the force or energy concept. The mathematical analysis can be performed through an analytical solution of the

DOI: 10.1201/9781003000020-1

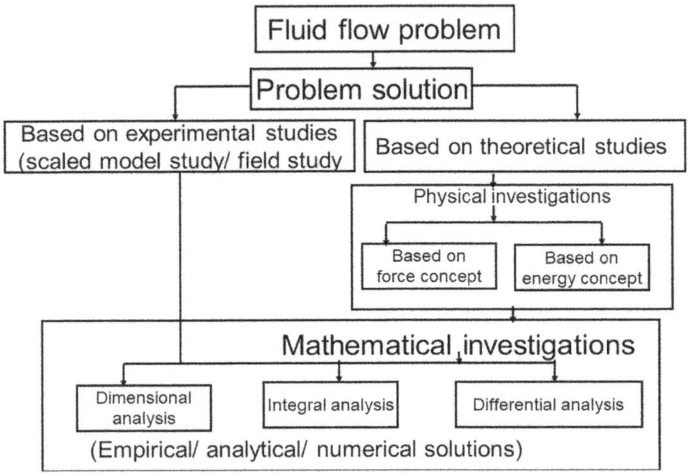

FIGURE 1.1 Schematic of fluid flow problem solution.

governing equation or through a numerical solution using a computer. Thus, the fluid flow analysis can be carried out analytically, experimentally, or numerically.

1.4 FLUID TURBULENCE

In nature, fluid flow turbulence is common and plays an important role in many engineering problems. Many of the geophysical processes such as river morphology, landscape evolution, atmospheric dynamics, and ocean currents are highly influenced by turbulent fluid flow. The turbulent flow by nature is irregular, random, and complex. The origin of turbulence with the transition from laminar to turbulent flow is of primary importance to many practical problems. Each turbulent flow is unique in its character, and hence it is very complex to predict the turbulent flow behavior. Although turbulent flow theories are reasonably developed in the past century, there is no exact solution to the turbulent flow equations. Thus, most of the times, we need to rely on experimental studies or computer models.

1.5 FLUVIAL HYDRAULICS

In general, hydraulics is a term used in engineering for studying liquids and how liquids function or flow. In hydraulics, various principles of hydrodynamics are directly used by engineers, applied scientists, and managers to study real practical problems of fluid flows. Fluvial hydraulics is the branch of hydraulics that studies the flow of water in river channels, canals, or streams. In fluvial hydraulics, the unsteady motion in turbulent conditions in natural channels exhibits its most complex form. The studies of fluvial hydraulics provide a sound qualitative and quantitative understanding of water and sediment flows in natural channels and rivers. For practical problems, this understanding is essential for modelling and predicting hydrologic and geomorphologic processes, erosion, sediment transport, water supply, water quality, habitat management, and flood hazards. The formation of river channels, flow turbulence, sediment transport, and deformations such as erosion and aggradation are studied in fluvial hydraulics.

1.6 SEDIMENT TRANSPORT

In nature, *Sediment transport* is the movement of solid particles due to the fluid motion in which particles are entrained. In fluvial processes, the mechanics of sediment transport deals with the processes of erosion, transportation, and deposition of sediment under the action of flowing water, wave, wind, and gravity. Depending on the nature of sediments, incipient motion, rolling, sliding, saltation, and suspension of individual sediment particles occur, and it depends on the physical properties and mineral composition of the particles, characteristics of the flow, and boundary conditions. In nature, sediment motion is the main cause of geomorphologic evolution, fluvial process, and environment change. In this process, erosion of land surfaces, transportation of eroded material, deposition of this material in lakes, rivers, and reservoirs happen. The sediment materials can be either non-cohesive or cohesive. In this book, mainly the transports of non-cohesive sediments are dealt with for which, incipient motion, rolling, sliding, saltation, and suspension of individual sediment particles are the main causes of transport phenomena. In nature, the sediment motion is the main cause of geo-morphological evolution, fluvial process, and environmental change.

1.7 SCOPE OF THE BOOK

This book "An Introduction to Advanced Fluid Dynamics and Fluvial Processes" is intended for engineering and applied science students, and faculties from introductory- to advanced-level courses in fluid dynamics and fluvial processes. This book is also intended for professionals and the scientific community working in the areas of fluid dynamics and fluvial processes. The state-of-the-art of this book can be visualized through a careful investigation of the development, starting from the theoretical to modern experimentation in the laboratories and fields using advanced technologies. The important features of this book are that from the beginning to most up-to-date research findings in fluid dynamics, turbulence, and physics of sediment transport are focused. This book is started with fundamentals of fluid dynamics applicable to open channel flow followed by turbulence, and its characteristics related to the sediment-laden flows and sediment transport. It may also serve the expert engineers, scientists, and researchers who need basic understanding of fluid mechanics, sediment transportation in both bedload and suspension for their professional and research work, as well as for more advanced level works. The merit of this book lies in its well-balanced exposure of physical concepts, mathematical treatments, validation of the models/theories, and experimentations using modern electronic gadgets within the scope. Thus, it is not surprising that this book is of interest to a wide circle of scientists, including engineers, fluid dynamists, hydrologists, applied statisticians, applied mathematicians, sedimentologists, geographers, and environmentalists for their development of judicious theoretical modelling that could be used for fluid flow dynamics and sediment transport processes in real river systems and in geological records. This book also describes the local scour problems around the hydraulic structures.

During the research work on interdisciplinary topics related to fluid mechanics, hydraulics, statistics and sedimentology at the Fluvial Mechanics Laboratory of Indian Statistical Institute (ISI), Kolkata, and Hydraulics Engineering Laboratory at IIT Bombay, Mumbai, the authors are encouraged by the responses of the interdisciplinary researchers, and so this book is an outcome of our teaching, theoretical and experimental research experiences on fluid dynamics, turbulence and sediment transport. Mostly, this book is designed for the final year Bachelor, Masters, research students and teachers, as well as for field investigators. Therefore, the primary attempt is to address fundamental aspects of fluid dynamics and basic concepts of sediment transport phenomena with experiments. In that respect, we tried to keep a level of detailed treatment for a proper understanding of turbulence and fluvial processes. In addition, descriptions in the book include fluid motions, sediment-fluid interactions, erosion and scouring, sediment suspension and bedload transportation, image processing for particle dynamics, and various problems of applied fluid mechanics in natural sciences. Throughout this book, we have tried categorically to describe the whole text, mathematical calculations, and data analyses for easy understanding to the interested readers.

1.8 HIGHLIGHTS OF THE BOOK

For professionals and students working in the area of fluid mechanics, hydraulics, and sedimentary environment, a fundamental understanding of fluid dynamics, turbulence, and fluvial processes are essential. The main highlight of this book is to provide a fundamental understanding of these topics and their advancements. Further, for construction of dams and reservoirs, the foundation of bridge piers, river management with respect to the bank erosion and siltation, saving of coastal and harbour areas, etc. are the most important examples for which the principles of fluid dynamics and fluvial processes are important, which are explained in this book. Moreover, the movement of various ships and barges in rivers and oceans associated with the field of naval architecture is based on the principles of fluid mechanics. In addition, soil erosion and deposition are mostly based on the principles of fluid dynamics related to the fluvial processes used in agriculture engineering and siltation management in rivers. Soil erosion of the river bank and bed is frequently observed from the head water to the mouth of any river. This is a challenge to human health, natural ecosystems, navigation and food security as well as to the socioeconomic development of countries. Moreover, the complex interaction of river bank sediments with the flow leads to erosion and deposition of sediment that results in an irregular topography of the river bank. The implications of erosion in the river environment experience multidimensional activities in terms of physical, ecological, navigational, and economical problems. These are all explained in this book.

The main highlights of this book include:

- Comprehensive treatment on fluid dynamics and fluvial process from the fundamentals to advanced level applications.
- Recent methodologies on the study of turbulent flow theories.
- Basic principles of fluvial processes with field examples.
- Well-balanced exposure of physical concepts, mathematical treatments, validation of the models/theories, and experimentations using modern electronic gadgets
- Treatment of various fluvial processes such as sediment-fluid interactions, erosion and scouring, sediment suspension and bed load transportation, image processing for particle dynamics, and various problems of applied fluid mechanics in natural sciences.

This book further provides the necessary contents for undergraduate, postgraduate and research levels courses on:

- Basic and advanced level fluid dynamics.
- Fluid turbulence – basic theories and experiments.
- Hydraulics and fluvial processes.
- Sediment transport – theory and practice.

1.9 BOOK CONTENT AND ITS USAGE

There are ten chapters in this book, starting with a brief introduction to the basic principles of fluid dynamics, fluvial processes of sediment transport, scope and highlights of this book. This introductory chapter leads to Chapter 2, which deals with the basic concepts and definitions of continuum mechanics, properties of fluids, laminar and turbulent flows, dimensional analysis, statistical analysis, open channel flows, hydraulics, and sediment transport. Chapter 3 presents a general description of fluid motion with conservations of mass, momentum and energy with examples. Chapters 4 deals with the concept of viscous flow theory with separation, vortex formation and drag. Chapter 5 deals with the approximate boundary layer equations and their solutions. Chapter 6 contains the fundamental concepts and classical theories of turbulent flows with Reynolds's decompositions in the Navier–Stokes equations. In addition, theories of turbulent kinetic energy equations, turbulence characteristics, turbulent burst, spectral analyses, isotropic and anisotropic turbulence, and turbulent length and time scales are discussed. Chapter 7 deals mostly with the instrumentations for the measurements of turbulent flows and bed form in open channels over waveform structures and around different shapes of piers, using several modern electronic gadgets, such as OTT current meter, laser Doppler anemometer (LDA), acoustic Doppler velocimeters (ADV), particle image velocimetry (PIV), hot wire anemometer (HWA), high-speed motion-scope camera (HSMC), ultra-sonic ranging system (URS), etc.; and in river flows using Electromagnetic current meters (MM511, MM527), Inter-Ocean current meters (S4), etc. to study the velocity distributions near the river bank area during the barge movements. Chapter 8 deals with the occurrence of grain-size distributions in nature, physical processes of sediment transport phenomena, including the particle's sizes and shapes, motion of sediment particles, critical shear stress, Hjulstrom and Shields criteria for particle movement, and different theoretical models. Chapter 9 gives bed load, saltation, suspension and total load, including their concepts of empirical formulations. In addition, characteristics of saltation of particles and image analysis of saltating motion using high-speed Motion-Scope Camera are focused. Chapter 10 consists of bedform migrations, formation of sand bars, flow over different bed forms and scour structures around the obstacles and piers.

In this book, Chapters 2–5 will be useful for professionals and students who look for advanced level topics of fluid dynamics. Chapter 6 can be referred for various theories of turbulent models with turbulent energy equations, production, dissipation and Kolmogorov's hypothesis. Chapter 7 can be referred for flow measurements and instrumentations using sophisticated electronic gadget. Chapters 8–10 can be referred for fluvial processes, sediment transport, scouring phenomena, and protection measures.

An extensive referencing of the available textbooks on the subject showed that there is no concise book deliberating the fluid dynamics principles with the fluvial processes so that the researchers, teachers, and students can refer to their daily needs on the subject. Accordingly, we hope that this book will be very useful and acceptable to all researchers, teachers, students, and practicing engineers in the areas of fluid dynamics and fluvial processes

1.10 SUMMARY

Some of the important points discussed in this chapter are summarized below.

- In fluid dynamics, we consider the forces acting on the fluid particles in motion with respect to one another that causes the acceleration of the fluid. In fluid dynamics, typical problems involve the calculation of various properties of the fluid, such as velocity, pressure, density, and temperature, as functions of space and time.
- The fluid flow analysis can be carried out analytically, experimentally, or numerically.
- The turbulent flow by nature is irregular, random, and complex. There are no exact solutions for turbulent flow problems and we need to rely on experimental studies or computer models.
- The studies of fluvial hydraulics provide a sound qualitative and quantitative understanding of water and sediment flows in natural channels and rivers. Sediment transport deals with the processes of erosion, transportation, and deposition of sediment under the action of flowing water, wave, wind, and gravity.
- Detailed discussions on various aspects of the available theoretical developments and applications on the topics of fluid dynamics, turbulence, fluvial processes, and sediment transport are elaborated in the subsequent chapters for interested readers.

2 Fundamental Fluid Properties and Definitions

2.1 INTRODUCTION

Fluid mechanics is a branch of physics concerned with the mechanics of fluid flows. The study of fluid flow has been described with the support of fundamental laws of mechanics (Schlichting, 1968). It has a wide range of applications in almost all branches of engineering fields including civil engineering, mechanical engineering, chemical engineering, aerospace engineering, agriculture engineering, ocean engineering, and others. Apart from the engineering profession, the principles of fluid mechanics are also used in all branches of sciences such as geophysics, astrophysics, physical chemistry, biomechanics, biomedical, meteorology, atmospherics, etc. It is one of the fastest rising basic sciences whose principles show applications in all aspects of daily life. For example, in civil engineering, the principles of fluid mechanics are used to design drainage systems, water networks, water-resisting structures such as dams and reservoirs, construction and foundation of bridge piers, river bank protection, river management, protection of coastal and harbor areas, etc. Moreover, designs of various types of ships, recreation boats, and barges or vessels for transportation in rivers and oceans associated with the field of naval architecture are based on the principles of fluid mechanics.

The study of fluid behavior can be divided into three categories: statics, kinematics, and dynamics. In the static case, the fluid elements are acted upon by the forces at rest. The static pressure forces on an immersed body in fluid are determined from the static analysis. In the case of kinematics of fluids, it deals with the geometry of fluid motion: study of translation, rotation, and deformation of fluid particles. This is useful for describing the motion of a particle and visualizing the flow patterns. In the dynamic case, the analysis involves considering the forces acting on the fluid particles in motion with respect to one another that cause acceleration of fluid. To determine the fluid flow pattern and its surroundings in motion, the dynamic analysis of a fluid in motion is required. In this chapter, the description of fundamental fluid properties and a fundamental definition of various terms used in fluid dynamics and fluvial hydraulics are briefly discussed.

2.2 CONTINUUM MECHANICS

Continuum mechanics deals with the motion of a fluid involving the mechanical behaviour of all discrete molecules which make up the fluid, where intermolecular cohesive forces induce the fluid to behave as a continuous mass of substance rather than as discrete particles. In gases, since the molecular motion varies, and the quantity of molecules is large, i.e., molecular density is high enough; the analysis of the molecular motion of the fluid is made using the average effects of all the molecules within the gas, which can be done with any degree of accuracy. If the molecular density is very high, the gas can be treated as continuous mass of fluid, called the continuum. There are two factors in continuum: the distance between the molecules indicating the molecular density and the time between the collisions. The distance between the molecules is characterized by mean free path which is a statistical average distance travel between two successive collisions. If the mean free path is very small compared to any characteristic length in the flow domain (i.e., the molecular density is very high) and the time of collisions is sufficiently small, the gas can be treated as continuous medium. If the mean free path is large compared to certain characteristic length, the gas cannot be considered continuous and must be

DOI: 10.1201/9781003000020-2

analyzed using molecular theory. In this case, a dimensionless parameter, known as Knudsen number is defined as $K_n = \lambda / L$, where λ is the mean free path and L is the characteristic length. Usually, when $K_n > 0.01$, the concept of continuum does not hold. However, when $K_n < 0.01$, the fluid is said to be continuum.

Continuum mechanics is the mathematical description of deformation and related stresses. It deals with the physical properties of solids and fluids which are independent of any particular coordinate system in which they are observed. The primary assumption adorned in the name is that the materials are assumed to be homogeneous, isotropic, continuous and independent of the coordinate system. This book primarily deals with continuous fluids.

2.3 FLUID PROPERTIES

Here some basic properties of the fluids in continuum are briefly discussed which are used in this book.

Pressure: The pressure at a point in the fluid is defined as the normal force exerted on the incremental area by the surrounding fluid particles. Pressure is determined as force per unit area. In SI (System International) unit system, the pressure unit is N/m^2.

Density: The density ρ_f of a fluid is its mass per unit volume. The density of fluid at a point is defined as the mass of fluid surrounding the point to the incremental volume surrounding the same point, i.e. the mass of fluid contained in a unit volume. The unit of density is kg/m^3.

Specific weight: The specific weight is closely related to the density, which represents the weight of the fluid per unit volume. Specific weight is defined as product of density and the acceleration due to gravity, which means the specific weight is given by $\gamma_f = \rho_f g$, where ρ_f is the fluid density, and g is the acceleration due to gravity. The unit of specific weight is N/m^3.

Specific volume: The specific volume of a fluid is the volume occupied by a unit mass of fluid. The unit of measurement is m^3/kg.

Specific gravity: The specific gravity of a fluid is the ratio of specific weight of fluid at actual conditions to the specific weight of pure water at a standard condition (101 kN/m^2 at 20°C).

Incompressible fluids: The theory of inviscid (zero viscosity) or viscous fluids in which the fluid density is assumed to be constant has a wide range of applications to the flow of liquids. An extensive mathematical theory has been developed for inviscid incompressible fluid which gives reliable results in the calculation of lift, drag, and wave motion for gas and water flows at low velocity, whereas for viscous incompressible fluids, it explains the phenomena of the viscous forces, separation of flow, and eddy viscosities.

2.4 PERFECT AND REAL FLUIDS

In the motion of perfect or ideal fluid in classical hydrodynamics, two contacting layers experience no shearing force but act on each other with normal force, which is the pressure only. This fluid is assumed to be inviscid and incompressible in nature; where no tangential force between adjacent fluid layers exists (Yuan, 1970). Perfect fluid does not offer any internal resistance to change its shape. The theoretical developments in the field of fluid mechanics are primarily based on the concept of a perfect fluid, which is frictionless and incompressible fluid. The theory of motion of perfect fluid is extremely mathematically developed and provides a suitable description of real motions in some cases, such as the motion of water waves or the formation of liquid jets in the air. On the other hand, the theory of perfect fluids fails to describe the drag of a body. Whereas in real fluids, there exist the tangential stress and the no-slip condition near the solid wall, which makes up the essential difference between *a perfect and real fluid*. The perfect fluid leads to a statement that a body moves uniformly through a fluid and extends to infinity experiences with no drag (dAlembert's paradox) (Yuan, 1970). This appalling result of a perfect fluid can be attributed to the fact that the inner layers of real fluid transmit tangential as well as normal stresses, this being also the case near a solid wall wetted by a fluid. This frictional force in real fluid is concerned with a property of fluid, which is called the viscosity of the fluid. Certain measurable quantities, like velocity, density, viscosity of fluid, etc., are present in the real fluid, which are the controlling factors.

Certain fluids that are of actual practical importance, such as water and air, have very small viscosity. In many instances, the motion of fluids with small viscosity agrees well with that of perfect fluid because, in most cases, the shearing stresses are very small. For this reason, the existence of viscosity is completely neglected in perfect fluids, which introduces an extensive simplification of the equations of motion. It is important to note that even in fluids with small viscosity; unlike in perfect fluids, the condition of no-slip near a solid boundary exists. The no-slip condition introduces in many cases in the laws of motion of perfect and real fluids. In particular, the drag in real and perfect fluids has its physical origin in terms of no-slip condition in the wall (Schlichting, 1968).

2.5 VISCOSITY

Viscosity is a measure of resistance to the movement of fluid flow layer relative to each other. It describes the internal friction of a moving fluid. A fluid with large viscosity resists motion because its molecular attraction gives a lot of internal friction, resulting in the deformation of fluid motion at different rates. The deformation occurs at different types of fluids. For example, the fluid flows between two plates (Figure 2.1), where the upper plate is moving with a velocity U and the lower plate is at rest. If the distance between the plates is denoted by h, and the pressure being constant throughout the fluid, then the fluid flow is confined between the boundaries as follows:

$$\begin{aligned} u &= U \text{ at } y = h \\ u &= 0 \text{ at } y = 0 \end{aligned} \tag{2.1}$$

The velocity distribution of the fluid between the plates is linear, so that fluid velocity is proportional to the distance y from the lower plate to the upper plate and we have (Som and Biswas, 1998):

$$u(y) = \frac{y}{h} U \tag{2.2}$$

To get the motion of fluid, it is necessary to apply a tangential force to the upper plate. This force is in equilibrium with the frictional force to the fluid. It is known that the force (per unit area) is proportional

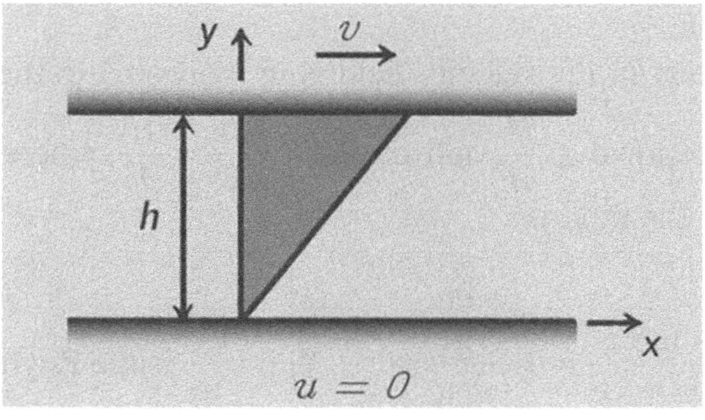

FIGURE 2.1 Effect of viscosity for fluid flow between two parallel plates.

to the velocity U of the upper plate and inversely proportional to the distance h. The frictional force τ per unit area is proportional to U/h, for which we substitute the velocity gradient $\frac{du}{dy}$ as

$$\tau \propto \frac{du}{dy} \text{ i.e., } \tau = \mu_f \frac{du}{dy} \tag{2.3}$$

where μ_f is the proportionality constant, known as *viscosity or dynamic viscosity or coefficient of viscosity*, which depends on the nature of the fluid. The parameter μ_f is small for thin fluid such as water or air, and large for oil or glycerin. This law is known as *Newton's law of viscosity or friction*. The dimension of viscosity is $[\text{kg/ms}]$. The laminar flow is represented by parallel layers sliding over the other without any exchange of mass between the layers. For turbulent flow, mixing between the layers takes place and the shear stress is given by

$$\tau = (\mu_f + \epsilon_t \rho_f) \frac{du}{dy} \tag{2.4}$$

where ϵ_t is the coefficient of eddy viscosity or turbulent diffusivity.

Equation (2.3) is also related to Hooke's law for an elastic solid body in which case the shearing stress is proportional to the shear strain as,

$$\tau = G \frac{d\xi}{dy} = G\gamma \tag{2.5}$$

where G is the modulus of shear, γ the change in angle between two lines which were originally at right angles, and ξ denotes the displacement in the direction of abscissa. *Kinematic viscosity* (v_f) is defined as the ratio of the coefficient of viscosity to the density (μ_f/ρ_f), which is another way to express the viscosity of fluid. The unit of v_f is m^2/s.

2.6 DIMENSIONLESS NUMBERS

In fluid mechanics, dimensionless numbers are useful to determine the fluid flow characteristics. Here several dimensionless numbers which arise from the fluid flows in different phenomena are discussed briefly.

Reynolds number (*Re*): This number is defined as the ratio of the inertial force to the viscous force.

$$Re = \frac{\text{Inertia force}}{\text{Viscous force}} = \text{constant}$$

The condition is that at all corresponding points in the flow, the ratio between the inertia force and the viscous force must be constant.

The velocity u at some point in the velocity field is proportional to the free stream velocity U, the velocity gradient $\frac{\partial u}{\partial x}$ is proportional to $\frac{U}{d}$, and similarly $\frac{\partial^2 u}{\partial y^2} \propto \frac{U}{d^2}$, where d is the characteristic linear dimension of the body. Then the ratio is

$$Re = \frac{\text{Inertia force}}{\text{Viscous force}} = \frac{\rho_f u \frac{\partial u}{\partial x}}{\mu_f \frac{\partial^2 u}{\partial y^2}} = \frac{\rho_f \frac{U^2}{d}}{\mu_f \frac{U}{d^2}} = \frac{Ud}{v_f} \text{ is the Reynolds number} \tag{2.6}$$

The Reynolds number is applicable to all types of flow, closed surface (e.g., pipe flow), as well as free surface flows.

Froude number (*Fr*): It is the ratio of the inertia force to the gravitational force of the fluid. The square root of this dimensionless ratio is known as Froude Number. It is defined as:

$$Fr = \sqrt{\frac{\text{Inertial force}}{\text{Gravitional force}}} = \frac{U}{\sqrt{gd}} \tag{2.7}$$

where g is the acceleration due to gravity. If $Fr < 1$, the fluid flow condition is said to be subcritical; if $Fr = 1$, critical; and if $Fr > 1$, supercritical. If $Fr > 1$, then the inertia effects outweigh the gravity effect. This dimensionless number Fr is very important to the flow with free surface, for example, in open channel flow, flow over dams or spillways.

Weber number (*We*): It is the ratio of inertia force to surface tension force. It is expressed as,

$$We = \frac{\text{inertia force}}{\text{surface tension force}} = \frac{\rho_f d\, U^2}{\sigma} \tag{2.8}$$

where σ is the surface tension of fluid (N/m). If the Weber number *We* is small, surface tension is larger and vice versa. Weber number is useful in analyzing fluid flows whenever there is an interface between two different fluids, especially for multiphase flows.

Euler number (*Eu*): The ratio of the pressure difference between the two different points to the inertia force is called the Euler number. The pressure force can be expressed in terms of its length as Δpd^2, so the Euler number *Eu* is written as:

$$Eu = \frac{\text{pressure force}}{\text{inertial force}} = \frac{\Delta pd^2}{\rho_f U^2 d^2} = \frac{\Delta p}{\rho_f U^2} \tag{2.9}$$

When the liquid flows through a pipe with governance of pressure and inertia forces, this Euler number describes the flow behaviour. It also plays a vital role in the study of cavitations, and to study the drag and lift forces.

Mach number: It is the ratio of the fluid velocity to the speed of sound in the same fluid in the same state. It is a very important dimensionless parameter useful in the analysis of compressible fluid flow. The ratio is defined as:

$$Mc = \frac{U}{c} \tag{2.10}$$

where c is the velocity of sound, which is used in the compressible fluid flow, which is not considered in this book.

Richardson number: This dimensionless number is the ratio of the buoyancy term to the flow shear term. The Richardson number is named after Lewis Fry Richardson. The flux Richardson number is traditionally defined as the ratio of the buoyancy flux to the production rate of turbulent kinetic energy. The flux Richardson number (often referred to as the mixing efficiency) is a widely used parameter in stably stratified turbulence which is intended to provide a measure of the amount of turbulent kinetic energy that is irreversibly converted to background potential energy due to turbulent mixing.

Rouse number: Rouse number $\left(z_* = \dfrac{v_0}{\kappa_0 u_*} \right)$ is defined as the ratio of settling velocity v_0 of sediment particle to the shear velocity u_* multiplied by von-Karman constant $\kappa_0 = 0.4$. This Rouse number designates the dimensionless settling velocity of particles.

Peclet number: Peclet number Pe $(= LU/D)$ measures the relative characteristic times of the diffusion process $\left(L^2/D\right)$ to the convection process (L/U), where U is the reference velocity, L is the characteristic length, and D is the molecular diffusion coefficient, which is constant.

Schmidt number: Schmidt number $(Sc = \upsilon_f/D)$ is the measure of the ratio of kinematic viscosity υ_f to the molecular diffusivity D.

Cauchy number: It is defined as the ratio between inertial and the compressibility force (elastic force) in a flow. It is used in compressible flows.

Strouhal number: It represents the ratio of inertial forces due to the local acceleration of the flow to the inertial forces due to the convective acceleration. It describes oscillating flow mechanisms.

2.7 STATISTICAL ANALYSIS

Most of the study in turbulent flow requires statistical or stochastic approach because the instantaneous motions are too complex to understand. The turbulent flow is irregular, random and complex in nature, which in general makes its manifestation to the fluids, gaseous or liquid flows. A slight change in the initial or boundary conditions may cause large changes in the solution at a given time and space; which is very difficult to anticipate without averaging. Therefore, simple statistical approaches are introduced to estimate the turbulent flow behaviour. The statistical analyses of turbulent flows are based on the idea of ensemble averaging.

The concept of an ensemble average is based on the existence of independent statistical events. Consider a number of individuals who are simultaneously flipping unbiased coins. If a value of one is assigned to a head and the value of zero is assigned to a tail, then the arithmetic mean of the numbers generated is defined as:

$$X_n = \frac{1}{n}\sum_{i=1}^{n} x_i \tag{2.11}$$

where n is the total number of flips. If all the coins are the same, it is not necessary whether one coin n time or n number of coins a single time is flipped. The key point is that they must all be *independent events* that mean, the probability of achieving a head or tail in a given flip must be completely independent, no matter what happens in all the other flips. The ensemble average of x is denoted by $\langle x \rangle$ or X and is defined as

$$X = \langle x \rangle = \frac{1}{n}\lim_{n\to\infty}\sum_{i=1}^{n} x_i \tag{2.12}$$

It is not possible to obtain the ensemble average experimentally because the infinite number of realizations is not possible. The most we can ever obtain is the arithmetic mean for the number of realizations. Therefore, the arithmetic mean can be referred to as the estimator for the true mean or ensemble mean. The time average is the average quantity of a single system over a time interval which is related to a real experiment. The ensemble average is the average quantity of many identical systems at a certain time. Some random processes have a property of ergodicity, which means that the time averaging is equivalent to ensemble average, which requires the process to be stationary, but the reverse is not true that the stationary process may not be ergodic. In general, x_i could be the realization of any random variable. The quantity $\langle x \rangle$ is defined as the ensemble average of it and is sometimes referred to as the expected value of the random variable x or simply mean.

For example, the velocity vector at a given point in space x and time t in a given turbulent flow can be considered to be a random variable, say $u_i(x, t)$. If there were a large number of identical experiments so that the $u_i^{(j)}(x, t)$ in each of them were identically distributed, then the ensemble average of $u_i^{(j)}(x, t)$ is given by

$$\langle u_i(x, t) \rangle = U_i(x, t) = \lim_{n\to\infty} \frac{1}{n} \sum_{j=1}^{n} u_i^j(x, t) \tag{2.13}$$

It is important to note that the ensemble average $U_i(x, t)$ will vary with independent variables x and t. Under certain conditions, it is seen that the ensemble average is the same as the averaging with time. Even when the time average is not meaningful, the ensemble average can be defined in stationary or periodic flow.

In turbulent flow, the fluid velocity at a fixed point in space changes with time in an irregular way and is characterized by random fluctuations. If the flow is unsteady, the mean velocity usually is meant by time-averaging. In describing the hydrodynamic quantity in the flow in a mathematical form, it is devised to separate the velocity components into a mean value (time-averaged value) and fluctuations. So, the time-averaged velocity components: stream-wise, bottom-normal (vertical) and transverse are denoted by $(\bar{u}, \bar{v}, \bar{w})$ and their fluctuations by (u', v', w') in x, y, and z directions, respectively, the following statistical relations for the instantaneous velocity components are defined as:

$$u = \bar{u} + u', \; v = \bar{v} + v', \; w = \bar{w} + w' \tag{2.14}$$

where the time-averaged velocity is obtained by integration at a fixed point in space given by

$$\bar{u} = \frac{1}{t_1} \int_{t_0}^{t_0+t_1} u dt \tag{2.15}$$

in which t_0 is any arbitrary time and time interval t_1 is the time over which the mean is taken. It is clear that the time t_1 should be sufficiently long time interval for reliable mean value and should be completely independent of time. It is easy to see that the average of fluctuations is zero, i.e., $\langle u_i' \rangle = 0$, average of square of the fluctuation is not zero, i.e., $\langle u_i'^2 \rangle \neq 0$, which is known as variance and is defined as

$$\langle u'^2 \rangle = \frac{1}{n} \sum_{i=1}^{n} (u_i - \bar{u})^2, \langle v'^2 \rangle = \frac{1}{n} \sum_{i=1}^{n} (v_i - \bar{v})^2, \langle w'^2 \rangle = \frac{1}{n} \sum_{i=1}^{n} (w_i - \bar{w})^2 \tag{2.16}$$

The root-mean-squares of the velocity components are defined as

$$\sigma_u = \sqrt{\overline{u'^2}} = \sqrt{\frac{1}{n} \sum_{i=1}^{n} (u_i - \bar{u})^2}, \sigma_v = \sqrt{\overline{v'^2}} = \sqrt{\frac{1}{n} \sum_{i=1}^{n} (v_i - \bar{v})^2}, \sigma_w = \sqrt{\overline{w'^2}} = \sqrt{\frac{1}{n} \sum_{i=1}^{n} (w_i - \bar{w})^2} \tag{2.17}$$

Here, the root-mean-squares of these quantities are the measure of the magnitudes of velocity fluctuations about the mean value, called the intensity of turbulence. Here are the three respective standard deviations of stream-wise (σ_u), bottom-normal (σ_v) and transverse (σ_w) velocity fluctuations. The m-order moment of the random variable is defined as:

$$\langle u^m \rangle = \lim_{n \to \infty} \frac{1}{n} \sum_{i=1}^{n} u_i^m \tag{2.18a}$$

The central moment is defined as:

$$\langle (u')^m \rangle = \lim_{n \to \infty} \frac{1}{n} \sum_{i=1}^{n} (u_i - \bar{u})^m \tag{2.18b}$$

For $m = 3$, it shows third-order moment. The third-order moments (skewness) provide the stochastic information relating to the flux of the Reynolds stresses. The skewness $\overline{u'^3}$ and $\overline{v'^3}$ define respectively the stream-wise flux of the stream-wise Reynolds normal stress $\overline{u'^2}$ and bottom-normal flux of the vertical Reynolds normal stress $\overline{v'^2}$, respectively. For $m = 4$, it shows the fourth-order moment. The third- and fourth-order moments show the skewness and kurtosis of the distribution, respectively. If the distribution is normal or Gaussian, the skewness (third-order moment) will be zero and kurtosis (fourth-order moment) will be 3. If the distribution is peaked less than the normal peak, i.e., kurtosis < 3, it is referred to as the leptokurtic distribution. If for a process the kurtosis > 3, the distribution becomes platykurtic (Mazumder and Das, 1992).

2.7.1 Probability

The frequency of occurrence of a given value from a finite number of observations of a random variable can be presented by dividing the range of possible values of the random variables into a number of slots or windows. Using all possible values from the random experiments, one window or slot can be arranged for each observation. For every observation, a count is made into the appropriate window. When all the observations are considered, the number of counts in each window is divided by the total number of observations. The result is termed as the histogram or frequency diagram. From the definition, it follows that the sum of the values of all the windows or slots is exactly one. The shape of the histogram or frequency diagram depends on the statistical distribution of the random variable but also depends on the total number of observations N and the size of the slot or window Δw. The histogram shows the relative frequency of occurrence of a given range in a given ensemble. A histogram or frequency diagram is shown in Figure 2.2. If the sample size is increased, the number of observations in each window also increases, implying less erratic in the diagram. That indicates the more representative of the actual probability of occurrence of the values of the variable as long as the window size is sufficiently small.

Let N be the number of observations of the random variable, Δw be the slot width or window width size, the histogram may be represented by the function $H_x(w, \Delta w, N)$, where $w \le x < w + \Delta w$ with Δw is the slot width. If N increases without any bound as the window size, $\Delta w \to 0$, the histogram divided by the window size goes to a limiting curve called the probability density function $F_x(w)$.

$$F_x(w) = \frac{\lim\limits_{\substack{N \to \infty \\ \Delta w \to 0}} H_x(w, \Delta w, N)}{\Delta w} \tag{2.19}$$

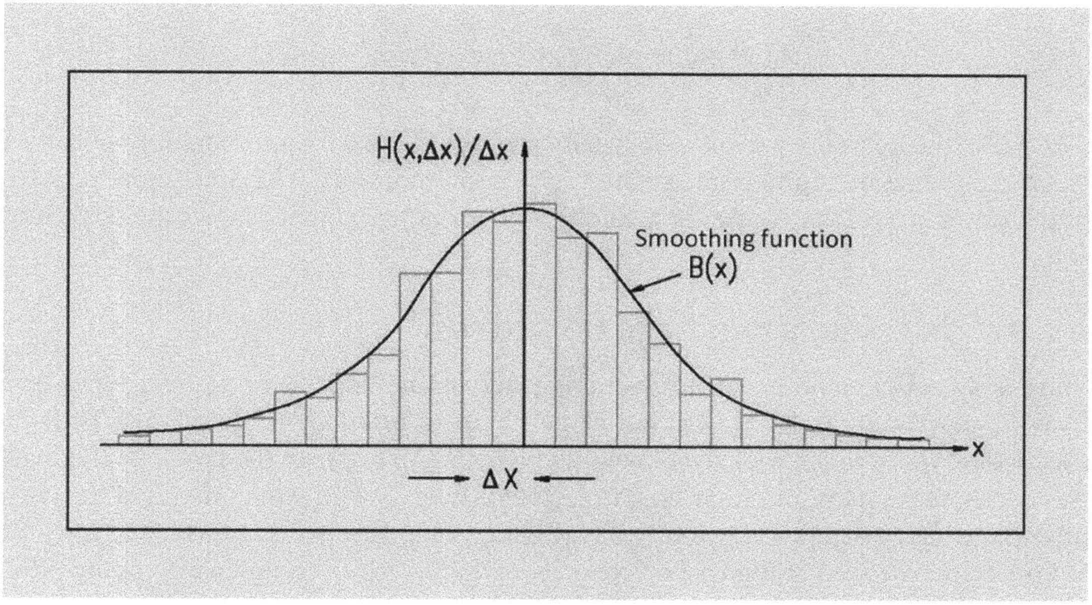

FIGURE 2.2 A typical frequency histogram.

As the window width $\Delta w \to 0$, so does the number of observations which fall into it. Thus, it is only when this number is divided by the window width that a meaningful limit is attained. The probability density function (pdf) has the following properties given below:

$$1)\ F_x(w) \geq 0 \text{ for } -\infty < w < \infty \tag{2.20a}$$

$$2)\ \text{Prob}\{w < x < w + \Delta w\} = F_x(w)dw \tag{2.20b}$$

$$3)\ \text{Prob}\{x < w\} = \int_{-\infty}^{w} F_x(w)dw \tag{2.20c}$$

$$4)\ \int_{-\infty}^{\infty} F_x(x)dx = 1 \tag{2.20d}$$

where the condition (1) simply states that negative probability is impossible, (2) indicates the probability of finding the observation in an interval around a certain value, (3) shows the probability that the observation is less than a prescribed value, and (4) corresponds to the requirement that the probability of an event that is certain to occur should be equal to unity (i.e., 1). Here $F_x(w)dw$ is the probability of random variable x assuming a value between w and $w + \Delta w$. The moment of the distribution can be computed by integrating the appropriate power of x over all possible values. The n-th moment of the distribution is given by:

$$\langle x^n \rangle = \int_{-\infty}^{\infty} w^n F_x(w)dw \tag{2.21}$$

If the probability density is given, all order moments can be obtained. The variance is determined as:

$$\text{Var}(x) = \langle (x - X)^2 \rangle = \int_{-\infty}^{\infty} (w - X)^2 F_x(w)\,dw \tag{2.22}$$

Equation (2.22) is the central moment which indicates the shape of the probability density function. Similarly, we can define the third- and fourth-order central moments. The third-order central moment represents the symmetry of the distribution, and the fourth-order moment shows the peakedness of the distribution.

2.8 CONCEPTS ON LAMINAR AND TURBULENT FLOWS

Flow of fluid is a continuous movement from one place to another. There are two types of real fluid flow: namely *laminar and turbulent*, which may be steady or unsteady (Figure 2.3). The laminar flow occurs when a fluid flows in parallel layers with no interruption between the layers, and no lateral mixing, i.e., in orderly movement. Laminar flow is a type of fluid flow in which the fluid travels smoothly or in regular paths and viscous force dominates (Figure 2.4a). However, in turbulent flow, the orderly pattern of flow ceases to exist at higher flow velocity (Figure 2.4b). The molecules in the fluid move in different directions and at different speeds. The fluid flows with highly irregular transverse motions, which lead to a complete scattering, random and fluctuating motion. The turbulent flow is characterized by eddies and swirling of fluid particles, which result in mixing (Hinze, 1959). This phenomenon often takes the form of vortices superimposed on the main flow. These turbulent motions are very common in nature. It occurs almost everywhere: in the oceans, in the atmosphere, in rivers, even in stars and galaxies. For example, when a fluid flows through a closed channel such as pipe or between two parallel plates, either of two types of flow may occur depending on the velocity and viscosity of the fluid: laminar flow or turbulent flow. Laminar flow tends to occur at lower velocities, below the threshold above which it becomes turbulent. In the laminar case, the velocity varies from zero at the walls due to no-slip condition to a maximum at the centre of the pipe or parallel plate channel (Hagen–Poiseuille flow) with cross-sectional parabolic, but for the turbulent case, due to the transfer of momentum in the transverse direction, the velocity becomes more uniform across the flow with zero velocity at the walls.

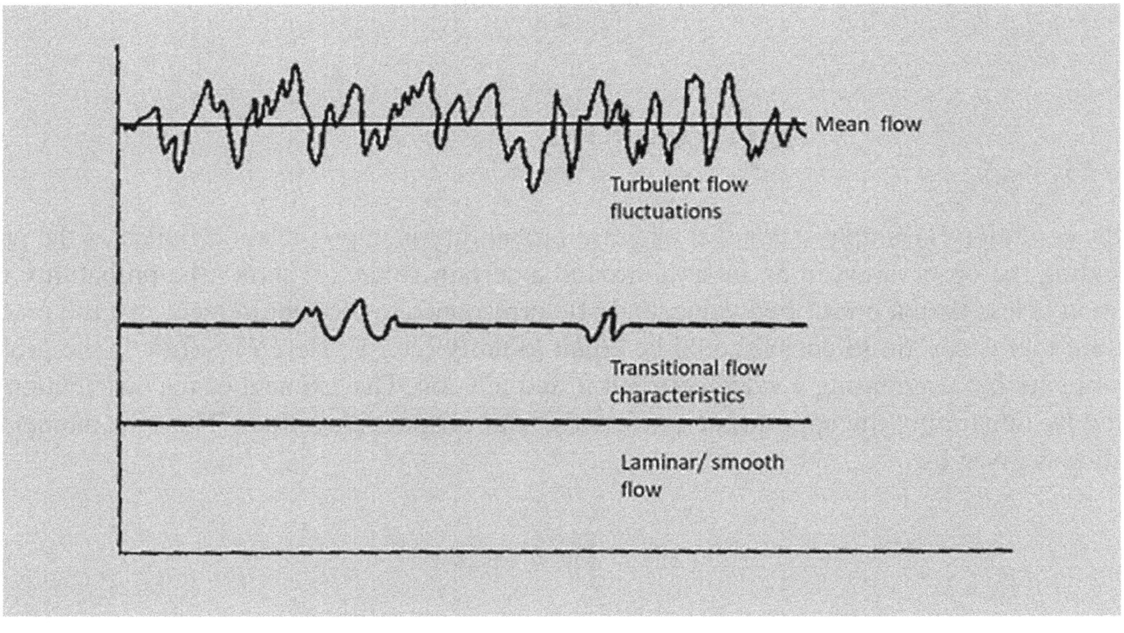

FIGURE 2.3 Typical characteristics of turbulent, transitional and laminar flows.

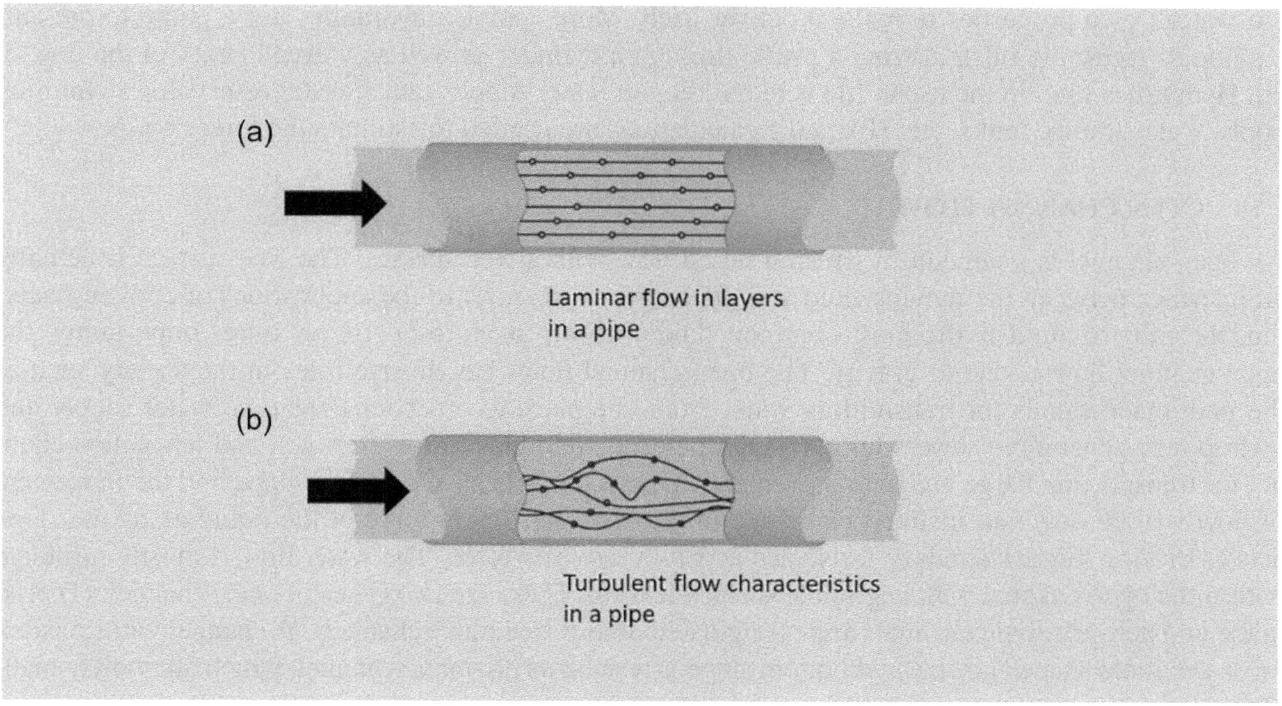

FIGURE 2.4 (a) Orderly laminar flow and (b) disorderly turbulent flow in a pipe.

It is also found that a flow through a pipe is laminar if the Reynolds Number (based on the diameter of the pipe) is less than 2,000 and is turbulent if it is greater than 4,000. Transitional flow prevails between these two limits.

Turbulence is a type of fluid flow in which the fluid experiences irregular fluctuations, or mixing, in contrast to laminar flow, in which the fluid moves orderly in smooth paths or layers. In turbulent flow, the speed of the fluid at a point is continuously changing in both magnitude and direction. Turbulent flow is visualized as an irregular swirl of motion called eddies. Usually, turbulence consists of many different sizes of eddies superimposed on each other. The relative strengths of these different scales of eddies define the turbulent spectrum.

2.9 HYDRAULICS AND HYDRODYNAMICS

Hydraulics is a branch of hydrodynamics. It deals with the conveyance of liquid through pipes and channels, generally as a source of mechanical force or control. Simple hydraulic systems include channels, aqueducts and irrigation systems that deliver water, using gravity to create water pressure. Hydraulics is a term used for studying liquids and how liquids function, but most people think of its use in engineering. In hydraulics, the principles of hydrodynamics could be directly used by engineers, applied scientists and managers to study some real practical problems of fluid flows. Therefore, realistic approaches were needed for understanding the practical problems. Most of the results from the practical problems were developed from empirical equations using curve fitting of data obtained from the experiments and field studies. In the early 20th century, the principles of hydrodynamics and hydraulics were essentially merged together from the work of L. Prandtl's boundary layer theory concept. In civil engineering, the properties and behaviour of fluid flow in different civil engineering applications, such as flow of water through canal for irrigation, flow through public supply pipelines, water drainage system and flow through open channel, are studied in hydraulics. Simple examples include calculations of fluid forces on a wall or a gate of a sluice channel, force exerted on a static fluid, etc. These systems essentially use water's own properties to make it deliver itself. More complex hydraulics use a pump to pressurize liquids (typically oils), moving a piston through a cylinder as well as valves to control the flow of oil. Hydraulics used in the urban life is in building of water supply, tanks, water reservoirs, swimming pools, water storage tanks, etc. Hydraulic calculations are needed to estimate the forces on the walls.

2.10 OPEN CHANNEL FLOW

An open channel is a conduit in which a liquid flow with a free surface. The free surface is actually an interface between the moving fluid and the constant pressure. In the applications of civil engineering, the moving fluid is the most common fluid with air at atmospheric pressure. Importantly, the open channel flow is due to gravity. The open-channel flows has diverse roles in the society such as the man-made canals for transmitting water from one place to other for irrigation, water supply and hydropower generation; sewers for carrying domestic and industrial wastewater, and navigation channel for transporting barge and ship movements. Open channels have a bottom slope and the movement of flow will be down an inclined plane due to gravity. The weight force of the liquid along the slope acts as driving force. Boundary resistance acts as a resisting force. The water flow is mostly turbulent within the open channel with negligible surface tension. There are two types of open-channel: (1) prismatic and non-prismatic channels and (2) rigid and mobile boundary channels. A channel with constant cross-sectional shape and size and bottom slope is termed as prismatic channel, which are mostly man-made channels. The channels with varying cross-sections are termed as non-prismatic, and most of the natural channels are non-prismatic. Rigid channels are those in which the boundary is not deformable, which means, the shape, bottom surface and roughness magnitudes are not the functions of flow parameters. The flow velocity and shear stress distribution will be such that no major scour, erosion or deposition takes place in the channels. The channel shape and roughness are essentially constant with respect to time. The mobile boundary channel means where the boundaries undergo deformation due to continuous erosion and deposition due to flow; subsequently, the flow carries considerable amounts of sediment through suspension. The mobile boundary channel is treated under the topic of sediment transport and sedimentation engineering, which attracts considerable attention to hydraulic engineers and their study constitutes a major area of multidisciplinary interest. The open channel flow is classified into steady and unsteady flows. In the steady-state flows, the flow parameters are constant with time, while in unsteady-state flows, the flow parameters change with time.

2.11 SEDIMENT TRANSPORT

Sediment transport is the movement of solid particles due to the fluid motion in which particles are entrained. The mechanics of sediment transport is a branch of fluid mechanics in which the processes of erosion, transportation and deposition of sediment take place under the action of flowing water, wave, wind and gravity. Microscopically, incipient motion, rolling, sliding, saltation and suspension of individual sediment particles depend on the physical properties, mineral composition of the particles and characteristics of the flow and boundary conditions. Macroscopically, sediment motion is the main direct cause of geomorphologic evolution, fluvial process and environment change. The processes of erosion of land surfaces, transportation of eroded material, deposition of this material in lakes and reservoirs depend on several factors. To understand the mechanism of sediment transport, some fundamental definitions and some characteristics of sediment and fluid flow, are needed. The study of sediment movement is not possible without considering the effects of fluid turbulence. In sediment transport, non-cohesive sediments such as silt, sand or gravel start to move after exceeding the threshold shear stress and the transportation can be classified into two broad categories: *bed load* and *suspended load* (Figure 2.5). The bed load is the movement of sediment particles very near the bed by sliding, rolling and saltation or hopping; and suspended load is the movement of the particles which are in suspension for appreciable period of time. The materials moving in suspension are being kept due to turbulent fluctuations in the fluid flow.

Many situations require the measurement of sediment concentration and particle size. Various industries require large amounts of sediment-free water. The planning of hydraulic structures such as dams, canals, etc. is practically impossible without sediment transport analysis. Sediment-free streams and reservoirs are highly valued for recreation purposes. Sediment deposition in stream or river channels can cause flooding. Therefore, the knowledge of sediment transport is practically very essential for the construction of bridge, piers, dams, reservoirs, and river management.

The transport of sediment particles in natural rivers, irrigation canals, pipes, and drainage ditches has fundamental importance and, for more than a century, has drawn a lot of attention to hydraulic engineers, coastal engineers, geologists, hydrologists, physical geographers, soil scientists, and other researchers in this domain. Numerous investigations related to sediment-laden turbulent flows in open channels have been undertaken to study the vertical velocity, bed-load and suspended sediment concentration profiles, and grain-size distributions. Here Figure 2.6 shows the flow pattern over the gravel bed in a river.

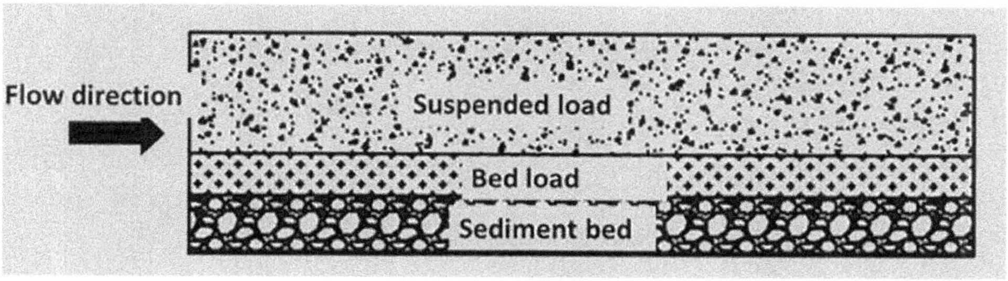

FIGURE 2.5 Sediment transport in a channel with sand bed, bed-load and suspension.

FIGURE 2.6 The flow behaviour in a river of gravel bed surface.

2.12 SUMMARY

Some of the important points discussed in this chapter are summarized below.

- In fluid mechanics, we consider the fluid motion in continuum. Continuum mechanics deals with the motion of a fluid involving the mechanical behavior of all discrete molecules which make up the fluid.
- Some of the important fluid properties such as fluid pressure, density, specific weight, specific volume, specific gravity, fluid compressibility, viscosity, temperature, shear stress, fluid velocity and acceleration govern the fluid motion and are very fundamental considerations in fluid dynamics and fluvial hydraulics.
- There are several important dimensionless numbers in fluid dynamics such as Reynolds number, Froude number, Euler number, Weber number, Mach number, Richardson number, Peclet number etc. By identifying the values of these numbers, we can identify the type of flow and various flow characteristics.
- In most of the fluid dynamics and fluvial hydraulics problems, the most important dimensionless numbers to identify the flow characteristic are Reynolds number and Froude number.
- In fluid dynamics and fluvial hydraulics, we use statistical analysis and probability theories to analyze turbulent flows and sediment transport due to its highly chaotic and random nature.
- Normal fluid flow can be laminar or turbulent and using the Reynolds number, we can identify the flow behavior.
- Hydraulics deals with the conveyance of liquid through a container such as pipes and channels, under the action of a force, mostly under controlled conditions. The fluvial hydraulics deals with transport of all types of fluids in channels or a conduit and the analysis of the flow characteristics.
- In fluvial hydraulics, the mechanics of sediment transport deals with the processes of erosion, transportation and deposition of sediment take place under the action of flowing water, wave, wind and gravity. In sediment transport mechanism, we need to understand the incipient motion, rolling, sliding, saltation and suspension of individual sediment particles that depend on the physical properties, mineral composition of the particles and characteristics of the flow and boundary conditions.

2.13 EXERCISE PROBLEMS

1. Explain the important fluid properties that govern the fluid dynamics and fluvial hydraulics.
2. Illustrate the important dimensionless numbers used in fluid mechanics with its definition and its applications.
3. Which are the most important dimensionless numbers used in the analysis of open channel and pipe flows?
4. What are the applications of statistical analysis and probability theories in fluid mechanics? Why it is essential for turbulent flow analysis?
5. Differentiate between laminar and turbulent flows?
6. In fluvial hydraulics, what are the important issues to be considered? In sediment transport, what are the important aspects to be considered in microscopically and macroscopically?

3 Elementary Fluid Kinematics and Dynamics

3.1 INTRODUCTION

In kinematics of fluids, various aspects of fluid motion are considered without concerning the actual forces and moments that cause the fluid motion. Thus, the kinematics of fluid motion deals with the geometry of the flow, which provides descriptions of the position, velocity, and acceleration of fluid particles in motion. The motion of a fluid can be analyzed using the same principles as those applied to the motion of a solid. However, there is a basic difference between the motion of a solid and the motion of a fluid. A solid body is compact and moves as one mass, and there is no relative motion between the particles. Therefore, we study the motion of the entire body. It is known that each particle of a fluid in motion has a definite value for its properties like density, velocity, acceleration, etc. As the fluid moves, the values of these properties will change from one position to another, from time to time. There are two approaches to describing fluid motion, namely, Lagrangian approach and Eulerian approach. In the Lagrangian approach, we study the velocity, acceleration, etc. of an individual fluid particle at every instant of time as the particle moves to different positions. In the Eulerian approach, we describe the flow, studying the velocity, acceleration, pressure, density, etc. at a fixed point in space. Due to the ease of application, Eulerian approach is most commonly adopted.

The dynamics of fluids deal with fluids in motion. The actual forces and moments in fluids are used to consider the various aspects of fluid motion. The dynamics of fluids are the foundation of the understanding of fluid flows in streams, pipes, and the subsurface. The fundamental principles of motion are applied to the flow of liquids or gases, namely, conservation of mass, conservation of momentum (known as Newton's second law of motion), and conservation of energy (from the first law of thermodynamics). The property of a fluid describes the resistance to motion under applied shear stress termed as viscosity, which may be a function of temperature. In this chapter, various aspects of kinematics of fluids, including the streamlines, streak lines, and path lines, are described with some examples. The concepts of vorticity, rotational, and irrotational flows are discussed. Further, the elementary fluid dynamic aspects, including conservation equations of mass, momentum, and energy, are discussed with examples.

3.2 GENERAL DESCRIPTIONS OF FLUID MOTION

Various types of flow in fluid mechanics are illustrated according to the fluid properties, which characterize the physical problems and their conditions. The geometry of the flow provides a description of the position, velocity, and acceleration of the system of fluid particles without considering the force and momentum, called as the *kinematics of fluid motion*. The rate of change of position at a point is called the velocity, and the rate of change of velocity is called as the acceleration of a fluid particle. To gain a complete picture of fluid flow, one must understand the position of every particle of fluid at every instant of time. The velocity and acceleration of a fluid particle are obtained from its change in position and velocity with time. The kinematics of flow illustrates the flow pattern. Once the flow pattern is recognized, the pressure and forces that act on a submerged body in the fluid can be determined. The motion of fluid particles can be identified using two important approaches (Ramsey, 1920): one is the Lagrangian approach, which describes the motion of each particle of fixed identity for all time along the path of the particle motion, i.e., the motion of a single fluid particle in the system, and the

DOI: 10.1201/9781003000020-3

other one is the Eulerian approach, which describes the motion of a fluid at every point with time at a particular location in a system without considering the individual particle. The Eulerian approach identifies a point or region in the system, i.e., a control volume is fixed at a fixed point and the particle passing through it is measured, whereas the Lagrangian approach follows the motion of an individual particle, which is used in the rigid-body dynamics. In most of the fluid flow applications, the Eulerian approach is generally used for analysis. If the fluid velocity at different points in space is independent of time, the fluid flow is called *steady*. On the other hand, if the velocity at the considered point is a function of time, then the flow is called *unsteady*. If an unsteady flow is in one reference space, it can be transferred to a steady flow in another reference space. For example, consider a uniform flow past an infinitely long cylinder at rest. If an observer at a fixed position looks the motion of the fluid to be steady, then the velocity at any point in space should remain invariant with time, i.e., there is no change with time; whereas when the cylinder moves with a uniform velocity with respect to the fluid at rest, the flow becomes unsteady with respect to the observer at the same point.

3.2.1 Lagrangian Approach

If the position vector $\vec{r} = (x,\ y,\ z)$ of a particle is a function of time t and the initial position vector

$$\vec{\xi} = (x_0,\ y_0,\ z_0) \text{ then } \vec{r} = F\left(\vec{\xi},\ t\right) \tag{3.1}$$

or,

$$x = F_1(x_0,\ y_0,\ z_0,\ t),\ y = F_2(x_0,\ y_0,\ z_0, t),\ z = F_3(x_0,\ y_0,\ z_0,\ t) \tag{3.2}$$

This set of equations gives the position of fluid particles at different times. The first and second-order partial derivatives with respect to time give the velocity and acceleration, respectively. The velocity and acceleration of a fluid particle at the position $\vec{\xi} = (x_0,\ y_0, z_0)$ can be obtained as:

$$u = \left(\frac{\partial x}{\partial t}\right)_\xi = \left(\frac{\partial F_1}{\partial t}\right)_\xi,\ v = \left(\frac{\partial y}{\partial t}\right)_\xi = \left(\frac{\partial F_2}{\partial t}\right)_\xi,\ w = \left(\frac{\partial z}{\partial t}\right)_\xi = \left(\frac{\partial F_3}{\partial t}\right)_\xi \tag{3.3}$$

and

$$\frac{\partial u}{\partial t} = \left(\frac{\partial^2 x}{\partial t^2}\right)_\xi,\ \frac{\partial v}{\partial t} = \left(\frac{\partial^2 y}{\partial t^2}\right)_\xi,\ \frac{\partial w}{\partial t} = \left(\frac{\partial^2 z}{\partial t^2}\right)_\xi \tag{3.4}$$

Here subscript ξ means the differentiation must be carried out at a fixed position $\vec{\xi} = (x_0,\ y_0,\ z_0)$. If the motion and trajectory of every fluid particle are known, it is possible to determine the behaviour of the fluid particle at any time.

3.2.2 Eulerian Approach

In the Eulerian approach, the individual fluid particles are not identified. A fixed position in space is chosen, and the velocity of the particle at that position is a function of time. Mathematically, the velocity of a particle at any point in space can be written as follows:

$$\vec{q} = \vec{f}(\vec{r},\ t) \tag{3.5}$$

where $\vec{q} = (u, v, w)$ is the velocity vector with the expressions as:

$$u = f_1(x, y, z, t),\ v = f_2(x, y, z, t),\ w = f_3(x, y, z, t) \tag{3.6}$$

The definite relationship between these two methods is obtained by using the sets of equations (3.3) for Lagrangian, and (3.6) for Eulerian:

$$\begin{aligned} \frac{dx}{dt} &= u(x, y, z, t) = f_1(x, y, z, t) \\ \frac{dy}{dt} &= v(x, y, z, t) = f_2(x, y, z, t) \\ \frac{dz}{dt} &= w(x, y, z, t) = f_3(x, y, z, t) \end{aligned} \tag{3.7}$$

Integration of equation (3.7) leads to three constants of integration, which are considered as the initial coordinates (x_0, y_0, z_0) of the fluid particle. Hence, the solutions of the set of equation (3.7) give the Lagrangian form of equations as:

$$x = F_1(x_0, y_0, z_0, t),\ y = F_2(x_0, y_0, z_0, t),\ z = F_3(x_0, y_0, z_0, t) \tag{3.8}$$

In principle, the Lagrangian approach can be derived from the Eulerian approach, using equation (3.7). The solutions of three simultaneous differential equation (3.7) are generally very difficult. In the Eulerian approach, the basic variables are the velocity components of fluid.

In general, the movement of a fluid element may consist of translation, rotation, and deformation. Let us consider the movement of a fluid element at a point P (x, y, z) with a velocity vector $\vec{q} = \vec{q}(\vec{r}, t)$, and at a small distance $\overrightarrow{dr}$ from P, the velocity is $\vec{q_1}$. The velocity vector $\vec{q_1}$ due to the movement of the fluid element from one point to another at small distance $\overrightarrow{dr}$ represents the translational velocity vector without change of shape, rigid rotation, and the rate of strain, i.e., rate of deformation. Any deformation of a continuum may be accomplished by two consecutive processes, which are independent from each other. The first one is a simple extension or compression, that is, normal strain. The second one is the shear strain, which measures the changes in the shape of the element. The normal strains are $\frac{\partial u}{\partial x}, \frac{\partial v}{\partial y}, \frac{\partial w}{\partial z}$, and the shearing strains are $\frac{\partial u}{\partial y} + \frac{\partial v}{\partial x}, \frac{\partial v}{\partial z} + \frac{\partial w}{\partial y}, \frac{\partial w}{\partial x} + \frac{\partial u}{\partial z}$.

3.3 STREAMLINES AND PATHLINES

The streamlines are the lines drawn through the fluid in such a way that they indicate the direction of the velocity of the fluid particles located on the lines at a particular instant of time. The velocity of fluid particle is always tangent to the streamline at that instant. Figure 3.1a shows a streamline ABC. The velocity at A is along the tangent to the streamline at A. The velocity at B is also along the tangent to the streamline at B. In case of two-dimensional flow, u and v are two velocity components along x and y directions. Figure 3.1b shows a number of streamlines AB, CD, and EF. For clarity, fluid particles on the streamline CD are also shown, and these are moving along CD around a cylinder. No fluid particle can flow across the streamlines; it can only move along the streamline. The streamline touches the cylinder at a point O. The point O is called stagnation point, where the velocity reduces to zero, and the pressure is maximum. For steady flow, particles move along fixed streamlines, whereas for an unsteady flow, direction of streamlines changes, and consequently the particles have different orientations with time.

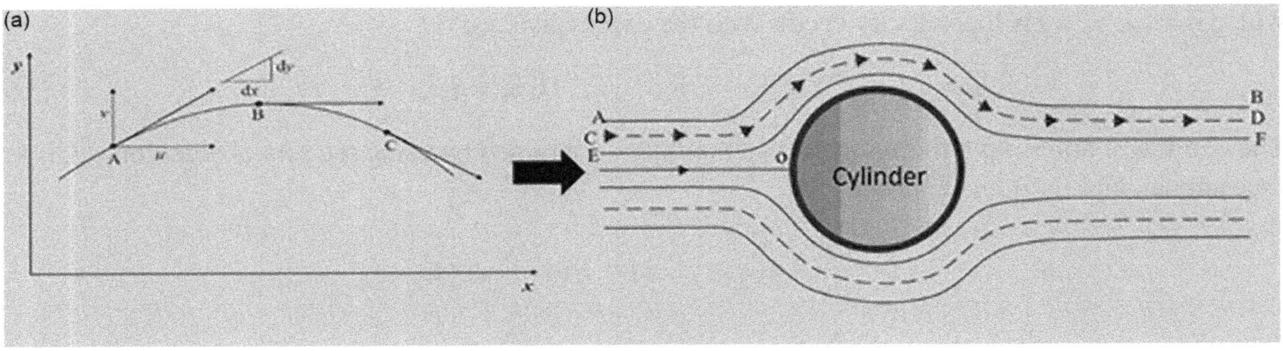

FIGURE 3.1 (a) Definition sketch of streamlines; (b) Typical streamlines around a cylinder.

The streamlines are used to represent the flow field during each instant of time. The Eulerian description gives a series of instantaneous streamlines for a state of motion, and the flow pattern changes with each instant of time.

Let the coordinates of a streamline be

$$d\vec{r} = \hat{i}dx + \hat{j}dy + \hat{k}dz \tag{3.9}$$

From the condition of streamline, we can write

$$\vec{q} \times d\vec{r} = 0 \tag{3.10}$$

at a given time. The differential equations of streamlines are as follows:

$$\frac{dx}{u} = \frac{dy}{v} = \frac{dz}{w} \tag{3.11}$$

For two-dimensional flow, the velocity of particles will have two components, u and v in the x and y directions, respectively. The particle velocity is always tangent to the streamline (Figure 3.1a). A circumferential region surrounded by a bundle of streamlines is called a *stream tube* (Figure 3.2). The fluid flows through a stream tube as if these are contained within a curved conduit AB. In two dimensions, the stream tube is formed between any two streamlines (Som and Biswas, 1998).

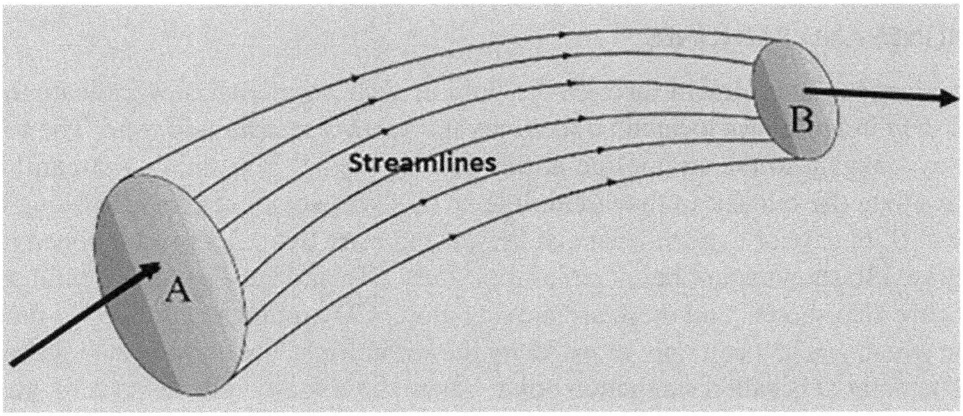

FIGURE 3.2 Concept of stream tube.

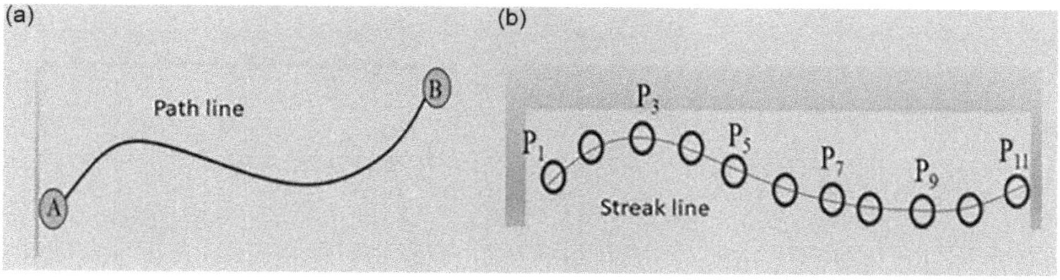

FIGURE 3.3 Definition sketch: (a) Path line; (b) Streak line.

There are two kinds of lines which are closely related to streamlines that are as follows: *pathlines* and *streaklines* (Hibbeler, 2015). The pathline of a fluid particle describes a path A to B through which the particle travels over a period of time (Figure 3.3a) and is given by equation (3.8). The pathline shows a path of a single fluid particle in a flow stream using a time-exposed photograph. In a steady flow, it is identical to the streamline passing through the point (x_0, y_0, z_0). The *streakline* is the line joining all fluid particles which pass through a given point in space. The streakline shows a path of many particles such as $P_1, P_2, \ldots P_{11}$ at an instant of time (Figure 3.3b), for example, the release of smoke in a gas or coloured dye in a liquid. In a steady flow, the streakline coincides with streamline, but in unsteady flow, all three lines (streamline, path line, and streak line) are, in general, different. Streak lines are very useful in flow visualization (Hibbeler, 2015). Timeline is generated by drawing a line through adjacent particles in flow at any instant of time.

Example 3.1

The components of velocity of a fluid particle are defined by $u = 4$ m/s and $v = 8t$ m/s, where t is the time in second. (a) Find the equation of path line of the fluid particle if it is released from the origin at time $t = 0$. (b) Find also the equation of streamlines for the particles at time $t = 2$ s.

a. **Equation of path line**: This problem is on Lagrangian concept because the velocity is only on function of time. The particle is released from origin (0, 0) at time $t = 0$. The velocity components $u = \frac{dx}{dt} = 4$ and $v = \frac{dy}{dt} = 8t$ (Figure 3.4). Integrating the equations, one gets as

$\int_0^x dx = 4\int_0^t dt$ and $\int_0^y dy = 8\int_0^t t\, dt$; which imply $x = 4t$ m and $y = 4t^2$ m

Eliminating t from the above two equations, one gets $y = \frac{x^2}{4}$. Therefore, the equation of path line is $y = \frac{x^2}{4}$, which is a parabola and describes the location of a particle (Figure 3.4).

b. **Equation of streamline**: To find the equation of streamline at time $t = 2$ second. The location of particle at time $t = 2$ second is $x = 8$ m, $y = 16$ m, and it has the velocity components $u = 4$ m/s and $v = 16$ m/s. (Figure 3.4). To obtain the equation of streamline at this instant, we have

$\frac{dy}{dx} = \frac{v}{u} = \frac{16}{4} = 4$, after integration, it gives $\int dy = \int 4dx$ or $y = 4x + C$

where C is the integrating constant determined from the conditions of (x, y). For $x = 8$, $y = 16$, the constant $C = -16$ m. So, the equation $y = 4x - 16$ is the equation of streamline for the problem and note that both streamline and path line have the same slope at (8 m, 16 m). This is to be expected since both must give the same direction for the velocity at time $t = 2$ s.

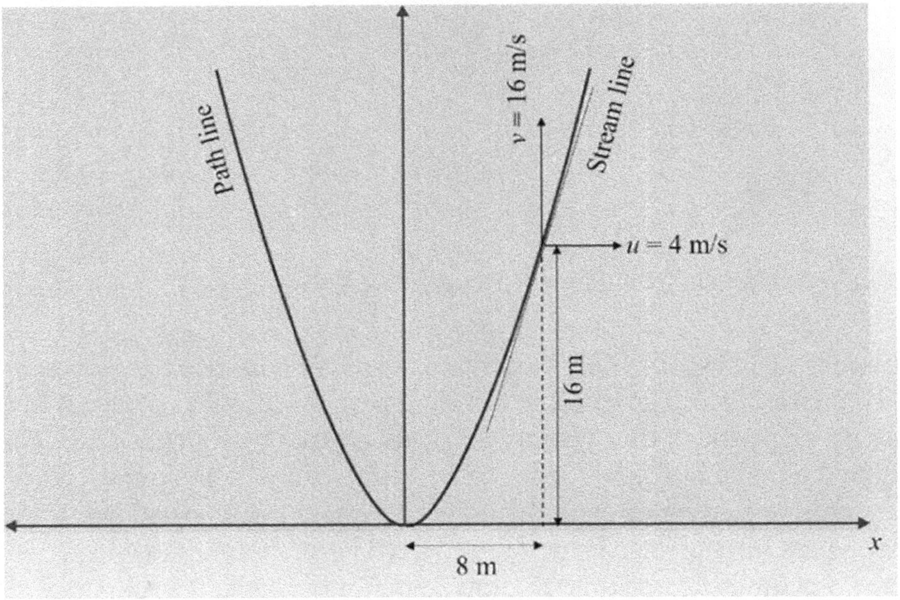

FIGURE 3.4 Parabola with location of particle.

3.4 VORTICITY

In fluid dynamics, vorticity is a vector quantity, and it shows a tendency of a fluid particle to rotate or circulate at a particular point (Aris, 1990; Yuan, 1970). Vorticity is a measure of local spin of the fluid element, which is defined as the curl of the velocity vector field. It shows how the velocity vector changes when fluid moves by an infinitesimal distance in a direction perpendicular to it. Mathematically, the vorticity of a three-dimensional flow is a pseudo-vector field, usually denoted by $\vec{\omega}$, defined as the curl or rotational of the velocity field $\vec{q}$ describing the continuum motion. In Cartesian coordinates, the vorticity $\vec{\omega}$ of the fluid is the curl of velocity vector $\vec{q}$, i.e.,

$$\vec{\omega} = \vec{\nabla} \times \vec{q} = \hat{i}\omega_x + \hat{j}\omega_y + \hat{k}\omega_z \tag{3.12}$$

where $\omega_x = \frac{\partial w}{\partial y} - \frac{\partial v}{\partial z}$, $\omega_y = \frac{\partial u}{\partial z} - \frac{\partial w}{\partial x}$, $\omega_z = \frac{\partial v}{\partial x} - \frac{\partial u}{\partial y}$

In a two-dimensional flow, where the velocity is independent of the z coordinate, the vorticity vector is always parallel to the z axis, and therefore can be expressed as a scalar field multiplied by a constant unit vector:

$$\vec{\omega} = \vec{\nabla} \times \vec{q} = \hat{k}\omega_z = \hat{k}\left(\frac{\partial v}{\partial x} - \frac{\partial u}{\partial y}\right) \tag{3.13}$$

where $\omega_z = \frac{\partial v}{\partial x} - \frac{\partial u}{\partial y}$

Vorticity measures the rotation of very small air parcels. A parcel has vorticity when it rotates on its axis as it moves along its path. A parcel that does not rotate on its axis is said to have zero vorticity.

The zero-vorticity flow is called the *irrotational flow*, means when the fluid element has no rotation. The equation is given by

$$\vec{\omega} = \vec{\nabla} \times \vec{q} = 0 \tag{3.14}$$

which means $\frac{\partial v}{\partial x} = \frac{\partial u}{\partial y}$. Similarly, for other two components, $\frac{\partial w}{\partial y} = \frac{\partial v}{\partial z}$ and $\frac{\partial u}{\partial z} = \frac{\partial w}{\partial x}$.

In other words, the vorticity of flow field is zero except at some singular points. For irrotational flow, a velocity potential ϕ is defined such that $\vec{q} = \nabla\phi$, where ϕ is a scalar function of $x,\ y,\ z$. From equation (3.14), one gets

$$\vec{\omega} = \nabla \times \vec{q} = \nabla \times \nabla\phi = 0, \tag{3.15}$$

Here the condition of irrotationality is satisfied for a potential flow. Hence the irrotational flow is called a potential flow. Since the velocity vector $\vec{q}$ is the gradient of the velocity potential ϕ, the streamline is perpendicular to the equi-potential line. The vorticity may be non-zero even when all particles are flowing along straight and parallel pathlines, if the flow speed varies across streamlines due to shear. For example, in the laminar flow within a pipe of constant cross section, all particles travel parallel to the axis of the pipe, with faster near the axis and practically zero near the walls (Figure 3.5). The vorticity will be zero on the axis, and maximum near the walls due to the largest shear at the walls. On the contrary, a flow may have zero vorticity even though fluid particles travel along curved trajectories. Irrotational vortices exist, in which most particles rotate about a straight axis at a speed that is inversely proportional to their distance from the axis.

A small bundle of continuum that does not straddle the axis will be rotated in one direction but sheared in the opposite sense, in such a way that their mean angular velocity about their center of mass is zero. For example, we could drop a chip of wood into a creek and watch its progress. The chip will move downstream with the flow of water, but it may or may not rotate as it moves downstream. If it does rotate, the chip has vorticity. In two ways, we can check the rotation of chip: (1) if the flow of water is moving faster on one side of the chip than the other, this is shear of the current; (2) if the bottom of the inlet is bent, the path has curvature. Figure 3.5a shows no change in velocity in fluid flow, indicating no vorticity (curl of the velocity vector is zero, i.e., $\nabla \times \vec{q} = 0$), whereas Figure 3.5b shows a change in velocity (non-uniform), and hence there is a gradient in velocity mostly near the boundary, so there is vorticity (i.e., $\nabla \times \vec{q} \neq 0$). Therefore, a uniform flow field (Figure 3.5a) with no velocity gradients is certainly an example of an irrotational flow. It is known that for an inviscid fluid there are no shearing stresses, only weight and pressure forces act on the fluid element. The weight acts through the center of gravity of the element, and pressure acts in the direction normal to the element surface, so neither of these forces can cause the element to rotate.

3.4.1 Relation between Vorticity and Angular Velocity

The vorticity has three distinct elements like components of a vector in Cartesian coordinates; the vorticity is treated as vector quantity (equation 3.12).

The angular velocities of the segments Δy and Δz about x-axis are as follows (Yuan, 1970):

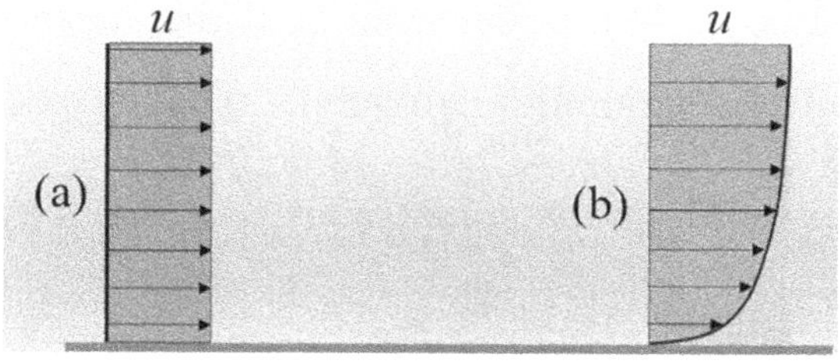

FIGURE 3.5 (a) Uniform velocity, vorticity = 0; (b). Non-uniform velocity, vorticity ≠ 0.

Angular velocity of segment $\Delta y = \dfrac{w + \left(\dfrac{\partial w}{\partial y}\right)\Delta y - w}{\Delta y} = \dfrac{\partial w}{\partial y}$, and angular velocity of segment $\Delta z = -\dfrac{v + \left(\dfrac{\partial v}{\partial z}\right)\Delta z - v}{\Delta z} = -\dfrac{\partial v}{\partial z}$

Therefore, the average angular velocity component Ω_x in the x-direction is

$$\Omega_x = \frac{1}{2}\left(\frac{\partial w}{\partial y} - \frac{\partial v}{\partial z}\right) \tag{3.16a}$$

Similarly, the average angular velocity components in y and z -directions may be obtained as

$$\Omega_y = \frac{1}{2}\left(\frac{\partial u}{\partial z} - \frac{\partial w}{\partial x}\right), \quad \Omega_z = \frac{1}{2}\left(\frac{\partial v}{\partial x} - \frac{\partial u}{\partial y}\right) \tag{3.16b,c}$$

Finally, the angular velocity vector $\vec{\Omega}$ of a fluid element in terms of velocity field is as follows:

$$\vec{\Omega} = \frac{1}{2}\left[\hat{i}\left(\frac{\partial w}{\partial y} - \frac{\partial v}{\partial z}\right) + \hat{j}\left(\frac{\partial u}{\partial z} - \frac{\partial w}{\partial x}\right) + \hat{k}\left(\frac{\partial v}{\partial x} - \frac{\partial u}{\partial y}\right)\right] \tag{3.17}$$

Combining equations (3.12) and (3.17), we have the following relation between the vorticity vector $\vec{\omega}$ and the angular velocity vector $\vec{\Omega}$ as follows:

$$\vec{\Omega} = \frac{1}{2}\vec{\omega} \tag{3.18}$$

Therefore, the vorticity vector $\vec{\omega}$ of a fluid element is the twice of angular velocity vector $\vec{\Omega}$ at that point. If a line is drawn so that the tangent to the line at each point in the direction of vorticity vector $\vec{\omega}$ at that point, the line is called the vortex line. If $\overrightarrow{dl}$ is the direction of vorticity vector, then

$$\vec{\omega} \times \overrightarrow{dl} = 0 \tag{3.19}$$

From this, equations of vortex lines are also obtained as follows:

$$\frac{dx}{\omega_x} = \frac{dy}{\omega_y} = \frac{dz}{\omega_z} \tag{3.20}$$

The equations of vortex lines (3.20) are similar to the equation of streamlines (3.11).

It may be noted that for a two-dimensional flow field in xy-plane, the vorticity components (ω_x, ω_y) will be always zero. By definition of two-dimensional flow, the velocity components u and v are not functions of z, and w is zero. The condition of irrotationality simply becomes $\omega_z = \dfrac{\partial v}{\partial x} - \dfrac{\partial u}{\partial y} = 0$, that means, $\dfrac{\partial v}{\partial x} = \dfrac{\partial u}{\partial y}$. So, it implies that both the components rotate with the same speed but in opposite directions so there is no rotation in the fluid element.

Example 3.2

For a certain two-dimensional flow, the velocity vector is given by equation $\vec{u} = \{6x^2y,\ 2(x^3 - y^3)\}$. What will be flow field, and identify whether this is rotational or irrotational?

Solution

For an irrotational flow, the vorticity vector $\vec{\omega} = \vec{\nabla} \times \vec{q} = \hat{i}\omega_x + \hat{j}\omega_y + \hat{k}\omega_z = 0$, where $\omega_x = \frac{\partial w}{\partial y} - \frac{\partial v}{\partial z}$, $\omega_y = \frac{\partial u}{\partial z} - \frac{\partial w}{\partial x}$, $\omega_z = \frac{\partial v}{\partial x} - \frac{\partial u}{\partial y}$

From the given problem, $u = 6x^2y$, $v = 2(x^3 - y^3)$, $w = 0$, so after differentiating, one gets

$$\omega_x = 0,\ \omega_y = 0,\ \omega_z = \frac{\partial v}{\partial x} - \frac{\partial u}{\partial y} = 6x^2 - 6x^2 = 0$$

Here all the vorticity components are zero. Thus, the flow field is irrotational.

3.5 MATERIAL DERIVATIVE AND ACCELERATION

The velocity of fluid at any point in the space can be written as

$$\vec{q} = \vec{f}(\vec{r}, t) \tag{3.21}$$

The velocity in x-direction at time t is in the form

$$u = f_1(x, y, z, t) \tag{3.22}$$

At an infinitesimal time Δt, the particle will move to the position $x + \Delta x,\ y + \Delta y,\ z + \Delta z$ with x-component of velocity $u + \Delta u$. The increments $\Delta x,\ \Delta y,\ \Delta z$ are

$$\Delta x = u\Delta t,\ \ \Delta y = v\Delta t,\ \ \Delta z = w\Delta t \tag{3.23}$$

We have $u + \Delta u = f_1(x + u\Delta t,\ \ y + v\Delta t,\ \ z + w\Delta t,\ \ t + \Delta t)$

$$= f_1(x, y, z, t) + \left(u\frac{\partial f_1}{\partial x}\Delta t + v\frac{\partial f_1}{\partial y}\Delta t + w\frac{\partial f_1}{\partial z}\Delta t + \frac{\partial f_1}{\partial t}\Delta t\right) + \cdots$$

Taking the limiting form, we get

$$\lim_{\Delta t \to 0}\frac{\Delta u}{\Delta t} = \frac{Du}{Dt} = \left(\frac{\partial}{\partial t} + u\frac{\partial}{\partial x} + v\frac{\partial}{\partial y} + w\frac{\partial}{\partial z}\right)u \tag{3.24}$$

The total derivative with respect to time

$$\frac{D}{Dt} = \frac{\partial}{\partial t} + u\frac{\partial}{\partial x} + v\frac{\partial}{\partial y} + w\frac{\partial}{\partial z} \tag{3.25}$$

Equation (3.25) is known as substantive/material derivative with respect to time. The first term at the right-hand side of equation (3.25) is the time variation of fluid particle at a fixed position and is called the local derivative and the last three terms are associated with change of positions of fluid particle, called as the convective derivative. In Cartesian co-ordinates (x, y, z) with the velocity components (u, v, w), the acceleration components are given by (Yuan, 1970),

$$A_x = \frac{\partial u}{\partial t} + u\frac{\partial u}{\partial x} + v\frac{\partial u}{\partial y} + w\frac{\partial u}{\partial z}$$

$$A_y = \frac{\partial v}{\partial t} + u\frac{\partial v}{\partial x} + v\frac{\partial v}{\partial y} + w\frac{\partial v}{\partial z} \tag{3.26}$$

$$A_z = \frac{\partial w}{\partial t} + u\frac{\partial w}{\partial x} + v\frac{\partial w}{\partial y} + w\frac{\partial w}{\partial z}$$

In cylindrical co-ordinate (r, θ, z) with velocity components (u_r, u_θ, u_z), the acceleration components are given by (Yuan, 1970):

$$a_r = \frac{\partial u_r}{\partial t} + u_r\frac{\partial u_r}{\partial r} + \frac{u_\theta}{r}\frac{\partial u_r}{\partial \theta} + u_z\frac{\partial u_r}{\partial z} - \frac{u_\theta^2}{r}$$

$$a_\theta = \frac{\partial u_\theta}{\partial t} + u_r\frac{\partial u_\theta}{\partial r} + \frac{u_\theta}{r}\frac{\partial u_\theta}{\partial \theta} + u_z\frac{\partial u_\theta}{\partial z} + \frac{u_r u_\theta}{r} \tag{3.27}$$

$$a_z = \frac{\partial u_z}{\partial t} + u_r\frac{\partial u_z}{\partial r} + \frac{u_\theta}{r}\frac{\partial u_z}{\partial \theta} + u_z\frac{\partial u_z}{\partial z}$$

The substantive/material derivative is used to determine the acceleration of a particle when the velocity field is known. It consists of two parts: local time derivative within the control volume, and the convective derivative due to the particle movement in and out of the control surface.

3.6 EQUATION OF CONTINUITY (CONSERVATION OF MASS)

Equation of continuity is a fundamental equation of fluid mechanics based on the conservation of mass of the fluid. Let u, v, w be the x, y, z components of velocity respectively with ρ_f is the density of the fluid at a point $A\ (x, y, z)$ in space at a given time t (Ramsey, 1920). Let us consider a parallelepiped with sides dx, dy and dz from a point $A(x, y, z)$. The mass flowing out of the parallelepiped in the x-direction is

$$\left(\rho_f u + \frac{\partial\, \rho_f u}{\partial x}dx\right)dydz - \rho_f u dydz = \frac{\partial\, \rho_f u}{\partial x}dxdydz \tag{3.28}$$

Similarly, in the y- and z-directions respectively as

$$\frac{\partial\, \rho_f v}{\partial y}dxdydz, \text{ and } \frac{\partial\, \rho_f w}{\partial z}dxdydz \tag{3.28a}$$

The total amount of mass flowing out at the parallelepiped is then

$$\left(\frac{\partial\, \rho_f u}{\partial x} + \frac{\partial\, \rho_f v}{\partial y} + \frac{\partial\, \rho_f w}{\partial z}\right)dxdydz \tag{3.29}$$

By the conservation of mass, the expression (3.29) must be equal to the time rate of change of mass in the parallelepiped, which consists of two points: one is the change at density of fluid in parallelepiped and other is the increase at mass due to a source σ_0 per unit volume.

That means

$$\left(\frac{\partial \rho_f u}{\partial x}+\frac{\partial \rho_f v}{\partial y}+\frac{\partial \rho_f w}{\partial z}\right)dxdydz=-\frac{\partial \rho_f}{\partial t}dxdydz+\sigma_0 dxdydz \tag{3.30}$$

which implies

$$\frac{\partial \rho_f}{\partial t}+\nabla\cdot\left(\rho_f\vec{q}\right)=\sigma_0 \tag{3.31}$$

where σ_0 is the source term. Here $\sigma_0=0$ except otherwise specified. Then

$$\frac{\partial \rho_f}{\partial t}+\nabla\left(\rho_f\vec{q}\right)=0$$

Since, $\rho_f\left(\nabla\cdot\vec{q}\right)+\left(\vec{q}\cdot\nabla\right)\rho_f=\nabla\left(\rho_f\vec{q}\right)$, the above equation can be written as:

$$\frac{\partial \rho_f}{\partial t}+\left(\vec{q}\cdot\nabla\right)\rho_f=-\rho_f\left(\nabla\cdot\vec{q}\right)$$

$$\frac{D\rho_f}{Dt}=-\rho_f\left(\nabla\cdot\vec{q}\right) \tag{3.32}$$

Equation (3.32) is equation of continuity of fluid mechanics with the term $\frac{D}{Dt}$ is known as the material derivative. In Cartesian coordinates, the above equation (3.32) shows for compressible fluid as:

$$\frac{\partial \rho_f}{\partial t}+\frac{\partial\left(\rho_f u\right)}{\partial x}+\frac{\partial\left(\rho_f v\right)}{\partial y}+\frac{\partial\left(\rho_f w\right)}{\partial z}=0 \tag{3.33a}$$

For incompressible fluid, the density ρ_f is constant and equation (3.32) becomes

$$\nabla\cdot\vec{q}=\frac{\partial u}{\partial x}+\frac{\partial v}{\partial y}+\frac{\partial w}{\partial z}=0 \tag{3.33b}$$

Equation (3.33a) represents the divergence of $\vec{q}$ and implies the flux flowing out of a unit volume in space. Equations (3.33a) and (3.33b) represent respectively equations of continuity for the flow of compressible fluid and that of incompressible fluid. For any closed surface drawn in a fluid, the increase in mass of fluid within the surface at any time interval must be equal to the excess of mass that flows into the volume through the surface over the mass that flows out.

In a cylindrical coordinate system (r, θ, z), the equation of continuity (3.32) can be written as (Granger, 1985):

$$\frac{\partial \rho_f}{\partial t}+\frac{\partial\left(\rho_f u_r\right)}{\partial r}+\frac{1}{r}\frac{\partial\left(\rho_f u_\theta\right)}{\partial \theta}+\frac{\partial\left(\rho_f u_z\right)}{\partial z}+\frac{\rho_f u_r}{r}=0 \tag{3.33c}$$

and in the case of incompressible fluid, equation (3.32) can be written as:

$$\frac{\partial u_r}{\partial r}+\frac{1}{r}\frac{\partial u_\theta}{\partial \theta}+\frac{\partial u_z}{\partial z}+\frac{u_r}{r}=0 \tag{3.33d}$$

In a spherical coordinate system (r, φ, θ), the equation of continuity (3.32) can be written as:

$$\frac{\partial \rho_f}{\partial t}+\frac{1}{r^2}\frac{\partial\left(r^2\ \rho_f u_r\right)}{\partial r}+\frac{1}{r\sin\varphi}\frac{\partial\left(\rho_f u_\varphi\ \sin\varphi\right)}{\partial \varphi}+\frac{1}{r\sin\varphi}\frac{\partial\left(\rho_f u_\theta\right)}{\partial \theta}=0 \tag{3.33e}$$

and for incompressible fluid, equation (3.32) can be written as:

$$\frac{1}{r}\frac{\partial\left(r^2 u_r\right)}{\partial r}+\frac{1}{\sin\varphi}\frac{\partial\left(u_\varphi\ \sin\varphi\right)}{\partial \varphi}+\frac{1}{\sin\varphi}\frac{\partial\left(u_\theta\right)}{\partial \theta}=0 \tag{3.33f}$$

All six above equations (3.33) are the equations of continuity for all three coordinate systems (Cartesian, cylindrical, and spherical) in compressible and incompressible fluids.

3.7 EQUATIONS OF MOTION (CONSERVATION OF MOMENTUM)

The equation of motion of fluid dynamics from the Eulerian point of view is the fundamental equation of fluid dynamics, known as Navier–Stokes equation of motion. Based on the Newton's second law of motion, that is, conservation of momentum as (Schlichting, 1968):

$$\text{Mass}\times\text{Acceleration}=\text{Force} \tag{3.34}$$

That is, $m\vec{f}=\vec{P}$, where $\vec{P}$ is the force, m is the mass and $\vec{f}$ is the acceleration. If this equation (3.34) is applied to a fluid element of unit volume, it can be written as:

$$\rho_f\frac{D\vec{q}}{Dt}=\vec{P} \tag{3.35}$$

Here $\vec{P}$ is the force per unit volume, ρ_f is the density and $\frac{D\vec{q}}{Dt}$ is the total acceleration given by equation (3.24), which gives the local acceleration $\frac{\partial\vec{q}}{\partial t}$ and the convective acceleration $(\vec{q}\cdot\nabla)\vec{q}$. It states that the total force acting on fluid mass enclosed in an arbitrary volume (Figure 3.6) fixed in space is equal to the time rate of change of linear momentum. Here the force $\vec{P}$ may be divided into two classes: surface force and body force. Since the fluid offers the resistance to the deformation of a velocity field, there are non-uniform distributions of stresses in the fluid motion. These stresses are the internal forces per unit area of the fluid. All the surface forces on the volume element determine the stress components. Here we are interested only to the isotropic Newtonian fluids, like gas, water, etc. In isotropic fluid, the relation between the stress tensor and the rate deformation tensor is the same in all directions. If the relation is linear, the fluid is called Newtonian fluid.

Let us consider a parallelepiped whose volume element is $dV=dxdydz$ (Figure 3.6). The following stresses acts on two normal surfaces along the x-axis as follows:

p_x and $p_x+\frac{\partial p_x}{\partial x}dx$, where the index x implies that the stress vector acts on surface element $dydz$ along x-direction. Similarly, other two terms are obtained from the surface elements $dxdz$ and $dxdy$ perpendicular to y and z, respectively. The total surface force per unit volume $dxdydz$ resulting from the stress is therefore:

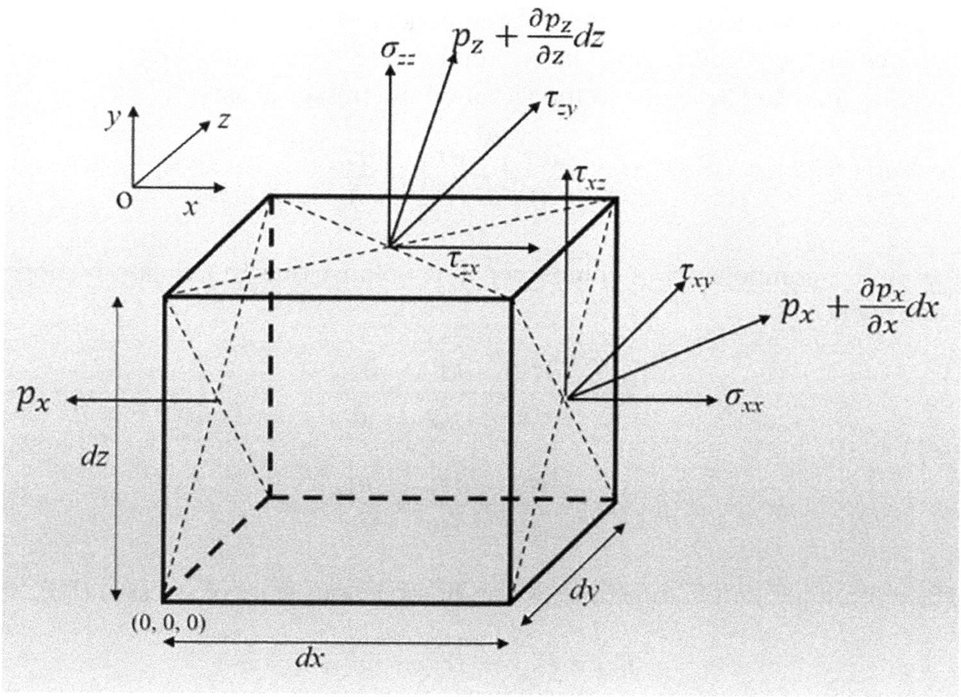

FIGURE 3.6 Typical control volume.

$$\vec{P} = \frac{\partial \overrightarrow{p_x}}{\partial x} + \frac{\partial \overrightarrow{p_y}}{\partial y} + \frac{\partial \overrightarrow{p_z}}{\partial z}.$$

Let $\overrightarrow{p_x}, \overrightarrow{p_y}$ and $\overrightarrow{p_z}$ are the respective forces per unit area on the surfaces *Oyz*, *Oxz* and *Oxy* of a small parallelepiped (Figure 3.6). Each of the forces $\overrightarrow{p_x}, \overrightarrow{p_y}, \overrightarrow{p_z}$ may be resolved into three components along x, y and z-axis (Schlichting and Gersten, 2000). These forces are

$$\overrightarrow{p_x} = \hat{i}\sigma_{xx} + \hat{j}\tau_{xy} + \hat{k}\tau_{xz}, \quad \overrightarrow{p_y} = \hat{i}\tau_{yx} + \hat{j}\sigma_{yy} + \hat{k}\tau_{yz}, \quad \overrightarrow{p_z} = \hat{i}\tau_{zx} + \hat{j}\tau_{zy} + \hat{k}\sigma_{zz} \tag{3.36a}$$

where σ_{xx}, σ_{yy} and σ_{zz} are the respective normal stresses perpendicular to the surface considered, and τ_{xy}, τ_{xz}, ... are the shearing stresses in the surface considered. For example, σ_{xx} is the normal stress acting on a plane whose normal is parallel to the x-axis and the stress itself is in the x-direction; and τ_{xy} denotes the component of shearing stress in the surface perpendicular to the x-axis and in the direction of y-axis. The state of stress is determined by nine scalar quantities and forms the stress tensor. The nine components of stress tensor together are called the stress matrix σ is given by

$$\sigma = \begin{pmatrix} \sigma_{xx} & \tau_{xy} & \tau_{xz} \\ \tau_{yx} & \sigma_{yy} & \tau_{yz} \\ \tau_{zx} & \tau_{zy} & \sigma_{zz} \end{pmatrix} \tag{3.36b}$$

The stress tensor and its matrix are symmetric: two tangential forces are equal, and their indices differ only by order. The equality in indices is shown by considering a parallelepiped in limiting condition.

If a parallelepiped is considered as a point A in macroscopic sense, equation (3.36a) gives the stresses at a point $A(x, y, z)$. The stress at a point in a fluid is a tensor of second order with nine components given in equation (3.36b). We may write the ij^{th} component of stress tensor as τ_{ij}, where $i,\ j = 1, 2, 3$. Considering the force on the parallelepiped in x-direction due to the stresses σ_{xx}, τ_{yx} and τ_{zx}, we get the force per unit volume in x-direction due to the variation of stresses is as:

$$X_{\text{vol}} = \frac{\partial \sigma_{xx}}{\partial x} + \frac{\partial \tau_{yx}}{\partial y} + \frac{\partial \tau_{zx}}{\partial z} \tag{3.37a}$$

Similarly, the y- and z-components of forces per unit volume due to the non-homogenous state of stresses are respectively

$$Y_{\text{vol}} = \frac{\partial \tau_{xy}}{\partial x} + \frac{\partial \sigma_{yy}}{\partial y} + \frac{\partial \tau_{zy}}{\partial z} \tag{3.37b}$$

$$Z_{\text{vol}} = \frac{\partial \tau_{xz}}{\partial x} + \frac{\partial \tau_{yz}}{\partial y} + \frac{\partial \sigma_{zz}}{\partial z} \tag{3.37c}$$

Hence, the surface force per unit volume due to non-homogenous state of stresses (Pai, 1956) is

$$\vec{P}_{\text{vol}} = \hat{i}X_{\text{vol}} + \hat{j}Y_{\text{vol}} + \hat{k}Z_{\text{vol}} \tag{3.38}$$

with the ith component P^i_{vol} as follows

$$P^i_{\text{vol}} = \frac{\partial \tau_{ij}}{\partial x^j} \tag{3.38a}$$

where the summation convention is used.

For an ideal or perfect fluid, the shearing stresses τ_{xy}, τ_{xz}, … are zero, and the pressure is taken as the value of normal stresses σ_{xx}, σ_{yy} and σ_{zz}, and these are negative values. Then, in an ideal fluid at a point in space, we have

$$\begin{aligned} \sigma_{xx} &= \sigma_{yy} = \sigma_{zz} = -p \\ \tau_{xy} &= \tau_{yx} = \tau_{xz} = \tau_{zx} = \cdots\cdots = 0 \end{aligned} \tag{3.39}$$

Equation (3.39) is a good approximation for the ideal fluid, particularly, for water and gaseous fluid. For ideal fluid, the conservation equation of momentum is called *Euler equation of motion*, which has no shearing forces. In Cartesian co-ordinates, the Euler's equations of motion including the body force $\vec{g}$ can be written as (Yuan, 1970):

$$\frac{\partial u}{\partial t} + u\frac{\partial u}{\partial x} + v\frac{\partial u}{\partial y} + w\frac{\partial u}{\partial z} = g_x - \frac{1}{\rho_f}\frac{\partial p}{\partial x} \tag{3.40a}$$

$$\frac{\partial v}{\partial t} + u\frac{\partial v}{\partial x} + v\frac{\partial v}{\partial y} + w\frac{\partial v}{\partial z} = g_y - \frac{1}{\rho_f}\frac{\partial p}{\partial y} \tag{3.40b}$$

$$\frac{\partial w}{\partial t} + u\frac{\partial w}{\partial x} + v\frac{\partial w}{\partial y} + w\frac{\partial w}{\partial z} = g_z - \frac{1}{\rho_f}\frac{\partial p}{\partial z} \tag{3.40c}$$

For real fluids or actual fluids, the shearing stresses are not zero, thus we have nine components of stress from equation (3.39). It can be shown that six shearing stress components can be reduced to three components from the equilibrium of moment on the element in the fluid. If the parallelepiped is sufficiently small, the body force may be neglected compared to the surface force. The equilibrium of moment about z-axis is

$$\tau_{xy} dydxdz = \tau_{yx} dxdzdy \Rightarrow \tau_{xy} = \tau_{yx} \tag{3.41}$$

Similarly, we have $\tau_{yz} = \tau_{zy}$ and $\tau_{zx} = \tau_{xz}$. Thus, we have the six stress components, like, σ_{xx}, σ_{yy} and σ_{zz} are the normal stresses, and τ_{xy}, τ_{yz} and τ_{xz} are the shearing stresses. Now the stress matrix σ has only six different stress components and is symmetric about the diagonal as:

$$\sigma = \begin{pmatrix} \sigma_{xx} & \tau_{xy} & \tau_{xz} \\ \tau_{xy} & \sigma_{yy} & \tau_{yz} \\ \tau_{xz} & \tau_{yz} & \sigma_{zz} \end{pmatrix} \tag{3.42}$$

From equations (3.37), (3.38), and (3.41), the surface force per unit volume is written as

$$\vec{P}_{vol} = \hat{i}\left(\frac{\partial \sigma_{xx}}{\partial x} + \frac{\partial \tau_{xy}}{\partial y} + \frac{\partial \tau_{xz}}{\partial z}\right) + \hat{j}\left(\frac{\partial \tau_{xy}}{\partial x} + \frac{\partial \sigma_{yy}}{\partial y} + \frac{\partial \tau_{yz}}{\partial z}\right) + \hat{k}\left(\frac{\partial \tau_{xz}}{\partial x} + \frac{\partial \tau_{yz}}{\partial y} + \frac{\partial \sigma_{zz}}{\partial z}\right) \tag{3.43}$$

Using equations (3.35) and (3.43) with $\vec{P}_{\text{vol}} = \vec{P}$ and neglecting the body force, one gets the equations as:

$$\rho_f \frac{Du}{Dt} = \frac{\partial \sigma_{xx}}{\partial x} + \frac{\partial \tau_{xy}}{\partial y} + \frac{\partial \tau_{xz}}{\partial z} \tag{3.44a}$$

$$\rho_f \frac{Dv}{Dt} = \frac{\partial \tau_{xy}}{\partial x} + \frac{\partial \sigma_{yy}}{\partial y} + \frac{\partial \tau_{yz}}{\partial z} \tag{3.44b}$$

$$\rho_f \frac{Dw}{Dt} = \frac{\partial \tau_{xz}}{\partial x} + \frac{\partial \tau_{yz}}{\partial y} + \frac{\partial \sigma_{zz}}{\partial z} \tag{3.44c}$$

Now it is useful to separate the pressure p from the normal stresses σ_{xx}, … as:

$$\sigma_{xx} = \tau_{xx} - p, \sigma_{yy} = \tau_{yy} - p, \sigma_{zz} = \tau_{zz} - p \tag{3.45}$$

Here the normal stresses are decomposed into the pressure p which is same in all directions and a departed stress τ_{xx}. Then equation (3.44) can be written as

$$\rho_f \frac{Du}{Dt} = -\frac{\partial p}{\partial x} + \frac{\partial \tau_{xx}}{\partial x} + \frac{\partial \tau_{xy}}{\partial y} + \frac{\partial \tau_{xz}}{\partial z} \tag{3.46a}$$

$$\rho_f \frac{Dv}{Dt} = -\frac{\partial p}{\partial y} + \frac{\partial \tau_{xy}}{\partial x} + \frac{\partial \tau_{yy}}{\partial y} + \frac{\partial \tau_{yz}}{\partial z} \tag{3.46b}$$

$$\rho_f \frac{Dw}{Dt} = -\frac{\partial p}{\partial z} + \frac{\partial \tau_{xz}}{\partial x} + \frac{\partial \tau_{yz}}{\partial y} + \frac{\partial \tau_{zz}}{\partial z} \tag{3.46c}$$

In vector form, if the body force components are taken in to account, then

$$\rho_f \frac{D\vec{q}}{Dt} = \vec{g} - \text{grad } p + \text{div } \tau \tag{3.47}$$

where $\vec{g}$ is the body force components, τ is the viscous stress tensor, which contains the departed stresses and these are symmetric. This equation of motion (3.47), in vector form, is known as the *Cauchy's equations* for conservation of momentum (Granger, 1985). From these equations, the well-known *Navier–Stokes equations* are derived, which is explained in Chapter 4.

3.8 BERNOULLI'S EQUATION (CONSERVATION OF ENERGY)

Let us consider the case of an inviscid or ideal fluid in which the stress tensor is given by equation (3.39) and the body force is the only gravitational force $\vec{g}$. The equation of motion in vector form is the Euler's equations as (Yuan, 1970):

$$\rho_f \frac{D\vec{q}}{Dt} = \rho_f \vec{g} - \nabla p = \rho_f \nabla \varphi_g - \nabla p \tag{3.48}$$

In Cartesian co-ordinates, the Euler's equations of motion can be written as (using $\vec{g} = \nabla \varphi_g$ for φ_g is a gravitational potential).

$$\frac{\partial u}{\partial t} + u\frac{\partial u}{\partial x} + v\frac{\partial u}{\partial y} + w\frac{\partial u}{\partial z} = \frac{\partial \varphi_g}{\partial x} - \frac{1}{\rho_f}\frac{\partial p}{\partial x} \tag{3.49a}$$

$$\frac{\partial v}{\partial t} + u\frac{\partial v}{\partial x} + v\frac{\partial v}{\partial y} + w\frac{\partial v}{\partial z} = \frac{\partial \varphi_g}{\partial y} - \frac{1}{\rho_f}\frac{\partial p}{\partial y} \tag{3.49b}$$

$$\frac{\partial w}{\partial t} + u\frac{\partial w}{\partial x} + v\frac{\partial w}{\partial y} + w\frac{\partial w}{\partial z} = \frac{\partial \varphi_g}{\partial z} - \frac{1}{\rho_f}\frac{\partial p}{\partial z} \tag{3.49c}$$

The set of equation (3.49) may be written in vector form as

$$\frac{\partial \vec{q}}{\partial t} + \nabla\left(\frac{1}{2}q^2\right) - \vec{q} \times \vec{\omega} = \nabla \varphi_g - \frac{1}{\rho_f}\nabla p \tag{3.50}$$

where $\vec{\omega}$ is the vorticity given by equation (3.12). For irrotational flow ($\vec{\omega} = 0$), putting $\varphi_g = -gz$, equation (3.50) can be written as:

$$\frac{\partial \vec{q}}{\partial t} + \nabla\left(\frac{1}{2}q^2\right) = -g\nabla z - \frac{1}{\rho_f}\nabla p \tag{3.51}$$

Since $\vec{q} = \nabla\phi$, here $\dfrac{\partial \vec{q}}{\partial t} = \dfrac{\partial(\nabla\phi)}{\partial t} = \nabla\dfrac{\partial \phi}{\partial t}$. Equation (3.51) gives

$$\nabla\left[\frac{\partial \phi}{\partial t} + \frac{1}{2}q^2 + gz + \frac{p}{\rho_f}\right] = 0 \tag{3.51a}$$

It may be recalled that the dot product of a gradient of a scalar and a differential length provides the differential change in the scalar in the direction of the differential length. That means,

$$\nabla\left(\frac{\partial\phi}{\partial t}\right)\cdot d\vec{r} = \frac{\partial\left(\frac{\partial\phi}{\partial t}\right)}{\partial x}dx + \frac{\partial\left(\frac{\partial\phi}{\partial t}\right)}{\partial y}dy + \frac{\partial\left(\frac{\partial\phi}{\partial t}\right)}{\partial z}dz = d\left(\frac{\partial\phi}{\partial t}\right)$$

Thus, multiplying equation (3.51) by $d\vec{r}$ in scalar and integrate over the whole space, we have

$$\frac{\partial\phi}{\partial t} + \frac{1}{2}q^2 + gz + \int\frac{\partial p}{\rho_f} = f(t) = B_0 \tag{3.52}$$

Equation (3.52) is known as Bernoulli's equation, where $f(t)$ is the arbitrary function of time and $f(t) = B_0$ is a constant along a streamline, which is known as Bernoulli's constant.

For incompressible and steady flow, equation (3.52) can be written as

$$\frac{1}{2}\frac{q^2}{g} + z + \frac{p}{\rho_f g} = \frac{B_0}{g} = \frac{1}{2}\frac{q_1^2}{g} + z_1 + \frac{p_1}{\rho_f g} \tag{3.53}$$

$$\Rightarrow \frac{q^2}{2} + gz + \frac{p}{\rho_f} = \text{constant} \tag{3.54}$$

Here subscript 1 refers to the value at a certain reference point. Equation (3.53) is known as *Bernoulli's equation*, which is widely used in hydraulics and fluid mechanics. Equation (3.54) is based on the assumption that no work between the fluid element and surrounding takes place. The first term of equation (3.54) signifies the kinetic energy per unit mass, second term for potential energy per unit mass and the third term for flow work per unit mass. Therefore, the sum of all three terms of the left-hand side of equation (3.54) is considered as the total mechanical energy per unit mass, which remains constant along a streamline for steady inviscid incompressible fluid flow. This equation is known as the Mechanical Energy Equation and is referred as Bernoulli's equation. The dimension of each term is energy per unit mass. Equation (3.54) can also be written in terms of energy per unit weight as:

$$\frac{q^2}{2g} + \frac{p}{\rho_f g} + z = \frac{B_0}{g}\text{(Constant)} \tag{3.55}$$

The energy per unit weight is termed as "head." From the above three terms, we may call the constant, that is, $\frac{B_0}{g}$ as the total head of the flow, because it is a sum of three heads (total energy per unit weight). The physical significance of these three terms on the left-hand side of equation (3.55) is as follows:

1. The term $\frac{q^2}{2g}$ is the velocity head of the fluid, which is dynamic pressure head of the fluid (kinetic energy per unit weight).
2. The term $\frac{p}{\rho_f g}$ is known as static pressure head (pressure energy or work per unit weight).
3. The term z is simply the elevation of the point considered (potential energy per unit weight).

When equation (3.55) is applied any two points, 1 and 2, located on the same streamline, then the equation can be written as

$$\frac{q_1^2}{2} + gz_1 + \frac{p_1}{\rho_f} = \frac{q_2^2}{2} + gz_2 + \frac{p_2}{\rho_f} \tag{3.56}$$

Equation (3.56) is written for the steady flow of an ideal fluid considering on the same streamline at two points. This equation can be considered as the principle of work and energy, as it applies to a fluid particle representing a unit mass of fluid. Equation (3.56) can also be written as

$$\left(\frac{q_2^2}{2} - \frac{q_1^2}{2}\right) = \left(\frac{p_2}{\rho_f} - \frac{p_1}{\rho_f}\right) + g(z_2 - z_1) \tag{3.57}$$

Equation (3.57) states the change in kinetic energy of the particle of unit mass equal to the work done by the pressure and the gravitation forces, as the fluid particles move from position 1 to 2. Here the pressure force is called the *flow work* and it is directed along the streamline; and the gravitational work is in the vertical direction (z). If the gravitational force is negligible, then equation (3.54) can be written as

$$\frac{1}{2}\rho_f q^2 + p = \text{constant} \tag{3.58}$$

The term $\frac{1}{2}\rho_f q^2$ has the unit as same as that of pressure. However, if the pressure exists, then only the fluid is in motion. Therefore, it can be defined as

$$\frac{1}{2}\rho_f q^2 = p_d \tag{3.59}$$

where p_d is called the dynamic pressure of the fluid.

3.9 APPLICATIONS OF THE BERNOULLI'S EQUATION

1. **Flow around a blunt body**

   Let us take an inviscid streamline flow around a blunt body (Figure 3.7). As the fluid flows around the blunt body, the energy of the fluid is transformed from one to another. Here let us consider two points 1 and 0 on the streamline A around the blunt body, then the Bernoulli equation (3.58) can be written as

$$\frac{1}{2}\rho_f q_1^2 + p_1 = \frac{1}{2}\rho_f q_0^2 + p_0 = \text{constant} \tag{3.60}$$

   However, on the streamline A, the flow is zero at the point O, i.e., $q_0 = 0$, because velocity is zero at the stagnation point (Figure 3.7). The pressure at point O is therefore called the stagnation pressure p_0 or total pressure, and is given by

$$p_0 = \frac{1}{2}\rho_f q_1^2 + p_1 \tag{3.61}$$

   Therefore, from equation (3.61), it is stated that for an inviscid incompressible fluid with negligible body force, the stagnation pressure p_0 is the equivalent to the sum of the dynamic pressure $\frac{1}{2}\rho_f q_1^2$ and the static pressure p_1, and which is constant throughout the fluid. If equation (3.54) is written as

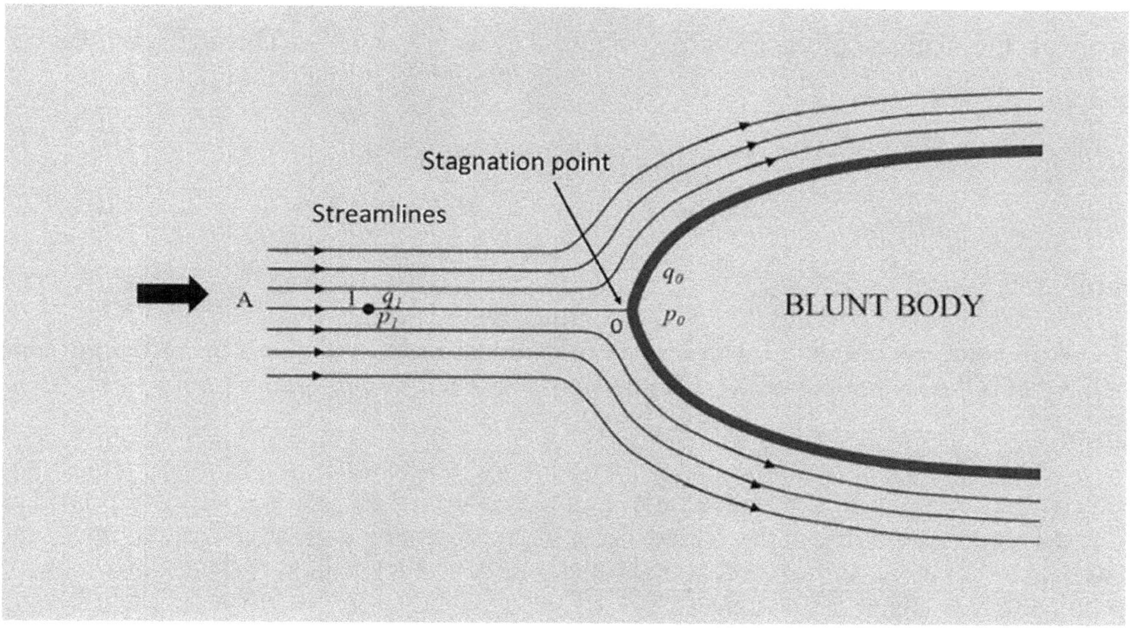

FIGURE 3.7 Streamlines along a blunt body.

$$\frac{q^2}{2g} + \frac{p}{\rho_f g} + z = \text{constant} \tag{3.62a}$$

Then, equation (3.62) is called the "head" equation because each term has the dimension in length.

2. **Flow through a constricted pipe**:
   Let us consider the flow of an inviscid fluid through the constriction in the pipe as shown in Figure 3.8.
   Here two points, 1 and 2, are considered at the respective levels h_1 and h_2 along the constricted pipe, where p_1, p_2 and q_1, q_2 are pressures and the velocities at those points. The static pressure $\frac{p}{\rho_f g}$ and the velocity head $\frac{q^2}{2g}$ vary with the position along the pipe. It is noticed that the total head stays always constant in agreement with equation (3.62), but component heads

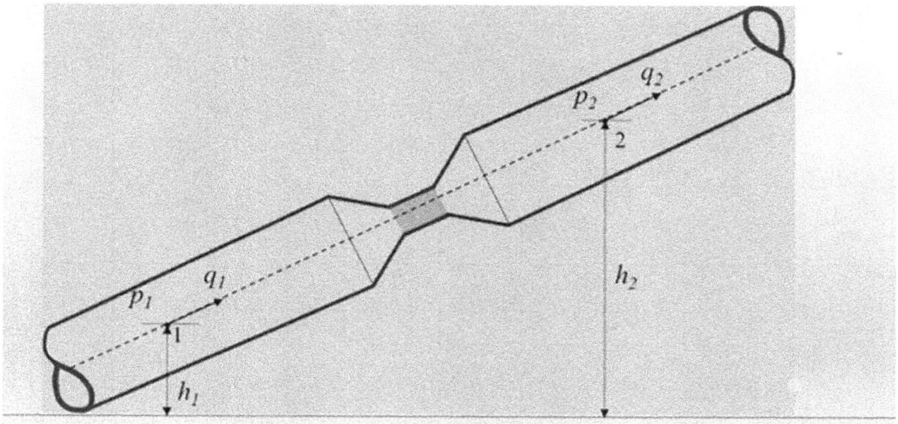

FIGURE 3.8 Flow through a constriction in a pipe.

vary with the distance along the pipe (Figure 3.8) as $\frac{q_1^2}{2g}$ and $\frac{q_2^2}{2g}$. Therefore, the Bernoulli's equation can be written as

$$\frac{q_1^2}{2g} + \frac{p_1}{\rho_f g} + h_1 = \frac{q_2^2}{2g} + \frac{p_2}{\rho_f g} + h_2 \tag{3.62b}$$

Example 3.3

A full tank of liquid is considered with a small orifice near its base (Figure 3.9). When the orifice is opened, what is the velocity of efflux of liquid from the tank?

Solution

Let the cross-sectional area of the tank be C_1 and that of orifice C_2. Let us consider a point 1 at the liquid surface and the point 2 at the orifice, like a jet, representing a streamline of the flow, with the velocities q_1 and q_2 at the surface and at the orifice respectively. Equation (3.54) for the points 1 and 2 can be written as (Figure 3.9) (Yuan, 1970)

$$gz_1 + \frac{p_1}{\rho_f} + \frac{q_1^2}{2} = gz_2 + \frac{p_2}{\rho_f} + \frac{q_2^2}{2} \tag{3.63}$$

Figure shows $z = z_1 - z_2$. From the continuity equation, the velocities q_1 and q_2 are related as

$$C_1 q_1 = C_2 q_2 \tag{3.64}$$

Substituting equations (3.64) in (3.63), one gets

$$q_2 = \sqrt{\frac{2}{\left(1 - \frac{C_2^2}{C_1^2}\right)} \left(\frac{p_1 - p_2}{\rho_f} + gz\right)} \tag{3.65}$$

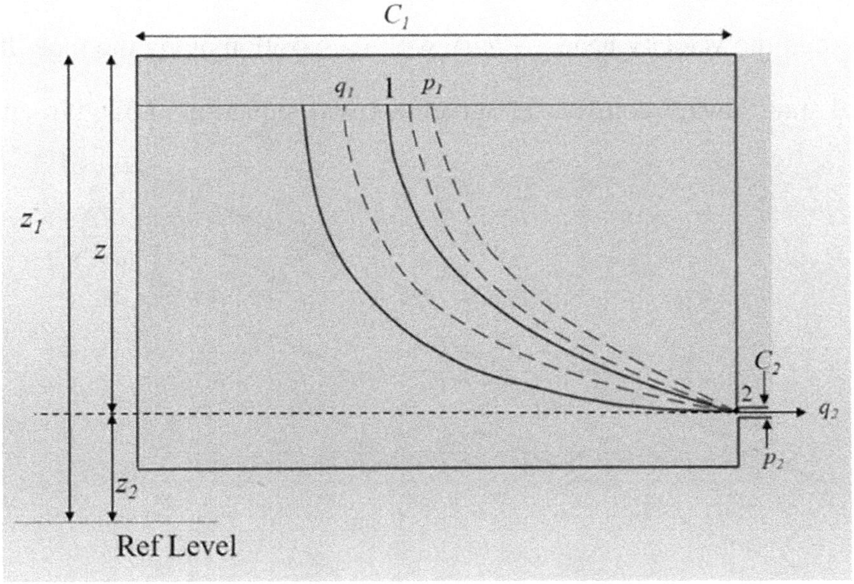

FIGURE 3.9 Liquid filled tank with a small orifice.

Equation (3.65) represents the velocity of liquid from the tank through the orifice. Now if $C_2 \ll C_1$, there are two cases may be considered.

1. If the tank is vented to the atmosphere or has an open surface so that $p_1 = p_2$ with $C_2 \ll C_1$, then equation (3.65) reduces to

$$q_2 = \sqrt{2gz} \tag{3.66}$$

Thus, the velocity of efflux from the vented tank is equal to that obtained by a fluid particle simply dropped from rest at the same height z, i.e., that of a rigid body falling freely from a height z. This is known as Torricelli's Theorem in the name of Italian Mathematician (1608–1647).

If for $C_2 \ll C_1$, $\frac{p_1 - p_2}{\rho_f} \gg gz$, then equation (3.65) becomes

$$q_2 = \sqrt{\frac{2(p_1 - p_2)}{\rho_f}} \tag{3.67}$$

Here the velocity q_2 of efflux depends on pressure difference at p_1 and p_2.

2. Considering the frictional effect into account, a correction factor is introduced as

$$q_2 = F_c\sqrt{2gz} \tag{3.68}$$

where F_c is the correction coefficient of discharge that can be determined experimentally, which varies from 0.95 to 1.0 for an opening type shown in Figure 3.9. For sharp-edged opening, the coefficient F_c is taken as low as 0.6.

Example 3.4

In example 3.3, if $p_1 = 0.1$ MN/m^2 and $p_2 = 0.2$ MN/m^2, the cross-sectional ratio $\frac{C_2}{C_1} = 0.01$ and elevation $z = 5$ m, then determine the velocity of the water jet.

Solution

According to equation (3.65), we get

$$q_2 = \sqrt{\frac{2}{1-(0.01)^2}\left(\frac{(0.2-0.1)\times 10^6}{1\times 10^3} + 9.81\times 5\right)} = \sqrt{298} = 17.3\,\text{m/s}$$

If $p_1 = p_2$, and $C_2 \ll C_1$, then discharge velocity is $q_2 = \sqrt{2(9.81)5} = 9.9\,\text{m/s}$. Experiments have shown that the actual jet velocity is somewhat smaller than the calculated velocity q_2 because the frictional effect is neglected.

Example 3.5

Water is put in tank of 1 m diameter and 5 m depth. A hole of 5 cm diameter is created at a point 4 m below the water surface. Determine the velocity through the hole.

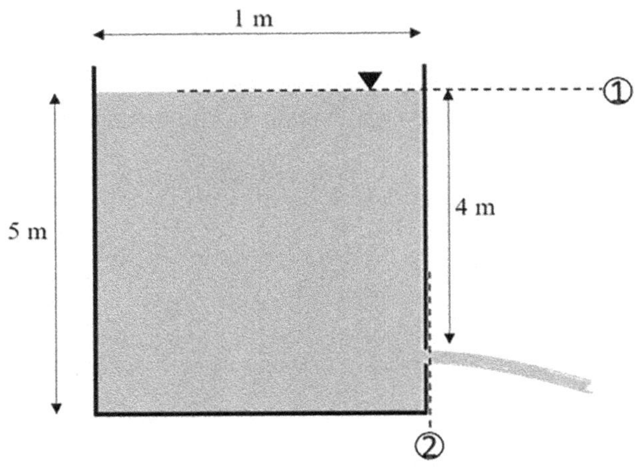

FIGURE 3.10 Water tank with a hole.

Solution

From the given problem (Figure 3.10), ignoring any loss of head, i.e., $E_1 = E_2$ and atmospheric pressures at point 1 and point 2 are same, so $p_1 = p_2$, and at at a point 1, velocity $q_1 = 0$. The velocity q_2 through the hole to be determined.

From the Bernoulli's equation, we get $\frac{p_1}{\rho_f g} + \frac{q_1^2}{2g} + z_1 = \frac{p_2}{\rho_f g} + \frac{q_2^2}{2g} + z_2$

From the above conditions,

$$z_1 = \frac{q_2^2}{2g} + z_2 \text{ or } \frac{q_2^2}{2g} = z_1 - z_2 = 4. \text{ Thus } q_2 = 8.86 \text{ m/s}.$$

Therefore, the velocity through the hole is 8.86 m/s.

Example 3.6

Water is entering in a pipe at a velocity 2 m/s and pressure of 300 kPa and coming out through a nozzle. Calculate the velocity of water coming out from the nozzle if the pipe is horizontal.

Solution

At the nozzle outlet in open air so the pressure at the section 2 will be atmospheric pressure (Figure 3.11). Therefore, atmospheric pressure $p_2 = 101300 \text{ Pa} = 101.3 \text{ kPa}$. As the pipe is horizontal, $z_1 = z_2$. From the Bernoulli's equation, we get

$$\frac{p_1}{\rho_f g} + \frac{q_1^2}{2g} + z_1 = \frac{p_2}{\rho_f g} + \frac{q_2^2}{2g} + z_2. \text{ Then } \frac{300 \times 10^3}{\rho_f} + \frac{2^2}{2} = \frac{101.3 \times 10^3}{\rho_f} + \frac{q_2^2}{2}.$$

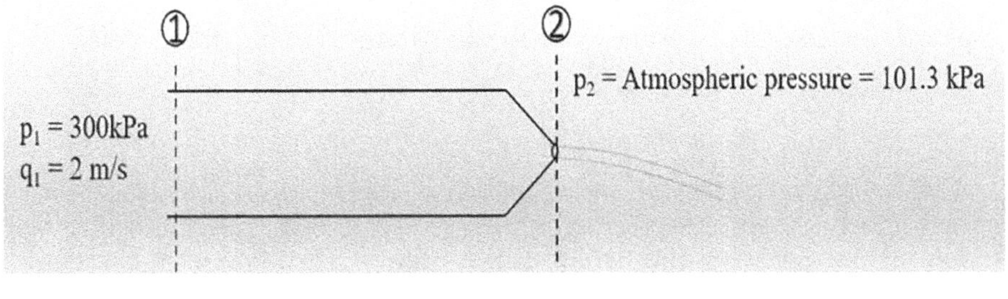

FIGURE 3.11 Pipe with a nozzle.

$$\frac{q_2^2}{2} = 302 - 101.3 \Rightarrow q_2 = 20.05 \text{ m/s}.$$

Therefore, the velocity of water coming out from the nozzle is 20.05 m/s.

3.9.1 Bernoulli's Equation with Head Loss

The derivation of Bernoulli's equation in a real fluid depends on the frictional work done by a moving fluid element. In many practical situations, Bernoulli's equation may be modified as:

$$\frac{q_1^2}{2g} + \frac{p_1}{\rho_f g} + z_1 = \frac{q_2^2}{2g} + \frac{p_2}{\rho_f g} + z_2 + h_l \quad (3.69)$$

where h_l is the frictional work done per unit weight of a fluid element moving from location 1 to location 2 along the streamline in the direction of flow. The term h_l is referred to as head loss from one location to other due to the loss of total mechanical energy per unit weight. The loss of energy from the location 1 to other is emerged due to the effect of frictional force. In fact, for inviscid fluid, the frictional work, $h_l = 0$, and the total mechanical energy is constant along the streamline.

Example 3.7

Water flows through a circular pipe. At one section the diameter is 0.4 m, static pressure is 270 kPa gauge, the velocity is 4 m/s with an elevation 10 m above the ground bed. The elevation of a section downstream is 0.0 m, the pipe diameter is 0.2 m. Find the gauge pressure at the downstream section. The frictional effect may be neglected. Assuming the density of water be 999 kg/m³.

Solution

From the equation of continuity, it is written as $A_1 q_1 = A_2 q_2$

Since A_1 and A_2 are the cross-sectional areas of the upstream and downstream sections and q_1 and q_2 are the corresponding velocities at the sections.

$A_1 = \pi r_1^2$, $A_2 = \pi r_2^2$, given that $r_1 = 0.2$ m, and $r_2 = 0.1$ m. $q_2 = \frac{A_1}{A_2} q_1 = \left(\frac{0.2}{0.1}\right)^2 \times 4 = 16$ m/s.

Applying Bernoulli's equation between upstream and downstream sections, we get $\frac{p_1}{\rho_f} + \frac{q_1^2}{2} + g z_1 = \frac{p_2}{\rho_f} + \frac{q_2^2}{2} + g z_2$, where p_1 is the pressure at upstream, $p_1 = 270$ kN/m² gauge (given)

$$p_1 + \frac{\rho_f}{2} q_1^2 + \rho_f g z_1 = p_2 + \frac{\rho_f}{2} q_2^2 + \rho_f g z_2.$$

Pressure gauge p_2 at the downstream to be determined $p_2 = p_1 + \frac{\rho_f}{2}\left(q_1^2 - q_2^2\right) + \rho_f g \left(z_1 - z_2\right)$

$$= 270 \times 10^3 + \frac{999}{2}\left(4^2 - 16^2\right) + 999 \times 9.81(10 - 0) = 270000 + \frac{999}{2} \times (-240) + 98001.9 = 270000 - 119880$$

$$+98001.9 = 248.122 \times 10^3 \text{ Pa gauge} = 248.12 \text{ kPa gauge}$$

3.9.2 Trajectory of a Free Jet

Consider a liquid jet (Figure 3.12). It is considered as a streamline. Here the atmospheric viscous effects are neglected. Since the entire jet is in the atmosphere, the pressure will be same everywhere on the

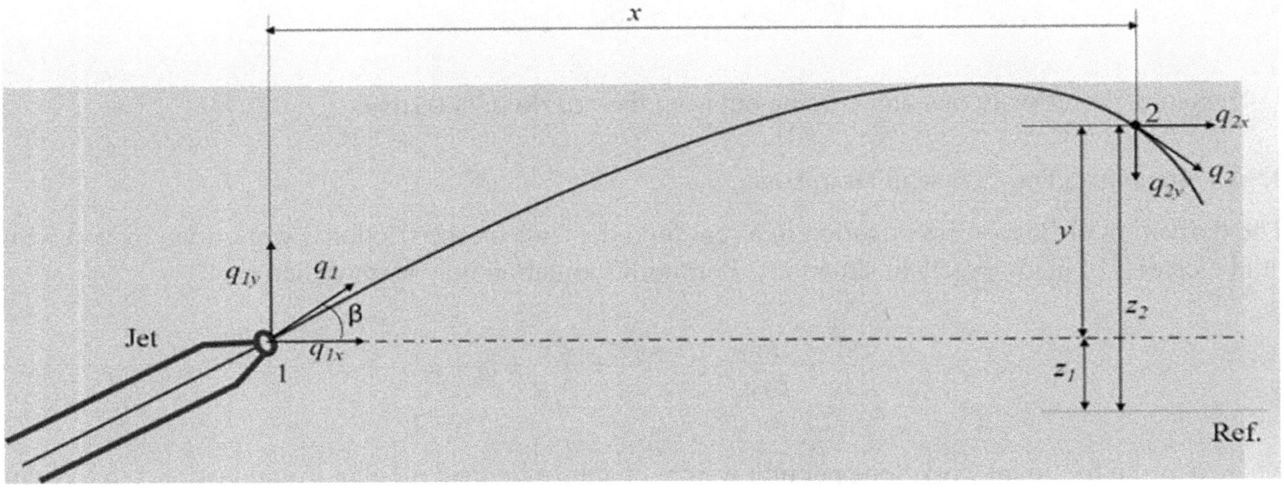

FIGURE 3.12 A typical free jet flow.

streamline, i.e., $p_1 = p_2$ are in the arbitrary points in the streamlines. Then the Bernoulli's equation between jet exit 1 and an arbitrary point 2 on the streamline is as follows:

$$gz_1 + \frac{q_1^2}{2} = gz_2 + \frac{q_2^2}{2} \tag{3.70}$$

Since $p_1 = p_2$, the pressure terms are cancelled from both sides. Now, one has to determine the velocity of the jet at any point on the trajectory as a function of jet flow rate Q, the horizontal distance x to the point 2, and the initial inclination β. Assuming the nozzle area is c_1, then the above Bernoulli's equation for the flow rate $Q = q_1c_1$ is written as follows:

$$\begin{aligned} q_2^2 &= 2g(z_1 - z_2) + q_1^2 \\ q_2^2 &= q_1^2 - 2gy \end{aligned} \tag{3.71}$$

where $y = z_2 - z_1$ is to be determined in terms of horizontal distance x and inclination β. The relations between the velocity components at points 1 and 2 are given by:

$$q_{2x} = q_{1x} \text{ and } q_{2y} = q_{1y} - gt$$

The coordinates at the point 2 are found as

$$x = q_{1x}t \text{ and } y = q_{1y}t - \frac{1}{2}gt^2$$

Eliminating t from the above two equations, one gets

$$y = \frac{q_{1y}}{q_{1x}}x - \frac{g}{2}\frac{x^2}{q_{1x}^2} \tag{3.72}$$

But, $\dfrac{q_{1y}}{q_{1x}} = \tan\beta$ and $q_{1x}^2 = q_1^2\cos^2\beta$,

Therefore, substituting y from equation (3.72) in (3.71), one gets

$$q_2^2 = \left(\frac{Q}{c_1}\right)^2 - 2g\left[\frac{q_{1y}}{q_{1x}}x - \frac{g}{2}\frac{x^2}{q_{1x}^2}\right]$$

$$= \left(\frac{Q}{c_1}\right)^2 - 2g\left[x\tan\beta - \frac{gx^2}{2q_1^2\cos^2\beta}\right]$$

$$q_2 = \sqrt{\left(\frac{Q}{c_1}\right)^2 - 2g\left[x\tan\beta - \frac{gx^2}{2q_1^2\cos^2\beta}\right]} \tag{3.73}$$

Equation (3.73) represents the velocity of the jet at any point (say at 2) on its trajectory or streamline.

3.9.3 Pitot Tube or Stagnation Tube

The Pitot tube is a device to measure the stagnation pressure, which is used to measure the velocity of a moving fluid in an *open channel* or *pipe* (Hibbeler, 2015). The total pressure or stagnation pressure is equal to the sum of the static pressure and the dynamic pressure on the streamline. A tube is bent at a right angle and immerged into the channel with a longitudinal axis aligned to the flow and checked the height to which the fluid elevated within the tube. The stagnation pressure can be determined using the Bernoulli's equation.

Consider two points 1 and 2 located on the horizontal streamline (Figure 3.13). The point 1 is located little upstream of the tube, where the velocity and pressure are q_1 and $p_1 = \rho_f g d$; and the point 2 is located at the opening of the tube, where the velocity is momentarily reduced to zero ($q_2 = 0$) due to its impact with the liquid within the tube. At the point 2, the liquid shows a static pressure, which reaches to the liquid level up to d, and a dynamic pressure, which forces the liquid additionally up to the level h above the liquid surface (Figure 3.13). The total pressure at the point 2 is $\rho_f g(d+h)$. Applying the Bernoulli's equation with the gravitational datum on the streamline, one gets

$$\frac{p_1}{\rho} + \frac{q_1^2}{2} + gz_1 = \frac{p_2}{\rho_f} + \frac{q_2^2}{2} + gz_2 \tag{3.74}$$

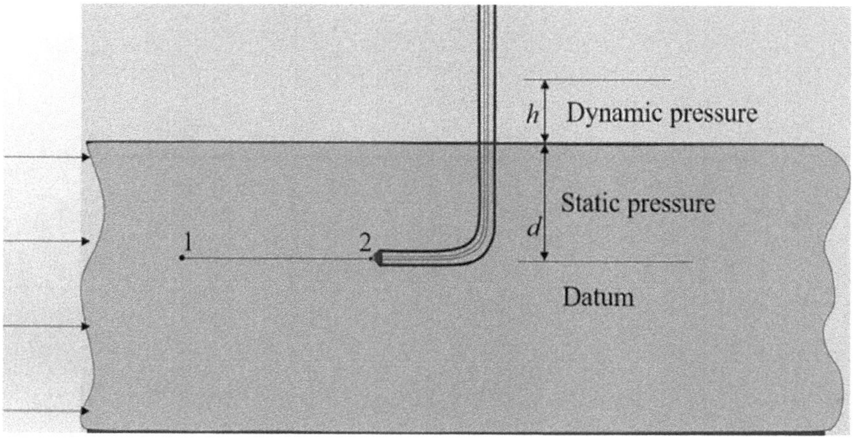

FIGURE 3.13 A typical pitot tube arrangement.

$$q_1 = \sqrt{2gh} \tag{3.75}$$

Here in the open channel flow, recording the level h on the Pitot tube, equation (3.75) is used to determine the velocity of the flow.

For flow through a *closed channel or pipe*, it is necessary to use both the Pitot tube and piezometer to determine the flow velocity. The piezometer measures the static pressure at a point 1 on the streamline (Figure 3.14). At the point located at 1, the hydrostatic pressure in the pipe or closed channel due to weight of the fluid is $\rho_f gd$ and the internal pressure in the pipe is $\rho_f gh$. The total pressure is at the location 1 is $\rho_f g(d+h)$, and the total pressure at the stagnation point 2 will be larger than the point 1 due to the additional dynamic pressure, which will be $\rho_f g(d+h+h_1)$, where h_1 is the height due to dynamic pressure. Therefore, applying the Bernoulli equation at the points 1 and 2 on the streamline using the measurements from the tube, the velocity q_1 at the point 1 can be obtained as follows:

$$\frac{\rho_f g(d+h)}{\rho_f} + \frac{q_1^2}{2} + 0 = \frac{\rho_f g(d+h+h_1)}{\rho_f} + 0 + 0 \tag{3.76}$$

which implies as $q_1 = \sqrt{2gh_1}$

In the closed channel or pipe, equation (3.76) is used to determine the velocity at a point 1 on the streamline, recording the levels from the Pitot tube and the piezometer.

Pitot-static tube is also used to determine the velocity of a flow in a closed conduit. It is constructed using two concentric tubes as shown in Figure 3.15. The stagnation pressure at the point 2 can be measured from a pressure tap at E of the inner tube. In the outer tube, there are several holes H at the downstream of the location 2. Here the outer tube acts like a piezometer, so that the static pressure can be measured from the pressure tap at C. Using two measured pressures at C and E, and using the Bernoulli's equation between two points 1 and 2, neglecting the elevation between C and E, one gets:

$$\frac{p_1}{\rho_f} + \frac{q_1^2}{2} + gz_1 = \frac{p_2}{\rho_f} + \frac{q_2^2}{2} + gz_2. \text{ Then, } \frac{p_c + \rho_f gh}{\rho_f} + \frac{q_1^2}{2} + 0 = \frac{p_E + \rho_f gh}{\rho_f} + 0 + 0$$

$$q_1 = \sqrt{\frac{2}{\rho_f}(p_E - p_C)} \tag{3.77}$$

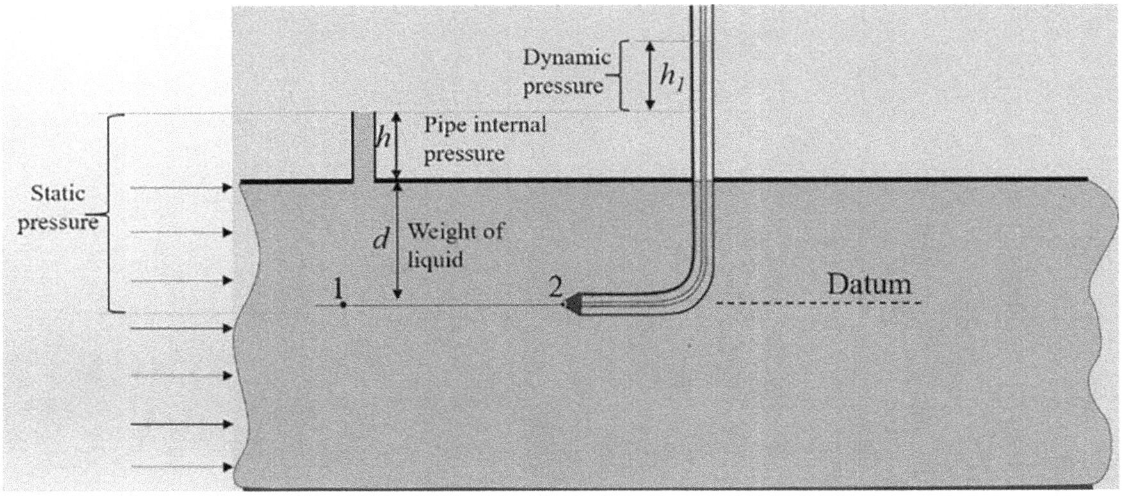

FIGURE 3.14 Typical pitot tube and piezometer arrangement.

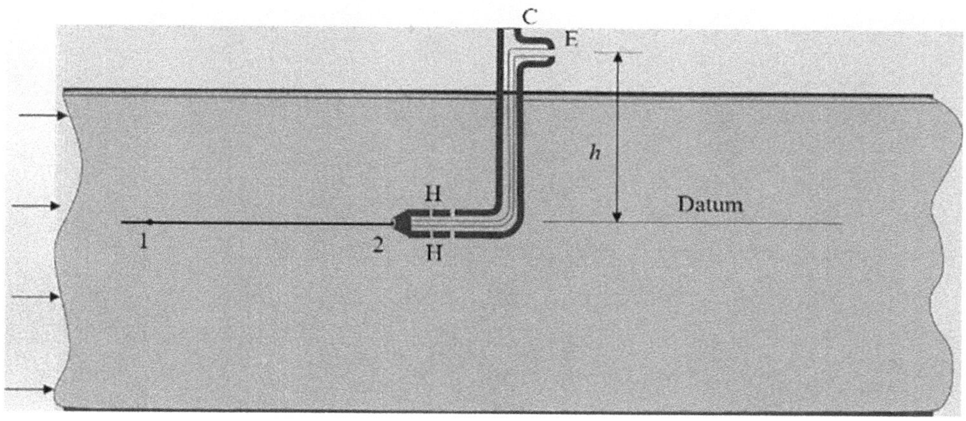

FIGURE 3.15 Arrangement of typical pitot-static tube.

The difference in pressures can be determined either by using a manometer attached to the outlets C and E, and measuring the differential height at the manometer liquid, or by pressure transducers.

3.9.4 Venturi Meter

Venturi meter is a device used to measure the *flow rate* of a fluid through a conduit. It consists of a conduit of constant diameter tapering to a smaller diameter and then again gradually expanding to its original diameter (Figure 3.16). The rings of the static pressure ports are placed at the entrance and exit of the tapering section. When the flow continues from larger to smaller diameter, the flow accelerates causing the higher velocity and the static pressure variation is produced on it, which provides an accurate measurement of the flow rate in the conduit.

Let C_1 and C_2 be the cross-sectional areas at the sections 1 and 2 shown in (Figure 3.16), q_1 and q_2 are the velocities at these sections respectively. Equation of continuity shows as follows:

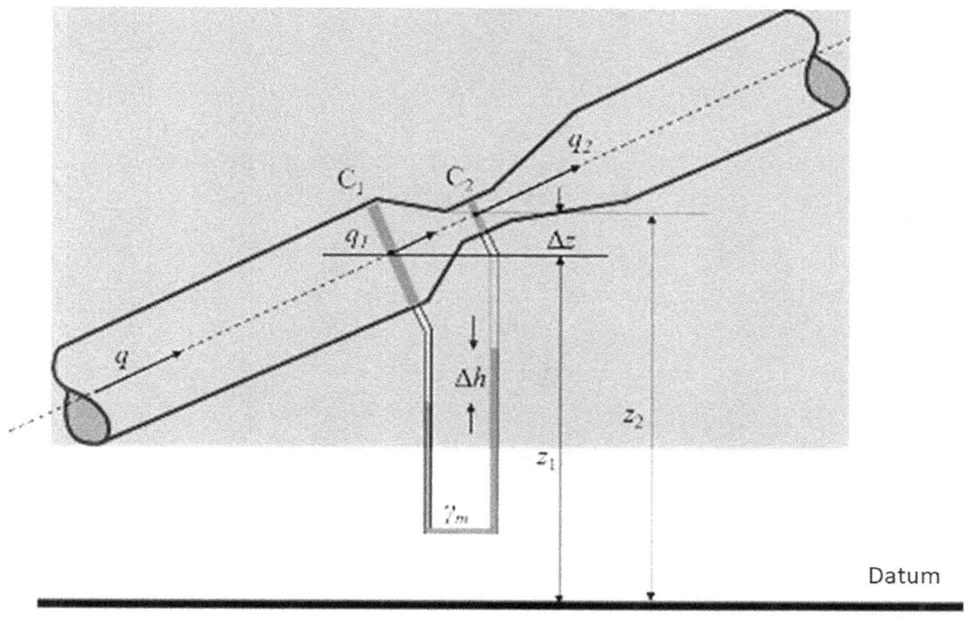

FIGURE 3.16 A typical venturimeter.

$$C_1 q_1 = C_2 q_2 \tag{3.78}$$

and the Bernoulli's equation is

$$\frac{p_1}{\rho_f} + \frac{q_1^2}{2} + g z_1 = \frac{p_2}{\rho_f} + \frac{q_2^2}{2} + g z_2 \tag{3.79}$$

From the above two equations, one gets velocity at the location 2 as:

$$q_2 = \sqrt{\frac{2\left[(p_1 - p_2)/\rho_f - g(z_2 - z_1)\right]}{1 - \left(\frac{C_2}{C_1}\right)^2}} \tag{3.80}$$

Multiplying the above equation by C_2, one gets theoretically the flow rate at either section as follows:

$$Q_{\text{theory}} = C_2 q_2 = C_2 \sqrt{\frac{2\left[(p_1 - p_2)/\rho_f - g(z_2 - z_1)\right]}{1 - \left(\frac{C_2}{C_1}\right)^2}} = C_1 q_1$$

$$= \frac{C_2}{\sqrt{1 - \left(\frac{C_2}{C_1}\right)^2}} \sqrt{2g\left[\frac{p_1 - p_2}{\rho_f g} - (z_2 - z_1)\right]} = \frac{C_2}{\sqrt{1 - \left(\frac{C_2}{C_1}\right)^2}} \sqrt{2g\left[\frac{\Delta p}{\rho_f g} - \Delta z\right]} \tag{3.81}$$

where $\Delta p = p_1 - p_2, \Delta z = z_2 - z_1$

Equation (3.81) represents theoretical flow rate at either section through the Venturi meter. If $z_2 = z_1 = 0$, equation (3.81) is written as

$$C_2 q_2 = \frac{C_2}{\sqrt{1 - \left(\frac{C_2}{C_1}\right)^2}} \sqrt{2\frac{\Delta p}{\rho_f}} \tag{3.82}$$

The static pressure difference $\Delta p = p_1 - p_2$ is often measured using a pressure transducer or a manometer. In terms of manometer deflection, equation (3.81) can be written as:

$$Q_{\text{theory}} = \frac{C_2}{\sqrt{1 - \left(\frac{C_2}{C_1}\right)^2}} \sqrt{2g\Delta h\left[\frac{\gamma_m}{\gamma} - 1\right]} \tag{3.83}$$

where $\Delta h = \left(\frac{\Delta p}{\gamma} - \Delta z\right)\left[\frac{\gamma}{\gamma_m - \gamma}\right]$ is the manometer deflection, $\gamma = \rho g$, and γ_m is the specific weight of the manometer liquid (Yuan, 1970). However, in practice the flow rate is somewhat less than the theoretical value because of the non-uniformity of velocity distribution. Therefore, one can write

$$Q_{\text{actual}} = \frac{C_d C_2}{\sqrt{1-\left(\frac{C_2}{C_1}\right)^2}}\sqrt{2g\Delta h\left[\frac{\gamma_m}{\gamma}-1\right]} \tag{3.84}$$

where $C_d = \frac{Q_{\text{actual}}}{Q_{\text{theoretical}}}$ is the co-efficient of discharge, which is determined experimentally for a given venture meter. To determine the velocity at the throat the velocity q_2 is given by

$$q_2 = C_d q_{2\ \text{theory}} \tag{3.85}$$

where $q_{2\ \text{theory}}$ is given by $\sqrt{\frac{2\left[\frac{\Delta p}{\rho_f} - g\Delta z\right]}{1-\left(\frac{C_2}{C_1}\right)^2}}$. Or, $q_2 = \frac{C_d}{\sqrt{1-\left(\frac{C_2}{C_1}\right)^2}}\sqrt{2\left[\frac{\Delta p}{\rho_f} - g\Delta z\right]}$

In order to use the venture meter due to non-uniformity of velocity, one should consider $q_2 = C_d q_{2\ \text{theory}}$, where $q_{2\ \text{theory}}$ is determined from the theoretical formula and C_d is the discharge coefficient or error correction.

The Venturi meter is used to measure the pipe flow rate, because the gradual expanding section remains the boundary layer separation to be a minimum, resulting a good pressure across the meter.

Example 3.8

Entrance diameter of a Venturi meter is 0.2 m and the throat diameter is 0.1 m. The center of throat is 0.5 m above the center of the entrance. Find the velocity of water at the throat when the pressure difference between the entrance and the throat is 30 kN/m² and the coefficient of discharge correction C_d is 0.97.

Solution

According to equation (3.81), one gets $C_2 q_2 = \frac{C_2}{\sqrt{1-\left(\frac{C_2}{C_1}\right)^2}}\sqrt{2g\left[\frac{\Delta p}{\rho_f g} - \Delta z\right]}$

The velocity at the throat q_2 is $q_2 = C_d q_{2\ \text{theory}} = C_d \sqrt{\frac{2\left[\frac{\Delta p}{\rho_f} - g\Delta z\right]}{1-\left(\frac{C_2}{C_1}\right)^2}}$

where $C_d = 0.97$ is given in the problem. $q_2 = 0.97\sqrt{\frac{2\left[\frac{\Delta p}{\rho_f} - g\Delta z\right]}{1-\left(\frac{C_2}{C_1}\right)^2}}$

$$\frac{\Delta p}{\rho_f} - g\Delta z = \frac{30\times 10^3}{1\times 10^3} - 9.81\times 0.5 = 25.1\ \text{m}^2/\text{s}^2.\ \frac{C_2}{C_1} = \left(\frac{0.1}{0.2}\right)^2 = 0.25$$

Therefore, $q_2 = 0.97\sqrt{\frac{2\times 25.1}{1-(0.25)^2}} = 7.1$ m/s

3.10 POTENTIAL FLOW

Few of the fluid mechanics problems can be approximated with the flow field which is incompressible and irrotational. A scalar function, called as a velocity potential, is introduced to the incompressible, inviscid fluid flows to satisfy the differential form of equation of continuity. The continuity equation is a first-order partial differential equation having three unknown velocity components in three dimensions. The velocity potential is such a function, using which the equation of continuity can be transformed in to the second-order linear partial differential equation of one unknown function, which is the velocity potential function ϕ. The derivatives of this function ϕ give the velocity components in the flow field. Substitution of velocity components in the equation of continuity presents the second-order differential equation, called the Laplace equation in two or three dimensions. The use of the velocity potential describing the fluid flow through the Laplace equation is popularly called as potential flow. The Laplace equation has many applications in different branches of science. In each application, there should be an equivalent idealized physical flow. The solution of this equation is important, because it is useful not only in fluid dynamics but also in other fields, like heat conduction equation, electromagnetic theory, wave equation, etc. (Granger, 1985). The derivations of Laplace equations for Cartesian, cylindrical and spherical coordinate systems are discussed in the following section.

3.10.1 Laplace Equation in Different Coordinate Systems

Using the differential operator in Cartesian coordinates as

$$\nabla\phi\cdot d\vec{r} = \vec{q}.d\vec{r} = \frac{\partial\phi}{\partial x}dx + \frac{\partial\phi}{\partial y}dy + \frac{\partial\phi}{\partial z}dz = d\phi = udx + vdy + wdz \tag{3.86}$$

where $u = \frac{\partial\phi}{\partial x}$, $v = \frac{\partial\phi}{\partial y}$, $w = \frac{\partial\phi}{\partial z}$.

If the gradient of the velocity potential for the velocity vector is substituted in the equation of continuity, the following second order partial differential equation is obtained as:

$$\nabla^2\phi = 0 \tag{3.87}$$

Or, in Cartesian coordinates,

$$\frac{\partial^2\phi}{\partial x^2} + \frac{\partial^2\phi}{\partial y^2} + \frac{\partial^2\phi}{\partial z^2} = 0 \tag{3.87a}$$

Equation (3.87) is known as Laplace equation, often called as velocity potential equation, and its solution ϕ is called harmonic function (Yuan, 1970). When the fluid is in contact with a solid surface at rest, the boundary conditions of Laplace equation must satisfy:

$$\vec{n}.\vec{q} = \frac{\partial\phi}{\partial n} = 0 \tag{3.88}$$

and the pressure should be continuous at a common boundary of fluids. Equation (3.88) is a kinematic relation which means that the contact of fluid and the surface must have the same normal velocity. If $\vec{q}$ and $\vec{q}_s$ are the velocities of fluid and solid surface, then

$$\vec{n}.\vec{q} = \vec{n}.\vec{q}_s, \Rightarrow \quad \vec{n}.(\vec{q} - \vec{q}_s) = 0 \tag{3.89}$$

When the velocity of solid surface $\vec{q}_s$ is zero, equation (3.89) reduces to equation (3.88), which means the fluid velocity is tangential to the surface in an inviscid fluid. If p_1 and p_2 are the pressures in two fluids, then

$$\vec{n}p_1.d\vec{A} = \vec{n}p_2.d\vec{A} \Rightarrow p_1 = p_2 \tag{3.90}$$

where $d\vec{A}$ is an infinitesimal surface area common to the fluids.

Similarly the Laplace equation in cylindrical coordinate system is derived as:

$$\vec{ds}.\vec{q} = \vec{ds}.\nabla\phi = \frac{\partial\phi}{\partial r}dr + \frac{\partial\phi}{\partial\theta}d\theta + \frac{\partial\phi}{\partial z}dz = d\phi \tag{3.91}$$

$$(dr,\ rd\theta,\ dz).(u_r,\ u_\theta,\ u_z) = u_r dr + u_\theta r d\theta + u_z dz \tag{3.92}$$

Comparing equations (3.91) and (3.92), one gets

$$u_r = \frac{\partial\phi}{\partial r}, \quad u_\theta = \frac{1}{r}\frac{\partial\phi}{\partial\theta}, \quad u_z = \frac{\partial\phi}{\partial z} \tag{3.93}$$

Using the velocity components from equation (3.93) to the equation of continuity (3.33b), one gets the Laplace equation in cylindrical coordinates as:

$$\nabla^2\phi = \nabla.\vec{q} = \frac{\partial^2\phi}{\partial r^2} + \frac{1}{r^2}\frac{\partial^2\phi}{\partial\theta^2} + \frac{\partial^2\phi}{\partial z^2} + \frac{1}{r}\frac{\partial\phi}{\partial r} = 0 \tag{3.94}$$

The Laplace equation in terms of spherical coordinate system $\left(dr,\ rd\varphi,\ r\sin\varphi d\theta\right)$ is also obtained as:

$$d\phi = \frac{\partial\phi}{\partial r}dr + \frac{\partial\phi}{\partial\varphi}d\varphi + \frac{\partial\phi}{\partial\theta}d\theta \tag{3.95}$$

$$(dr,\ rd\varphi,\ r\sin\varphi d\theta).(u_r,\ u_\varphi,\ u_\theta) = u_r dr + u_\varphi r d\varphi + u_\theta r\sin\varphi d\theta$$

One gets from above two equations as:

$$u_r = \frac{\partial\phi}{\partial r}, \quad u_\varphi = \frac{1}{r}\frac{\partial\phi}{\partial\varphi}, \quad u_\theta = \frac{1}{r\sin\varphi}\frac{\partial\phi}{\partial\theta} \tag{3.96}$$

Using the velocity components from equation (3.96) to the equation of continuity (3.33b), one gets the Laplace equation in spherical coordinates as:

$$\nabla^2\phi = \nabla.\vec{q} = \frac{1}{r^2}\frac{\partial}{\partial r}\left(r^2\frac{\partial\phi}{\partial r}\right) + \frac{1}{r^2\sin\varphi}\frac{\partial}{\partial\varphi}\left(\sin\varphi\frac{\partial\phi}{\partial\varphi}\right) + \frac{1}{r^2\sin^2\varphi}\frac{\partial^2\phi}{\partial\theta^2} = 0 \tag{3.97}$$

The general analytic method of solutions of above Laplace's equations has been attempted by many mathematicians during last centuries (Yuan, 1970). The solutions of these equations are important, because it is useful not only in fluid dynamics but also in other fields, like heat conduction equation, electromagnetic theory, water wave theory etc. Since it is a linear equation, the superposition of different solutions of velocity potential is valid. A more detailed clarification on potential flows can be found in many fluid mechanics basic books, such as, Granger (1985), Yuan (1970), Ojha et al. (2010), Hibbeler (2015) etc.

3.11 SUMMARY

Some of the important aspects discussed in this chapter are summarized below:

- Kinematics only takes care of variables, like displacement, velocity, acceleration, deformation and rotation of fluid elements.
- The motion of fluid particle is identified by two approaches: Lagrangian and Eulerian approaches. The Lagrangian approach follows the motion of an individual particle, whereas Eulerian approach follows a point or a region in the fluid flow.
- A streamline is a line imagined within the flow for which the tangent at any point shows velocity vector at that point. The velocity of fluid particle is always tangent to the streamline at that instant.
- A path line of a fluid particle describes a path through which the particle travels over a period of time. In a steady flow, it is identical to the streamline. A *streak line* is a line joining all fluid particles which pass through a given point in space.
- Vorticity is defined for rotating flow, that means $\vec{\omega} = \vec{\nabla} \times \vec{q}$. If the vorticity $\vec{\omega}$ is zero, then the flow is called irrotational flow, and if it is non-zero, then the flow is rotational flow.
- The components of vorticity vector $\vec{\omega}$ of a fluid particle are as: $\omega_x = \frac{\partial w}{\partial y} - \frac{\partial v}{\partial z}$, $\omega_y = \frac{\partial u}{\partial z} - \frac{\partial w}{\partial x}$, $\omega_z = \frac{\partial v}{\partial x} - \frac{\partial u}{\partial y}$
- For irrotational flow, the velocity potential is defined. The components of velocity in terms of velocity potential in Cartesian coordinates are as: $u = \frac{\partial \phi}{\partial x}$, $v = \frac{\partial \phi}{\partial y}$, $w = \frac{\partial \phi}{\partial z}$
- The fundamental equations of fluid mechanics, known as continuity equation based on the conservation of mass, momentum and energy equations based of Newton's second law of motion (conservation of momentum), are derived.
- For ideal fluid, the conservation equation of momentum is called *Euler equation of motion*, which has no shearing forces, only normal stresses exist.
- For an ideal fluid flow of an incompressible fluid, Bernoulli's equation states that the total energy consists of kinetic energy, pressure energy and datum energy at any point of the fluid is constant. That means, $\frac{q_1^2}{2g} + \frac{p_1}{\rho_f g} + z_1 = \frac{q_2^2}{2g} + \frac{p_2}{\rho_f g} + z_2$, where $\frac{q_1^2}{2g}$ is the kinetic energy per unit weight = kinetic head, $\frac{p_1}{\rho_f g}$ is the pressure energy per unit weight = pressure head, and z_1 is the datum energy per unit weight = datum head.
- For real or practical fluid flow, Bernoulli's equation states that there should be a loss of energy due to friction, which is given by: $\frac{q_1^2}{2g} + \frac{p_1}{\rho_f \rho g} + z_1 = \frac{q_2^2}{2g} + \frac{p_2}{\rho_f g} + z_2 + h_l$, where h_l is the loss of energy between two sections 1 and 2.
- The explicit forms of Laplace equation in Cartesian, cylindrical and spherical coordinates are presented. Here, $\nabla^2 \phi = 0$ is the Laplace equation, where ∇^2 is the Laplace operator, and ϕ is the velocity potential.

3.12 EXERCISE PROBLEMS

1. What do you mean by the Lagrangian and Eulerian approaches to represent the flow field? How do you differentiate between these two?
2. What is irrotational flow? How do you distinguish between rotational and irrotational flows? Explain with example.
3. What is the physical significance of equation of continuity? Write down the equations of continuity for compressible and incompressible fluid flows. How do you distinguish between the two?
4. What is the difference between the Euler's equation of motion and Navier-Stokes equations? What are the physical significances of these equations? On what basis the equations are derived?
5. Write down the Bernoulli's equation. What is the physical significance of this equation? What is the basis of deriving this equation?
6. What will be effect in Bernoulli's equation in ideal and real fluid flows?
7. What is potential flow? Write down the Laplace equation for Cartesian coordinate system. What is the basic idea of deriving the Laplace equation?
8. A pipe is lying on the ground. Water is coming out of it at a speed of 6 m/s. If the pipe mouth is lifted to a height of 1 m above the ground, what is the speed of water to come out?
9. A 60 mm pipe is connected to a reservoir. Water level in the reservoir is 5 m from the ground level. The water is discharged through the pipe in the air at the ground level. Determine the discharge.
10. A curved solid body is moving through still water at a speed 6 m/s produces a water velocity of 4 m/s at a point 1 m ahead from the nose. What is the difference in pressure between the nose of the body and the point 1 m ahead?

4 Basic Concepts of Viscous Fluid Flows

4.1 INTRODUCTION

Viscous flow in fluid dynamics is a major class of flow where there is resistance in motion. The viscosity of a fluid articulates its resistance to shearing flows, in which neighboring fluid layers are in relative motion. A simple example of such a shearing flow is a planar Couette flow, where a fluid is contained between two infinitely large plates, one is fixed and the other one is in parallel motion at constant speed. Although viscosity plays a role in general flows, it is easy to define and visualize in a Couette flow. If the speed of the top plate is low enough, then in a steady state, the fluid particles move parallel to it, and their speed varies from the bottom to the top. Each layer of fluid moves faster than the one just below it, and friction between them gives rise to a force resisting their relative motion. The role of viscosity near the boundary is focused in this chapter with some examples, like flow through a pipe or channel, similarity principles, viscous boundary layers, some exact solutions of Navier–Stokes equations, Stokes's flows, etc.

4.2 FLOW THROUGH A PIPE (HAGEN–POISEUILLE FLOW)

When the fluid moves due to the pressure gradient in the direction of x-axis along the pipe, and the pressure may be regarded as constant perpendicular to the pipe axis, is known as Hagen–Poiseuille flow. Here the flow through a straight pipe of the circular cross-section of diameter 2R in which the fluid velocity at the wall is zero and reaches a maximum value on the axis, $y=0$ (Figure 4.1). Due to the frictional effect, individual layers act on each other with a shearing stress which is proportional to the velocity gradient $\frac{du}{dy}$. Hence, a fluid particle is accelerated by the pressure gradient and retarded by the frictional force. Here the inertia forces are absent because along every streamline the velocity remains constant. The flow of this kind is known as laminar flow. To establish the condition of equilibrium, we consider a coaxial fluid cylinder of length l and radius y (Figure 4.1). The condition of equilibrium in x direction requires that the pressure force $(p_1 - p_2)\pi\ y^2$ acting on the faces of the cylinder is equal to the shear $2\pi y l \tau$ acting on the circumferential area. So, we have the shear stress τ as:

$$\tau = \frac{p_1 - p_2}{l}\frac{y}{2} \tag{4.1}$$

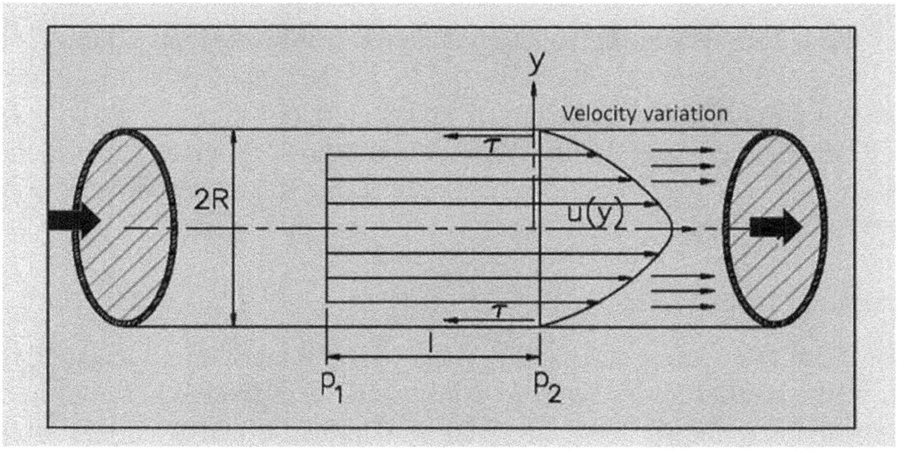

FIGURE 4.1 Laminar flow through a pipe.

DOI: 10.1201/9781003000020-4

From the law of friction, we can write as:

$$\tau = -\mu_f \frac{du}{dy} \tag{4.2}$$

Equation (4.1) leads to $\frac{du}{dy} = -\frac{p_1 - p_2}{\mu_f l}\frac{y}{2}$ and on integration, one gets

$$u(y) = \frac{p_1 - p_2}{\mu_f l}\left(C - \frac{y^2}{4}\right) \tag{4.3}$$

Here the constant of integration C is obtained from the no-slip condition at the wall:

$$u = 0 \quad \text{at} \quad y = R \tag{4.4}$$

so that $C = \frac{R^2}{4}$. Therefore,

$$u(y) = \frac{p_1 - p_2}{4\mu_f l}\left(R^2 - y^2\right) \tag{4.5}$$

The maximum velocity u_m on the axis becomes

$$u_m = \frac{p_1 - p_2}{4\mu_f l} R^2 \tag{4.6}$$

The volume Q flowing through a section per unit time

$$Q = 2\pi \frac{p_1 - p_2}{4\mu_f l}\left[\int_0^R R^2 y dy - \int_0^R y^2 y dy\right] = \frac{\pi R^4 (p_1 - p_2)}{8\mu_f l} \tag{4.7}$$

Equation (4.7) states that the volume rate of flow is proportional to the first power of pressure drop per unit length $\frac{(p_1 - p_2)}{l}$ and to the fourth power of the radius of the pipe. If $\bar{u} = \frac{Q}{\pi R^2}$ is introduced, then

$$p_1 - p_2 = 8\mu_f \frac{l}{R^2}\bar{u} \tag{4.8}$$

This is known as the Hagen–Poiseuille equation of laminar flow through a pipe. This flow is generated due to only pressure differences in the pipe. If $p_1 = p_2$, there is no fluid flow in the pipe. This type of laminar flow exists in reality only for relatively small radii and small flow velocities. The character of the fluid velocity changes completely for larger radii and flow velocities because the pressure drops ceases to be proportional to first power of $\bar{u}$ (equation 4.8), but it comes approximately proportional to $\bar{u}^2$, which will be discussed in detail later.

4.3 SIMILARITY PRINCIPLES

In Hagen–Poiseuille laminar flow, every fluid particle moves under the influence of pressure and frictional forces, where the inertial forces are zero everywhere. In a diverged or converged channel, fluid particles are acted upon by inertia force in addition to pressure and frictional forces. Now we are interested in the flow conditions under which the different fluids about two geometrically comparable bodies show geometrically similar streamlines. The motions of geometrically similar streamlines are called dynamically similar flows. For two different flows about two geometrically comparable bodies like spheres, the following situation should be fulfilled: for all geometrically similar points, the forces acting on a fluid particle must bear a fixed ratio at every instant of time.

The condition of similarity is fulfilled at all corresponding points for the inertial and viscous forces when the ratio of inertial and viscous forces is the same. In a motion parallel to x-axis, the inertia force per unit volume has the magnitude of $\rho_f \frac{Du}{Dt}$, where ρ_f is the density of fluid; u is the component of velocity in x-direction and $\frac{D}{Dt}$ indicates the substantive derivative. For steady flow, the inertia force will be like $\rho_f u \frac{\partial u}{\partial x}$, where $\frac{\partial u}{\partial x}$ is the change in velocity with respect to the position x. Thus, the inertia force per unit volume is $\rho_f u \frac{\partial u}{\partial x}$ and the viscous force per unit volume is $\frac{\partial \tau}{\partial y}\left(= \mu_f \frac{\partial^2 u}{\partial y^2}\right)$. The ratio between the inertia force and viscous force is a constant as:

$$\frac{\text{Inertia force}}{\text{Viscous force}} = \frac{\rho_f u \frac{\partial u}{\partial x}}{\mu_f \frac{\partial^2 u}{\partial y^2}} = \text{constant} \tag{4.9}$$

The condition is that at all corresponding points, the ratio of the inertia force to the viscous force must be constant.

The velocity u at some point in the velocity field is proportional to the free stream velocity U, the velocity gradient $\frac{\partial u}{\partial x}$ is proportional to $\frac{U}{d}$, and similarly $\frac{\partial^2 u}{\partial y^2} \propto \frac{U}{d^2}$, where d is the characteristic linear dimension of the body. Then the ratio is

$$\frac{\text{Inertia force}}{\text{Viscous force}} = \frac{\rho_f u \frac{\partial u}{\partial x}}{\mu_f \frac{\partial^2 u}{\partial y^2}} = \frac{\rho_f\, U^2/d}{\mu_f U^2/d^2} = \frac{Ud}{\upsilon_f} \tag{4.10}$$

with $\upsilon_f = \frac{\mu_f}{\rho_f}$ is the kinematic viscosity, dimension is cm^2/s. Therefore, the condition of similarity is fulfilled if the Reynolds number $\frac{Ud}{\upsilon_f}$ has the same value in both flows. The quantity $\frac{Ud}{\upsilon_f}$ is a dimensionless number because it is the ratio of the two forces, which is known as *Reynolds number.* Thus, two flows are similar when the Reynolds number $\frac{Ud}{\upsilon_f}$ is equal for both. This principle was first articulated by Reynolds (1883) in connection to the flow through pipes and is known as the *Reynolds Principle of similarity.* The Reynolds number is dimensionless and can be at once confirmed by considering the units of the physical parameters. Instead of considering the condition of dynamic similarity, the method of indices can also be used, which is the combination of velocity, linear dimension of a body, density, and viscosity (Schlichting, 1968).

4.3.1 Dimensional Analysis

The dimensional analysis is important to reduce the number of physical variables in the experimental investigations. All physical phenomena can be expressed in terms of basic physical dimensions. In fluid mechanics, usually mass M, length L, time T, and temperature θ are the basic dimensions, and these may be known as $MLT\theta$ system. Sometimes the force F is used instead of mass M to obtain the set as $FLT\theta$. The relationships between the dependent and independent variables are studied in terms of basic dimensions to determine the physical phenomena. There are several methods to reduce the number of variables to obtain the dimensionless parameters which are very important from the fluid dynamics point of view. The new group of dimensionless parameters obtained from the number of variables will be independent of the system of units chosen to use. This analysis called dimensional analysis is used in a variety of physical problems. There are two common methods: (i) Raleigh's method and (ii) Buckingham *Pi* theorem (Ojha et al., 2010). The method of Buckingham *Pi* theorem is discussed in this section because it is a simple one generally used for fluid mechanics. The following steps are used for the dimensional analysis (Ojha et al., 2010):

1. Let us consider $m = 3$ independent fundamental units, such as length L, time T, and mass M or force F.
2. There are n physical quantities in a phenomenon whose dimensional quantities may be expressed in m fundamental units.
3. The dimensional quantity A_0 (dependent variable) may be related to the independent dimensional quantities $A_1, A_2, A_3, \ldots.\ A_{n-1}$ by

$$A_0 = KA_1^{y_1} A_2^{y_2} A_3^{y_3} \ldots A_{n-1}^{y_{n-1}} \tag{4.11}$$

   where K is the dimensionless quantity, and $y_1, y_2, y_3, \ldots y_{n-1}$ are the integer exponents.

4. Equation (4.11) is the independent of system of units chosen and is dimensionally homogeneous, that is, the quantities on both sides of the equation have the same dimensions.

4.3.2 Derivations of Some Dimensionless Numbers

Let the position of a point in the space around the geometrically similar bodies be indicated by the coordinates (x, y, z), then the ratios $\frac{x}{d}, \frac{y}{d}$ and $\frac{z}{d}$ are its dimensionless coordinates, where d is a nominal distance. Similarly, the velocity components are also made dimensionless by the free stream velocity U, thus $\frac{u}{U}, \frac{v}{U}, \frac{w}{U}$ are dimensionless velocity components and the pressure p and shear stress τ are made dimensionless by referring them to the double of the dynamic head, i.e., $\rho_f U^2 : \frac{p}{\rho_f U^2}, \frac{\tau}{\rho_f U^2}$.

For two geometrically similar systems with *equal Reynolds number* $\frac{Ud}{\upsilon_f}$, the dimensionless quantities $\frac{u}{U}, \frac{v}{U}, \frac{w}{U} \ldots \frac{p}{\rho_f U^2}, \frac{\tau}{\rho_f U^2}$ depend only on the dimensionless coordinates $\frac{x}{d}, \frac{y}{d}, \frac{z}{d}$. If the two systems are geometrically similar, but not dynamically, i.e., if the *Reynolds numbers are different,* the dimensionless quantities under coordinates must also depend on the characteristic quantities U, d, ρ_f, μ_f of two systems. In principle, the physical laws are independent of systems of units. The dimensionless quantities $\frac{u}{U}, \frac{v}{U}, \ldots \frac{p}{\rho_f U^2}, \frac{\tau}{\rho_f U^2}$ can only be dependent on a dimensionless combination of U, d, μ_f, ρ_f, which is unique as the Reynolds number $\frac{Ud}{\upsilon_f}$. Therefore, for two geometrically similar systems which have different Reynolds numbers, the dimensionless quantities $\frac{u}{U}, \frac{v}{U}, \ldots \frac{\tau}{\rho_f U^2}$ can only depend on $\frac{x}{d}, \frac{y}{d}, \frac{z}{d}$ and Reynolds number $\frac{Ud}{\upsilon_f}$. Dimensionless analysis concludes that for a geometrically similar system, this coefficient can only be dependent on the group of parameters U, d, ρ_f, μ_f i.e., on Reynolds number $Re = \frac{\rho_f U d}{\mu_f}$.

It is important to note that the Reynolds similarity principle is valid only if the underlying assumptions are satisfied, i.e., if forces acting in the flow are due to friction and inertia only. Other important dimensionless numbers are discussed for some specific applications.

Froude number: The ratio of the inertia force $\rho_f U^2 d^2$ to the weight of the fluid $\rho_f g d^3$ gives U^2/gd. The square root of this dimensionless ratio is known as Froude Number. It is defined as:

$$Fr = \sqrt{\frac{\text{Inertial force}}{\text{Gravitational force}}} = \frac{U}{\sqrt{gd}} \tag{4.12}$$

William Froude, a naval architect, who studied the surface waves produced by the motion of a ship proposed this number. He indicated the relative importance of inertial force and gravitational force in fluid motion. If $Fr > 1$, then the inertia effects outweigh the gravity effect. This dimensionless number Fr is very important to the flow with free surface, for example, in open channel flow, flow over dams or spillways.

Mach number: It is the ratio of the fluid velocity to the speed of sound in the same fluid in the same state. It is an important dimensionless parameter useful in the analysis of compressible fluid flow. The ratio is defined as:

$$Mc = \frac{U}{c} \tag{4.13}$$

where c is the velocity of sound. For perfect gas, the velocity of sound is $\sqrt{k_s R_{gc} T}$, where k_s is the specific heat ratio, R_{gc} is the gas constant, and T is the absolute temperature. Hence the Mach number Mc can be computed by

$$Mc = \frac{U}{\sqrt{k_s R_{gc} T}} \tag{4.14}$$

Based on the Mach number Mc, the flow is defined as several kinds, such as incompressible flow $(Mc \ll 1)$, subsonic flow $(Mc < 1)$, transonic flow $(Mc \approx 1)$, Sonic flow $(Mc = 1)$, supersonic flow $(Mc > 1)$, and hypersonic flow $(Mc \gg 1)$ (Ojha et al., 2010). Here we restrict ourselves in this book to only the incompressible fluid flow $(Mc \ll 1)$.

Example 4.1

If x is the distance from the initial height of a falling body near a surface, w is the weight of the body, g is the acceleration due to gravity, and t represents the time of falling the body, find the relation of the distance x as a function of w, g, and t, using Buckingham Pi theorem.

Solution

From Buckingham Pi theorem, the fundamental units of force F, length L and time T are involved with the four physical quantities $A_0 = x$, $A_1 = w$, $A_2 = g$, and $A_3 = t$. Here $m = 3$ and $n = 4$ (following steps 1 and 2). Here we assume a relation of the form as:

$$x = f(w, g, t) = Kw^{y_1} g^{y_2} t^{y_3} \quad (1)$$

where K is an arbitrary dimensionless constant. The relationship can be written as: $[x] = K[w]^{y_1} [g]^{y_2} [t]^{y_3}$

$$\text{Or, } F^0 L^1 T^0 = K(F)^{y_1} \left(LT^{-2}\right)^{y_2} (T)^{y_3} = KF^{y_1} L^{y_2} T^{y_3 - 2y_2} \quad (2)$$

Equating the power from both sides of equation (1), one gets
$y_1 = 0$ for F, $y_2 = 1$ for L, and $y_3 - 2\ y_2 = 0$ for T
Solving the above equations, one gets from equation (1) as:

$$x = Kw^0 g^1 t^2 = Kgt^2 \quad (3)$$

From the mechanics, the constant K is given by $\frac{1}{2}$. So, the equation is $x = \frac{1}{2} gt^2$, where $\frac{1}{2}$ does not come from the dimensional analysis.

Example 4.2

A fluid of density ρ_f and viscosity μ_f is flowing through a tube of diameter d with a velocity U. Establish the Reynolds number of the flow from the physical quantities using the dimensional analysis Buckingham Pi theorem.

Solution

From the problem, it is realized that the flow is a function of all four quantities density ρ_f, viscosity μ_f, diameter d and velocity U; but the fundamental dimensions (M, L, T) are involved with the four physical quantities. We will choose ρ_f, viscosity μ_f, and velocity V, we have $m = 3$. Let us assume a relation of the form as:

$$d = f(\rho_f, \mu_f, V) = K\rho_f^{y_1} \mu_f^{y_2} U^{y_3} \quad (4)$$

where K is an arbitrary dimensionless constant. The relation can be written as: $[d] = [\rho_f]^{y_1} [\mu_f]^{y_2} [U]^{y_3}$
Or, $M^0 L^1 T^0 = \left(M^{y_1} L^{-3y_1}\right)\left(M^{y_2} L^{-y_2} T^{-y_2}\right)\left(L^{y_3} T^{-y_3}\right) = M^{y_1 + y_2} L^{-3y_1 - y_2 + y_3} T^{-y_2 - y_3}$
Equating the power from both sides of the above equation, one gets

$$y_1 + y_2 = 0, -3y_1 - y_2 + y_3 = 1, -y_2 - y_3 = 0$$

Solving the above equations, one gets $y_1 = -1$, $y_2 = 1$, $y_3 = -1$

$$\text{Therefore, } d = K\rho_f^{-1} \mu_f^1 U^{-1} \quad (5)$$

$K = \dfrac{Ud}{\mu_f/\rho_f}$ is the dimensionless number known as Reynolds number. Using this idea, dimensionless numbers can be derived.

4.4 DRAG AND LIFT FORCES

A fluid flowing past the surface of a body exerts a force on it. It makes no difference whether the fluid flows past a stationary body or the body moves through a stationary volume of fluid. When a fluid flows with a steady and uniform free stream velocity U, and it comes across a body of a curved surface, then the fluid exerts both viscous tangential force and normal pressure force on the surface of the body (Figure 4.2a). The resultant force exerted on the body which is parallel to the direction of free stream

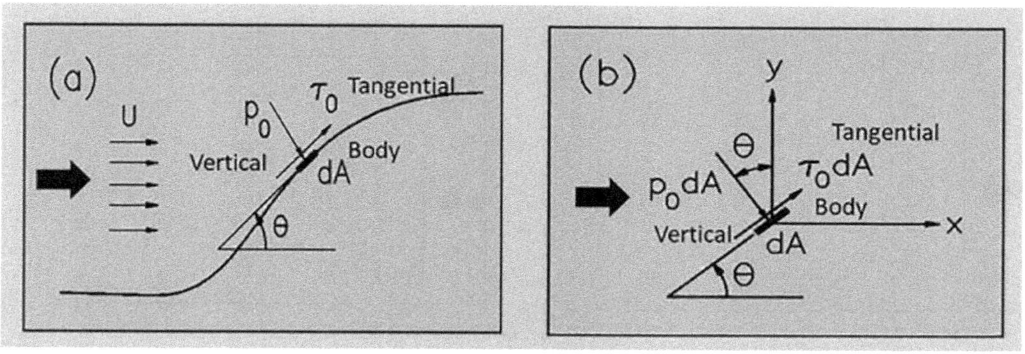

FIGURE 4.2 (a) Tangential and normal pressure forces, (b) Horizontal and vertical pressure forces.

flow (U) is called the *drag*. The lift is the component of this force that acts perpendicular to the free stream velocity U. Lift is always accompanied by a drag force, which is the component of the surface force parallel to the flow direction. For an element dA on the surface, the forces are resolved into the horizontal and vertical components (Figure 4.2b). Integrating over the entire surface of the body, the drag force is given by

$$F_D = \int \tau_0 \cos\theta dA + \int p_0 \sin\theta dA \tag{4.15}$$

and the lift force is given by

$$F_L = \int \tau_0 \sin\theta dA - \int p_0 \cos\theta dA \tag{4.16}$$

If the body shape, the distributions of the tangential stress τ_0 and the pressure p_0 along the surface are known, these integrations can be carried out. These distributions are often difficult to obtain either experimentally or theoretically.

The total drag of a body is made up of two components: frictional drag and pressure drag. The relative properties of these two depend on the shape of the body and the flow conditions. It is noted that the drag and lift forces are the results of a combination of the viscous stress and the pressure acting on the surface of the body. For example, only viscous shear drag will be produced on the plate, when the plate surface is aligned with the flow (Figure 4.3a). To achieve both drag and lift, the plate must be oriented at an angle to the flow (Figure 4.3b). A curved or irregular body surface can only be subjected to drag and lift forces. When the flow is perpendicular to the plate, which acts as a bluff body, the pressure drag is created (Figure 4.3c). The pressure drag is caused by changes in the momentum of the fluid. The resultant shear drag is negligible in this case because the shear stresses act equally up and down of the front surface of the plate. Note that the lift is also zero because no effect produces in the vertical direction perpendicular to the flow.

4.4.1 Drag and Lift Analysis

The total force consists of forces due to shear stresses and normal pressure acting on the surface of the body. The dimensionless quantities are important to describe the total force, which acts on the body. The dimensional quantities are L, ρ_f, μ_f, g, U, where L is the length of the body, ρ_f is the density, μ_f is the kinematic viscosity, g is the acceleration due to gravity, and U is the free-stream velocity. Using these quantities, drag and lift forces can be written as:

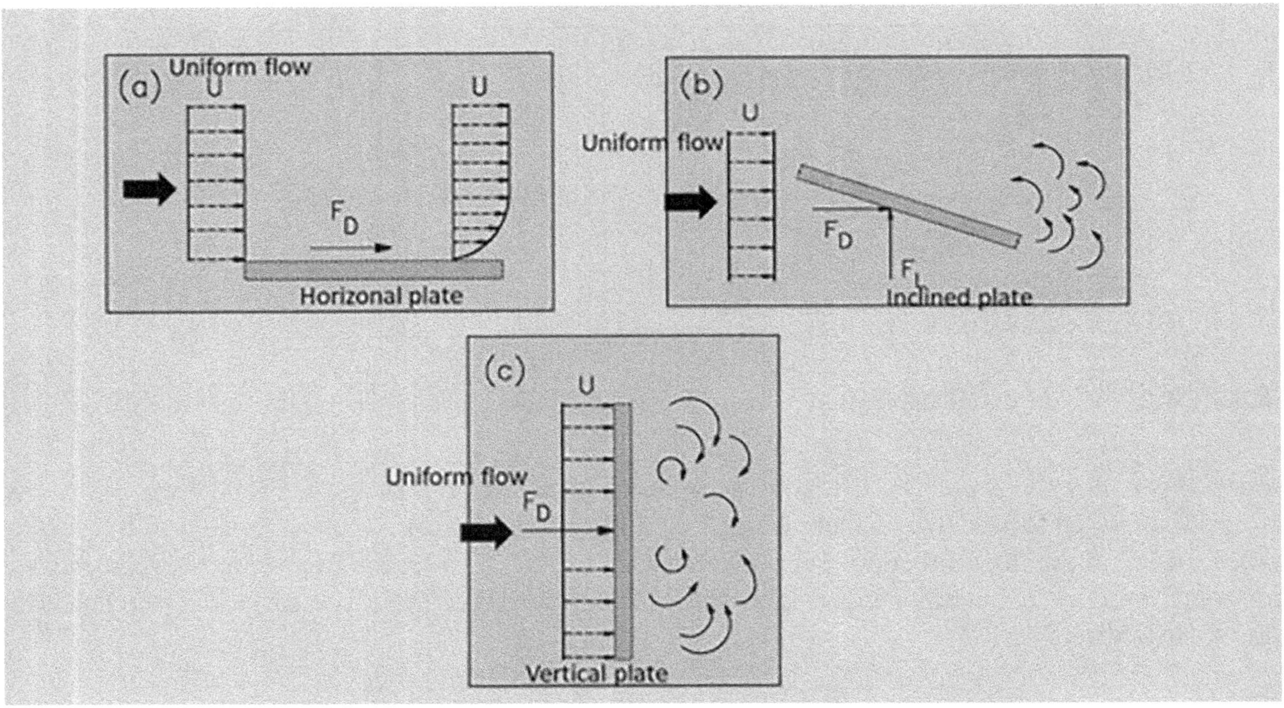

FIGURE 4.3 (a) Drag force aligned to the flow, (b) Drag and lift forces when plate is oriented with the flow, (c) Drag force when plate is perpendicular to the flow. (Modified from Hibbeler, 2015.)

$$F_D = f_1(L, \rho_f, \mu_f, g, U) \tag{4.17}$$

and

$$F_L = f_2(L, \rho_f, \mu_f, g, U) \tag{4.18}$$

Using the dimensionless analysis, it can be shown as:

$$\frac{F_D}{\rho_f L^2 U^2} = f_3\left(\frac{\rho_f LU}{\mu_f}, \frac{U^2}{Lg}\right) = f_3(Re, Fr) \tag{4.19}$$

$$\frac{F_L}{\rho_f L^2 U^2} = f_4\left(\frac{\rho_f LU}{\mu_f}, \frac{U^2}{Lg}\right) = f_4(Re, Fr) \tag{4.20}$$

Here *Re* is Reynolds number and *Fr* is Froude number. Instead of choosing $\rho_f U^2$ as the reference, if we choose the stagnation pressure $\frac{\rho_f U^2}{2}$, the dimensionless coefficients for the lift and drag can be written as:

$$C_D = \frac{F_D}{\frac{1}{2}\rho_f U^2 A} \text{ and } C_L = \frac{F_L}{\frac{1}{2}\rho_f U^2 A} \tag{4.21}$$

where *A* is the surface area of the body.

Therefore,

$$C_D = f_3(Re, Fr) \text{ and } C_L = f_4(Re, Fr) \tag{4.22}$$

Here the dimensionless Froude number Fr is important only in the flow of liquid with the free surface; otherwise, the dimensionless lift and drag coefficients are dependent on the Reynolds number Re, that is,

$$C_D = f_3(Re) \text{ and } C_L = f_4(Re) \tag{4.23}$$

where f_3 and f_4 are the functions of Reynolds number Re representing the coefficients of lift and drag, respectively. The conclusion from the similarity principle with respect to the Reynolds number is only valid as long as the gravitational and elastic forces are not considered.

The aerodynamic force is the force exerted on a body by the air in which the body is immersed and is due to the relative motion between the body and the air. The aerodynamic force arises from two cases: normal force due to pressure on the surface of the body, and the friction force due to the viscosity of the air, and also as skin friction. For example, lift associated with the wing of the aircraft. It is widely generated by the streamline bodies such as propellers, kites, helicopters, turbines, hydrofoils in water, etc.

4.5 VISCOUS BOUNDARY LAYERS

As the fluid flows through say a pipe or a channel or a plate, at the wall the fluid sticks to it, which means the frictional force delays the motion of the fluid in a thin layer near the wall. In that thin layer, the fluid velocity increases from zero at the wall (no-slip condition) to its full value which corresponds to external frictionless or free-stream flow. This layer is known as the boundary layer. There is a thin layer δ near the wall where the velocity is considerably smaller than at a larger distance from it. The thickness of this boundary layer augments along the plate in a downstream direction (Figure 4.4).

In front of the leading edge of the plate, the velocity distribution is uniform. With increasing the distance from the leading edge along downstream, the thickness of the retarded layer increases continuously, and the thin layer $\delta(x)$ is an increasing function of x; because the increasing quantity of fluid

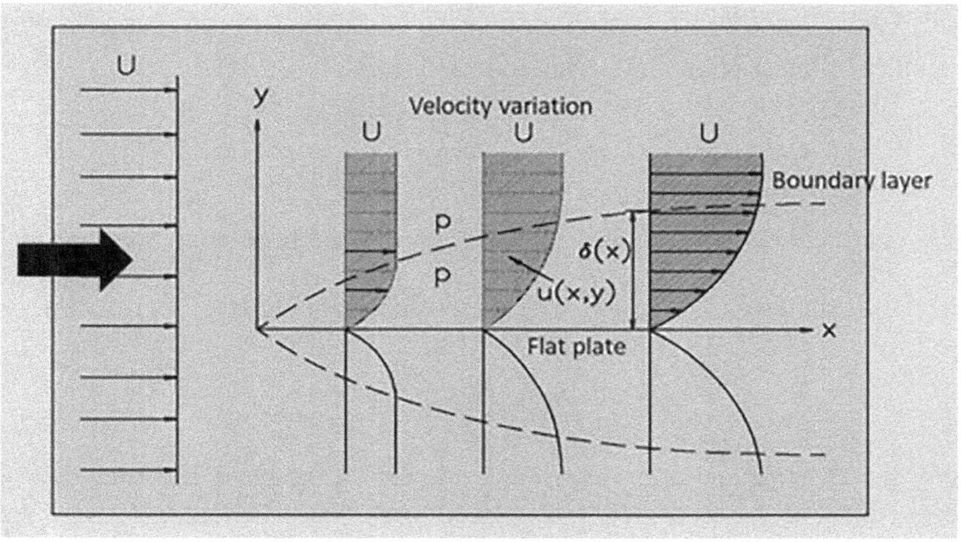

FIGURE 4.4 Boundary layer over a flat plate at zero incidences. (Modified from Schlichting and Gersten, 2000.)

becomes affected. The thickness of the boundary layer decreases with decreasing viscosity. On the other hand, even with the small viscosities for large Reynolds numbers, the shearing stress $\tau = \mu_f \dfrac{du}{dy}$ within the boundary layer is substantial due to the large velocity gradient across the flow, whereas outside the boundary layer, they are very small. The physical picture suggests that the flow field in the case of small viscosity can be divided into two regions: thin boundary layer region near the wall, in which frictional force must be considered and the region outside the boundary layer, the force due to friction is very small and may be neglected, and the theory of perfect or ideal fluid shows a good approximation there.

The decelerated fluid particles in the boundary layer do not remain in the thin layer, which stick to the boundary along the whole wetted length of the wall. In some cases, the boundary layer increases its thickness considerably in the downstream and the flow in the boundary layer becomes reversed, which causes the decelerated fluid particles to be forced outward; and consequently, the boundary layer separates from the wall. So, there is boundary layer separation. The occurrence of separation is always associated with the formation of vortices having large energy losses in the wake of the body. It occurs primarily near blunt bodies, such as circular cylinders and spheres. Behind such a body there exists a region of strongly decelerated flow, called wakes, in which the pressure distribution moves away considerably from that in a frictionless fluid.

4.5.1 Estimation of Boundary Layer Thickness

The boundary layer thickness that is not separated can be easily predicted. The friction force is neglected with respect to inertia force outside the boundary layer due to negligible viscosity. In the boundary layer, the inertial and viscous forces are of comparable order of magnitude. The inertia force per unit volume is $\rho_f u \dfrac{\partial u}{\partial x}$. For a plate of length l, the gradient $\dfrac{\partial u}{\partial x}$ is proportional to $\dfrac{U}{l}$, where U denotes the free-stream velocity outside the boundary layer. Hence the inertia force is of order $\rho_f \dfrac{U^2}{l}$. The friction force per unit volume is $\dfrac{\partial \tau}{\partial y}$ which is equal to $\mu_f \dfrac{\partial^2 u}{\partial y^2}$ for laminar flow. The gradient $\dfrac{\partial u}{\partial y}$ in the direction perpendicular to the wall is of order $\dfrac{U}{\delta}$, so that the friction force per unit volume is $\dfrac{\partial \tau}{\partial y} = \mu_f \dfrac{\partial^2 u}{\partial y^2} \sim \mu_f \dfrac{U}{\delta^2}$. From the condition of equality of the friction and inertia forces, one gets

$$\mu_f \frac{U}{\delta^2} \sim \rho_f \frac{U^2}{l} \Rightarrow \delta \sim \sqrt{\frac{\mu_f l}{\rho_f U}} \sim \sqrt{\frac{\upsilon_f l}{U}} \tag{4.24}$$

Blasius (1908) found that the proportionality factor is approximately equal to 5. Hence for Laminar flow in the boundary layer, we have $\delta = 5\sqrt{\dfrac{\upsilon_f l}{U}}$. The dimensionless boundary layer thickness is as follows:

$$\frac{\delta}{l} = 5\sqrt{\frac{\upsilon_f}{Ul}} = \frac{5}{\sqrt{R_l}} \tag{4.25}$$

where R_l denotes the Reynolds number related to the length of the plate. It is seen that boundary layer thickness is proportional to $\sqrt{\upsilon_f}$ and $\sqrt{l}$. If l is replaced by the variable distance x from the leading edge of the plate, it is seen that $\delta \propto \sqrt{x}$. On the other hand, the relative boundary layer thickness δ/l reduces with increasing Reynolds number R_l so that in the limiting case of frictionless flow with $R_l \to \infty$, the boundary layer thickness vanishes.

To estimate the shearing stress τ_0 at the wall and the total drag, we use Newton's law of friction at the wall ($y = 0$) as:

$$\tau_0 = \mu_f \left.\frac{\partial u}{\partial y}\right|_{y=0} \tag{4.26}$$

Here $\left.\frac{\partial u}{\partial y}\right|_0 \sim \frac{U}{\delta}$, then one gets $\tau_0 \sim \mu_f \frac{U}{\delta}$. Substituting the value of δ from equation (4.25), shearing stress τ_0 is rewritten as:

$$\tau_0 \sim \mu_f\, U \sqrt{\frac{\rho_f U}{\mu_f l}} \sim \sqrt{\frac{\mu_f \rho_f U^3}{l}} \tag{4.27}$$

The frictional shear stress near the wall is proportional to $U^{3/2}$. The dimensionless shearing stress with reference to $\rho_f U^2$, we have

$$\frac{\tau_0}{\rho_f U^2} \sim \sqrt{\frac{\mu_f}{\rho_f U l}} \sim \frac{1}{\sqrt{R_l}} \tag{4.27a}$$

The dimensionless shearing stress depends only on Reynolds number. The total drag D on the plate is equal to $bl\tau_0$, where b denotes the width of the plate, i.e., $D \sim bl\tau_0$. Therefore, the total drag is as:

$$D \sim b\sqrt{\rho_f \mu_f U^3 l} \tag{4.28}$$

The laminar frictional drag is seen to be proportional to $U^{3/2}$ and $l^{1/2}$. The dimensionless drag coefficient C_D of a plate in parallel laminar flow is given by

$$C_D = \frac{D}{\frac{1}{2}\rho_f U^2 bl} \sim \sqrt{\frac{\mu_f}{\rho_f U l}} = \frac{1}{\sqrt{R_l}}$$

$$\Rightarrow C_D = \frac{1.328}{\sqrt{R_l}} \tag{4.29}$$

where the constant of proportionality is 1.328 from Blasius's solution (1908).

The meaning of boundary layer thickness is, to a certain extent, arbitrary thickness because the transition of velocity in the boundary layer takes place asymptotically to the outside of the boundary layer. The velocity in the boundary layer attains a value, which is very close to the external velocity U at a small distance from the wall. That distance from the wall is called the *boundary layer thickness* δ where the velocity differs by 1% from the external velocity. Instead of boundary layer thickness, a more realistic meaningful definition of the thickness known as displacement thickness δ_1 is used (Figure 4.5) (Ojha et al., 2010):

$$U\delta_1 = \int_0^{\delta} \left(U - u(y)\right) dy \tag{4.30}$$

The displacement thickness in the boundary layer indicates the distance by which the external streamlines are shifted owing to the development of the boundary layer and also indicates the decrease in total

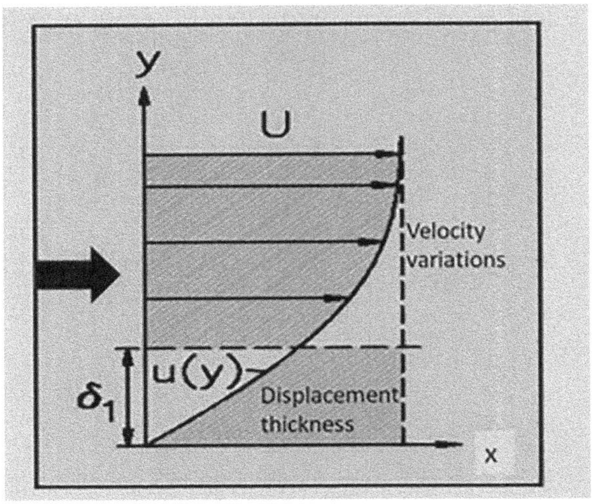

FIGURE 4.5 Displacement thickness δ_1 of the boundary layer.

flow rate due to friction. In the case of parallel flow past a flat plate at zero incidence, the displacement thickness δ_1 is about one-third of the boundary layer thickness δ.

4.5.2 Flow Separation and Vortex Formation

The development of the boundary layer over a flat plate at zero incidence in parallel flow is particularly straightforward and simple because the static pressure remains constant in the whole fluid flow (Figure 4.4). Since the velocity remains constant at the outside of the boundary layer, the same is applied to the pressure because of the frictionless flow, where Bernoulli's equation remains valid. So, the pressure remains constant over the whole width of the boundary layer at a given distance x. Thus, the pressure within the boundary layer is equal to the pressure outside the boundary layer at the same distance x. The appearance of boundary layer separation is closely associated with the pressure distribution in the boundary layer. *In the case of the motion past a flat plate, the pressure remains constant throughout the boundary layer. No flow separation takes place in the boundary on a flat plate, because there is no backflow* (Figure 4.4). Figure 4.4 shows that there is no boundary layer separation, i.e., no occurrence of backflow over the flat plate. For the case of an arbitrary body, the pressure outside the boundary layer varies along the wall, when the outer pressure is impressed on the boundary layer.

Here we consider the flow over a circular cylinder in frictionless symmetric flow to explain the boundary layer separation (Figure 4.6) (Schlichting and Gersten, 2000). In Figure 4.6, due to pressure drop, the fluid particles are accelerated on the upstream half of the cylinder from the point P_D to P_E, while the flow is decelerated on the downstream half of the cylinder from P_E to P_F due to rise of pressure. This happens due to the transformation of energy. When the flow is started up the motion, initially it is nearly frictionless and remains so as long as the boundary layer remains very thin. The pressure energy is transformed into kinetic energy, for a particle moving from P_D to P_E in the outer layer, and while moving from P_E to P_F, the kinetic energy is transformed into pressure (Schlichting, 1968) so that the particle at P_F has the same velocity as it had at P_D.

As shown in Figure 4.6, a fluid particle that moves in the vicinity of the wall of the cylinder in the boundary layer remains under the impact of the same pressure field as that existing outside. The fact is that the external pressure is impressed in the boundary layer. Due to the large frictional force, in the thin boundary layer, a particle consumes much of its kinetic energy on its path from the point P_D to P_E so that the remainder is too small to overcome the *pressure hill* from P_E to P_F. Such a particle cannot move far into the region of increasing pressure between P_E to P_F and its motion is eventually curtailed.

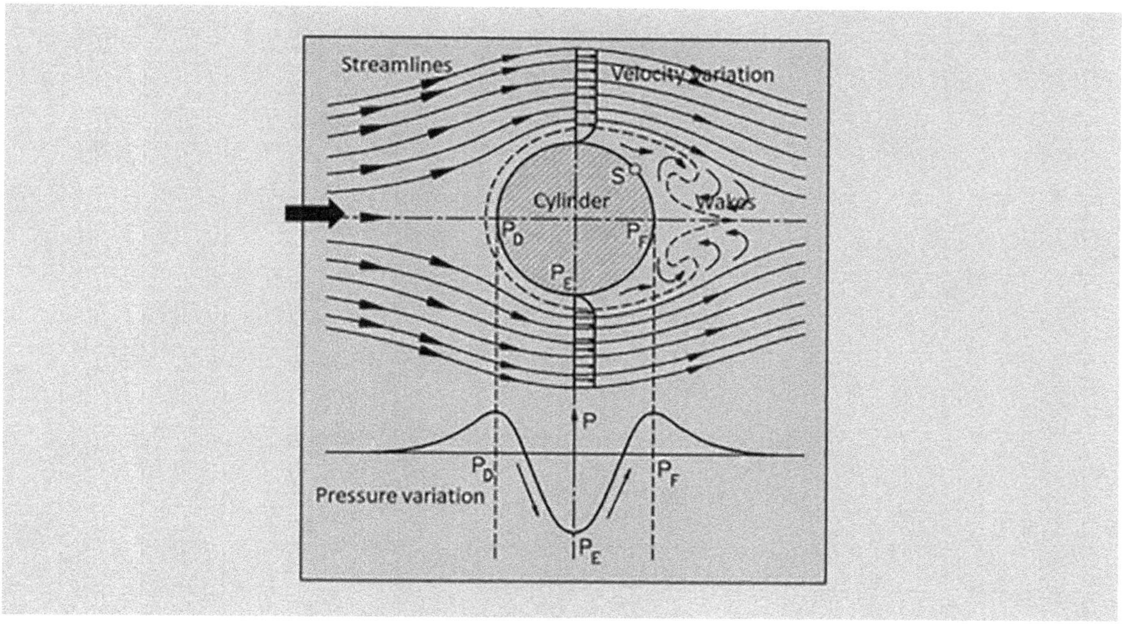

FIGURE 4.6 Boundary layer separation and formation of vortex past a cylinder, with point of separation. (Modified from Schlichting and Gersten, 2000.)

Here separation of flow occurs, as the external or adverse pressure causes it to move in the backward direction.

4.5.2.1 Separation of Flow

The boundary layer separation exists always in the regions where there is an adverse pressure gradient and the possibility of occurrence of separation increases in the case of steep pressure curves, i.e., behind the blunt bodies. At the point of separation, one streamline intersects the wall at a definite angle, and the point of separation itself is determined by the condition that the velocity gradient normal to the wall is zero at the boundary. Therefore, the separation of flow is defined as:

$$\left.\frac{\partial u}{\partial y}\right|_{\text{wall}} = 0 \tag{4.31}$$

which means the velocity gradient normal to the wall is zero at point S in the wall and the streamlines in the boundary layer near the separation point are shown (Figure 4.7). As a result, the backflow close to the wall occurs, and due to the backflow, a strong thick boundary layer takes place with the mass transporting away into the outer flow. The precise location of the point of separation can be determined only with the aid of an exact calculation, i.e., by the integration of the boundary layer equations. Figure 4.7 shows the streamlines in the boundary layer around the point of separation (Yuan, 1970).

The separation of flow, as described for the circular cylinder, can also occur in a highly divergent channel. In front of the throat due to a decrease of pressure in the direction of flow, the velocity is very high, and the flow adheres to the walls completely like a frictionless fluid. However, behind the throat, the boundary layer becomes separated from the walls because of large divergent of channel, and consequently vortices are formed (Schlichting, 1968). At a large distance from the body along downstream, it is possible to distinguish a regular pattern of vortices which moves alternatively clockwise and anticlockwise (Figure 4.8), known as a *Karman Vortex Street* (Schlichting, 1968). In general, it forms in the flow behind a blunt body. The vortex street moves with a velocity u, which is smaller than the flow

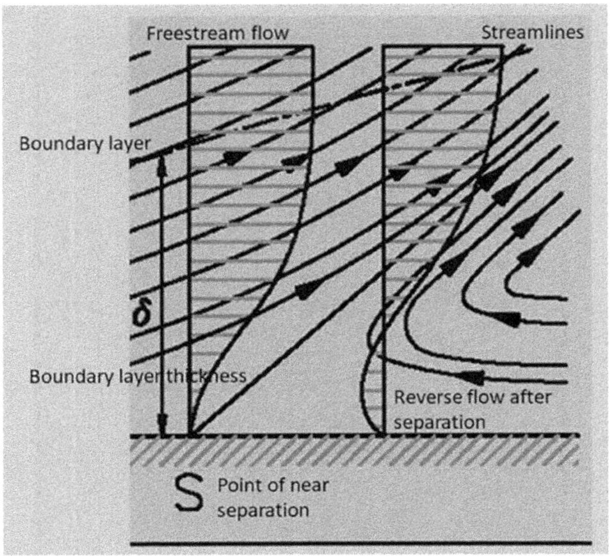

FIGURE 4.7 Boundary layer flow close to separation point S. (Modified from Schlichting and Gersten, 2000.)

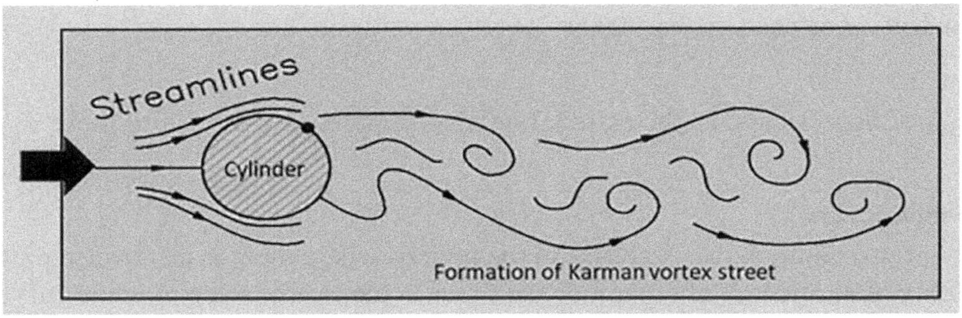

FIGURE 4.8 Sketch of Karman Vortex Street past a cylinder.

velocity U in front of the body. The kinetic energy contained in the velocity field of the vortex street must be continually created as the body moves through the fluid.

4.6 BASIC EQUATIONS OF LAMINAR FLOW (NAVIER–STOKES EQUATIONS)

To obtain a complete and primary understanding of the fluid dynamics problems, the general governing equations of continuity, momentum and energy need to be known. All the fluid motions, whether laminar or turbulent, are governed by those dynamical equations. The complete solution for those equations is not possible because of intractable mathematical difficulties due to high nonlinearity and complexity. However, these equations are solvable subject to great oversimplification in some individual cases.

For a Newtonian fluid (fluids with shear stress proportional to the velocity gradient), the momentum and continuity equations can be written using Cartesian tensor notations as:

$$\rho_f \frac{Du_i}{Dt} = \rho_f X_i + \frac{\partial \sigma_{ij}}{\partial x_j} \tag{4.32}$$

$$\frac{D\rho_f}{Dt} + \rho_f \nabla . u_j = 0 \tag{4.33}$$

where $\frac{D}{Dt} = \frac{\partial}{\partial t} + u_j \frac{\partial}{\partial x_j}, \nabla \equiv \frac{\partial}{\partial x_j}$, t is the time, x_i is the Cartesian coordinates, u_i is the x_i component of fluid velocity and X_i is the body force (such as gravity force), ρ_f is the density of the fluids. The quantity σ_{ij} is the stress on the fluid and is given by

$$\sigma_{ij} = -p\delta_{ij} + \mu_f \left(\frac{\partial u_i}{\partial x_j} + \frac{\partial u_j}{\partial x_i} \right) \tag{4.34}$$

The above relation is the constitutive equation for a Newtonian fluid. Here p is the fluid pressure, μ_f is the viscosity, and δ_{ij} is the Kronecker delta defined as

$$\begin{aligned} \delta_{ij} &= 1 \text{ for } i = j \\ &= 0 \text{ for } i \neq j \end{aligned} \tag{4.35}$$

In these equations, Einstein's summation convention is used, namely there is a summation over the repeated indices (in three-dimensional space, sum from 1 to 3). Here we have employed a Cartesian coordinate system with $x_1 = x$-axis in the direction of flow, $x_2 = y$-axis in the cross- stream (span-wise) direction, and $x_3 = z$-axis (vertical) perpendicular to the flow. In reality, equation (4.32) contains three equations for $i = 1, 2, 3$ and $j = 1, 2, 3$. These three equations are derived just from Newton's second law of motion written for a continuum in a spatial (or Eulerian) frame of reference. These equations are known as the *equations of motion* of fluid particles, which are known as the *Navier–Stokes equations* and it interprets the conservation of momentum. Equation (4.33) represents the equation of conservation of mass and is known as *equation of continuity*. For an incompressible fluid, equations (4.32) and (4.33) in the absence of body forces are simplified to

$$\rho_f \frac{Du_i}{Dt} = -\frac{\partial p}{\partial x_i} + \mu_f \frac{\partial^2 u_i}{\partial x_j^2}, \quad \frac{\partial u_j}{\partial x_j} = 0 \tag{4.36}$$

There are four unknowns $u_i (i = 1, 2, 3)$ and p, and four equations and hence the above system is in closed form mathematically. These equations are solvable when the boundary and initial conditions are specified. In case of viscous fluid, the no-slip conditions on the solid boundary must be satisfied, i.e., on the wall, both the normal and tangential components of the velocity vanish.

The Navier–Stokes and continuity equations for an incompressible fluid flow in three-dimensional Cartesian coordinate system, including the body forces, are written as (Schlichting, 1968):

$$\rho_f \left(\frac{\partial u}{\partial t} + u\frac{\partial u}{\partial x} + v\frac{\partial u}{\partial y} + w\frac{\partial u}{\partial z} \right) = g_x - \frac{\partial p}{\partial x} + \mu_f \left(\frac{\partial^2 u}{\partial x^2} + \frac{\partial^2 u}{\partial y^2} + \frac{\partial^2 u}{\partial z^2} \right) \tag{4.37a}$$

$$\rho_f \left(\frac{\partial v}{\partial t} + u\frac{\partial v}{\partial x} + v\frac{\partial v}{\partial y} + w\frac{\partial v}{\partial z} \right) = g_y - \frac{\partial p}{\partial y} + \mu_f \left(\frac{\partial^2 v}{\partial x^2} + \frac{\partial^2 v}{\partial y^2} + \frac{\partial^2 v}{\partial z^2} \right) \tag{4.37b}$$

$$\rho_f \left(\frac{\partial w}{\partial t} + u\frac{\partial w}{\partial x} + v\frac{\partial w}{\partial y} + w\frac{\partial w}{\partial z} \right) = g_z - \frac{\partial p}{\partial z} + \mu_f \left(\frac{\partial^2 w}{\partial x^2} + \frac{\partial^2 w}{\partial y^2} + \frac{\partial^2 w}{\partial z^2} \right) \tag{4.37c}$$

$$\frac{\partial u}{\partial x} + \frac{\partial v}{\partial y} + \frac{\partial w}{\partial z} = 0 \tag{4.37d}$$

where the set of equation (4.37a–c) represents equations of motion and equation (4.37d) represents the equation of continuity. If the vector notation is used, equation (4.37a–d) can be written as:

$$\rho_f \frac{D\vec{q}}{Dt} = \vec{g} - \text{grad } p + \mu_f \nabla^2 \vec{q} \tag{4.38}$$

$$\nabla.\vec{q} = 0 \tag{4.39}$$

where $\vec{g}$ is the body force components, ∇^2 is the Laplace operator, $\nabla^2 = \frac{\partial^2}{\partial x^2} + \frac{\partial^2}{\partial y^2} + \frac{\partial^2}{\partial z^2}$. By deleting the viscous term $\mu_f \nabla^2 \vec{q}$, one can get Euler's equations of motion (3.40).

The system of equation (4.37) can also be rewritten as:

$$\rho_f \left(\frac{\partial u}{\partial t} + \frac{\partial u^2}{\partial x} + \frac{\partial uv}{\partial y} + \frac{\partial uw}{\partial z} \right) = g_x - \frac{\partial p}{\partial x} + \mu_f \left(\frac{\partial^2 u}{\partial x^2} + \frac{\partial^2 u}{\partial y^2} + \frac{\partial^2 u}{\partial z^2} \right) \tag{4.40a}$$

$$\rho_f \left(\frac{\partial v}{\partial t} + \frac{\partial vu}{\partial x} + \frac{\partial v^2}{\partial y} + \frac{\partial vw}{\partial z} \right) = g_y - \frac{\partial p}{\partial y} + \mu_f \left(\frac{\partial^2 v}{\partial x^2} + \frac{\partial^2 v}{\partial y^2} + \frac{\partial^2 v}{\partial z^2} \right) \tag{4.40b}$$

$$\rho_f \left(\frac{\partial w}{\partial t} + \frac{\partial wu}{\partial x} + \frac{\partial wv}{\partial y} + \frac{\partial w^2}{\partial z} \right) = g_z - \frac{\partial p}{\partial z} + \mu_f \left(\frac{\partial^2 w}{\partial x^2} + \frac{\partial^2 w}{\partial y^2} + \frac{\partial^2 w}{\partial z^2} \right) \tag{4.40c}$$

$$\frac{\partial u}{\partial x} + \frac{\partial v}{\partial y} + \frac{\partial w}{\partial z} = 0 \tag{4.40d}$$

Equation (4.40) are non-linear second-order differential equations and there is no general method to solve the equations. The solutions of the above equations are fully determined physically when the boundary and initial conditions are specified. In the case of viscous fluids, the conditions of no-slip on the solid boundaries must be prescribed, i. e., on the wall both the normal and tangential components of velocity must vanish, that is,

$$v_n = 0 \text{ and } v_t = 0 \text{ on the solid walls} \tag{4.41}$$

Similarly, the Navier–Stokes equations along with the equation of continuity can also be written in cylindrical polar coordinate systems (r, θ, z) for further reference (Schlichting, 1968). If r, θ, z denote the radial, tangential, and axial coordinates of a three-dimensional system, and u_r, u_θ, u_z are the velocity components in respective directions, then for the incompressible fluid flow, equations of motion can be written as:

$$\rho_f \left(\frac{Du_r}{Dt} - \frac{u_\theta^2}{r} \right) = X_r - \frac{\partial p}{\partial r} + \mu_f \left(\nabla^2 u_r - \frac{u_r}{r^2} - \frac{2}{r^2} \frac{\partial u_\theta}{\partial \theta} \right) \tag{4.42a}$$

$$\rho_f \left(\frac{Du_\theta}{Dt} - \frac{u_r u_\theta}{r} \right) = X_\theta - \frac{1}{r} \frac{\partial p}{\partial \theta} + \mu_f \left(\nabla^2 u_\theta + \frac{2}{r^2} \frac{\partial u_r}{\partial \theta} - \frac{u_\theta}{r} \right) \tag{4.42b}$$

$$\rho_f \frac{Du_z}{Dt} = X_z - \frac{\partial p}{\partial z} + \mu_f \nabla^2 u_z \tag{4.42c}$$

$$\frac{\partial u_r}{\partial r} + \frac{1}{r}\frac{\partial u_\theta}{\partial \theta} + \frac{\partial u_z}{\partial z} + \frac{u_r}{r} = 0 \tag{4.42d}$$

where, $\frac{D}{Dt} = \frac{\partial}{\partial t} + u_r \frac{\partial}{\partial r} + \frac{u_\theta}{r}\frac{\partial}{\partial \theta} + u_z \frac{\partial}{\partial z}$ and $\nabla^2 = \frac{\partial^2}{\partial r^2} + \frac{1}{r}\frac{\partial}{\partial r} + \frac{1}{r^2}\frac{\partial^2}{\partial \theta^2} + \frac{\partial^2}{\partial z^2}$

These are fully non-linear differential equations. However, there are simple physical problems which enable one to simplify equations (4.40)–(4.42) into linear differential equations with corresponding boundary conditions. Few exact solutions of the Navier–Stokes equations in some special cases are possible and discussed in the Section 4.8.

4.7 VORTICITY TRANSPORT EQUATIONS

In two-dimensional unsteady flow in the x, y plane neglecting the body forces, equations (4.37) can be written as:

$$\left(\frac{\partial u}{\partial t} + u\frac{\partial u}{\partial x} + v\frac{\partial u}{\partial y}\right) = -\frac{1}{\rho_f}\frac{\partial p}{\partial x} + \nu_f\left(\frac{\partial^2 u}{\partial x^2} + \frac{\partial^2 u}{\partial y^2}\right) \tag{4.43a}$$

$$\left(\frac{\partial v}{\partial t} + u\frac{\partial v}{\partial x} + v\frac{\partial v}{\partial y}\right) = -\frac{1}{\rho_f}\frac{\partial p}{\partial y} + \nu_f\left(\frac{\partial^2 v}{\partial x^2} + \frac{\partial^2 v}{\partial y^2}\right) \tag{4.43b}$$

$$\frac{\partial u}{\partial x} + \frac{\partial v}{\partial y} = 0 \tag{4.43c}$$

If we introduce the vorticity vector, $\nabla \times \vec{q}$, which reduces to one component about the z-axis for two-dimensional flow as:

$$\frac{1}{2}\nabla \times \vec{q} = \Omega_z = \Omega = \frac{1}{2}\left(\frac{\partial v}{\partial x} - \frac{\partial u}{\partial y}\right) \tag{4.44}$$

If the flow is irrotational, then $\nabla \times \vec{q} = 0$. Eliminating the pressure from equation (4.43a and b), one gets

$$\frac{\partial \Omega}{\partial t} + u\frac{\partial \Omega}{\partial x} + v\frac{\partial \Omega}{\partial y} = \nu_f\left(\frac{\partial^2 \Omega}{\partial x^2} + \frac{\partial^2 \Omega}{\partial y^2}\right) \tag{4.45}$$

$$\text{Or, } \frac{D\Omega}{Dt} = \nu_f \nabla^2 \Omega \tag{4.46}$$

Equation (4.46) is known as vorticity transfer equation. The variation of vorticity Ω depends on the local and convective terms, which is the rate of dissipation of vorticity through frictional effects. Here the vorticity equation (4.46) together with the equation of continuity (4.43c) forms a set of equations of two unknown components u and v. These can be possible to form one equation with one unknown variable, introducing stream function variable $\psi(x, y)$ as:

$$u = \frac{\partial \psi}{\partial y};\ v = -\frac{\partial \psi}{\partial x} \tag{4.47}$$

which satisfies the continuity equation (4.43c). Using equation (4.47), equation (4.44) becomes

$$\Omega = -\frac{1}{2}\nabla^2\psi \tag{4.48}$$

and using equation (4.48) the vorticity transport equation (4.45) contains only one unknown ψ, which is written as:

$$\frac{\partial \nabla^2\psi}{\partial t} + \frac{\partial \psi}{\partial y}\frac{\partial \nabla^2\psi}{\partial x} - \frac{\partial \psi}{\partial x}\frac{\partial \nabla^2\psi}{\partial y} = \nu_f \nabla^4\psi \tag{4.49}$$

The vorticity equation (4.49) contains inertial terms on the left-hand side and the frictional term on the right-hand side. This equation shows a fourth-order non-linear partial differential equation in stream function, which is very difficult to solve. Equation (4.49) was solved by Jenson (1959) for a low Reynolds number for the case of flow around a sphere. The resulting patterns of streamlines on the vorticity in the flow field are obtained for different Reynolds numbers ranging from 5 to 40 (Schlichting, 1968).

4.8 SOME EXACT SOLUTIONS OF NAVIER–STOKES EQUATIONS

The exact solution of Navier–Stokes equations is possible for certain particular cases, e.g., parallel flow between two flat plates, Couette flow, flow through a tube or an annular pipe, etc. Let us consider a parallel flow case in which all fluid particles travel along the x-direction. For parallel flow,

$$u \neq 0,\ v = w = 0 \tag{4.50}$$

$$\frac{\partial u}{\partial x} + \frac{\partial v}{\partial y} + \frac{\partial w}{\partial z} = 0 \Rightarrow \frac{\partial u}{\partial x} = 0 \Rightarrow u = u(y, z, t) \tag{4.51}$$

Neglecting the inertial terms and the body forces from the Navier–Stokes equation (4.37), one gets the equation in the x-direction as:

$$\frac{\partial u}{\partial t} = -\frac{1}{\rho_f}\frac{\partial p}{\partial x} + \nu_f\left(\frac{\partial^2 u}{\partial y^2} + \frac{\partial^2 u}{\partial z^2}\right) \tag{4.52}$$

and y- and z-directions

$$0 = \frac{1}{\rho_f}\frac{\partial p}{\partial y} = -\frac{1}{\rho_f}\frac{\partial p}{\partial z} \tag{4.53}$$

where the pressure varies in the x-direction only.

4.8.1 Parallel Flow through a Straight Channel

Let us consider the flow between two parallel straight plates named as plane Poiseuille flow, which constitutes a simple class of motion (Figure 4.9). Consider a steady, two-dimensional laminar flow of a viscous incompressible fluid flowing between two fixed parallel straight plates separated by a distance $2b$.

Let x be the direction of flow, y be the direction perpendicular to the flow and width of the plates parallel to the z-direction, which is assumed to be large compared to $2b$ between two plates; hence the flow may be treated as two-dimensional $\left(\frac{\partial}{\partial z} = 0\right)$. The plates are long enough in the x-direction for the flow to be parallel so that the velocity components v and w are zero everywhere. The equation

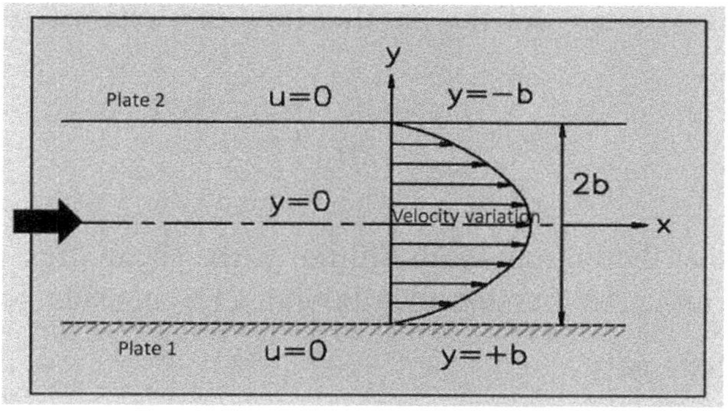

FIGURE 4.9 Parallel flow with parabolic velocity distribution (plane Poiseuille flow).

of continuity then reduces to $\frac{\partial u}{\partial x}=0$ giving $u=u(y)$. In case of steady and parallel flow through the channel, the components are: $u(y)$, $v=w=0$. Here the flow is only due to pressure gradient $\frac{\partial p}{\partial x}$, so the Navier–Stokes equation (4.52) can be written as

$$\frac{\partial p}{\partial x}=\mu_f\frac{\partial^2 u}{\partial y} \tag{4.54}$$

where μ_f is the coefficient of viscosity of the fluid, and the velocity u is a function of y alone.

The no-slip boundary conditions are

$$u=0 \text{ at } y=b \text{ and } y=-b \tag{4.55}$$

Equation (4.54) is solved by integrating using the above boundary conditions (4.55) and is written as:

$$u=\frac{1}{2\mu_f}\frac{\partial p}{\partial x}y^2+C_1y+C_2 \tag{4.56}$$

where C_1 and C_2 are the constants. Applying the boundary conditions (4.55), the constants are determined and the solution for velocity u as:

$$u=-\frac{1}{2\mu_f}\frac{\partial p}{\partial x}\left(b^2-y^2\right) \tag{4.57}$$

which u is the parabolic velocity profile. At $y=0, u=u_{max}$, this yields the maximum velocity u_{max} as follows:

$$u_{max}=-\frac{b^2}{2\mu_f}\frac{\partial p}{\partial x} \tag{4.58}$$

The average velocity u_{av} is given by

$$u_{av}=-\frac{2}{6\mu_f}\frac{dp}{dx}b^2 \tag{4.59}$$

$$\text{So, } u_{max}=\frac{3}{2}u_{av} \tag{4.60}$$

The shear stress at the walls ($y = \pm b$) for the parallel flow in a channel can be determined from the velocity gradient as

$$\tau_{xy} = -2\mu_f \frac{u_{\max}}{b} \tag{4.61}$$

This shear stress is at $y = b$ at the bottom wall. Similarly, the shear stress at the upper wall at $y = -b$ can be determined. To determine the pressure distribution, let us consider the average velocity equation (4.59) as $u_{av} = -\frac{2}{6\mu_f}\frac{dp}{dx}b^2$, one gets

$$\Delta p = \frac{3\mu_f u_{av}}{b^2}\Delta x \tag{4.62}$$

where $\Delta p = p_1 - p_2$, and $\Delta x = x_2 - x_1$. The head loss h_l in the length Δx is given by $h_l = \frac{3\mu_f u_{av}}{b^2}\Delta x$

Example 4.3

A fluid of viscosity six poise flows between two parallel plates separated by a distance of 20 mm apart. If the pressure drop in length 1.5 m is 0.4×10^4 N/m^2, find the rate of flow of fluid between the plates. The width of plate can be considered as 1 m.

Solution

From the given problem, $\mu_f = 0.6$ Ns/m^2; $2b = 20$ mm; $b = 10$ mm $= 0.01$ m; $\Delta p = 0.4 \times 10^4$ N/m^2.

Therefore $u_{av} = -\frac{1}{3\mu_f}\frac{dp}{dx}b^2 = \frac{1}{3\mu_f}\frac{\Delta p}{\Delta x}b^2 = \frac{1}{3 * 0.6}\frac{0.4 \times 10^4}{1.5}0.01^2 = 0.148$ m/s

Rate of flow of fluid between the plates = $u_{av} \times area = 0.148 \times 1 \times 0.02 = 2.96 \times 10^{-3}$ cumecs.

4.8.2 Couette Flow

Couette flow is the purely shearing flow between two parallel straight plates (Figure 4.10), lower plate at $y = 0$ is at rest and the other one at $y = b$ is moving with a uniform velocity U, and there is no pressure gradient $\frac{\partial p}{\partial x}$ with respect to x, and $v = w = 0$.

The differential equation subject to the boundary conditions is

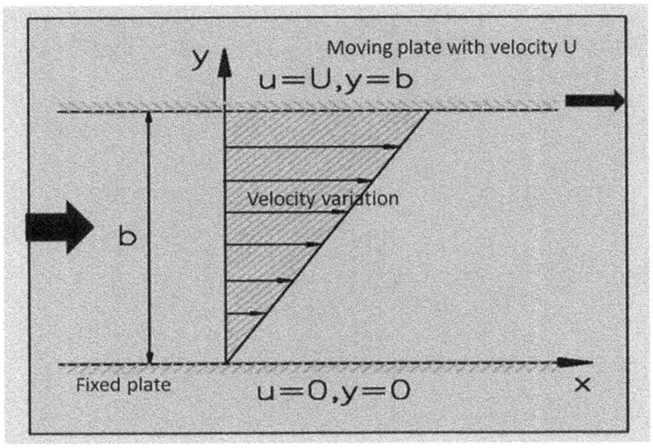

FIGURE 4.10 Velocity distribution between two parallel plates (Couette flow).

$$0 = \frac{\partial^2 u}{\partial y^2} \tag{4.63}$$

with coefficient viscosity $\mu_f \neq 0$, and the boundary conditions are

$$u = 0 \text{ at } y = 0, u = U \text{ at } y = b \tag{4.64}$$

The solution of equation (4.63) subject to boundary equation (4.64) is

$$u = U\,\frac{y}{b} \tag{4.65}$$

Equation (4.65) represents a linear function of *y*, and the gradient of velocity is proportional to the uniform velocity U and inversely proportional to the distance between the plates. This problem is very important in lubrication mechanics.

4.8.3 Generalized Couette Flow

The generalized Couette flow is known, when the flow is due to pressure gradient $\frac{\partial p}{\partial x}$ and uniform movement of the plate (Figure 4.11) with a velocity U, the differential equation subject to the boundary conditions are:

$$\frac{\partial p}{\partial x} = \mu_f \frac{\partial^2 u}{\partial y^2} \tag{4.66}$$

$$u = 0 \text{ at } y = 0 \text{ and } u = U \text{ at } y = b \tag{4.67}$$

The solution of equation (4.66) is given by

$$u = \frac{1}{2\mu_f}\frac{\partial p}{\partial x} y^2 + C_1 y + C_2 \tag{4.68}$$

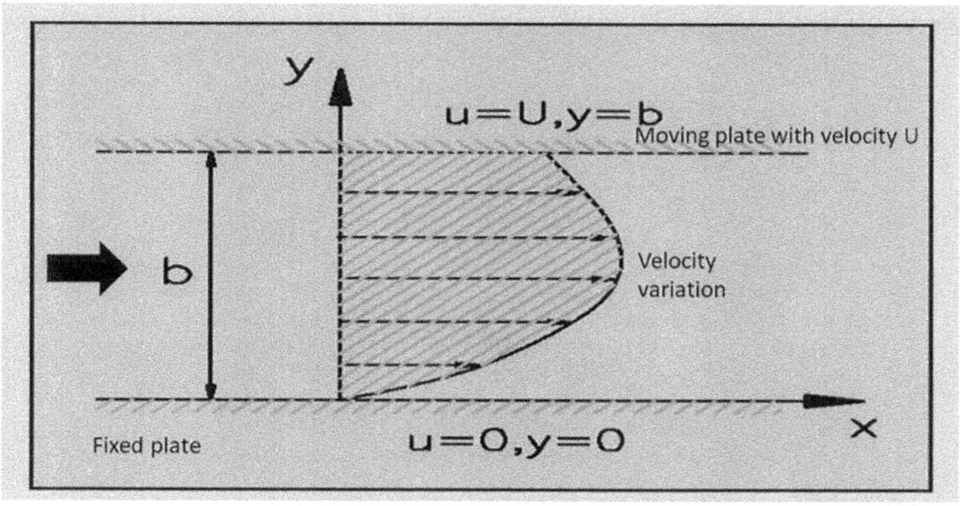

FIGURE 4.11 Generalized Couette flow between two parallel plates.

and using boundary conditions to determine the constants C_1 and C_2, one gets the solution, after simplification, as:

$$u = \frac{y}{b} + P\frac{y}{b}\left(1 - \frac{y}{b}\right) \tag{4.69}$$

where $P = -\frac{b^2}{2\mu_f U}\left(\frac{\partial p}{\partial x}\right)$ is the dimensionless pressure gradient. This is known as Generalized Couette Flow (Schlichting, 1968; Yuan, 1970), which is also known as Couette-Poiseuille flow between two parallel flat walls. If the pressure gradient $P = 0$, then $\frac{u}{U} = \frac{y}{b}$, which is a simple Couette flow with a linear variation. For the quantitative description of generalized Couette flow, equation (4.69) is plotted against height y/b for different values of P in Figure 4.12. When $P > 0$, negative pressure gradient $\left(-\frac{\partial p}{\partial x}\right)$ in the direction of motion generates the flow, i.e., for the pressure drop in the direction of motion of the upper plate, the velocity is positive over the whole width of the channel. When $P < 0$, positive or adverse pressure gradient $\left(\frac{\partial p}{\partial x}\right)$ in the direction of motion, occurs backflow, where the dragging action of the faster nearer layers on the fluid close to the moving wall is not enough to overcome the influence of the adverse pressure gradient. Figure 4.13 shows the schematic diagrams of velocity profiles for generalized Couette flow with different pressure gradients P.

The average velocity of the generalized Couette flow between two parallel plates is given by the expression as:

$$u_{\text{av}} = \left(\frac{1}{2} + \frac{P}{6}\right)U \tag{4.70}$$

Here if $P = 0$, it shows the average velocity is $U/2$ for simple Couette flow, which increases linearly with U from zero value at the stationary plate. When $P = -3$, $u_{av} = 0$, which means there is no flow across the passage between two plates because the influence of the adverse pressure gradient is balanced with the dragging force due to viscosity. Using equation (4.69) for velocity u, the volumetric flow Q between two parallel plates per unit width is given by:

$$Q = \int_0^b u\,dy = \left(\frac{1}{2} + \frac{P}{6}\right)Ub \tag{4.71}$$

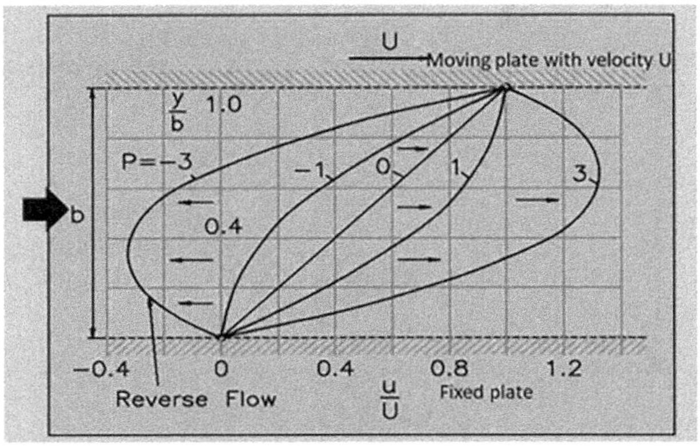

FIGURE 4.12 Generalized Couette flow for various values of pressure gradient P between two parallel flat plates (Modified from Schlichting, 1968.)

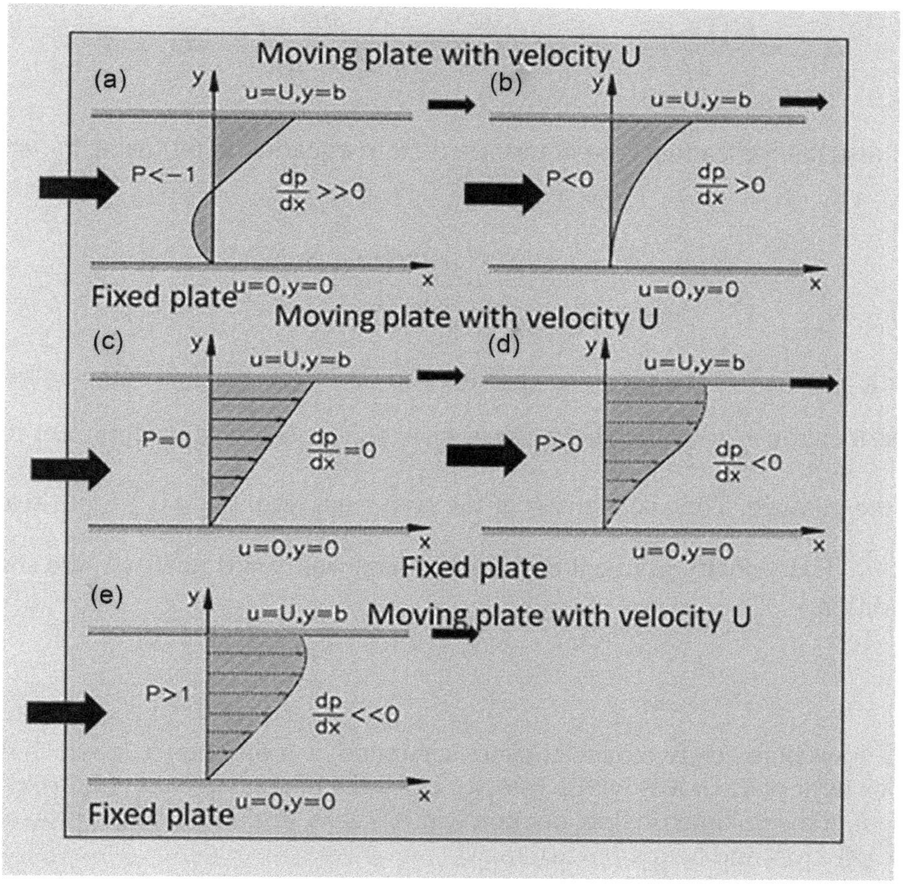

FIGURE 4.13 (a–e) Schematic diagrams of velocity profiles for generalized Couette flow with different pressure gradients P. (Modified from Som and Biswas, 1998.)

The location of maximum or minimum velocity in the channel is determined by differentiating equation (4.69) with respect to y and putting equal to zero, we get

$$\frac{du}{dy} = \frac{U}{b} + \frac{PU}{b}\left(1 - 2\frac{y}{b}\right) = 0 \tag{4.72}$$

Then for maximum or minimum velocity $\dfrac{du}{dy} = 0$, which gives

$$\frac{y}{b} = \frac{1}{2} + \frac{1}{2P} \tag{4.73}$$

The maximum velocity for $P = 1$ occurs at $\dfrac{y}{b} = 1$ and equals to U from equation (4.69). For $P > 1$, the maximum velocity occurs at a location $\dfrac{y}{b} < 1$, which means for $P > 1$, the fluid particles attain a velocity higher than that of moving plate at a location somewhere below the moving plate. On the other hand, if $P = -1$, the minimum velocity occurs at $\dfrac{y}{b} = 0$. For $P < -1$, the minimum velocity occurs at $\dfrac{y}{b} > 0$, which means the occurrence of backflow near the fixed plate. Therefore, the maximum and minimum velocities are determined by putting the value of y in equation (4.69) as:

$$u_{\max} = \frac{U(1+P)^2}{4P} \text{ for } P \geq 1 \text{ and } u_{\min} = \frac{U(1+P)^2}{4P} \text{ for } P \leq -1 \tag{4.74}$$

The shear stress distribution can also be determined from equation (4.69) using $\tau_{yx} = \mu_f \frac{du}{dy}$, which is given by

$$\tau_{yx} = \mu_f \frac{U}{b} + \frac{\mu_f UP}{b}\left(1 - \frac{2y}{b}\right) \tag{4.75}$$

For simple Couette flow, when the pressure gradient $P = 0$, the shearing stress remains constant across the plates. At the location $\frac{y}{b} = \frac{1}{2}$, the shear stress $\tau_{yx} = \mu_f \frac{U}{b}$, which is independent of the pressure P at the centre of the passage. The shear stress at the stationary wall $\left(\frac{y}{b} = 0\right)$ is positive for $P > -1$ and negative for $P < -1$. The velocity gradient at the stationary wall $y = 0$ for $P = 1$, the shear stress is also zero for this condition.

Example 4.4

A viscous fluid flows between two parallel plates separated by a distance b, in which the lower plate is fixed and the upper plate moves with a velocity U. Find a relation between the velocity U and the pressure gradient, if the volumetric flow per unit width is zero, and also find the pressure gradient for the velocity $U = 0.2$ m/s and height $b = 20$ cm.

Solution

From the given problem, the volumetric flow Q per unit width is given by the equation as: $Q = \int_0^b u dy = \left(\frac{1}{2} + \frac{P}{6}\right) Ub$ from equation (4.71). According to the problem, if the volumetric flow $Q = 0$, we get

$\left(\frac{1}{2} + \frac{P}{6}\right) Ub = 0$ for $U \neq 0$ and $b \neq 0$, then $\left(\frac{1}{2} + \frac{P}{6}\right) = 0$, which implies $P = -3$.

Thus, $\frac{\partial p}{\partial x} = \frac{6\mu_f}{b^2} U$.

If $U = 0.2$ m/s, and $b = 20$ mm, then the pressure gradient is $\frac{\partial p}{\partial x} = 30\mu_f$, with μ_f is the dynamic viscosity of the flowing fluid (Figure 4.14).

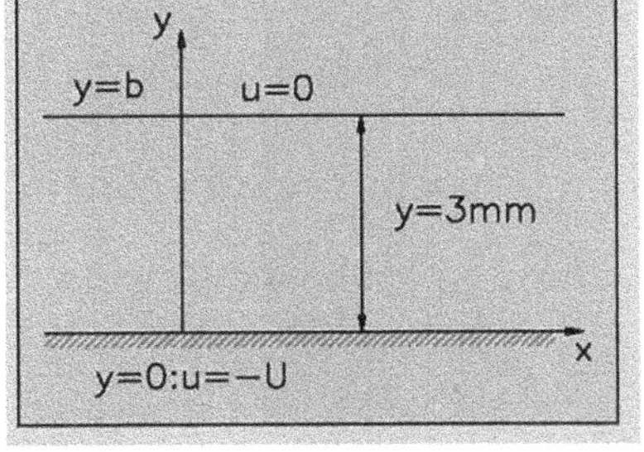

FIGURE 4.14 Example 4.4.

Example 4.5

Water flows in a laminar condition between two parallel flat plates separated by a distance of 3 mm, where the lower plate moves to the left from right at a speed of 0.3 m/s. and determine the pressure gradient for which the net flow will be zero across the section ($\mu_f = 4.7\times10^{-4}\,\text{Ns/m}^2$at 60°C).

Solution

From the Navier–Stokes equation and the problem given, the equation can be written as $\frac{dp}{dx} = \mu_f \frac{d^2u}{dy^2}$ and the given boundary conditions are $u = 0$ at $y = b = 3\text{ mm}$; $u = -U$ at $y = 0$.

The solution of the equation subject to the boundary conditions is $u = \frac{1}{2\mu_f}\frac{dp}{dx}y^2 + C_1 y + C_2$; where C_1 and C_2 are the integrating constants determined using the conditions. So, $C_1 = \frac{U}{h} - \frac{1}{2\mu_f}\frac{\partial p}{\partial x}b$, $C_2 = -U$. Therefore $u = U\left(\frac{y}{b} - 1\right) + \frac{1}{2\mu_f}\frac{dp}{dx}by\left(\frac{y}{b} - 1\right)$.

Discharge $Q = \int_0^b u\,dy = -\frac{1}{12\mu_f}\frac{dp}{dx}b^3 - \frac{Ub}{2}$. According to the problem, if $Q = 0$, then

$$\frac{dp}{dx} = -\frac{6U\mu_f}{b^2} = \frac{-6\times0.3\times4.7\times10^{-4}}{(0.003)^2} = -94\ \text{N/m}^2.$$

4.8.4 Steady Flow Through Pipes

The steady-state laminar flow through a circular pipe or tube is known as Hagen–Poiseuille flow. Let us consider a laminar flow through a long straight pipe of circular cross-section with axial symmetry (Figure 4.15), similar to the plane Poiseuille flow, with no body force. Let z is along the axial direction and r be the radial coordinate measured from the z-axis. Here the velocity components u_r along the radial direction and u_θ along tangential direction are zero. Under these flow conditions, the equation of continuity (equation 4.42d) reduces to

$$\frac{\partial u_z}{\partial z} = 0,\ \ u_r = u_\theta = 0$$

Or,

$$u_z = u_z(r) \tag{4.76}$$

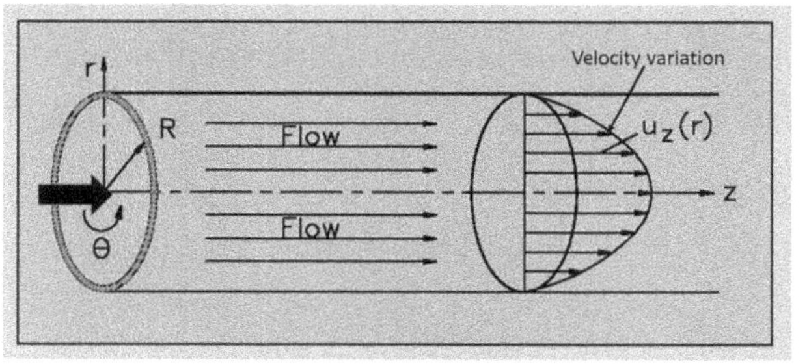

FIGURE 4.15 Flow through a pipe (Poiseuille flow). (Modified from Yuan, 1970.)

From the Navier–Stokes equations in cylindrical coordinates for Hagen–Poiseuille flow, equations can be written as $\frac{\partial p}{\partial r} = \frac{\partial p}{\partial \theta} = 0$, and $p = p(z)$ only. Then the equation of motion for axial direction along the z-axis becomes

$$\mu_f \left(\frac{d^2 u_z}{dr^2} + \frac{1}{r} \frac{du_z}{dr} \right) = \frac{dp}{dz} \tag{4.77}$$

Equation (4.77) is the balance of the shearing force and the pressure force, so that no acceleration takes place. In the present case, we have shown that $u_z = u_z(r)$ and $p = p(z)$, and the pressure is a linear function of z. To solve this equation (4.77), two boundary conditions are necessary. The first boundary condition is found from the symmetry of the flow, which is

$$r = 0, \quad \frac{du_z}{dr} = 0 \tag{4.78}$$

and then the second boundary condition is the no-slip condition at the wall, i.e.,

$$r = R, u_z = 0 \tag{4.79}$$

Under these conditions, by integrating equation (4.77), the solution is obtained as follows:

$$u_z = \frac{1}{4\mu_f} \frac{dp}{dz} r^2 + A \ln r + B \tag{4.80}$$

where A and B are the integrating constants determined from the boundary conditions. From the first boundary condition, one gets $A = 0$, and the second boundary condition, B is obtained

$$B = -\frac{1}{4\mu_f} \frac{dp}{dz} R^2 \tag{4.81}$$

Substitution of A and B into equation (4.80) gives the axial velocity distribution of the Hagen-Poiseuille flow through a pipe as:

$$u_z = -\frac{R^2}{4\mu_f} \frac{dp}{dz} \left[1 - (r/R)^2 \right] \tag{4.82}$$

which has the form of a paraboloid of revolution. Equation (4.82) can only occur when the flow reaches a fully developed state and remains laminar. Once the velocity distribution becomes parabolic, it remains unchanged from there as in the flow direction; hence it is called fully developed flow. As long as the Reynolds number $R_e = 2u_z R / \upsilon_f$ has a value less than the critical Reynolds number (~2300), the laminar flow can be maintained. Equation (4.82) is also obtained from the condition of equilibrium in the axial direction that the shearing stress acting on the circumferential area of the cylinder is equal to the pressure force acting on the cross-section of the cylinder. Equation (4.82) is the same as equation (4.3), the only difference is in the pressure term: $\frac{p_1 - p_2}{l} = -\frac{dp}{dz}$, where l is the length of fluid cylinder.

4.8.4.1 Maximum and Average Velocities

The maximum velocity occurs at the centre of the pipe, where $r = 0$ in equation (4.82). This can be obtained as

$$(u_z)_{\max} = -\frac{R^2}{4\mu_f} \frac{dp}{dz}, \tag{4.83}$$

where $\frac{dp}{dz} < 0$. The average velocity $(u_z)_{av}$ can be obtained as

$$(u_z)_{av} = \frac{1}{\pi R^2}\int_0^{2\pi}\int_0^R u_2\ r\ dr\ d\theta = -\frac{R^2}{8\mu_f}\frac{dp}{dz}. \tag{4.84}$$

The volumetric flow is as given by:

$$Q = \pi R^2 u_z = \frac{\pi}{8\mu_f} R^4\left(-\frac{dp}{dz}\right) \tag{4.85}$$

Shearing stress τ_{rz} is $\tau_{rz} = -\mu_f \dfrac{du_z}{dr}, (\tau_{rz})_R = -\dfrac{R}{2}\dfrac{dp}{dz} = 2\mu_f \dfrac{(u_z)_{max}}{R}$, where $(\tau_{rz})_R$ is the shearing stress at the wall. The frictional coefficient C_f for laminar flow through a circular pipe is given by

$$C_f = \frac{\tau_0}{\frac{1}{2}\rho_f (u^2{}_z)_{av}} \tag{4.86}$$

$$\Rightarrow C_f = \frac{16}{Re} \tag{4.87}$$

where $Re = \dfrac{(u_z)_{av}(2R)}{\upsilon_f}$

4.8.5 Flow Between Two Co-Axial Cylinders

The velocity distribution of a viscous fluid flows steadily parallel to the axis in the annular space between two co-axial cylinders (Figure 4.16) of radii R_1 and R_2 is given by equation (4.80) as:

$$u_z = \frac{1}{4\mu_f}\frac{dp}{dz} r^2 + A_1 \ln r + B_1 \tag{4.88}$$

with boundary conditions as:

$$u_z = 0 \text{ at } r = R_1 \text{ and } R_2 \tag{4.89}$$

Determining the constants of integrations A_1 and B_1 using the boundary conditions, the velocity distribution in the annular space is given by (Yuan, 1970) as:

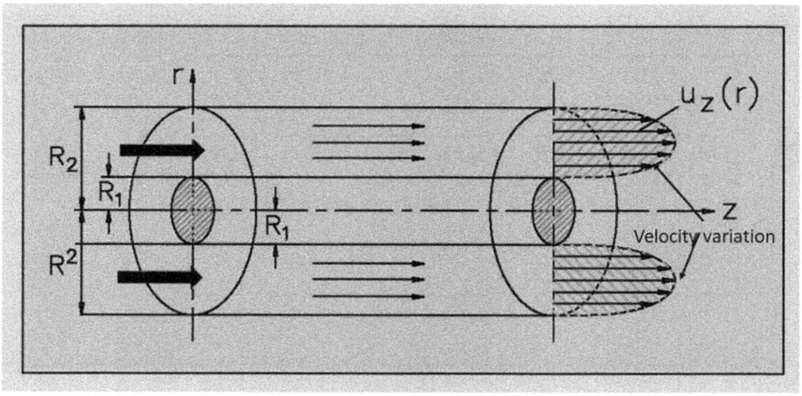

FIGURE 4.16 Flow between two co-axial cylinders. (Modified from Yuan, 1970.)

$$u_z = -\frac{1}{4\mu_f}\frac{dp}{dz}\left[\left(R_1^2 - r^2\right) + \frac{n^2-1}{\ln n}\ln\frac{r}{R_1}\right] \tag{4.90}$$

where $n = R_2/R_1$. The rate of volumetric flow for this case is as follows:

$$Q = \int_0^{2\pi}\int_0^{nR_1} u_z\ r\ dr\ d\theta = -\frac{\pi R_1^4}{8\mu_f}\frac{dp}{dz}\left[\left(n^4-1\right) - \frac{\left(n^2-1\right)^2}{\ln n}\right] \tag{4.91}$$

The average velocity in the annulus is determined from equation (4.91) as:

$$u_{z\,\mathrm{av}} = -\frac{R_1^2}{8\mu_f}\frac{dp}{dz}\left[\left(n^2+1\right) - \frac{n^2-1}{\ln n}\right] \tag{4.92}$$

The shear stresses at the wall of the inner and outer cylinders respectively are

$$\left(\tau_{rz}\right)_{R_1} = \left(\mu_f\frac{du_z}{dr}\right)_{R_1} = -\frac{R_1}{4}\frac{dp}{dz}\left(\frac{n^2-1}{\ln n} - 2\right) \tag{4.93}$$

$$\left(\tau_{rz}\right)_{R_2} = -\left(\mu_f\frac{du_z}{dr}\right)_{R_2} = -\frac{R_1}{4}\frac{dp}{dz}\left(2n - \frac{n^2-1}{n}\frac{1}{\ln n}\right) \tag{4.94}$$

It is seen from the above two equations (4.93) and (4.94) that the shearing stresses at both walls are positive; however, the velocity gradient at the wall of the outer cylinder is negative.

4.8.6 Flow Between Two Concentric Rotating Cylinders

Let us consider the flow between two concentric rotating cylinders of radii R_1 and R_2 respectively (Figure 4.17). Here the flow is assumed to be only peripheral so that the tangential component of velocity u_θ exists. Let ω_1 and ω_2 be the steady angular velocities of inner and outer cylinders respectively. According to the equation of continuity (4.42d) in cylindrical co-ordinates, it becomes

$$\frac{\partial u_\theta}{\partial \theta} = 0 \text{ or } u_\theta = u_\theta(r) \tag{4.95}$$

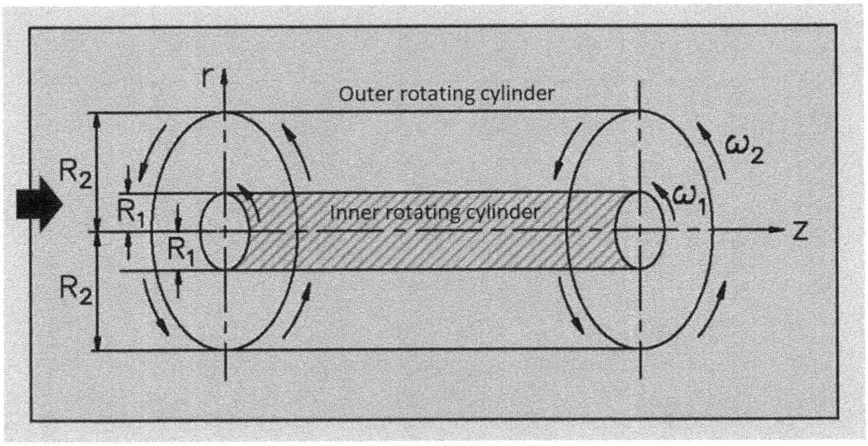

FIGURE 4.17 Flow between two concentric rotating cylinders. (Modified from Yuan, 1970.)

The Navier–Stokes equations (equation 4.42) for the present case reduce to (Yuan, 1970):

$$\rho_f \frac{u_\theta^2}{r} = \frac{dp}{dr} \tag{4.96}$$

$$\frac{d^2 u_\theta}{dr^2} + \frac{d}{dr}\left(\frac{u_\theta}{r}\right) = 0 \tag{4.97}$$

Since u_θ is a function of r only, it follows from axial symmetry that the pressure p in the above equation must be either a function of r or a constant. Hence, the term $\frac{dp}{d\theta}$ does not appear in equation (4.97). The integration of equation (4.97) gives

$$\frac{du_\theta}{dr} + \frac{u_\theta}{r} = 2c_1$$

Here $2c_1$ is considered for convenience. Rewriting the equation in the following form,

$$\Rightarrow \frac{1}{r}\frac{d}{dr}(ru_\theta) = 2c_1 \Rightarrow \frac{d}{dr}(ru_\theta) = 2rc_1 \tag{4.98}$$

After integration, the equation gives

$$u_\theta = c_1 r + \frac{c_2}{r} \tag{4.99}$$

where c_1 and c_2 are the constants of integration to be determined.

The boundary conditions of the flow are as follows:

$$u_\theta = R_1\omega_1 \text{ at } r = R_1; u_\theta = R_2\omega_2 \text{ at } r = R_2 \tag{4.100}$$

Using the boundary conditions (4.100), the solution given by (4.99) can be written as:

$$u_\theta = \frac{1}{R_2^2 - R_1^2}\left[\left(\omega_2 R_2^2 - \omega_1 R_1^2\right)r - \frac{R_1^2 R_2^2}{r^2}(\omega_2 - \omega_1)\right) \tag{4.101}$$

Substituting equations (4.101) in (4.96), the radial pressure from the peripheral motion is obtained (Yuan, 1970) as:

$$p = p_1 + \frac{\rho_f}{\left(R_1^2 - R_2^2\right)^2}\left[\begin{array}{c}\left(\omega_2 R_2^2 - \omega_1 R_1^2\right)\left(\frac{r^2 - R_1^2}{2}\right) \\ -2R_1^2 R_2^2(\omega_2 - \omega_1)\left(\omega_2 R_2^2 - \omega_1 R_1^2\right)\ln\frac{r}{R_1} \\ -\frac{R_1^4 R_2^4}{2}(\omega_2 - \omega_1)^2\left(\frac{1}{r^2} - \frac{1}{R_1^2}\right)\end{array}\right] \tag{4.102}$$

where $p = p_1$ at $r = R_1$.

The shearing stress can be determined as

$$\tau_{r\theta} = \mu_f\left[r\frac{d}{dr}\left(\frac{u_\theta}{r}\right)\right] \tag{4.103}$$

Substituting equations (4.102) in (4.103), one gets:

$$\tau_{r\theta} = \frac{2\mu_f}{\left(R_2^2 - R_1^2\right)} \frac{R_1^2 R_2^2}{r^2} (\omega_2 - \omega_1) \tag{4.104}$$

The shear stress at the wall of the outer cylinder is given by

$$(\tau_{r\theta})_{r=R_2} = 2\mu_f(\omega_2 - \omega_1)\frac{1}{\left(\frac{R_2}{R_1}\right)^2 - 1} \tag{4.105}$$

Similarly, the shear stress at the wall of the inner cylinder is obtained by using $r = R_1$.

4.9 UNSTEADY MOTION OF A FLAT PLATE (STOKES'S FIRST PROBLEM)

Let us consider an unsteady motion of a flat plate in an infinite fluid medium. The motion of the plate is started impulsively from rest. Consider an infinite long plate in an infinite fluid initially at rest and then suddenly given a constant velocity U in the x-direction (Figure 4.18). This problem is often called Stokes' first Problem (Yuan, 1970). Since the stream limes are parallel to the plate, two components of velocity perpendicular to the plate are zero, and all the derivatives of u are zero except those in y and t. The Navier–Stokes equations without body force in this case becomes

$$\frac{\partial u}{\partial t} = \upsilon_f \frac{\partial^2 u}{\partial y^2} \tag{4.106}$$

All three pressure gradients in equation (4.38) are zero because the plate is situated in an infinite fluid where the pressure is constant everywhere. This equation is also like heat conduction equation and diffusion equation. The initial and boundary conditions are

$$u = 0, \ \ t = 0 \text{ for all } y \tag{4.107}$$

$$\left.\begin{matrix} u = U \ \text{ at } \ y = 0 \\ u = 0 \ \text{ at } \ y = \infty \end{matrix}\right\} \text{when } t > 0 \tag{4.108}$$

Equation (4.106) is a partial differential equation that can be transformed into an ordinary differential equation by the principle of similarity. The principle of similarity of the velocity profile can be written as

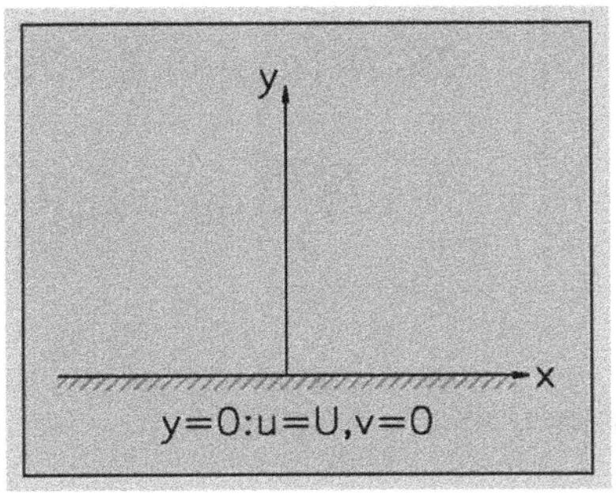

FIGURE 4.18 Unsteady flow past an infinite plate suddenly from rest.

$$\frac{u}{U} = f(\eta) \tag{4.109}$$

where

$$\eta = \frac{1}{2}\upsilon_f^n y^\alpha t^\beta \tag{4.110}$$

with n, α, β are the numbers to be determined and the factor $\frac{1}{2}$ is inserted for convenience. Substituting equations (4.109) and (4.110) in the partial differential equation (4.106), one obtains

$$\frac{\partial u}{\partial t} = \frac{\partial u}{\partial \eta}.\frac{\partial \eta}{\partial t} = U\frac{\partial f}{\partial \eta}\left(\frac{\beta}{2}\upsilon_f^n y^\alpha t^{\beta-1}\right) = U\beta t^{-1}\eta\frac{\partial f}{\partial \eta}$$

$$\frac{\partial u}{\partial y} = \frac{\partial u}{\partial \eta}.\frac{\partial \eta}{\partial y} = U\frac{\partial f}{\partial \eta}\left(\frac{\alpha}{2}\upsilon_f^n y^{\alpha-1} t^\beta\right)$$

$$\frac{\partial^2 u}{\partial y^2} = \frac{\partial}{\partial \eta}\left[U\frac{\partial f}{\partial \eta}\left(\frac{\alpha}{2}\upsilon_f^n y^{\alpha-1} t^\beta\right)\right]\frac{\partial \eta}{\partial y}$$

$$= U\left[\left(\frac{\alpha^2}{4}\upsilon_f^{2n} y^{2(\alpha-1)} t^{2\beta}\right)\frac{\partial^2 f}{\partial \eta^2} + \frac{\alpha}{2}(\alpha-1)\upsilon_f^n y^{\alpha-2} t^\beta \frac{\partial f}{\partial \eta}\right]$$

$$\text{Or}, U\beta\, t^{-1}\eta\frac{\partial f}{\partial \eta} = U\left[\left(\frac{\alpha^2}{4}\upsilon_f^{2n+1} y^{2(\alpha-1)} t^{2\beta}\right)\frac{\partial^2 f}{\partial \eta^2} + \frac{\alpha}{2}(\alpha-1)\upsilon_f^{n+1} y^{\alpha-2} t^\beta \frac{\partial f}{\partial \eta}\right] \tag{4.111}$$

To make it an ordinary differential equation, the power of t and y on both sides of the equation must be identical. Hence

$$\alpha = 1, \beta = -\frac{1}{2}, n = -\frac{1}{2} \tag{4.112}$$

Then the above equation reduces to

$$\frac{d^2 f}{d\eta^2} + 2\eta\frac{\partial f}{\partial \eta} = 0 \tag{4.113}$$

with similarity parameter

$$\eta = \frac{y}{2\sqrt{\upsilon_f t}} \tag{4.114}$$

The solution of equation (4.113) is $f(\eta) = c_1\int_0^\eta e^{-\eta^2} d\eta + c_2$.

Using the boundary conditions (4.108) and similarity parameter (4.114), the constant c_1 and c_2 are derived as follows:

$$c_2 = 1, \quad c_1 = -\frac{1}{\int_0^\infty e^{-\eta^2}\, d\eta} = -\frac{2}{\sqrt{\pi}} \tag{4.115}$$

The velocity distribution equation (4.109) is therefore

$$f(\eta) = \frac{u}{U} = 1 - \frac{2}{\sqrt{\pi}} \int_0^\eta e^{-\eta^2}\, d\eta = 1 - \operatorname{erf}\eta \tag{4.116}$$

The integral in equation (4.116) is called the error function or probability integral $(\operatorname{erf}\eta)$. This function is available in many mathematics books (Kreyszig, 1985). The velocity distribution given by equation (4.116) is plotted in Figure 4.19a in terms of similarity parameter η.

It is seen from Figure 4.18 that the distance y from the plate at $u = U$ extends asymptotically to infinity. To define a viscous layer from the surface of the plate to a distance, $y = \delta$, where the viscous stresses are negligible, a constant $(\tau_{xy})_{y=0}$ is assumed (Figure 4.19b). This shows that

$$(\tau_{yx})_{y=0} = -\mu_f \left(\frac{du}{\partial y}\right)_{y=0} = \mu_f \frac{U}{\delta} \tag{4.117}$$

$$\text{or,} \quad \frac{\delta}{2\sqrt{\upsilon_f t}} = -\frac{1}{\left(\dfrac{\partial f}{\partial \eta}\right)_{\eta=0}} = \frac{1}{\left[\dfrac{2}{\sqrt{\pi}}\right]\left[e^{-\eta^2}\right]_{\eta=0}} = \frac{\sqrt{\pi}}{2} \tag{4.118}$$

Therefore, the thickness of the boundary layer δ is

$$\delta = \sqrt{\pi}\sqrt{\upsilon_f t} \tag{4.119}$$

We define the mean boundary layer thickness δ_m so that the total quantity of fluid is the same as the actual quantity. At any given time, the total quantity of fluid moving above the moving plate is

$$\int_0^\infty u dy \tag{4.120}$$

If $\delta_m U$ represents the actual quantity of fluid moving above the moving plate, then

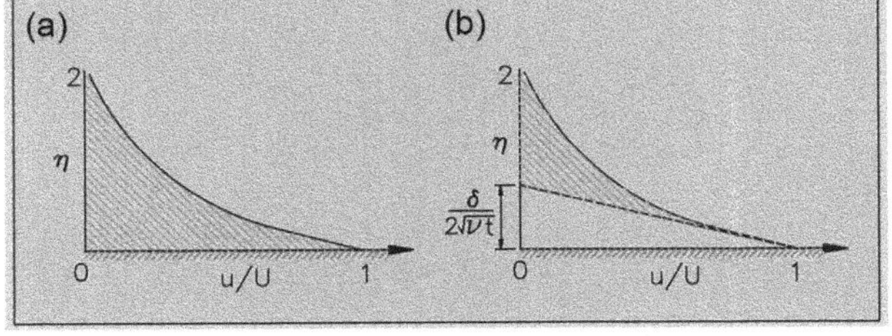

FIGURE 4.19 Velocity profiles (a) against similarity parameter η, and (b) against similarity parameter η with a viscous layer $y = \delta$ from the plate, where the viscous stresses are negligible. (Modified from Yuan, 1970.)

$$U\delta_m = \int_0^\infty u dy \tag{4.121}$$

Solving for δ_m, we have

$$\delta_m = 2\sqrt{\upsilon_f t}\int_0^\infty f(\eta)d\eta = 2\sqrt{\upsilon_f t}\int_0^\infty [1-\text{erf}(\eta)]d\eta \tag{4.122}$$

The integral in equation (4.122) can be evaluated by integration by parts

$$\int_0^\infty \text{erf}\,\eta d\eta = \int_0^\infty \varphi(\eta)d\eta = \lim_{m\to\infty}[\eta\varphi(n)]_0^m - \int_0^\infty \eta\varphi'(\eta)d\eta$$

But $\varphi(\infty) = \frac{2}{\sqrt{\pi}}\int_0^\infty e^{-\eta^2}d\eta = \frac{2}{\sqrt{\pi}}\left(\frac{\sqrt{\pi}}{2}\right) = 1;\ \varphi'(\eta) = \frac{2}{\sqrt{\pi}}\frac{d}{d\eta}\int_0^\eta e^{-\eta^2}d\eta = \frac{2}{\sqrt{\pi}}e^{-\eta^2}$. Hence,

$$\int_0^\infty \varphi(\eta)d\eta = \lim_{m\to\infty}[m] - \frac{2}{\sqrt{\pi}}\int_0^\infty \eta e^{-\eta^2}d\eta = \lim_{m\to\infty}\left[m + \frac{1}{\sqrt{\pi}}\left(e^{-m^2}-1\right)\right] \tag{4.123}$$

Substituting (4.123) into (4.122), the mean boundary layer thickness δ_m is as follows

$$\delta_m = 2\sqrt{\upsilon_f t}\lim_{m\to\infty}\left\{m - \left[m + \frac{1}{\sqrt{\pi}}\left(e^{-m^2}-1\right)\right]\right\} = \frac{2}{\sqrt{\pi}}\sqrt{\upsilon_f t} \tag{4.124}$$

The value of boundary layer thickness δ is within the range

$$\frac{2}{\sqrt{\pi}}\sqrt{\upsilon_f t} < \delta < \sqrt{\pi}\sqrt{\upsilon_f t} \tag{4.125}$$

It may be concluded that the thickness of boundary layer δ is proportional to the square root of the kinematic viscosity and time.

For *Couette flow*, similar to the formation of boundary layer over an unsteady motion of a flat plate in an infinite fluid (Stokes's first problem), when the bottom plate moves in a direction parallel to another flat plate at rest separated by a distance h from the bottom plate. This problem is known as the unsteady Couette flow, which follows the same differential equation (4.106) as Stokes's first problem with only difference in the initial and boundary conditions as follows (Schlichting, 1968):

$$\begin{aligned} &u = 0 \text{ for all } y \text{ if } 0 \le y \le h \text{ for } t = 0 \\ &u = U \text{ at } y = 0, u = 0 \text{ at } y = h \text{ for } t > 0 \end{aligned} \tag{4.126}$$

The solution of equation (4.106) subject to the initial and boundary conditions (4.126) can be written in the form of a series solution of complementary error functions as

$$\frac{u}{U} = \sum_{n=0}^{\infty}\text{erf}\,c\,[2n\,\eta_1 + \eta] - \sum_{n=0}^{\infty}\text{erf}\,c\,[2(n+1)\,\eta_1 - \eta] \tag{4.127}$$

where $\eta_1 = \dfrac{h}{2\sqrt{\upsilon_f t}}$ is the dimensionless distance between the two plates.

4.9.1 Unsteady Flow due to an Oscillating Flat Plate (Stokes's Second Problem)

Consider an infinitely extended flat plate carrying out harmonic oscillations in its own plane (Figure 4.20). This problem was first treated by G.G. Stokes (1851) and later by Lord Rayleigh (1911). Let x be the coordinate parallel to the direction of motion and y-axis be perpendicular to the plate (Schlichting and Gersten, 2000). Because of no-slip condition at the plate, the fluid velocity must be equal to the motion of the plate, which is given by

$$y = 0 : u(0,t) = U \cos \omega t \tag{4.128}$$

where ω is the frequency of oscillation and t is the time. The differential equation together with the above boundary condition (4.128) is known from the theory of conduction equation.

$$\frac{\partial u}{\partial t} = \upsilon_f \frac{\partial^2 u}{\partial y^2} \tag{4.129}$$

The solution of this equation can be obtained as

$$u(y, t) = Ue^{-\eta_s} \cos(\omega t - \eta_s) \tag{4.130}$$

with $\eta_s = y\sqrt{\dfrac{\omega}{2\upsilon_f}}$. The velocity profile $u(y, t)$ has the form of a damped harmonic oscillation, the amplitude of which is $U\exp(-\eta_s)$. That means, the velocity distribution oscillates with an amplitude $U\exp(-\eta_s)$ decreasing outwards, where the layer at a distance y from the wall has a phase lag of $y\sqrt{\dfrac{\omega}{2\upsilon_f}}$ compared to the wall motion. The velocity distribution is shown in Figure 4.21 for different instants of time. Here two layers of fluid oscillate each other in phase at a distance $2\,\pi\sqrt{\dfrac{2\upsilon_f}{\omega}}$. This distance is considered as wavelength of the oscillation, which is called *depth of penetration* of the viscous wave. The layer which is carried by the wall has a thickness of order ~ $\sqrt{\dfrac{\upsilon_f}{\omega}}$, which decreases with decreasing kinematic viscosity υ_f and with increasing frequency ω. Velocity distribution close to an oscillating wall is known as Stokes's second problem. The boundary layer thickness can be obtained as

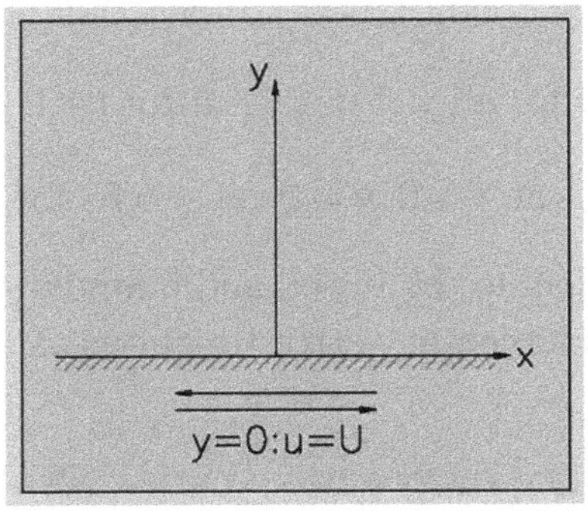

FIGURE 4.20 Unsteady flow due to an oscillating flat plate.

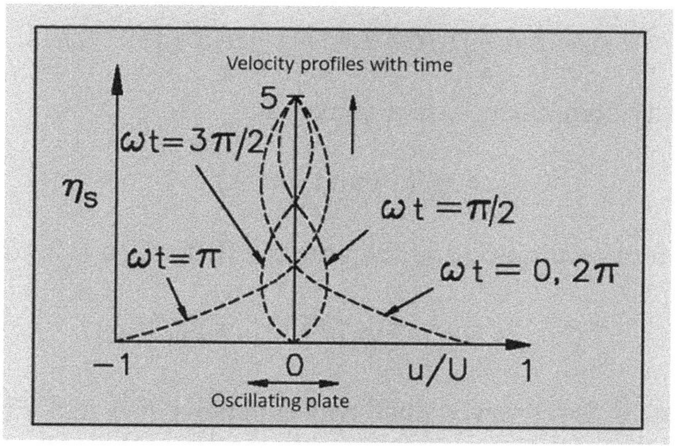

FIGURE 4.21 Velocity distribution close to an oscillating plate for different instants of time.

$\delta_s = 4.6\sqrt{\frac{2\upsilon_f}{\omega}}$. Equation (4.130) represents an analogous solution of the thermal energy equation and is presented in Figure 4.21.

4.10 STAGNATION POINT FLOWS

In the two-dimensional flow problem, let us now consider some exact solutions in which some inertial terms are retained, so that the non-linear equation has to be considered. The first simple example of the flow is the stagnation point flow in a plane surface, i.e., *in two-dimensional flow* (Figure 4.22). The velocity distribution of the frictionless flow in the neighborhood of the stagnation point at $x = y = 0$ is given by

$$U = ax \text{ and } V = -ay \tag{4.131}$$

where a is a constant.

Here the flow is coming from the y-axis and impinges on the flat wall placed at $y = 0$, and separates into two streams on the wall and leaves in both directions x and y. This simple flow is called a *plane potential flow.* Here the viscous flow must stick to the wall, whereas the potential flow glides along the wall with the pressure, given by Bernoulli's equation. If p_0 is the stagnation pressure, and p is the pressure at an arbitrary point, we have in potential flow $p_0 = \frac{1}{2}\rho_f U^2$ and $p = \frac{1}{2}\rho_f V^2$. Then

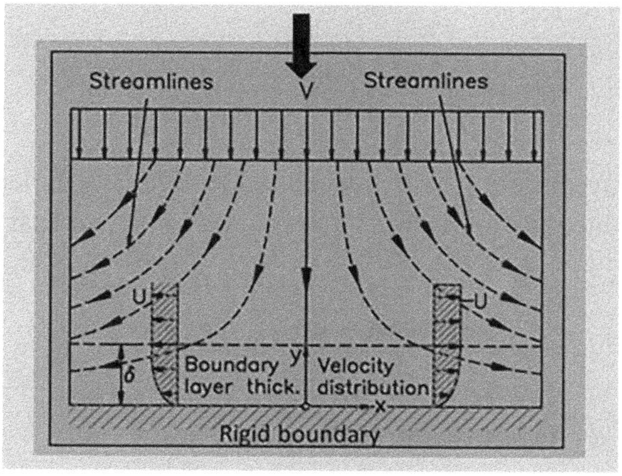

FIGURE 4.22 Steady two-dimensional stagnation point flow at a rigid boundary. (Modified from Schlichting, 1979.)

$$p_0 - p = \frac{1}{2}\rho_f(U^2 + V^2) = \frac{1}{2}\rho_f a^2(x^2 + y^2) \tag{4.132}$$

For viscous flow, the velocity components are assumed as:

$$u = xf'(y) \text{ and } v = -f(y) \tag{4.133}$$

which satisfies the equation of continuity, and the pressure difference is assumed as:

$$p_0 - p = \frac{1}{2}\rho_f a^2\left[x^2 + F(y)\right] \tag{4.134}$$

From the above equations, it is observed that two functions $f(y)$ and $F(y)$ are to be determined. Substituting the above equations (4.133) and (4.134) in two-dimensional Navier–Stokes equations, one gets two ordinary differential equations for f and F as (Batchelor, 1993):

$$f'^2 - ff'' = a^2 + \upsilon_f f''' \tag{4.135}$$

$$ff' = \frac{1}{2}aF' - \upsilon_f f'' \tag{4.136}$$

The boundary conditions for f and F are obtained from the conditions $u = v = 0$ at the wall $x = y = 0$, where the stagnation pressure $p = p_0$; and at a large distance from the wall $y \to \infty$ the velocity $u = U = ax$. Thus, in the functional form:

$$y = 0: f = 0,\ f' = 0,\ F = 0 \tag{4.137a}$$

$$y \to \infty: f' = a. \tag{4.137b}$$

Equations (4.135) and (4.137) are two non-linear ordinary differential equations for functions $f(y)$ and $F(y)$, which determine the velocity and pressure distributions. Here the non-linear equation (4.135) has to be solved first for $f(y)$ and then $F(y)$ has to be determined from the second equation (4.136) using $f(y)$ from the first. To solve equation (4.135) numerically, it is convenient to assume variables as:

$$f(y) = A\varphi(\eta),\ \eta = \alpha y \tag{4.138}$$

Thus, one gets from equation (4.135) as

$$\alpha^2 A^2\left(\varphi'^2 - \varphi\varphi''\right) = a^2 + \upsilon_f\ A\ \alpha^3\varphi''' \tag{4.139}$$

where prime denotes the differentiation with respect to η. The coefficients of the equations become identically equal to unity, if one puts $\alpha^2 A^2 = a^2$ and $\upsilon_f A\alpha^3 = a^2$, so that

$$f(y) = \sqrt{a\upsilon_f}\ \varphi(\eta),\ \eta = y\sqrt{\frac{a}{\upsilon_f}} \tag{4.140}$$

The differential equation for $\varphi(\eta)$ has the form as

$$\varphi''' + \varphi\varphi'' - \varphi'^2 + 1 = 0 \tag{4.141}$$

with the boundary conditions as:

$$\eta = 0: \varphi = 0, \varphi' = 0; \eta = \infty: \varphi' = 1 \tag{4.142}$$

The velocity component parallel to the wall is as follows:

$$\frac{u}{U} = \frac{1}{a} f'(y) = \varphi'(\eta) \tag{4.143}$$

The solution of the differential equation (4.141) subject to the boundary condition (4.142) is known as *Hiemenz flow* (Schlichting and Gersten, 2000); and it is improved by Howarth (1935). The function $\varphi'(\eta)$ increases linearly from $\eta = 0$ and then tends approximately to unity. It is observed that at $\eta = 2.4$, one gets $\varphi' = 0.99$, which reached with an accuracy of 1%. If it is considered for the corresponding distance from the wall, we have the boundary layer thickness δ as:

$$\delta = \eta_\delta \sqrt{\frac{\upsilon_f}{a}} = 2.4\sqrt{\frac{\upsilon_f}{a}} \tag{4.144}$$

which shows that the layer influenced by viscosity is small at low kinematic viscosity, which is proportional to $\sqrt{\upsilon_f}$, that means, $\delta \propto \sqrt{\upsilon_f}$, and the pressure gradient $\frac{\partial p}{\partial y}$ is proportional to $\rho_f a\sqrt{\upsilon_f a}$ and also very small for low kinematic viscosity υ_f. Further, it is worthwhile to note that the dimensionless velocity distribution $\frac{u}{U}$ and the boundary layer thickness from equation (4.144) are independent of x along the wall.

In three-dimensional flow, in a similar way the stagnation point flow can also be obtained as an exact solution of the Navier–Stokes equations for the axially symmetrical case. A fluid flow impinges on a wall at right angles to it and it flows away radially in all directions. For this case, one can use the cylindrical coordinates (r, θ, z) and put the derivatives of velocity components with respect to axial coordinate θ and time t are zero. The solution for the potential flow with the stagnation point $r = z = 0$ is:

$$U = ar, V = 0, \text{ and } W = -2az \tag{4.145}$$

$$p_0 - p = \frac{1}{2}\rho_f(U^2 + W^2) = \frac{1}{2}\rho_f a^2(r^2 + 4z^2) \tag{4.146}$$

which satisfy the Navier–Stokes equation without no-slip condition at the wall ($z = 0$). Here a is a constant, and p_0 is a stagnation pressure. To satisfy the no-slip condition, the corresponding viscous solution may be written as:

$$u = rf'(z); w = -2f(z), \tag{4.147a}$$

$$p_0 - p = \frac{1}{2}\rho_f a^2\left[r^2 + F(z)\right], \tag{4.147b}$$

with boundary conditions as follows:

$$z = 0: u = w = 0, p = p_0; z = \infty: u = ar \tag{4.148}$$

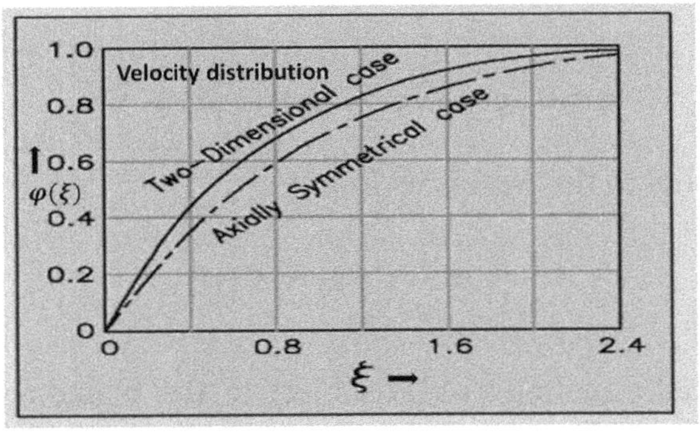

FIGURE 4.23 Velocity distributions near a stagnation point. (Modified from Schlichting and Gersten, 2000.)

Substituting the equations for velocities and pressure from equation (4.147) into the Navier–Stokes equations, we get

$$f'^2 - 2ff'' = a^2 + \upsilon_f f''' \tag{4.149}$$

$$2ff' = \frac{1}{4}a^2 F' - \upsilon_f f'' \tag{4.150}$$

with boundary conditions for f and F as:

$$z = 0 : f = f' = 0, F = 0; z = \infty : f' = a \tag{4.151}$$

Using the transformations, $\xi = z\sqrt{\frac{a}{\upsilon_f}}$, $f(z) = \sqrt{a\upsilon_f}\varphi(\xi)$. the differential equation (4.149) reduces to

$$\varphi''' + 2\varphi\varphi'' - \varphi'^2 + 1 = 0 \tag{4.152}$$

with the boundary conditions as:

$$\eta = 0 : \varphi = \varphi' = 0; \xi = \infty : \varphi' = 1 \tag{4.153}$$

This equation is identical to equation (4.141) except by a factor 2 in the second term. Then for $F(\xi)$, one gets $F(\xi) = \varphi^2 + \varphi'$. The solution of equation (4.152) subject to the boundary conditions (4.153) is given in Schlichting and Gersten (2000) using the method of series expansion. Figure 4.23 shows the plots of two-dimensional and three-dimensional stagnation point flow. The function $\varphi(\xi)$ for three-dimensional stagnation point flow differs slightly from that of two-dimensional case.

4.11 LOW REYNOLDS NUMBER FLOWS (HIGH VISCOSITY)

The low Reynolds number flows are those in which the inertial force plays a non-significant role in the field. The Reynolds number is characterized only by the viscosity and density and is defined by $Re = \frac{LV\rho_f}{\mu_f}$, where L is the length scale of the body, U is the characteristic velocity, and ρ_f and μ_f are the density and viscosity of the fluid. In the low Reynolds number flows, the value of Re provides a rough estimate of the relative importance of inertial and viscous forces. When Reynolds number (Re) is small, the effect of inertial force is negligible compared to viscosity.

4.11.1 Creeping Flows

For low Reynolds number, the approximate solutions of the Navier–Stokes equations are valid in the limiting case when the viscous forces are considerably greater than the inertial forces. The inertial forces are proportional to the square of the velocity, whereas the viscous forces are the first power of velocity. So, the viscous forces are dominant when the velocity is small, which may happen when the Reynolds number *Re* is very small. When the inertia force terms are omitted from the equations of motion, the resulting solutions are valid approximately for Reynolds number $Re \ll 1$. This can be seen from the Navier–Stokes equations if the equations are made dimensionless form, when the inertial force terms are multiplied by a factor $Re = \dfrac{LV\rho_f}{\mu_f}$. However, the motion at very low Reynolds number is sometimes called the *creeping flow*. When the inertial force terms are neglected, the Navier–Stokes equation (4.40) for incompressible fluid are written as:

$$0 = -\nabla p + \mu_f \nabla^2 \vec{q} \tag{4.154}$$

$$\nabla.\vec{q} = 0 \tag{4.155}$$

These two equations (4.154) and (4.155) are the basic equations for creeping flow. In the Cartesian coordinate system, these equations can be written as:

$$0 = -\frac{\partial p}{\partial x} + \mu_f \left(\frac{\partial^2 u}{\partial x^2} + \frac{\partial^2 u}{\partial y^2} + \frac{\partial^2 u}{\partial z^2} \right) \tag{4.156a}$$

$$0 = -\frac{\partial p}{\partial y} + \mu_f \left(\frac{\partial^2 v}{\partial x^2} + \frac{\partial^2 v}{\partial y^2} + \frac{\partial^2 v}{\partial z^2} \right) \tag{4.156b}$$

$$0 = -\frac{\partial p}{\partial z} + \mu_f \left(\frac{\partial^2 w}{\partial x^2} + \frac{\partial^2 w}{\partial y^2} + \frac{\partial^2 w}{\partial z^2} \right) \tag{4.156c}$$

$$\frac{\partial u}{\partial x} + \frac{\partial v}{\partial y} + \frac{\partial w}{\partial z} = 0 \tag{4.156d}$$

These systems of equations are supplemented by both the normal and tangential components of velocity, which must vanish at the walls, i.e., the no-slip conditions:

$$v_n = 0 \text{ and } v_t = 0 \text{ on the solid walls} \tag{4.157}$$

If the divergence of equation (4.154) is taken, it becomes

$$\nabla.\nabla p = \nabla^2 p = \mu_f \nabla.\left(\nabla^2 \vec{q}\right) = \mu_f \nabla^2 \left(\nabla.\vec{q}\right) = 0 \tag{4.158}$$

Equation (4.158) indicates that the pressure p satisfies the Laplace equation; hence for a very slow motion the pressure is a harmonic. Equations (4.154), (4.155), and (4.158) may be applied to the problem of a steady uniform flow around a sphere at rest. This problem was first solved by Stokes (1856) and is often referred to as Stokes law. The pressure field in the creeping motion satisfies the potential equation and the pressure p (x, y, z) is potential function. For two-dimensional creeping motion, the stream function ψ is defined by $u = \dfrac{\partial \psi}{\partial y}$ and $v = -\dfrac{\partial \psi}{\partial x}$. From equation (4.156a,b), pressure is eliminated by differentiating equation (4.156a) with respect to x and equation (4.156b) with respect to y, and adding the both together, the stream function ψ must satisfy the equation as:

$$\nabla^4 \psi = 0 \tag{4.159}$$

The stream function of plane creeping motion is thus a bi-potential or bi-harmonic function. The only difference in these two equations (4.49) and (4.159) is due to the presence of unsteady inertial terms in equation (4.49). The simplified form of equation (4.159) is possible to satisfy as many boundary conditions as for the case of vorticity transfer equation (4.49) because of the order of the differential equation is the same for both equations. The flow described by equation (4.159) for very low Reynolds number (tends to zero) is known as the *Creeping motions*, where there is no inertial force. The creeping flow theory may be applied to different problems like:

1. Parallel flow past a sphere;
2. In the fully developed duct flow, where the inertia terms are neglected;
3. The hydrodynamic theory of lubrication, the flow is in narrow tube but variable passages;
4. The Hele-Shaw flow; and
5. Flow through a porous media, i.e., the groundwater movement.

4.11.2 Stokes's Law and Applications

In particular, the solutions of creeping flow equation (4.159) were obtained by Stokes (1851) for the case of sphere and by Lamb (1945) for the case of a circular cylinder. The Stokes solution was valid for falling of a spherical particle in air, water or oil, where the velocities were so small that inertia can be neglected. Much later, Stokes's idea of creeping flow motion was described by Prandtl (1935) for the case of parallel flow past a sphere, using equation (4.156). Solutions of the set of equation (4.156) subject to the boundary conditions equation (4.157) for the case of a sphere placed in a parallel flow of uniform velocity along the x-axis, including the pressure distributions were given by Prandtl (1935). He also determined the shear stress and pressure distributions over the surface of the sphere. The integration of the pressure distribution and the shearing stress over the surface of the sphere resulted in the total drag F_D (Schlichting, 1968) and is given by

$$F_D = 3\pi\mu_f dU \tag{4.160}$$

where μ_f is the viscosity of the fluid, d is the diameter of the sphere, and U is the uniform velocity along x-axis. Equation (4.160) is a well-known *Stokes equation* for the drag of a sphere. From the equation it is noted that the drag is proportional to the first power of velocity. It may be seen that one-third of the drag of a sphere is due to pressure distribution and that rest two-thirds of the drag are due to shear stress. Here the drag coefficient C_D for the sphere is determined as the ratio of the total drag F_D from equation (4.160) to the dynamic head in the projected area of the sphere and is given by

$$C_D = \frac{F_D}{\frac{1}{2}\pi\rho_f d^2 U^2} = \frac{24}{Re} \tag{4.161}$$

For Stokes equation, the drag coefficient C_D applies to the only $Re < 1$. Oseen (1910) considered the part of the inertia terms to improve the Stokes solution, considering small perturbation in the velocity components, subject to the same boundary conditions in the Navier–Stokes equations. A little improvement was there in the expression as:

$$C_D = \frac{24}{Re}\left(1 + \frac{3}{16}Re\right) \tag{4.162}$$

which improves up to Reynolds number $Re = 5$.

4.11.3 Hydrodynamic Lubrication

For very low Reynolds numbers, another example of flow in which the effect of viscosity is predominant is the hydrodynamic lubrication that is very important from a practical point of view. In particular, for measurement of viscosity, the flow through a capillary tube is used for determining the viscosity of fluids. A relation between the rate of volumetric flow and the pressure gradient is verified experimentally by Hagen, and later by Poiseuille to measure the viscosity of fluid passing through a parallel plate channel. Since some distance from the entry section is required for fluid to be fully developed, the parabolic velocity distribution through the capillary tube, the pressure is recorded to check the fully developed flow. If the radius and length of the tube are known, the difference in pressure between two points along the axial distance is determined, and hence the volume of flow per unit time is measured. The coefficient of viscosity μ_f is calculated from the equation as:

$$\mu_f = \frac{\pi R^4}{8Q}\frac{p_1 - p_2}{l} = \frac{\pi R^4}{8Q} g\rho_f\left(\frac{h}{l}\right) \tag{4.163}$$

where g is the acceleration due to gravity, ρ_f is the density of the fluid, p_1 and p_2 are respective pressure at the different points, R is the radius of the capillary tube and $p_1 - p_2 = \rho gh$ with h is the depth of the fluid.

Whenever there is a rubbing or sliding action between two metallic contacts, the movement will be easy if there is some light oil or viscous oil in between the metallic bodies; otherwise, dissipation of mechanical energy will lead to frictional effect and heat generation. Therefore, the lubricated bearing is the most important application in the laminar viscous fluid flows. For example, let us consider a flow between two parallel sliding walls, the pressure gradient $\frac{dp}{dx}$ in the direction of flow is no longer constant, but there is no pressure gradient $\frac{dp}{dy}$. The equation of resulting motion comes to the generalized Couette flow as:

$$\frac{dp}{dx} = \mu_f \frac{\partial^2 u}{\partial y^2} \tag{4.164}$$

with boundary conditions as:

$$u = -U \text{ at } y = 0, u = 0 \text{ at } y = b; \ p = p_0 \text{ at } x = 0 \text{ and at } x = 1 \tag{4.165}$$

Solving equation (4.164) subject to the boundary conditions (4.165), the velocity, discharge and pressure distributions are obtained (Ojha et al., 2010; Pani, 2016).

The application of Couette flow is illustrated in the study of lubrication, using slipper bearings. The combination of Poiseuille flow due to pressure gradient and Couette flow due to plate movement is known as the generalized Couette flow. The velocity distribution becomes

$$u = -U\left(1 - \frac{y}{h}\right) - \frac{h^2}{2\mu_f}\frac{dp}{dx}\frac{y}{h}\left(1 - \frac{y}{h}\right) \tag{4.166}$$

Here first term represents the flow due to the plate movement with a uniform velocity U and the second term due to the pressure gradient $\frac{dp}{dx}$. If the pressure gradient is zero, equation (4.169) reduces to the Couette flow, and if the plate movement is zero, it reduces to Poiseuille flow. The rate of volumetric flow Q in every section is obtained from the equation of continuity as:

$$Q = \int_0^h u dy = \text{constant}$$

$$= -\frac{Uh}{2} - \frac{h^3}{12\mu_f}\frac{dp}{dx} \Rightarrow \frac{dp}{dx} = -\frac{12\mu_f}{h^3}\left(\frac{uh}{2} + Q\right) \tag{4.167}$$

For further details on lubrication using the slipper bearing, readers may refer to other textbooks such as Som and Biswas (1998), Ojha et al. (2010), Pani (2016) etc.

Another example of low Reynolds flow is flow through a homogeneous porous media with fine-grained sand. Within the tiny pores of sand grain materials, the movement of the fluid is very low. The relationship between the velocity, pressure gradient and the body forces are controlled solely by the high viscous shear stress acting on the fluid space, where the convective accelerations have no effect, i.e., the inertial forces are neglected. So, the laminar flow for very low Reynolds number can be applied (Ojha et al., 2010).

4.12 SUMMARY

Some of the important aspects discussed in this chapter are summarized below:

- Principles of similarity and study of dimensionless parameters are very important in the solution of fluid dynamics problems. The dimensionless numbers and similarity principles can be effectively used in many flow analyses.
- Boundary layer is defined over a smooth flat surface, where the effect of viscosity is very important. The concepts of boundary layer at high Reynolds number are divided into two regions: one is called the boundary layer at the wall, where the viscosity effect is important, and other one at the outer region of the boundary layer, where the viscosity effect is neglected, called the inviscid flow layer or potential flow.
- When the flow past a circular cylinder or a sphere, the boundary layer separation occurs at a point due to the adverse pressure gradient and consequently, the vortices are also formed behind the body.
- The fundamental equations of fluid dynamics in laminar flow are derived from the conservation of mass and linear momentum. These equations are called the continuity and Navier–Stokes equations $\left(\nabla.\vec{q} = 0;\ \rho_f \frac{D\vec{q}}{Dt} = \vec{g} - \text{grad}\, p + \mu_f \nabla^2 \vec{q}\right)$. There are no exact solutions to the full Navier–Stokes equations. However, some simplified solutions of these equations, like Poiseuille flow, Couette flow, co-axial flow concentric rotating flow, etc. are possible and described with applications and examples.
- Unsteady motions of flat plate in an infinite fluid with an impulsively started from a rest, known as Stokes first problem, and the motion with harmonic oscillation, known as Stokes second problem.
- In fluid dynamics, stagnation point flow represents the flow of a fluid in the immediate neighborhood of a solid surface. It has many applications in engineering problems.
- If the Reynolds number is very low (for high viscosity), the inertial forces can be

neglected. In that case, for Low Reynolds number flow such as creeping flow, hydrodynamic lubrication flow, porous media flow are important examples.

- For Stokes flow past a sphere for low Reynolds numbers, the total drag F_D (Schlichting, 1968) is given by $F_D = 3\pi\mu_f dU$, where ρ_f is the viscosity of the fluid, d is the diameter of the sphere, and U is the uniform velocity along the x-axis. This equation is the well-known *Stokes equation* for the drag of a sphere.

4.13 EXERCISE PROBLEMS

1. What do you mean by boundary layer thickness? Show that shear stress at the wall τ_0 is proportional to the power 3/2 of free stream velocity U, and find the total drag on the wall. Under what conditions does separation of flow occur in the boundary layer?
2. Derive the velocity expression for generalized Couette flow between two parallel plates, where one plate is at rest and the other plate is moving with a uniform velocity. Show the schematic diagram of positive and negative velocity gradient in the flow.
3. State the Stokes first problem in the fluid flow. Show that the solution of the Stokes first problem is $u = U_\infty(1 - \text{erf}\,\eta)$, where U_∞ is the motion of a plate started impulsively from rest in an infinite fluid, erf η error function and η is a similarity variable. Describe the Stokes's Second Problem.
4. What is the difference between the Stokes first and second problems in the fluid flow?
5. A fluid flows at 0.001 m^3/s through a narrow region between two smooth plates separated by 10 mm apart. Determine the pressure gradient acting on the fluid.
6. A viscous fluid flows between two parallel plates separated by a distance of 0.015 m, in which the lower plate is fixed and the upper plate moves with a velocity of 1 m/s. The dynamic viscosity of the fluid is 0.8 pose. Calculate the velocity distribution, discharge and shear stress on the upper plate. The pressure drop in 100 m distance is 50 kPa.

5 Boundary Layer Theory

5.1 INTRODUCTION

In fluid dynamics, the boundary layer (BL) concept for viscous fluid flow was introduced by Prandtl (1904). The influence of viscosity at a high/moderate Reynolds number is confined to a very thin layer in the immediate neighborhood of the solid wall, where the viscous stresses are very important. Although the layer is very thin, the physics of flow is very important within the layer with the no-slip condition, where the mean flow velocity tends to zero at the wall. The velocity gradient normal to the surface within the layer is very large compared to the gradient of this component outside the layer, where the free stream velocity exists. Many important fluids, such as water, air, and others, have very small viscosity while considering high Reynolds number flows. As a result, BL approximations also hold true in these cases. The BL theory has been well developed for many engineering problems. The BL can be a laminar BL or a turbulent BL. In general, if the Reynolds number is less than 5×10^5, then the BL is considered laminar and otherwise, a turbulent BL. In this chapter, BL approximations specifically, the laminar BL, for various problems and solutions are discussed.

5.2 BOUNDARY LAYER EQUATIONS AND APPROXIMATIONS

Let us consider a two-dimensional flow of fluid with very low viscosity past a submerged slender cylindrical body (Figure 5.1a). At the leading stagnation point, the thickness of the BL is zero, and it grows slowly toward the rear of the body. Apart from the immediate neighborhood of the body, the velocity is of the order of the free stream velocity U. Within the very thin BL thickness δ, a large velocity gradient exists, i.e., the velocity increases from zero at the wall to the free stream velocity U, i.e., potential flow outside the BL δ. The streamline pattern and the velocity distribution deviate from those of frictionless flow. The transition from zero velocity at the wall to the free stream velocity with a full magnitude at some distance from the wall takes place in a very thin layer, called boundary layer thickness (Figure 5.1a). The velocity gradients perpendicular to the wall are much larger than the gradients parallel to the wall, therefore, some terms in the Navier–Stokes equations can be neglected. According to the theory, the flow field can be identified into two different regions: (i) BL region dominated by viscosity, described by the momentum and energy equations and (ii) outer region described by the inviscid flow. Although the division between the two is not very sharp, clear distinction between the regions is defined as follows (Schlichting, 1968):

1. A very thin layer just attached to the body where the velocity gradient $\frac{\partial u}{dy}$ normal to the wall is very high within the layer. A very small viscosity μ_f can play an important role there, since the shear stress $\tau = \mu_f \frac{\partial u}{dy}$ can attain a significant value.
2. Outside the BL, there is no large velocity gradient so viscosity is not important. This region is called frictionless or potential region, dominated by inviscid flow.

In general, it is stated that the thickness of BL decreases with decrease in viscosity or more generally, BL decreases with the increase of Reynolds number. We know the BL thickness is proportional to the square root of kinematic viscosity, i.e., $\delta \sim \sqrt{\nu_f}$.

DOI: 10.1201/9781003000020-5

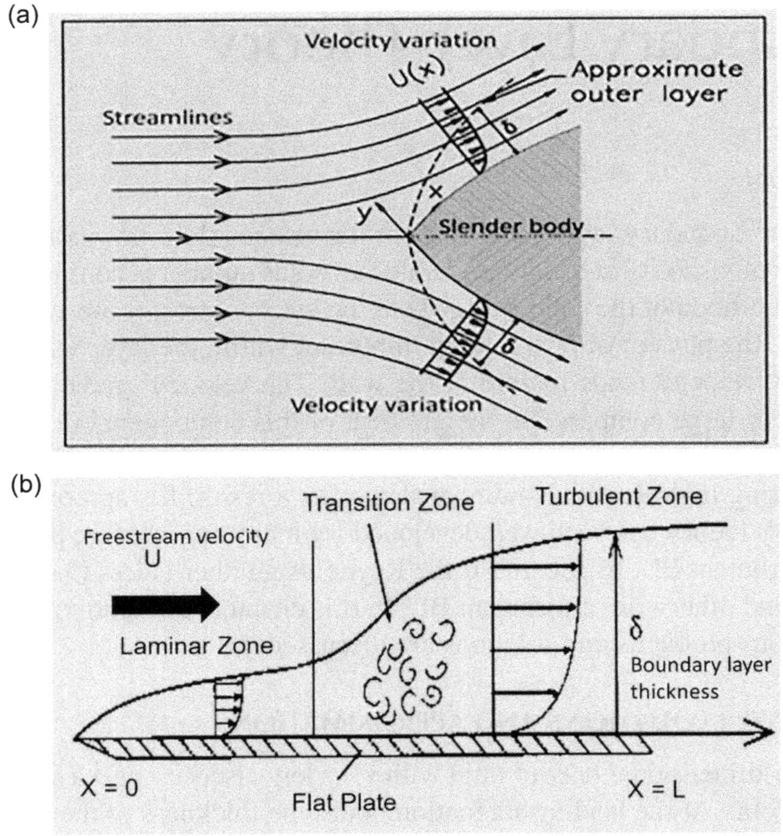

FIGURE 5.1 (a) Boundary layer flow along a thin slender cylinder. (Modified from Schlichting, 1979.) (b) Schematic diagram of boundary layer growth over a thin flat plate.

For the simplification of Navier–Stokes equations, it is assumed that the BL thickness is very small compared to the unspecified linear dimension of the body l, i.e., $\delta \ll l$.

Now if we use U as the free stream velocity, and l is the characteristic length of body, then the relation $\delta \sim \sqrt{\nu_f}$ leads to

$$\Rightarrow \delta^* = \frac{\delta}{l} \sim \frac{1}{\sqrt{Re}}, \text{ since } \delta \sim \sqrt{\frac{\nu_f l}{U}},$$

where $Re = \dfrac{Ul}{\nu_f}, \Rightarrow \delta^* \to 0$ as $Re \to \infty$.

Considering the two-dimensional flow over a flat plate (Figure 5.1b) with x-axis is along the plate and y-axis is perpendicular to it, we can write the dimensionless Navier–Stokes equations and equation of continuity with characteristic length l and free stream velocity U without body force for two-dimensional flow as follows (Yuan, 1970):

$$\frac{\partial u}{\partial t} + u\frac{\partial u}{\partial x} + v\frac{\partial u}{\partial y} = -\frac{\partial p}{\partial x} + \frac{1}{Re}\left(\frac{\partial^2 u}{\partial x^2} + \frac{\partial^2 u}{\partial y^2}\right) \tag{5.1}$$

$$\frac{\partial v}{\partial t} + u\frac{\partial v}{\partial x} + v\frac{\partial v}{\partial y} = -\frac{\partial p}{\partial y} + \frac{1}{Re}\left(\frac{\partial^2 v}{\partial x^2} + \frac{\partial^2 v}{\partial y^2}\right) \tag{5.2}$$

The equation of continuity is

$$\frac{\partial u}{\partial x}+\frac{\partial v}{\partial y}=0 \tag{5.3}$$

where $Re=\dfrac{Ul}{\upsilon_f}$ is the Reynolds number assumed to be relatively large. It is ensured that the dimensionless derivative $\dfrac{\partial u}{\partial x}$ does not exceed unity in the region under consideration. The pressure is made dimensionless using $\rho_f U^2$ and the time by l/U. According to the exact solution of the unsteady motion of a flat plate, it was found that the thickness of the BL δ is proportional to the square root of the kinematic viscosity υ_f which is very small.

The dimensionless momentum equations respectively in the x- and y-directions are as follows:

$$\frac{\partial u}{\partial t}+u\frac{\partial u}{\partial x}+v\frac{\partial u}{\partial y}=-\frac{\partial p}{\partial x}+\frac{1}{Re}\left(\frac{\partial^2 u}{\partial x^2}+\frac{\partial^2 u}{\partial y^2}\right)$$

$$1 \quad 1\;1 \quad \delta\;\frac{1}{\delta} \qquad\qquad \delta^2 \quad 1 \quad \frac{1}{\delta^2} \tag{5.4}$$

$$\frac{\partial v}{\partial t}+u\frac{\partial v}{\partial x}+v\frac{\partial v}{\partial y}=-\frac{\partial p}{\partial x}+\frac{1}{Re}\left(\frac{\partial^2 v}{\partial x^2}+\frac{\partial^2 v}{\partial y^2}\right)$$

$$\delta \quad 1\;\delta \quad \delta\;1 \qquad\qquad \delta^2 \quad \delta \quad \frac{1}{\delta} \tag{5.5}$$

The dimensionless continuity equation is

$$\frac{\partial u}{\partial x}+\frac{\partial u}{\partial y}=0$$

$$1 \quad 1$$

The boundary conditions are as

$$u=v=0 \text{ for } y=0 \text{ and } u=U \text{ for } y\to\infty \tag{5.6}$$

If $Re\to\infty$, the equations reduce to those for inviscid fluids and also describes the potential flow with the uniform free stream velocity. Since the friction forces play an important role in this layer, friction terms in the equations describing the flow are not neglected. The order of magnitude of various terms in the above equations of motions and continuity is indicated under the respective terms. The order of magnitude of the individual terms is estimated to drop the small terms and thus to achieve the desired simplification of equations. The length x and the velocity u have the order of magnitude O (1), so $\dfrac{\partial u}{\partial x}$ is of order 1, and from the equation of continuity that equally $\dfrac{\partial v}{\partial y}$ is of the order 1. Since at the wall v=0, and for $Re\to\infty$, i.e., $\delta\to 0$, the continuity equation should not change, that means $v=O(\delta)$. Thus, $\dfrac{\partial v}{\partial x}$ and $\dfrac{\partial^2 v}{\partial x^2}$ are also $O(\delta)$; and the term $\dfrac{\partial^2 u}{\partial x^2}$ is of O (1). The local acceleration term $\dfrac{\partial u}{\partial t}$ has the same order of magnitude as convective acceleration term $u\dfrac{\partial u}{\partial x}$. That means, there is no sudden acceleration due to strong pressure waves.

In fact, some of the viscous terms must be of the same order of magnitude as the inertia terms, at least in the immediate neighborhood of the wall. Hence, some of the second derivatives of velocity must be very large near the wall. Since the velocity component u parallel to the wall increases from zero at the wall to the value of order 1 in the free stream across the BL thickness δ, one gets $\frac{\partial u}{\partial y} \sim \frac{1}{\delta}$ and $\frac{\partial^2 u}{\partial y^2} \sim \frac{1}{\delta^2}$, whereas $\frac{\partial v}{\partial y} \sim \frac{\delta}{\delta} \sim 1$, and $\frac{\partial^2 v}{\partial y^2} \sim \frac{\delta}{\delta^2} \sim \frac{1}{\delta}$. If these values are substituted in equations (5.4) and (5.5), it follows from equation (5.4) that the viscous terms in the BL can be of the same order of magnitude as inertia terms only if the Reynolds number Re is order of $\frac{1}{\delta^2}$, that means, $\frac{1}{Re} = \delta^2$. The equation of continuity remains unchanged for very large Reynolds number. Equation (5.4) can be simplified by neglecting $\frac{\partial^2 u}{\partial x^2}$ with respect to $\frac{\partial^2 u}{\partial y^2}$. Here equation (5.5) shows that $\frac{\partial p}{\partial y}$ is of the order δ. The pressure increases across the BL which would be obtained by integrating equation (5.5) is of order δ^2 which is very small, indicating that the pressure normal to the boundary is practically constant; and assumed equal to that at the outer edge of the BL where it is determined by the potential flow. This pressure is known as the impressed pressure on the BL by the outer flow, which is a known function that depends on x and t, as far as the BL flow is concerned.

5.2.1 Prandtl's Boundary Layer Equations

Let us consider the laminar BL growth over a flat plate (Figure 5.1b). According to Prandtl, the flow can be considered into two regions, (i) Region outside the BL and (ii) Region within BL. At the outer edge of the BL, the velocity u parallel to the flow becomes equal to the free stream velocity $U(x, t)$. Since there is no velocity gradient, the viscous terms vanish for large Re, and thus for outer flow, one gets pressure equation in dimensional form as:

$$\frac{\partial U}{\partial t} + U\frac{\partial U}{\partial x} = -\frac{1}{\rho_f}\frac{\partial p}{\partial x} \tag{5.7}$$

In the case of steady flow, the equation is simplified for the pressure $\frac{\partial p}{\partial x}$ as:

$$U\frac{\partial U}{\partial x} = -\frac{1}{\rho_f}\frac{\partial p}{\partial x}, \tag{5.8}$$

which may be written as the usual form of Bernoulli's equation $p + \frac{1}{2}\rho_f U^2 = \text{constant}$. The boundary conditions at the outer flow of the BL are nearly the same as frictionless flow.

In the region of BL region, Prandtl (1935) derived the BL equations using the Navier–Stokes equations, in dimensional form as:

$$\frac{\partial u}{\partial t} + u\frac{\partial u}{\partial x} + v\frac{\partial u}{\partial y} = -\frac{1}{\rho_f}\frac{\partial p}{\partial x} + \nu_f\frac{\partial^2 u}{\partial y^2} \tag{5.9}$$

$$0 = -\frac{\partial p}{\partial y} \tag{5.10}$$

$$\frac{\partial u}{\partial x} + \frac{\partial v}{\partial y} = 0$$

subject to the boundary conditions

$$u = v = 0 \text{ for } y = 0, \text{ and } u = U(x,t) \text{ for } y \to \infty \tag{5.11}$$

Equation (5.10) shows that the pressure p is independent of y, which is constant across the cross-section of the BL. The pressure at the edge of the BL is determined by the inviscid flow. The pressure p depends on x and t. At the edge of the BL, the longitudinal velocity u passes over the velocity in the outer flow $U(x, t)$, i.e., $u = U(x, t)$. Eliminating the pressure from the Prandtl's BL equations, one gets the equations as

$$\frac{\partial u}{\partial t} + u\frac{\partial u}{\partial x} + v\frac{\partial u}{\partial y} = \frac{\partial U}{\partial t} + U\frac{\partial U}{\partial x} + \nu_f \frac{\partial^2 u}{\partial y^2} \tag{5.12}$$

$$\frac{\partial u}{\partial x} + \frac{\partial v}{\partial y} = 0$$

with the boundary conditions

$$y = 0: u = 0,\ v = 0; y \to \infty : u = U(x, t). \tag{5.13}$$

In case of steady flow, the equation is simplified further, where the pressure depends only on x and the equation can be written as

$$u\frac{\partial u}{\partial x} + v\frac{\partial u}{\partial y} = -\frac{1}{\rho_f}\frac{\partial p}{\partial x} + \nu_f \frac{\partial^2 u}{\partial y^2}$$

$$\frac{\partial u}{\partial x} + \frac{\partial v}{\partial y} = 0 \tag{5.14}$$

with the boundary conditions

$$y = 0: u = 0,\ v = 0; y \to \infty : u = U(x) \tag{5.15}$$

To eliminate the pressure from the above equation, we use

$$U\frac{\partial U}{\partial x} = -\frac{1}{\rho_f}\frac{\partial p}{\partial x} \tag{5.16}$$

In the dimensional form, eliminating the pressure gradient, one can write the equation as:

$$u\frac{\partial u}{\partial x} + v\frac{\partial u}{\partial y} = U\frac{dU}{dx} + \nu_f \frac{\partial^2 u}{\partial y^2}$$

$$\frac{\partial u}{\partial x} + \frac{\partial v}{\partial y} = 0 \tag{5.17}$$

with the boundary conditions

$$y = 0: u = 0,\ v = 0; y \to \infty : u = U(x) \tag{5.15}$$

5.2.2 Some Physical Properties of Boundary Layers

The solutions of BL equations are used to determine the important physical parameters like wall shear stress and displacement thickness. Here these are briefly explained.

5.2.2.1 Skin-Friction Coefficient

For dimensionless wall shear stress, the skin-friction coefficient is introduced as:

$$C_f(x) = \frac{\tau_\omega(x)}{\frac{\rho_f}{2}U^2} \tag{5.18}$$

where $\tau_\omega(x)$ is shear stress defined by

$$\tau_\omega = \mu_f \left.\frac{\partial u}{\partial y}\right)_{\text{wall}}$$

$$C_f = \frac{2\mu_f \left.\frac{\partial u}{\partial y}\right)_{\text{wall}}}{\rho_f U^2} \tag{5.19}$$

The skin-friction coefficient is determined from the velocity gradient at the wall $(y = 0)$.

5.2.2.2 Separation Point

When a flow past a body, due to adverse pressure gradient, the separation of flow at the BL occurs and the point of separation is defined as

$$\left.\frac{\partial u}{\partial y}\right|_{y=0} = 0 \tag{5.20}$$

The point at which wall shear stress vanishes is called the separation of flow at that point, which is due to adverse pressure gradient existing along the wall. That means, the retarded fluid particles cannot penetrate too far into the region of increased pressure owing to their small kinetic energy. Thus, the BL is deflected sideway from the wall, separates from it, and moves into the main stream. The point of separation occurs only due to adverse pressure gradient i.e., $\frac{\partial p}{\partial x} > 0$; and the BL equations are valid only up to the point of separation (Figure 5.2). A shape of streamlines near the point of separation is shown in Figure 5.3; and the velocity distributions around the point of separation are shown in Figure 5.4 for different conditions. The short distance beyond the point of flow separation, the BL becomes so thick that the derivations of BL equations are no longer valid.

From the BL equation (5.14), we get for $u = 0,\ v = 0$ at $y = 0$

$$0 = -\frac{1}{\rho_f}\frac{\partial p}{\partial x} + \nu_f \frac{\partial^2 u}{\partial y^2} \text{ at } y = 0 \Rightarrow \mu_f \frac{\partial^2 u}{\partial y^2} = \frac{\partial p}{\partial x} \text{ at } y = 0.$$

$$\text{If } \frac{\partial p}{\partial x} > 0, \frac{\partial^2 u}{\partial y^2} > 0 \text{ at } y = 0; \text{ if } \frac{\partial p}{\partial x} < 0, \frac{\partial^2 u}{\partial y^2} < 0 \text{ at } y = 0$$

Then again, if we differentiate with respect to y, $\frac{\partial^3 u}{\partial y^3} = 0$ at $y = 0$

If $\frac{\partial p}{\partial x} < 0$ for decreasing pressure, i.e., accelerated flow, we have $\left.\frac{\partial^2 u}{\partial y^2}\right)_{\text{wall}} < 0$.

$\therefore \frac{\partial^2 u}{\partial y^2} < 0$ over the whole width of the BL (Figure 5.5).

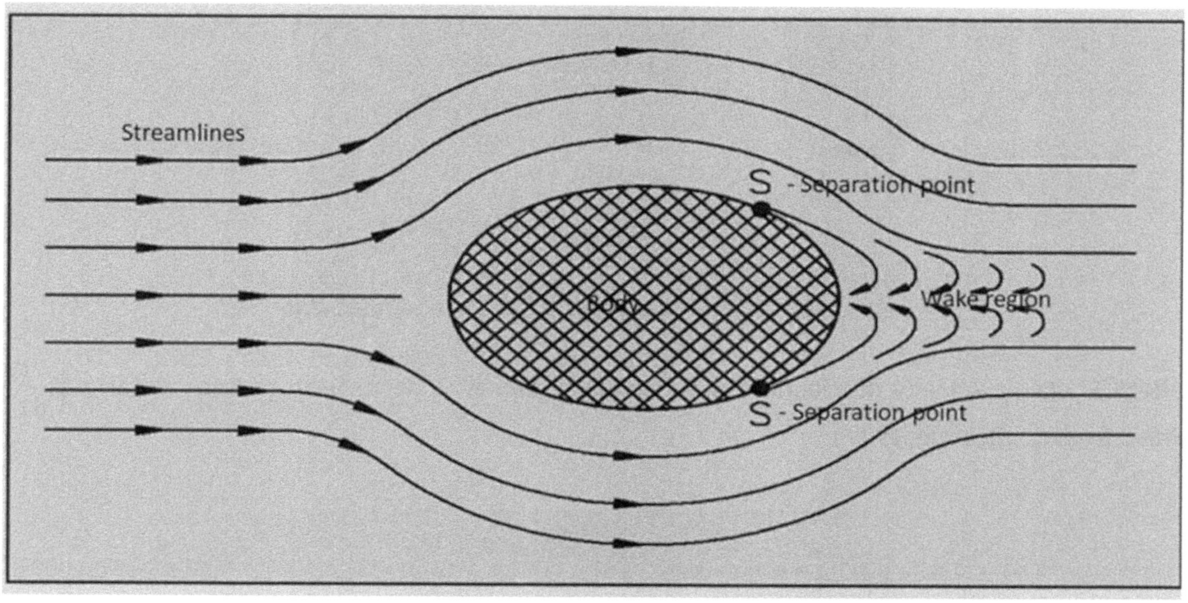

FIGURE 5.2 Boundary layer separation over a body with separation points S. (Modified from Schlichting, 1979.)

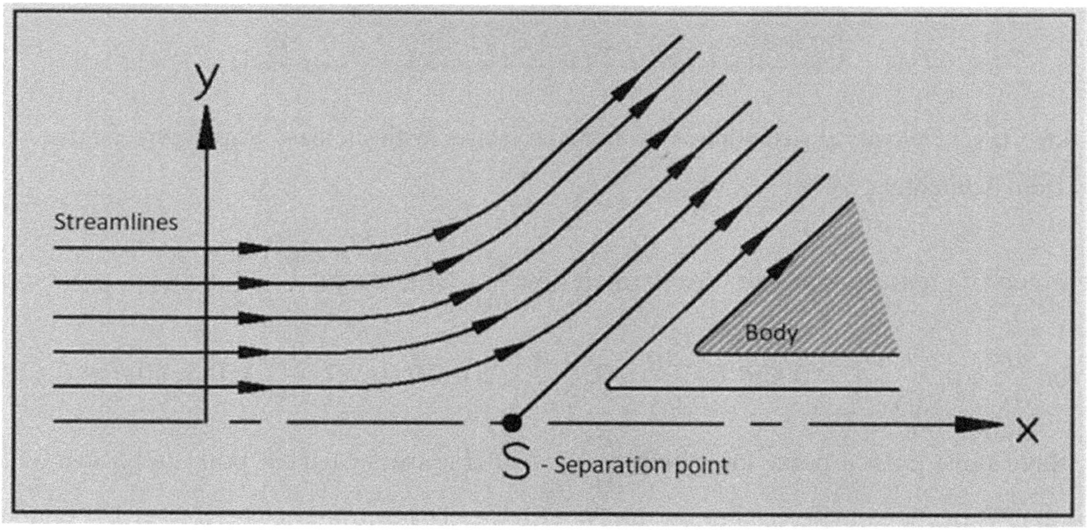

FIGURE 5.3 Stream line flow near a point of separation S. (Modified from Schlichting, 1979.)

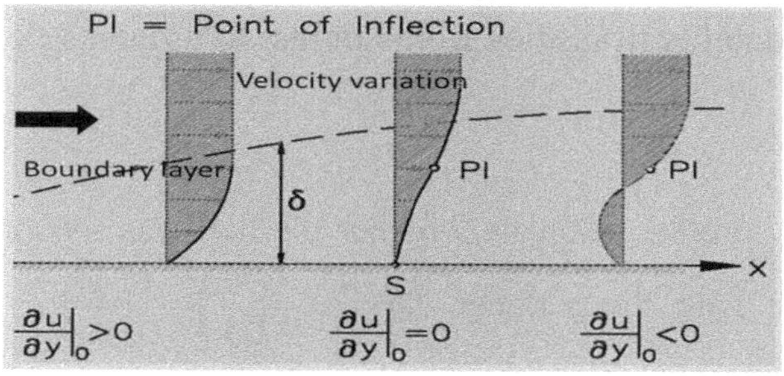

FIGURE 5.4 Velocity distributions near the point of separation S with point of inflection PI. (Modified from Schlichting, 1979.)

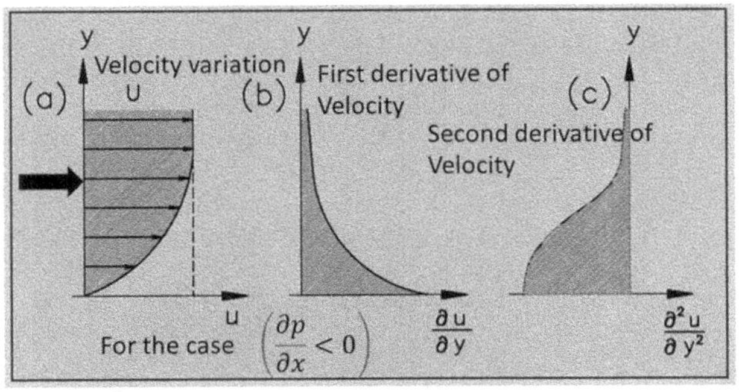

FIGURE 5.5 (a–c) Velocity distributions in a boundary layer with decrease in pressure gradient $\left(\frac{\partial p}{\partial x} < 0\right)$. (Modified from Schlichting, 1979.)

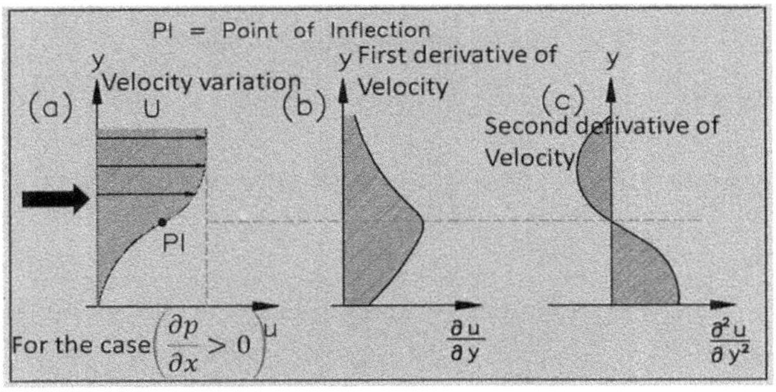

FIGURE 5.6 (a–c) Velocity distributions in a boundary layer with increase in pressure gradient $\left(\frac{\partial p}{\partial x} > 0\right)$. (Modified from Schlichting, 1979.)

In the region of pressure increase (decelerated flow $\frac{\partial p}{\partial x} > 0$), we get $\left.\frac{\partial^2 u}{\partial y^2}\right)_0 > 0$.

Since for $\frac{\partial p}{\partial x} < 0$, $\left.\frac{\partial^2 u}{\partial y^2}\right)_0 < 0$ and for $\frac{\partial p}{\partial x} > 0$, $\left.\frac{\partial^2 u}{\partial y^2}\right)_0 > 0$. However, $\frac{\partial^2 u}{\partial y^2} < 0$ at a large distance from the wall, there must exist a point for which $\left.\frac{\partial^2 u}{\partial y^2}\right)_0 = 0$ (Figure 5.6). That point is known as the point of inflection (PI) of the velocity profile in the BL (Figure 5.6). In the region of retarded potential flow, the velocity profile in the BL always displays a point of inflexion. Since the velocity profile at the point of separation and with a zero tangent with a point of inflection, it follows that the separation can only occur when the potential flow is retarded. The existence of point of inflection in the velocity profile in the BL is important for its stability (transition from laminar to turbulent flow).

Example 5.1

For the following velocity profiles, evaluate whether the flow has separated or is on the verge of separation or will attach with the surface?. (a) $\frac{u}{U} = \frac{5}{3}\left(\frac{y}{\delta}\right) - \frac{1}{3}\left(\frac{y}{\delta}\right)^3$; (b) $\frac{u}{U} = -3\left(\frac{y}{\delta}\right) + 2\left(\frac{y}{\delta}\right)^2$; (c) $\frac{u}{U} = -4\left(\frac{y}{\delta}\right)^2 - 3\left(\frac{y}{\delta}\right)^3$?.

Solution

a. $\frac{u}{U} = \frac{5}{3}\left(\frac{y}{\delta}\right) - \frac{1}{3}\left(\frac{y}{\delta}\right)^3$; $\frac{\partial u}{\partial y} = \frac{5U}{3}\frac{1}{\delta} - \frac{U}{3}3\left(\frac{y}{\delta}\right)^2\frac{1}{\delta}$

At $y=0$, $\frac{\partial u}{\partial y} = \frac{5U}{3\delta}$, which is positive. Hence flow will not separate and will remain attached.

b. $\frac{u}{U} = -3\left(\frac{y}{\delta}\right) + 2\left(\frac{y}{\delta}\right)^2$; $\frac{\partial u}{\partial y} = -3U\left(\frac{1}{\delta}\right) + 4u\left(\frac{y}{\delta}\right)\frac{1}{\delta}$.

At $y=0, \frac{\partial u}{\partial y} = -3\frac{U}{\delta}$. Flow has separated.

c. $\frac{u}{U} = -4\left(\frac{y}{\delta}\right)^2 - 3\left(\frac{y}{\delta}\right)^3$; $\frac{\partial u}{\partial y} = 4U \times 2\left(\frac{y}{\delta}\right)\frac{1}{\delta} - 3U \times 3\left(\frac{y}{\delta}\right)^2 \times \frac{1}{\delta}$

At $y=0$, $\frac{\partial u}{\partial y} = 0$

Flow is on the verge of separation.

5.3 BOUNDARY LAYER ALONG A FLAT PLATE

Consider the flow of an incompressible viscous fluid past a thin flat plate which is placed in the direction of a uniform velocity U (Figure 5.7). The plate is of infinite in length. The problem is one of the two-dimensional motions which can be analyzed using the Prandtl BL equations (Prandtl, 1935).

Let O is the origin at the leading edge, x-axis along the flow and y-axis normal to the flow. Under the BL approximations, this can be formulated as

$$u\frac{\partial u}{\partial x} + v\frac{\partial u}{\partial y} = \nu_f \frac{\partial^2 u}{\partial y^2} \tag{5.21}$$

$$\frac{\partial u}{\partial x} + \frac{\partial v}{\partial y} = 0 \tag{5.22}$$

Boundary conditions are

$$\begin{aligned} y = 0 &: u = v = 0; \\ y = \delta&, u = U \end{aligned} \tag{5.23}$$

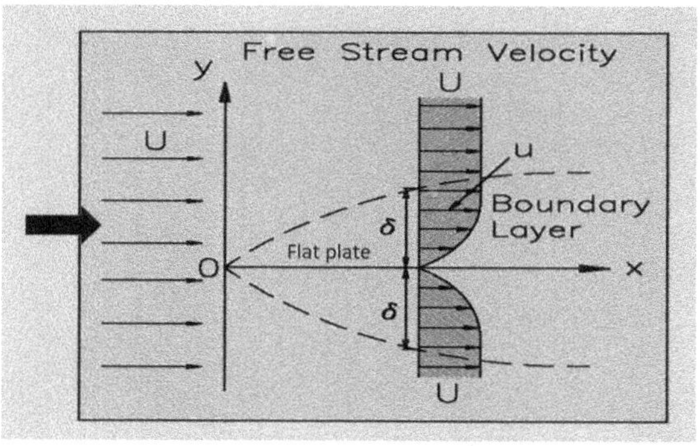

FIGURE 5.7 Boundary layer flow past a flat plate. (Modified from Schlichting, 1979.)

The characteristic parameters in these problems are U, ν, x, y, i.e. the problem is defined completely by these parameters. In accordance with the law of similarity, the velocity profile may be assumed as (Yuan, 1970; Schlichting and Gersten, 2000):

$$\frac{u}{U} = F(x, y, \nu, U) = F(\eta) \tag{5.24}$$

According to the exact solution of the unsteady motion of a flat plate, we found that (refer Chapter 4, Section 4.9)

$$\delta \sim \sqrt{\nu_f t} \sim \sqrt{\frac{\nu_f x}{U}} \tag{5.25}$$

Here the time t is the time required for a fluid particle to travel a distance x with velocity U. Hence the nondimensional distance η may be expressed as

$$\eta = \frac{y}{\delta} = \frac{y}{\sqrt{\frac{\nu_f x}{U}}} \tag{5.26}$$

which is in agreement with equation (5.24)

The stream function ψ can now be obtained in terms of velocity equation (5.24) as

$$\psi = \int u dy = \int UF(\eta) d\eta \left(\sqrt{\frac{\nu_f x}{U}} \right) = \sqrt{U\nu_f x} \int F(\eta) d\eta = \sqrt{U\nu_f x} f(\eta) \tag{5.27}$$

Now

$$u = \frac{\partial \psi}{\partial y} = \frac{\partial \psi}{\partial \eta} \frac{\partial \eta}{\partial y} = \sqrt{U\nu_f x} f'(\eta) \left(\frac{\nu_f x}{U} \right)^{-1/2}$$

$$u = Uf'(\eta) \tag{5.28}$$

$$\nu = -\frac{\partial \psi}{\partial x},$$

where $\psi = \sqrt{U\nu_f x} f(\eta)$.

Now,

$$\psi = \sqrt{U\nu_f x} f(\eta) \tag{5.29}$$

$$\frac{\partial \psi}{\partial x} = f(\eta) \cdot \frac{1}{2} \sqrt{\nu_f U} \frac{1}{\sqrt{x}} - \frac{1}{2} \sqrt{U\nu_f x} \frac{y}{\sqrt{\frac{\nu_f}{U}} x\sqrt{x}} f'(\eta)$$

$$\frac{\partial \psi}{\partial x} = \frac{1}{2} \sqrt{\frac{U\nu_f}{x}} \left[f(\eta) - \eta f'(\eta) \right]$$

$$\nu = -\frac{\partial \psi}{\partial x} = \frac{1}{2} \sqrt{\frac{U\nu_f}{x}} \left[\eta f'(\eta) - f(\eta) \right] \tag{5.30}$$

$$\frac{\partial u}{\partial x}=\frac{\partial^2 \psi}{\partial x\,\partial y}=-\frac{U}{2}\frac{\eta}{x}f''(\eta) \tag{5.31}$$

$$\frac{\partial u}{\partial y}=U\sqrt{\frac{U}{\nu_f x}}f''(\eta) \tag{5.32}$$

$$\frac{\partial^2 u}{\partial y^2}=\frac{U^2}{\nu_f x}f'''(\eta) \tag{5.33}$$

where $\frac{\partial \eta}{\partial y}=\sqrt{\frac{U}{\nu_f x}}$, $\frac{\partial \eta}{\partial x}=-\frac{1}{2}\frac{y}{x}\sqrt{\frac{U}{\nu_f x}}=-\frac{1}{2}\frac{\eta}{x}$.

Substituting the above equation in (5.21), we get

$$-\frac{U^2}{2}\frac{\eta}{x}ff''+\frac{U^2}{2x}[\eta f'-f]f''=\frac{U^2}{x}f''' \tag{5.34}$$

After simplification, one gets as

$$2f'''+ff''=0 \tag{5.35}$$

This is **Blasius's equation (1908)**. Boundary conditions are in terms of f and η as follows:

$$\left.\begin{aligned}&\eta=0: f=0,\ f'=0\\ &\eta=\infty: f'=1\end{aligned}\right| \tag{5.36}$$

Equation (5.35) is a third-order nonlinear differential equation and no closed-form solution has been found. Blasius (1908) obtained the solution in the form of power series expansion about $\eta=0$ as (Yuan, 1970):

$$f(\eta)=A_0+A_1\eta+\frac{A_2}{2!}\eta^2+\frac{A_3}{3!}\eta^3+\cdots \tag{5.37a}$$

$$f'(\eta)=A_1+A_2\eta+\frac{A_3}{2!}\eta^2+\frac{A_4}{3!}\eta^3+\cdots \tag{5.37b}$$

$$f''(\eta)=A_2+A_3\eta+\frac{A_4}{2!}\eta^2+\frac{A_5}{3!}\eta^3+\cdots \tag{5.37c}$$

$$f'''(\eta)=A_3+A_4\eta+\frac{A_5}{2!}\eta^2+\frac{A_6}{3!}\eta^3+\cdots \tag{5.37d}$$

Applying equation (5.37) for f, f', f'' and f''' into equation (5.35), one gets

$$2A_2+(2A_4)\eta+\left(A_2^2+2A_5\right)\frac{\eta^2}{2!}+(4A_2A_3+2A_6)\frac{\eta^3}{3!}+\cdots=0 \tag{5.38}$$

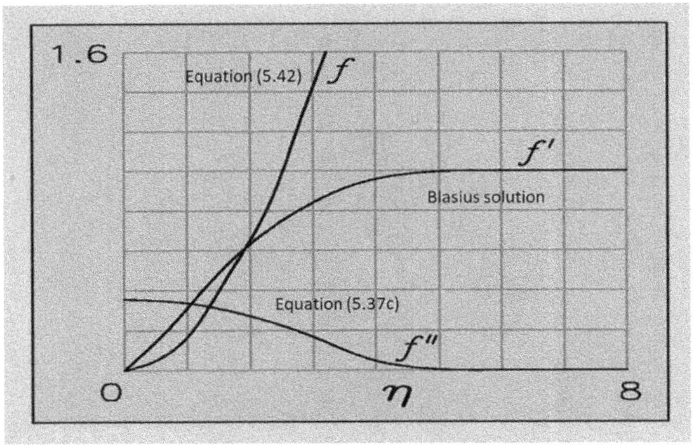

FIGURE 5.8 The functions f, f' and f'' shown against η along the flat plate. (Modified from Yuan, 1970.)

Applying boundary conditions (5.36) at $\eta = 0$ to equation (5.37), we get

$$A_0 = 0,\ A_1 = 0 \tag{5.39}$$

Equating the coefficients of various powers at η must be identically equal to zero. Hence

$$\left.\begin{aligned} &A_3 = A_4 = A_6 = A_7 = 0 \\ &A_5 = -\frac{1}{2}A_2^2,\ A_8 = \frac{11}{4}A_2^3 \end{aligned}\right\} \tag{5.40}$$

Using equations (5.39) and (5.40) into (5.37), the series of f in terms of η and A_2 is obtained as:

$$f = \frac{A_2}{2!}\eta^2 - \frac{1}{2}\frac{A_2^2}{5!}\eta^5 + \frac{1}{4}\frac{11A_2^3}{8!}\eta^8 - \frac{1}{8}\cdot\frac{375A_2^4}{11!}\eta^{11} + \cdots \tag{5.41}$$

Equation (5.41) satisfies the two boundary conditions at $\eta = 0$, and the constant A_2 is determined from the boundary condition at $\eta = \infty : f' = 1$.

Equation (5.41) can be rewritten as follows:

$$f = A_2^{\frac{1}{3}}\left[\frac{\left(A_2^{\frac{1}{3}}\eta\right)^2}{2!} - \frac{1}{2}\frac{\left(A_2^{\frac{1}{3}}\eta\right)^5}{5!} + \frac{1}{4}\frac{\left(A_2^{\frac{1}{3}}\eta\right)^8}{8!} - \frac{1}{8}\frac{375\left(A_2^{\frac{1}{3}}\eta\right)^{11}}{11!} + \cdots\right] = A_2^{\frac{1}{3}}\,F\left(A_2^{\frac{1}{3}}\eta\right) \tag{5.42}$$

The application of boundary condition at $\eta = \infty$ to the equation, we get

$$\lim_{\eta\to\infty}\left[A_2^{\frac{2}{3}}F'\left(A_2^{\frac{1}{3}}\eta\right)\right] = f'(\infty) = 1$$

$$A_2 = \left(\frac{1}{\lim\limits_{\eta\to\infty} F'(\eta)}\right)^{\frac{3}{2}} \tag{5.43}$$

The value of A_2 can be determined numerically from (5.43). Howarth (1938) found $A_2 = 0.3321$. The functions f, f', f'' are plotted in Figure 5.8 (Yuan, 1970).

5.3.1 Shear Stress and Boundary Layer Thickness

The shearing stress on the surface of the plate can be determined from the results of Blasius solution as follows:

$$\tau_0 = \mu_f \left.\frac{\partial u}{\partial y}\right)_{y=0} = \frac{\mu_f U f''(0)}{\sqrt{\frac{\nu_f x}{U}}} = \mu_f \frac{UA_2}{\sqrt{\frac{\nu_f x}{U}}} = \frac{0.332}{\sqrt{\text{Re}}}\rho U^2 \tag{5.44}$$

where $Re = \frac{Ux}{\nu_f}$, and using $A_2 = 0.332$.

The local skin-friction coefficient or frictional drag coefficient is

$$C_f = \frac{\tau_0}{\frac{1}{2}\rho_f U^2} = \frac{0.664}{\sqrt{Re}} \tag{5.45}$$

Total frictional force per unit width for one side of the plate of length l is given by

$$F = \int_0^l \tau_0 dx = 0.664\ \rho_f U^2 \sqrt{\frac{\nu_f l}{U}} \tag{5.46}$$

From equation (5.46), it is reported that the frictional force F is proportional to $U^{\frac{3}{2}}$, i.e., $F \propto U^{\frac{3}{2}}$. The average skin-friction coefficient in this case is

$$C_F = \frac{F}{\frac{1}{2}\rho_f l U^2} = \frac{0.664\ \rho_f U^2 \sqrt{\frac{\nu_f l}{U}}}{\frac{1}{2}\rho_f U^2 l} = \frac{1.328}{\sqrt{Re}} \tag{5.47}$$

where $Re = \frac{Ul}{\nu_f}$.

According to the boundary condition, the velocity in the BL does not reach the value of free stream velocity until $y \to \infty$. Hence this theory does not give good estimation of BL thickness. However, at a certain finite value of η the velocity in the BL asymptotically merges into the free stream velocity. If an arbitrary limit of the BL at $y = 0.9975$ is used, the thickness of the BL can be found to be

$$\delta = 5.65\sqrt{\frac{\nu_f x}{U}} \tag{5.48}$$

Displacement thickness: Since the definition above of the BL thickness is somewhat arbitrary, a more physically meaningful definition of the thickness δ is introduced as

$$U\delta^* = \int_0^\delta (U - u)\, dy \tag{5.49}$$

where δ^* is called the displacement thickness in a BL, where the free stream velocity is displaced outward due to decrease in velocity. The right-hand side of equation (5.49) indicates the reduction of total flow rate due to the frictional effect and the term in the left-hand side signifies the potential flow moved from the wall. The distance by which potential velocity is moved outwards due to the reduction of velocity in the BL is

$$\delta^* = \int_0^{\delta}\left(1-\frac{u}{U}\right)dy \tag{5.50}$$

Physically, the displacement thickness may be considered as the transverse distance by which the external free stream is effectively displaced due to BL formation.

Using $\frac{u}{U}$ and η from equations (5.28) and (5.26) into (5.50), one gets

$$\delta^* = \sqrt{\frac{v_f x}{U}}\int_0^{\delta}(1-f')d\eta = \sqrt{\frac{v_f x}{U}}\lim_{\eta\to\infty}\left[\eta - f(\eta)\right] = 1.721\sqrt{\frac{v_f x}{U}} \tag{5.51}$$

From the analogy of displacement thickness, a momentum thickness is also defined in accordance with the momentum law. This is recognized by evaluating the loss of momentum flow as a consequence of the wall friction in the BL to the momentum flow in the absence of the BL. Thus

$$\rho_f \theta U^2 = \rho_f \int_0^{\delta} u(U-u)dy \tag{5.52}$$

The term on the left-hand side indicates the momentum flow in absence of BL, and the term on the right-hand side indicates the loss of momentum flow due to the wall friction.

Momentum thickness: The momentum thickness is defined as

$$\theta = \int_0^{\delta}\frac{u}{U}\left(1-\frac{u}{U}\right)dy \tag{5.53}$$

Physically the momentum thickness may be considered as the transverse distance by which the boundary should be displaced to compensate for the reduction in momentum due to the formation of BL.

Again using $\frac{u}{U}$ and η, we can evaluate numerically the value of θ for a flat plate as follows:

$$\theta = \sqrt{\frac{vx}{U}}\int_0^{\infty} f'(1-f')d\eta = 0.664\sqrt{\frac{v_f x}{U}} \tag{5.54}$$

Shape factor: The ratio of displacement thickness to momentum thickness is called as shape factor ($H = \delta^*/\theta$) and it is useful in BL analysis.

Example 5.2

Water flows over a flat plate at a free stream velocity of 0.15 m/s. There is no pressure gradient and laminar BL is 6 mm thick. Assume a sinusoidal velocity profile $\frac{U}{U_\infty} = \sin\frac{\pi}{2}\left(\frac{y}{\delta}\right)$. Calculate the local wall shear stress and skin-friction coefficient with $\mu_f = 1.02\times\frac{10^{-3}\ \text{gm}}{\text{ms}}$, $\rho_f = 1000\ \text{kg/m}^3$.

Solution

Shear stress is given by: $\tau = \mu_f \dfrac{du}{dy} = \dfrac{\mu_f U}{\delta} \cdot \dfrac{\partial\left(\frac{u}{U}\right)}{\partial \eta}$, where $\eta = \dfrac{y}{\delta}$

$$\tau = \frac{\mu_f U}{\delta} \frac{\pi}{2} \cos\left(\frac{\pi}{2} \cdot \frac{y}{\delta}\right) = \frac{1.57\mu_f U}{\delta} \cos\left(\frac{\pi}{2} \cdot \frac{y}{\delta}\right)$$

Now, $\tau_w = \tau|_{y=0} = \dfrac{1.57\mu_f U}{\delta}$. Therefore $\tau_w = \dfrac{1.57 \times 1.02 \times 10^{-3} \times 0.15}{6 \times 10^{-3}} = 0.04\ \text{N/m}^2$.

$$c_f = \frac{\tau_w}{\frac{1}{2}\rho_f U^2} = \frac{2 \times 0.04}{1000 \times (0.15)^2} = 3.5 \times 10^{-3}.$$

5.4 BOUNDARY LAYER WITH PRESSURE GRADIENT ON A SURFACE

We have discussed the BL along a flat plate with zero pressure gradient based on a similar velocity profile. Now let us consider BL with pressure gradient. Here the velocity at outside the BL is assumed to be proportional to a power of a distance along the wall (Yuan, 1970). When the flow past a wedge and the velocity of the potential flow is given by

$$U(x) = U_1 x^m \tag{5.55}$$

where U_1 is a constant. The coordinate x is measured from the stagnation point O and the wedge angle is denoted by $\pi\beta$ (Figure 5.9).

The fundamental BL equation for steady state replacing the pressure gradient by inertial force due to free stream velocity U is:

$$u\frac{\partial u}{\partial x} + v\frac{\partial u}{\partial y} = U\frac{\partial U}{\partial x} + \nu_f \frac{\partial^2 u}{\partial y^2} \tag{5.56}$$

$$\frac{\partial u}{\partial x} + \frac{\partial v}{\partial y} = 0$$

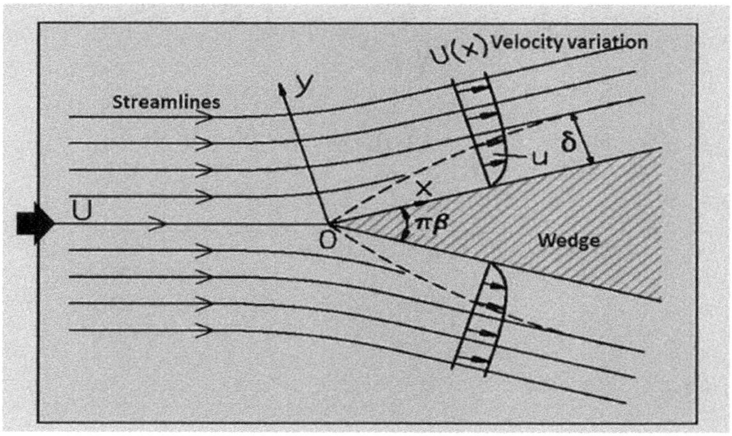

FIGURE 5.9 Flow past a wedge submerged in a low viscous fluid. (Modified from Yuan, 1970.)

subject to the boundary conditions as:

$$y=0: u=v=0$$

$$y=\infty: u=U(x) \tag{5.57}$$

Using the stream function ψ, we get

$$\frac{\partial\psi}{\partial y}\frac{\partial^2\psi}{\partial x\partial y}-\frac{\partial\psi}{\partial x}\frac{\partial^2\psi}{\partial y^2}=U\frac{\partial U}{\partial x}+v_f\frac{\partial^3\psi}{\partial y^3} \tag{5.58}$$

with boundary conditions

$$y=0: \frac{\partial\psi}{\partial x}=\frac{\partial\psi}{\partial y}=0$$

$$y=\infty: \frac{\partial\psi}{\partial y}=U(x) \tag{5.59}$$

The solution of the above equation can be obtained if the transformation is made as:

$$\eta_1 \sim \frac{y}{\sqrt{\frac{v_f x}{U}}}=\frac{cy}{\sqrt{\frac{v_f x}{U}}}$$

$$=cy\frac{1}{\sqrt{\frac{v_f x}{U_1 x^m}}}=yc\sqrt{\frac{U_1}{v}}x^{(m-1)/2} \tag{5.60}$$

(Using $U(x)=U_1x^m$ from equation 5.55), and c is the arbitrary constant to be determined. The stream function ψ is obtained by integrating the continuity equation, i.e.,

$$\psi=\int u\,dy=\frac{U_1x^m}{c\sqrt{\frac{U_1}{v}}x^{(m-1)/2}}\int F(\eta_1)\,d\eta_1$$

$$\text{since } \frac{u}{U(x)}=F(\eta_1) \text{ or, } u=U_1x^mF(\eta_1)$$

$$\psi=\frac{1}{c}\sqrt{v_fU_1}\,x^{\frac{m+1}{2}}f(\eta_1) \tag{5.61}$$

where $f(\eta_1)=\int F(\eta_1)\,d\eta_1$.

Similarly, derivatives of ψ with respect to x and y are transformed as

$$
\begin{aligned}
&\frac{\partial \psi}{\partial y} = \frac{\partial \psi}{\partial \eta_1}\frac{\partial \eta_1}{\partial y} = Uf'(\eta_1) \\
&\frac{\partial \psi}{\partial x} = c\sqrt{\frac{\nu_f U}{x}}\left[f + \frac{c^2-1}{c^2}\eta_1 f'\right] \\
&\frac{\partial^2 \psi}{\partial x \partial y} = (2c^2-1)\frac{U}{x}\left[f' - \frac{c^2-1}{2c^2-1}\eta_1 f''\right] \\
&\frac{\partial^2 \psi}{\partial y^2} = cU\sqrt{\frac{U}{\nu_f x}}f'' \\
&\frac{\partial^3 \psi}{\partial y^3} = U\left(c^2 U/\nu_f x\right)f'''
\end{aligned} \tag{5.62}
$$

Using the relations (5.62) in equation (5.58), one gets

$$
Uf'\left[(2c^2-1)\frac{U}{x}\left(f' - \frac{c^2-1}{2c^2-1}\eta_1 f''\right)\right] - \left[c\sqrt{\frac{\nu_f U}{x}}\left(f + \frac{c^2-1}{c^2}\eta_1 f'\right)\right]\cdot\left[cU\sqrt{\frac{U}{\nu_f x}}f''\right]
$$

$$
= \nu_f U\left(c^2\frac{U}{\nu_f x}f'''\right) + m\frac{U^2}{x}
$$

after simplification, one gets

$$
f''' + ff'' - \left(\frac{2c^2-1}{c^2}\right)f'^2 + \frac{m}{c^2} = 0 \tag{5.63}
$$

Coefficient of the last two terms can be equal if $2c^2 - 1 = m$, or $c = \sqrt{\frac{1+m}{2}}$.

Equation (5.63) can be rewritten as

$$
\frac{\partial^3 f}{\partial \eta_1^3} + f\frac{\partial^2 f}{\partial \eta_1^2} - \beta\left[\left(\frac{\partial f}{\partial \eta_1}\right)^2 - 1\right] = 0 \tag{5.64}
$$

where $\beta = \frac{2m}{m+1}$

with boundary conditions:

$$
\left.
\begin{aligned}
&\eta_1 = 0 : f = 0,\ f' = 0 \\
&\eta_1 \to \infty : \frac{\partial f}{\partial \eta_1} = 1
\end{aligned}
\right| \tag{5.65}
$$

Now for $m = 0$, equation (5.64) reduces to $f''' + ff'' = 0$, which is the Blasius Equation for the case of flow over a flat plate.

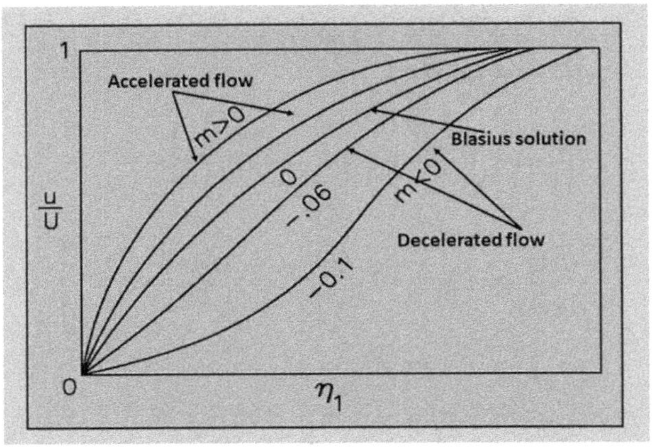

FIGURE 5.10 Velocity profiles u/U against η_1 for different values of power m. (Modified from Yuan, 1970.)

For $m = 1$, the equation reduces to

$$f''' + ff'' - f'^2 + 1 = 0, \tag{5.66}$$

$$\text{when } U(x) = U_1 x \text{ where } m = 1$$

which is the flow in the vicinity of a stagnation point in a plane with wedge angle π with $\beta = 1$. In this case the velocity potential is linearly proportional to the distance along with the wall. If $m > 0, \beta$ is also positive, the flow is accelerated and the velocity profiles have no point of inflection. On the other hand, if $m < 0, (\beta < 0)$, the flow is decelerated with adverse pressure gradient and a point of inflection occurs in the velocity profile. Separation of flow occurs for $\beta = -0.199$, i.e. for $m = -0.091$, indicating the laminar BL is capable to support only a very small deceleration without separation of flow occurring (Figure 5.10).

5.5 MOMENTUM INTEGRAL THEOREM FOR THE BOUNDARY LAYER FLOW

This is the most useful method named as von-Karman Pohlhausan method, based on the integral theorem (Schlichting, 1968). The basic concept of this method is that the solutions of the BL equation satisfy the differential equation only on average. The solution may not satisfy at every point (x, y), but the momentum integral equation and the boundary conditions must be satisfied. The integral equation is obtained by integrating the BL equations with respect to y or by the momentum law.

5.5.1 The von-Karman Integral Relations

The BL equations are

$$\left.\begin{aligned} \frac{\partial u}{\partial t} + u\frac{\partial u}{\partial x} + v\frac{\partial u}{\partial y} = -\frac{1}{\rho_f}\frac{\partial p}{\partial x} + \nu_f\frac{\partial^2 u}{\partial y^2} \\ \frac{\partial u}{\partial x} + \frac{\partial v}{\partial y} = 0 \end{aligned}\right| \tag{5.67}$$

Integrating this equation with respect to y from $y = 0$ to $y = \delta(x)$ the outer edge of the BL, we have (Yuan, 1970)

$$\frac{\partial}{\partial t}\int_0^\delta u\,\partial y+\int_0^\delta u\frac{\partial u}{\partial x}\partial y+\int_0^\delta v\frac{\partial u}{\partial y}\partial y=-\frac{1}{\rho_f}\int_0^\delta \frac{\partial p}{\partial x}dy+\nu_f\int_0^\delta \frac{\partial^2 u}{\partial y^2}\partial y \tag{5.68}$$

Now

$$\int_0^\delta v\frac{\partial u}{\partial y}\partial y=\int_0^\delta \frac{\partial}{\partial y}(uv)\partial y-\int_0^\delta u\frac{\partial v}{\partial y}dy=uv|_0^\delta-\int_0^\delta u\frac{\partial v}{\partial y}dy=U\int_0^\delta \frac{\partial v}{\partial y}dy-\int_0^\delta u\frac{\partial v}{\partial y}dy$$

Since $uv|_0^\delta=u(\delta)v(\delta)-u(0)v(0)=U\int_0^\delta \frac{\partial v}{\partial y}dy$

$$\text{Therefore, }\int_0^\delta v\frac{\partial u}{\partial y}dy=U\int_0^\delta \frac{\partial v}{\partial y}dy-\int_0^\delta u\frac{\partial v}{\partial y}dy=-U\int_0^\delta \frac{\partial u}{\partial x}dy+\int_0^\delta u\frac{\partial u}{\partial x}dy \tag{5.69}$$

Then

$$\frac{\partial}{\partial t}\int_0^\delta u\,dy+\int_0^\delta u\frac{\partial u}{\partial x}dy-U\int_0^\delta \frac{\partial u}{\partial x}dy+\int_0^\delta u\frac{\partial u}{\partial x}dy=-\frac{1}{\rho_f}\int_0^\delta \frac{\partial p}{\partial x}dy+\frac{1}{\rho_f}\int_0^\delta \frac{\partial}{\partial y}\left(\mu\frac{\partial u}{\partial y}\right)dy$$

$$\frac{\partial}{\partial t}\int_0^\delta u\,dy+\int_0^\delta \frac{\partial}{\partial x}\left(u^2\right)dy-U\int_0^\delta \frac{\partial u}{\partial x}dy=-\frac{1}{\rho_f}\int_0^\delta \frac{\partial p}{\partial x}dy+\frac{1}{\rho_f}\mu\frac{\partial u}{\partial y}\bigg|_0^\delta$$

$$=-\frac{1}{\rho_f}\int_0^\delta \frac{\partial p}{\partial x}dy-\frac{1}{\rho_f}\tau_0 \tag{5.70}$$

where $\tau_0=\mu_f\left.\frac{\partial u}{\partial y}\right|_{y=0}$

In equation (5.70), as given below, we can show that $\int_0^\delta \frac{\partial}{\partial x}\left(u^2\right)dy=\frac{d}{dx}\int_0^\delta u^2dy-U^2\frac{d\delta}{dx}$.

Using Leibnitz Integral rule (Kreyszig, 1985):

$$\boxed{\frac{d}{dx}\int_{a(x)}^{b(x)} f(x,t)dt=f(x,b(x))\cdot b'(x)-f(x,a(x))\cdot a'(x)+\int_{a(x)}^{b(x)}\frac{\partial}{\partial x}f(x,t)dt}$$

This formula can be derived using the fundamental theorem of integral calculus. Using this theorem,

$$\frac{d}{dx}\int_0^{\delta(x)} u^2 dy = u^2(x,\delta(x))\delta'(x) - u^2(x,0)\cdot 0 + \int_0^{\delta}\frac{\partial}{\partial x}u^2 dy$$

$$\frac{d}{dx}\int_0^{\delta} u^2 dy = U^2\frac{d\delta}{dx} - 0 + \int_0^{\delta}\frac{\partial u^2}{\partial x}dy \tag{5.71}$$

$$\int_0^{\delta}\frac{du^2}{dx}dy = \frac{d}{dx}\int_0^{\delta}u^2 dy - U^2\frac{d\delta}{dx}$$

$$\text{Similarly, } U\int_0^{\delta}\frac{\partial u}{\partial x}dy = U\frac{d}{dx}\int_0^{\delta}u dy - U^2\frac{d\delta}{dx} \tag{5.72}$$

$$\left.\begin{aligned} \int_0^{\delta}\frac{\partial u^2}{\partial x}dy &= \frac{d}{dx}\int_0^{\delta}u^2 dy - U^2\frac{d\delta}{dx} \\ U\int_0^{\delta}\frac{\partial u}{\partial x}dy &= U\frac{d}{dx}\int_0^{\delta}u dy - U^2\frac{d\delta}{dx} \end{aligned}\right\} \tag{5.73}$$

Then using (5.73), equation (5.70) can be written as

$$\frac{\partial}{\partial t}\int_0^{\delta}u dy + \frac{d}{dx}\int_0^{\delta}u^2 dy - U\frac{d}{dx}\int_0^{\delta}u dy = -\frac{1}{\rho_f}\int_0^{\delta}\frac{\partial p}{\partial x}dy - \frac{1}{\rho_f}\tau_0 \tag{5.74}$$

Equation (5.74) is the one form of von-Karman Integral relation. This is known as the momentum integral equation of the BL.

Now if we consider

$$\frac{\partial U}{\partial t} + U\frac{\partial U}{\partial x} = -\frac{1}{\rho_f}\frac{dp}{dx} \tag{5.75}$$

where U is the free stream velocity in the potential flow just outside the BL, then Euler equation reduced to the above equation.

Substituting equations (5.75) in (5.74) in steady flow

$$\frac{d}{dx}\int_0^{\delta}u^2 dy - U\frac{d}{dx}\int_0^{\delta}u dy - \frac{dU}{dx}\cdot\delta U = -\frac{\tau_0}{\rho_f} \tag{5.76}$$

or,

$$\frac{d}{dx}\int_0^{\delta}u^2 dy - U\frac{d}{dx}\int_0^{\delta}u dy - \frac{dU}{dx}\cdot\int_0^{\delta}U dy = -\frac{\tau_0}{\rho_f} \tag{5.77}$$

Since $\delta U = \int_0^\delta U dy$. Now

$$\frac{d}{dx}\int_0^\delta U u dy = \frac{d}{dx}\left[U\int_0^\delta u dy\right] = U\frac{d}{dx}\int_0^\delta u dy + \int_0^\delta u dy\frac{dU}{dx}$$

$$U\frac{d}{dx}\int_0^\delta u dy = \frac{d}{dx}\int_0^\delta U u dy - \int_0^\delta u dy.\frac{dU}{dx}$$

$$\frac{d}{dx}\int_0^\delta u^2 dy - \frac{d}{dx}\int_0^\delta U u dy + \frac{dU}{dx}\int_0^\delta u dy - \frac{dU}{dx}\int_0^\delta U dy = -\frac{\tau_0}{\rho_f}$$

$$\frac{d}{dx}\int_0^\delta \left(u^2 - Uu\right)dy + \frac{dU}{dx}\int_0^\delta (u-U)dy = -\frac{\tau_0}{\rho_f}$$

$$\frac{d}{dx}\int_0^\delta u(U-u)dy + \frac{dU}{dx}\int_0^\delta (U-u)dy = \frac{\tau_0}{\rho_f} \tag{5.78}$$

Taking limit $\delta \to \infty$, one gets

$$\frac{d}{dx}\int_0^\infty u(U-u)dy + \frac{dU}{dx}\int_0^\infty (U-u)dy = \frac{\tau_0}{\rho_f} \tag{5.79}$$

Using displacement and momentum thicknesses into equation (5.79), it is obtained

$$\frac{d}{dx}\int_0^\infty \frac{U^2}{U}u\left(1-\frac{u}{U}\right)dy + \frac{dU}{dx}\int_0^\infty U\left(1-\frac{u}{U}\right)dy = \frac{\tau_0}{\rho_f}$$

If $\theta = \int_0^\infty \frac{u}{U}\left(1-\frac{u}{U}\right)dy$ is the momentum thickness of the BL, and

$\delta^* = \int_0^\infty \left(1-\frac{u}{U}\right)dy$ is the displacement thickness, then $\boxed{\frac{d}{dx}\left(U^2\theta\right) + U\frac{dU}{dx}\delta^* = \frac{\tau_0}{\rho_f}}$

After simplification, one gets

$$\frac{d\theta}{dx} + \left(2\theta + \delta^*\right)\frac{1}{U}\frac{dU}{dx} = \frac{\tau_0}{\rho_f U^2} \tag{5.80}$$

Equation (5.80) is an ordinary differential equation for the BL thickness. Based on this equation (5.80), it is necessary to find a suitable expression for the velocity function in the BL. The velocity function must satisfy the no-slip condition of the wall, and the conditions at the point where the potential flow solution exists. The equation (5.80) can be written as

$$\boxed{U^2\frac{d\theta}{dx}+\left(2\theta+\delta^*\right)U\frac{dU}{dx}=\frac{\tau_0}{\rho_f}} \tag{5.81}$$

$$\text{If } \frac{dp}{dx}=0,\ U\frac{dU}{dx}=0$$

$$\Rightarrow \frac{d\theta}{dx}=\frac{\tau_0}{\rho_f U^2} \tag{5.82}$$

where τ_0 is the shearing stress. This is the momentum integral equation for the flat plate, based on the von-Karman integral relation. The momentum thickness can be written as

$$\frac{\theta}{\delta}=\int_0^1 \frac{u}{U}\left(1-\frac{u}{U}\right)d\eta \tag{5.83}$$

where $\eta=\frac{y}{d}$. The shearing stress τ_0 on the surface of the plate is expressed in the form as,

$$\tau_0=\mu_f\frac{\partial u}{\partial y}_{y=0}=\mu_f\frac{U}{\delta}\left[\frac{\partial}{\partial\eta}\left(\frac{u}{U}\right)\right] \tag{5.84}$$

The local frictional coefficient is given by

$$C_f=\frac{\tau_0}{\frac{1}{2}\rho_f U^2} \tag{5.85}$$

Substituting the value of τ_0 in the above equation, one gets the frictional coefficient.

Example 5.3

Air moves over a flat plate with a uniform free stream velocity of 10 m/s. At a position 15 cm away from the front edge of the plate, what is the BL thickness, using a parabolic profile in the BL?. For air, $v_f=1.5\times10^{-5}\ \text{m}^2/\text{s}$ & $\rho_f=0.23\ \text{kg/m}^3$.

Solution

For a parabolic profile, let us take

$$\frac{u}{U}=a+by+cy^2 \tag{1}$$

The boundary conditions are

$$\left.\begin{array}{r} y=0: u=0 \\ y=\delta: u=U,\ \dfrac{\partial u}{\partial y}=0 \end{array}\right| \tag{2}$$

Evaluating the constants, we get

$$\frac{u}{U}=2\left(\frac{y}{\delta}\right)-\left(\frac{y}{\delta}\right)^2=2\eta-\eta^2 \tag{3}$$

$$\tau_w = \mu_f \left.\frac{du}{dy}\right|_{y=0} = \frac{\mu_f U}{\delta}.\left.\frac{\partial\left(\frac{u}{U}\right)}{\partial\eta}\right|_{\eta=0}, \text{ where } \eta = \frac{y}{\delta} \tag{4}$$

$$= \frac{\mu_f U}{\delta}.\left.\frac{d\left(2\eta-\eta^2\right)}{\partial\eta}\right|_{\eta=0} = \frac{2\mu_f U}{\delta}$$

Now $\theta = \int_0^\infty \frac{u}{U}\left(1-\frac{u}{U}\right)dy,\ \eta = \frac{y}{\delta} \Rightarrow y = \delta\eta = \int_0^\infty \frac{u}{U}\left(1-\frac{u}{U}\right)\delta d\eta$

$$\frac{d\theta}{dx} = \frac{d\delta}{dx}\int_0^1 \frac{u}{U}\left(1-\frac{u}{U}\right)d\eta \tag{5}$$

Applying momentum integral equation (5.82),

$$\tau_w = \rho_f U^2 \frac{d\theta}{dx} = \rho_f U^2 \frac{d\delta}{dx}\int_0^1 \frac{u}{U}\left(1-\frac{u}{U}\right)d\eta \tag{6}$$

Combining equations (4) and (6), one gets

$$\frac{2\mu_f U}{\delta} = \rho_f U^2 \frac{d\delta}{dx}\int_0^1 \left(2\eta-\eta^2\right)\left(1-2\eta+\eta^2\right)d\eta \text{ or } \frac{2\mu_f U}{\delta\rho_f U^2} = \frac{d\delta}{dx}\int_0^1 \left(2\eta-5\eta^2+4\eta^3-\eta^4\right)d\eta$$

$$\frac{2\mu_f}{\delta\rho_f U} = \frac{2}{15}\frac{d\delta}{dx}$$

$$\delta d\delta = \frac{15\mu_f}{\rho_f U}dx \tag{7}$$

Then, $\frac{\delta^2}{2} = \frac{15\mu_f}{\rho_f U}x + c$ at $x = 0 : \delta = 0, \Rightarrow c = 0.\ \delta = \sqrt{\frac{30\mu_f}{\rho_f U}x}$

or, $\frac{\delta}{x} = \sqrt{\frac{30\nu_f}{Ux}} = \frac{5.48}{\sqrt{Re}}$. In this problem: $Re = \frac{10\times15\times10^{-2}}{1.5\times10^{-5}} = 1\times10^5$

$\delta = \frac{5.48}{\sqrt{R_{ex}}}\times15\text{ cm} = 0.259\text{ cm};\ \delta = 2.59\text{ mm}.$

5.6 APPLICATIONS OF THE MOMENTUM INTEGRAL EQUATION TO BOUNDARY LAYERS

5.6.1 von-Karman – Pohlhausen Method for Non-Zero Pressure Gradient

The solution of BL equation discussed above reveals that the velocity distribution in the BL is proportional to the ratio y/δ, that is, velocity distribution $\propto y/\delta$. In addition, velocity distribution must satisfy the following boundary conditions,

$$\left.\begin{aligned} y=0&:u=0,\ v=0 \\ y=\delta&:u=U,\ \frac{\partial u}{\partial y}=0 \end{aligned}\right| \tag{5.86}$$

The point of inflection occurs in the velocity profile if the flow is decelerated, and the separation of flow begins when the velocity gradient $\left.\frac{\partial u}{\partial y}\right|_{y=0}=0$.

With this information, Pohlhausen introduced a fourth-degree polynomial of velocity function in terms of dimensionless distance from the wall $\eta = y/\delta$ as follows (Yuan, 1970):

$$\frac{u}{U}=f(\eta)=a_1\eta+a_2\eta^2+a_3\eta^3+a_4\eta^4 \tag{5.87}$$

in the range $0\le\eta\le 1$, where $\eta=\frac{y}{\delta}$. The constants a_1, a_2, a_3, a_4 are to be determined from the boundary conditions:

$$y=0:u=0,\ v=0,\ \frac{\partial^2 u}{\partial y^2}=\frac{1}{\mu_f}\frac{dp}{\partial x}=-\frac{U}{\nu_f}\frac{dU}{dx} \tag{5.88a}$$

$$y=\delta:u=U,\ \frac{\partial u}{\partial y}=0,\ \frac{\partial^2 u}{\partial y^2}=0 \tag{5.88b}$$

First two conditions of equation (5.88a) are satisfied identically by the velocity equation (5.87). Third condition of (5.88a) is obtained from the Prandtl BL equation at $y=0$. For an accelerated flow $\frac{dp}{dx}<0$, and $\frac{dU}{dx}>0$, we have $\left.\frac{\partial^2 u}{\partial y^2}\right|_{y=0}<0$ (Figure 5.5 a–c) and also $\frac{\partial^2 u}{\partial y^2}<0$ over the entire thickness of the BL (Figure 5.5c). For decelerated flow $\frac{\partial p}{\partial x}>0$, and $\frac{dU}{dx}<0$, the curvature of velocity at the wall is positive, i.e. $\left.\frac{\partial^2 u}{\partial y^2}\right)_0>0$ (Figure 5.6a–c); but due to the nature of the velocity profile, $\frac{\partial^2 u}{\partial y^2}<0$ occurs within the BL thickness. Therefore, there exists a point for which $\frac{\partial^2 u}{\partial y^2}=0$. This point is called a point of inflection at the velocity profile (Figure 5.6 a–c). From equation (5.87), we have for $\frac{\partial^2 u}{\partial y^2}=0$.

$$\frac{\partial^2 u}{\partial y^2}=2a_2+6a_3\eta+12a_4\eta^2=0 \tag{5.89}$$

which has two roots having two inflection points in the BL. One inflection point I_1 is located near the wall and other one I_2 is located at the upper region of the BL (Yuan, 1970). The inflection point I_2 at the upper layer is physically unacceptable, so to remove the inflection point I_2, we impose the following condition, $\frac{\partial^2 u}{\partial y^2}=0$ at $y=\delta$.

Finally, the constants a_1, a_2, a_3, a_4 are determined from the conditions are

$$\eta = 0: \frac{\partial^2 u}{\partial y^2} = -\frac{U}{\nu_f}\frac{dU}{dx} \text{ or } a_2 = -\frac{\delta^2}{2\nu_f}\frac{dU}{dx} = -\frac{\Delta}{2},$$

where $\Delta = \frac{\delta^2}{\nu_f}\frac{dU}{dx}$, which depends on the pressure gradient of the flow.

$$\left.\begin{aligned} \eta = 1: \frac{u}{U} = 1 \text{ or } a_1 - \frac{\Delta}{2} + a_3 + a_4 = 1 \\ \eta = 1: \frac{\partial u}{\partial y} = 0 \text{ or } a_1 - \Delta + 3a_3 + 4a_4 = 0 \\ \eta = 1: \frac{\partial^2 u}{\partial y^2} = 0 \text{ or } -\Delta + 6a_3 + 12a_4 = 0 \end{aligned}\right\} \quad (5.90)$$

Solving for a_1, a_2, a_3, a_4, we get

$$a_1 = 2 + \frac{\Delta}{6}, a_2 = -\frac{\Delta}{2}, a_3 = -2 + \frac{\Delta}{2}, a_4 = 1 - \frac{\Delta}{6} \quad (5.91)$$

Hence the expression for the velocity profile which satisfies the boundary conditions is

$$\frac{u}{U} = f(\eta) = F_1(\eta) + \Delta F_2(\eta) \quad (5.92)$$

with $F_1(\eta) = 2\eta - 2\eta^3 + \eta^4$ and $F_2(\eta) = \frac{1}{6}(\eta - 3\eta^2 + 3\eta^3 - \eta^4) = \frac{\eta}{6}(1-\eta)^3$.

From equation (5.92), the slope and the curvature of the velocity in the BL are obtained respectively as:

$$\frac{df}{d\eta} = (2 - 6\eta^2 + 4\eta^3) + \frac{\Delta}{6}\left((1-\eta)^2(1-4\eta)\right) \quad (5.93)$$

$$\frac{d^2 f}{d\eta^2} = (1-\eta)\left[12\left(\frac{\Delta}{6} - 1\right)\eta - \Delta\right] \quad (5.94)$$

The velocity gradient at $\eta = 0$

$$\left.\frac{df}{d\eta}\right|_{\eta=0} = 2 + \frac{\Delta}{6} \quad (5.95)$$

For separation point, the velocity gradient is zero at the boundary, i.e.,

$$\left.\frac{df}{d\eta}\right|_{\eta=0} = 0 \Rightarrow 2 + \frac{\Delta}{6} = 0$$
$$\Rightarrow \Delta = -12 \quad (5.96)$$

Therefore, the separation occurs for $\Delta = -12$.

The condition of zero curvature of the velocity is as $\frac{d^2 f}{d\eta^2} = 0$

$$\Rightarrow \eta = \frac{\Delta}{12}\left[\Delta/6 - 1\right] \tag{5.97}$$

It is seen that for $\Delta \le 12, \eta \ge 1$ and for $\Delta > 12, \eta < 1$.

Hence for $\Delta > 12$, the inflection point occurs within $\eta = 1.0$, i.e., the velocity profile in the BL becomes greater than the velocity in the potential flow. This is not physically realistic; therefore, the parameter Δ is limited to the range

$$-12 \le \Delta \le 12 \tag{5.98}$$

If $\Delta = 0$, the velocity profile corresponds to the Blasius solution.

Using equation (5.92), one gets the displacement thickness of the BL as:

$$\frac{\delta^*}{\delta} = \int_0^1 \left[1 - F_1(\eta) - \Delta F_2(\eta)\right] d\eta = \frac{3}{10} - \frac{\Delta}{120} \tag{5.99}$$

Similarly, the momentum thickness is as:

$$\frac{\theta}{\delta} = \int_0^1 \left[F_1(\eta) + \Delta F_2(\eta)\right]\left[1 - F_1(\eta) - \Delta F_2(\eta)\right] d\eta$$

$$= \frac{37}{315} - \frac{\Delta}{945} - \frac{\Delta^2}{9072} \tag{5.100}$$

Let us consider the momentum integral equation (5.80),

$$\frac{d\theta}{dx} + \left(2\theta + \delta^*\right)\frac{1}{U}\frac{dU}{dx} = \frac{\tau_0}{\rho_f U^2}$$

Multiplying the above equation by $\frac{U\theta}{\nu_f}$, one gets

$$\frac{U\theta}{\nu_f}\frac{d\Delta}{dx} + \left(2 + \frac{\delta^*}{\theta}\right)\frac{dU}{dx}.\frac{\theta^2}{\nu_f} = \frac{\tau_0}{\mu_f}\frac{\theta}{U} \tag{5.101}$$

with $\frac{dU}{dx} = \frac{\Delta \nu_f}{\delta^2}$

Equation (5.101) is called as the momentum integral equation of BL.

Now shear stress is

$$\tau_0 = \mu_f \left.\frac{du}{dy}\right|_0 = \frac{\mu_f U}{\delta}\left(2 + \frac{\Delta}{6}\right) \tag{5.102}$$

The Karman momentum integral equation is as given in equation (5.80)

$$\frac{d\theta}{dx} + \left(2\theta + \delta^*\right)\frac{1}{U}\frac{dU}{dx} = \frac{\tau_0}{\rho_f U^2}$$

which implies as

$$U^2 \frac{d\theta}{dx} + \left(2\theta + \delta^*\right) U \frac{dU}{dx} = \frac{\tau_0}{\rho_f}$$

5.6.2 For Zero Pressure Gradient

When $\dfrac{dp}{dx} = 0, \text{ then } U\dfrac{dU}{dx} = 0$

$$\Rightarrow \frac{d\theta}{dx} = \frac{\tau_0}{\rho_f U^2} \tag{5.103}$$

for zero pressure gradient.

We choose a velocity profile as a third-degree polynomial in the form

$$\frac{u}{U} = a_0 + a_1\eta + a_2\eta^2 + a_3\eta^3 \tag{5.104}$$

In order to determine the four constants a_0, a_1, a_2, a_3 we shall prescribe the following boundary conditions:

$$\begin{aligned} &\eta = 0: \ \frac{u}{U} = 0, \\ &\eta = 0: \ \frac{\partial^2 u/U}{\partial \eta^2} = 0 \\ &\eta = 1: \ \frac{u}{U} = 1 \\ &\eta = 1: \ \frac{\partial u/U}{\partial \eta} = 0 \end{aligned} \tag{5.104a}$$

From the above conditions, one gets

$$a_0 = 0, a_2 = 0, a_1 + 3a_3 = 0, a_1 + a_3 = 1.$$

Finally, one gets the following values for the coefficients as:

$$a_0 = 0, a_1 = \frac{3}{2}, a_2 = 0, a_3 = -\frac{1}{2}.$$

Using the above values of the constants a_0, a_1, a_2, a_3 in the velocity profile (5.104), one gets the velocity profile as:

$$\frac{u}{U} = \frac{3}{2}\eta - \frac{1}{2}\eta^3 \tag{5.105}$$

For flow over a flat plate, $\dfrac{dp}{dx} = 0,$

Hence, $U\dfrac{dU}{dx} = 0.$

Now momentum thickness θ is given by

$$\theta = \int_0^\delta \frac{u}{U}\left(1 - \frac{u}{\mathrm{U}}\right)dy = \frac{39}{280}\delta \tag{5.106}$$

Wall shear stress is defined as

$$\tau_0 = \mu_f \left.\frac{du}{dy}\right|_{y=0} = \frac{3\mu_f U}{2\delta} \tag{5.107}$$

Substituting θ and τ_0 in equation (5.103), one gets

$$\frac{39}{280}\frac{d\delta}{dx} = \frac{3\mu_f U}{2\delta\rho_f U^2}$$

Using the condition at leading edge $x = 0,\ \delta = 0$, integration gives as:

$$\delta^2 = \frac{280}{13}\frac{\nu_f x}{U}$$

Hence, the BL thickness is,

$$\delta = 4.64\sqrt{\frac{\nu_f x}{U}} = \frac{4.64x}{Re}. \tag{5.108}$$

Here Re is the Reynolds number at x- distance from origin.

This is the BL thickness δ in a flat plate. Although the method is an approximate one, the result is found to be reasonably accurate. The accuracy depends on the order of the velocity profile.

Instead of the third-order polynomial, we take a fourth-order polynomial as:

$$\frac{u}{U} = a_0 + a_1\eta + a_2\eta^2 + a_3\eta^3 + a_4\eta^4 \tag{5.109}$$

We need one additional boundary condition as:

$$\text{at } y = \delta : \frac{\partial^2 u}{\partial y^2} = 0 \ \text{ or at } \eta = 1 : \frac{\partial^2 u/U}{\partial \eta^2} = 0 \tag{5.110}$$

Using the conditions (5.104) and (5.110), we get the coefficients as:

$$a_0 = 0,\ a_1 = 2,\ a_2 = 0,\ a_3 = -2,\ a_4 = 1,$$

Finally, the velocity profile is given by

$$\frac{u}{U} = 2\eta - 2\eta^3 + \eta^4 \tag{5.111}$$

and correspondingly the BL thickness δ is given by

$\delta = \frac{5.83x}{\sqrt{Re}}$, which is close to the exact value.

5.7 BOUNDARY LAYER FOR ENTRY FLOW IN A DUCT

Let us consider a steady uniform flow entering in a duct or pipe. The duct can be considered as two parallel plates kept at distance D apart. The growth of BL has considerable effect on the flow through the duct or pipe (Som and Biswas, 1998; Muralidhar and Biswas, 2015). The BL starts growing on the wall at the entrance of the pipe. Gradually it becomes thicker in the downstream. The flow becomes fully developed when the BLs from the wall meet at the axis of the pipe. The velocity gradient is steeper at the wall, causing a higher value of shear stress as compared to a developed flow. Momentum flux across any section is higher than that typically at the inlet due to the change in the shape of the velocity profile. Arising out of these, an additional pressure drop is brought about at the entrance region as compared to the pressure drop in the fully developed region. The velocity profile is nearly rectangular at the entrance, and it gradually changes to a parabolic profile at the fully developed region. Before the BLs from the periphery meet at the axis, there prevails a core region which is uninfluenced by the viscosity. Since the volume flow must be the same for every section and the boundary-layer thickness increases in the flow direction, the inviscid core accelerates, and there is a corresponding fall in pressure.

Entrance length: It can be shown that for laminar incompressible flows, the velocity profile approaches the parabolic profile through a distance L from the entry of the pipe of diameter D (Figure 5.11). This is known as entrance length and is given by (Som and Biswas, 1998; Muralidhar and Biswas, 2015) as: $\frac{L}{D} \approx 0.05Re$, where $Re = \frac{U_{av}D}{\nu_f}$. The entry length constitutes both the inlet zone and the filled zone. For a Reynolds number of 2,000, this distance, the entrance length is about 100 pipe-diameters. For turbulent flows, the entrance region is shorter, since the turbulent BL grows faster (Figure 5.11).

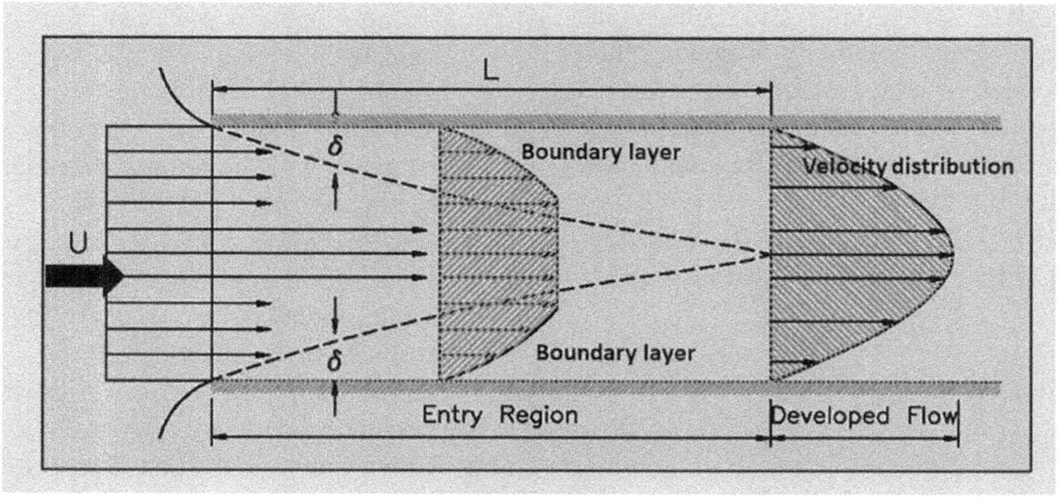

FIGURE 5.11 Development of boundary layer in the entrance region of a duct.

5.8 SUMMARY

Some of the important aspects discussed in this chapter are summarized below:

- The BL can be laminar or turbulent. If the Reynolds number is less than 5×10^5, then the BL is considered as laminar and otherwise, turbulent BL.
- The BL is identified by the BL thickness (δ) which is defined as the height at which the mean velocity (u) reaches 99% of the free stream velocity U.
- For BL, the displacement thickness is given by: $\delta^* = \int_0^\delta \left(1-\frac{u}{U}\right)dy$
- For BL, the momentum thickness is given by: $\theta = \int_0^\delta \frac{u}{U}\left(1-\frac{u}{U}\right)dy$
- Prandtl proposed the BL theory in 1908 and according to him, flow be divided into two regions, that are: (1) Region of BL which lies in immediate neighborhood of solid boundary and within which about 99% of velocity change takes place, velocity gradient normal to solid boundary is very large where Navier–Stokes Equations are valid. (2) Region outside BL, which constitutes the main body of flowing fluid has velocities of the order of free stream velocity U. The flow in this zone can be considered is governed by potential flow (ideal fluid) theory.
- Prandtl (1910) derived the BL equations using the Navier–Stokes equations, in 2D:

$$\frac{\partial u}{\partial t}+u\frac{\partial u}{\partial x}+v\frac{\partial u}{\partial y}=-\frac{1}{\rho_f}\frac{\partial p}{\partial x}+\nu_f\frac{\partial^2 u}{\partial y^2};$$

$$0=-\frac{\partial p}{\partial y};\frac{\partial u}{\partial x}+\frac{\partial v}{\partial y}=0$$

subject to the boundary conditions $u=v=0$ for $y=0$, and $u=U\,(x,t)$ for $y\to\infty$.

- When a flow past a body, due to adverse pressure gradient, the separation of flow at the BL occurs and the condition for flow separation $\left.\frac{\partial u}{\partial y}\right|_{y=0}=0$.
- von-Karman Momentum Integral Equation of BL is given as:

$$\frac{\partial}{\partial t}\int_0^\delta u\,dy+\frac{d}{dx}\int_0^\delta u^2dy-U\frac{d}{dx}\int_0^\delta u\,dy$$

$$=-\frac{1}{\rho_f}\int_0^\delta \frac{\partial p}{\partial x}dy-\frac{1}{\rho_f}\tau_0$$

From this equation, we can get a relation $\frac{d}{dx}\left(U^2\theta\right)+U\frac{dU}{dx}\delta^*=\frac{\tau_0}{\rho_f}$

- The momentum integral equation for the flat plate, based on the von-Karman integral relation is $\frac{d\theta}{dx}=\frac{\tau_0}{\rho_f U^2}$ where τ_0 is the shearing stress, U is free stream velocity and θ is momentum thickness.

5.9 EXERCISE PROBLEMS

1. How do you derive the two-dimensional BL equations under the approximations? Show the schematic diagrams for positive and negative pressure gradients with point of inflection. What is the condition of point of inflection in the BL flow?
2. Derive the Prandtl BL equations along a flat plate with boundary conditions; and the equation can be written as $2f'''+ff''=0$ using the law of similarity principles.
3. What are the physical significances of displacement thickness and momentum thickness of BL flow?
4. Derive the Blasius equation of BL with pressure gradient on a flat surface.
5. Derive the von-Karman momentum integral equation of the BL thickness for steady flow.

6. Air at 20°C flows parallel to a flat plate at free stream velocity $U = 2\,\text{m/s}$. Calculate BL thickness and local shear friction coefficient at $x = 3\,\text{m}$ from the leading edge of the plate. Use Blasius solution. Kinematic viscosity = $1.533 \times 10^{-4}\,\text{m}^2/\text{s}$.

7. Assume the velocity distribution in the BL is given by $\frac{u}{U} = \left(\frac{y}{\delta}\right)^{1/7}$, calculate displacement thickness and momentum thickness in terms of δ, if at a certain section, the free stream velocity observed to be 15 m/s. Take density of air, $\rho_f = 1.226\ \text{kg/m}^2$.

6 Turbulent Flow Analysis

6.1 INTRODUCTION

The turbulent flow is common in nature and has an important role in several geophysical processes related to a variety of phenomena such as river morphology, landscape modeling, atmospheric dynamics, and ocean currents. The origin of turbulence with the transition from laminar to turbulent flow is of primary importance to the whole science of fluid flow. The first experimental results on this complex problem were recognized by Reynolds (1883) in relation to flow through a straight pipe, feeding into it a thin thread of liquid dye, and the experiment is known as Reynold's dye experiment. Even after 506 years (Leonardo da Vinci around 1510, see Gad-El-Hak, 2000), turbulence studies are still in their infancy. Thus, we are still discovering how turbulence behaves. Though turbulent flow theories are reasonably developed, there is no analytical solution to equations. Thus, most of the time, we need to rely on experimental studies or computer models. We do have a crude, practical, working understanding of many turbulent phenomena but certainly not a comprehensive theory, and there is nothing that provides predictions of the accuracy demanded by designers. In this chapter, the basic concepts of turbulence flows, various theories, and solution methodologies are described in detail.

6.2 LAMINAR AND TURBULENT FLOWS

We have already discussed the basics of laminar flow in Chapter 4. Transition from laminar flow to turbulent flow is of fundamental importance in the field of fluid dynamics. The flows of real fluids vary from laminar flows, which display a characteristic feature of turbulence or chaotic nature. When velocity increases or Reynolds number *Re* is increased, internal flows and boundary layers formed at the solid body experience an amazing transition from laminar to turbulence. In a flow of low Reynolds number through a straight pipe of uniform cross-section with smooth walls, every fluid particle moves with a uniform velocity along a straight path. Viscous flow slows down near the wall in relation to those in the external core. The flow is well ordered, and particles travel along neighboring layers in laminar flow (Figure 6.1a).

However, examination shows that this orderly pattern of flow comes to an end when velocity is increased and at higher Reynolds numbers, and that a strong mixing of all particles appears (Figure 6.1b). As long as the flow is laminar, the injected thin thread of liquid dye maintains sharply defined layers all along the stream. As soon as the flow becomes turbulent, the thread gets disseminated into the stream and the fluid gets uniformly colored even at a short distance downstream. An ancillary motion at the right angle to the pipe flow results in mixing in the flow. The pattern of stream lines at a fixed point is subjected to continuous fluctuations, and the ancillary motion causes an exchange of momentum in the transverse direction because each particle substantially retains its forward momentum while the mixing takes place. As a result, the velocity distribution over the cross-section is considerably more uniform in turbulent flow than that in laminar flow (Figure 6.2). In the velocity distributions, the mass flow is the same for both cases. According to Hagen–Poiseuille in laminar flow, the velocity distribution over the cross-section is parabolic (Figure 6.2a), but in turbulent flow, owing to the transfer of momentum along the transverse direction, the distribution gets more uniform in the core region of the pipe or channel (Figure 6.2b). At a given point in it, the velocity and pressure are not constant in time but exhibit very irregular with high-frequency fluctuations. Reynolds (1883) was the first to investigate the conditions in transition from laminar to turbulence. He discovered the law of similarity which states that transition

DOI: 10.1201/9781003000020-6

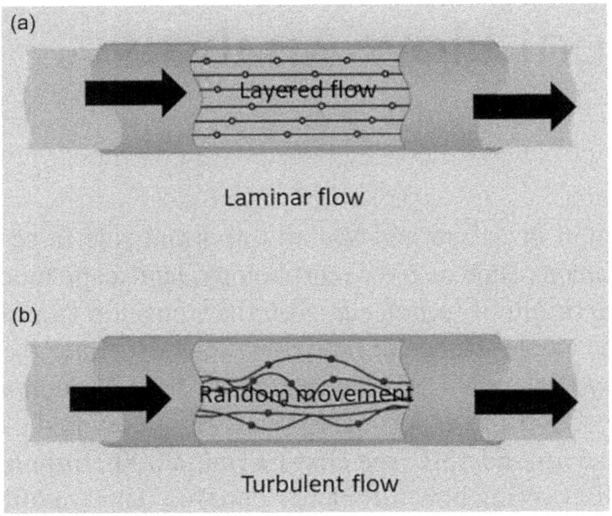

FIGURE 6.1 (a) Laminar flow, and (b) Turbulent flow.

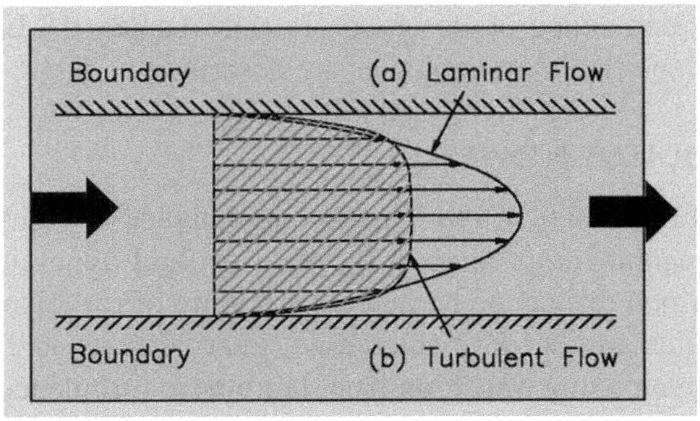

FIGURE 6.2 Velocity distributions: (a) Laminar, and (b) Turbulent flow.

from laminar to turbulent flow always occurs at nearly the same Reynolds number $Re = \frac{\bar{u}d}{\nu_f}$ where $\bar{u} = \frac{Q}{A}$ is the mean flow velocity, Q = volume rate of flow, A = cross-sectional area, ν_f is the kinematic viscosity and d is flow depth. For pipe flows, the transition occurs approximately for

$$Re_{cr} = \left(\frac{\bar{u}d}{\nu_f}\right)_{\text{crit}} = 2300 \tag{6.1}$$

The flow for which the Reynolds number, $Re < Re_{cr}$, is supposed to be laminar, and for $Re > Re_{cr}$, the flow is supposed to be turbulent. The value of Re_{cr} depends strongly on condition of the pipe inlet. Reynolds (1883) thought that the critical Reynolds number Re_{cr} will increase, if the disturbances in the flow before the pipe decrease. Later, the fact was confirmed by many researchers, like Barnes and Coker (1905) and Schiller (1922), who reached the value of critical Reynolds number Re_{cr} up to 20,000. Subsequently, Ekman (1910) achieved in maintaining the laminar flow up to $Re_{cr} = 40,000$, arranging an inlet experimentally free from the disturbance. Transition from laminar to turbulence is escorted by an apparent change in the law of resistance. In laminar flow, the longitudinal pressure gradient which maintains the motion proportional to the first power of mean velocity, by contrast for turbulent flow; this pressure gradient becomes nearly proportional to the square of mean velocity, that is,

$$-\frac{\partial p}{\partial x} \propto \bar{u} \text{ for laminar flow, } -\frac{\partial p}{\partial x} \propto \overline{u^2} \text{ for turbulent flow.} \quad (6.2)$$

6.2.1 Transition from Laminar to Turbulent Flow

The process from a laminar flow to the turbulent is known as laminar-turbulent transition (Figure 6.3a). The main parameter characterizing transition is the flow velocity or Reynolds number. The flow Reynolds number is defined as the inertial force by viscous forces. This process applies to any fluid flow and is most often used in the context of boundary layer flow. A boundary layer is a thin region of fluid flow that is developed by the presence of a boundary. The boundary layer forms when a flow past an impermeable wall with no-slip condition at the wall, which results in a large shear and viscous force close to the wall. There is a critical Reynolds number for which the boundary layer changes from laminar to turbulent flow. The boundary layer flow over a flat surface begins as laminar, and then under certain circumstances, the flow will have a transition from laminar to turbulent (Figure 5.1b) (Reynolds, 1883, 1885). The transition occurs mainly by a sudden increase in boundary layer thickness. In laminar flow, the dimensionless boundary layer thickness remains constant at approximately five for critical Reynolds number, when the flow exceeds the critical Reynolds number, a sudden increase in boundary layer thickness is visible. From Figure 6.3b, it is observed that a small ball retains its original shape when the ball flows in the uniform flow outside the boundary layer. Once the ball enters the laminar boundary layer, the ball gets distorted like ellipsoid because of the velocity gradient within the laminar boundary layer flow near the leading edge (Figure 6.3b) (Munson et al., 2006). At some distance along the flow near the leading edge, the laminar boundary layer becomes turbulent, and the fluid ball becomes fully distorted due to random and irregular nature of the turbulent flow. This transition from laminar to turbulent boundary layer flow occurs at a critical Reynolds number. The value of the critical Reynolds number is about (3–5) × 10^5 regarded as a lower limit. The transition involves a remarkable change in the shape of the velocity distribution curve, which is the transition from laminar to turbulent. The Reynolds number is based on the length along a flat plate in a parallel flow at zero incidence. The changes in the typical velocity profiles in the laminar, transitional, and turbulent flow are shown in Figure 6.4 (Munson et al., 2006) for different distances along the plate at a fixed external velocity U. It is observed that the turbulent velocity profiles have a larger velocity gradient at the wall and generate a larger boundary layer thickness with respect to the laminar boundary layer. The laminar boundary layer profile in the fully developed flow calculated by Blasius (1908) changes to the fully developed turbulent

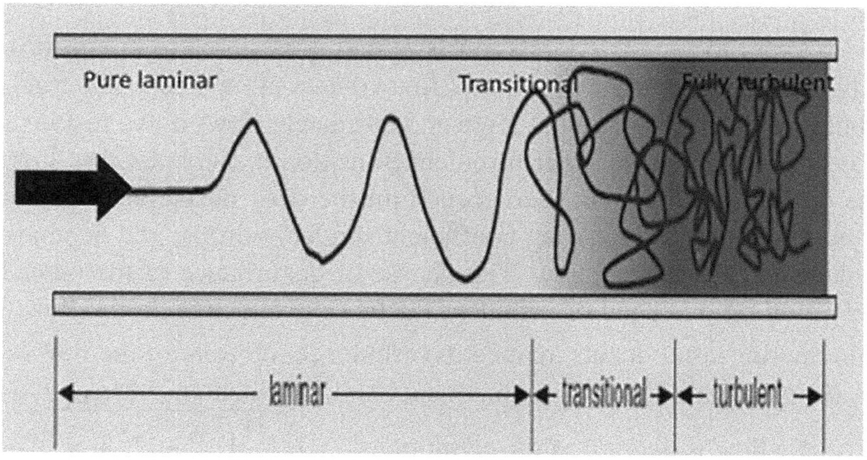

FIGURE 6.3A Schematic strictures of laminar to turbulent flow.

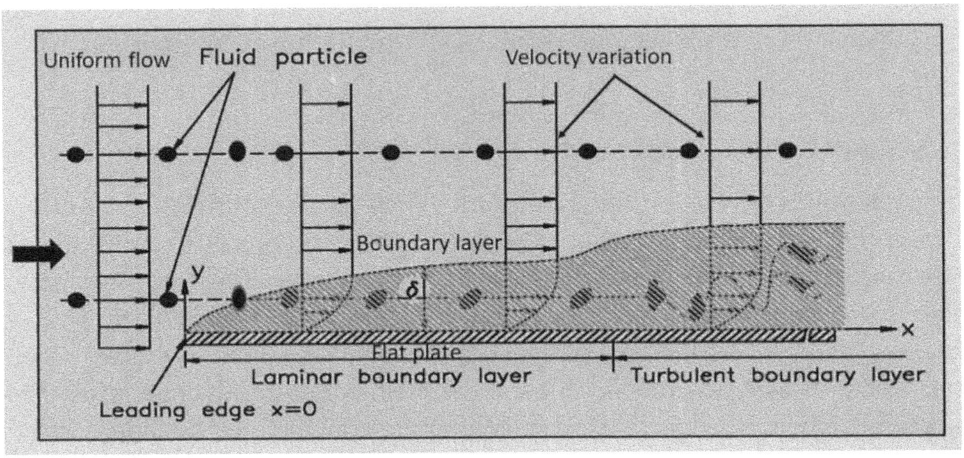

FIGURE 6.3B Boundary layer formation. (Modified from Munson et al., 2006.)

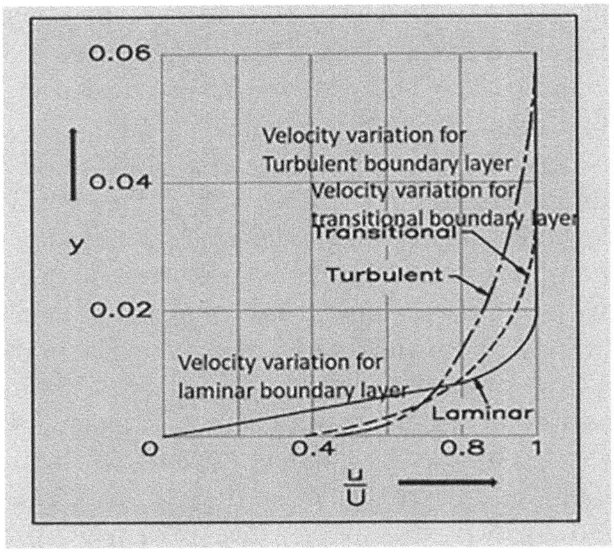

FIGURE 6.4 Boundary layer profiles over a flat plate for laminar, transitional and turbulent flows. (Modified from Munson et al., 2006.)

flow. The velocity external to the boundary layer is U and is assumed to be uniform until interacting with the solid body at the leading edge or point of first contact at the body. The fluid continues to flow over the surface, but is affected by frictional force and ultimately slows down inside the boundary layer.

A very important parameter for laminar-turbulent transition in the boundary layer is the degree of disturbance of the outer flow. This has been seen from the drag measurements at sphere, where the critical Reynolds number at which the drag coefficient suddenly drops, and depends very strongly on the perturbation intensity of the outer flow. The degree of disturbance of the outer flow can be measured quantitatively by time-averaged of turbulent fluctuating velocities in the flow. The time-average of three velocity fluctuating components, named as turbulence intensity of the flow is given by $\overline{u'^2}$, $\overline{v'^2}$, $\overline{w'^2}$, where prime denotes the fluctuating components and bars denote averaging. The dimensionless turbulence intensity of a flow is defined as the quantity as: $T_{in} = \sqrt{\frac{1}{3}\left(\overline{u'^2} + \overline{v'^2} + \overline{w'^2}\right)} / U$, where U is the free stream velocity or basic flow. A measure of the intensity of fluctuations is the turbulence intensity and can be written as: $T_{in} = \sqrt{\frac{2}{3}k}/U$, where $k = \frac{1}{2}\left(\overline{u'^2} + \overline{v'^2} + \overline{w'^2}\right)$ is the kinetic energy. In general, in a wind tunnel flow there is so-called isotropic flow at some distance from the screen. In isotropic flow, the average fluctuating velocity is the same in all three coordinate directions: $\overline{u'^2} = \overline{v'^2} = \overline{w'^2}$. In this case, the stream-wise velocity fluctuation alone can be used for turbulence intensity, that is, $T_{in} = \sqrt{\overline{u'^2}}/U$. It is important to note that the laminar-turbulent transition is greatly dependent on the turbulence intensity of the outer flow because in practice the outer flow is not completely free from turbulence.

6.2.2 Intermittent Flows

Intermittent nature is commonly observed in fluid flows that occur in a fraction of time as laminar and turbulent flow, one after another at an irregular interval. By this we mean that the flow is sometime laminar and sometime turbulent. A flow field (or a signal) is called intermittent, if the turbulent activity stops from time to time and starts again with a laminar flow in an alternative nature, for example, turbulent wake of a bullet, cloud in clear sky, plumes, exhaust stream, dust storms with sharp irregular boundary, etc. The fraction of time is a statistical function of the distance from the centerline of the wake or plume. Comprehensive studies on the process of transition reveal that in a certain range of critical Reynolds number Re_{cr}, the flow becomes "intermittent" which means that it changes with time from laminar to turbulent alternatively. This variation of velocity with time shows a random sequence in orderly and disorderly flow, which represents the results of velocity measurements (Figure 6.5). In highly turbulent flows, intermittency is seen in the irregular dissipation of kinetic energy.

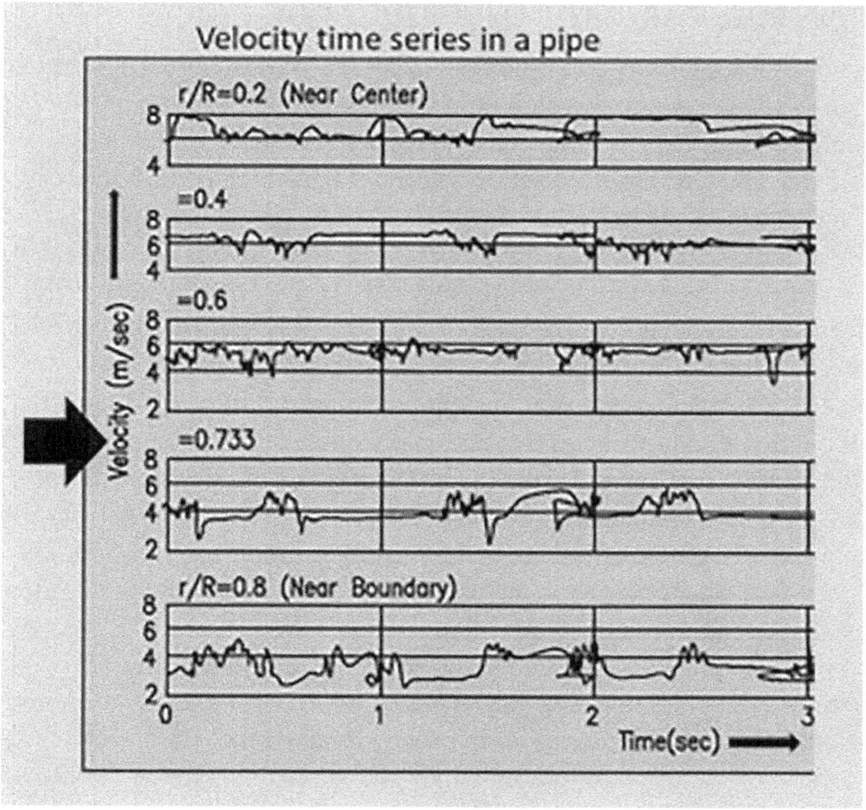

FIGURE 6.5 Typical velocity time series in the pipe flow close to laminar-turbulent transition for various distances from the center to the boundary. (Modified from Rotta, 1956.)

Rotta (1956) performed the measurements at different distances along the radius of the pipe. His measurement in velocity shows the periods of laminar and turbulent flow succeed each other in a random sequence. At a position closer to the center line of the pipe, the velocity in the laminar intervals exceeds the time-averaged velocity of turbulent intervals; while at the position close to the boundary wall, conditions are reversed. There is time series of consecutive intervals of turbulent and laminar flows. The intervals of low turbulence activity are interrupted irregularly by strong turbulence (Lohse and Grossmann, 1993). In the region of intermittent flow, the velocity distribution changes from the fully developed laminar flow to the corresponding fully developed turbulent flow. This physical nature of flow can be described with the aid of the intermittency factor γ, which is defined as the fraction of time during which the flow is turbulent at a given point. Hence, $\gamma = 1$ corresponds to the turbulent flow, and $\gamma = 0$ corresponds to continuous laminar flow or outside the turbulent flow. For pipe flow, the intermittency factor γ is plotted for various Reynolds numbers with respect to axial distance x (Figure 6.6a). At a fixed Reynolds number, the factor γ increased with axial distance x continuously. From the figure, it is observed that how γ is changing with the Reynolds number. The intermittency factors were also obtained by Klebanoff (1954) and Corrsin and Kistler (1954) for the boundary layer over the flat plate. The equation was expressed as:

$$\gamma = \frac{1}{2}\left\{1 - \text{erf } 5\left[\left(\frac{y}{\delta}\right) - 0.78\right]\right\} \tag{6.3}$$

where y/δ is the dimensionless boundary layer thickness. Figure 6.6b shows the intermittency factor γ varying with y/δ represented by equation (6.3).

6.3 TRANSITION IN BOUNDARY LAYER OVER SOLID BODY

Flow in a boundary over a solid body can undergo transition from laminar to turbulent. The occurrence of transition over a solid body was deliberated much later than that of the transition in the pipe flow. The transition in a boundary layer over a solid body is affected by many parameters, such as the pressure distribution in the external flow, wall roughness, and the nature of disturbances in the free flow. For the occurrence of transition in the boundary layer over a solid body, the spheres or circular cylinders, flat plate at zero incidence, slender bodies (airfoils or streamline bodies) are considered as solid bodies. The drag coefficient of sphere or cylinder drops suddenly at a Reynolds number *Re* about (3–5) $\times 10^5$. The sudden drop of drag coefficient for the case of sphere is a consequence of transition in the boundary layer (Eiffel, 1912). This transition causes the separation point to move downstream. For a flat plate at zero incidences, the process of transition is rather simpler to understand than that on a blunt body. The transition in the boundary layer becomes most noticeable by a sudden raise in the boundary layer thickness and in the shear stress near the wall. The dimensionless boundary layer thickness becomes constant for laminar flow and is approximately equal to five for the Reynolds number up to 3.2×10^5. At the Reynolds number *Re* greater than 3.2×10^5, a sudden increase in boundary layer thickness is noticeable and an identical phenomenon in shear stress is also observed. The sudden increase in these quantities indicates the transitional flow from laminar to turbulent. Based on this, the critical Reynolds number for the flat plate is approximately 3.2×10^5 along the axial direction of the plate from the leading edge. Transition on a flat plate makes a large change in the resistance to flow, and hence in the skin friction. In the laminar flow, the skin friction is proportional to the 1.5 power of velocity, whereas in the turbulent flow, the power increases to about 1.85 and shown by Froude (1872), who performed the towing experiments with plates at a very high Reynolds number. It is important to note that the pressure gradient along the wall influences the position of the point of transition of the boundary layer. For accelerated flow (decrease in pressure) the boundary layer remains laminar, whereas even in small increase

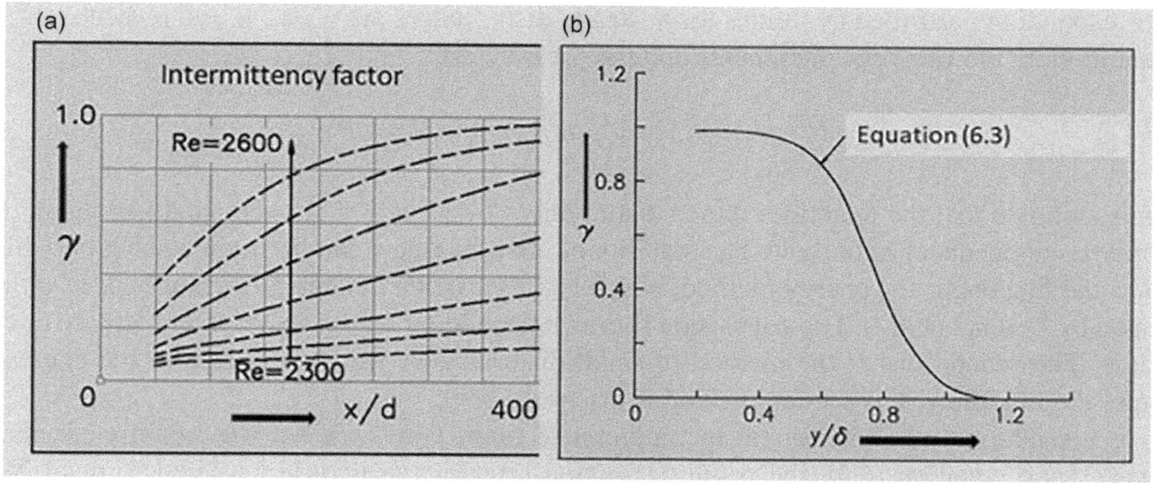

FIGURE 6.6 (a) Intermittency factor γ versus *x/d* for different Reynolds number *Re* (Modified Schlichting and Gersten, 2000), and (b) Intermittency factor γ varies with y/δ. (Modified from Cebeci, 2004.)

in pressure, the flow may be in transition. Therefore, the reduction of skin friction on the slender bodies like aero-foils, streamline bodies, etc. is possible only by shifting the point of transition downstream as far as possible, which may be achieved by a suitable choice of the shape with corresponding pressure gradient. The location of the point of transition and the scale of skin friction may be controlled by sucking the boundary layer.

6.4 PRINCIPLES OF STABILITY THEORY

The theoretical investigations of stability are based on the assumption that the laminar flows are affected by certain small disturbances, which arise mostly in the entry flow. The stability theory depends on laminar to turbulent transition. In the case of pipe flow, these disturbances may originate due to the inlet flow, wall roughness and irregularities in the external flows. The nature of such disturbances in time is noticeable, when they superimposed in the main flow. The question is whether these disturbances augment or die out with time, bearing in mind their exact forms still remain to be studied. If the disturbances decompose with time, the main flow is considered to be stable; on the other hand, if it increases with time, the flow is unstable, and there exists a possibility of transition from laminar to a turbulent flow. In this way, the theory of stability is established, and its purpose is to locate the critical Reynolds number Re_{cr} for the main flow. The principle of this stability theory was developed by Reynolds (1883), who invented that the laminar flow pattern represents a possible type of flow which is a solution of the differential equations of fluid dynamics, but becomes unstable beyond a specific limit and changes to the turbulent flow. Much work had been studied on the mathematical foundations of Reynolds's hypothesis for last many decades such as Reynolds (1883), Rayleigh (1887, 1895, 1911). Details of experimental verification on the stability theory are available in Schlichting (1968), Lin (1955), and Stuart (1956). There was a remarkable agreement between the theory and the experiments.

6.4.1 Method of Small Disturbances

To study the stability theory of laminar flow, the fluid motion is decomposed into a basic flow and a disturbance superimposed on the basic flow. Here the stability of basic flow constitutes the subject of study. Let the basic flow considered as steady be described by the Cartesian velocity components U, V, W and its pressure P. The basic flow is the solution of Navier–Stokes equations or boundary layer equations.

The corresponding perturbed quantities for unsteady disturbances are u', v', w', p' respectively. In the resultant motion, the velocity components and the pressure are

$$u = U + u', \quad v = V + v', \quad w = W + w', \quad p = P + p' \tag{6.4}$$

Here it is assumed that the quantities due to disturbances u', v', w', p' are very small compared to the respective components of main flow. The stability of disturbed flow can be studied using two different methods: the first one is the energy method, which is the variation of energy of disturbances with time, developed by Lorentz (1907). The conclusion is drawn depending on the decrease or increase of energy with time. The second one is the method of small disturbances, which is related to the equations of motion of the fluid flow. The second method is discussed here.

Let us consider a two-dimensional incompressible main flow with equally two-dimensional disturbance. Using equation (6.4), the resulting motion satisfies the two-dimensional form of Navier–Stokes equations. We further simplify the problem that the basic velocity U depends only on y, i.e., $U = U(y)$, $V = W = 0$ for flow through a channel of parallel walls with a gap width b. We also came across earlier the same flow problem named as parallel flow in Chapter 4. This flow can be considered with enough accuracy at a sufficient distance from inlet section. As far as the pressure in the main flow is concerned, it is necessary to assume a dependence of pressure on x and y, i.e., $P(x, y)$ because the pressure gradient generates the flow. If the basic flow is two-dimensional and parallel, the velocity distribution must be either linear or parabolic. A two-dimensional disturbance, which is a function of time and space, is superimposed on the basic flow $U(y)$. The resulting motion is described as

$$U = U(y) + u', v = v', w = 0, p = P + p' \tag{6.5}$$

where $u'(t, x, y)$, $v'(t, x, y)$, $p'(t, x, y)$ are the disturbances of velocity components and pressure respectively. Here it is assumed that the basic flow $U(y)$ is obtained from solution of Navier–Stokes equations and it is required that equation (6.5) must also satisfies the Navier–Stokes equations (equation 4.40). It is also assumed that the superimposed perturbed velocity components from equation (6.5) are considered to be very small, so that all the quadratic terms in the perturbed components are neglected with respect to the linear terms. Substituting the expressions (6.5) in Navier–Stokes equation for two-dimensional, incompressible unsteady flow and neglecting quadratic terms in the disturbances of velocity components, one gets as follows:

$$\frac{\partial u'}{\partial t} + U\frac{\partial u'}{\partial x} + v'\frac{dU}{dy} + \frac{1}{\rho_f}\frac{\partial P}{\partial x} + \frac{1}{\rho_f}\frac{\partial p'}{\partial x} = \upsilon_f\left(\frac{d^2U}{dy^2} + \nabla^2 u'\right) \tag{6.6a}$$

$$\frac{\partial v'}{\partial t} + U\frac{\partial v'}{\partial x} + \frac{1}{\rho_f}\frac{\partial P}{\partial y} + \frac{1}{\rho_f}\frac{\partial p'}{\partial y} = \upsilon_f \nabla^2 v' \tag{6.6b}$$

$$\frac{\partial u'}{\partial x} + \frac{\partial v'}{\partial y} = 0 \tag{6.6c}$$

where $\nabla^2 = \dfrac{\partial^2}{\partial x^2} + \dfrac{\partial^2}{\partial y^2}$

If it is considered that the basic flow itself satisfies the Navier–Stokes equations, the above equations are simplified as

$$\frac{\partial u'}{\partial t}+U\frac{\partial u'}{\partial x}+v'\frac{dU}{dy}+\frac{1}{\rho_f}\frac{\partial p'}{\partial x}=\upsilon_f\nabla^2u' \tag{6.7a}$$

$$\frac{\partial v'}{\partial t}+U\frac{\partial v'}{\partial x}+\frac{1}{\rho_f}\frac{\partial p'}{\partial y}=\upsilon_f\nabla^2v' \tag{6.7b}$$

$$\frac{\partial u'}{\partial x}+\frac{\partial v'}{\partial y}=0 \tag{6.7c}$$

There are three equations for u', v' and p'. Boundary conditions are $u' = v' = 0.$ at the walls (no-slip condition), and the pressure p' is easily eliminated from the above two equations (6.7a, b). Therefore, together with equation of continuity, there are two equations for u' and v', which give the closed-form solution. It may be questioned why the variation of basic velocity $U(y)$ with respect to x and the normal component of velocity V are neglected? In this context, Pretsch (1941) showed that those equations containing the terms are not important for the stability of a boundary layer. Details are explained in Cheng (1953).

It is assumed that the basic laminar flow $U(y)$ in the x-direction is influenced by a disturbance which is composed of number of waves propagating in the x-direction. As the perturbation is two-dimensional; it is possible to introduce a stream function $\psi(x,y,t)$, and thus satisfying the equation of continuity (6.7c). Let us assume the stream function considering a single oscillation of the disturbance in the form

$$\psi(x,y,t)=\phi(y)e^{i(\alpha x-\beta t)} \tag{6.8}$$

where $\phi(y)$, the amplitude of fluctuation is a function of y like $u(y)$, α is real quantity, $\lambda = 2\pi/\alpha$ is the wavelength of disturbance, $\beta = \beta_r + i\beta_i$ with β_r is the circular frequency of the partial oscillation, and β_i is the amplification factor, determines the degrees of amplification or damping. The disturbances are damped if $\beta_i < 0$, and the laminar mean flow is stable, whereas if $\beta_i > 0$, instability sets in. Using the stream function ψ from equation (6.8), one gets:

$$u'=\frac{\partial\psi}{\partial y}=\phi'(y)e^{i(\alpha x-\beta t)} \tag{6.9}$$

$$v'=-\frac{\partial\psi}{\partial x}=-i\alpha\phi(y)e^{i(\alpha x-\beta t)} \tag{6.10}$$

Substituting equations (6.9) and (6.10) in the Navier–Stokes equation (6.7), and after eliminating the pressure from the equations, the following fourth-order, ordinary differential equation of the amplitude $\phi(y)$ is given by

$$(U-c)(\phi''-\alpha^2\phi)-U''\phi=-\frac{i}{\alpha Re}(\phi''''-2\alpha^2\phi''+\alpha^4\phi) \tag{6.11}$$

with $c=\dfrac{\beta}{\alpha}=c_r+ic_i$, where c_r is the velocity of propagation of the wave in the x-direction, and c_i determines the degree of damping. Here $Re=\dfrac{U_m b}{\upsilon_f}$ denotes the Reynolds number with the maximum velocity U_m of the basic flow through parallel walls, and primes denote the differentiation with respect to the dimensionless coordinate y/b. The terms on the left-hand side of equation (6.11) represent the inertial terms and those on the right-hand side from viscous terms of the equations of motion. The boundary conditions represent the components of perturbed velocity and must vanish at the wall ($y = 0$) and at a large distance from the wall, which are given by:

$$\left.\begin{aligned} y = 0 &: u' = v' = 0;\ \phi = 0, \phi' = 0 \\ y \to \infty &: u' = v' = 0; \phi = 0, \phi' = 0 \end{aligned}\right| \tag{6.12}$$

Equation (6.11) is the fundamental differential equation for the disturbance (stability equation), which is the point of departure of stability theory of the laminar flows. This equation is commonly known as the *Orr-Sommerfeld equation*. The theory of stability of parallel flows is appropriate to the flow which is basically parallel to one direction, such as boundary layer flows.

It is important to note that the disturbances superimposed on a two-dimensional flow pattern need not be two-dimensional; if a complete analysis of stability equation is to be achieved. The issue was resolved by Squire (1933) by assuming the disturbances which were also periodic in z-direction. The problem of stability theory has been converted to the eigenvalue problem of equation (6.11) subject to the boundary conditions (6.12), which is not considered here in detail. For interested readers, more details are available in Schlichting (1968)

6.5 FUNDAMENTALS OF TURBULENT FLOWS

In general, fluid flow is classified into two categories: laminar flow and turbulent flow. When the flow velocity is low, the flow is a layered flow where layers of fluid move with different velocities without exchange of fluid particles perpendicular to the flow direction. This type of orderly moving flow is called *laminar flow*. Figure 6.7a shows a laminar velocity profile through a pipe, which shows a parabolic profile with zero values at the boundaries. If the flow velocity is increased, the flow picture changes drastically, where the flow is characterized by a highly irregular, random, and fluctuating motion. This disorderly moving flow is called *turbulent flow* through a pipe, where Figure 6.7b shows a turbulent velocity profile with random fluctuations in velocity data.

Most of the fluid flows which occur in nature and in practical applications are turbulent. The turbulent motion in fluid flow is generated, when an irregular fluctuation (mixing or eddy motion) is superimposed on the main stream, consisting of irregular swirls of motion, called eddies. That flow is an irregular and complex flow in which various physical quantities show a random variation with time and space, so that the statistically distinct averages can be differentiated. The flow is so complex that it

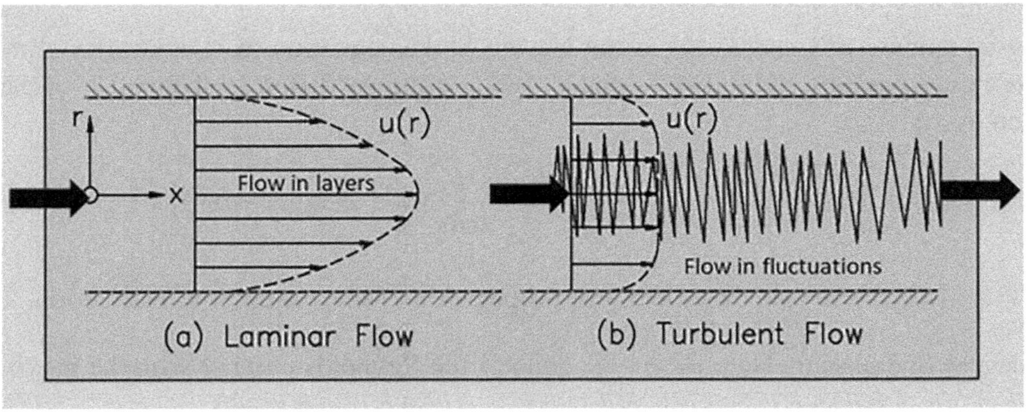

FIGURE 6.7 Typical velocity profiles in a pipe for laminar and turbulent flows. (Modified from Som and Biswas, 1998.)

seems to be inaccessible to mathematical treatment, but the resulting mixing motion is very essential for the course of flow and for the equilibrium of forces. At large Reynolds number, there is a continuous transfer of energy from the main flow to the large eddies. However, that energy is dissipated by the smaller and smaller eddies and the process occurs in a narrow strip at the immediate neighborhood of the wall inside the boundary layer. The turbulence is a state of fluid motion which characterizes random and chaotic three-dimensional vortex motion (Schlichting and Gersten, 2000). As the turbulent flow is irregular, apparently random (chaotic) and complex, till today no analytical solution exists. We are still looking for approximate solutions and empirical relations how turbulence behaves, in many respects. The source of turbulence and the associated transition from laminar to turbulent flow are of fundamental importance for the whole science of Fluid Mechanics. Turbulent flow has always been a challenge for scientists that have an important role in several complex problems in fluid flows.

We shall discuss problems related to *fully developed turbulent flow*. In this context, we are required to confine ourselves to the consideration of time-averaged of turbulent motion. The most important feature of turbulent flow consists of the fact that both the velocity and pressure at a fixed point in space always change with time, but they perform very irregular fluctuations with high frequency. The amplitudes of the fluctuations are completely irregular. The lumps of fluid which execute such fluctuations in the direction of flow and right angles to it do not follow a single molecule as assumed in the kinetic theory of gas; they are macroscopic fluid balls or lumps of varying size called *eddies* (Figure 6.8), which has the crucial influence to the whole motion. The turbulence flow can be interpreted as a population of many eddies (vortices) of different sizes and strengths with a random appearance to the flow. Two variables, like characteristic diameter of eddy and their orbital velocities, play an important role. Since the turbulent flow consists of many eddies of varying size and speed, they assume values in a certain range. That motion spans a wide range of scales ranging from a macro-scale to a micro-scale at which energy is dissipated by viscosity. The interaction among eddies of various sizes transfers energy from the larger eddies to the smaller ones. This process of transfer of energy from larger sizes of eddies to smaller sizes of eddies is known as the energy cascade (Kolmogorov, 1941).

Such fluid balls, eddies or lumps are clearly visible to the flow. The sizes of such fluid balls, which continually agglomerate and break up, determine the *scale of turbulence*; their sizes are determined by the external flow conditions, for example, by the mesh size or honeycomb through which flow can pass. A time-average of turbulent motion is performed because a complete theoretical formulation is almost impossible owing to the complexity of fluctuations in flow. Accordingly, certain theoretical principles are established which allow to predict the turbulence parameters generated from the experimental observations. Moreover, in certain cases, prediction from theoretical principles under certain assumptions shows a good agreement with the experimental observations. This article is focused on the influence of velocity fluctuations on the mean flow, the development of equations of turbulent motion, and semi-empirical relations developed for the calculation of turbulent flow using the mixing length concept

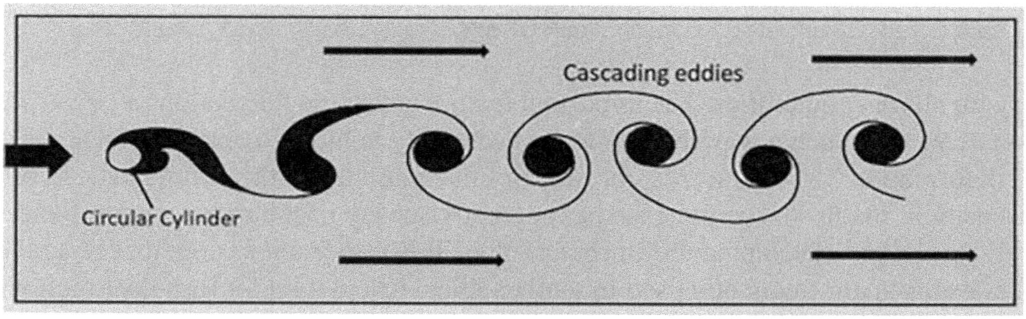

FIGURE 6.8 Development of macroscopic balls or eddies or lumps of flow due to a cylinder.

due to Prandtl (1925). Turbulence can be categorized as follows: *Homogeneous turbulence* – which has the same configuration qualitatively everywhere in the flow field. *Isotropic turbulence* – which has no directional preference in terms of statistical features, and the perfect disorder, persists. *Anisotropic turbulence* – the statistical features have directional preference and there is a mean velocity gradient.

6.5.1 Time-Averaged Motion and Fluctuations

In turbulent flow, both velocity and pressure at a fixed point in space do change with time at a highly irregular way, and are characterized by random fluctuations of high frequency. Such fluctuations in the fluid flow illustrate like macroscopic fluid balls or lumps of varying small size. In describing the physical quantities of a turbulent flow in mathematical form, it is therefore convenient to separate the velocity components and pressure into a mean and fluctuations. Denoting the time-average of velocity components by $(\bar{u}, \bar{v}, \bar{w})$ and their fluctuations by (u', v', w') in x, y and z directions respectively, and denoting $\bar{p}$ and p' to be the mean and fluctuations of pressure, we can write the following relations for the instantaneous velocity components and pressure as:

$$u = \bar{u} + u', v = \bar{v} + v', w = \bar{w} + w', p = \bar{p} + p' \tag{6.13}$$

The time-averaged velocity is found at a fixed point in space is given by

$$\bar{u} = \frac{1}{t_1} \int_{t_0}^{t_0+t_1} u dt \tag{6.14}$$

It is understood that the mean values are taken over a sufficiently long-time interval t_1, to be completely independent of time. That means, in a discrete way the stream-wise mean velocity $\bar{u}$ is defined as:

$$u_{\text{mean}} = \bar{u} = \lim_{n \to \infty} \frac{1}{n} \sum_{i=1}^{n} u_i \tag{6.14a}$$

where n is the total number of observations, and similarly, for other two components. The technique for decomposing the instantaneous motion from the hydrodynamics quantities is referred to as Reynolds' decomposition. By definition, the time-averages of the fluctuations are as follows:

$$\overline{u'} = 0, \overline{v'} = 0, \overline{w'} = 0, \overline{p'} = 0 \tag{6.15}$$

which means,

$$\overline{u'} = \lim_{n \to \infty} \frac{1}{n} \sum_{i=1}^{n} u'_i = 0 \tag{6.15a}$$

and similarly for all other quantities. The important feature is that the fluctuations u', v', w' influence the mean motion $\bar{u}$, $\bar{v}$, $\bar{w}$ in such a way that the latter eventually exhibit an apparent increase in the resistance to the deformation. In other words, the velocity fluctuations are the manifestations of increase in apparent viscosity in the flow. The increase in apparent viscosity of the mean motion focuses the main concept of all theoretical development of turbulent flow. It is important to note that several operational rules on time-averages are frequently used in mathematical calculation for turbulent motion. If f and g are two dependent variables whose mean values are to be formed and s denotes any one of the independent variables x, y, z, t, then the following time-averaging rules are applied as:

$$\overline{\overline{f}} = \overline{f}, \quad \overline{f+g} = \overline{f} + \overline{g}$$

$$\overline{\overline{f} \cdot g} = \overline{f} \cdot \overline{g} \tag{6.16}$$

$$\overline{\frac{\partial f}{\partial s}} = \frac{\partial \overline{f}}{\partial s}, \quad \overline{\int ds} = \int \overline{f} ds$$

Let us now assume an elementary area $d\Omega$ in a turbulent stream whose velocity components are u, v, w. The directions y and z are in the plane of $d\Omega$, and the normal to the area is taken as parallel to x-axis. The mass of fluid passing through this area in time dt is given by $\rho_f u \cdot d\Omega \cdot dt$ and the momentum flux in the x-direction is $dJ_x = \rho_f u^2 d\Omega dt$, and the corresponding momentum fluxes in y- and z-directions respectively are $dJ_y = \rho_f uvd\Omega dt$, and $dJ_z = \rho_f uwd\Omega dt$, where ρ_f is the fluid density.

Considering the density ρ_f is constant, the time-averages for the fluxes of momentum per unit time are

$$\overline{dJ_x} = d\Omega \rho_f \overline{u^2}, \quad \overline{dJ_y} = d\Omega \rho_f \overline{uv}, \quad \overline{dJ_z} = d\Omega \rho_f \overline{uw} \tag{6.17}$$

Now $u^2 = (\overline{u} + u')^2 = \overline{u}^2 + 2\overline{u}u' + u'^2$, applying the operational rules, one gets

$$\overline{u^2} = \overline{u}^2 + \overline{u'^2} \tag{6.18}$$

since $\overline{\overline{u}u'} = 0 = \frac{1}{t_1} \int_{t_0}^{t_0+t_1} \overline{u}u' dt = \frac{\overline{u}}{t_1} \int_{t_0}^{t_0+t_1} u' dt = 0.$

Similarly, it is easy to show that

$$\overline{u \cdot v} = \overline{u} \cdot \overline{v} + \overline{u'v'}, \quad \overline{u \cdot w} = \overline{u} \cdot \overline{w} + \overline{u'w'} \tag{6.19}$$

Therefore, using equation (6.18), the expressions for momentum fluxes per unit time are as follows:

$$\overline{dJ_x} = d\Omega \cdot \rho_f \left(\overline{u}^2 + \overline{u'^2}\right), \overline{dJ_y} = d\Omega \cdot \rho_f \left(\overline{u} \cdot \overline{v} + \overline{u'v'}\right), \overline{dJ_z} = d\Omega \cdot \rho_f \left(\overline{u} \cdot \overline{w} + \overline{u'w'}\right) \tag{6.20}$$

These quantities are the rate of change of momentum fluxes with the dimension of forces on the elementary area $d\Omega$, and dividing them by area $d\Omega$, one gets the term as $\frac{\overline{dJ_x}}{d\Omega}$, the force per unit area i.e., stress, and similarly for other components of stresses $\frac{\overline{dJ_y}}{d\Omega}$ and $\frac{\overline{dJ_z}}{d\Omega}$. Since the flux of momentum through an area per unit time is always equivalent to an equal and opposite force exerted on the area by the surroundings, it can be considered that the area under consideration, which is normal to the x-axis is acted upon by stresses, such as,

$$-\rho_f \left(\overline{u}^2 + \overline{u'^2}\right) \text{ in the } x \text{ - direction}$$

$$-\rho_f \left(\overline{uv} + \overline{u'v'}\right) \text{ in the } y \text{ - direction} \tag{6.21}$$

$$-\rho_f \left(\overline{uw} + \overline{u'w'}\right) \text{ in the } z \text{ - direction}$$

In equation (6.21), the first one is the normal stress, and the other two are the shearing stresses. Therefore, the superposition of fluctuations on the mean motion offers three additional stresses and is designated as

$$\sigma'_x = -\rho_f \overline{u'^2}, \;\; \tau'_{yx} = -\rho_f \overline{u'v'}, \;\; \tau'_{xz} = -\rho_f \overline{u'w'} \tag{6.22}$$

acting on the elementary surface, where σ'_x is the normal stress and τ'_{yx} and τ'_{xz} are the shearing stresses. These terms are known as *apparent or virtual or Reynolds stresses* of turbulent flow, and must be added to the stresses caused by steady laminar flow. Similarly, other two areas normal to the two remaining axes y and z form corresponding stresses; and hence there is a set of complete stress tensor of turbulent flow.

It is seen that the time-averages of product of velocity fluctuations are different from zero. The transport of x-momentum through a surface normal to y-axis is represented as the stress component $\tau'_{xy} = \tau'_{yx} = -\rho_f \overline{u'v'}$. For example, let us consider a mean flow $\overline{u} = \overline{u}(y)$, $\overline{v} = \overline{w} = 0$ with $\frac{d\overline{u}}{dy} > 0$. We see that the mean product $\overline{u'v'}$ is different from zero, which emerges a very important quantity, when multiplied by the density ρ_f is known as the Reynolds shear stress term. It is a correlation between two velocity components at a point. If these two quantities are not related to each other, the correlation is zero, but actually these two quantities are physically correlated. The particles which travel upwards due to turbulent fluctuation $v' > 0$ arrive at a layer y from a region where a smaller mean velocity $\overline{u}$ overcomes (Figure 6.9). Since they do preserve their original velocity $\overline{u}$, they give rise to a negative velocity $u' < 0$ in a layer y, which means, it is low momentum fluid moving to a region of high momentum. Conversely, the particles, which arrive from the above layer with negative $v' < 0$ give rise to a positive $v' > 0$ in it, i.e., the high momentum fluid are moving to a region of low momentum. On an average, a positive v' is mostly associated with a negative u' and a negative v' is mostly associated with positive u'. That means, a positive $v' > 0$ has a high probability that negative $u' < 0$, and if $v' < 0$, it has high probability that $u' > 0$. Therefore, the time-average $\overline{u'v'}$ is not only different from zero but also negative, i.e., the shear stress term $\overline{u'v'} < 0$. In this case, the turbulent shearing stress $\tau_{xy} = -\rho_f \overline{u'v'}$ is positive, thus it has the same sign as the relevant laminar shear stress $\tau = \mu_f \frac{du}{dy}$. The term $\overline{u'v'}$ is the turbulent vertical advection of stream-wise turbulent momentum or simply the vertical flux of stream-wise momentum. Now if the velocity gradient $\frac{d\overline{u}}{dy} < 0$, then the shear stress term $\overline{u'v'} > 0$. Therefore, the sign of momentum flux represents whether the flux induces a net increase or decrease in momentum. The momentum flux terms are frequently referred to as the Reynolds stresses because they play an analogous role in the governing equations to the viscous stress. It can be also articulated as the correlation between the two velocity-fluctuating components u' and v' at a given point.

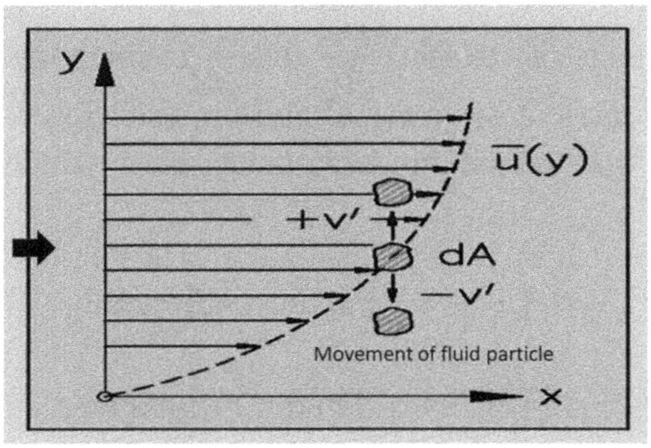

FIGURE 6.9 Momentum transport due to turbulent velocity fluctuation. (Modified Schlichting and Gersten, 2000.)

In the next section, the well-known Navier–Stokes equation will be modified for turbulent flow using the Reynolds decomposition to obtain Reynolds stress equations, which are widely used in the mathematical formulation of turbulent flow.

6.5.2 Reynolds Equations and Reynolds Stresses

As we know the turbulent motion consists of the sum of the mean and fluctuating flows, the sum is introduced into the Navier–Stokes equations, and time-averaging of the resulting equations are formed. The time-averaged equations provide a noticeable insight into the characteristics of turbulent motion with its prime parameters. The Navier–Stokes equations for an incompressible fluid flow, neglecting the body forces, in a three-dimensional Cartesian coordinate system along with the equation of continuity are written as (Schlichting and Gersten, 2000):

$$\rho_f \left\{ \frac{\partial u}{\partial t} + u\frac{\partial u}{\partial x} + v\frac{\partial u}{\partial y} + w\frac{\partial u}{\partial z} \right\} = -\frac{\partial p}{\partial x} + \mu_f \nabla^2 u \tag{6.23a}$$

$$\rho_f \left\{ \frac{\partial v}{\partial t} + u\frac{\partial v}{\partial x} + v\frac{\partial v}{\partial y} + w\frac{\partial v}{\partial z} \right\} = -\frac{\partial p}{\partial y} + \mu_f \nabla^2 v \tag{6.23b}$$

$$\rho_f \left\{ \frac{\partial w}{\partial t} + u\frac{\partial w}{\partial x} + v\frac{\partial w}{\partial y} + w\frac{\partial w}{\partial z} \right\} = -\frac{\partial p}{\partial z} + \mu_f \nabla^2 w \tag{6.23c}$$

$$\frac{\partial u}{\partial x} + \frac{\partial v}{\partial y} + \frac{\partial w}{\partial z} = 0 \tag{6.23d}$$

where ∇^2 is the Laplace's operator.

With the aid of physical argument caused by the turbulent fluctuations, the expressions for Reynolds stress equations are derived in a more general form directly from the Navier–Stokes equations. If the averaging procedure is carried out on the above equation (6.23), using the Reynolds decomposition and time-averaging rules (6.16), the following system of equations is obtained:

$$\rho_f \left(\bar{u}\frac{\partial \bar{u}}{\partial x} + \bar{v}\frac{\partial \bar{u}}{\partial y} + \bar{w}\frac{\partial \bar{u}}{\partial z} \right) = -\frac{\partial \bar{p}}{\partial x} + \mu_f \nabla^2 \bar{u} - \rho_f \left[\frac{\partial \overline{u'^2}}{\partial x} + \frac{\partial \overline{u'v'}}{\partial y} + \frac{\partial \overline{u'w'}}{\partial z} \right] \tag{6.24a}$$

$$\rho_f \left(\bar{u}\frac{\partial \bar{v}}{\partial x} + \bar{v}\frac{\partial \bar{v}}{\partial y} + \bar{w}\frac{\partial \bar{v}}{\partial z} \right) = -\frac{\partial \bar{p}}{\partial y} + \mu_f \nabla^2 \bar{v} - \rho_f \left[\frac{\partial \overline{u'v'}}{\partial x} + \frac{\partial \overline{v'^2}}{\partial y} + \frac{\partial \overline{v'w'}}{\partial z} \right] \tag{6.24b}$$

$$\rho_f \left(\bar{u}\frac{\partial \bar{w}}{\partial x} + \bar{v}\frac{\partial \bar{w}}{\partial y} + \bar{w}\frac{\partial \bar{w}}{\partial z} \right) = -\frac{\partial \bar{p}}{\partial z} + \mu_f \nabla^2 \bar{w} - \rho_f \left[\frac{\partial \overline{u'w'}}{\partial x} + \frac{\partial \overline{v'w'}}{\partial y} + \frac{\partial \overline{w'^2}}{\partial z} \right] \tag{6.24c}$$

$$\frac{\partial \bar{u}}{\partial x} + \frac{\partial \bar{v}}{\partial y} + \frac{\partial \bar{w}}{\partial z} = 0 \tag{6.24d}$$

From equations (6.23d) and (6.24d), one gets

$$\frac{\partial u'}{\partial x} + \frac{\partial v'}{\partial y} + \frac{\partial w'}{\partial z} = 0 \tag{6.24e}$$

The terms which are linear in the turbulent components such as e.g., $\frac{\partial u'}{\partial t}$ and $\frac{\partial^2 u'}{\partial x^2}$ vanish in view of (6.15). It is seen that the time-averaged velocity and fluctuating components, each satisfy the equation of continuity for incompressible flow. Here,

$$\nabla \vec{q} = 0, \ \nabla \vec{q'} = 0, \tag{6.25}$$

where $\vec{q} = (u, v, w)$. The above set of equation (6.24a, b, c) is known as the Reynolds averaged Navier–Stokes (RANS) equations with equations of continuity (6.24d, e). The left-hand side of equation (6.24a, b, c) is identical to the steady-state of Navier–Stokes equations, where the velocity components (u, v, w) are replaced by their time-averages ($\bar{u}$, $\bar{v}$, $\bar{w}$) and the same is true for the pressure and viscous terms on the right-hand side. In addition, the equations contain terms due to turbulent fluctuations of the velocity component, known as stress components, and these can be constructed as the components of stress tensor. Comparing the right-hand side terms of equation (4.37) of Chapter 4 with equation (6.24a, b, c) of the this chapter, the additional terms on the right-hand side of equation (6.24a, b, c) can be interpreted as the components of Reynolds stress tensor. The resulting surface force per unit volume due to these terms is considered as:

$$\rho_f\left(\bar{u}\frac{\partial \bar{u}}{\partial x} + \bar{v}\frac{\partial \bar{u}}{\partial y} + \bar{w}\frac{\partial \bar{u}}{\partial z}\right) = -\frac{\partial \bar{p}}{\partial x} + \mu_f \nabla^2 \bar{u} - \rho_f\left[\frac{\partial \sigma'_{xx}}{\partial x} + \frac{\partial \tau'_{xy}}{\partial y} + \frac{\partial \tau'_{xz}}{\partial z}\right] \tag{6.26a}$$

$$\rho_f\left(\bar{u}\frac{\partial \bar{v}}{\partial x} + \bar{v}\frac{\partial \bar{v}}{\partial y} + \bar{w}\frac{\partial \bar{v}}{\partial z}\right) = -\frac{\partial \bar{p}}{\partial y} + \mu_f \nabla^2 \bar{v} - \rho_f\left[\frac{\partial \tau'_{yx}}{\partial x} + \frac{\partial \sigma'_{yy}}{\partial y} + \frac{\partial \tau'_{yz}}{\partial z}\right] \tag{6.26b}$$

$$\rho_f\left(\bar{u}\frac{\partial \bar{w}}{\partial x} + \bar{v}\frac{\partial \bar{w}}{\partial y} + \bar{w}\frac{\partial \bar{w}}{\partial z}\right) = -\frac{\partial \bar{p}}{\partial z} + \mu_f \nabla^2 \bar{w} - \rho_f\left[\frac{\partial \tau'_{zx}}{\partial x} + \frac{\partial \tau'_{zy}}{\partial y} + \frac{\partial \sigma'_{zz}}{\partial z}\right] \tag{6.26c}$$

Comparing the right-hand sides of equations (6.24) and (6.26), the Reynolds stress tensor due to velocity fluctuations can be written as:

$$\begin{pmatrix} \sigma'_{xx}\tau'_{xy}\tau'_{xz} \\ \tau'_{yx}\sigma'_{yy}\tau'_{yz} \\ \tau'_{zx}\tau'_{zy}\sigma'_{zz} \end{pmatrix} = -\begin{pmatrix} \rho_f\overline{u'^2} & \rho_f\overline{u'v'} & \rho_f\overline{u'w'} \\ \rho_f\overline{v'u'} & \rho_f\overline{v'^2} & \rho_f\overline{v'w'} \\ \rho_f\overline{w'u'} & \rho_f\overline{w'v'} & \rho_f\overline{w'^2} \end{pmatrix} \tag{6.27}$$

where $\sigma'_{xx}, \sigma'_{yy}$ and σ'_{zz} are normal stresses and the others τ'_{xy}, τ'_{xz}, and τ'_{yz} are the shearing stresses. The component τ'_{xy} of the shear stress tensor is the same as obtained from the momentum equation (6.22). These additional stresses are known as apparent or virtual stresses of turbulent flow or Reynolds stresses. They are due to turbulent fluctuations and are given by the time-averaged values of the quadratic fluctuating components. Since these stresses are added to the ordinary viscous stresses in the laminar flow and have a similar influence on the flow, it is often called eddy viscosity. Thus, the total stresses are the sum of the ordinary viscous stresses and the apparent turbulent stresses so that

$$\sigma_x = -p + 2\mu_f \frac{\partial \bar{u}}{\partial x} - \rho_f \overline{u'^2} \tag{6.28}$$

$$\tau_{xy} = \mu_f\left(\frac{\partial \bar{u}}{\partial y} + \frac{\partial \bar{v}}{\partial x}\right) - \rho_f \overline{u'v'} \tag{6.29}$$

and other components of total stresses. These apparent stresses form a symmetrical second-order tensor; so, there are six components of Reynolds stress. In general, apparent turbulent stresses predominate over the viscous stresses, so the latter can often be dropped away from near the wall. In Reynolds stress equations, there are ten unknowns (namely, three components of mean velocity, the pressure and six components of Reynolds stress) and only four equations; hence the system is not in closed form. This is referred to as a closure problem as the Reynolds averaged Navier–Stokes (RANS) equations are not a closed form set of equations, whereas in the case of laminar flow of incompressible fluid, there are four unknowns (namely, three velocity components and the pressure) and four equations. Therefore, the system is in closed form. The closure problem of RANS equations is solved by developing models for the new terms based on resolved quantities.

6.5.3 Boundary Conditions for Velocity Components

The boundary conditions for the velocity components of these differential equations are the same as in ordinary laminar flow, namely they all vanish at solid walls (no-slip conditions). All the components of turbulent velocity must vanish at the walls, and they are very small in the immediate neighborhood. It follows that all the components of apparent stress tensor vanish at the solid walls and the only stresses which act near them are viscous stresses of laminar flow as they do not vanish there. Furthermore, it is seen that in the immediate neighborhood of a wall, the apparent stresses are small compared with viscous stresses, and it follows that in every turbulent flow, there exists a very thin layer next to the wall which behaves like one in the laminar motion, known as the *Laminar Sub-layer.* Its velocity is so small that the viscous forces dominate over the inertia forces. The thickness of the laminar sub-layer is so small, so that it is very difficult to observe under experimental conditions. Thus, the turbulence may not exist in it.

6.5.4 Turbulent Boundary Layer Equations

In the case of two-dimensional flows ($\overline{w} = 0$), incompressible, turbulent flow over a flat surface, assuming the thickness of turbulent boundary layer to be much smaller than the axial length, the main turbulent equations of motion (6.24) can be modified by the turbulent boundary layer approximations as (Schlichting, 1968):

$$\overline{u}\frac{\partial \overline{u}}{\partial x} + \overline{v}\frac{\partial \overline{u}}{\partial y} = -\frac{1}{\rho_f}\frac{d\overline{p}}{dx} + \frac{1}{\rho_f}\frac{\partial}{\partial y}\left(\mu_f \frac{\partial \overline{u}}{\partial y} - \rho_f \overline{u'v'}\right) \tag{6.30a}$$

$$\frac{\partial \overline{u}}{\partial x} + \frac{\partial \overline{v}}{\partial y} = 0 \tag{6.30b}$$

For laminar flow, $\frac{\partial p}{\partial y}$ is of $O(\delta)$, but that for turbulent flow, it is of $O(1)$. Consequently, the pressure variation across the boundary layer is of $O(\delta)$, so that in comparison with the stream-wise pressure gradient $\frac{\partial p}{\partial x}$, the pressure gradient across the boundary layer $\frac{\partial p}{\partial y}$ is still small and can be neglected within the boundary-layer approximations. Therefore, the y-component of momentum boundary layer equation vanishes.

If we compare this turbulent boundary layer equation (6.30) with laminar boundary layer, then it follows as:

a. Velocity components and pressure u, v, p are replaced by $\overline{u}$, $\overline{v}$, $\overline{p}$ (time-averaged).

b. The inertial terms and pressure term remain unchanged, and the viscous term $\nu_f \dfrac{\partial^2 u}{\partial y^2}$ must be replaced by $\dfrac{1}{\rho_f}\dfrac{\partial}{\partial y}\left(\mu_f \dfrac{\partial \bar{u}}{\partial y} - \rho_f \overline{u'v'}\right)$.

The last term on the right-hand side of equation (6.30a) is equivalent to the force per unit volume $\dfrac{\partial \tau_l}{\partial y}$ for laminar flow must be replaced by $\dfrac{1}{\rho_f}\dfrac{\partial}{\partial y}(\tau_l + \tau_t)$ in which $\tau_l\left(= \mu_f \dfrac{\partial \bar{u}}{\partial y}\right)$ is viscous shear stress for laminar flow, and $\tau_t\left(= -\rho_f \overline{u'v'}\right)$ is the apparent turbulent shear stress.

For a fully developed turbulent flow over the flat surface (Figure 6.3b), the inertial effects are insignificant at the near-wall region. Therefore, equation (6.30a) can be written as

$$\frac{\partial}{\partial y}\left(\mu_f \frac{\partial \bar{u}}{\partial y} - \rho_f \overline{u'v'}\right) = \frac{\partial \tau}{\partial y} = 0 \tag{6.31}$$

After integration with respect to y, it gives

$$\mu_f \frac{\partial \bar{u}}{\partial y} - \rho_f \overline{u'v'} = \text{constant} = \tau \tag{6.32}$$

where τ is the total shear stress, is composed of two parts, viscous shear stress τ_l and Reynolds shear stress τ_t. It is known that the shear stress at the wall is purely viscous due to the existence of laminar sub-layer. At the wall boundary $y = 0$, the shear stress $\tau = \tau_w$. Therefore, equation (6.32) can be written as:

$$\mu_f \frac{\partial \bar{u}}{\partial y} - \rho_f \overline{u'v'} = \tau_w \tag{6.33}$$

which shows that the shear stress is constant over the entire depth, and equal to the wall shear stress τ_w.

Multiplying equation (6.33) by $\dfrac{\partial \bar{u}}{\partial y}$ and integrating from $y = 0$ to h, one gets

$$\int_0^h -\overline{u'v'}\frac{\partial \bar{u}}{\partial y} dy + \int_0^h \nu_f \left(\frac{\partial \bar{u}}{\partial y}\right)^2 dy = u_*^2 \int_0^h \frac{\partial \bar{u}}{\partial y} dy = u_*^2 \bar{u}_m \tag{6.34}$$

where $\bar{u}_m$ is the maximum velocity. The term $u_*^2 \bar{u}_m$ indicates the work of the mean flow against the bed shear stress τ_w/ρ_f. Equation (6.34) implies that mean-flow energy is lost by the following two mechanisms: turbulence generation or production and direct viscous dissipation of mean energy. The turbulence generation or production produces turbulent fluctuations associated mainly with large-scale eddies. The turbulent energy of the large-scale eddies is then transferred to smaller-scale eddies through a cascade process, and it is finally dissipated into heat by molecular viscosity. As the Reynolds number increases, the direct dissipation of mean-flow energy is significant only near the wall.

6.6 CLASSICAL TURBULENCE MODELING USING SEMI-EMPIRICAL THEORIES

Because of the extremely complicated nature of turbulent flow, it is very difficult to achieve a complete understanding of the turbulence mechanism. The exact relationship between the Reynolds stresses and the mean flow quantities is not yet known. To understand the mean velocity in a fully developed open channel turbulent flow, many semi-empirical investigations have been carried out for the last few decades. The methods of calculation developed so far are based on empirical hypotheses which endeavor to set up a relationship between the Reynolds stresses from the mixing motion and mean velocity. Here some of the velocity distribution laws for open-channel turbulent flow over smooth and rough boundaries are discussed; though none of them is adequate to investigate the turbulent flow successfully. The following three basic semi-empirical theories proposed by Boussinesq (1877), Prandtl (1925), von-Karman (1930), and others are discussed in this section:

6.6.1 Boussinesq's Eddy Viscosity Hypothesis

The idea of mixing motion due to fluctuating velocity components was first developed by Boussinesq (1877, 1896) to introduce a mixing coefficient A_τ analogous with the coefficient of viscosity μ_f in Stokes's law (1851) for laminar flow. The shear stress for laminar flow is given by $\tau_l = \mu_f \dfrac{d\bar{u}}{dy}$, and the Reynolds shear stress for turbulent flow is given by

$$\tau_t = -\rho_f \overline{u'v'} = A_\tau \frac{d\bar{u}}{dy} \tag{6.35}$$

If the shear stresses for laminar and turbulent flows are compared, the mixing coefficient A_τ for turbulent flow corresponds to the viscosity μ_f for laminar flow. Here the mixing coefficient A_τ is called apparent or virtual or eddy viscosity, but this assumption has some drawback because the *eddy viscosity A_τ is not a property of fluid, it is a property of flow, but not like viscosity μ_f, which is a property of fluid.* The apparent (virtual) kinematic viscosity is defined as $\epsilon_t = A_\tau/\rho_f$, which is analogous to the kinematic viscosity $\nu_f = \mu_f/\rho_f$. Therefore, the equations for shearing stress are given by $\tau_l = \rho_f \nu_f \dfrac{d\bar{u}}{dy}$ for laminar flow and $\tau_t = \rho_f \epsilon_t \dfrac{d\bar{u}}{dy}$ for turbulent flow. In the case of incompressible, two-dimensional turbulent boundary layer equation (6.30a) including equation of continuity are obtained as:

$$\begin{aligned} &\bar{u}\frac{\partial \bar{u}}{\partial x} + \bar{v}\frac{\partial \bar{u}}{\partial y} = -\frac{1}{\rho_f}\frac{\partial \bar{p}}{\partial x} + \frac{\partial}{\partial y}\left(\left(\nu_f + \epsilon_t\right)\frac{\partial \bar{u}}{\partial y}\right) \\ &\frac{\partial \bar{u}}{\partial x} + \frac{\partial \bar{v}}{\partial y} = 0 \end{aligned} \tag{6.36}$$

which correspond to the equations for laminar boundary layer flow with identical boundary conditions. Here the eddy viscosity ϵ_t is not known, that is the property of flow depending on random fluctuations in fluid flow. However, it is necessary to find out the empirical relations or models between the eddy viscosity ϵ_t and the mean velocity $\bar{u}$. The thickness of the viscous sub-layer is approximately equal to 0.1%–1% of the total thickness of boundary layer. The mean velocity is determined from the laminar shear stress τ_l and is given by $\tau_l = \rho_f \nu_f \dfrac{d\bar{u}}{dy}$ implies $\nu_f \dfrac{d\bar{u}}{dy} = u_*^2$, using turbulent shear stress $\tau_t = 0$ at $y = 0$. After integration, the dimensionless mean-velocity $\bar{u}^+ = y^+$. This is valid for $y^+ < 5$, which is known as the hydraulically smooth region. In the turbulent part of the boundary layer, where laminar shear stress τ_l is small compared to the turbulent shear stress τ_t and can be neglected. Equation (6.35) reduces to $-\overline{u'v'} = u_*^2 = \epsilon_t \dfrac{d\bar{u}}{dy}$ which will be determined later. Here the eddy viscosity ϵ_t is connected with the transport of momentum from one layer to another. This transport is determined by the diffusion of fluid particles due to turbulence (Sumer, 2002).

6.6.2 Prandtl's Mixing Length Hypothesis

To develop the turbulence models, Prandtl (1925) assumed a simplified representation on the velocity fluctuations from the movement of individual fluid elements displaced due to the fluctuations by a mean distance l, called the *mixing length*, which is perpendicular to the main flow retaining their momentum. The mixing length may be represented as the mean free path in kinetic theory of gas. Empirical relationships between the mixing coefficient and mean velocity were assumed.

In developing the hypothesis, the simplest case of parallel flow in which velocity varies only along transverse to the flow direction, i.e., from streamline to streamline, is considered. The principal direction of flow is assumed parallel to x-axis and $\bar{u}=\bar{u}(y)$; $\bar{v}=0$; $\bar{w}=0$. As the fluid passes over the flat plate in turbulent motion, fluid particles coalesce into lumps which move bodily in both longitudinal and transverse directions, retaining their momentum parallel to x-axis. It is assumed that such a lump of fluid, which comes from a layer at $y_1 - l$ with velocity, $\bar{u}(y_1-l)$ is displaced over a distance l in the transverse direction (Figure 6.10). This distance l is known as *Prandtl's mixing length*. As the lump of fluid preserves its original momentum, its velocity in the new layer at y_1 is smaller than the velocity overcoming there. The difference in velocity is then given by:

$$\Delta u_1 = \bar{u}(y_1) - \bar{u}(y_1 - l) \approx l\left(\frac{d\bar{u}}{dy}\right)_1 \tag{6.37}$$

The last term is obtained by expressing the function $\bar{u}\,(y_1-l)$, in a Taylor series and neglecting all higher-order terms, that is,

$$\bar{u}(y_1 - l) = \bar{u}(y_1) - l\left.\frac{d\bar{u}}{dy}\right|_{y_1} + \frac{l^2}{2}\left.\frac{d^2\bar{u}}{dy^2}\right|_{y_1} + \cdots \tag{6.38}$$

for the transverse motion $v' > 0$ Similarly, a lump of fluid which arrives at y_1 from the lamina at $y_1 + l$ possesses a velocity $\bar{u}\,(y_1 + l)$ which exceeds that around it, the difference being

$$\Delta u_2 = \bar{u}(y_1 + l) - \bar{u}(y_1) \approx l\left(\frac{d\bar{u}}{dy}\right)_1 \tag{6.39}$$

for negative transverse velocity $v' < 0$. Hence the time-average of the absolute value of this fluctuation is obtained as

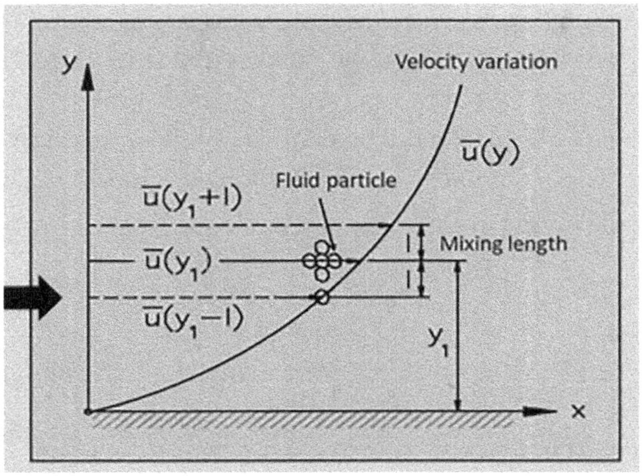

FIGURE 6.10 Prandtl's mixing length concept. (Modified from Schlichting, 1979.)

$$\left|\overline{u'}\right| = \frac{1}{2}\left(|\Delta u_1| + |\Delta u_2|\right) = l\left|\left(\frac{d\bar{u}}{dy}\right)_1\right| \tag{6.40}$$

Equation (6.40) leads to a physical interpretation that, the mixing length l is a distance in the transverse direction which is covered by an accumulation of fluid particles traveling with its original mean velocity to make the difference between its velocity and the velocity in the new lamina equal to the mean transverse fluctuation in turbulent flow. Here Prandtl's concept of mixing is analogous to the mean free path of the kinetic theory of gases. The main difference is that the latter deals with the microscopic motion of the molecules, whereas the present concept is on the macroscopic motion of lump of fluid particles. Following the argument of Prandtl, it is assumed that the transverse component v' is of the same order of magnitude as u' and it is as follows:

$$\left|\overline{v'}\right| = \text{const.}\left|\overline{u'}\right| = \text{const.}\, l\frac{d\bar{u}}{dy} \tag{6.41}$$

To investigate the value of $\overline{u'v'}$, it follows that the lumps of fluid which arrive at layer y_1 upward from below with $v' > 0$, give rise mostly to $u' < 0$, so that their product is $u'v' < 0$; and lumps with $v' < 0$ downwards from above are mostly associated with $u' < 0$ and the product is again negative. Thus, the time-average of $u'v'$ is different from zero and negative. Hence, we assume $\overline{u'v'} = -K\left|\overline{u'}\right|\left|\overline{v'}\right|$ with $0 < K < 1$ and it appears to be identical with the correlation factor. Then we obtain

$$\overline{u'v'} = -Kl^2\left(\frac{d\bar{u}}{dy}\right)^2 \tag{6.42}$$

Here K is constant included with the unknown mixing length l and we write

$$\overline{u'v'} = -l^2\left(\frac{d\bar{u}}{dy}\right)^2 \tag{6.43}$$

Consequently, the turbulent shearing stress is given by

$$\tau_t = -\rho_f\overline{u'v'} = \rho_f l^2\left(\frac{d\bar{u}}{dy}\right)^2 \tag{6.44}$$

This yields the turbulent model of the mixing length

$$\tau_t = \rho_f l^2\left|\frac{d\bar{u}}{dy}\right|\frac{d\bar{u}}{dy} \tag{6.45}$$

The absolute value has been taken to ensure that negative of $\frac{d\bar{u}}{dy}$ implies the negative of τ_t. This is known as *Prandtl's mixing length hypothesis* (Schlichting, 1979) which is very important in the calculation of turbulent flows. This equation is fruitfully applied to study the turbulent motion along the wall of pipe, channel, or plate, and to the problem of free turbulent jet flow.

Now comparing the expression in equation (6.45) with that of Boussinesq hypothesis equation (6.35), one gets the following expression for the apparent viscosity A_τ as:

$$A_\tau = \rho_f l^2\left|\frac{d\bar{u}}{dy}\right|, \tag{6.46}$$

and for the virtual kinematic viscosity ϵ_t, one gets

$$\epsilon_t = l^2 \left| \frac{d\bar{u}}{dy} \right|. \tag{6.46a}$$

where $\epsilon_t = A_\tau / \rho_f$. It is known from the experiments that turbulent drag is roughly proportional to the square of velocity, and the same is found from equation (6.46), if the mixing length l is assumed to be independent of the velocity. In fact, *the mixing length l is not a property of the fluid, but it is just a length.* In flows over a smooth wall, the mixing length l must vanish at the wall itself; and in flow over the rough wall, the mixing length l near the wall must tend to the value of the same order of magnitude as the solid protrusions.

6.6.3 von-Karman Similarity Hypothesis

The expression of mixing length l due to Prandtl's hypothesis is limited to the flow of shallow layers since the mixing length l is of the same order of magnitude as the distance above the reference surface, but the value of mixing length l due to von-Karman is allowed to depend on the space coordinate. von-Karman's hypothesis is that the turbulent fluctuations are similar at all points in the fluid flow, i.e., they differ only from point to point by time and length scales, following the similarity hypothesis. The time and length scales are replaced by time and velocity units. The basic hypothesis is that in a moving system with mean velocity, the fluctuations at all points differ only in length and time scales. The mixing length l is a function of distribution of mean motion of turbulent flow, but it is related only to the distance y from the wall. With reference to the moving coordinate system, the velocity components $u = \bar{u} + u'$, $v = 0 + v'$, after expanding by Taylor series in the neighborhood of a point at P, may be written as

$$u = \left.\frac{\partial \bar{u}}{\partial y}\right|_P \cdot y + \left.\frac{\partial^2 \bar{u}}{\partial y}\right|_P \cdot \frac{y^2}{2} + + \cdots + u', \; v = v' \tag{6.47}$$

According to similarity hypothesis, the relations are given by

$$x = l\xi, \; y = l\eta, \; u' = Au_1', \; v' = Av_1' \tag{6.48}$$

where the quantities ξ, η, u_1', v_1' for the amplitude of disturbance and l and A for length and velocity scales. Substituting these relations in the above equations, one gets

$$u = \left.\frac{\partial \bar{u}}{\partial y}\right|_P \cdot l\eta + \left.\frac{d^2 \bar{u}}{dy^2}\right|_P \cdot \frac{l^2\eta^2}{2} + \cdots + Au_1', \; v = Av_1' \tag{6.49}$$

The similarity conditions require

$$l \sim \frac{\left.\frac{d\bar{u}}{dy}\right|_P}{\left.\frac{d^2\bar{u}}{dy^2}\right|_P} \text{ and } A \sim l \left.\frac{d\bar{u}}{dy}\right|_P \tag{6.50}$$

von-Karman suggested that the mixing length l takes the form as

$$l = \kappa_0 \frac{d\bar{u}}{dy} \bigg/ \frac{d^2\bar{u}}{dy^2} \tag{6.51}$$

where κ_0 is the von-Karman constant generally taken as 0.4. This equation designates that the mixing length l is a local function and depends only on the velocity distribution in the neighborhood of a particular point. The turbulent shearing stress (6.45) in this case is written as

$$\tau_t = \rho_f \kappa_0^2 \left(\frac{\frac{d\bar{u}}{dy}}{\frac{d^2\bar{u}}{dy^2}} \right)^2 \left(\frac{d\bar{u}}{dy} \right)^2 = \rho_f \kappa_0^2 \frac{\left(\frac{d\bar{u}}{dy} \right)^4}{\left(\frac{d^2\bar{u}}{dy^2} \right)^2} \tag{6.52}$$

This equation is known as von-Karman's law of friction. Both von-Karman law of friction and Prandtl's law are applied to the problem of finding the velocity distribution in a rectangular channel. These laws are called the *Universal laws* of fundamental importance to consider the velocity distributions in rectangular as well as circular channels.

Example 6.1

In a pipe of 0.3 m diameter, turbulent flow of water occurs and the velocity profile is observed to be approximately u = 1.5 + ½ log (y), where velocity is in m/s and distance y in m from the wall. Analytically, the shear stress was obtained as 14 N/m² at a point = 0.1 m. For the given conditions, estimate the values of turbulent viscosity, mixing length and turbulent constant.

Solution

The velocity profile is given as: u = 1.5+½ log (y). Therefore, $du/dy = 1/(2y)$; $d^2u/dy^2 = 1/(2y^2)$
At y = 0.05 m, $du/dy = 1/(2y) = 1/(2*0.1)= 5\ s^{-1}$; $d^2u/dy^2 = 1/(2y^2)= 1/(2*0.1^2) = 50$

a. From the Boussinesq relation for turbulent shear stress and eddy viscosity, $\tau_t = \mu_f \frac{du}{dy}$. Therefore, $14 = \mu_f * 5;\ \mu_f = 2.8\ \text{Ns/m}^2$.

b. From Prandtl's mixing length equation, $\tau_t = \rho_f l^2 \left|\frac{d\bar{u}}{dy}\right| \frac{d\bar{u}}{dy}$. Therefore, $14 = 1000 \times l^2 \times 5^2$ Thus, the mixing length is: 0.0234 m.

c. From von-Karman's law of friction, $\tau_t = \rho_f \kappa_0^2 \frac{\left(\frac{d\bar{u}}{dy} \right)^4}{\left(\frac{d^2\bar{u}}{dy^2} \right)^2}$.

d. Therefore, $14 = 1000\ \kappa_0^2 \frac{(5)^4}{50^2}$; Thus, turbulent constant is: 0.236.

6.7 VELOCITY DISTRIBUTIONS OVER THE FLAT SURFACE

Experimental results show that the turbulent velocity profiles, such as pipe flows, open channel flows and boundary layer flows, over a smooth boundary are difficult to describe the flow phenomena in the entire boundary layer. Therefore, it is necessary to treat the turbulent boundary layer as a composite layer consisting of inner and outer regions, even for the flow over the flat plate (Coles, 1956). The *inner region* being defined where the turbulence is directly affected by the bed surface and the *outer region* where the flow is only indirectly affected by the bed through its shear stress (Figure 6.11), in which $u^+ = \bar{u}/u_*$ and $y^+ = u_* y/v_f$ are the inner variables, which is linear. For further clarification, Figure 6.12 is also provided. The inner region can again be divided into a viscous sub-layer, a buffer layer or transitional layer and an overlap layer or fully turbulent layer. Since the variation from the inner region to the outer region is gradual, the overlap is also a part of the outer region (Kundu, 1990). In transitional region (buffer layer), both laminar and turbulent shear stresses are important. Prandtl's mixing length and Boussinesq's eddy viscosity apply to the fully turbulent region. Flow in the buffer region is in a state of transition. Thus, the outer region can also be further divided into the overlap layer and a wake layer (Nezu and Nakagawa, 1993).

In a broad sense, the flow layer over a flat plate can be divided into three layers:

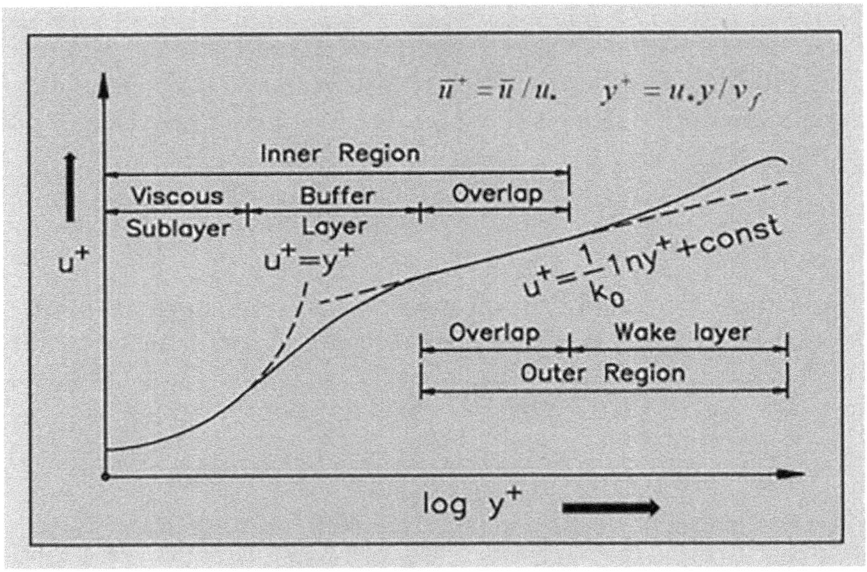

FIGURE 6.11 Sketch of velocity profiles in open channel. (Modified from Nezu and Nakagawa, 1993.)

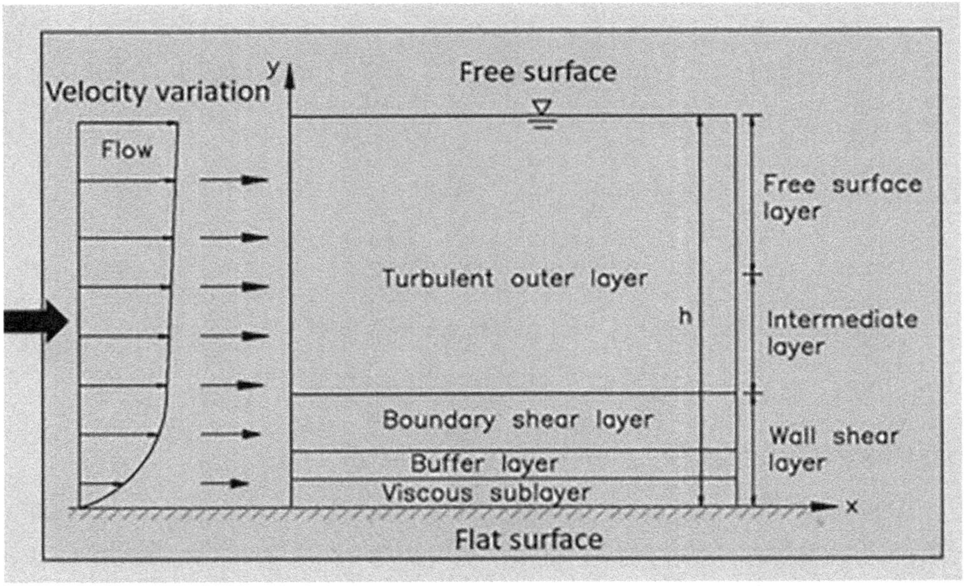

FIGURE 6.12 Categorization of flow in different layers.

1. **Wall shear layer** ($y/h < 0.2$): This is the inner layer of a fully developed turbulent flow closest to the channel boundary. The length and velocity scales are $\frac{v_f}{u_*}$ (smooth bed) or d (rough-bed), and u_* respectively, where v_f is the kinematic viscosity of water and u_* is the friction or shear velocity and d is the roughness height. The layer is considered as a turbulent bursting horizontal near-boundary flow region, where the turbulent kinetic energy (TKE) production rate exceeds the TKE dissipation rate. In this layer, the law of the wall is conserved. The boundary roughness plays a significant role in defining the velocity distribution in the viscous sub-layer, which was first observed by Nikuradse in pipe flow (Nikuradse, 1933). He introduced the concept of equivalent roughness, k_s called Nikuradse's equivalent roughness or equivalent sand roughness (Schlichting and Gersten, 2000). Based on the experimental findings on roughness, the flow regime is classified as hydraulically smooth (shear Reynolds number, $R_s = \frac{u_* k_s}{\upsilon_f \leq 5}$), hydraulically rough ($R_s \geq 70$) and hydraulically transitional flow ($5 < R_s < 70$).
2. **Free surface layer** ($0.6 < y/h \leq 1$): This is the layer closest to the free surface of the channel of a fully developed turbulent flow. The length scale is defined by the flow depth h or the boundary layer thickness δ and the velocity scale by maximum stream-wise velocity $U_{\max}$. The TKE dissipation rate exceeds the TKE production rate within this layer. Therefore, the TKE is transmitted from the wall shear layer to this layer by a TKE diffusion process.
3. **Intermediate layer** ($0.2 \leq y/h \leq 0.6$): This layer is influenced by the composite characteristics of the wall shear layer and the free surface layer. The length and velocity scale are y and $\sqrt{\frac{\tau}{\rho_f}}$, respectively (where τ is the total shear stress, and ρ_f is the fluid density). Viscous effects are negligible in the layer. The TKE production rate equals the TKE dissipation rate.

In brief, the flow domain in the wall shear turbulence can be divided into four layers: (1) viscous sub-layer, (2) *buffer layer or transitional region,* (3) *fully turbulent region or overlap or intermediate layer,* and (4) *wake layer or outer region.* In the viscous sub-layer, the stresses are mainly viscous, since the turbulent fluctuations and mean velocities become zero at the wall. The viscous region very-near the wall is not uniform with time or distance along the wall. The nonuniform thickness near the wall is time-averaged, and that thickness is denoted by y_o called viscous sub-layer. Thus, for $y \leq y_0$, the flow may be assumed to be viscous. In the region $y > y_0$, the effect of the viscosity on the flow decreases gradually with distance from the wall, and it goes to a region where the flow is completely turbulent, and the effect of viscosity is negligibly small. The intermediate or turbulent region or transitional region is defined as where the total shear stress is partly viscous and partly turbulent. If an average distance y_t from the wall is considered, beyond which the flow is fully turbulent, the transitional region is indicated as $y_0 < y \leq y_t$. Both the thicknesses of the viscous sub-layer and the transitional layer are quite small compared to that of fully turbulent region. In the outer region, the turbulent boundary layer contains 80%–90% of the boundary layer thickness. The velocity profile in each layer is reviewed below:

6.7.1 Linear Law in the Viscous Sub-Layer in the Inner Layer

We assume that the velocity profile near the bed can be expressed as a Taylor series, i.e.

$$\bar{u} = (\bar{u})_{y=0} + \left(\frac{d\bar{u}}{dy}\right)_{y=0} \cdot y + \frac{1}{2}\left(\frac{d^2\bar{u}}{dy^2}\right)_{y=0} \cdot y^2 + \cdots \tag{6.53}$$

where $\bar{u}$ is the time-mean velocity in the flow direction; and y is the distance from the bed. The no-slip condition at the bed requires that

$$(\bar{u})_{y=0} = 0 \tag{6.54}$$

According to Newton's friction law, $\left(\dfrac{d\bar{u}}{dy}\right)_{y=0}$ is related to the bed shear stress τ_0, i.e.

$$\left(\frac{d\bar{u}}{dy}\right)_{y=0} = \frac{u_*^2}{\nu_f} \tag{6.55}$$

in which $u_* = \sqrt{\tau_0/\rho_f}$ is the friction velocity with ρ_f is the water density. Substitution of (6.54) and (6.55) into equation (6.53) and neglecting the higher-order terms yield the velocity profile near the bed as

$$\bar{u}^+ = y^+ \tag{6.56}$$

in which $\bar{u}^+ = \bar{u}/u_*$ and $y^+ = u_* y/\nu_f$ are the inner variables, which are linear. Experiments (Schlichting, 1979) show that the above equation is valid in the range of $0 \le y^+ \le 5$. Here y^+ is a kind of Reynolds number based on the wall distance y. The quantities u_* and ν_f are called the inner flow parameters or wall parameters (Figure 6.12).

Again, consider a boundary layer flow in an open channel. The flow in the channel is driven by gravity force $g_x = gS$, where S is the slope of the channel. The x-component of Reynolds equation (6.30a) with total shear stress τ is written as

$$\frac{\partial}{\partial y}\left(\mu_f \frac{\partial \bar{u}}{\partial y} - \rho_f \overline{u'v'}\right) = \frac{d\tau}{dy} = -\rho_f g_x S = -\gamma S \tag{6.57}$$

where γ is the specific weight of the fluid. Integration of (6.57) gives

$$\tau = -\gamma S y + \text{constant} \tag{6.58}$$

Considering the boundary condition at the free surface $y = h$ as shear stress τ equal to zero, the constant is as γSh. Then the shear stress τ is given by

$$\tau = \gamma Sh\left(1 - \frac{y}{h}\right) \tag{6.59}$$

At the wall $y = 0$, the wall shear stress $\tau_0 = \gamma Sh$. Hence, the shear stress τ is given by

$$\tau = \tau_0\left(1 - \frac{y}{h}\right) \tag{6.60}$$

Therefore, shear stress close to the wall is ~ τ_0. This thin layer is known as the constant stress layer. The velocity close to the wall is given by *the law of wall* as Equation (6.56).

6.7.2 Velocity Distribution from Prandtl's Mixing Length Theory

A velocity distribution law can be deduced from Prandtl's hypothesis for the turbulent shearing stress equation (6.45), which is

$$\tau_t = \rho_f l^2 \left|\frac{d\bar{u}}{dy}\right| \frac{d\bar{u}}{dy} \tag{6.61}$$

If a turbulent flow over a smooth flat wall is considered, according to Prandtl's hypothesis in the neighborhood of the wall, the mixing length l must be proportional to the distance y from the wall with a velocity $\bar{u}(y)$; and l must vanish at the wall because the transverse motions are inhibited by its presence. Therefore, it is assumed the simplest relation between l and y is given by

$$l = \kappa_0 y \tag{6.62}$$

where κ_0 is the proportionality constant, known as von-Karman constant and is determined from the experiments. This assumption is reasonable because the turbulent shearing stress at the wall is zero due to the disappearance of the fluctuations. Therefore, using equation (6.62), Prandtl's turbulent shearing stress is given by:

$$\tau = \rho_f \kappa_0^2 y^2 \left(\frac{d\bar{u}}{dy} \right)^2 \tag{6.63}$$

An additional assumption was also introduced by Prandtl that the shearing stress should be constant at the wall, i.e., $\tau = \tau_0$, where τ_0 denotes the shearing stress at the wall. Introducing the friction velocity $u_{*0} = \sqrt{\tau_0 / \rho}$, one gets

$$\frac{d\bar{u}}{dy} = \frac{u_{*0}}{\kappa_0 y} \tag{6.64}$$

On integrating

$$\bar{u} = \frac{u_{*0}}{\kappa_0} \ln y + C \tag{6.65}$$

where C is the integrating constant to be determined from the condition at the wall. This equation is valid only in the neighborhood of the wall because of the assumption that $\tau =$ constant at the wall. The constant of integration C is determined from the condition that the turbulent velocity distribution must join the laminar velocity distribution in the immediate neighborhood of the wall, where the laminar and turbulent shearing stresses are of the same order of magnitude. Therefore, the constant C is determined from the condition that $\bar{u} = 0$ at a certain distance $y = y_0$ from the wall. Using the dimensional argument, we find that the distance y_0 is proportional to the ratio $\frac{v_f}{u_{*0}}$, and the velocity is obtained as:

$$\bar{u} = \frac{u_{*0}}{\kappa_0} (\ln y - \ln y_0) = \frac{u_{*0}}{\kappa_0} \ln \frac{y}{y_o} \tag{6.66}$$

The distance y_0 is of the order of magnitude of the laminar sub-layer thickness. We can write the distance y_0 as

$$y_0 \propto \frac{v_f}{u_{*0}} = \frac{\beta v_f}{u_{*0}} \tag{6.67}$$

where β is the proportionality constant. Substituting equation (6.67) for y_0 in equation (6.66), one gets,

$$\frac{\bar{u}}{u_{*0}} = \frac{1}{\kappa_0} \ln \eta - \frac{1}{\kappa_0} \ln \beta \tag{6.68}$$

where $\eta = y u_{*0} / v_f$ is the dimensionless distance. Equation (6.68) is the dimensionless, logarithmic, universal velocity-distribution law, and shows that the velocity is a function of the dimensionless wall distance $\eta = yu_{*0}/\nu_f$, which is a Reynolds number based on the wall distance y and the friction velocity u_{*0} at the wall. Equation (6.68) contains two empirical constants κ_0 and β, where κ_0 is the universal constant, which is independent of the nature of the wall (rough or smooth) $\kappa_0 = 0.4$. The second constant β depends on the nature of the wall surface, whether the wall surface is rough or smooth. Hence, the universal velocity distribution law equation (6.68) has the form as:

$$\frac{\bar{u}}{u_{*0}} = A_1 \ln \eta + c \tag{6.69}$$

where $A_1 = \kappa_0^{-1}$, and $c = -\kappa_0^{-1} \ln \beta$ are the constants. The constant c is to be determined from the nature of the wall. It may be mentioned here that value of β is 0.111 for smooth surface. Equation (6.69) is derived for the case of flat wall (rectangular open channel flow), which also holds its fundamental importance for flows through circular pipes, with the expectation that equation (6.69) leads to a good agreement with experimental data. Nikuradse (1930) performed the experiments and it was shown that excellent agreement was obtained not only for points near the wall but also for the whole region up to axis of the pipe, and the numerical values of the constants are found to be as $A_1 = 2.5$, $c = -\left(\kappa_0^{-1}\right)\ln \beta = 5.5$ for smooth surface. The universal velocity distribution law for very large Reynolds numbers *for smooth surface* has the form as

$$\left.\begin{aligned} \frac{\bar{u}}{u_{*0}} &= 2.5 \ln \eta + 5.5 \\ \frac{\bar{u}}{u_{*0}} &= 5.75 \log \eta + 5.5 \end{aligned}\right\} \tag{6.70}$$

This equation is valid only in regions, where the laminar shear stress can be neglected in comparison to that of turbulent flow. In the immediate neighborhood of the wall, where the turbulent shear stress decreases to zero and the laminar shear stress is prominent, this law may deviate. The layer in which the log-law is satisfied is called the *logarithmic layer.* The lower level of the log-law layer is often taken as $\eta = 70$ in the literature. The layer of fluid which lies within the region of the viscous sub-layer and log-law layer is called the buffer layer $5 < \eta < 70$, where both viscosity and turbulence effects are equally important. To satisfy the logarithmic law, we need to meet these two conditions, such as, (1) $y << h$ should be in the constant stress layer $\sim \tau_0$, where h is the water depth for open channel flow, the radius for pipe flow, or the boundary-layer thickness over the flat wall, and (2) y should be very large so that $\frac{d\bar{u}}{dy}$ is independent of viscosity. Therefore, $\eta > 30$ to 70 is necessary. These two conditions should be satisfied for fitting of the logarithmic law equation (6.70) of velocity distribution. Reichardt (1951) extended this measurement for very small distances from the wall in the channel or pipe flow (Figure 6.13). The curve (A) in Figure 6.13 corresponds to laminar flow for which $\tau_0 = \mu_f \bar{u}/y$, where $\tau_0 = \rho_f u_{*o}^2$. Then one gets $\frac{\bar{u}}{u_{*o}} = \frac{yu_{*o}}{\nu_f}$, or $\frac{\bar{u}}{u_{*0}} = \eta$ for laminar flow. The curve (B) represents the transition from the laminar sub-layer to the turbulent boundary layer. The curve (C) in Figure 6.13 represents the turbulent flow over the boundary layer. A power law velocity distribution, known as 1/7-th power velocity-distribution law (Schlichting, 1979), derived from Blasius's resistance formula is given by $\frac{\bar{u}}{u_{*0}} = 8.74\ \eta^{\frac{1}{7}}$, which was compared with the Nikuradse's experiments (curve D). To obtain a better agreement with the measured

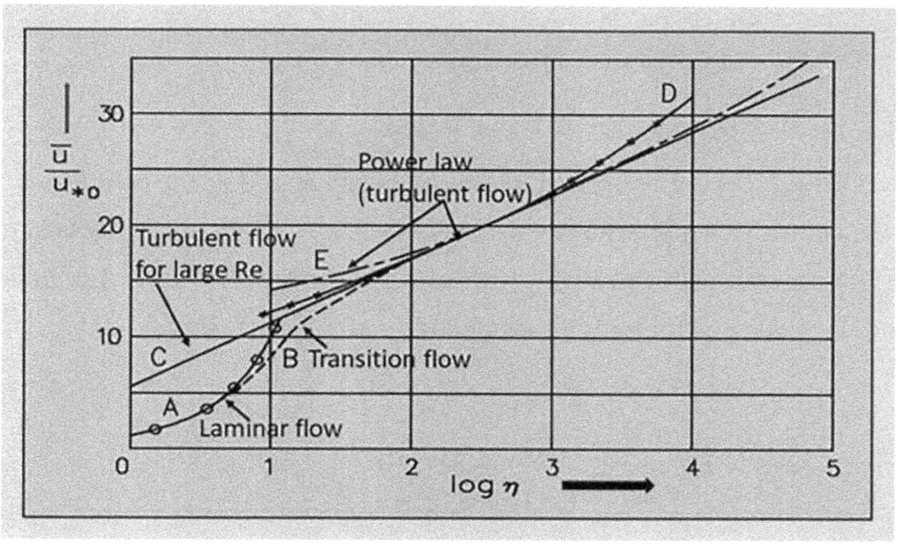

FIGURE 6.13 Velocity variation for various conditions: Curve (A) corresponds to the laminar flow. Curve (B) for the transition from laminar sub-layer to turbulent boundary layer, curve (C) for turbulent flow over boundary layer for large Reynolds number, and curves (D) and (E) represent power law. (Modified from Schlichting, 1979.)

data, the exponent should be little less instead of 1/7 th by 1/10 th, which comes to $\frac{\bar{u}}{u_{*0}} = 11.5\ \eta^{\frac{1}{10}}$ for higher Reynolds number (curve E).

From this, it can be seen that for values $\eta < 5$, the contribution from turbulent friction may be neglected compared to the laminar friction. In the range $5 < \eta < 70$, both contributions (laminar and turbulent frictions) are of the same order of magnitude, whereas for $\eta > 70$, the laminar contribution is negligible compared to the turbulent friction. Therefore

a. purely laminar friction $\eta < 5$
b. laminar-turbulent friction $5 < \eta < 70$ (6.71)
c. purely turbulent friction $\eta > 70$.

Hence, the thickness of the laminar sub-layer is seen to be equal to

$$\frac{u_* \delta_l}{\nu_f} \approx 5 \tag{6.72}$$

It may be mentioned here that the turbulence from the main body of the flow can enter into the thin layer, called viscous sub-layer. The effect of turbulence may be visible in the form of turbulent fluctuations within the viscous sub-layer.

Alternatively, even though equation (6.65) is valid near the wall, an attempt is made to use it for the whole region up to the depth $y = h$ analogous with that of von Karman's log-law. Since at $y = h, \bar{u} = \bar{u}_{max}$, we have

$$\frac{\bar{u}_{max} - \bar{u}}{u_{*0}} = \frac{1}{\kappa_0} In\frac{h}{y} = -\frac{1}{\kappa_0} \ln\frac{y}{h} \tag{6.73}$$

where y is the distance from the wall. Equation (6.73) is known as universal velocity-defect law due to Prandtl (Schlichting, 1979). This velocity law is comparable to von-Karman logarithmic velocity law (described later). The only difference between these two laws is that von-Karman assumed a linear shear stress distribution with mixing length given by equation (6.50), whereas Prandtl assumed a constant shear stress with mixing length $l \sim y$ given by equation (6.62).

6.7.3 Velocity-Distribution from von-Karman's Hypothesis

If the pressure gradient along x-axis is assumed to be constant and shearing stress τ is a linear function of width of the channel, then it follows $\tau = \tau_0 \frac{y}{h}$, where τ_0 denotes shear stress at the wall, and h is the half-width of the channel with center at $y = 0$. To determine the velocity-distribution law using the von-Karman similarity rule equation (6.52) to equation $\tau = \tau_0 \frac{y}{h}$, one gets as:

$$\frac{\tau_0}{\rho_f}\frac{y}{h} = \kappa_0^2 \frac{\left(\frac{d\bar{u}}{dy}\right)^4}{\left(\frac{d^2\bar{u}}{dy^2}\right)^2} \tag{6.74}$$

After simplification of the above equation and using the friction velocity at the wall $u_* = \sqrt{\frac{\tau_0}{\rho}}$, the above equation is obtained as

$$\left(\frac{d\bar{u}}{dy}\right)^2 = \frac{u_*}{\kappa_0}\sqrt{\frac{y}{h}}\left(\frac{d^2\bar{u}}{dy^2}\right) \tag{6.75}$$

Now integrating twice and determination of integrating constants from the condition $\bar{u} = \bar{u}_{\max}$ at $y = 0$, it is obtained as:

$$\bar{u} = \bar{u}_{\max} + \frac{1}{\kappa_0}\sqrt{\frac{\tau_0}{\rho_f}}\left[\ln\left\{1 - \sqrt{\frac{y}{h}}\right\} + \sqrt{\frac{y}{h}}\right] \tag{6.76}$$

The equation can be written in dimensionless form as:

$$\frac{\bar{u}_{\max} - \bar{u}}{u_*} = -\frac{1}{\kappa_0}\left[\ln\left\{1 - \sqrt{\frac{y}{h}}\right\} + \sqrt{\frac{y}{h}}\right] \tag{6.77}$$

This is the universal velocity distribution law derived by von-Karman (1930). From equation (6.77), it is observed that at the wall $y = h$, the velocity goes to infinity indicating breakdown of assumption, where the molecular friction is neglected compared to apparent turbulent friction because in the neighborhood of the wall turbulent flow goes to the laminar sub-layer. Moreover, it is remarkable to note that the universal velocity-distribution law does not contain either roughness or the Reynolds number explicitly. Therefore, the velocity-distribution law is known as *velocity defect law* (Figure 6.14).

6.7.4 Velocity Distribution from Cole's Law of Wake Strength for the Whole Layer

A classic logarithmic velocity distribution in a pipe flow with zero-pressure gradient boundary layer was first developed by Prandtl (1925), using mixing length hypothesis (that is called Prandtl mixing length theory). Later on, von-Karman (1930) attempted to establish the universal velocity distribution law based on the similarity rule (Schlichting, 1979). Keulegan (1938) suggested that the logarithmic velocity distribution may be applied to the open channel flow throughout the depth. For fully developed open-channel flow, the vertical mean velocity in the inner region ($y \leq 0.2h$, where y is the distance

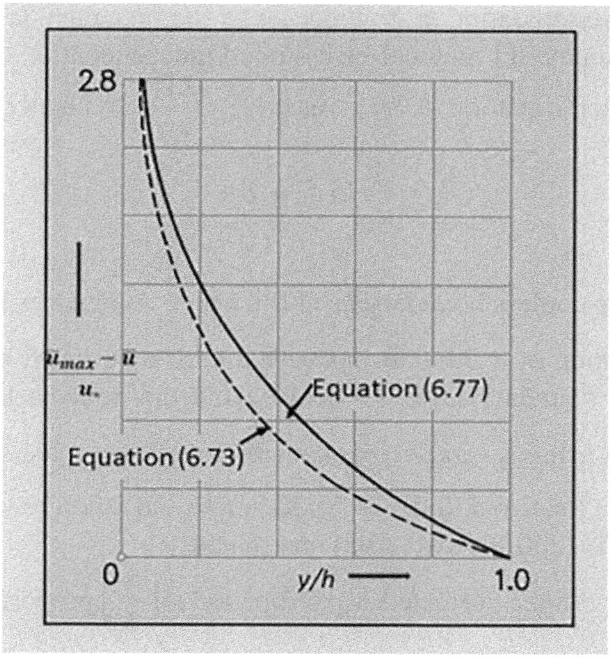

FIGURE 6.14 Velocity distributions for turbulent flow: Curve (1) for equation (6.77), and Curve (2) for equation (6.73). (Modified from Schlichting, 1979.)

from the bed and h is the flow depth) shows log-law. But this log-law deviates at the outer region. This deviation cannot be always modified by adjusting the von-Karman constant κ_0 or by the additive constant present in the log law. The departure from the logarithmic velocity profiles was observed with the distance from the boundary. This phenomenon was first noticed by Laufer (1954) in his experiment for pipe flow in the outer region. He found that the experimental data near a pipe axis systematically deviate from the log law. Subsequently, in his experiments Coles (1956) pointed out that similar deviations exist near the upper boundary of all boundary layer flows including open-channel flows; and he confirmed this behaviour for boundary layers and suggested the law of the wake. Coles (1956) proposed a remarkable mean-velocity distribution in both inner and outer regions using a concept of correction function called law of wake strength and is given by

$$\frac{\bar{u}}{u_*} = \frac{1}{\kappa_0} \ln \frac{y u_*}{\nu_f} + B + \omega\left(\frac{y}{\delta}\right) \tag{6.78}$$

where $\omega\left(\frac{y}{\delta}\right)$ is the correction function to the velocity profile. The correction function $\omega\left(\frac{y}{\delta}\right)$ is an empirical fit to the measured velocity profiles proposed by Hinze (1975) to describe the wake function by the following equation as:

$$\omega\left(\frac{y}{\delta}\right) = \frac{2\Pi}{\kappa_0} \sin^2\left[\left(\frac{\pi}{2}\right)\left(\frac{y}{\delta}\right)\right] \tag{6.79}$$

where Π is the Coles parameter expressing the strength of the wake function does not depend on y/δ. This Coles parameter Π is related to the local friction coefficient $c_f = 2u_*^2/U^2$, where U is the velocity at the edge of the boundary layer $y = \delta$. Equation (6.78) is known as the *law of the wake*, which expresses the similarity with wake flow. Experimental evidence suggests that this law is of universal character, which is applicable to flows with and without pressure gradient. For a boundary layer with zero pressure-gradient, the parameter Π was assumed independent of x and he obtained the parameter values as $\Pi = 0.55,\ B = 5.1$. If the separation of flow occurs in the boundary layer due to adverse pressure gradient $dp/dx > 0$, the parameter Π may not be assumed independent of x. At the edge of the boundary layer $y = \delta$, the expression Equation (6.79) gives $\omega(1) = \frac{2\Pi}{\kappa_0}$, and hence equation (6.78) reduces to

$$\frac{U}{u_*} = \frac{1}{\kappa_0}\ln\delta^+ + B + \frac{2\Pi}{\kappa_0} \tag{6.80}$$

where $\delta^+ = \frac{\delta u_*}{\nu_f}$ is the dimensionless boundary layer thickness. Again, it is important to note that equation (6.78) with the substitution of (6.79) does not give a derivative $d\bar{u}/dy = 0$ at the center line of the pipe or at the edge of the boundary layer $y = \delta$. To avoid this drawback, the Coles wake function $\omega\left(\frac{y}{\delta}\right)$ has been corrected using several expressions by many researchers such as, Granville (1976), Coleman and Alanso (1983), Nezu and Rodi (1986), Kironoto and Graf (1994), Lyn (2000), Ghoshal and Mazumder (2005), Guo et al. (2005), Absi (2011) and others.

The velocity expression with the corrected wake function $\omega\left(\frac{y}{\delta}\right)$ proposed by Granville (1976) is as follows:

$$\frac{\bar{u}}{u_*} = \frac{1}{\kappa_0}\ln\eta + B + \frac{1}{\kappa_0}\left[\Pi\left(1 - \cos\pi\frac{y}{\delta}\right) + \left(\left(\frac{y}{\delta}\right)^2 - \left(\frac{y}{\delta}\right)^3\right)\right] \tag{6.81}$$

The expression (6.81) is applicable for the region $\eta \geq 30$. However, this equation can be extended to the region $0 \leq \eta \leq 30$ by the following equations given by Thompson (1965) as:

$$\frac{\bar{u}}{u_*} = \frac{yu_{*o}}{\nu_f} = \eta \quad \text{for } \eta \leq 5 \tag{6.82a}$$

$$\frac{\bar{u}}{u_*} = 1.0828 - 0.414\ \ln\ \eta + 2.2661(\ln\ \eta)^2 - 0.324(\ln\ \eta)^3 \quad \text{for } 5 \leq \eta \leq 30 \tag{6.82b}$$

Combining equation (6.82) with equation (6.81), the velocity distribution is obtained for whole layer of flow over the flat surface. Using equation (6.81), the displacement boundary thickness δ^* from equation (5.50 of Chapter 5) and the momentum thickness θ from equation (5.53 of Chapter 5) are computed. The displacement boundary layer thickness δ^* is as:

$$\frac{\delta^*}{\delta} = \int_0^1 \frac{U - \bar{u}}{u_*}dy^+ = \frac{u_*}{\kappa_0 U}\left(\frac{11}{12} + \Pi\right) \tag{6.83}$$

and the momentum thickness θ^* is defined as

$$\frac{\theta^*}{\delta} = \int_0^1 \frac{\bar{u}}{U}\left(1 - \frac{\bar{u}}{U}\right)dy \tag{6.84}$$

Details calculations are available in Cebeci (2004).

Now let us compute the velocity-defect law, using Cole's wake strength expression (6.81). Subtracting equation (6.80) from equation (6.78), one gets a formula for modified velocity-defect distribution, using Coles wake strength parameter as:

$$\frac{U-\bar{u}}{u_*} = -\frac{1}{\kappa_0}\ln\left(\frac{y}{\delta}\right) + \frac{\Pi}{\kappa_0}\left(2-\omega\left(\frac{y}{\delta}\right)\right) \tag{6.85}$$

Of course, by setting $\Pi = 0$, equation (6.85) turns into the usual log-law. Therefore, Π describes the deviation from the log law in the outer region. Cole's wake function appears to be the most reasonable extension of this law for the outer region. Figure (6.15) shows the calculated and experimental data of Klebanoff (1954) and shows good agreement.

Coleman (1981, 1986) introduced a modified form of logarithmic law as follows:

$$\frac{\bar{u}}{u_*} = \left[\frac{2.303}{\kappa_0}\ln\eta + B\right] - \frac{\Delta\bar{u}}{u_*} + \frac{\Pi}{\kappa_0}\omega\left(\frac{y}{\delta}\right) \tag{6.86}$$

where B is the integrating constant, $\frac{\Delta\bar{u}}{u_*}$ is the channel roughness velocity reduction function. The term in the square bracket is the original log-law of the wall resulting from the Prandtl-Karman derivation. The velocity reduction and augmentation terms in the above equation, with the proper choice of numerical values of κ_0, A, and Π, will describe the entire velocity profile inside the boundary layer thickness δ in an open channel flow, except for the viscous sub-layer at the channel bed.

Nezu and Rodi (1986) re-examined the law of the wall and the velocity defect law, using their measured velocity data by LDA because of the application of the log-law to the open channels without detailed verification. As the Reynolds number becomes larger, the deviation from the log-law cannot be neglected in the outer region (say for $y/h > 0.6$). It was impossible to measure the mean velocity at the free surface using the LDA method because of the wakes generated near the free surface. The LDA laser beams were interrupted by small surface waves. This deviation can be expressed well by Cole's wake strength function which involves a Reynolds-number dependent parameter Π. The calculated values of Π are obtained from the best-fitted curve with the data of the outer region. Subsequently, they also modified the distributions of eddy viscosity ϵ_t and mixing length l, using Cole's wake strength parameter Π (for details, Nezu and Nakagawa, 1993). For flows with zero pressure gradients, the value of parameter Π is equal to 0.55 provided the momentum thickness Reynolds number should be greater than 5,000. In the equilibrium boundary layer, Π is constant depending on the strength of pressure gradient. In nonequilibrium boundary layers, the profile parameter Π becomes a function of x.

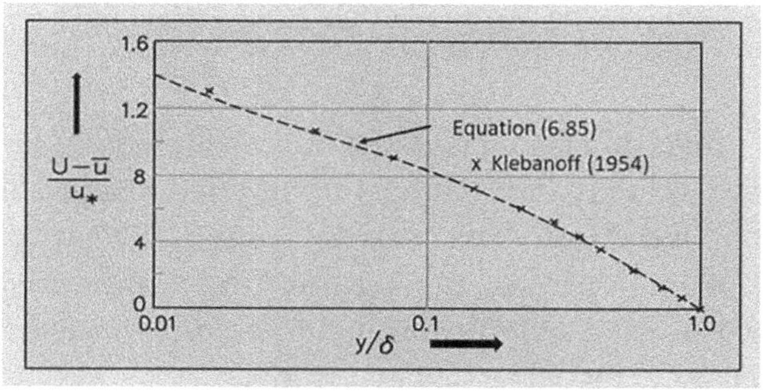

FIGURE 6.15 Velocity-defect law according to Cole's wake strength expression equation (6.85) with the experimental data of Klebanoff (1954).

Guo (1998) showed that the wake correction to the lag-law is not sufficient to match a velocity profile throughout the flow depth, particularly at the free surface. A boundary correction due to modified shear stress from the wind velocity at the water surface has been imposed on the log-wake law to obtain the velocity profile and the modified log-wake law is given by

$$\frac{U-\bar{u}}{u_*}=\frac{1}{\kappa_0}\ln\frac{y}{\delta}+\frac{2\Pi}{\kappa_0}\cos^2\left(\frac{\pi}{2}\frac{y}{\delta}\right)-\left[\frac{1}{\kappa_0}-\frac{1}{u_*}\left(\frac{d\bar{u}}{d\frac{y}{\delta}}\right)_{\frac{y}{\delta}=1}\right]\left(1-\frac{y}{\delta}\right) \tag{6.87}$$

The modified log-wake law has been applied to fit many experimental data and it shows a good agreement near the free surface. Guo et al. (2005) stated that the mean velocity $\bar{u}$ follows only an asymptotic to the free stream velocity U at the so-called boundary layer edge δ, i.e., $\bar{u}\to U$ at $y\to\delta$, and the velocity gradient $d\bar{u}/dy=0$ at $y\to\delta$ as a good approximation to satisfy the velocity at the boundary. To satisfy the zero velocity gradients at the boundary one must modify the boundary correction factor with a cubic term $-\frac{(y/\delta)^3}{3\kappa_0}$ in the velocity expression. Combining the cubic term $-\frac{(y/\delta)^3}{3\kappa_0}$ with equations (6.78) and (6.79), Guo et al. (2005) suggested the modified log-wake law velocity equation as:

$$\frac{\bar{u}}{u_*}=\frac{1}{\kappa_0}\ln\eta+B+\frac{2\Pi}{\kappa_0}\sin^2\left\{\left(\frac{\pi}{2}\right)\left(\frac{y}{\delta}\right)\right\}-\frac{1}{3\kappa_0}\left(\frac{y}{\delta}\right)^3 \tag{6.88}$$

This modified log-wake law (6.88) agrees well with the experimental data from the overlap region to the outer region of the boundary layer. This equation is different from the conventional log-law because of the zero-velocity gradient at the boundary, and the constant B in equation (6.88) accounts for the effect of the Reynolds number. To eliminate the effect of Reynolds number in (6.88), the free stream velocity U at the edge of the boundary $y\to\delta$ should be introduced to the modified log-wake law (6.88). Equation (6.88) is written as:

$$\frac{U}{u_*}=\frac{1}{\kappa_0}\ln\frac{\delta u_*}{v_f}+B+\frac{2\Pi}{\kappa_0}-\frac{1}{3\kappa_0} \tag{6.89}$$

Subtracting equation (6.88) from equation (6.89), the velocity-defect law from the modified log-wake law equation is given by:

$$\frac{U-\bar{u}}{u_*}=-\frac{1}{\kappa_0}\ln\left(\frac{y}{\delta}\right)+\frac{2\Pi}{\kappa_0}\cos^2\frac{\pi}{2}\frac{y}{\delta}-\frac{1}{3\kappa_0}\left(1-\left(\frac{y}{\delta}\right)^3\right) \tag{6.90}$$

Here both κ_0 and Π are the universal constants under the assumption of large Reynolds number.

Absi (2009, 2011) proposed velocity distributions in an open channel flow based on RANS equations using a *dip phenomenon* at the free surface as well as the Coles wake strength correction. In 3D open-channel flows with secondary currents, the log-wake law was unable to predict the velocity-dip-phenomenon. Absi (2009) suggested a suitable formula to predict the velocity distribution by adding a dip correction parameter (α) to the log-wake law and is given by:

$$\frac{\bar{u}}{u_*}=\frac{1}{\kappa_0}\ln\frac{y}{y_0}+B+\frac{2\Pi}{\kappa_0}\sin^2\left[\left(\frac{\pi}{2}\right)\left(\frac{y}{\delta}\right)\right]+\frac{\alpha}{\kappa_0}\ln\left(1-\frac{y}{\delta}\right) \tag{6.91}$$

This equation (6.91) is referred to as the simple dip-modified-log-wake law, which reverts to the log-wake law for $\alpha = 0$ in two dimensional open-channel flows. Subsequently, Absi (2011) assumed maximum velocity $U = U_{\max}$ at a dip distance $y = y_{\text{dip}}$ below the water surface and also obtained dip correction parameter $\alpha = \frac{1}{y_{\text{dip}}} - 1$, that means, $y_{\text{dip}} = (\alpha + 1)^{-1}$, using gradient of the velocity is equal to zero at the dip distance y_{dip}, and correspondingly the velocity distribution becomes as:

$$\frac{\bar{u}}{u_*} = \frac{1}{\kappa_0}\ln\frac{y}{y_0} + \frac{2\Pi}{\kappa_0}\sin^2\left[\left(\frac{\pi}{2}\right)\left(\frac{y}{\delta}\right)\right] + \frac{\alpha}{\kappa_0}\ln\left(1-\frac{y}{\delta}\right) - \frac{\alpha\pi\Pi}{\kappa_0}\int_{\frac{y_0}{\delta}}^{\frac{y}{\delta}} \frac{\frac{y}{\delta}}{1-\frac{y}{\delta}} \sin\pi\left(\frac{y}{\delta}\right) d\left(\frac{y}{\delta}\right) \quad (6.92)$$

Equation (6.92) is referred as the *full dip-modified-log-wake law*, which is different from other equations because of the consideration of the dip phenomenon occurred below the free surface. In determining model parameters, Tominaga and Nezu (1992) experimentally showed that B is about 5.29 for sub-critical flow while it decreases with Froude number for supercritical flow. As for the wake strength Π, Coles (1956) suggested a value of 0.55 for flat plate boundary layers. However, the previous investigators proposed the various values of Π. Coleman (1981, 1986) obtained $\Pi = 0.19$; Nezu and Rodi (1986) found $\Pi = 0.2$; Kirkgoz (1989) reported a value of $\Pi = 0.1$; Cardoso et al. (1989) observed $\Pi = -0.077$ in a flow over smooth bed; Kironoto and Graf (1994) stated that $\Pi = -0.08 \sim 0.04$ for flows over gravel bed; and Wang and Plate (1996) got $\Pi = -0.06 \sim 0.2$.

Velocity-dip phenomenon in an open channel flow is defined below the water surface at a height $y = \delta$, where the mean velocity is maximum. Above that height, velocity reduces due to the frictional effect with surface. In the narrow open-channel flow, where *aspect ratio* defined by the channel width to flow depth is less than five and near side walls or corner zones even for wide open channels (Vanoni, 1946), the maximum velocity occurs below the free surface producing the velocity-dip-phenomenon. This phenomenon, which was reported more than a century ago (Francis, 1878; Stearns, 1883), was observed both in open-channels and rivers. It is related to secondary currents generated in three-dimensional (3D) open-channel flows (Imamoto and Ishigaki, 1986; Wang and Cheng, 2005). Yang et al. (2004) provided an insight into the understanding of the velocity profile in the open channel with dip phenomenon based on the analysis of the RANS equations. They proposed an empirical model with dip-correction parameter, which is valid for narrow and wide channels. This law involves two logarithmic distances, one from the bed (i.e., the log law), and the other from the free surface with the dip-correction parameter. Bonakdari et al. (2008) proposed a sigmoid model for velocity profile in the outer region which was valid for both wide and narrow open channel flow with smooth boundaries, using the Reynolds averaged Navier–Stokes (RANS) equations. The proposed model is able to predict time-averaged primary velocity in the outer region of the turbulent boundary layer for both narrow and wide-open channels. This model is based on the knowledge of the aspect ratio and which involves a parameter depending on the position of the maximum velocity.

In open channel flow, for *aspect ratio* of the channel width to flow depth less than a certain value about six, the effects of side walls of a flume intrude near the free surface and develop a dip phenomenon in the velocity profiles (Yang et al., 2004). Furthermore, Yang et al. (2004, 2005) showed from the analysis of several researchers' velocity data (Yassin, 1953; Coleman, 1981; Knight et al., 1984 and others) that the velocity-dip phenomenon appeared for narrow channel even for smaller aspect ratio around 2.0. Cardoso et al. (1989) attributed the dip phenomenon due to weak secondary currents. According to Long et al. (1990) and Tachie et al. (2004), a two-dimensional flow in the central portion of the flume was achieved with a width/depth ratio ($\approx$2). Later on, Absi (2011) proposed an equation to predict the velocity-dip-phenomenon, i.e., the maximum velocity below the free surface, based on the

Reynolds-Averaged Navier–Stokes (RANS) equations and a log-wake modified eddy viscosity distribution. He reported that the velocity dip phenomenon, where the maximum velocity appeared below the water surface, occurred even if the aspect ratio was less than five.

6.7.5 Velocity Distribution Due to van Driest's Damping Concept to the Wall

Several models have been proposed for turbulent shear stress to describe the mean velocity distribution. van Driest (1956) proposed a remarkable expression for a mixing length l, modifying the expression of Prandtl's mixing length:

$$l = \kappa_0 y\left[1-\exp(-y/A)\right] \tag{6.93}$$

where κ_0 is the von-Karman constant (= 0.4), A is a damping–length constant defined as $A_* v_f / u_*$, and A is restricted to the incompressible turbulent boundary layer with zero pressure gradient. Parameter A is referred to as Van Driest damping parameter to be determined. If $y = 0$, mixing length $l \to 0$, that means, no turbulence mixing at the wall, and for large y, the mixing length $l \to xy$. Using Prandtl's mixing length equation (6.44) together with van Driest's mixing length equation (6.93), the total shear stress equation (6.34) can be written as:

$$(\kappa_0 y)^2\left[1-\exp\left(-\frac{y}{A}\right)\right]^2\left(\frac{d\bar{u}}{dy}\right)^2 + v_f\frac{d\bar{u}}{dy} = u_*^2 \tag{6.94}$$

In dimensionless form, equation (6.94) can be written as

$$\frac{du^+}{dy^+} = \frac{-1+(1+4a)^{\frac{1}{2}}}{2a} \tag{6.95}$$

where $a = (\kappa_0 y^+)^2\left[1-\exp\left(-\frac{y^+}{A_*}\right)\right]^2$, $y^+ = \frac{yu_*}{v_f}$, $A_* = 26$. Considering $\kappa_0 = 0.4$, and comparing the model with the experimental data at Reynolds number greater than 5,000, van Driest determined the value of the damping constant A_* as 26. After integration of the differential equation (6.95) with the condition $u^+ = 0$ at $y^+ = 0$, the normalized stream-wise mean velocity ($\bar{u}/u_*$) is found to be the van Driest (1956) velocity profiles as:

$$u^+ = 2\int_0^{y^+}\frac{dy^+}{1+\left\{1+4\kappa_0^2 y^{+2}\left[1-\exp\left(-y^+/A_*\right)\right]^2\right\}^{\frac{1}{2}}} \tag{6.96}$$

where $u^+ = \left(\frac{\bar{u}}{u_*}\right)$ and $y^+ = \frac{yu^*}{v_f}$ are the dimensionless stream-wise velocity and distance respectively. As y^+ approaches zero, equation (6.96) approaches the familiar linear equation $u^+ = y^+$ for the viscous sub-layer. This relation extends up to y^+ equal to about five. The velocity profile (6.96) approaches the logarithmic velocity profile for large y^+ values which applies above the buffer region and is given by

$$u^+ = \left(\frac{1}{\kappa_0}\right)\ln y^+ + 5 \tag{6.97}$$

where $\kappa_0(=0.4)$ is the von-Karman constant. It is observed from Figure 6.16 that the viscous sub-layer exists for $y^+ < 5$, buffer layer $y^+ < 30$, and beyond the logarithmic velocity distribution is approached. Equation (6.96) shows a continuous velocity distribution in the inner region of the turbulent boundary layer from the viscous sub-layer to the turbulent flow region through the transitional or buffer region. Therefore, under these conditions, the mean velocity shown by equation (6.96) agrees quite well with the experimental data of Laufer (1954), Klebanoff (1954), and Wieghardt (1944). It is remarkable that the van Driest velocity profile agrees well with the transitional of the buffer region $(5 < y^+ < 70)$. It is seen from the figure that for small values of $y^+ < 5$, the van Driest profile goes to linear velocity distribution in the viscous sub-layer, while for large values of $y^+ > 70$, it goes to the logarithmic velocity distribution in the log layer. The van Driest velocity profile agrees reasonably well with the measured velocity data including the buffer layer $5 < y^+ < 70$. Once the mixing length l and the mean velocity distribution $\bar{u}$ are known, the mixing co-efficient or eddy viscosity A_t from equation (6.46) and the turbulent shear stress τ_t from equation (6.35) can be obtained. Over the years the algebraic forms of eddy viscosity and mixing length have been explored for their accuracy with a range of turbulent shear flows, such as van Driest (1956), Cebeci and Smith (1968), Patankar and Spalding (1968), Cebeci and Chang (1978), Nezu and Rodi (1986), and others.

Cebeci and Chang (1978) proposed a mixing length similar to that of van Driest (1956) for transitional and rough surfaces and is given by

$$l_m = \kappa_0\left(y^+ + \Delta y^+\right)\left[1-\exp\left(-\frac{\left(y^+ + \Delta y^+\right)}{A_*}\right)\right] \tag{6.98}$$

where y^+ is the distance from the wall where the velocity $\bar{u}$ is zero at the top level of roughness elements; and Δy^+ is a coordinate displacement or shift, which is a function of equivalent sand roughness parameter $k_s^+(=u_* k_s/\nu_f)$ or roughness Reynolds number, and is given by:

$$\Delta y^+ = 0.9\left(\frac{\nu_f}{u_*}\right)\left[\sqrt{k_s^+} - k_s^+ \exp\left(-\frac{k_s^+}{6}\right)\right] \quad \text{for } 4.535 < k_s^+ < 2000 \tag{6.99}$$

where the lower limit of 4.535 corresponding to the upper bound of the hydraulically smooth region. The distance $y^+ + \Delta y^+$ is the distance measured from the theoretical wall. Similar to the case of the smooth wall, the velocity $\bar{u}$ can be obtained from the model equation (6.96) of velocity using mixing length l_m from (6.98) and is given by

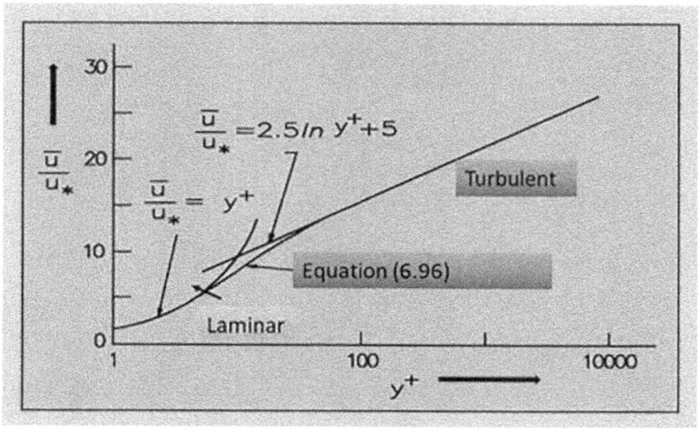

FIGURE 6.16 Velocity distribution using van Driest (1956) equation. (Modified from Sumer, 2002.)

$$u^+ = 2\int_0^{y^+} \frac{dy^+}{1+\left\{1+4\kappa_0^2\left(y^+ + \Delta y^+\right)^2\left[1-\exp\left(-\frac{y^+ + \Delta y^+}{A_*}\right)\right]^2\right\}^{1/2}} \tag{6.100}$$

equation (6.100) represents the velocity profiles including the smooth-wall profile and completely hydraulically rough wall. In Figure 6.17, it is also shown three distinct regions designated by A, B. and C and the momentum takes in different regions due to three different phenomena. Let the thickness of region A indicated as "Δy," may be called as "false layer" on the top of the rough surface, and region B shows the combined viscosity and turbulence, and region C shows only the turbulence. The flow over the roughness of the "false layer" is considered to be identical with that of a smooth wall. Therefore, the velocity over the false layer is as:

$$u^+ + u^+(\Delta y) = u_*\left[\frac{1}{\kappa_0}\ln\left(y^+ + \Delta y^+\right) + f(0)\right] \tag{6.101}$$

where $f(0)$ is the constant for the smooth wall, u^+ is the actual velocity, and $u^+(\Delta y)$ is the velocity at the false layer Δy. Considering the large value of y implying that $y^+ + \Delta y^+ \approx y^+$, the above equation becomes

$$u^+ = u_*\left[\frac{1}{\kappa_0}\ln\left(y^+\right) + f(0)\right] - u^+\left(\Delta y^+\right) \tag{6.102}$$

Here the velocity of the false layer $u^+\left(\Delta y^+\right)$ can be calculated from equation (6.96) replacing y by Δy, and is given by Rotta (1962) as

$$u^+(\Delta y) = 2u_*\int_0^{y^+} \frac{dy^+}{1+\left\{1+4\kappa_0^2\left(\Delta y^+\right)^2\left[1-\exp\left(-\frac{\Delta y^+}{A_*}\right)\right]\right\}^{1/2}} \tag{6.103}$$

The variation of equation (6.103) with respect to the roughness agrees fairly well with the expression (6.99) given by Cebeci and Chang (1978). The coordinate shift Δy^+ may be interpreted as the distance of the theoretical wall from the top of the roughness elements.

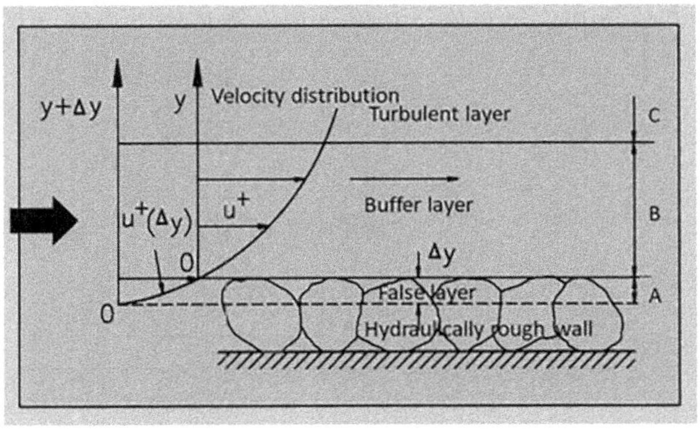

FIGURE 6.17 Flow over a rough wall. (Modified from Rotta, 1962.)

The mean velocity distribution for turbulence modeling across the entire section is proposed by Spalding (1961) for engineering applications. Here he proposed the law of wall equation (6.56) for viscous sub-layer in its inverse form as follows (Hinze, 1973):

$$y^+ = u^+ + C\left[\exp\left(\kappa_0 u^+\right) - 1 - \kappa_0 u^+ - \frac{\left(\kappa_0 u^+\right)^2}{2!} - \frac{\left(\kappa_0 u^+\right)^3}{3!} - \frac{\left(\kappa_0 u^+\right)^4}{4!}\right] \tag{6.104}$$

where the constant $C = \exp(-\kappa_0 B)$ with B from the constant of log-law. This equation (6.104) reduces to the linear distribution for small values of u^+, i.e. for small y^+, and to the logarithmic distribution at large $u^+$and y^+. Spalding suggested von-Karman constant $\kappa_0 = 0.4$ and $C = 0.111$, which corresponds to $B = 5.5$.

The velocity distribution (6.100) is valid for near the wall, but when it is applied to the entire cross-section, the predicted velocity profiles are apparently smaller than the measured data. This difference is corrected by Cole's wake strength function (6.79) defined earlier and it is adopted by Finley et al. (1966) and then Coleman and Alanso (1983). When the mean velocity equation (6.100) is added to Cole's function (6.79), the derivative of the additive velocity distribution with respect to y is not zero at the free boundary or at the centerline of a pipe. To avoid this drawback, Finley et al. (1966) modified Cole's wake function, and the final form of velocity distribution is as follows (Coleman and Alanso, 1983):

$$u^+ = \int_0^{y^+} \frac{2dy^+}{1 + \left\{1 + 4\kappa_0^2\left(y^+ + \Delta y^+\right)^2\left[1 - \exp\left(-\frac{y^+ + \Delta y^+}{A_*}\right)\right]^2\right\}^{1/2}} + \left(\frac{y}{\delta}\right)^2\left(1 - \frac{y}{\delta}\right) + \frac{2\Pi}{\kappa_0}\left(\frac{y}{\delta}\right)^2\left[3 - 2\left(\frac{y}{\delta}\right)\right] \tag{6.105}$$

Equation (6.105) is a universal model for the velocity distribution in the inner and outer regions of bounded shear flow. The equation represents the velocity profile over both smooth and rough boundaries in the coordinates used in Figure 6.18.

The second term on the right in equation (6.105) forces $du^+/dy^+ \to 0$ at $y^+ \to \delta^+$. The second term is the velocity growth, which varies with parameter Π. Equation (6.105) is clear-cut that a value of u^+ can be computed for a given value of y^+, if a value of k_s^+ can be specified in calculating the roughness shift Δy^+. Values of δ^+ and Π can also be specified for calculating the velocity in outer region. The values of these three parameters are not known, and in this case, the model represented by equation (6.105) is a parametric, so we need to predict the values of the parameters k_s^+, δ^+, and Π. For standard Nikuradse's sand-grain roughness, $k_s^+ (= u_* k_s / \nu_f)$, in which k_s is the particle size of the uniform sand comprising boundary roughness. Here k_s^+ is valid for the computation of velocity profiles over the beds of close-packed spheres with a planar bed of uniform sand. Hey (1979) has reviewed several methods for computing this roughness. Generally, in open channel flow, the boundary layer thickness, δ^+, is frequently assumed to be equal to the total flow depth, so that the parameter δ^+ could be replaced by y^+. However, in the majority of open channel flows, the boundary layer thickness δ^+ is actually considered as 80%–90% of depth *y*. Here Cole's wake strength parameter Π basically depends on two factors named as the driving pressure gradient and the degree of intermittency of the turbulence in the outer region. In gradually varied flows, the converging flow due to favorable pressure gradient tends to the lower values of Π, whereas for adverse pressure gradient, high values of Π show. Similarly, if the turbulence in the outer region is relatively strong with a low degree of intermittency, the values of Π will be low, while if the turbulence in the outer region is weak and intermittent, the value of Π will

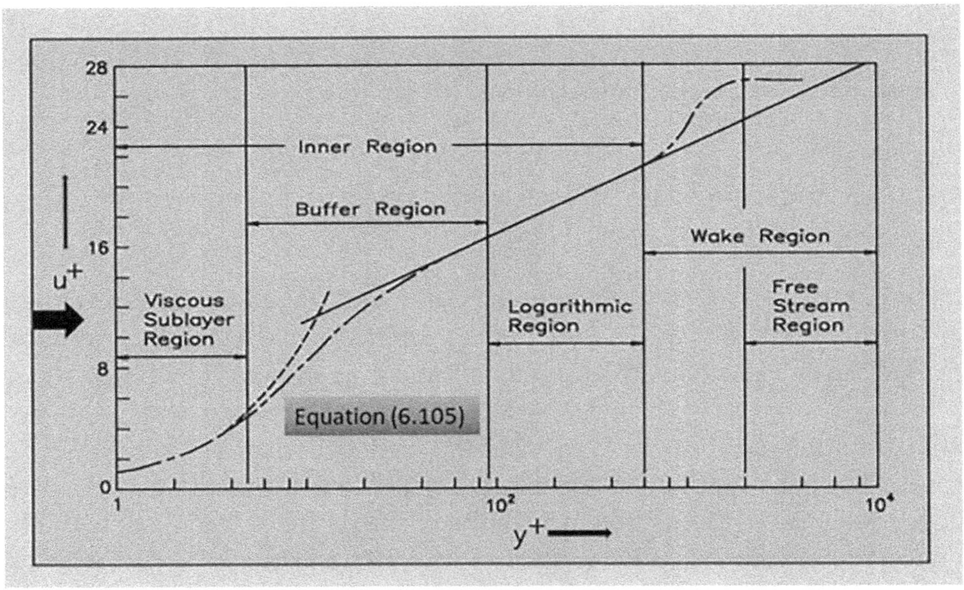

FIGURE 6.18 Velocity distribution in the inner and outer regions of bounded shear flow. (Modified from Coleman and Alanso, 1983.)

be high (Coleman and Alanso, 1983). The value of Π is determined within a range from 0 to 0.55. The significance of Π value is best understood by converting equation (6.105) to the velocity defect form. Considering maximum velocity U at the top of the boundary $y^+ \rightarrow \delta^+$, then from equation (6.105) the velocity-defect law can be written as (Coleman and Alanso, 1983):

$$U^+ - u^+ = -\frac{2.303}{\kappa_0}\log\left(\frac{y}{\delta}\right) + \frac{2\Pi}{\kappa_0} - \left(\frac{y}{\delta}\right)^2\left(1-\frac{y}{\delta}\right) - \frac{2\Pi}{\kappa_0}\left(\frac{y}{\delta}\right)\left[3-2\left(\frac{y}{\delta}\right)\right] \quad (6.106)$$

which is valid over the logarithmic zone and the outer region for flow over both smooth and rough boundaries with value of $\kappa_0 = 0.41$.

Guo et al. (2005) extended his proposed modified log-wake law equation (6.88) for zero pressure gradient in the turbulent boundary layer flow including the van Driest's mixing length model (6.93) where the damping factor varies with the Reynolds number. They suggested that the modified log-wake law model using van Driest's damping factor is applied to the entire boundary layer flow to the open-channel turbulence. Though the modified log–wake law equation (6.88) indeed agrees with the experimental data in the outer region, it is notable that the law equation (6.88) can be extended to the inner region by applying van Driest's mixing-length model (Schlichting, 1979) for the law of the wall. Therefore, replacing the logarithmic law from equation (6.88) with van Driest's mixing-length model equation (6.96), one can get a complete velocity profile equation for the entire boundary layer flow and is given by:

$$\frac{\bar{u}}{u_*} = \int_0^{y^+} \frac{2dy^+}{1+\left\{1+4\kappa_0^2\left(y^+\right)^2\left[1-\exp\left(-\frac{y^+}{A_*}\right)\right]^2\right\}^{1/2}} + \frac{2\Pi}{\kappa_0}\sin^2\left\{\left(\frac{\pi}{2}\right)\left(\frac{y}{\delta}\right)\right\} - \frac{1}{3\kappa_0}\left(\frac{y}{\delta}\right)^3 \quad (6.107)$$

where $\kappa_0 = 0.4$, $\Pi = 0.7577$, $y^+ = \frac{u_* y}{\nu}$, and $A_* = 26$ .The modified equation (6.107) is valid for the entire boundary layer velocity profile, including the inner and outer regions. This model is verified with 70 recent observed velocity profiles, and showed excellent agreement. In brief, this extension of the modified log–wake law equation (6.107) satisfies all boundary conditions at the wall and at the boundary layer edge, and connects to the constant potential velocity smoothly. This profile shows excellent agreement not only to the mean velocity profiles but also for the skin friction factor.

The displacement boundary layer thickness δ^* given by equation (6.83) and the momentum thickness θ given by equation (6.84) are two important parameters that can be estimated from the proposed modified log-wake law equation (6.107). The displacement thickness δ^* is defined as

$$\frac{\delta^*}{\delta} = \int_0^1 \frac{U - \bar{u}}{u_*} dy^+ = \frac{u_*}{\kappa_0 U}\left(\frac{3}{4} + \Pi\right) \tag{6.108}$$

and the momentum thickness θ is defined as

$$\frac{\theta}{\delta} = \int_0^1 \frac{\bar{u}}{U}\left(1 - \frac{\bar{u}}{U}\right) dy^+ = \alpha\left(\frac{u_*}{U}\right)^2 + \beta\left(\frac{u_*}{U}\right) \tag{6.109}$$

where $\alpha = -26.538$, and $\beta = 3.7693$. The details of calculations are available in Guo et al. (2005).

For outer region, the convective terms of the momentum equation (6.30a) should be considered because the convective term is of the same order of magnitude as the shear stress gradient term at a greater distance from the wall, and in that region, τ_l is quite small compared to τ_t, and can be neglected. Therefore, the momentum boundary layer equation (6.30a), neglecting the pressure gradient term, is written as

$$\bar{u}\frac{\partial \bar{u}}{\partial x} + \bar{v}\frac{\partial \bar{u}}{\partial y} = \frac{1}{\rho_f}\frac{\partial \tau_t}{\partial y}, \; y_o < y < \delta \tag{6.110}$$

where $\tau_t = -\rho_f \overline{u'v'}$, and y_o is the distance from the wall. The momentum equation (6.30a) for the outer region of turbulent boundary layer is the same as the momentum equation for laminar boundary layer except using the eddy-viscosity term $\tau_t = \epsilon_t \frac{\partial \bar{u}}{\partial y}$. If ϵ_t is constant, then it is similar to kinematic viscosity ν_f. If the proper variables are chosen, a close similarity to the laminar and turbulent velocity profiles in the outer region can be shown, and the well-known Blasius equation for laminar boundary layer flow is obtained, but it is important to note that in the similarity variable, instead of kinematic viscosity ν_f, the constant eddy-viscosity ε_t should be considered (Cebeci, 2004).

6.7.6 Velocity Distributions over Rough Surface

The abovementioned discussions pertain strictly to the turbulent flow structures over a smooth bed in open-channel/pipe flow. For the flow along a smooth wall, the concept of a viscous sub-layer is introduced. Now the effect of wall roughness on the mean velocity distribution will be discussed. The velocity-defect law equation (6.77) is valid for both smooth and rough surfaces. In the case of flow over a rough wall, the influence of roughness is only in the inner region of the flow, and hence its velocity distribution is affected by the rough wall, particularly, when the average height of the roughness elements becomes greater than the thickness of the viscous sub-layer. Turbulent flows over rough beds are important in hydraulic engineering because almost all river beds are composed of sand grains and complicated bed configurations like ripples and dunes. Therefore, many Researchers have investigated on flow resistance and friction laws over fixed rough bed configurations with roughness k_s in open channel flows.

Most of these smooth-bed relations can, however, be applied to flow over rough walls with a little modification. Knowledge of mean velocity and turbulence characteristics over fixed sand-grain beds is, however, limited. Nikuradse (1933) used the equivalent sand roughness k_s for his systematic experiments in pipe flows. For a rough bed composed of uniform sand grains attached densely to the wall, he found that the sand diameter itself can be used for roughness k_s. For most roughness, the equivalent sand roughness k_s can be determined from the *friction law derived from the log-law.* For others, one can determine the value of k_s *from the mean velocity distribution* in the region, where it coincides with the log-law of the wall region. However, these two methods of determining k_s do not necessarily give the same result. For example, it may be difficult to determine the equivalent sand roughness k_s for irregular surfaces. If the roughness size k_s is linked to the length scale, the turbulent structure in the wall region over the rough beds may be composed of multiple length scales such as k_s and ν_f/u_*. Such turbulent flows are complex and they do not easily can be solved by analytical investigation. Hence, a consideration into a single scale turbulent flow should be effected by using the ratio $k_s^+ = u_* k_s/\nu_f$ of the two length scales. Here k_s^+ is known as roughness Reynolds number based on the roughness length k_s and the friction velocity u_*. According to Nikuradse's experiments, the effects of roughness element k_s^+ are usually classified into three categories:

a. Hydraulically smooth bed ($k_s^+ < 5$), where the roughness height is smaller than the thickness of the viscous sub-layer. The effect of protrusions or roughness is covered by a viscous sub-layer; there is no direct effect of protrusions to the main flow.
b. Incompletely rough bed or transition regime ($5 \le k_s^+ \le 70$), where the effects of protrusions are partly outside the viscous sub-layer and the resistances are due to mainly the form drag by the protrusions in the boundary layer.
c. Completely rough bed ($k_s^+ > 70$), where only the effect of roughness element dominates on the flow. All the protrusions come outside the laminar sub-layer, and the largest part of resistance to flow is due to the form drag, and hence the law of resistance becomes quadratic.

Roughness effects disappear if the bed is hydraulically smooth because of the viscous sub-layer, whereas viscous effects disappear in the case of completely rough beds because the roughness elements penetrate the fully turbulent logarithmic layer. An incompletely rough bed is the transition between (a) and (b) and it is affected by both viscosity and roughness.

The logarithmic law for velocity distribution equation (6.66), in the case of hydraulically smooth wall, remains valid for rough wall, except that the constant length scale y_0 must be replaced by a different numerical value, depending on the nature of the surface (smooth or rough). Here, let the length scale $y_0 = \frac{\beta k_s}{k_s^+}$, that means, $\frac{y_0}{k_s} = \frac{\beta}{k_s^+} = f(k_s^+)$ is a function of dimensionless roughness element. For a completely rough wall, the constant y_0 is determined from Nikuradse's measurements by the condition that $\bar{u} = 0$ at a certain distance $y = y_0$, where roughness length y_0 is the function of k_s, u_{*0}, and ν_f, that means, $y_0 = k_s \mathrm{f}\left(k_s^+\right)$, where $\mathrm{f}\left(k_s^+\right) = \frac{1}{30}$ from the Nikuradse's experiments. Thus, the velocity distribution law from equation (6.66) can be written as

$$\frac{\bar{u}}{u_{*0}} = \frac{1}{\kappa_0}\ln\left(\frac{y}{y_0}\right) = \frac{1}{\kappa_0}\ln\left(\frac{y}{k_s}\right) - \frac{1}{\kappa_0}\ln f\left(k_s^+\right) \tag{6.111}$$

which is the logarithmic law for flow over a completely rough wall. For a completely rough regime, the velocity distribution can be written as:

$$\frac{\bar{u}}{u_{*0}} = \frac{1}{\kappa_0}\ln\left(\frac{y}{k_s}\right) + B_2 \tag{6.112}$$

where $B_2 = -\frac{1}{\kappa_0}\ln f\left(k_s^+\right)$. Generally speaking, B_2 is a function of roughness Reynolds number $k_s^+ = u_{*0}k_s / \nu_f$. From Nikuradse's measurements for a completely rough regime, the constant $B_2 = 8.5$. So equation (6.112) can be written as:

$$\left.\begin{aligned} \frac{\bar{u}}{u_{*0}} &= 2.5\ln\left(\frac{y}{k_s}\right) + 8.5 \\ \frac{\bar{u}}{u_{*0}} &= 5.75\log\left(\frac{y}{k_s}\right) + 8.5 \end{aligned}\right\} \tag{6.112a}$$

Again, $\frac{y_0}{k_s} = f(k_s^+) = \frac{1}{30}$. The logarithmic equation (6.112) for completely rough wall is compared with the experimental data of Grass (1971) (Figure 6.19). It is observed from the figure that near the rough boundary, the velocity distribution does not satisfy properly. The universal velocity distribution law equation (6.70) for very large Reynolds numbers for the smooth surface can be written in the form as

$$\frac{\bar{u}}{u_{*0}} = 2.5\ln\frac{y}{k_s} + 2.5\ \ln k_s^+ + 5.5 \tag{6.113}$$

Comparing equations (6.112) and (6.113), the function B_2 for hydraulically smooth is written as

$$B_2 = -\frac{1}{\kappa_0}\ln f\left(k_s^+\right) = 2.5\ln k_s^+ + 5.5 \tag{6.114}$$

The function B_2 is plotted against the dimensionless roughness element k_s^+ from hydraulically smooth to completely rough regime through the transitional region in Figure 6.20. From the figure, it is observed that the value of $k_s^+ = 5$ corresponds to hydraulically smooth walls, the range within $5 < k_s^+ < 70$ corresponds to transitional regime and $k_s^+ > 70$ corresponds to the completely rough regime.

The wall roughness is influenced by several factors such as the height and shape of roughness element, and packing set up. The roughness determined by Nikuradse with sand was the maximum density because the sand grains were glued to the wall surface in such a way that the sand particles were

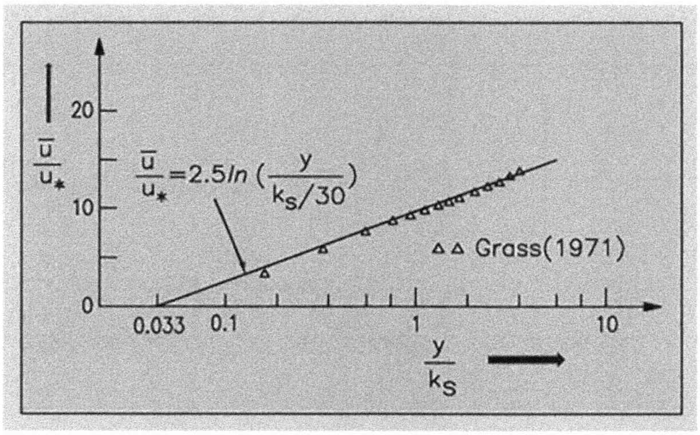

FIGURE 6.19 Velocity distribution over rough wall. Triangles indicate observed data of Grass (1971). (Modified from Sumer, 2002.)

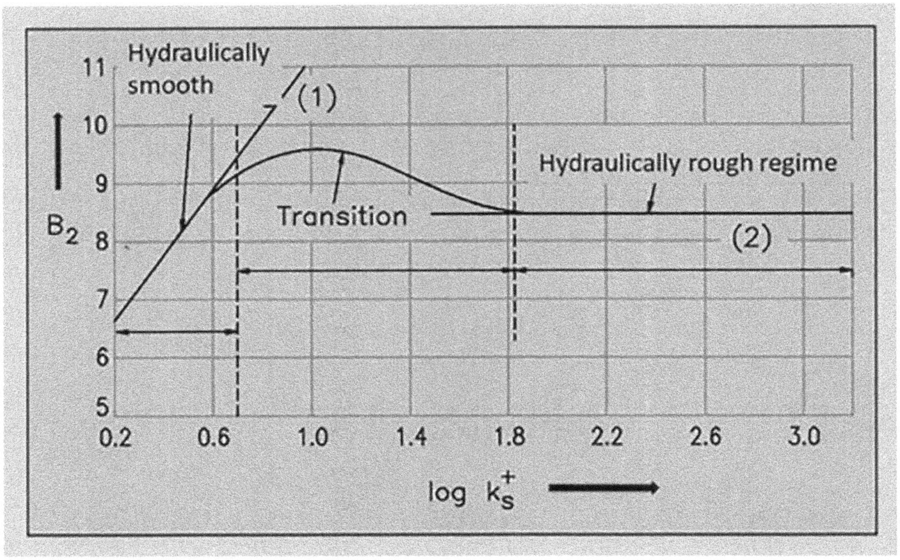

FIGURE 6.20 Plots of B_2 vs log (k_s^+). Line (1) represents hydraulically smooth, and line (2) represents log-law in completely rough region $B_2 = 8.5$. (Modified from Cebeci, 2004.)

as close as to each other, which was different from other loose roughness. To consider the roughness in different situations, there should be a common baseline for consideration of roughness heights. The basic idea of Nikuradse's experiments was the equivalent sand roughness k_s. The Nikuradse's equivalent sand roughness k_s was defined from the log-law distribution across the depth. The measured velocity was plotted in semi-log scale, and y-intercept of the line representing the log-law will be the roughness $k_s/30$, which is Nikuradse's equivalent sand roughness reported in the literature for various situations of roughness. Various values of equivalent sand roughness are reported by several researchers (Kamphuis, 1974; Bayazit, 1983; Fredsoe et al., 2000; Schlichting and Gersten, 2000; and others).

To determine the roughness of the wall, the exact location of the rough wall is not well defined. It is difficult to guess where should be the measurements of roughness height: at the top of the roughness elements, at the base, or at somewhere in between. Therefore, there is a concept of *theoretical wall*, which is defined as the location from which the vertical height y distance is measured (Figure 6.21). For convenience, there should be a change in coordinate y^1.

The theoretical bed starts at a distance y_l from the base surface, then the vertical coordinate will be as: $y = y^1 - y_1$. The logarithmic law can be written as

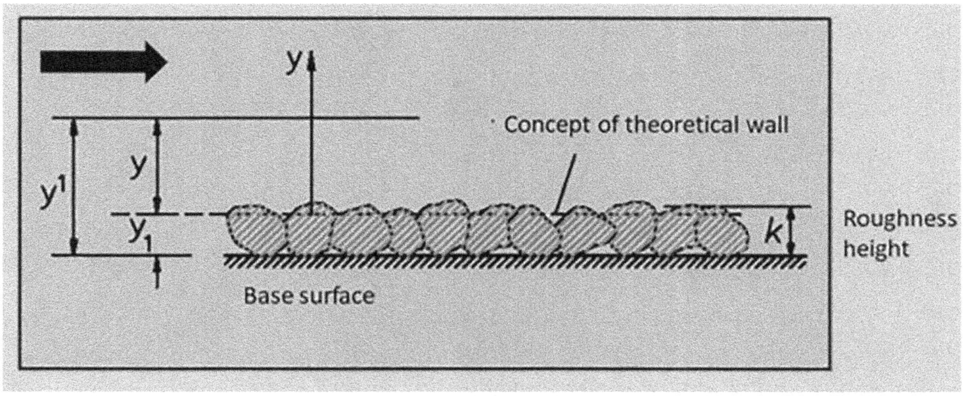

FIGURE 6.21 Theoretical wall over rough bed.

$$\frac{\bar{u}}{u_{*0}} = \frac{1}{\kappa_0}\ln\left(\frac{y^1 - y_1}{y_0}\right) = \frac{1}{\kappa_0}\ln\left(\frac{30\left(y^1 - y_1\right)}{k_s}\right) \tag{6.115}$$

Here the velocity $\bar{u}\left(y^1\right)$ is the function of friction velocity u_{*0}, roughness parameter k_s, and the location of theoretical wall y_1, and of course κ_0. The log-law velocity distribution is determined from equation (6.115).

Here the velocity profile near a rough surface is explained briefly. If the roughness height y_0 is smaller than the average height of the surface asperities, the velocity $\bar{u}(y)$ will be zero somewhere within the asperities, where local flow generates into small vortices between the peaks of the rough surface (Figure 6.22). Due to the peaks in rough surface, flow separation occurs. The negative values at the rough surface predicted by the logarithmic profile are not physically realized in the figure. It is important to note that the roughness height is not the average height of bumps or peaks on the surface but is equal to a small fraction of it, about one-tenth (Garratt, 1992).

6.7.7 Universal Velocity Distribution Laws Through Pipes

Since the simpler assumption on the mixing length $l = \kappa y$ does not seem to be suitable for the whole pipe diameter, it is preferable to deduce the dependence of mixing length on distance directly from experiment and then apply Prandtl hypothesis to calculate the velocity distribution from linear shearing stress distribution. Prandtl's hypothesis on the shear stress is as follows:

$$\tau = \rho_f l^2 \left(\frac{d\bar{u}}{dy}\right)^2$$

$$\frac{d\bar{u}}{dy} = \frac{1}{l}\sqrt{\frac{\tau}{\rho_f}} \tag{6.116}$$

The linear shear stress distribution is

$$\tau = \tau_0\left(1 - \frac{y}{R}\right) \tag{6.117}$$

where R is the radius of the pipe with zero to the center-line. Combining equations (6.116) and (6.117), it is now possible to calculate the variation of mixing length with y/R directly together with the measured velocity $u(y)$ as:

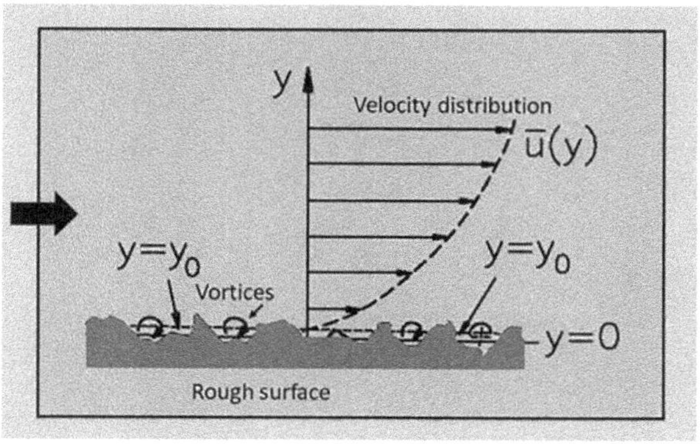

FIGURE 6.22 Velocity profile in the vicinity of a rough wall.

$$\tau_0\left(1-\frac{y}{R}\right)=\rho_f l^2\left(\frac{d\bar{u}}{dy}\right)^2 \tag{6.118}$$

The calculation was carried out by Nikuradse, who obtained the remarkable result which represents the variation of mixing length over the diameter of the pipe for the case of smooth pipes and it is seen that it is independent of the Reynolds number. This function can be represented by the empirical relation:

$$\frac{l}{R}=0.14-0.08\left(1-\frac{y}{R}\right)^2-0.06\left(1-\frac{y}{R}\right)^4 \tag{6.119}$$

In the neighborhood of the wall, the equation can be written as:

$$l=0.4y-0.44\frac{y^2}{R^2}+\ldots \tag{6.120}$$

which shows that Prandtl's hypothesis, $l=\kappa_0 y$, is confirmed for small distances from the wall, with $\kappa_0=0.4$. It can be shown that the variation of mixing length with wall distance given by equations (6.119) and (6.120) remains valid for both rough and smooth pipes.

The variation of mixing length with wall distance remains valid for rough as well as smooth pipes (Figure 6.23). Nikuradse (Schlichting, 1979) measured velocity on pipe artificially roughened with sand of different grain sizes and found that the statement is confirmed. For the sake of simplicity, the expression for mixing length can be written as

$$l=\chi y f\left(\frac{y}{R}\right) \tag{6.121}$$

where $f\left(\frac{y}{R}\right)\to 1$ for $\frac{y}{R}\to 0$

Now the velocity distribution calculated using the mixing length equation (6.120) will be valid for rough and smooth pipes. Now, from equation (6.118), one obtains

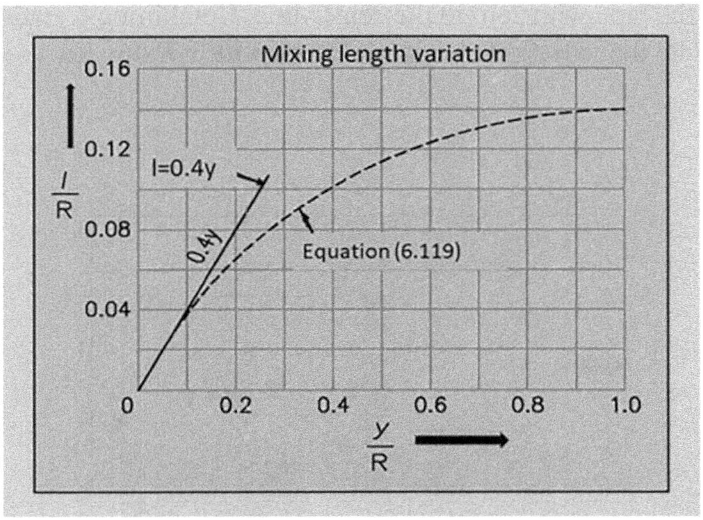

FIGURE 6.23 Variation of mixing length in smooth and rough pipes. Equation (6.119) represents dashed line; and the solid line represents Prandtl mixing length. (Modified from Schlichting, 1979.)

$$\frac{d\bar{u}}{dy} = \frac{1}{l}\sqrt{\frac{\tau}{\rho_f}} = \frac{1}{l}\sqrt{\frac{\tau_0\left(1-\frac{y}{R}\right)}{\rho_f}} = \frac{u_*}{\kappa_0}\frac{\sqrt{\left(1-\frac{y}{R}\right)}}{yf\left(\frac{y}{R}\right)} \tag{6.122}$$

by integration

$$\bar{u} = \frac{u_*}{\kappa_0}\int_{\frac{y_0}{R}}^{\frac{y}{R}} \frac{\sqrt{\left(1-\frac{y}{R}\right)}}{\frac{y}{R}f\left(\frac{y}{R}\right)} d\left(\frac{y}{R}\right) \tag{6.123}$$

Here the lower limit of integration y_0 is of the order of the laminar sub-layer thickness, where the velocity $\bar{u} = 0$ and y_0 is proportional to ν_f / u_*. Thus $\frac{y_0}{R} = F\left(\frac{u_* R}{\nu_f}\right)$.

The maximum velocity u_{max} in the center of the pipe can be deduced from equation (6.123) above, which is given by

$$\bar{u}_{max} = \frac{u_*}{\kappa_0}\int_{\frac{y_0}{R}}^{1} \frac{\sqrt{\left(1-\frac{y}{R}\right)}}{\frac{y}{R}f\left(\frac{y}{R}\right)} d\left(\frac{y}{R}\right) = \frac{u_*}{\kappa_0}\int_{\frac{y_0}{R}}^{y/R} \frac{\sqrt{\left(1-\frac{y}{R}\right)}}{\frac{y}{R}f\left(\frac{y}{R}\right)} d\left(\frac{y}{R}\right) + \frac{u_*}{\kappa_0}\int_{y/R}^{1} \frac{\sqrt{\left(1-\frac{y}{R}\right)}}{\frac{y}{R}f\left(\frac{y}{R}\right)} d\left(\frac{y}{R}\right)$$

$$\bar{u}_{max} = \bar{u} + \frac{u_*}{\kappa_0}F\left(\frac{y}{R}\right), \quad \text{where } F\left(\frac{y}{R}\right) = \int_{\frac{y}{R}}^{1} \frac{\sqrt{\left(1-\frac{y}{R}\right)}}{\frac{y}{R}f\left(\frac{y}{R}\right)} d\left(\frac{y}{R}\right)$$

$$\bar{u}_{max} - \bar{u} = \frac{u_*}{\kappa_0}F\left(\frac{y}{R}\right) \tag{6.124}$$

Thus, this equation is directed to the universal velocity-distribution law, which is valid for rough as well as for smooth pipes; and the function $F\left(\frac{y}{R}\right)$ is valid for both cases. It may be noted the above form of the velocity-distribution law was first derived by Stanton and Pannel (1914). The explicit expression of $F\left(\frac{y}{R}\right)$ can be, however, written as the velocity distribution law for smooth pipe, and given by

$$\bar{u}_{max} - \bar{u} = \frac{u_*}{\kappa_0}\ln\left(\frac{R}{y}\right) = 5.75u_* \log\left(\frac{R}{y}\right)$$

$$\text{or,} \quad \frac{\bar{u}_{max} - \bar{u}}{u_*} = 5.75\log\frac{R}{y} \tag{6.125}$$

This equation (6.125) due to Prandtl is plotted in Figure (6.24) and agrees with the experimental results for smooth and rough pipes. At the same time, the universal velocity-distribution law can be determined from the von-Karman similarity law (6.77) as:

$$\frac{\bar{u}_{max} - \bar{u}}{u_*} = -\frac{1}{\kappa_0}\left[\ln\left\{1-\sqrt{1-\frac{y}{h}}\right\} + \sqrt{1-\frac{y}{h}}\right] \tag{6.126}$$

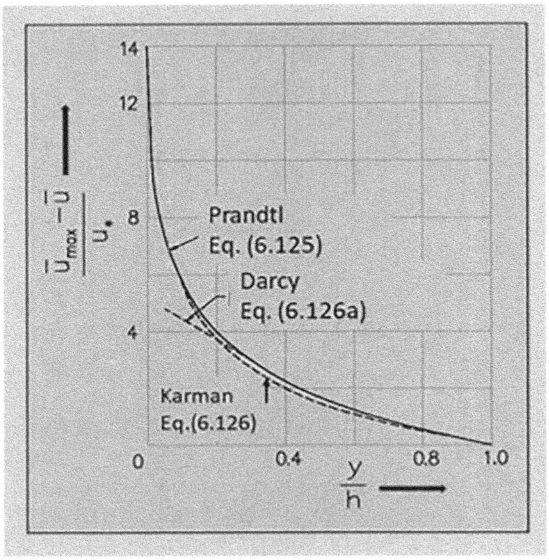

FIGURE 6.24 Universal velocity distributions by Prandtl (equation 6.125); Karman and Darcey (equation 6.126a). (Modified from Schlichting, 1979.)

Equation (6.126) is also plotted in Figure 6.24 and agrees well with the experimental data, if $\kappa_0 = 0.36$ is considered. It is meaningful to point out that both equations (6.125) and (6.126) for two-dimensional channel flows agree well with the experimental data trend, even for the case of pipe flow with axial symmetry. It can be mentioned here that in the laminar flow the velocity distribution is parabolic in both cases. It is noted that the Darcy's empirical equation (Schlichting, 1968) on the velocity distribution agrees well at all points except near the wall $y/h < 0.25$. The Darcy equation is written as:

$$\frac{\bar{u}_{max} - \bar{u}}{u_*} = 5.08\left(1 - \frac{y}{h}\right)^{\frac{3}{2}} \tag{6.126a}$$

Example 6.2

In a meteorological station, the wind velocity was measured at two points 3 m and 6 m above the ground, the values obtained being 2.0 m/s and 2.3 m/s respectively. Compute friction/shear velocity u_{*0}. What is the probable laminar sub-layer thickness for the problem considered? Estimate the velocity at 9 m above the ground level from the given data? Assume that the boundary is smooth and take kinematic viscosity $\nu = 0.145$ Stokes. The turbulent constant κ_0 (Karman's constant) can be considered as 0.4.

Solution

From equation (6.66), the velocity distribution for smooth boundary is given by

$$\bar{u} = \frac{u_{*0}}{\kappa_0}(\ln y - \ln y') \tag{1}$$

Taking ground level as origin, at $y = 3$ m; $\bar{u}_x = 2m/s$ and at $y = 6$ m; $\bar{u}_x = 2.3m/s$.

$$\text{Hence, } 2 = \left(\frac{u_{*0}}{\kappa_0}\right)[\log_e 3 - \log_e y'] \tag{2}$$

$$2.3 = \left(\frac{u_{*0}}{\kappa_0}\right)[\log_e 6 - \log_e y'] \tag{3}$$

a. Solving (2) and (3) we can get, $u_{*0} = 0.173\ \text{m/s}$
b. To find the shear stress, $\tau_0 = \rho u_{*0}^2$ (4)

The laminar sub-layer thickness can be obtained from Reichardt's observation that the laminar sub-layer extends up to $\delta_l \approx 5\frac{v_f}{u_*}$. Taking $y' = \delta_l$,

Thickness of laminar sub layer $= 5 * 0.145/(100*100)*(1/0.173) = 4.19\times10^{-4}\ \text{m}$

However, from equation (2), putting the value of u_* we get $y' = 0.029\ \text{m}$; $y' \sim \delta_l$, but at present:

$\delta_l = 4.19\times10^{-4}$ and $y' = 2.9\times10^{-2}$.

c. To find the velocity at 9 m height, at $y = 9\ \text{m}$; $\bar{u}_x = (0.173/0.4)\,[\log_e 9 - \log_e 0.029]$

Therefore, the velocity at 9 m height is $\bar{u}_x = 2.48\ \text{m/s}$

6.8 UNIVERSAL RESISTANCE LAW

Let us consider the fully developed turbulent flow through a straight pipe of circular cross-section with length L and radius R. The flow in the pipe is not acted upon by inertial forces. We write down the condition of equilibrium between the forces due to shear stress acting on circumferential area as $2\pi RL\tau$ and the pressure difference as $(p_1 - p_2)\pi R^2$ on the end faces in the form

$$2\pi RL\tau = (p_1 - p_2)\pi R^2$$

$$\Rightarrow \tau = \frac{p_1 - p_2}{L}\frac{R}{2} \tag{6.127}$$

which is equally valid for laminar and turbulent motion, where R is the radius of the pipe. In the present case, the shear stress τ denotes the sum of laminar and turbulent shear stress, therefore shear stress distribution over the cross-section is linear which occurs highest value at the wall, and is represented by τ_0, and that can be written as

$$\tau_0 = \frac{p_1 - p_2}{L}\frac{R}{2} \tag{6.128}$$

The shear stress at the wall τ_0 is determined directly by measuring the pressure gradient along the pipe. The relation between the pressure gradient and the mean velocity of flow is normally represented by introducing a resistance coefficient λ of pipe flow. It is known that the pressure gradient proportional to the dynamic head i.e., to the square of the mean velocity of flow, according to equation $\frac{dp}{dx} \sim \bar{u}^2$ fits turbulent flow very well, but for laminar flow $\frac{dp}{dx} \sim \bar{u}$. Thus, for laminar flow, the parameter λ ceases to be constant. Therefore, for turbulent flow, one can write as:

$$-\frac{dp}{dx} = \frac{\lambda}{d}\frac{\rho_f}{2}\bar{u}^2 = \frac{p_1 - p_2}{L} \tag{6.129}$$

where $d = 2R$ denotes the diameter of the cross-section. Using equations (6.128) and (6.129), one gets

$$\tau_0 = \frac{1}{8}\lambda\rho_f\bar{u}^2 \tag{6.130}$$

Blasius (1910) made a critical review of the accessible results with a dimensionless form and established the following relation in agreement with Reynolds's law of similarity, which is an empirical relation as:

$$\lambda = 0.3164\left(\frac{\bar{u}d}{v_f}\right)^{-\frac{1}{4}} = \frac{0.3164}{Re^{0.25}} \tag{6.131}$$

This is the valid frictional resistance formula for smooth circular pipe, known as Blasius formula, where Re is the Reynolds number determined from the mean velocity and the diameter of the pipe.

6.8.1 Universal Resistance Law through Smooth Pipes at Large Reynolds Numbers

A pipe-resistance formula is now derived from the universal logarithmic velocity distribution law. The log-law distribution equation (6.125) was derived under the condition that the laminar friction was negligible compared to the turbulent friction, which meant that it could be extrapolated for any large Reynolds numbers that may be true for resistance law, according to Prandtl (1933). On integration of equation (6.125) over the cross-section area, one obtains the mean velocity as

$$\bar{u} = \bar{u}_{\max} - 3.75u_* \tag{6.132}$$

From equation (6.130), one gets

$$\lambda = 8\left(\frac{u_*}{\bar{u}}\right)^2 \tag{6.133}$$

From the universal velocity distribution law $\frac{\bar{u}}{u_*} = 2.5\ln\frac{yu_*}{v_f} + 5.5$

At center of the pipe $y = R, \bar{u} = \bar{u}_{\max}$

$$\bar{u}_{\max} = u_*\left[2.5\ln\frac{Ru_*}{v_f} + 5.5\right]$$

which combined with equation (6.132), one gets

$$\bar{u} = u_*\left\{2.5\ln\frac{Ru_*}{v_f} + 1.75\right\} \tag{6.134}$$

We introduced the Reynolds number from

$$\frac{Ru_*}{v_f} = \frac{1}{2}\frac{\bar{u}d}{v_f}\cdot\frac{u_*}{\bar{u}} = \frac{\bar{u}d}{v_f}\frac{\sqrt{\lambda}}{4\sqrt{2}}$$

One gets from (6.133) and (6.134) as:

$$\lambda = \frac{8}{\left\{2.5\ln\left(\frac{\bar{u}d}{\nu_f}\cdot\sqrt{\lambda}\right) - 2.5\ln 4\cdot\sqrt{2} + 1.75\right\}^2} = \frac{1}{\left\{2.035\log\left(\frac{\bar{u}d}{\nu_f}\cdot\sqrt{\lambda}\right) - 0.91\right\}^2}$$

$$\text{or, } \frac{1}{\sqrt{\lambda}} = 2.035\log\left(Re\cdot\sqrt{\lambda}\right) - 0.91 \tag{6.135}$$

where $Re = \frac{\bar{u}d}{\nu_f}$, is the Reynolds number. This is the universal law of friction for a smooth pipe that should give a straight line if $\frac{1}{\sqrt{\lambda}}$ is plotted against $\log\left(Re\sqrt{\lambda}\right)$, which agrees well with the experimental data (Figure 6.25).

This phenomenon concurs extremely well with experimental data, where the results of measurements of many authors have been plotted. The continuous line in Figure 6.25 is represented by the equation

$$\frac{1}{\sqrt{\lambda}} = 2.0\log\left(\frac{\bar{u}d}{\nu_f}\cdot\sqrt{\lambda}\right) - 0.8\,(\text{smooth}) \tag{6.136}$$

This is the universal law of friction due to Prandtl for smooth pipes, and the dashed line in Figure 6.25 represents the Blasius resistance formula (6.131) as: $\lambda = 0.3164/Re^{0.25}$, which agrees well with universal resistance equation (6.136) up to $Re = 10^5$, but at higher values of Reynolds number, the Blasius equation deviates from the measurements.

For the pipe of smooth noncircular cross-section, the velocity distributions and laws of friction were determined for the pipes of rectangular, triangular, trapezoidal, and circular pipes with notches (Schiller, 1922; Nikuradse, 1930). Here, introducing the hydraulic diameter d_h instead of pipe diameter d, a coefficient of resistance to the turbulent flow is given by

$$\frac{\lambda}{d_h}\frac{\rho_f}{2}\bar{u}^2 = \frac{p_1 - p_2}{L} \tag{6.137}$$

where $d_h = \frac{4A}{C}$ is the hydraulic diameter with A is the cross-sectional area and C be the wetted perimeter. For circular cross-section, the hydraulic diameter d_h is equal to diameter of the pipe. Equation (6.137) gives the plots of resistance λ against Reynolds number Re for several cross-sectional shapes. For turbulent flow, the results are well presented by the law of circular pipes, whereas for noncircular

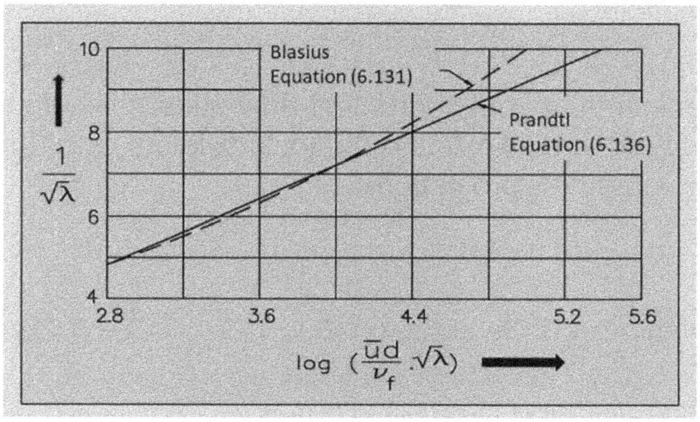

FIGURE 6.25 Law of friction for smooth pipe. Continuous line represents Prandtl (equation 6.136), and dashed line for Blasius (equation 6.131). (Modified from Schlichting, 1979.)

pipes when reduced to hydraulic diameter results deviate, which depend on the shape of the duct cross-section. The velocity distributions in such pipes or ducts are different, mostly for the corners of the pipes. In all pipes of noncircular cross-section, there exist secondary flows. The secondary flows continuously transport momentum from the center to the corners and generate high vortices.

6.8.2 Universal Resistance Law for Rough Pipes at Large Reynolds Numbers

The resistance formula for a rough pipe can be determined in a similar manner as was done for the case of smooth pipes. We begin with mean velocity as;

$$\bar{u} = \bar{u}_{\max} - 3.75u_*$$

Substituting equation (6.112) with $\bar{u} = \bar{u}_{\max}$ at $y = R$ in the above equation, one gets

$$\frac{\bar{u}}{u_*} = 2.5\ln\left(\frac{R}{k_s}\right) + 4.75 \tag{6.138}$$

From equation (6.133), one can write as:

$$\lambda = 8\left(\frac{u_*}{\bar{u}}\right)^2 = 8\left[2.5\ln\left(\frac{R}{k_s}\right) + 4.75\right]^{-2}$$

This equation may be written as

$$\lambda = \left[2\log\left(\frac{R}{k_s}\right) + 1.68\right]^{-2} \tag{6.139}$$

This is the resistance formula for completely rough flow first derived by von-Karman (1930) from the similarity law, which agrees well with Nikuradse's experiment with a little change in the constant with 1.74 instead of 1.68. Therefore, the resistance formula for the completely rough regime is as:

$$\lambda = \left[2\log\left(\frac{R}{k_s}\right) + 1.74\right]^{-2} \tag{6.140}$$

It is noted that equation (6.140) may be applied to the pipes of noncircular cross-section using the hydraulic radius instead of pipe diameter.

Example 6.3

In a pipeline of 20 cm diameter, fully rough turbulent flow takes place. In the pipeline, at a location, the center line velocity is found to be 3 m/s and the local velocity at mid-radius is 2.8 m/s. Calculate the flow rate and height of roughness projections.

Solution

In the case of rough turbulent flow, from equation (6.125), $\frac{\bar{u}_{\max} - \bar{u}}{u_*} = 5.75\log\frac{R}{y}$. From the given data, $\frac{3 - 2.8}{u_*} = 5.75\log\frac{R}{R/2} = 1.731$. Therefore, $u_* = 0.115$ m/s.

The mean velocity is related to $u_{\max}$ as, $\bar{u} = \bar{u}_{\max} - 3.75u_*$.

Therefore, mean velocity is: 2.57 m/s.

a. The flow rate through the pipe = Area of cross section × Average velocity. Thus $Q = \pi r^2 * 2.57 = \pi \times 0.1^2 \times 2.57 = 0.0807\ \text{m}^3/\text{s}$.

b. To find the roughness height, $\frac{\bar{u}}{u_{*0}} = 5.75\log\left(\frac{y}{k_s}\right) + 8.5$; Putting for centerline velocity, $\frac{3}{0.115} = 5.75\log\left(\frac{R}{k_s}\right) + 8.5$. That is, $\log\left(\frac{R}{k_s}\right) = 3.0586$. Therefore, $\left(\frac{R}{k_s}\right) = 1144.45$

Thus, the roughness height is: 8.74×10^{-5}.

6.9 DERIVATIONS OF ENERGY EQUATIONS

The turbulent energy equations are obtained from the Reynolds stress equations. The time-average of velocity components and pressure of the turbulent flow satisfy the equations as the velocity components of laminar flow (Navier–Stokes equations), where the friction forces of the laminar flow are adjoined by additional stresses, known as the stress tensor, which are called the apparent stresses of the turbulent flow. These additional stresses are generated due to turbulent fluctuations and are specified by the time-average of quadratic fluctuation terms.

6.9.1 Kinetic Energy Equation for Laminar Flow

In tensor form, the continuity equation for incompressible fluid, using summation convention, is:

$$\frac{\partial u_i}{\partial x_i} = 0 \tag{6.141}$$

and the Navier–Stokes equation is

$$\rho_f\left(\frac{\partial u_i}{\partial t} + u_j\frac{\partial u_i}{\partial x_j}\right) = \rho_f g_i + \frac{\partial}{\partial x_j}(\sigma_{ij}) \tag{6.142}$$

Here t is the time, g_i is the gravity forces and x_i is the Cartesian Coordinates $(i, j = 1, 2, 3)$ and σ_{ij} is the stress quantity of the fluid and is given by

$$\sigma_{ij} = -p\delta_{ij} + \mu_f\left(\frac{\partial u_i}{\partial x_j} + \frac{\partial u_j}{\partial x_i}\right) \tag{6.143}$$

with δ_{ij} is the Kronecker delta defined as:

$$\begin{aligned} \delta_{ij} &= 1 \text{ for } i = j \\ &= 0 \text{ for } i \neq j \end{aligned} \tag{6.144}$$

In these equations, Einstein summation convention is used, and there is a summation over the repeated indices (in three-dimensional space, sum from 1 to 3). Using the continuity equation (6.141), the Navier–Stokes equation (6.142) can be written as (Schlichting, 1979; Sumer, 2002):

$$\rho_f\frac{\partial u_i}{\partial t} + \frac{\partial}{\partial x_j}(\rho_f u_i u_j) = \rho_f g_i + \frac{\partial}{\partial x_j}\sigma_{ij} \tag{6.145}$$

After averaging each term, one gets

$$\rho_f\frac{\partial \overline{u_i}}{\partial t} + \frac{\partial}{\partial x_j}(\rho_f \overline{u_i u_j}) = \rho_f \overline{g_i} + \frac{\partial}{\partial x_j}\overline{\sigma_{ij}} \tag{6.146}$$

Using the Reynolds decomposition and equation (6.143), equation (6.146) can be written as:

$$\rho_f\left(\frac{\partial \overline{u_i}}{\partial t}+\overline{u_j}\frac{\partial \overline{u_i}}{\partial x_j}\right)=\rho_f\overline{g_i}+\frac{\partial}{\partial x_j}\left(\overline{\sigma_{ij}}\right)-\frac{\partial}{\partial x_j}\left(\rho_f\overline{u_i'u_j'}\right) \tag{6.147}$$

where $\overline{\sigma_{ij}}=-\overline{p}\delta_{ij}+\mu_f\left(\frac{\partial \overline{u_i}}{\partial x_j}+\frac{\partial \overline{u_j}}{\partial x_i}\right)$, and $-\rho_f\overline{u_i'u_j'}$ is the usual Reynolds stress terms with nine stress components as (Schlichting and Gersten, 2000):

$$-\rho_f\overline{u_i'u_j'}=\begin{bmatrix}-\rho_f\overline{u_1'u_1'}-\rho_f\overline{u_1'u_2'}-\rho_f\overline{u_1'u_3'}\\-\rho_f\overline{u_2'u_1'}-\rho_f\overline{u_2'u_2'}-\rho_f\overline{u_2'u_3'}\\-\rho_f\overline{u_3'u_1'}-\rho_f\overline{u_3'u_2'}-\rho_f\overline{u_3'u_3'}\end{bmatrix} \tag{6.148}$$

Now expanding equation (6.147), using $\overline{\sigma_{ij}}$, one gets

$$\rho_f\left(\frac{\partial \overline{u_i}}{\partial t}+\overline{u_\alpha}\frac{\partial \overline{u_i}}{\partial x_\alpha}\right)=\rho_f\overline{g_i}+\frac{\partial}{\partial x_j}\left[-p\delta_{ij}+\mu_f\left(\frac{\partial \overline{u_i}}{\partial x_j}+\frac{\partial \overline{u_j}}{\partial x_i}\right)\right]-\frac{\partial}{\partial x_\alpha}\left(\rho_f\overline{u_i'u_\alpha'}\right)$$

$$=\rho_f\overline{g_i}-\frac{\partial \overline{p}}{\partial x_i}+\mu_f\frac{\partial}{\partial x_\alpha}\left(\frac{\partial \overline{u_i}}{\partial x_\alpha}+\frac{\partial \overline{u_\alpha}}{\partial x_i}\right)-\frac{\partial}{\partial x_\alpha}\left(\rho_f\overline{u_i'u_\alpha'}\right)$$

$$\text{Now, } \mu_f\frac{\partial}{\partial x_\alpha}\left(\frac{\partial \overline{u_i}}{\partial x_\alpha}+\frac{\partial \overline{u_\propto}}{\partial x_i}\right)=\mu_f\left(\frac{\partial^2 \overline{u_i}}{\partial x_\alpha\partial x_\alpha}+\frac{\partial^2 \overline{u_\alpha}}{\partial x_\alpha\partial x_i}\right)$$

$$=\mu_f\frac{\partial^2 \overline{u_i}}{\partial x_\alpha\partial x_\alpha}+\mu_f\frac{\partial}{\partial x_i}\left(\frac{\partial \overline{u_\alpha}}{\partial x_\alpha}\right)$$

$$=\mu_f\frac{\partial^2 \overline{u_i}}{\partial x_\alpha\partial x_\alpha},\text{ since }\frac{\partial \overline{u_\alpha}}{\partial x_\alpha}=0\text{ from (6.141).}$$

Therefore, equation (6.147) can be rewritten as:

$$\rho_f\left(\frac{\partial \overline{u_i}}{\partial t}+\overline{u_\alpha}\frac{\partial \overline{u_i}}{\partial x_\alpha}\right)=\rho_f\overline{g_i}-\frac{\partial \overline{p}}{\partial x_i}+\mu_f\frac{\partial^2 \overline{u_i}}{\partial x_\alpha\partial x_\alpha}-\frac{\partial}{\partial x_\alpha}\left(\rho_f\overline{u_i'u_\alpha'}\right) \tag{6.149}$$

Equation (6.149) is known as **Reynolds stress equation** in tensor notation, using Reynolds decomposition. Now, we have the equation of continuity (6.141) and three equations of motion (6.149). Therefore, we have four equations, but ten unknowns, including six components of Reynolds stress $\left(-\rho_f\overline{u_i'u_j'}\right)$ given by equation (6.148) due to symmetry, and hence the system is not in closed form. This problem is known as the closure problem of turbulence.

Derivation of energy equation for laminar flow

In laminar flow, the Navier–Stokes equation is written in tensor notation as:

$$\frac{\partial u_i}{\partial t}+u_\alpha\frac{\partial u_i}{\partial x_j}=g_i-\frac{1}{\rho_f}\frac{\partial p}{\partial x_i}+\nu_f\frac{\partial^2 u_i}{\partial x_\alpha\partial x_\alpha} \tag{6.150}$$

Multiplying both sides of (6.150) by u_j, one gets

$$u_j \frac{\partial u_i}{\partial t} + u_j u_\alpha \frac{\partial u_i}{\partial x_\alpha} = u_j g_i - \frac{1}{\rho_f} u_j \frac{\partial p}{\partial x_i} + \nu_f u_j \frac{\partial^2 u_i}{\partial x_\alpha \partial x_\alpha} \tag{6.151}$$

The first term of the left-hand side of equation (6.151) is

$$u_j \frac{\partial u_i}{\partial t} = \frac{\partial}{\partial t}(u_i u_j) - u_i \frac{\partial u_j}{\partial t} \tag{6.152}$$

Substituting (6.152) in (6.151), we get, neglecting g_i,

$$\frac{\partial}{\partial t}(u_i u_j) - u_i \frac{\partial u_j}{\partial t} + u_j u_\alpha \frac{\partial u_i}{\partial x_\alpha} = -u_j \frac{1}{\rho_f} \frac{\partial p}{\partial x_i} + \nu_f u_j \frac{\partial^2 u_i}{\partial x_\alpha \partial x_\alpha} \tag{6.153}$$

$$u_i \frac{\partial u_j}{\partial t} = \frac{\partial}{\partial t}(u_i u_j) + u_j u_\alpha \frac{\partial u_i}{\partial x_\alpha} + u_j \frac{1}{\rho_f} \frac{\partial p}{\partial x_i} - \nu_f u_j \frac{\partial^2 u_i}{\partial x_\alpha \partial x_\alpha} \tag{6.154}$$

Substituting $\frac{\partial u_j}{\partial t}$ from (6.150) in (6.152) and putting $i = j$, one gets

$$u_j \frac{\partial u_i}{\partial t} = \frac{\partial}{\partial t}(u_i u_j) - u_i \left[-\frac{1}{\rho_f} \frac{\partial p}{\partial x_j} + \nu_f \frac{\partial^2 u_j}{\partial x_\alpha \partial x_\alpha} - u_\alpha \frac{\partial u_j}{\partial x_\alpha} \right] \tag{6.155}$$

Now subtracting (6.151) from (6.155), one gets

$$\frac{\partial}{\partial t}(u_i u_j) - u_i \left[-\frac{1}{\rho_f} \frac{\partial p}{\partial x_j} + \nu_f \frac{\partial^2 u_j}{\partial x_\alpha \partial x_\alpha} - u_\alpha \frac{\partial u_j}{\partial x_\alpha} \right] = -u_j u_\alpha \frac{\partial u_j}{\partial x_\alpha} - u_j \frac{1}{\rho_f} \frac{\partial p}{\partial x_i} + \nu_f u_j \frac{\partial^2 u_i}{\partial x_\alpha \partial x_\alpha} \tag{6.156}$$

$$\frac{\partial}{\partial t}(u_i u_j) + \frac{1}{\rho_f}\left(u_i \frac{\partial p}{\partial x_j} + u_j \frac{\partial p}{\partial x_i} \right) + \left(u_i u_\alpha \frac{\partial x_j}{\partial x_\alpha} + u_j u_\alpha \frac{\partial u_i}{\partial x_\alpha} \right) - \nu_f \left(u_i \frac{\partial^2 u_j}{\partial x_\alpha \partial x_\alpha} + u_j \frac{\partial u_i}{\partial x_\alpha \partial x_\alpha} \right) = 0 \tag{6.156a}$$

Putting $j = i$ in (6.156a), one gets

$$\frac{\partial}{\partial t}(u_i u_i) + 2u_i \frac{1}{\rho_f} \frac{\partial p}{\partial x_i} + 2u_i u_\alpha \frac{\partial u_i}{\partial x_\alpha} - 2\nu_f u_i \frac{\partial^2 u_i}{\partial x_\alpha \partial x_\alpha} = 0$$

$$\text{or, } \frac{\partial}{\partial t}\left(\frac{\rho_f}{2} u_i u_i \right) + \rho_f u_i u_\alpha \frac{\partial u_i}{\partial x_\alpha} = -u_i \frac{\partial p}{\partial x_i} + \mu_f u_i \frac{\partial^2 u_i}{\partial x_\alpha \partial x_\alpha} \tag{6.157}$$

Here $\mu_f u_i \frac{\partial^2 u_\alpha}{\partial x_i \partial x_\alpha} = 0 \Rightarrow \mu_f u_i \frac{\partial}{\partial x_i}\left(\frac{\partial u_\alpha}{\partial x_\alpha} \right) = 0,$

$\because \frac{\partial u_\alpha}{\partial x_\alpha} = 0$ from the equation of continuity.

Adding $\mu_f u_i \frac{\partial^2 u_\alpha}{\partial x_i \partial x_\alpha}$ term on right sides of (6.157), we get

$$\underset{(1)}{\frac{\partial}{\partial t}\left(\frac{\rho_f}{2} u_i u_i \right)} + \underset{(2)}{\rho_f u_i u_\alpha \frac{\partial u_i}{\partial x_\alpha}} = \underset{(3)}{-u_i \frac{\partial p}{\partial x_i}} + \underset{(4)}{\mu_f u_i \frac{\partial^2 u_i}{\partial x_\alpha \partial x_\alpha}} + \underset{(5)}{\mu_f u_i \frac{\partial^2 u_\alpha}{\partial x_i \partial x_\alpha}} \tag{6.158}$$

Simplifying (6.158) term by term,

Term 2 of equation (6.158):

$$\rho u_i u_\alpha \frac{\partial u_i}{\partial x_\alpha} = \frac{\partial}{\partial x_\alpha}(\rho_f u_i u_\alpha u_i) - \rho_f u_i \frac{\partial}{\partial x_\alpha}(u_i u_\alpha) = \frac{\partial}{\partial x_\alpha}(\rho_f u_i u_\alpha u_i) - \rho_f u_i u_\alpha \frac{\partial u_i}{\partial x_\alpha}$$

Therefore, $\rho_f u_i u_\alpha \dfrac{\partial u_i}{\partial x_\alpha} = \dfrac{\partial}{\partial x_\alpha}\left(\dfrac{\rho_f}{2} u_i u_i u_\alpha\right)$

Term 3: $u_i \dfrac{\partial p}{\partial x_i} = \dfrac{\partial}{\partial x_i}(\rho_f u_i)$, using $\dfrac{\partial u_i}{\partial x_i} = 0$

Terms 4 and 5:

$$\mu_f u_i \frac{\partial^2 u_i}{\partial x_\alpha \partial x_\alpha} + \mu_f u_i \frac{\partial^2 u_\alpha}{\partial x_i \partial x_\alpha} = \mu_f u_i \frac{\partial}{\partial x_\alpha}\left(\frac{\partial u_i}{\partial x_\alpha}\right) + \mu_f u_i \frac{\partial}{\partial x_\alpha}\left(\frac{\partial u_\alpha}{\partial x_i}\right) = \mu_f u_i \frac{\partial}{\partial x_\alpha}\left(\frac{\partial u_i}{\partial x_\alpha} + \frac{\partial u_\alpha}{\partial x_i}\right)$$

$$= \mu_f \frac{\partial}{\partial x_\alpha}\left[u_i\left(\frac{\partial u_i}{\partial x_\alpha} + \frac{\partial u_\alpha}{\partial x_i}\right)\right] - \mu_f\left(\frac{\partial u_i}{\partial x_\alpha} + \frac{\partial u_\alpha}{\partial x_i}\right)\frac{\partial u_i}{\partial x_\alpha}$$

Putting the simplified terms in (6.158), one gets

$$\frac{\partial}{\partial t}\left(\frac{\rho}{2} u_i u_i\right) + \frac{\partial}{\partial x_\alpha}\left(\frac{\rho_f}{2} u_i u_i u_\alpha\right) = -\frac{\partial}{\partial x_i}(\rho_f u_i) + \mu_f \frac{\partial}{\partial x_\alpha}\left\{u_i\left(\frac{\partial u_i}{\partial x_\alpha} + \frac{\partial u_\alpha}{\partial x_i}\right)\right\} - \mu_f\left(\frac{\partial u_i}{\partial x_\alpha} + \frac{\partial u_\alpha}{\partial x_i}\right)\frac{\partial u_i}{\partial x_\alpha} \tag{6.159}$$

Equation (6.159) can be written as:

$$\frac{\partial}{\partial t}\left(\frac{\rho_f}{2} u_i^2\right) + \frac{\partial}{\partial x_\alpha}\left(\frac{\rho_f}{2} u_\alpha u_i^2\right) + \frac{\partial}{\partial x_i}(p u_i) - \mu_f \frac{\partial}{\partial x_\alpha} u_i\left(\frac{\partial u_i}{\partial x_\alpha} + \frac{\partial u_\alpha}{\partial x_i}\right) = -\mu_f\left(\frac{\partial u_i}{\partial x_\alpha} + \frac{\partial u_\alpha}{\partial x_i}\right)\frac{\partial u_i}{\partial x_\alpha}$$

$$\Rightarrow \frac{\partial k}{\partial t} + \frac{\partial}{\partial x_\alpha}(u_\alpha k) + \frac{\partial}{\partial x_\alpha}(p u_\alpha) - \frac{\partial}{\partial x_\alpha}\mu_f u_i\left(\frac{\partial u_i}{\partial x_\alpha} + \frac{\partial u_\alpha}{\partial x_i}\right) = -\mu_f\left(\frac{\partial u_i}{\partial x_\alpha} + \frac{\partial u_\alpha}{\partial x_i}\right)\frac{\partial u_i}{\partial x_\alpha}$$

$$\Rightarrow \frac{\partial k}{\partial t} + \frac{\partial}{\partial x_\alpha}\left[k u_\alpha + p u_\alpha - \mu_f u_i\left(\frac{\partial u_i}{\partial x_\alpha} + \frac{\partial u_\alpha}{\partial x_i}\right)\right] = -\mu_f\left(\frac{\partial u_i}{\partial x_\alpha} + \frac{\partial u_\alpha}{\partial x_i}\right)\frac{\partial u_i}{\partial x_\alpha} = -\phi \tag{6.160}$$

where $k = \dfrac{1}{2}\rho_f u_i^2 = \dfrac{\rho_f}{2}\left(u^2 + v^2 + w^2\right)$ is the kinetic energy per unit volume of fluid, and

$$\phi = \mu_f\left(\frac{\partial u_i}{\partial x_\alpha} + \frac{\partial u_\alpha}{\partial x_i}\right)\frac{\partial u_i}{\partial x_\alpha} \tag{6.160a}$$

is the energy dissipation per unit volume and per unit time.
Equation (6.160) is the **kinetic energy equation for laminar flow** in tensor form.
Equation (6.160) can be written as (Sumer, 2002):

$$\underset{(1)}{\frac{\partial k}{\partial t}} + \underset{(2)}{\frac{\partial}{\partial x_\alpha}\left(u_\alpha (k + p)\right)} - \underset{(3)}{\mu_f \frac{\partial}{\partial x_\alpha}\left[u_i\left(\frac{\partial u_i}{\partial x_\alpha} + \frac{\partial u_\alpha}{\partial x_i}\right)\right]} = \underset{(4)}{-\mu_f\left(\frac{\partial u_i}{\partial x_\alpha} + \frac{\partial u_\alpha}{\partial x_i}\right)\frac{\partial u_i}{\partial x_\alpha}} = -\phi \tag{6.161}$$

The meanings of various terms in (6.161) are given as follows:-

(1) The total change in kinetic energy per unit mass and time,
(2) The change in convective transport of the pressure and kinetic energy per unit mass and time, or work done per unit mass and time by total dynamic pressure.
(3) This term represents the work done per unit mass and time by viscous stresses.
(4) This term represents dissipation per unit mass.

Here the term (2) can be interpreted as the work done per unit mass and time by the total dynamic pressure $p+k$ with $k=\frac{\rho_f}{2}\left(u^2+v^2+w^2\right)$ or $k=\frac{\rho_f}{2}u_iu_i$. This term may also be interpreted as the change in transport of the energy $\frac{p}{\rho_f}+\frac{1}{2}u_iu_i$ per unit mass through convection by velocity u_j.

6.9.2 Kinetic Energy Equation for Fluctuating Flow

This is determined from the instantaneous momentum and the continuity equations. Considering the Reynolds decomposition of instantaneous flow and pressure in the Navier–Stokes equation, we get (Schlichting, 1979; Sumer, 2002):

$$\rho_f\left[\frac{\partial}{\partial t}(u_i+u_i')+(u_j+u_j')\frac{\partial}{\partial x_j}(u_i+u_i')\right]=-\frac{\partial}{\partial x_i}(P+p)+2\mu_f\frac{\partial}{\partial x_j}(S_{ij}+S_{ij}') \tag{6.162}$$

where $S_{ij}=\frac{1}{2}\left(\frac{\partial u_i}{\partial x_j}+\frac{\partial u_j}{\partial x_i}\right), S_{ij}'=\frac{1}{2}\left(\frac{\partial u_i'}{\partial x_j}+\frac{\partial u_j'}{\partial x_i}\right)$ and u_i' are the fluctuating components of velocity and u_i are the mean velocity components. P is the mean pressure and p is the fluctuating pressure. From equation (6.162), we get

$$\rho_f\left(\frac{\partial u_i}{\partial t}+\frac{\partial u_i'}{\partial t}+u_j\frac{\partial u_i}{\partial x_j}+u_j\frac{\partial u_i'}{\partial x_j}+u_j'\frac{\partial u_i}{\partial x_j}+u_j'\frac{\partial u_i'}{\partial x_j}\right)$$

$$=-\frac{\partial}{\partial x_i}(P+p)+2\mu_f\frac{\partial}{\partial x_j}(S_{ij}+S_{ij}') \tag{6.163}$$

Multiplying equation (6.163) by u_i' and averaging with respect to time, equation (6.163) results as:

$$\rho_f\left[\overline{u_i'\frac{\partial u_i}{\partial t}}+\overline{u_i'\frac{\partial u_i'}{\partial t}}+\overline{u_i'u_j\frac{\partial u_i}{\partial x_j}}+\overline{u_i'u_j\frac{\partial u_i'}{\partial x_j}}+\overline{u_i'u_j'\frac{\partial u_i}{\partial x_j}}+\overline{u_i'u_j'\frac{\partial u_i'}{\partial x_j}}\right]=-\overline{u_i'\frac{\partial}{\partial x_i}(P+p)}+2\mu_f\overline{u_i'\frac{\partial}{\partial x_j}(S_{ij}+S_{ij}')} \tag{6.164}$$

Using the rules of averaging: $\overline{u_i}=u_i, \overline{u_i'}=0,\ \overline{u_iu_i'}=\overline{u_i}\,\overline{u_i'}=0,\ \overline{u_i'u_i'}\neq 0,\ \overline{u_i'u_j'}\neq 0,$

$$\overline{S_{ij}'}=\frac{1}{2}\left(\frac{\partial\overline{u_i'}}{\partial x_j}+\frac{\partial\overline{u_j'}}{\partial x_i}\right)=0,\ \overline{S_{ij}}=\frac{1}{2}\left(\frac{\partial\overline{u_i}}{\partial x_j}+\frac{\partial\overline{u_j}}{\partial x_i}\right)\neq 0$$

Let us simplify all the terms of equation (6.164) individually,

$$Term\ 1: \overline{u_i'\frac{\partial u_i}{\partial t}}=\overline{u_i'}\frac{\partial\overline{u_i}}{\partial t}=0\cdot\frac{\partial\overline{u_i}}{\partial t}=0$$

$$Term\ 2: \overline{u_i' \frac{\partial u_i'}{\partial t}} = \frac{\partial}{\partial t}\left(\overline{u_i' u_i'}\right) - \overline{u_i' \frac{\partial u_i'}{\partial t}}$$

$$\Rightarrow \frac{\partial}{\partial t}\left(\overline{u_i' u_i'}\right) = 2\overline{u_i' \frac{\partial u_i'}{\partial t}} \Rightarrow u_i' \frac{\partial u_i'}{\partial t} = \frac{\partial}{\partial t}\left(\frac{1}{2}\overline{u_i' u_i'}\right) = \frac{\partial k}{\partial t},$$

$$\text{where } k = \frac{1}{2}\overline{u_i' u_i'} = \frac{1}{2}\overline{u_i'^2},\ k = \frac{\rho_f}{2} u_i'^2,$$

$$Term\ 3: \overline{u_i' u_j \frac{\partial u_i}{\partial x_j}} = \overline{u_i' u_j} \frac{\partial \overline{u_i}}{\partial x_j} = \overline{u_i'}\,\overline{u_j} \frac{\partial \overline{u_i}}{\partial x_j} = 0, \text{ since } \overline{u_j'} = 0$$

$$Term\ 4: \overline{u_i' u_j \frac{\partial u_i'}{\partial x_j}}$$

$$\overline{u_j u_i' \frac{\partial u_i'}{\partial x_j}} = \overline{u_j}\,\overline{u_i' \frac{\partial u_i'}{\partial x_j}} = \overline{u_j} \frac{\partial}{\partial x_j}\left(\frac{1}{2}\overline{u_i' u_i'}\right) = \overline{u_j} \frac{\partial k}{\partial x_j},$$

$$\text{where } k = \frac{\rho}{2} u_i'^2$$

$$Term\ 5: \overline{u_i' u_j' \frac{\partial u_i}{\partial x_j}} = \overline{u_i' u_j'} \frac{\partial u_i}{\partial x_j} = \overline{u_i' u_j'} \frac{1}{2}\left(\frac{\partial u_i}{\partial x_j} + \frac{\partial u_j}{\partial x_i}\right) = \overline{u_i' u_j'} S_{ij}$$

$$\text{where } S_{ij} = \frac{\partial u_i}{\partial x_j} = \frac{1}{2}\left(\frac{\partial u_i}{\partial x_j} + \frac{\partial u_j}{\partial x_i}\right)$$

$$Term\ 6: \overline{u_i' u_j' \frac{\partial u_i'}{\partial x_j}} = \frac{\partial}{\partial x_j}\left(\overline{u_i' u_j' u_i'}\right) - \overline{u_i' \frac{\partial}{\partial x_j} u_i' u_j'}$$

$$= \frac{\partial}{\partial x_j}\left(\overline{u_i' u_j' u_i'}\right) - \overline{u_i'\left[u_i' \frac{\partial u_j'}{\partial x_j} + u_j' \frac{\partial u_i'}{\partial x_j}\right]}$$

$$= \frac{\partial}{\partial x_j}\left(\overline{u_i' u_j' u_i'}\right) - \overline{u_i' u_i' \frac{\partial u_j'}{\partial x_j}} - \overline{u_i' u_i' \frac{\partial u_i'}{\partial x_j}}$$

$$= \frac{\partial}{\partial x_j}\left(\overline{u_i' u_j' u_i'}\right) - 0 - \overline{u_i' u_j' \frac{\partial u_i'}{\partial x_j}}$$

$$\text{or, } 2\overline{u_i' u_j' \frac{\partial u_i'}{\partial x_j}} = \frac{\partial}{\partial x_j}\left(\overline{u_i' u_j' u_i'}\right)$$

$$\text{Therefore, } \overline{u_i' u_j' \frac{\partial u_i'}{\partial x_j}} = \frac{\partial}{\partial x_j}\left(\frac{1}{2}\overline{u_i' u_j' u_i'}\right)$$

$$\text{Term } 7: \overline{-u_i' \frac{\partial}{\partial x_i}(P+p)}$$

$$= -\frac{\partial}{\partial x_i}\left[\overline{u_i'(P+p)}\right] - \overline{(P+p)\frac{\partial u_i'}{\partial x_i}} = -\frac{\partial}{\partial x_i}\left(\overline{u_i'(P+p)}\right) - 0$$

$$= -\frac{\partial}{\partial x_i}\left(\overline{u_i' p}\right), \text{ since } \overline{u_i' P} = \overline{u_i'} \cdot \overline{P} = 0$$

$$\text{Term } 8: \overline{2\mu_f u_i' \frac{\partial}{\partial x_j}\left(S_{ij} + \mathcal{S}_{ij}'\right)}$$

$$= 2\mu_f \overline{u_i' \frac{\partial}{\partial x_j}\left(S_{ij}\right)} + 2\mu_f \overline{u_i' \frac{\partial}{\partial x_j}\left(\mathcal{S}_{ij}'\right)} = 2\mu_f \overline{u_i'}\frac{\partial}{\partial x_j}\overline{S_{ij}} + 2\mu_f \overline{u_i' \frac{\partial}{\partial x_j}\left(\mathcal{S}_{ij}'\right)}$$

$$= 0 + 2\mu_f \overline{u_i' \frac{\partial}{\partial x_j}\left(\mathcal{S}_{ij}'\right)}$$

$$= 2\mu_f \overline{u_i' \frac{\partial}{\partial x_j}\left(\mathcal{S}_{ij}'\right)} = 2\mu_f \left[\frac{\partial}{\partial x_j}\left(\overline{u_i' \mathcal{S}_{ij}'}\right) - \overline{\mathcal{S}_{ij}' \frac{\partial u_i'}{\partial x_j}}\right]$$

$$= 2\mu_f \left[\frac{\partial}{\partial x_j}\left(\overline{u_i' \mathcal{S}_{ij}'}\right) - \overline{\mathcal{S}_{ij}' \mathcal{S}_{ij}'}\right], \text{ since } \frac{\partial u_i'}{\partial x_j} = \mathcal{S}_{ij}'$$

Now, substituting all the above-average terms in equation (6.164), one gets:

$$0 + \frac{\partial k}{\partial t} + 0 + u_j\frac{\partial k}{\partial x_j} + \rho_f \overline{u_i' u_j'}S_{ij} + \frac{\partial}{\partial x_j}\left(\frac{\rho_f}{2}\overline{u_i' u_j' u_i'}\right) = -\frac{\partial}{\partial x_i}\left(\overline{u_i' p}\right) + 2\mu_f\left[\frac{\partial}{\partial x_j}\left(\overline{u_i' \mathcal{S}_{ij}'}\right) - \overline{\mathcal{S}_{ij}' \mathcal{S}_{ij}'}\right]$$

$$\Rightarrow \frac{\partial k}{\partial t} + u_j\frac{\partial k}{\partial x_j} + \rho_f \overline{u_i' u_j'}S_{ij} + \frac{\partial}{\partial x_j}\left(\frac{\rho_f}{2}\overline{u_i' u_j' u_i'}\right) = -\frac{\partial}{\partial x_i}\left(\overline{u_i' p}\right) + 2\mu_f\frac{\partial}{\partial x_j}\left(\overline{u_i' \mathcal{S}_{ij}'}\right) - 2\mu_f \overline{\mathcal{S}_{ij}' \mathcal{S}_{ij}'}$$

$$\Rightarrow \frac{\partial k}{\partial t} + u_j\frac{\partial k}{\partial x_j} = -\frac{\partial}{\partial x_j}\left[\overline{u_j' p} + \frac{\rho_f}{2}\overline{u_i' u_j' u_i'} - 2\mu_f \overline{u_i' \mathcal{S}_{ij}'}\right] - \rho_f \overline{u_i' u_j'}S_{ij} - 2\mu_f \overline{\mathcal{S}_{ij}' \mathcal{S}_{ij}'}$$

$$\Rightarrow \frac{\partial k}{\partial t} + \frac{\partial}{\partial x_j}\left[ku_j + \overline{u_j' p} + \frac{\rho_f}{2}\overline{u_i' u_j' u_i'} - 2\mu_f \overline{u_i' \mathcal{S}_{ij}'}\right] = -\rho_f \overline{u_i' u_j'}S_{ij} - 2\mu_f \overline{\mathcal{S}_{ij}' \mathcal{S}_{ij}'}$$

$$\Rightarrow \frac{\partial k}{\partial t} + \frac{\partial}{\partial x_j}\left[ku_j + \overline{u_j' p} + \frac{\rho_f}{2}\overline{u_i' u_j' u_i'} - 2\mu_f \overline{u_i' \mathcal{S}_{ij}'}\right] = -\rho_f \overline{u_i' u_j'}\frac{\partial \overline{u_i}}{\partial x_j} - 2\mu_f \overline{\mathcal{S}_{ij}' \mathcal{S}_{ij}'} \tag{6.165}$$

since $S_{ij} = \dfrac{\partial \overline{u_i}}{\partial x_j}$

$$\text{or, } \frac{\partial k}{\partial t}+\frac{\partial}{\partial x_j}\left[ku_j+\overline{u_j'p}+\frac{\rho_f}{2}\overline{u_i'u_j'u_i'}-2\mu_f\overline{u_i'\left(\frac{\partial u_i'}{\partial x_j}+\frac{\partial u_j'}{\partial x_i}\right)}\right]=-\rho_f\overline{u_i'u_j'}\frac{\partial \overline{u_i}}{\partial x_j}-2\mu_f\overline{S_{ij}'S_{ij}'} \quad (6.166)$$

Again, the 8th term of Equation (6.164) can be written as:

$$2\mu_f\overline{u_i'\frac{\partial}{\partial x_j}(S_{ij}+S_{ij}')}=2\mu_f\,\overline{u_i'\frac{\partial}{\partial x_j}S_{ij}}+2\mu_f\overline{u_i'\frac{\partial S_{ij}'}{\partial x_j}}$$

$$=2\mu_f\overline{u_i'}\frac{\partial}{\partial x_j}S_{ij}+2\mu_f\overline{u_i'\frac{\partial S_{ij}'}{\partial x_j}}$$

$$=0+2\mu_f\overline{u_i'\frac{\partial S_{ij}'}{\partial x_j}},\ \text{since } S_{ij}'=\frac{1}{2}\left(\frac{\partial u_j'}{\partial x_j}+\frac{\partial u_j'}{\partial x_i}\right)$$

$$=\overline{\not{2}\mu_f u_i'\frac{\partial}{\partial x_j}\frac{1}{\not{2}}\left(\frac{\partial u_i'}{\partial x_j}+\frac{\partial u_j'}{\partial x_i}\right)}=\mu_f\overline{u_i'\frac{\partial^2 u_i'}{\partial x_j\partial x_j}}+\mu_f\overline{u_i'\frac{\partial^2 u_j'}{\partial x_j\partial x_i}}$$

$$=\mu_f\overline{u_i'\frac{\partial^2 u_i'}{\partial x_j\partial x_j}}+\mu_f\overline{u_i'}\frac{\partial}{\partial x_i}\left(\frac{\partial \overline{u_j}}{\partial x_j}\right)=\mu_f\overline{u_i'\frac{\partial^2 u_j'}{\partial x_j\partial x_j}}+0,\ \text{because } \frac{\partial u_j'}{\partial x_j}=0$$

$$\text{Therefore, } 2\mu_f\overline{u_i'\frac{\partial}{\partial x_j}(S_{ij}+S_{ij}')}=\mu_f\overline{u_i'\frac{\partial^2 u_i'}{\partial x_j\partial x_j}} \quad (6.167)$$

Replacing the 8th term of equation (6.164) with (6.167), again equation (6.165) can be written as:

$$\frac{\partial k}{\partial t}+\frac{\partial}{\partial x_j}\left[ku_j+\overline{pu_j'}+\frac{\rho}{2}\overline{u_i'u_j'u_i'}\right]=-\rho_f\overline{u_i'u_j'}\,S_{ij}+\mu_f\,\overline{u_i'\frac{\partial^2 u_i'}{\partial x_j^2}} \quad (6.168)$$

As the term $u_i'\dfrac{\partial u_i'}{\partial x_i\partial x_j}=0$, this term is added to the above equation (6.168), one gets

$$\frac{\partial k}{\partial t}+\frac{\partial}{\partial x_j}\left(ku_j+\overline{pu_j'}+\frac{\rho_f}{2}\overline{u_i'u_j'u_i'}\right)=-\rho_f\overline{u_i'u_j'}\,S_{ij}+\mu_f\overline{u_i'\frac{\partial^2 u_i'}{\partial x_j\partial x_j}}+\mu_f\overline{u_i'\frac{\partial^2 u_i'}{\partial x_i\partial x_j}} \quad (6.169)$$

The sum of the last two terms of (6.169) can be written as

$$\mu_f u_i'\frac{\partial}{\partial x_j}\left(\frac{\partial u_i'}{\partial x_j}+\frac{\partial u_j'}{\partial x_i}\right)=\mu_f\frac{\partial}{\partial x_j}\left[u_i'\left(\frac{\partial u_i'}{\partial x_j}+\frac{\partial u_j'}{\partial x_i}\right)\right]-\mu_f\left(\frac{\partial u_i'}{\partial x_j}+\frac{\partial u_j'}{\partial x_i}\right)\frac{\partial u_i'}{\partial x_j}$$

Using the above expression of the last two terms in equation (6.169), we get

$$\frac{\partial k}{\partial t}+\frac{\partial}{\partial x_j}\left[ku_j+\overline{pu_j'}+\frac{\rho_f}{2}\overline{u_i'u_j'u_i'}-\mu_f\overline{u_i'\left(\frac{\partial u_i'}{\partial x_j}+\frac{\partial u_j'}{\partial x_i}\right)}\right]=-\rho_f\overline{u_i'u_j'}\,S_{ij}-\mu_f\overline{\left(\frac{\partial u_i'}{\partial x_j}+\frac{\partial u_j'}{\partial x_i}\right)\frac{\partial u_i'}{\partial x_j}}$$

$$\text{or, } \frac{\partial k_f}{\partial t}+\frac{\partial}{\partial x_j}\left[k_f u_j+\overline{pu_j'}+\frac{\rho_f}{2}\overline{u_i'u_j'u_i'}-\mu_f\overline{u_i'\left(\frac{\partial u_i'}{\partial x_j}+\frac{\partial u_j'}{\partial x_i}\right)}\right]=-\phi-\rho_f\overline{u_i'u_j'}\frac{\partial u_i}{\partial x_j} \quad (6.170)$$

where $k_f = \frac{1}{2}\rho_f \overline{u_i' u_j'} = \frac{\rho}{2}\left(u'^2 + v'^2 + w'^2\right)$ is the kinetic energy per unit volume of fluid for fluctuating flow, and $\phi = \mu_f \left(\frac{\partial u_i'}{\partial x_j} + \frac{\partial u_j'}{\partial x_i}\right)\frac{\partial u_i'}{\partial x_j}$ is the viscous dissipation of the turbulent energy per unit volume of fluid and per unit time. *Equation (6.170) is the turbulent kinetic energy equation for the fluctuating flow.*

Equation (6.170) can be rewritten as:

$$\underset{(1)}{\frac{\partial k_f}{\partial t}} + \underset{(2)}{u_j \frac{\partial k_f}{\partial x_j}} = -\underset{(3)}{\frac{\partial}{\partial x_j}\overline{u_j'(p+k_f)}} - \underset{(4)}{\rho_f \overline{u_i' u_j'} \frac{\partial u_i}{\partial x_j}} + \underset{(5)}{\mu_f \frac{\partial}{\partial x_j}\overline{u_i'\left(\frac{\partial u_i'}{\partial x_j} + \frac{\partial u_j'}{\partial x_i}\right)}} - \underset{(6)}{\mu_f \overline{\left(\frac{\partial u_i'}{\partial x_j} + \frac{\partial u_j'}{\partial x_i}\right)\frac{\partial u_i'}{\partial x_j}}} \quad (6.171)$$

The meanings of the terms of equation (6.171) are as follows:

The sum of the two terms (1) and (2) on the left-hand side of equation (6.171) represents the change in kinetic energy of fluctuating flow per unit mass and time with the convective transport by the mean motion. Term (3) of the right-hand side of equation (6.171) is the convective diffusion by turbulence of the total mechanical energy or work by the total dynamic pressure of turbulence. Term (4) represents the work of deformation of the mean motion by the turbulence stresses. Term (5) represents the work by the viscous shear stresses of turbulent motion, and the term (6) represents the viscous dissipation by the turbulent motion per unit mass and the time.

6.9.3 Kinetic Energy Equation for the Mean Flow

Similarly, the kinetic energy equation for the mean flow can be obtained from the Reynolds Stress equation (6.149), multiplying by the mean flow u_i, we get the Reynolds equation as (Sumer, 2002):

$$\rho_f u_i \frac{\partial u_i}{\partial t} + \rho_f u_i u_j \frac{\partial u_i}{\partial x_j} = -u_i \frac{\partial p}{\partial x_i} + u_i \mu_f \frac{\partial}{\partial x_j}\left(\frac{\partial u_i}{\partial x_j}\right) - u_i \frac{\partial}{\partial x_j}\left(\rho_f \overline{u_i' u_j'}\right)$$

$$\text{or, } \frac{\partial}{\partial t}\left(\frac{\rho_f}{2}u_i^2\right) + \frac{\partial}{\partial x_j}u_j\left(\frac{\rho_f}{2}u_i^2\right) = -\frac{\partial}{\partial x_i}(p u_i) - \frac{\partial}{\partial x_j}\left(\rho_f \overline{u_i' u_j'} u_i\right) + \rho_f \overline{u_i' u_i'} \frac{\partial u_i}{\partial x_j}$$

$$+\mu_f u_i \frac{\partial}{\partial x_j}\left(\frac{\partial u_i}{\partial x_j}\right) + \mu_f u_i \frac{\partial^2 u_j}{\partial x_i \partial x_j} \quad (6.172)$$

The last term $\mu_f u_i \frac{\partial^2 u_i}{\partial x_i \partial x_j}$ of (6.172) is added because this term is zero due to equation of continuity.

Therefore, equation (6.172) can be written as:

$$\frac{\partial}{\partial t}\left(\frac{\rho_f}{2}u_i^2\right) + \frac{\partial}{\partial x_j}\left(u_i\left(\frac{\rho_f}{2}u_i^2 + p\right)\right) = \rho_f \overline{u_i' u_j'} \frac{\partial u_i}{\partial x_j} - \frac{\partial}{\partial x_j}\left(\rho_f \overline{u_i' u_j'} u_i\right) + \mu_f u_i \frac{\partial}{\partial x_j}\left(\frac{\partial u_i}{\partial x_j} + \frac{\partial u_j}{\partial x_i}\right)$$

$$\text{or, } \underset{(1)}{\frac{\partial}{\partial t}\left(\frac{\rho_f}{2}u_i^2\right)} + \underset{(2)}{\frac{\partial}{\partial x_j}\left[u_j\left(\frac{\rho_f}{2}u_i^2 + p\right)\right]} = -\underset{(3)}{\left(-\rho_f \overline{u_i' u_j'} \frac{\partial u_i}{\partial x_j}\right)} + \underset{(4)}{\frac{\partial}{\partial x_j}\left(-\rho_f \overline{u_i' u_j'} u_i\right)} + \underset{(5)}{\mu_f \frac{\partial}{\partial x_j} u_i\left(\frac{\partial u_i}{\partial x_j} + \frac{\partial u_j}{\partial x_i}\right)}$$

$$-\underset{(6)}{\mu_f\left(\frac{\partial u_i}{\partial x_j} + \frac{\partial u_j}{\partial x_i}\right)\frac{\partial u_i}{\partial x_j}} \quad 6.173)$$

Equation (6.173) is the balance equation for kinetic energy of the mean flow. Here out of six terms in equation (6.173), the terms 1, 2, 5, and 6 have the same meaning as the corresponding terms in the kinetic energy equation (6.161) for laminar flow through concerning the mean motion. Here only 3rd and 4th terms are due to turbulence stresses per unit mass and time. These two terms are the work of deformation by the turbulence stresses per unit mass and time. It is known that $-\rho_f \overline{u_i' u_j'}$ for $i \neq j$ is the shear stress. Here it is important to note that equation (6.161) is the kinetic energy equation for laminar flow, equation (6.170) is kinetic energy equation for turbulent fluctuating flow, and equation (6.173) is the balance equation for kinetic energy of mean motion.

Finally, the kinetic energy equation for laminar flows from (6.161) is as:

$$\frac{\partial k}{\partial t}+\frac{\partial}{\partial x_j}\left[u_j k+u_j p-\mu_f u_i\left(\frac{\partial u_i}{\partial x_j}+\frac{\partial u_j}{\partial x_i}\right)\right]=-\mu_f\left(\frac{\partial u_i}{\partial x_j}+\frac{\partial u_j}{\partial x_i}\right)\frac{\partial u_i}{\partial x_j} \tag{6.161a}$$

where $k=\frac{\rho_f}{2}\left(u^2+v^2+w^2\right)$ is the kinetic energy for laminar flow.

The *kinetic energy equation for fluctuating flows* from (6.171) is as follows:

$$\frac{\partial k_f}{\partial t}+\frac{\partial}{\partial x_j}\left[\overline{\kappa_f u_j+u_j' p-\frac{\rho_f}{2}\overline{u_i' u_j' u_i'}-\mu_f u_i'\left(\frac{\partial u_i'}{\partial x_j}+\frac{\partial u_j'}{\partial x_i}\right)}\right]=-\rho_f\overline{u_i' u_j'}\frac{\partial u_i}{\partial x_j}-\mu_f\overline{\left(\frac{\partial u_i'}{\partial x_j}+\frac{\partial u_j'}{\partial x_i}\right)\frac{\partial u_i'}{\partial x_j}} \tag{6.171a}$$

where $k_f=\frac{\rho_f}{2}\overline{u_i' u_j'}=\frac{\rho_f}{2}\left(u'^2+v'^2+w'^2\right)$ is the kinetic energy equation for the fluctuating components.

The *kinetic energy equation for the mean flow* from (6.173) is as follows:

$$\frac{\partial k_m}{\partial t}+\frac{\partial}{\partial x_j}\left[k_m u_j+u_j p+\rho_f\overline{u_i' u_j' u_i}-\mu_f u_i\left(\frac{\partial u_i}{\partial x_j}+\frac{\partial u_j}{\partial x_i}\right)\right]=-\left(-\rho_f\overline{u_i' u_j'}\frac{\partial u_i}{\partial x_j}\right)-\mu_f\left(\frac{\partial u_i}{\partial x_j}+\frac{\partial u_j}{\partial x_i}\right)\frac{\partial u_i}{\partial x_j} \tag{6.173a}$$

where $k_m=\frac{\rho_f}{2}\left(u_1^2+u_2^2+u_3^2\right)$ is the kinetic energy for the mean flow.

It is noticed that the first term of the right-hand side of both equations (6.171a) and (6.173a) has an opposite sign which means, the term $\left(-\rho_f\overline{u_i' u_j'}\right)\frac{\partial u_i}{\partial x_j}$ is common in both the equations with an opposite sign. This term is the work of deformation by the turbulence stresses per unit of mass and time, which gives the positive contribution to the turbulence kinetic energy, and the negative contribution to the kinetic energy of the mean motion. This implies that the term is an energy gain (source) for turbulence and it is an energy loss (sink) for the mean flow or vice versa. Therefore, this is referred to as turbulence production. Through the extraction of energy from the mean motion, it transfers this energy to the turbulent motion (Sumer, 2000).

Equation (6.165) is a transport equation, in which the viscous term consists of two different effects, such as sink/source and transport effects. It behaves as a sink as it describes the rate of dissipation of turbulent energy to heat as well as it describes the rate of transport of turbulent energy by viscosity. These two viscous terms of (6.165) is replaced by the 8th term of equation (6.164), which is equation (6.167) as:

$$2\mu_f\overline{u_i'\frac{\partial}{\partial x_i}\left(S_{ij}+S_{ij}'\right)}=\mu_f\overline{u_i'\frac{\partial^2 u_i'}{\partial x_j\,\partial x_j}} \tag{6.167}$$

Substitution of (6.167) in the last term of (6.164) gives equation (6.168).

As $\overline{u_i' \dfrac{\partial^2 u_i'}{\partial x_i \partial x_j}} = \overline{u_i' \dfrac{\partial}{\partial x_j}\left(\dfrac{\partial u_i'}{\partial x_i}\right)} = 0$, one can write

$$\mu_f \overline{u_i' \frac{\partial^2 u_i'}{\partial x_j \partial x_j}} = \overline{\mu_f u_i' \frac{\partial^2 u_i'}{\partial x_j \partial x_j} + \mu_f u_i' \frac{\partial^2 u_j'}{\partial x_i \partial x_j}}$$

$$= \overline{\mu_f u_i' \frac{\partial}{\partial x_j}\left(\frac{\partial u_i'}{\partial x_j}\right) + \mu_f u_i' \frac{\partial}{\partial x_j}\left(\frac{\partial u_j'}{\partial x_i}\right)} = \mu_f \overline{u_i' \frac{\partial}{\partial x_j}\left(\frac{\partial u_i'}{\partial x_j} + \frac{\partial u_j'}{\partial x_i}\right)}$$

$$= \mu_f \frac{\partial}{\partial x_j}\left\{\overline{u_i'\left(\frac{\partial u_i'}{\partial x_j} + \frac{\partial u_j'}{\partial x_i}\right)}\right\} - \mu_f \overline{\left(\frac{\partial u_i'}{\partial x_j} + \frac{\partial u_j'}{\partial x_i}\right)\frac{\partial u_i'}{\partial x_j}}$$

$$= \mu_f \frac{\partial}{\partial x_j}\left(\overline{u_i' \frac{\partial u_i'}{\partial x_j}}\right) + \mu_f \frac{\partial}{\partial x_j}\left(\overline{u_i' \frac{\partial u_j'}{\partial x_i}}\right) - \mu_f \overline{\left(\frac{\partial u_i'}{\partial x_j} + \frac{\partial u_j'}{\partial x_i}\right)\frac{\partial u_i'}{\partial x_j}}$$

$$= \mu_f \frac{\partial^2 k}{\partial x_j^2} + \mu_f \overline{\frac{\partial u_j'}{\partial x_i} \cdot \frac{\partial u_i'}{\partial x_j}} - \mu_f \overline{\left(\frac{\partial u_i'}{\partial x_j} + \frac{\partial u_j'}{\partial x_i}\right)\frac{\partial u_i'}{\partial x_j}}, \text{ where } \kappa = \frac{1}{2} u_i'^2$$

Finally, the expression for viscous term as:

$$\mu_f \overline{u_i' \frac{\partial^2 u_i'}{\partial x_j \partial x_j}} = \mu_f \left[\frac{\partial^2 k}{\partial x_j^2} + \overline{\frac{\partial u_j'}{\partial x_i} \cdot \frac{\partial u_i'}{\partial x_j}}\right] - \mu_f \overline{\left(\frac{\partial u_i'}{\partial x_j} + \frac{\partial u_j'}{\partial x_i}\right)\frac{\partial u_i'}{\partial x_j}} = \mu_f \frac{\partial^2 \overline{k}}{\partial x_j^2} - \mu_f \overline{\frac{\partial u_i'}{\partial x_j} \cdot \frac{\partial u_j'}{\partial x_i}} \tag{6.174}$$

The first term on the right-hand side of (6.174) is the full transport term and second term is the dissipation term. Equation (6.174) represents the physical behavior of the viscous term in the turbulent kinetic energy equation for fluctuating flow.

The turbulence-energy equation (6.171) can be read as:

$$\frac{\partial k_f}{\partial t} + \frac{\partial}{\partial x_j}\left[k_f u_j + \overline{u_i' p} + \frac{\rho_f}{2}\overline{u_i' u_j' u_i'}\right] = -\rho_f \overline{u_i' u_j'} \frac{\partial u_i}{\partial x_j} + \mu_f \frac{\partial}{\partial x_j}\overline{u_i'\left(\frac{\partial u_i'}{\partial x_j} + \frac{\partial u_j'}{\partial x_i}\right)} - \mu_f \overline{\left(\frac{\partial u_i'}{\partial x_j} + \frac{\partial u_j'}{\partial x_i}\right)\frac{\partial u_i'}{\partial x_j}}$$

$$= -\rho_f \overline{u_i' u_j'} \frac{\partial u_i}{\partial x_j} + \mu_f \overline{\left(\frac{\partial u_i'}{\partial x_j} + \frac{\partial u_j'}{\partial x_i}\right)\frac{\partial u_i'}{\partial x_j}} + \mu_f \overline{u_i' \frac{\partial^2 u_i'}{\partial x_j \partial x_j}} + \mu_f \overline{u_i' \frac{\partial^2 u_j'}{\partial x_j \partial x_i}} - \mu_f \overline{\left(\frac{\partial u_i'}{\partial x_j} + \frac{\partial u_j'}{\partial x_i}\right)\frac{\partial u_i'}{\partial x_j}}$$

$$= -\rho_f \overline{u_i' u_j'} \frac{\partial u_i}{\partial x_j} + \mu_f \overline{u_i' \frac{\partial^2 u_i'}{\partial x_j \partial x_j}} + \mu_f \overline{u_i' \frac{\partial^2 u_j'}{\partial x_j \partial x_i}}$$

$$= -\rho_f \overline{u_i' u_j'} \frac{\partial u_i}{\partial x_j} + \mu_f \overline{u_i' \frac{\partial^2 u_i'}{\partial x_j \partial x_j}}, \quad \because \mu_f \overline{u_i' \frac{\partial^2 u_j'}{\partial x_j \partial x_i}} = 0 \text{ from the equation of continuity.}$$

Using (6.174) for the term $\mu_f \overline{u_i' \dfrac{\partial^2 u_i'}{\partial x_j \partial x_j}}$, one gets:

$$\frac{\partial k_f}{\partial t} + \frac{\partial}{\partial x_j}\left[k_f u_j + \overline{u_i' p} + \frac{\rho_f}{2}\overline{u_i' u_j' u_i'}\right] = -\rho_f \overline{u_i' u_j'} \frac{\partial u_i}{\partial x_j} + \mu_f \frac{\partial^2 \overline{k}}{\partial x_j^2} - \mu_f \overline{\frac{\partial u_i'}{\partial x_j} \frac{\partial u_j'}{\partial x_i}} \tag{6.175}$$

It must be noted that two viscous terms in equation (6.175) do not have the same meaning as the viscous terms in equation (6.171).

Looking at the turbulence kinetic energy equation (6.171a), one can visualize the flow of energy. Due to the interaction between the mean motion and the turbulent motion, energy is extracted from the mean motion through work of deformation by the turbulence stresses i.e., termed as turbulence production, converted into turbulence energy which ultimately is converted through work of deformation by viscous stresses in the turbulent motion into heat. The term $-\rho_f \overline{u_i'u_j'}\frac{\partial u_i}{\partial x_j}$ in equation (6.171a) should be positive, which means, this term is an energy gain for turbulence, and therefore it is an energy loss for the mean flow in equation (6.173a). It implies that turbulence gains its energy from the mean flow and the rate at which the energy is extracted from the mean flow is $\left(-\rho_f \overline{u_i'u_j'}\right)\frac{\partial u_i}{\partial x_j}$. The turbulence is generated only when there exists a velocity gradient $\frac{\partial u_i}{\partial x_j}$ in the flow, when there is no gradient of velocity, no turbulence will be there. In this case, any field of turbulence introduced into the flow will be dissipated (Sumer, 2002).

The process of transport, production and dissipation can be expressed as:

$$\frac{Dk}{Dt} + \nabla \cdot T' = P + \in \tag{6.176}$$

where the first term of the left-hand side is the mean flow material derivative of k, the second term is the turbulent transport of turbulent kinetic energy (TKE), the third term is the production of TKE, and the fourth term is the TKE dissipation.

Assuming the density and viscosity both constant, the full form of the turbulent kinetic energy (TKE) equation in tensor form is as:

$$\underset{(1)}{\frac{\partial k}{\partial t}} + \underset{(2)}{\bar{u}_j\frac{\partial k}{\partial x_j}} = \underset{(3)}{-\frac{1}{\rho_f}\frac{\partial \overline{u_i'p'}}{\partial x_i}} \underset{(4)}{-\frac{1}{2}\frac{\partial \overline{u_i'u_j'u_i'}}{\partial x_i}} + \underset{(5)}{\nu_f\frac{\partial^2 k}{\partial x_j^2}} \underset{(6)}{- \overline{u_i'u_j'}\frac{\partial \bar{u}_i}{\partial x_j}} \underset{(7)}{- \nu_f\overline{\left(\frac{\partial u_i'}{\partial x_j}\right)\left(\frac{\partial u_j'}{\partial x_j}\right)}} \underset{(8)}{- \frac{g}{\rho_f}\overline{u_j'S_{ij}}} \tag{6.177}$$

Each of these terms represents the process mentioned below.

(1) Local diffusion, (2) advection, (3) pressure diffusion, (4) turbulent transport, (5) molecular viscous transport, (6) turbulent production, (7) turbulent energy dissipation, and (8) buoyancy flux. By examining these phenomena, the turbulence kinetic energy budget for a particular flow can be found.

6.9.4 Equations for Turbulent Kinetic Energy in Cartesian Coordinate System

The transport equation of turbulent kinetic energy describes how the mean flow promotes the kinetic energy into turbulence. The balance of the kinetic energy of the velocity fluctuation is crucial for understanding the physical processes in turbulent fluctuations, especially in turbulence modeling. Defining the turbulent kinetic energy k with velocity fluctuations, the balance of the quantity can be represented as:

$$k = \frac{1}{2}\overline{q^2} = \frac{1}{2}\overline{\left(u'^2 + v'^2 + w'^2\right)} \tag{6.178}$$

$$\text{with } q^2 = u'^2 + v'^2 + w'^2$$

The equation of turbulent kinetic energy (TKE) is derived from the Navier–Stokes equations in Cartesian coordinates. Detailed derivations are available in the literature (Tennekes and Lumley, 1972;

Schlichting and Gertten, 2000; Kumar, 2016). For unsteady flows with constant physical properties, TKE equation is presented as:

$$\frac{\partial k}{\partial t}+\bar{u}\frac{\partial k}{\partial x}+\bar{v}\frac{\partial k}{\partial y}+\bar{w}\frac{\partial k}{\partial z}=-\left[\overline{u'^2}\frac{\partial \bar{u}}{\partial x}+\overline{u'v'}\frac{\partial \bar{u}}{\partial y}+\overline{u'w'}\frac{\partial \bar{u}}{\partial z}+\overline{u'v'}\frac{\partial \bar{v}}{\partial x}+\overline{v'^2}\frac{\partial \bar{v}}{\partial y}+\overline{v'w'}\frac{\partial \bar{v}}{\partial z}+\overline{u'w'}\frac{\partial \bar{w}}{\partial x}+\overline{v'w'}\frac{\partial \bar{w}}{\partial y}+\overline{w'^2}\frac{\partial \bar{w}}{\partial z}\right]$$

$$-\frac{1}{\rho_f}\left[\overline{u'\frac{\partial p'}{\partial x}}+\overline{v'\frac{\partial p'}{\partial y}}+\overline{w'\frac{\partial p'}{\partial z}}\right]-\frac{1}{2}\left[\frac{\partial \overline{u'^3}}{\partial x}+\frac{\partial \overline{u'^2v'}}{\partial y}+\frac{\partial \overline{u'^2w'}}{\partial z}+\frac{\partial \overline{u'v'^2}}{\partial x}+\frac{\partial \overline{v'^3}}{\partial y}+\frac{\partial \overline{v'^2w'}}{\partial z}+\frac{\partial \overline{u'w'^2}}{\partial x}+\frac{\partial \overline{v'w'^2}}{\partial y}+\frac{\partial \overline{w'^3}}{\partial z}\right]$$

$$+\nu_f\left[\frac{\partial^2\overline{\frac{1}{2}u'^2}}{\partial x^2}+\frac{\partial^2\overline{\frac{1}{2}u'^2}}{\partial y^2}+\frac{\partial^2\overline{\frac{1}{2}u'^2}}{\partial z^2}+\frac{\partial^2\overline{\frac{1}{2}v'^2}}{\partial x^2}+\frac{\partial^2\overline{\frac{1}{2}v'^2}}{\partial y^2}+\frac{\partial^2\overline{\frac{1}{2}v'^2}}{\partial z^2}+\frac{\partial^2\overline{\frac{1}{2}w'^2}}{\partial x^2}+\frac{\partial^2\overline{\frac{1}{2}w'^2}}{\partial y^2}+\frac{\partial^2\overline{\frac{1}{2}w'^2}}{\partial z^2}\right]$$

$$-\nu_f\left[\overline{\left(\frac{\partial u'}{\partial x}\right)^2}+\overline{\left(\frac{\partial u'}{\partial y}\right)^2}+\overline{\left(\frac{\partial u'}{\partial z}\right)^2}+\overline{\left(\frac{\partial v'}{\partial x}\right)^2}+\overline{\left(\frac{\partial v'}{\partial y}\right)^2}+\overline{\left(\frac{\partial v'}{\partial z}\right)^2}+\overline{\left(\frac{\partial w'}{\partial x}\right)^2}+\overline{\left(\frac{\partial w'}{\partial y}\right)^2}+\overline{\left(\frac{\partial w'}{\partial z}\right)^2}\right] \quad (6.179)$$

Using $K^2 = u'^2 + v'^2 + w'^2$ for steady flow condition, $\frac{\partial k}{\partial t}=0$, equation (6.179) is rearranged, and can be written as:

$$\rho_f\left(\bar{u}\frac{\partial k}{\partial x}+\bar{v}\frac{\partial k}{\partial y}+\bar{w}\frac{\partial k}{\partial z}\right)=-\left[\frac{\partial}{\partial x}\left\{\overline{u'\left(p'+\frac{\rho}{2}K^2\right)}\right\}+\frac{\partial}{\partial y}\left\{\overline{v'\left(p'+\frac{\rho}{2}K^2\right)}\right\}+\frac{\partial}{\partial z}\left\{\overline{w'\left(p'+\frac{\rho}{2}K^2\right)}\right\}\right]$$

$$+\mu_f\left[\frac{\partial^2}{\partial x^2}\left(k+\overline{u'^2}\right)+\frac{\partial^2}{\partial y^2}\left(k+\overline{v'^2}\right)+\frac{\partial^2}{\partial z^2}\left(k+\overline{w'^2}\right)+2\left(\frac{\partial^2\overline{u'v'}}{\partial x\partial y}+\frac{\partial^2\overline{v'w'}}{\partial y\partial z}+\frac{\partial^2\overline{w'u'}}{\partial z\partial x}\right)\right]$$

$$-\rho_f\left[\overline{u'^2}\frac{\partial \bar{u}}{\partial x}+\overline{u'v'}\frac{\partial \bar{v}}{\partial x}+\overline{u'w'}\frac{\partial \bar{w}}{\partial x}+\overline{u'v'}\frac{\partial \bar{u}}{\partial y}+\overline{v'^2}\frac{\partial \bar{v}}{\partial y}+\overline{v'w'}\frac{\partial \bar{w}}{\partial y}+\overline{u'w'}\frac{\partial \bar{u}}{\partial z}+\overline{v'w'}\frac{\partial \bar{v}}{\partial z}+\overline{w'^2}\frac{\partial \bar{w}}{\partial z}\right]$$

$$-\mu_f\left[2\overline{\left(\frac{\partial u'}{\partial x}\right)^2}+2\overline{\left(\frac{\partial v'}{\partial y}\right)^2}+2\overline{\left(\frac{\partial w'}{\partial z}\right)^2}+\overline{\left(\frac{\partial u'}{\partial y}+\frac{\partial v'}{\partial x}\right)^2}+\overline{\left(\frac{\partial u'}{\partial z}+\frac{\partial w'}{\partial x}\right)^2}+\overline{\left(\frac{\partial v'}{\partial z}+\frac{\partial w'}{\partial y}\right)^2}\right] \quad (6.180)$$

The term on the left-hand side of equation (6.180) is the convection term, and on the right side of equation (6.180), there are four third brackets like the symbol [...] that contain the terms are respectively known as turbulent diffusion, viscous diffusion, turbulence production and dissipation. Combining the terms viscous diffusion and dissipation (second and fourth terms) of equation (6.180), one gets

$$\mu_f\left[\frac{\partial^2 k}{\partial x^2}+\frac{\partial^2 k}{\partial y^2}+\frac{\partial^2 k}{\partial z^2}\right]-\mu_f\left[\overline{\left(\frac{\partial u'}{\partial x}\right)^2}+\overline{\left(\frac{\partial v'}{\partial x}\right)^2}+\overline{\left(\frac{\partial w'}{\partial x}\right)^2}+\overline{\left(\frac{\partial u'}{\partial y}\right)^2}+\overline{\left(\frac{\partial v'}{\partial y}\right)^2}+\overline{\left(\frac{\partial w'}{\partial y}\right)^2}+\overline{\left(\frac{\partial u'}{\partial z}\right)^2}+\overline{\left(\frac{\partial v'}{\partial z}\right)^2}+\overline{\left(\frac{\partial w'}{\partial z}\right)^2}\right]$$

(6.181)

The second term under third bracket is called pseudo-dissipation (Schlichting and Gersten, 2000). It is also called only dissipation. The energy equation here describes the balance between four contributions to the energy budget of turbulent fluctuations, such as convection, diffusion, production, and dissipation. The turbulent diffusion is composed of two parts: viscous and turbulent. The terms within the last third bracket of equation (6.180) is dissipation rate which is always positive, the dissipation term is an energy sink because of negative sign. In contrast, the turbulence production is generally positive. If the production and dissipation are much larger than the remaining factors in the flow, one can say an equilibrium region, since the production is approximately equal to dissipation. There exist some turbulent flows where the turbulence production is negative, that means, the energy flows reverse from the fluctuations to the mean flow. However, the change in turbulent energy due to convection is compensated by turbulence production (energy source), dissipation (energy sink) and energy transport (diffusion). The dissipation means a change from turbulent kinetic energy to internal energy.

At very large Reynolds numbers, turbulent flows have a locally isotropic structure in nature (Kolmogorov, 1941), excluding the region close to the walls and edges. Isotropic turbulence means that there is no particular direction of fluctuations in a small neighborhood close to point. Since the dissipation occurs in the region of smallest eddies, it is assumed as isotropic turbulence (Kolmogorov, 1941). In this case, $\overline{u'^2} = \overline{v'^2} = \overline{w'^2}$, and $\overline{u'v'} = \overline{u'w'} = \overline{v'w'} = 0$, the dissipation term can be simplified (Hinze, 1973) as $15\mu_f \overline{\left(\frac{\partial u'}{\partial x}\right)^2}$.

6.9.5 Turbulent Energy Equation for Boundary Layer Flows

The plane turbulent boundary layer equations from equation (6.30) with constant physical properties as:

$$\bar{u}\frac{\partial \bar{u}}{\partial x} + \bar{v}\frac{\partial \bar{u}}{\partial y} = -\frac{1}{\rho_f}\frac{d\bar{p}}{dx} + \frac{1}{\rho_f}\frac{\partial}{\partial y}\left(\mu_f\frac{\partial \bar{u}}{\partial y} - \rho_f\overline{u'v'}\right) \tag{6.30a}$$

$$\frac{\partial \bar{u}}{\partial x} + \frac{\partial \bar{v}}{\partial x} = 0 \tag{6.30b}$$

Under the boundary layer approximation, the y-component of the momentum boundary layer equation vanishes. Since the turbulent kinetic energy (TKE) equation (6.180) is used for turbulent flow modeling, the TKE equation for turbulent boundary layer flow is written in a simplified form as:

$$\underbrace{\rho_f\left(\bar{u}\frac{\partial k}{\partial x} + \bar{v}\frac{\partial k}{\partial y}\right)}_{(1)} = \underbrace{\mu_f\frac{\partial^2 k}{\partial y^2}}_{(2)} - \underbrace{\frac{\partial}{\partial y}\left\{\overline{v'\left(p' + \frac{\rho}{2}K^2\right)}\right\}}_{(3)} - \underbrace{\rho_f\overline{u'v'}\frac{\partial \bar{u}}{\partial y}}_{(4)} - \underbrace{\rho_f\left(\overline{u'^2} - \overline{v'^2}\right)\frac{\partial \bar{u}}{\partial x}}_{(5)} - \underbrace{\mu_f[\cdots]}_{(6)}. \tag{6.182}$$

The terms on the right-hand side of equation (6.182) are as: (2) viscous diffusion, (3) turbulent diffusion, (4 and 5) production and (6) dissipation. Here there are two production terms, in which the term (5) production $\rho_f\left(\overline{u'^2} - \overline{v'^2}\right)\frac{\partial \bar{u}}{\partial x}$ is neglected compared to the term (4) $\rho_f\overline{u'v'}\frac{\partial \bar{u}}{\partial y}$. Generally, the production term is considered as $-\rho_f\overline{u'v'}\frac{\partial \bar{u}}{\partial y}$ in turbulent modeling. Therefore, equation (6.182) is considered for fully developed internal turbulent flows. Ultimately, equation (6.182) can be written as:

$$\underbrace{\rho_f\left(\bar{u}\frac{\partial k}{\partial x} + \bar{v}\frac{\partial k}{\partial y}\right)}_{(1)} = \underbrace{\mu_f\frac{\partial^2 k}{\partial y^2}}_{(2)} - \underbrace{\frac{\partial}{\partial y}\left\{\overline{v'\left(p' + \frac{\rho}{2}K^2\right)}\right\}}_{(3)} - \underbrace{\rho_f\overline{u'v'}\frac{\partial \bar{u}}{\partial y}}_{(4)} - \underbrace{\mu_f[\cdots]}_{(6)}. \tag{6.183}$$

where $\rho_f\left(\overline{u'^2} - \overline{v'^2}\right)\frac{\partial \bar{u}}{\partial x} \approx 0$

Couette flow: Let us consider the turbulent Couette flow as a case study for the turbulent boundary layer flow to determine the kinetic energy balance equations for mean motion and turbulent fluctuations. The turbulent Couette flow is defined as the flow between two parallel plates separated by a distance $2H$ apart. The origin is chosen at the lower fixed plate along x-axis and y is the vertical axis from the lower plate. The upper plate at $y = 2H$ moves at a constant velocity $2u_{2H}$, where u_H is the velocity at the center line $y = H$. Under the boundary layer approximations, the balance force for the fully developed turbulent Couette flow is written (from equation 6.30a) as:

$$\frac{\partial}{\partial y}\left(\mu_f \frac{\partial \bar{u}}{\partial y} - \rho_f \overline{u'v'}\right) = \frac{\partial \tau}{\partial y} = 0 \tag{6.184}$$

After integration with respect to y, it gives

$$\mu_f \frac{\partial \bar{u}}{\partial y} - \rho_f \overline{u'v'} = \text{constant} = \tau \tag{6.185}$$

where τ is the total shear stress, is composed of two parts, viscous shear stress τ_l and Reynolds shear stress τ_t. It is known that the shear stress at the wall is purely viscous due to the existence of laminar sub-layer. At the wall boundary $y = 0$, the shear stress $\tau = \tau_w$. Therefore, equation (6.185) is as:

$$\mu_f \frac{\partial \bar{u}}{\partial y} - \rho_f \overline{u'v'} = \tau_w \tag{6.186}$$

which shows that the shear stress is constant over the entire depth, and equal to the wall shear stress τ_w. Equation (6.186) can be written as:

$$\nu_f \frac{\partial \bar{u}}{\partial y} - \overline{u'v'} = \frac{\tau_w}{\rho_f} = u_*^2 \tag{6.187}$$

Multiplying equation (6.187) by $\frac{\partial \bar{u}}{\partial y}$, one gets

$$u_*^2 \frac{\partial \bar{u}}{\partial y} = \nu_f \left(\frac{\partial \bar{u}}{\partial y}\right)^2 - \overline{u'v'} \frac{\partial \bar{u}}{\partial y} \tag{6.188}$$

Equation (6.188) is written in dimensionless form as

$$\underset{(1)}{\frac{\partial u^+}{\partial y^+}} = \underset{(2)}{\left(\frac{\partial u^+}{\partial y^+}\right)^2} + \underset{(3)}{\tau_t^+ \frac{\partial u^+}{\partial y^+}} \tag{6.189}$$

where $u^+ = \frac{\bar{u}}{u_*}$, $y^+ = \frac{y u_*}{\nu_f}$, $\tau_t^+ = \frac{-\rho_f \overline{u'v'}}{\tau_w}$ are the dimensionless quantities. Equation (6.189) is interpreted as the energy balance equation of the mean flow. The term (1) on the left-hand side is the energy supply, and first term (2) on the right-hand side is the direct dissipation and second term (3) is the turbulence production or generation due to fluctuations. The nature of different terms is shown in Figure (6.26).

To determine the energy balance equations for the turbulent fluctuations in the Couette flow, one can use the kinetic energy equation (6.183), which reduces to the following:

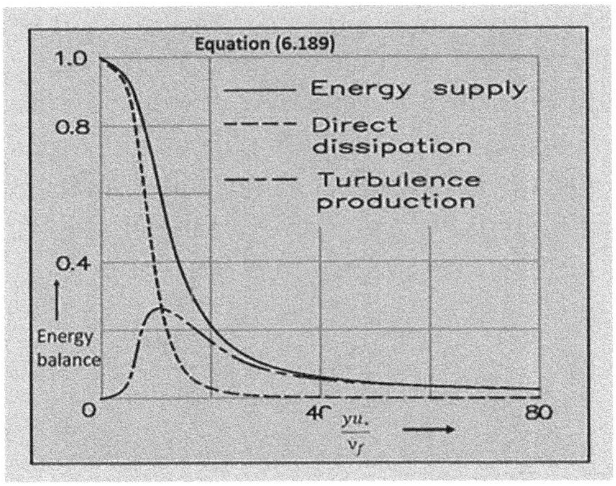

FIGURE 6.26 Typical energy balance equation for the mean motion in the wall layer (equation 6.189). (Modified from Schlichting and Gersten, 2000.)

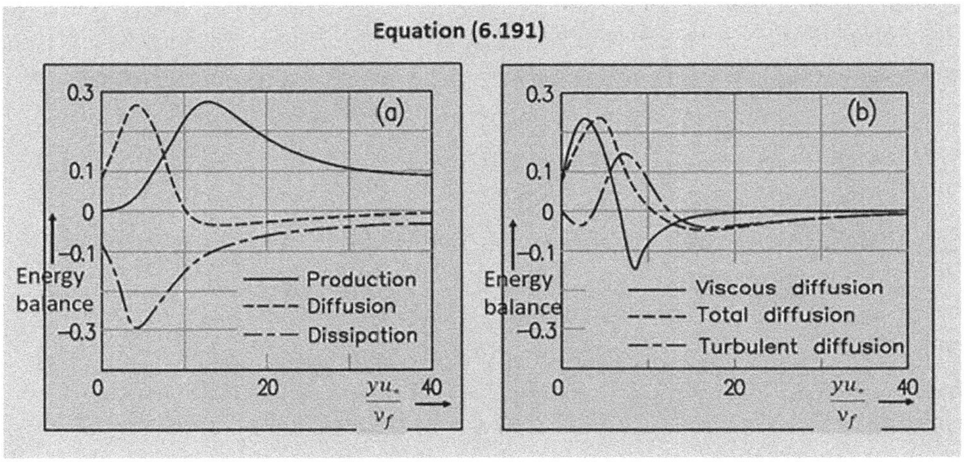

FIGURE 6.27(a, b) Typical universal energy balance of turbulent fluctuations in the wall layer corresponds to equation (6.191). (Modified from Schlichting and Gersten. 2000.)

$$0 = \mu_f \frac{\partial^2 k}{\partial y^2} - \frac{\partial}{\partial y}\left\{\overline{v'\left(p' + \frac{\rho_f}{2}K^2\right)}\right\} - \rho_f \overline{u'v'}\frac{\partial \bar{u}}{\partial y} - \mu_f[\cdots]. \tag{6.190}$$

Equation (6.190) is written in the dimensionless form as (Schlichting and Gersten. 2000):

$$\underset{(1)}{\frac{d^2k^+}{dy^{+2}}} + \underset{(2)}{\frac{dB^+}{dy^+}} + \underset{(3)}{\tau_t^+ \frac{du^+}{dy^+}} - \underset{(4)}{\in^+} = 0 \tag{6.191}$$

with dimensionless quantities $k^+ = \dfrac{k}{u_*}$, $B^+ = -\dfrac{\overline{v'\left(p' + \frac{\rho_f}{2}K^2\right)}}{u_*\tau_w}$, $\in^+ = \dfrac{\mu_f[\cdots]}{u_*^4}$

Here term (1) represents viscous diffusion, (2) turbulent diffusion, (3) turbulence production, and (4) turbulent dissipation. The curves of these terms corresponding to equation (6.191) are individually shown in Figure 6.27(a, b)

6.10 MEAN FLOWS AND TURBULENCE CHARACTERISTICS

In this section, the mean flow and turbulence characteristics in boundary layer flow are defined and explained in detail. The most important characteristic of turbulent flow is the fact that velocity and pressure fluctuate at a point in a random manner. The mixing in turbulent flow is more due to these fluctuations. Turbulent flow can be generated by frictional forces at the confining solid walls, and the flow of layers of fluids with different velocities over one another. The generation of turbulence may be referred in two ways: wall turbulence and free turbulence. The turbulence generated and continuously affected by fixed walls is designated as wall turbulence, and the turbulence generated by two adjacent layers of fluids in the absence of walls is termed as the free turbulence. The viscosity effect on turbulence is to make the flow more homogeneous and less dependent on direction. Therefore, the turbulence is categorized as follows: homogeneous, isotropic, and anisotropic turbulence.

Homogeneous turbulence is defined as the same structure quantitatively in all parts of the flow field. The term homogeneous turbulence implies that the velocity fluctuations in the system are random but the average turbulent characteristics are independent of the position in the fluid, i.e., invariant to axis translation. In homogeneous turbulence, the root mean square (rms) values of u', v', and w' can all be different, but each value must be constant over the entire turbulent flow field. It is important to note that even if rms fluctuations of any component, say u' are constant over the entire field, the instantaneous values of u necessarily differ from the point to point at any instant. Isotropic turbulence is defined as the statistical features have no directional preference and perfect disorder persists. The velocity fluctuations are independent of the axis of reference, i.e., invariant to the axis of rotation and reflection. In such cases, the gradient of mean velocity does not exist; the mean velocity is either zero or constant throughout. Anisotropic turbulence is defined as the statistical features have directional preference and mean velocity has a gradient.

Turbulent flow is diffusive and dissipative. In general, turbulence brings about better mixing of a fluid and produces an additional diffusion agent. Such diffusion is termed as eddy-diffusion. At a large Reynolds number, there exists a continuous transport of energy from the free stream to the large eddies. Then, from the large eddies, smaller eddies are continuously formed. Near the wall, the smallest eddies destroy themselves in dissipating energy, i.e., converting kinetic energy of eddies into intermolecular energy.

Turbulent flows always occur at high Reynolds numbers. They are formed owing to the complex interaction between the viscosity and the inertia in the momentum equations. If the Reynolds number is high enough, the flow becomes turbulent and fully irregular and random, and that is characterized by the irregular movement of fluid particles. Therefore, the turbulent flows are usually described statistically because the deterministic approach is very difficult. Turbulent flows are always chaotic, but not all chaotic flows are turbulent. According to Tennekes and Lumley (1972), turbulence cannot preserve itself; rather it depends on the surrounding environmental source to gain energy. All turbulent flows vary both spatially and temporally (Pope, 2000). The key characteristics of turbulent flow are fluid transport, mixing and dissipation.

For deeper understanding of mechanism of turbulent flow, due to irregular and random nature with certain stationary properties, the statistical approach is viable. To develop the empirical or semi-empirical relations from the turbulent flow using phenomenological ideas, it is necessary to measure the pressure and velocity components with fluctuations from the experimental work of turbulent flows. Reichardt (1933, 1938, 1939) carried out such measurements in a wind tunnel with a rectangular test section 1m wide and 24.4 cm high, with the aid of hot-wire anemometers. He studied the variation of mean velocity components, intensities of turbulence and shear stress components from his experimental data. The measurement of turbulent flow is quite important for most practical applications; particularly, the actual measurement of fluctuating components of velocity and pressure is essential for deeper understanding of the mechanism of turbulent flow. The turbulent boundary layer flow over a surface in a flume or channel is statistically characterized by mean velocity, turbulence intensity, Reynolds shear stress, turbulent kinetic energy (TKE), eddy viscosity, skewness, and kurtosis, TKE production, TKE dissipation, and energy spectra. Details of all parameters of turbulent flow are described in detail in the proceeding subsections.

6.10.1 Mean Flows

The turbulent motions associated with eddies are approximately random, and these are characterized by statistical concepts. In turbulent flow, the fluid velocity at a fixed point in space changes with time in an irregular way and is characterized by random fluctuations. Such fluctuations in the fluid flows demonstrate like macroscopic fluid balls or lumps of varying tiny size called eddies. The size of eddies is determined by the external conditions of flow, passing through the mesh of a honeycomb placed at the entry of the flow. If the flow is unsteady, the mean velocity usually is meant by time-averaging. In describing the hydrodynamic quantity in the flow in a mathematical form, it is devised to separate the velocity components into a mean value (time-averaged value) and fluctuations. Therefore, the time-averaged velocity components: stream-wise, vertical, and transverse are denoted by $(\overline{u}, \overline{v}, \overline{w})$ and their fluctuations by (u', v', w') in x, y and z directions respectively, the following similar statistical relations (equations 6.13, 6.14, and 6.15) for the instantaneous velocity components are defined as:

$$u = \overline{u} + u', \; v = \overline{v} + v', \; w = \overline{w} + w' \tag{6.192}$$

where the time-averaged velocity is found through integration at a fixed point in space given by

$$\overline{u} = \frac{1}{t_1} \int_{t_0}^{t_0+t_1} u \, dt \tag{6.193}$$

in which t_0 is any arbitrary time and time internal t_1 is the time over which the mean is taken. It is clear that the time t_1 should be sufficiently long time internal for reliable mean value and should be completely independent of time. The mean velocity is the velocity in a fluid motion which can be understood from the statistical definition of velocity in a fluid flow. In theory, the velocity record is continuous and the mean value can be evaluated through integration. However, in practice the measured velocity is a series of discrete points. Therefore alternatively, in a discrete way the time-averaged stream-wise $\overline{u}$, wall-normal $\overline{v}$ and transverse $\overline{w}$ velocities are defined as

$$\overline{u} = \lim_{n \to \infty} \frac{1}{n} \sum_{i=1}^{n} u_i, \; \overline{v} = \lim_{n \to \infty} \frac{1}{n} \sum_{i=1}^{n} v_i, \; \overline{w} = \lim_{n \to \infty} \frac{1}{n} \sum_{i=1}^{n} w_i \tag{6.194}$$

where n is the total number of observations. This averaging is known as the time averaging. There are other kinds of averaging, such as ensemble averaging, moving averaging, etc.

The technique for decomposing the instantaneous value from the hydrodynamic quantity is referred to as the Reynolds' decomposition (Schlichting, 1968). The mean value of a certain quantity in fluid flow depends on the number of observations or samples. If the observations are more, the accuracy of the mean value will be statistically better. The observations can be made only after reaching a statistically steady velocity field or equilibrium condition of flow, which means, the simulation is integrated in time to obtain the representative time-averaged properties. The randomness and chaotic nature of the velocity field make the instantaneous values of fluid flow almost ineffective for understanding to work with. Therefore, the state of equilibrium is necessary. The state of equilibrium condition of flow at a point in a location was confirmed from the repeated samplings of instantaneous velocity data for a particular interval of time; say 60 s using electronic gadgets, like ADV, LDA, or PIV. If the mean and variance of instantaneous velocity data of that particular time interval at a point show almost the same at that point, the stable state of flow is confirmed. Therefore, the instantaneous velocity data should be continuously collected using electronic gadgets for different lengths of time period. The collection of velocity data should be maintained for an appropriate time interval to reproduce the turbulent coherent structures in the desired experiments, and to satisfy the ergodicity condition with reference to the velocity mean and variance (Bendat and Piersol, 2000; Barman et al., 2016; Chatterjee et al., 2018). However, when the time averaging is done, the velocity field becomes predictable and suitable for work. If we are looking at an object in motion over a period of time, we are able to determine the mean velocity of the object for any sub-interval of time as well as over the entire period of time. Whenever the velocity data are collected from the flume experiments or fields, using any electronic gadget, these data are instantaneous and time dependent with too many fluctuations with or without high spikes. During the process of laboratory or field data, the high spikes or out liars of raw data should be taken care before analyzing the data for prediction, even to determine the mean value, which is most important. The collected velocity data are generally affected by the Doppler noise associated with the measuring system and thus the noise has to be removed before the analysis of turbulence parameters. The raw velocity data collected from the laboratory or fields should be processed to remove noise using a phase space threshold despiking technique described by Goring and Nikora (2002) and implemented in the win-ADV software. The data should be analyzed systematically for any point. The effects of large noise should be removed by minimizing the possible aliasing effect near the Nyquist frequency. It should be remembered that about 2%–4% of the raw data signals would be excluded from the main data. Such excluded data signals should be replaced by a polynomial interpolation technique for better results, which is important.

For fully developed turbulent flow in a channel, the transverse velocity and the bottom normal velocity components should be approximately zero throughout the depth of the flow (Figure 6.28). In addition, the flow should be independent of the x-coordinate (longitudinal direction). If the flow is fully developed (i.e., $\partial \bar{u}/\partial x = 0$), it follows that $\bar{v} = 0$ from the equation of continuity. Therefore, the inertial terms vanish (Schlichting and Gersten, 2000). For hydraulically rough surface, the stream-wise mean velocity ($\bar{u}/u_*$) is found to follow the log law as:

$$\frac{\bar{u}}{u_*} = \frac{1}{\kappa_0} \ln\left(\frac{y}{y_0}\right) \tag{6.195}$$

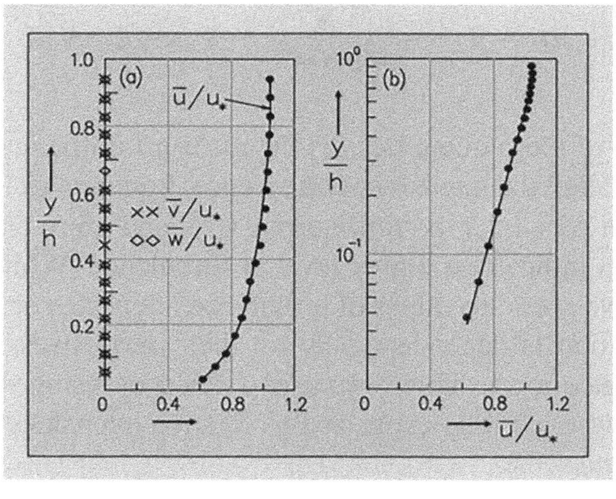

FIGURE 6.28 Fully developed flow: (a) velocity profiles in linear scale, (b) velocity profile in log-scale. (Data verified from the author's laboratory.)

where κ_0 is the von-Karman constant, u_* is the friction or shear velocity and y_0 is the equivalent bed roughness (Nikuradse roughness) with coefficient of regression $R^2 \approx 1.0$. Here, u_* (cm/s) is the shear velocity determined from the shear stress profile or log-law, while y_0 value is determined from the log-law adjusting from the data (Figure 6.28). It is important to note that in summary the mean velocity distributions in open-channel flows can be predicted as a log-law in the wall region $y/h < 0.2$, and as log-wake law in the outer region $y/h > 0.2$ with the von-Karman constant $\kappa_0 = 0.41$ and the integral constant within the range 5.0–5.3 irrespective of flow. These constants are universal for any type of flow, like open-channel flow over smooth surface, boundary layers, pipe flows and closed-channel flows (Nezu and Nakagawa, 1993). But the mean flows over the rough surface, the measurements were performed by many researchers. Tominaga et al. (1989) performed experiments in open channel flow over rough and smooth surfaces and they proved that the von-Karman constant κ_0 is also universal constant irrespective of roughness size with the integral constant being rather different for roughness size. For further details, see Monin and Yaglom (1973).

6.10.2 Turbulence Intensities and Reynolds Shear Stresses

6.10.2.1 Turbulence Intensities

Turbulence intensity is defined as ratio of the root-mean-square (rms) of the velocity fluctuations to the mean flow velocity and commonly given as percentage. Time-averaged values of velocity fluctuations are zero; it follows:

$$\overline{u'} = 0, \overline{v'} = 0, \overline{w'} = 0, \tag{6.196}$$

That means,

$$\overline{u'} = \lim_{n\to\infty} \frac{1}{n} \sum_{i=1}^{n} u_i' = 0 \tag{6.197}$$

and similarly for all other quantities. The important feature is that the fluctuations u', v', w' influence the mean motion $\bar{u}$, $\bar{v}$, $\bar{w}$ in such a way that the latter eventually display an apparent increase in the resistance to the deformation. However, the mean of the squares of the fluctuating components is not zero, but all should be positive. The root-mean-squares of the velocity components are defined as

$$\sigma_u = \sqrt{\overline{u'^2}} = \sqrt{\frac{1}{n}\sum_{i=1}^{n}(u_i - \bar{u})^2}, \sigma_v = \sqrt{\overline{v'^2}} = \sqrt{\frac{1}{n}\sum_{i=1}^{n}(v_i - \bar{v})^2}, \sigma_w = \sqrt{\overline{w'^2}} = \sqrt{\frac{1}{n}\sum_{i=1}^{n}(w_i - \bar{w})^2} \tag{6.198}$$

Here, the root-mean-squares of these quantities are the measure of the magnitudes of velocity fluctuations about the mean value, called the intensity of turbulence. Equation (6.198) contains three respective standard deviations of stream-wise (σ_u), bottom-normal (σ_v) and transverse (σ_w) velocity fluctuations. The larger standard deviation indicates a higher level of turbulence. While studying the turbulent dissipation, it is important to have accurate values of turbulence intensities or velocity fluctuations because the turbulent energy dissipation is dependent only on their derivatives. The intensity of turbulence shows the degree of turbulence in the flow, which is defined by using the fluctuating component of velocity in all directions. These are also expressed as relative intensities by three quantities (stream-wise, bottom-normal, and transverse):

$$u^+ = \frac{\sigma_u}{u_*}, v^+ = \frac{\sigma_v}{u_*}, w^+ = \frac{\sigma_w}{u_*} \tag{6.199}$$

A stationary turbulent flow is characterized by a constant stream-wise mean velocity with zero vertical and transverse velocities and constant values of variances $\overline{u'^2}$, $\overline{v'^2}$, $\overline{w'^2}$. The true velocity at a point at any instant is never known, but at least certain mean quantities can be identified. One such measure is the relative measure of turbulence σ in a flow, which is normalized by friction or shear velocity u_* as:

$$\sigma = \frac{1}{u_*}\sqrt{\frac{(\overline{u'^2} + \overline{v'^2} + \overline{w'^2})}{3}} \tag{6.200}$$

If the turbulence is isotropic, all three quantities are the same, that is, $\overline{u'^2} = \overline{v'^2} = \overline{w'^2}$. The isotropic turbulence is developed in a wind tunnel by placing a uniform grid across a duct. A few mesh lengths downstream, the flow becomes essentially isotropic in nature. This quantity σ is directly related to the kinetic energy of the turbulence. The relative turbulence intensity components (u^+, v^+, w^+) against y/h in the flow along a smooth flat surface (Cebeci, 2004) are shown in Figure (6.29).

All three components of turbulence intensities stream-wise (σ_u), bottom-normal (σ_u) and transverse (σ_w) were first measured by Nakagawa et al. (1975) for open channel flows using hot-film anemometers. The normalized turbulence intensities ($u^+ = \sigma_u / u_*, v^+ = \sigma_v / u_*, w^+ = \sigma_w / u_*$) are the function of independent variable (y/h) (Figure 6.29). These quantities were later investigated for different values of Reynolds numbers and Froude numbers in smooth open-channel flows. The expressions were compared with standard hot-wire data for pipe and closed channel flows. Nakagawa et al. (1975) proposed a power-law distribution for turbulence intensity $u^+ = \sigma_u / u_*$ in the energy equilibrium region, using log-law. On the basis of a phenomenological consideration, the exponential laws of the normalized turbulence intensity components u^+, v^+ and w^+ gave better fits in the intermediate region $0.1 < y/h < 0.6$ than the power laws proposed by Nakagawa et al. (1975). For comparison, see the standard pipe-flow data of Laufer (1954) and Clark (1968). Nezu and Rodi (1986) proposed the following universal semi-analytical relationship for normalized turbulence intensities u^+, v^+ and w^+ as

$$u^+ = \frac{\sigma_u}{u_*} = B_u \exp\left(-C_u \frac{y}{h}\right) \tag{6.201}$$

$$v^+ = \frac{\sigma_v}{u_*} = B_v \exp\left(-C_v \frac{y}{h}\right) \tag{6.202}$$

$$w^+ = \frac{\sigma_w}{u_*} = B_w \exp\left(-C_w \frac{y}{h}\right) \tag{6.203}$$

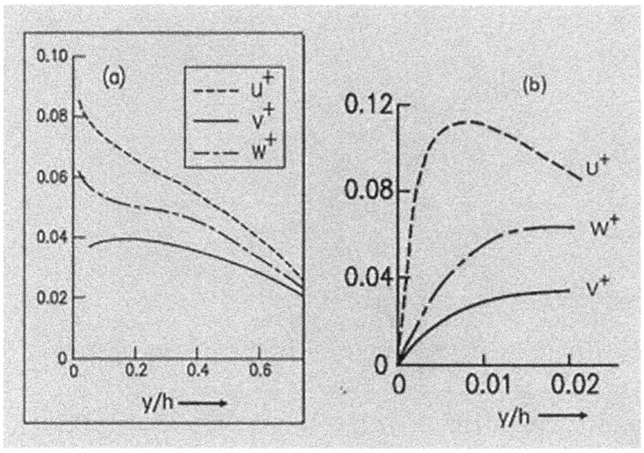

FIGURE 6.29 (a) Turbulence intensities in linear scale (all three components). (b) represents the magnified version for small values of y/h. (Modified from Cebeci, 2004.)

where B_u, B_v, B_w, C_u, C_v, and C_w are the dimensionless coefficients determined from the observed data. Here the coefficients B_u, B_v, B_w are respectively given by 2.30, 1.27, 1.63; and for all the coefficients C_u, C_v and C_w, the value is approximately 1.00, which are proved to be independent of the Reynolds and Froude numbers. It may be mentioned here that LDA data collected by Nezu and Rodi (1986) for $u^+ = \sigma_u / u_*$ show the values of $B_u = 2.26$ and $C_u = 0.88$, which are slightly different from the previous values. This difference is due to the experimental scattered data. For practical purpose, the semi-empirical exponential functions for the turbulence intensities are useful for prediction throughout the depth irrespective of Reynolds and Froude number except near the wall. Laufer (1950, 1951) studied all three components of turbulence intensity from their measurements, and found that all the intensities relative to the mean velocity were minimal at the center of the channel and increased towards the wall until a maximum point was reached; and then rapidly dropped to zero value at the wall. The stream-wise intensity varied from 3% to 12% of mean velocity value at the center portion of the channel, while the transverse and bottom normal intensities were almost the same and much smaller than that of stream-wise intensity. Their variation was about 2%–5% of the mean velocity.

The effect of wall roughness on the turbulence intensities u^+, v^+ and w^+ was measured by Grass (1971), using hydrogen bubble technique, and by Nezu (1977) using hot-film anemometers. Wall roughness is equivalent to sand roughness and it is classified into three types of roughness, such as hydraulically smooth, incompletely rough, and completely rough (mentioned previously). If y/h is greater than 0.3, there is not effect of roughness on turbulence intensities, and consequently, the observed values match well with semi-empirical relations (6.201–6.203); and observed that they are independent of the wall roughness. Raupach (1981) pointed out that the effect of wall roughness element on turbulence intensities occurs primarily near the wall $y/h < 0.3$ of the boundary layer flows; and thus, the region is called the roughness layer. The values of u^+ decreases gradually with roughness k_s^+, but the intensities v^+ and w^+ are less affected by roughness. The peak value of stream wise intensity u^+ is about 2.8 over the smooth surface, while it is 2.0 over a completely rough surface. Though the turbulent production in flow over the smooth surface occurs mainly in the buffer layer in the range $10 < y^+ < 30$, this layer completely or partly disappears over the rough surface as it merges with the roughness elements (Nezu and Nakagawa, 1993). Wei and Willmarth (1989) suggested the effect of Reynolds number $Re = \dfrac{hu_*}{\nu_f}$ on the turbulence quantities u^+, and v^+ for the smooth bed surface (Figure 6.30). Near the wall for small distance, the effect of Re disappears. However, for large distance, the Reynolds number Re significantly influence the turbulence quantities. With an increase

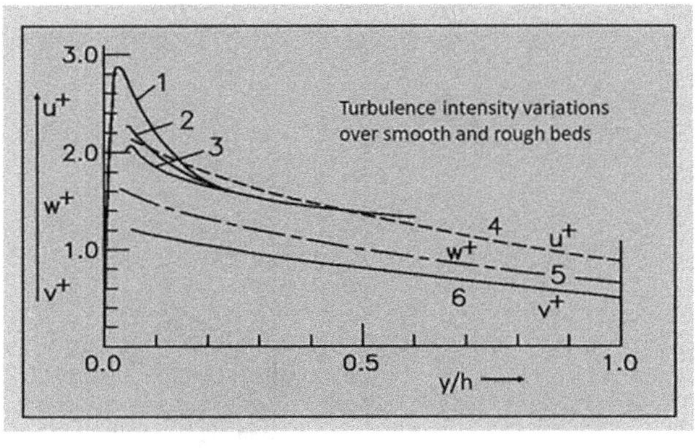

FIGURE 6.30 Turbulence intensity variations over smooth and rough beds as a function of y/h. Curves (1–3) represent intensities respectively for roughness k_s=0, 2.1, 8.5. Curves (4–6) represent respectively normalized stream-wise, bottom-normal, and transverse turbulence intensities, given by equations (6.201–6.203). (Modified from Nezu and Nakagawa, 1993.)

in *Re*, both intensities increase. This effect is due to increase in flow depth, when *Re* increases. Kironoto (1993) observed that the turbulence intensities, u'/u_* and v'/u_*, depend also on the pressure gradients. An acceleration of the mean flow is accompanied by a reduction of the turbulence intensities (damping); when compared with the uniform-flow profiles, the turbulence intensities become more concave. The opposite tendency was observed in decelerating flows; the turbulence intensities increase, and the profiles become convex.

6.10.2.2 Reynolds Shear Stresses

In fluid dynamics, the Reynolds stress is the component of the total stress tensor in a fluid flow obtained from the averaging over the Navier–Stokes equations to account for turbulent fluctuations in fluid momentum. Reynolds (1885) showed that a critical characteristic of turbulence was the capacity to transfer the momentum through eddy motion generated from the fluid flows. The time-averaged Reynolds shear stress (RSS) components (τ_{xy}, τ_{yz}, and τ_{xz}) over three respective planes at each measuring point are defined as:

$$\tau_{xy} = -\rho_f \overline{u'v'} = -\rho_f \frac{1}{n}\sum_{i=1}^{n}(u_i - \overline{u})(v_i - \overline{v}) \tag{6.204}$$

$$\tau_{yz} = -\rho_f \overline{v'w'} = -\rho_f \frac{1}{n}\sum_{i=1}^{n}(u_i - \overline{v})(w_i - \overline{w}) \tag{6.205}$$

$$\tau_{xz} = -\rho_f \overline{u'w'} = -\rho_f \frac{1}{n}\sum_{i=1}^{n}(u_i - \overline{u})(w_i - \overline{w}) \tag{6.206}$$

where n is the total number of observations or samples at each measuring point and p_f is the density of the fluid. These stresses are called "apparent stresses" of turbulent flow and must be added to the steady flow stresses to the laminar flow. These covariance terms or Reynolds shear stress terms arise due to momentum transfer of fluctuating velocity (Pope, 2000). The momentum flux terms are usually referred to as the Reynolds stress terms as they play an analogous role in the governing equations to the viscous stress. The sign in the momentum flux terms indicates the increase or decrease in momentum depending on the positive or negative respectively. It may be noted that the negative shear stress at the boundary leads to outward flux of momentum, and the positive shear stress leads to inward flux of momentum. There are six independent Reynolds stress terms: three normal stress terms $-\rho_f \overline{u'^2}, -\rho_f \overline{v'^2}, -\rho_f \overline{w'^2}$, and three shear stress terms $-\rho_f \overline{u'v'}, -\rho_f \overline{v'w'}, -\rho_f \overline{u'w'}$. These Reynolds stress terms are normalized by the square of friction velocity u_*, that means, normalized shear stress $\tau_{uv} = -\overline{u'v'}/u_*^2$. Here the shear stress $-\rho_f \overline{u'v'}$ indicates the vertical advection of stream-wise turbulent momentum, or simply the vertical flux of stream wise momentum, similarly for other shear stress terms. It is important to note that the Reynolds stress components are directly estimated, if the three-dimensional velocity components with fluctuations at any point with time are known. For example, measurements of the Reynolds shear stress $-\rho_f \overline{u'v'}$ in water flow were first reported by McQuivey and Richardson (1969). Nakagawa et al. (1975) used simultaneous measurements of stream wise $u(t)$ and vertical $v(t)$, using dual-sensor hot-film probe to determine the Reynolds shear stress $-\rho_f \overline{u'v'}$. The following expression for the Reynolds shear stress $-\rho_f \overline{u'v'}$ is obtained from the linear shear stress and the log-wake law given by:

$$-\frac{\overline{u'v'}}{u_*^2} = \left(1 - \frac{y}{h}\right) - \frac{1}{\kappa_0 Re}\left(\frac{h}{y} + \pi\,\Pi\,\mathrm{Sin}\left(\pi\frac{y}{h}\right)\right) \tag{6.207}$$

If there is no variation of velocity in the transverse or z-direction, the shear stress $-\rho_f \overline{u'w'}$ should be zero, and also if there is no mean flow velocity in yz-plane, the shear stress $-\rho_f \overline{u'w'}$ should be zero. Furthermore, the Reynolds stress tensor is symmetric with respect to the diagonal. If it is assumed like this, there will be only four independent Reynolds stresses, and these are the three diagonal components (Figure 6.31a), and one off-diagonal component of shear stresses, namely, $-\rho_f \overline{u'v'}$ (Figure 6.31b) (Sumer, 2002).

6.10.2.3 Correlation Coefficients

As is seen from the Reynolds stresses, the correlations or time-averages of the products of fluctuating quantities play a vital role in describing the turbulent flows. The correlations of different fluctuating quantities of different velocity components at the same point at different times (auto-correlation) or at the same time at different positions (space correlation) are of also important to know the turbulence behaviours. The coefficient of correlation R_{uv} is a statistical measure of structural consistency of the turbulence between two variables u' and v'. To quantify the efficiency of turbulent mixing, the correlation coefficient (R_{uv}) between two variables is defined as follows:

$$R_{uv} = -\frac{\overline{u'v'}}{\sigma_u \sigma_v} \tag{6.208}$$

where σ_u and σ_v are the turbulence intensities along u and v directions. A correlation coefficient of the Reynolds shear stress also indicates the degree of similarity of turbulence, which is very important for describing the turbulent flows. The value of the correlation coefficient R_{uv} increases monotonically with y/h in the wall region and decreases in the free surface region; and also, R_{uv} remains nearly constant about 0.4–0.5 in the intermediate region (Nezu, 1977; Nezu and Nakagawa, 1993; Singh et al., 2007) at the flat surface (Figure 6.32). The comparative studies for boundary layers, pipe flows and the open-channel flows indicate that distribution of correlation coefficient R_{uv} is universal, and it is independent of the mean flow and wall roughness. Substituting equations (6.201) and (6.202) for σ_u and σ_v and equation (6.207) for shear stress $-\overline{u'v'}$, the empirical relation for R_{uv} can be estimated as:

$$R_{uv} = -\frac{\overline{u'v'}}{\sigma_u \sigma_v} = \frac{\left(1-\frac{y}{h}\right) - \frac{1}{\kappa_0 Re}\left(\frac{h}{y} + \pi \Pi \sin\left(\pi \frac{y}{h}\right)\right)}{2.92 \exp\left(-\frac{2y}{h}\right)} \tag{6.209}$$

Equation (6.209) agrees well with the experimental values except in the vicinity of the wall.

The correlation coefficients between the values of stream-wise fluctuations at two different points in space can also be determined. The scales of turbulence and micro-scales at different

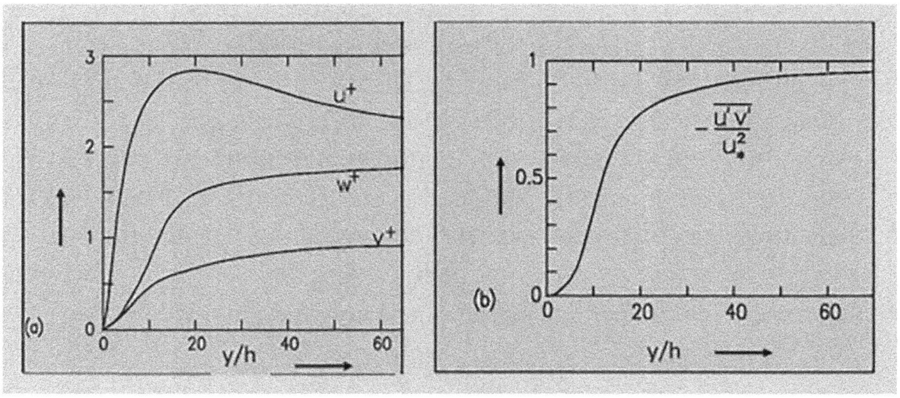

FIGURE 6.31 Schematic of (a) dimensionless turbulence intensities, and (b) shear stress. (Modified from Sumer, 2002.)

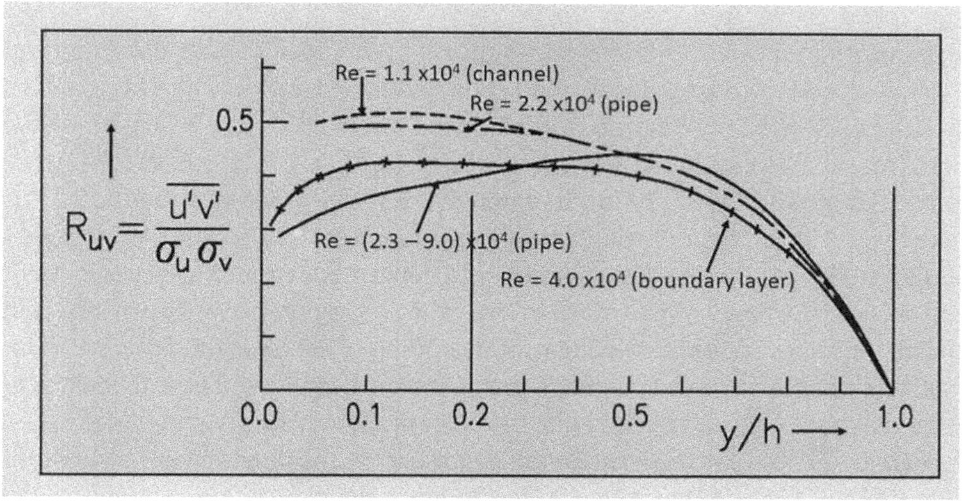

FIGURE 6.32 Trend of correlation coefficient for Reynolds shear stress against *y*/*h*. (Modified from Nezu and Nakagawa, 1993.)

points across the channel can be obtained from these measurements. The scales of turbulence in the central region are independent of Reynolds number and depend only on channel width, while the micro-scales depend on the Reynolds number. The distributions of the dimensionless Reynolds shear stress decrease, becoming concave, and compared with the one in uniform flow in accelerating flows, and increase, becoming convex in decelerating flows (Kironoto, 1993). Close to the wall, the Reynolds (total) stress distribution can be explained analytically. The distributions of the correlation coefficient R_{uv} are universal; being essentially independent of the pressure-gradient parameter.

6.10.3 Turbulence Bursting Process

The ratio of the inertial force to the viscous force is the Reynolds number. In the open channel laminar flow, the Reynolds number defined by the hydraulic radius is less than $Re = 500$. If the Reynolds number *Re* exceeds this limit 500, the turbulent burst appears intermittently in the flow. With increase in *Re*, the frequency and duration of the bursts also increase until $Re > O$ (1000), at which point the turbulence is fully persistent. During the years 1960–1970, the experimental work has shown that the nature of flow pattern near the wall in turbulent boundary layer is repetitive, and the flow at the wall becomes quasi-cyclic process, which is a deterministic sequence of events that occurs randomly in space and time. Kline et al. (1967) first discovered the idea of bursting phenomenon to describe the turbulence features based on the transfer of momentum between the turbulent and laminar regions near a boundary. Rao et al. (1971) described the "bursting" phenomenon in a turbulent boundary layer. Taking into consideration of quasi-cyclic process of flow pattern and the basic flow features due to bursting, Offen and Kline (1973, 1975) proposed a model explaining the complete cycle of events. Review of literature can be found in Laufer (1975), Hinze (1975), Sumer and Oguz (1978), Raupach (1981), Grass (1983), Nezu and Nakagawa (1993) and Mazumder (2000). According to Offen and Kline's (1973, 1975) investigations, the flow consists of a transverse change of low-speed and high-speed streak regions, which form locally and momentarily in the boundary. The boundary layer separation causes due to the adverse pressure gradient, similar to which, a low-speed wall streak forms subject to separation, which causes the coherent, and consequently the low-speed fluid entrains into the main body of the fluid. The low-speed fluid moving towards the main body is called the *ejection event*. This ejected fluid grows in size and is convected downstream with coherent entity for a certain period of time, and finally at some time ($t = T$), that coherent structure breaks up. At the same time, some fluid returns back and intrudes to the wall with a high-speed streak, and spread out sideways. This returned high-speed fluid streak is called *sweep* or *in-rush* event. This way, two such neighbouring high-speed streaks unite and slow down there, and again a new low-speed fluid streak is formed. The complete time period from the formation of low-speed streak at time $t = 0$ to the formation of high-speed streak at time $t = T_1$ is known as the *bursting period*. The occurrence of events from low-speed streak to high-speed one is continuous process in a repetitive manner (Sumer and Oguz, 1978). It is noticed that this quasi-cyclic process, known as *bursting process*, takes place in the case of rough wall as well (Grass, 1971; Grass et al., 1991).

Fulgosi et al. (2003) defined a *coherent structure* as a connected, large-scale turbulent fluid mass with a phase-correlated vorticity throughout its spatial extent. The separation between coherent and non-coherent motion is of crucial importance to obtain a better understanding of the transport processes. On a smooth-wall turbulent boundary layer, stream-wise coherent structures have been linked to ejections and sweeps, which are responsible for moving slow-moving fluid into the outer layer and bringing high momentum fluid into the wall region, respectively. These events generate the major part of the drag and are well correlated with heat and mass transfer fluxes (Lam and Banerjee, 1992). The key issue is to define a suitable criterion that identifies boundaries, topology, and dynamics in the spatial and temporal extent of these structures.

There are several ways to describe and quantify the *turbulent structures*. High vorticity magnitude is a possible candidate for vortex identification in free-shear flows. However, the vortex contours can be extremely complex, and the structures that are visible in plots of vorticity depend largely on the choice of a low-level threshold (McIlwain and Pollard, 2002). Besides, in the presence of a boundary the mean-shear creates a residual vorticity, which is uncorrelated with the vorticity caused by the coherent motions. Another commonly used method is the so-called *Quadrant Analysis* which is used to analyse the generation mechanisms of the turbulent structures or process of bursting. Turbulent structures play an important role in heat and momentum transfer, sediment transport, and dispersion of contaminants. Identification of coherent vortical structures is not only useful for understanding turbulent motion, but crucial in the development of viable turbulence models for complex flows.

6.10.4 Quadrant Threshold Technique

The quadrant analysis is originally developed to classify the contributions of turbulent events to the Reynolds shear stress τ from each quadrant of instantaneous values on the $u'v'$- plane (say). The turbulent boundary layer is directly associated with intermittent large-scale coherent structures. These arrangements are quasi-periodic and occupy the entire depth of flow and are classified randomly by ejection and sweep events. The intermittent contributions to shear stress are associated with this ejection and sweep events. These events characterize the kinematics of structures in the vertical plane. The normalized shear stress τ_{dim} is sought to examine the turbulence over the structures with respect to turbulent events.

Researchers (Corino and Brodkey, 1969; Lu and Willmarth, 1973; Clifford et al., 1993; Katul et al., 2006; Ojha and Mazumder, 2008; Mazumder et al., 2009; Maity and Mazumder, 2017) attempted to determine quantitative results about the structure of Reynolds shear stress from the velocity data. Following Raupach (1981) the turbulent events are defined by four quadrants (Figure 6.33): (Q_i^{xy}, $i = 1, 2, 3, 4$) with outward interaction ($i = 1$, $u'>0$, $v'>0$), when high-speed fluid streak moves far away from the bottom boundary; ejections ($i = 2$, $u'<0$, $v'>0$), for low-speed fluid streak moving far away from the bottom boundary; inward interactions ($i = 3$, $u'<0$, $v'<0$), for low-speed fluid moving towards the bottom boundary; and sweeps ($i = 4$, $u'>0$, $v'<0$), occurring when high-speed fluid moves toward the bottom boundary, also called *in-rush* event (Nakagawa and Nezu, 1977). Contributions of turbulent events to the total Reynolds shear stress originate from four different quadrants in the (u, v)-plane. A powerful tool for investigating the nature and mechanisms of turbulent processes is the method of conditional sampling. The quadrant technique and the conditional averaged method are used to examine the characteristics of bursting events. The ejection is a sudden outward motion of low-speed fluid from

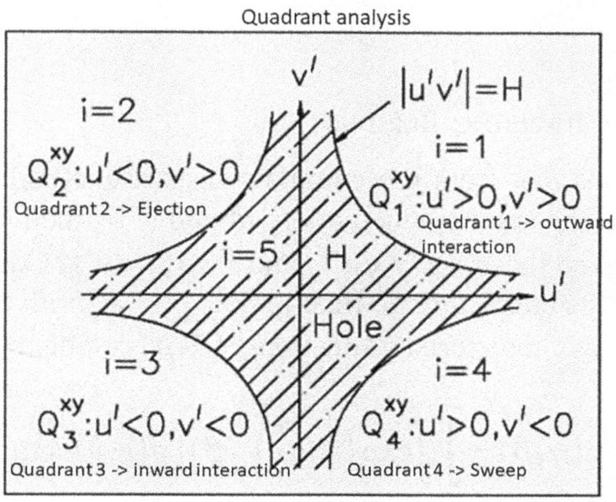

FIGURE 6.33 Quadrant analysis for turbulent events. (Modified from Raupach, 1981.)

the bottom boundary associated with the values of $u'v'$ in the second quadrant ($u'<0$, $v'>0$). Similarly, a sweep phase is the motion of high-speed fluid moving toward bottom boundary associated with the amplitude values of $u'v'$ in the fourth quadrant ($u'>0$, $v'<0$). Two-dimensional quadrant analyses are performed to study the effects of different events on the Reynolds shear stress for various values of flow parameters. At any point in a stationary flow, the contributions of the turbulent events to the Reynolds shear stress from quadrant i, excluding a hyperbolic hole region H, is given by

$$\langle u'v' \rangle_{i,H} = \lim_{T\to\infty} \int_0^T u'(t)v'(t)I_{i,H}\left[u'(t),v'(t)\right]dt \tag{6.210}$$

where square brackets denote a conditional average and the indicator function $I_{i,H}$ is defined as

$$I_{i,H}(u',v') = \begin{cases} 1, \text{ if } (u'v') \text{ is in the ith quadrant and } |u'v'| \geq H\left|\overline{u'v'}\right|, \\ 0, \text{ otherwise} \end{cases} \tag{6.211}$$

The parameter H is a threshold level in the Reynolds shear stress signals. The contributions to each quadrant of the extreme value of shear stress were determined (Nezu and Nakagawa, 1993; Clifford et al., 1993). Here the stress fraction is given by

$$S_{i,H} = \frac{\langle u'v' \rangle_{i,H}}{\overline{u'v'}} \tag{6.212}$$

By definition, $S_{i,H}>0$ if i is even (ejections and sweeps) and $S_{i,H}<0$ if i is odd (outward and inward interactions). Moreover, $S_{1,0}+S_{2,0}+S_{3,0}+S_{4,0}=1$ for threshold level (hole size) $H=0$. The intensities of stress fraction $|S_{i,H}|$ for all events decrease with increasing the vertical distance from the near-wall, that agree reasonably well with Fer et al. (2004). This increment near the boundary is probably linked with relatively high turbulence level. The overall trend shows that the contributions of ejections and sweeps (Q_2 and Q_4) to the shear stress $|S_{i,H}|$ are much higher than that of the outward and inward interactions (Q_1 and Q_3) for any H for all these parameter values (Ojha et al., 2019). It is interesting to note that the sum of contributions of all quadrant events decreases from the bottom level to height $y = 1.0\,\text{cm}$ for parameter $H > 0$ (Figure 6.33).

6.10.5 Theoretical Model for Turbulence Bursting

To estimate the stress fractions $S_{i.H}$ from the cumulant discard method (Nakagawa and Nezu, 1977; Raupach, 1981), we normalize the velocity fluctuations u' and v' by their standard deviation (root mean square r.m.s) in each direction so that $\hat{u} = u' / \sqrt{\overline{u'^2}}$, and $\hat{v} = v' / \sqrt{\overline{v'^2}}$. Considering the joint probability function of $\hat{u}$ and $\hat{v}$ by $p(\hat{u},\hat{v})$ its characteristic function by $\chi(\alpha,\beta)$ with the moment of $\overline{\hat{u}^s\hat{v}^t}$ by m_{st} and corresponding cumulant q_{st}, the characteristic function $\chi(\alpha,\beta)$ can be defined as:

$$\chi(\alpha,\beta) = \int_{-\infty}^{+\infty}\int_{-\infty}^{+\infty} \exp\{i(\hat{u}\alpha + \hat{v}\beta)\}\, p(\hat{u},\hat{v})\, d\hat{u}\, d\hat{v} \tag{6.213}$$

where α and β are its arguments. Here m_{st} and q_{st} correspond respectively to the coefficients in Taylor series expansions of $\chi(\alpha,\beta)$ and $\ln \chi(\alpha,\beta)$; and hence the relationship between the moments m_{st} and cumulants q_{st} are successively obtained. Using and inverse transform of equation (6.213) in which the terms of $\chi(\alpha,\beta)$ less than fourth order are taken into account, $p(\hat{u},\hat{v})$ can be written as:

$$p(\hat{u},\hat{v}) = \frac{1}{2\pi}\int_{-\infty}^{+\infty}\int_{-\infty}^{+\infty} \exp\{-i(\hat{u}\alpha + \hat{v}\beta)\}\, \chi(\alpha,\beta)\, d\alpha\, d\beta \tag{6.214}$$

$$p(\hat{u},\hat{v}) = \phi(\hat{u},\hat{v})\left[1 + \sum_{s+t=3}^{4} \frac{q_{st}}{s!t!} H_{st}(\hat{u},\hat{v})\right] \tag{6.215}$$

$$\text{where, } \phi(\hat{u},\hat{v}) \equiv \frac{1}{2\pi\sqrt{1-R^2}} \exp\left\{-\frac{\hat{u}^2 + 2R\hat{u}\hat{v} + \hat{v}^2}{2(1-R^2)}\right\} \tag{6.216}$$

Here $\phi(\hat{u},\hat{v})$ is the Gaussian distribution for two variables and $H_{st}(\hat{u},\hat{v})$ is a Hermite polynomial of order $(s+t)$. Equation (6.215) represents a joint probability density distribution of the Gram-Charlier type in bivariate case.

Normalizing the Reynolds shear stress $\tau_n\left(= \frac{u'v'}{\overline{u'v'}} = -\frac{\hat{u}\hat{v}}{R}\right)$ and using the change of variables, the probability distribution $P_{\tau_n}(\tau_n)$ of τ_n is given by

$$P_{\tau_n}(\tau_n) = \int_{-\infty}^{+\infty} \frac{R}{|\hat{u}|}\, p\left(\hat{u}, -\frac{R\tau_n}{\hat{u}}\right) d\hat{u}$$

$$= \frac{R}{\pi\sqrt{1-R^2}} \exp\left(\frac{R^2\tau_n}{1-R^2}\right) \int_0^{x} \exp\left\{-\frac{\hat{u}^2 + R^2\left(\frac{\tau_n}{\hat{u}}\right)^2}{2(1-R^2)}\right\} \times \left[1 + \sum_{s+t=3}^{4} \frac{q_{st}}{s!t!}\frac{1}{2}\left\{H_{st}\left(\hat{u}, -\frac{R\tau_n}{\hat{u}}\right) + H_{st}\left(-\hat{u}, \frac{R\tau_n}{\hat{u}}\right)\right\}\right] \frac{d\hat{u}}{\hat{u}} \tag{6.217}$$

where $-R = \left(\frac{\overline{u'v'}}{\sigma_u \sigma_v}\right)$ is the correlation coefficient.

Let $p_{Q_i}(\tau_n)(i=1, 2, 3, 4)$ denote the probability distribution of the i th event, the total probability distribution $p_{\tau_n}(\tau_n)$ is the sum of all events as:

$$p_{\tau_n}(\tau_n)=p_{Q1}(\tau_n)+p_{Q2}(\tau_n)+p_{Q3}(\tau_n)+p_{Q4}(\tau_n) \tag{6.218}$$

The probability density function of normalized shear stress is

$$p_{\tau_n}(\tau_n)=2p_G(\tau_n) \tag{6.219}$$

$$\text{with, } p_G(\tau_n)=\frac{R}{2\pi}\exp(R\xi)\frac{K_0(|\xi|)}{\sqrt{1-R^2}},\quad \xi=\frac{R\tau_n}{(1-R^2)} \tag{6.220}$$

where K_0 is the zeroth order modified Bessel function of second kind.

The contribution to the Reynolds stress $S_{i,H}$ corresponding to each event can be represented by

$$\begin{aligned} S_{i,H} &= \int_H^{\infty} \tau_n p_{Q_i}(\tau_n)d\tau_n > 0 \ \ (i=2, 4) \\ S_{i,H} &= \int_{-\infty}^{-H} \tau_n p_{Q_i}(\tau_n)d\tau_n < 0 \ \ (i=1, 3) \end{aligned} \tag{6.221}$$

Then we define the relation for ΔS_H and Δl_H

$$\Delta S_H = \Delta S_{4,H} + \Delta S_{2,H} \tag{6.222}$$

$$\Delta l_H = \Delta S_{3,H} + \Delta S_{1,H} \tag{6.223}$$

Equations (6.222) and (6.223) represent the downward and upward fluxes of momentum respectively. The ratio of upward to downward flux of momentum is given by

$$E_H = \frac{\Delta l_H}{\Delta S_H} \tag{6.224}$$

which is a measure of relative dominance of downward and upward momentum fluxes for different threshold parameter H.

6.10.6 Distribution of Eddy Viscosity

The turbulent flow is often described by eddy viscosity as a local property of flow; whereas the molecular viscosity is a property of fluid. The closure models of the Reynolds shear stresses normally use an eddy viscosity hypothesis based on analogy between the molecular and turbulent motions. The turbulence eddies are thought of as lumps of fluid, like molecules, they collide each other and exchange momentum (Chatterjee et al., 2018). One of the most striking features of turbulence is the generation of turbulent eddies and self-similar behavior over a range of scales (Batchelor and Townsend, 1949). These eddies play an important role in river turbulence, which affects morpho-dynamics and sediment transport (Nikora and Goring, 2000). The corresponding mixing length due to the movement of turbulent eddies behaves like molecular mean free path derived from kinetic theory of gas (Markatos, 1986). Since the Reynolds shear stress and the mean velocity gradient are known at different locations along the flow for fully developed flow over a flat surface, the eddy viscosity for 2-D flows according to Boussinesq's concept is defined as

$$\nu_t^+ = \frac{-\overline{u'v'}}{\dfrac{\partial \bar{u}}{\partial y}} \tag{6.225}$$

The eddy viscosity is the characteristics of turbulence, closely correlated to the Reynolds shear stress. The normalized eddy viscosity ν_t with depth (*y*/*h*) for the fully developed flow region is written from the log-law as

$$\nu_t = \frac{\nu_t^+}{u_* h} = \kappa_0 \frac{y}{h}\left(1 - \frac{y}{h}\right) \tag{6.226}$$

In a fully developed flow with logarithmic velocity profile, the distribution of normalized eddy viscosity ν_t shows a standard parabolic profile with respect to the vertical axis (*y*/*h*), and approximately linear near the bottom as suggested by Nezu and Rodi (1986) using experimental data. Nezu and Rodi (1986) evaluated the normalized eddy viscosity ν_t with depth (*y*/*h*) using his collected LDA data, and are plotted in Figure (6.34). The experimental data collected in open channel flow by Jobson and Sayre (1970b), Ueda et al. (1977) and the data collected from closed channel by Hussain and Reynolds (1975) are included in Figure (6.34). They obtained an empirical equation as follows:

$$\nu_t = \kappa_0\left(1 - \frac{y}{h}\right)\left[\frac{h}{y} + \pi \Pi \sin\left(\pi \frac{y}{h}\right)\right]^{-1} \tag{6.227}$$

where Π is the Coles wake parameter, which has a significant effect on the eddy viscosity near the mid-depth of the channel. It is important to note that the eddy viscosity ν_t is not zero at the symmetrical axis of the closed channel flow, while it approaches to zero at the free surface in open channel flow. The results suggest significant differences in turbulent structure between the central region of the closed channel and the free surface region of the open channel.

The idea of employing an eddy viscosity assumption to the closure problem of turbulence is used to determine the Reynolds stress analogous to that resulting from viscosity, the Reynolds stress tensor can be read as

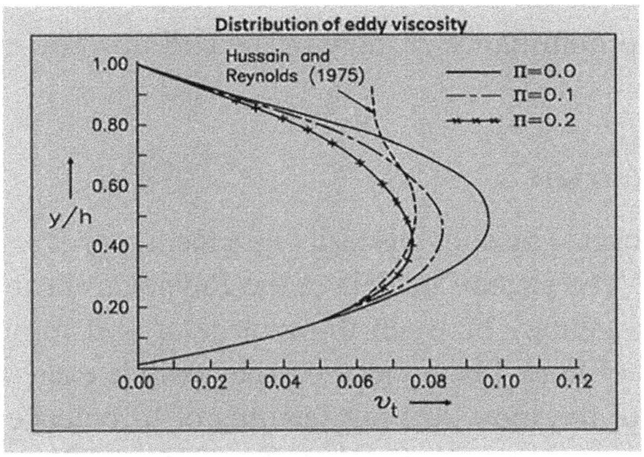

FIGURE 6.34 Distribution of eddy viscosity *u*/*t* vs. vertical distance *y*/*h* in open-channel flow. (Modified from Nezu and Nakagawa, 1993.)

$$-\overline{u_i u_j} = -\frac{1}{3}\overline{q^2}\delta_{ij} + \nu_t\left(\frac{\partial u_i}{\partial x_j} + \frac{\partial u_j}{\partial x_i}\right) \tag{6.228}$$

where $\overline{q^2} = 3u'^2 = \overline{u_i u_i}$ is the total mean-squared turbulent velocity. In equation (6.228), a term analogous to the pressure in the usual stress tensor for a viscous fluid, which can be termed as real pressure in the Reynolds stress equations. Therefore, this equation is used in practice with an eddy viscosity ν_t, which is positive, may vary with position and time and must be specified to the Reynolds stress for definite prediction.

The negative values of ν_t may also be observed, which imply that the eddy motion is transferring energy from smaller scales to the larger scales, known as inverse cascading. Negative values of ν_t indicates higher degree of anisotropy in the flow (Sivashinsky and Yakhot, 1985; Sivashinsky and Frenkel, 1992). It happens when the flow undergoes high frequency periodic oscillations. The eddy viscosity generated by rapidly oscillating eddies may be negative, thus reducing the total (molecular and eddy) viscosity of the system. At high Reynolds numbers, the eddy viscosity might even become negative promoting the spontaneous formation of large-scale eddies (Sivashinsky and Frenkel, 1992).

6.10.7 Mixing Length

In fluid dynamics, the mixing length model is a method to describe momentum transfer by Reynolds stresses within a boundary layer by means of eddy viscosity. Mixing length is defined as that distance in the transverse direction which must be covered by a lump of fluid particle travelling with its original mean velocity to make the difference between its velocity and the velocity of the new layer equal to the mean transverse fluctuation in the turbulent flow. OR simply it can be defined as the average distance that a small mass of fluid travels before it exchanges its momentum with another mass of fluid. Prandtl proposed that when there is mixing between two fluid elements; there is complete exchange of momentum. The model was developed by Prandtl in the early 20th century. He had reservations about the concept describing as only a rough approximation, but it has been used in numerous fields.

The mixing length theory of turbulent motion was originally developed by Prandtl (Schlichting and Gersten, 2000) to study the vertical velocity profile in a fully developed channel flow. Prandtl's hypothesis about turbulent motion was proposed that the typical values of the fluctuating velocity components in the x and y directions, are each proportional to $\frac{d\overline{u}}{dy}$, where l is the mixing length (discussed in Section 6.6.2 in Prandtl mixing length hypothesis). Prandtl used the simplest form of mixing length for clear water as $l = \kappa_0 y$, which was used to determine the well-known log-law. Prandtl provided another form of mixing length l as:

$$l = \kappa_0 y\sqrt{1 - \frac{y}{h}} \tag{6.229}$$

for clear water flow. According to van Driest (1956), $l = \kappa_0 y$ is supposed to represent fully developed turbulent flow near a wall and such fully developed turbulent flow occurs only beyond a distance sufficiently remote from the wall that the eddies themselves are not damped in turn by the nearness of the wall. Near the wall, the damping factor was assumed as $1 - \exp(-y^+/A_*)$, where $y^+ = yu_*/\nu_f$, $A_* = 26$ as a constant. van Driest (1956) proposed a remarkable expression for a mixing length l^+, modifying the expression of Prandtl's mixing length:

$$l^+ = \kappa_0 y^+\left[1 - \exp\left(-\frac{y^+}{A_*}\right)\right] \tag{6.230}$$

where κ_0 is the von-Karman constant (= 0.4), A_* is a damping constant, and is restricted to the incompressible turbulent boundary layer with zero pressure gradient. The parameter A_* is referred to as Van Driest damping parameter. If $y^+ = 0$, mixing length $l^+ \rightarrow 0$, that means, no turbulence mixing as the wall is approached, and for large y^+, the mixing length $l^+ \rightarrow \kappa_0 y^+$. A comparative study of several mixing length formulae near the wall, modifying the Van Driest damping parameter, is focused by Launder and Priddin (1973). Galbraith et al. (1977) proposed a modified mixing length in the wall region of the turbulent boundary layer flow. Nezu and Rodi (1986) suggested the mixing length l from their LDA data is as follows:

$$\frac{l}{h} = \kappa_0 \sqrt{\left(1 - \frac{y}{h}\right)} \left[\frac{h}{y} + \pi \Pi \sin\left(\pi \frac{y}{h}\right)\right]^{-1} \tag{6.231}$$

As the free surface is approached, l falls to zero in contrast to the pipe flow. However, this depends again on whether the log-wake law is valid at the free surface (Figure 6.35). Granville (1989) proposed a modified Van Driest mixing length formula for turbulent boundary layer flow with pressure gradients. Umeyama and Gerritsen (1992) proposed an excellent theoretical model for velocity distribution in sediment-laden flow based on new mixing length hypothesis, that depends on sediment concentration. They verified that the mixing length for highly concentrated flow is smaller than that for low concentrated flow. Grifoll and Giralt (2000) formulated the mixing length consistently with the available information of the turbulent viscosity profile near the wall and proposed a mixing length equation valid for momentum and heat transfer across the turbulent wall-bounded flows. Buschmann and Gad-elHak (2005) proposed a mixing-length approach to study the mean velocity profiles in turbulent boundary layer flow.

Obermeier (2006) revisited the Prandtl's mixing length model of Reynolds stresses in fully developed flows. It is known that the Prandtl's mixing length model fails to describe the channel flow correctly from the wall to the center, and the Reynolds stresses very close to the wall. To overcome these shortfalls, Obermeier (2006) replaced the characteristic mixing length by different mixing lengths for velocity fluctuations along and normal to the wall, respectively. The modified model describes the mean velocity and all Reynolds stresses, which agree well with the experimental data and with the data obtained from the numerical simulation. Crimaldi et al. (2006) proposed a formula of mixing length

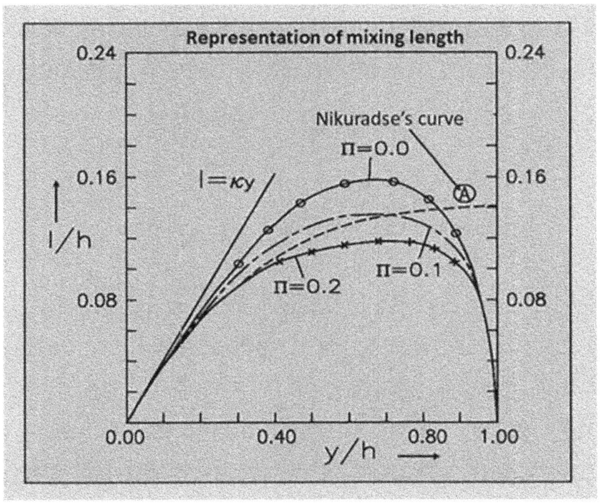

FIGURE 6.35 Mixing length: Continuous line for $l = k_0 y$, continuous line with circle for $\Pi = 0.0$, dash-dot line for $\Pi = 0.1$, dash-cross line for $\Pi = 0.2$, and dashed line indicates Nikuradse's curve. (Modified from Nezu and Rodi, 1986; Nezu and Nakagawa, 1993.

over the rough bed, modifying the Van Driest's momentum mixing length formula. They introduced geometry of roughness elements to describe the effect of roughness on the flow. They defined a roughness Reynolds number as $k_s^+ = \dfrac{k_s u_*}{\nu_f}$. For $k_s^+ < k_{ss}^+$, where $k_{ss}^+ = 5$, the flow is said to be dynamically smooth because the roughness element within the viscous sub-layer, for $k_s^+ > k_{Rs}^+$, where $k_{Rs}^+ = 60$, there is no effect of roughness in the flow, and for $k_{ss}^+ < k_s^+ < k_{Rs}^+$, the flow is within the transitional region. The Van Driest mixing length is modified, considering the effect of roughness, and is given by:

$$l^+ = \kappa_0 y^+ \left[1 - \frac{k_{Rs}^+ - k_s^+}{k_{Rs}^+ - k_{ss}^+} \exp\left(-\frac{y^+}{A_*}\right)\right] \tag{6.232}$$

When $k_s^+ = k_{ss}^+$, equation (6.232) matches with Van Driest equation (6.230), and for $k_s^+ = k_{Rs}^+$, equation (6.232) matches with Prandtl mixing length $l^+ = ky^+$. For intermediate value of k_s^+, equation (6.232) applies a linear interpolation of the Van Driest exponential factor, where the linear interpolation is chosen. Ghoshal and Mazumder (2006) proposed a modified mixing length to investigate the velocity and concentration distributions in sediment-mixed mixed fluid. Recently, Kundu et al. (2018) reexamined the mixing length in an open channel turbulent flow.

6.10.8 Turbulent Kinetic Energy (TKE)

Turbulent kinetic energy (TKE) represents the extraction of energy from the mean flow by the motion of turbulent eddies. Over a flat bed, TKE is largest near the boundary where turbulence production is highest. Turbulence *kinetic energy* (TKE) is the mean *kinetic energy* per unit mass associated with eddies in *turbulent* flow and arises as a result of friction-induced shear. This energy is transferred down the turbulence energy cascade, and ultimately dissipating as a result of viscous forces. The kinetic energy is produced in and transferred from the mean flow to fluctuating flow and lost by a heat increase through energy cascade (Tennekes and Lumley, 1972). Physically, the *turbulence kinetic energy* is characterized by measured root-mean-square (RMS) velocity fluctuations, which calculates the intensity of turbulence. Turbulent energy continuously drives eddies from the greatest to the least by fluid shear (mechanical) and thermals (buoyant). The equation for the *turbulent kinetic energy* is obtained from Navier–Stokes equation for incompressible flows, after modifying the equations by Reynolds decompositions of velocity and pressure. In RANS equations, the turbulence kinetic energy can be calculated based on the closure method, i.e., a turbulence model. The total kinetic energy is simply the sum of the kinetic energy of the mean and turbulent flows. The kinetic energy of the mean and turbulent parts of the flow per unit mass respectively can be expressed as:

$$MKE = \frac{1}{2}\left(\bar{u}^2 + \bar{v}^2 + \bar{w}^2\right) \tag{6.233}$$

$$TKE = \frac{1}{2}\left(\overline{u'^2} + \overline{v'^2} + \overline{w'^2}\right) \tag{6.234}$$

where *MKE* is the mean kinetic energy of the flow per unit mass, and *TKE* is the turbulent kinetic energy of the flow per unit mass. The kinetic energy may be normalized by u_*^2. The instantaneous kinetic energy *k*(*t*) is the sum the mean kinetic energy and the turbulent kinetic energy, that means, *k*(*t*) = MKE+TKE. Pope (2000) defined TKE to be half the trace of Reynolds stress tensor, proportional to the sum of the three normal stresses. The turbulent velocity component is the difference between the instantaneous velocity and the average velocity $u' = u - \bar{u}$, whose mean and variance respectively are as:

$$\overline{u'} = \frac{1}{T}\int_0^T \left(u(t) - \bar{u}\right)dt = 0, \text{ and } \overline{u'^2} = \frac{1}{T}\int_0^T \left(u(t) - \bar{u}\right)^2 dt \geq 0. \tag{6.235}$$

TKE can be produced by fluid shear, friction, or buoyancy, or through external forcing at low-frequency eddy scales. Turbulence kinetic energy is then transferred down the turbulence energy cascade, and is dissipated by viscous forces at the Kolmogorov length scale. The TKE distribution can be divided into production, diffusion, and dissipation through heat due to viscous forces. TKE rises maximum value near the boundary ($y/h \sim 0.1$) and this generally remains constant or it decreases with increase in y/h. At low aspect ratio of the flume, the TKE may show the bimodality in the TKE distribution. The rate of turbulent kinetic energy is $\frac{\partial K}{\partial t}$, where $K = \frac{1}{2}\overline{u_i'u_j'}$ the turbulent kinetic energy is defined by equation (6.234).

6.10.9 Coefficients of Skewness and Kurtosis

Higher-order correlations contain important information on turbulent statistics. The gradients of the higher-order correlation terms represent the turbulent diffusion terms in both turbulent kinetic energy and Reynolds stress equations (Agelinchaab and Tachie, 2006). The third order correlation terms $\overline{u'^3}$ and $\overline{u'^2v'}$ physically signify respectively the transport of $\overline{u'^2}$ in the stream-wise and wall-normal directions; similarly, $\overline{u'v'^2}$ and $\overline{v'^3}$ represent the transport of $\overline{v'^2}$ in the respective directions (Agelinchaab and Tachie, 2006). The third-order moments preserve their sign, positive or negative, providing useful stochastic information on the distribution of the velocity fluctuations with respect to the time-averaged velocity. Mathematically, the set of third-order moments can be specified as (Raupach, 1981; Sarkar and Mazumder, 2018b): $M_{ab} = \overline{\tilde{u}^a\tilde{v}^b}$ with $a + b = 3$, where

$$\tilde{u} = u'/\left(\overline{u'u'}\right)^{1/2}, \text{and } \tilde{v} = v'/\left(\overline{v'v'}\right)^{1/2} \tag{6.236}$$

The skewness of stream wise velocity fluctuation u' is given by $M_{30} = \overline{\tilde{u}^3} = \frac{\overline{u'^3}}{\sigma_u^3}$, which refers to the stream wise flux of Reynolds normal stress $\overline{u'^2}$ along x-direction, and the skewness of vertical velocity fluctuation v' is $M_{03} = \overline{\tilde{v}^3} = \frac{\overline{v'^3}}{\sigma_v^3}$, signifies the Reynolds normal stress $\overline{v'^2}$ along y-direction, where σ_u and σ_v are the standard deviations along u and v directions. Apart from this, the third order correlations are $M_{21} = \overline{\tilde{u}^2\tilde{v}}$, defining the advection of $\overline{u'^2}$ in y-direction, and $M_{12} = \overline{\tilde{u}\tilde{v}^2}$, characterizing the advection of $\overline{v'^2}$ in x-direction. In general, the gradients of triple correlation of velocity represent the diffusion of turbulent kinetic energy and Reynolds shear stress. The third-order moments M_{30} and M_{03} provide a measure of skewness, which is an important criterion to determine the degree of symmetry of the distribution of velocity fluctuations about the mean. The positive or negative value of skewness represents the useful statistical information on the shape of the velocity distributions, in terms of right or left skewed respectively. When the coefficient of skewness = 0, it leads to a symmetric distribution. The coefficients of skewness, S_u, S_v, and S_w along x, y, and z directions, respectively are given as follows:

$$S_u = \frac{\overline{u'^3}}{\sigma_u^3}, S_v = \frac{\overline{v'^3}}{\sigma_v^3}, S_w = \frac{\overline{w'^3}}{\sigma_w^3} \tag{6.237}$$

where σ_u, σ_v, and σ_w are the turbulent intensities (standard deviation) in x, y, and z directions, respectively. From the definitions of skewness of the velocity fluctuations along the directions of u, v, and w, the third-order moments may be interpreted as the fluxes of the normal stresses of u', v', and w'

respectively. The third-order correlations characterize the temporal behavior of the velocity fluctuations, and hence it is directly related to the turbulent coherent structures because of their changes of signs (Simpson et al., 1981; Gad-El-Hak and Bandyopadhyay, 1994). The third-order moment also helps in understanding the mechanism of the transport of Reynolds shear stress and the characterization of turbulent events (Lacey and Rennie, 2012). The negative value of S_u indicates the low-speed fluid streaks, i.e., occurrence ejection event; and positive value of S_u indicates the high-speed fluid streaks, i.e., occurrence of sweep event. The positive value of S_v emphasizes a strong ejection event, while the negative value of S_v indicates a strong sweep. More precisely, non-zero skewness addresses the acceleration versus deceleration or sweep versus ejection. Similarly, for positive and negative values of skewness S_w indicate the transverse flux.

The fourth-order moment may be referred to as the coefficient of kurtosis with respect to the x, y and z directions and are defined as

$$F_u = \frac{\overline{u'^4}}{\sigma_u^{\ 4}} - 3;\ F_v = \frac{\overline{v'^4}}{\sigma_u^{\ 4}} - 3;\ F_w = \frac{\overline{w'^4}}{\sigma_u^{\ 4}} - 3 \tag{6.238}$$

The coefficient of kurtosis is an important criterion to measure the degree of peakedness or flatness of the distribution. It is important to note that the coefficient of kurtosis greater than zero signifies the peakedness, representing highly characterized turbulent event, while the value less than zero indicates to flatness. The fourth-order moments i.e., coefficient of kurtosis (flatness factor), reflects the level of intermittency of turbulence (Nikora et al., 2002).

If the statistical moments such as mean, variance, skewness and flatness are known, the probability density function (PDF) is presented, which includes information of all the statistical moments (Aberle and Nikora, 2006). If the PDF follows the Gaussian distribution, the PDFs of fluctuating velocities follow uni-modal distribution with symmetric shape, otherwise it will be asymmetric. The deviation from the Gaussian distribution is commonly attributed to large-scale flow features leading to symmetry breaking. As explained by Raupach (1981), coefficient of kurtosis attains the value much higher than the value for Gaussian probability distribution and is equal to three. However, it is important to note that the value of the flatness factor greater than three signifies a relationship with a peak signal characteristic of intermittent turbulent events, while the value of the flatness factor less than three indicates to flat characteristic (Agelinchaab and Tachie, 2006). The coefficients of skewness and kurtosis are used to examine the deviations from the Gaussian distribution. If the distributions are exactly Gaussian, both the coefficients should zero. If the coefficient of kurtosis (flatness factor) is negative, indicates the leptokurtic distribution; if the coefficient is positive, indicates the platykurtic distribution. If the coefficient is zero, the distribution shows the Gaussian peak (Mazumder and Das, 1992).

6.10.10 Turbulent Kinetic Energy Flux

The turbulent kinetic energy fluxes in x, y and z directions, are presented in the following section. The normalized turbulence kinetic energy fluxes (f_{ku}, f_{kv}, f_{kw}) are defined as (Raupack, 1981; Maity and Mazumder, 2014):

$$f_{ku} = \frac{0.5\left(\overline{u'u'u'} + \overline{u'v'v'} + \overline{u'w'w'}\right)}{u_*^3} \tag{6.239}$$

$$f_{kv} = \frac{0.5\left(\overline{v'u'u'} + \overline{v'v'v'} + \overline{v'w'w'}\right)}{u_*^3} \tag{6.240}$$

$$f_{kw} = \frac{0.5\left(\overline{w'u'u'} + \overline{w'u'u'} + \overline{w'w'w'}\right)}{u_*^3} \tag{6.241}$$

The negative and positive values of the flux f_{ku} lead to the transport of energy in the backward and forward directions respectively; and these are related to the ejection–sweep character of the Reynolds stress. Balachandar and Bhuiyan (2007) have shown that the contribution to the stream-wise energy flux shows negative values for flat smooth surfaces. The positive values of f_{kw} indicate upward transport of TKE that is produced from the bottom. The negative values of f_{ku} indicate that transport of TKE occurs in the opposite direction of longitudinal mean velocity. This is probably because kinetic energy fluctuations are sustained by extraction of energy from the mean flow. The positive value of energy flux f_{kw} leads to the occurrence of upward transport of energy flux. The negative and positive values of vertical TKE flux lead to transport of energy in the downward and upward directions respectively, and these are related to the sweep-ejection characters of the Reynolds shear stress (Tachie et al., 2004; Maity and Mazumder, 2014, 2017).

6.10.11 Turbulent Kinetic Energy Budget

The turbulent kinetic energy (TKE) budget for two-dimensional uniform open-channel flow as proposed by Nezu and Nakagawa (1993) is,

$$-\overline{u'v'}\frac{\partial \bar{u}}{\partial y} = \frac{1}{\rho_f}\cdot\frac{\partial(\overline{p'v'})}{\partial y} + \frac{\partial f_{kv}}{\partial y} - \upsilon_f\frac{\partial^2 k}{\partial y^2} + \varepsilon \tag{6.242}$$

In fully developed turbulent flow, the viscous diffusion rate ($\upsilon_f \cdot \partial^2 k/\partial y^2$) is negligible due to high Reynolds numbers. Moreover, in the intermediate region, TKE is transported from the near-bed region to the near-free surface maintaining dynamic equilibrium, in which $-\overline{u'v'} \cdot \partial\bar{u}/\partial y \approx \varepsilon$. As observed from equation (6.242) turbulent production is the sum of pressure energy diffusion, TKE diffusion and dissipation rate. Our interest is on the large-scale eddies because these eddies are responsible for most of the turbulent production in the mean flow. Therefore, only the turbulent production term is analyzed from the energy budget equation. One can write a TKE budget equation that is a sum of the production and loss terms. If the production terms are larger than the loss terms, TKE will increase and the boundary layer becomes more turbulent. If the loss terms are larger than the production terms, TKE will decrease and the boundary layer becomes less turbulent.

6.10.12 Turbulent Kinetic Energy Production

Kinetic energy is produced by and transferred from the mean flow to the turbulent flow. In the flow, kinetic energy is generated from the mean flow through the large eddies present in it. This transfer of energy is governed by the dynamics of large eddies generated from the turbulent flow, which contribute the most to the production (Tennekes and Lumley, 1972) and that transfers the energy from larger eddies to smaller through scales, i.e., from the larger to smaller. Ultimately this energy cascades down to the smallest scales and is dissipated to heat. This cascading process is carried on through ever smaller eddies, until eventually dissipation; that is, the transfer from mechanical energy to internal energy occurs in the smallest eddies. The turbulent production involves interactions of the Reynolds shear stress with mean velocity gradients and is analyzed from the energy budget equation. Mathematically, for steady flow with non-varying fluid properties, the equation for the turbulence production T_p is written based on three-dimensional flows, which is taken from the turbulent energy equation (Schlichting and Gersten, 2000) as:

$$T_p = -\rho_f \overline{u'^2}\left(\frac{\partial \bar{u}}{\partial x}\right) - \rho_f \overline{u'v'}\frac{\partial \bar{v}}{\partial x} - \rho_f \overline{u'w'}\frac{\partial \bar{w}}{\partial x} - \rho_f \overline{u'v'}\frac{\partial \bar{u}}{\partial y} - \rho_f \overline{v'^2}\left(\frac{\partial \bar{v}}{\partial y}\right) - \rho_f \overline{v'w'}\frac{\partial \bar{w}}{\partial y} - \rho_f \overline{u'w'}\frac{\partial \bar{u}}{\partial z}$$
$$-\rho_f \overline{v'w'}\frac{\partial \bar{v}}{\partial z} - \rho_f \overline{w'^2}\left(\frac{\partial \bar{w}}{\partial z}\right) \qquad (6.243)$$

In two-dimensional flows, the turbulence production T_p is written as:

$$T_p = -\rho_f \overline{u'^2}\left(\frac{\partial \bar{u}}{\partial x}\right) - \rho_f \overline{v'^2}\left(\frac{\partial \bar{v}}{\partial y}\right) - \rho_f \overline{u'v'}\left[\frac{\partial \bar{u}}{\partial y} + \frac{\partial \bar{v}}{\partial x}\right] \qquad (6.244)$$

Using the equation of continuity, this equation can be rewritten as:

$$T_p = -\rho_f \left(\overline{u'^2} - \overline{v'^2}\right)\left(\frac{\partial \bar{u}}{\partial x}\right) - \rho_f \overline{u'v'}\left[\frac{\partial \bar{u}}{\partial y} + \frac{\partial \bar{v}}{\partial x}\right] \qquad (6.245)$$

Here the term $\rho_f \left(\overline{u'^2} - \overline{v'^2}\right)\frac{\partial \bar{u}}{\partial x}$ is negligible compared to the term $\rho_f \overline{u'v'}\left[\frac{\partial \bar{u}}{\partial y} + \frac{\partial \bar{v}}{\partial x}\right]$. Generally, the production term is considered as $-\rho_f \overline{u'v'}\frac{\partial \bar{u}}{\partial y}$ in turbulent modeling due to boundary layer. The magnitude of turbulence energy production or generation T_p (cm^2/sec^3) by vertical shear in a two-dimensional and uniform flow is defined as (Nezu and Nakagawa, 1993; Venditti and Bennett, 2000, Schlichting and Gersten, 2000; Sarkar and Mazumder, 2014):

$$T_p = -\overline{u'v'}\frac{\partial \bar{u}}{\partial y} \qquad (6.246)$$

During the flow reversal event, the negative turbulent production in the boundary layer turbulence is also observed by Gayan and Sarkar (2011), indicating the energy transfer from the velocity fluctuations to the mean flow.

6.10.13 Kinetic Energy Dissipation

To sustain turbulent flow, a constant source of energy is required because turbulence dissipates rapidly as the kinetic energy is converted into internal energy by viscous shear stress. Turbulence causes the formation of eddies of many different length scales. Most of the kinetic energy of the turbulent motion is contained in large-scale structures. The large eddies are unstable and break up, transferring their energy to somewhat smaller eddies. These smaller eddies undergo a similar break-up process and transfer their energy to yet smaller eddies and continue until the Reynolds number $Re(l) = u(l)l / \nu_f$ is sufficiently small that the eddy motion is stable. The turbulent dissipation is the rate at which the turbulent kinetic energy is transformed into internal energy, which is associated with the turbulent eddies in a fluid flow. The turbulent energy is absorbed by breaking down the eddies into smaller and smaller eddies until it is ultimately converted into heat by viscous forces. The rate of dissipation of energy in a fluid at any instant depends only on the viscosity and on the instantaneous distribution of velocity. In turbulent flows, the mechanical energy is transformed into internal energy through dissipation due to viscosity and turbulent fluctuations. In a flow with local equilibrium, the dissipation would be equal to the production of turbulent kinetic energy. The *dissipation rate of turbulent kinetic energy* is defined as $\varepsilon = \nu_f \left(\frac{\partial u'}{\partial x}\right)^2$, where u' is the fluctuating velocity in longitudinal direction, and ν_f is kinematic viscosity. The dissipation rate depends on the viscosity and the velocity gradient in the turbulent eddies. As the mean flow velocity is greater than the turbulent fluctuations, Taylor's frozen hypothesis (Taylor, 1935; Lohse and Grossmann, 1993; Sarkar and Mazumder, 2018b) is used to simplify the dissipation rate by $x = \bar{u}t$, where $\bar{u}$ denotes the mean velocity and t is the time. Hence, the turbulent dissipation rate can be written as $\varepsilon \propto \left(\frac{\partial u'}{\partial t}\right)^2$. The dissipation rate ε is estimated (Tennekes and Lumley, 1972; Hinze, 1975; Krogstad and Antonia, 1999) with a constant of proportionality as

$$\varepsilon = \frac{15\nu_f}{\bar{u}^2} \overline{\left(\frac{\partial u'}{\partial t}\right)^2} \tag{6.247}$$

Here the constant of proportionality is 15 in equation (6.247) for the isotropic turbulence. If, the statistical properties of the velocity field can be expressed by the correlation R_y between the values of u at two points y apart in the direction of y, it must be possible to deduce from them the rate of dissipation of energy. This would be in general a complicated analysis, but the problem can be much simplified if the field of turbulent flow is assumed to be isotropic.

6.11 ISOTROPIC TURBULENCE

In *isotropic turbulence* the average value of any function of the velocity components, defined in relation to a given set of axes, is unaltered if the axes of reference are rotated in any manner. There is a strong tendency to isotropy in turbulent motion. It has been shown by Fage and Townend (1934), for instance, that the average values of the three components of velocity in the central region of a pipe of square section are nearly equal to one another. In the atmosphere the same phenomenon has been observed; though, as might be expected, the vertical components are smaller near the ground than the horizontal ones, this inequality decreases with height above the ground. The assumption of isotropy immediately introduces many simplifications both into the statistical demonstration of turbulence and into the expression for the mean rate of dissipation of energy.

Taylor (1935) introduced a concept of simple type of turbulence, which is statistically isotropic turbulence. In isotropic turbulence, the mean value of any function of velocity components and their space derivatives is unaltered by any rotation or reflection of the axes of references. Isotropic turbulence by its definition is always homogeneous. In isotropic turbulence fluctuations are independent of the direction of reference. Thus, in particular, $\overline{u'^2} = \overline{v'^2} = \overline{w'^2}$, and $\overline{u'v'} = \overline{u'w'} = \overline{v'w'} = 0$. At very large Reynolds numbers, the turbulent flows are locally isotropic in nature that has been shown by Kolmogorov (1941). The isotropy is excluded only regions near the walls and edges. It means that the fluctuations in a close neighborhood of a point possess no particular direction, i.e., they are isotropic. Since the dissipation occurs in the region of the smallest eddies, dissipation equation can be simplified under the assumption of isotropic turbulence.

The general expression for the rate of dissipation (Taylor, 1935) is

$$\varepsilon = \mu_f \left[2\overline{\left(\frac{\partial u'}{\partial x}\right)^2} + 2\overline{\left(\frac{\partial v'}{\partial y}\right)^2} + 2\overline{\left(\frac{\partial w'}{\partial z}\right)^2} + \overline{\left(\frac{\partial u'}{\partial y} + \frac{\partial v'}{\partial x}\right)^2} + \overline{\left(\frac{\partial u'}{\partial z} + \frac{\partial w'}{\partial x}\right)^2} + \overline{\left(\frac{\partial v'}{\partial z} + \frac{\partial w'}{\partial y}\right)^2} \right] \tag{6.248}$$

Making the assumption that the turbulence is statistically isotropic, the relations

$$\overline{\left(\frac{\partial u'}{\partial x}\right)^2} = \overline{\left(\frac{\partial v'}{\partial y}\right)^2} = \overline{\left(\frac{\partial w'}{\partial z}\right)^2} \tag{6.249}$$

$$\overline{\left(\frac{\partial u'}{\partial y}\right)^2} = \overline{\left(\frac{\partial u'}{\partial z}\right)^2} = \overline{\left(\frac{\partial v'}{\partial x}\right)^2} = \overline{\left(\frac{\partial v'}{\partial z}\right)^2} = \overline{\left(\frac{\partial w'}{\partial x}\right)^2} = \overline{\left(\frac{\partial w'}{\partial y}\right)^2} \tag{6.250}$$

$$\overline{\frac{\partial v'}{\partial x}\frac{\partial u'}{\partial y}} = \overline{\frac{\partial w'}{\partial y}\frac{\partial v'}{\partial z}} = \overline{\frac{\partial u'}{\partial z}\frac{\partial w'}{\partial x}} \tag{6.251}$$

Adding the above expressions, equations (6.249–6.251), equation (6.248) can be written as

$$\varepsilon = 6\mu_f\overline{\left(\frac{\partial u'}{\partial x}\right)^2} + 6\mu_f\overline{\left(\frac{\partial u'}{\partial y}\right)^2} + 6\mu_f\overline{\frac{\partial v'}{\partial x}\frac{\partial u'}{\partial y}} \tag{6.252}$$

It can be shown that the terms in equation (6.252) are all related to one another so that if the value of one is known, the other two are known. The conditions of statistical isotropy and the equation of continuity therefore lead to the relationship

$$\overline{\left(\frac{\partial u'}{\partial x}\right)^2} = -2\overline{\frac{\partial u'}{\partial x}\frac{\partial v'}{\partial y}} \tag{6.253}$$

or

$$\frac{\overline{\dfrac{\partial u'\partial v'}{\partial x\partial y}}}{\sqrt{\overline{\left(\dfrac{\partial u'}{\partial x}\right)^2}}\sqrt{\overline{\left(\dfrac{\partial v'}{\partial y}\right)^2}}} = -\frac{1}{2} \tag{6.254}$$

There is a definite correlation coefficient between $\frac{\partial u'}{\partial x}$ and $\frac{\partial v'}{\partial y}$ equal to –1/2. Through a long process of using linear algebra, the possible terms in (6.252) are linearly related to each other, and that

$$2\overline{\left(\frac{\partial u'}{\partial x}\right)^2} = \overline{\left(\frac{\partial u'}{\partial y}\right)^2} \tag{6.255}$$

The derivation is shown in Taylor (1935). In particular, the dissipation may be expressed in terms of $\overline{\left(\frac{\partial u'}{\partial y}\right)^2}$. Finally, substitutions of these relations into (6.252) give the dissipation equation for isotropic turbulence as

$$\varepsilon = 15\mu_f\overline{\left(\frac{\partial u'}{\partial x}\right)^2} \tag{6.256}$$

Taylor's frozen hypothesis (Taylor, 1935; Lohse and Grossmann, 1993) is used to simplify the dissipation rate by $x = \overline{u}t$, where $\overline{u}$ denotes the mean velocity and t is the time. Hence, the turbulent dissipation rate ε can be written as $\varepsilon \propto \left(\frac{\partial u'}{\partial t}\right)^2$. The dissipation rate ε is estimated (Tennekes and Lumley, 1972; Hinze, 1975) with a constant of proportionality as

$$\varepsilon = \frac{15\nu_f}{\bar{u}^2}\overline{\left(\frac{\partial u'}{\partial t}\right)^2} \tag{6.257}$$

Equation (6.248) can be written as

$$\varepsilon = 15\frac{\mu_f \overline{u'^2}}{\lambda^2} \tag{6.258}$$

where λ is a measure of the diameter of the smallest eddies which are responsible for the dissipation of energy.

6.11.1 Spectral Analysis

Turbulence spectra provide information on temporal or spatial scales of velocity fluctuations, from which we can determine the frequency or wave number sub-range in the region where turbulence is isotropic. A common way to characterize the variability across a wide range of scales is through computing the power spectral density, using Fast Fourier transform (FFT) algorithm converting time-domain signals to the frequency domain. In turbulent flows, the flow structure contains eddies of varying sizes: the region occupied by larger eddies contains smaller eddies, and the later eddies also contain even smaller eddies and so on. The eddies of size l have a characteristic velocity $u(l)$ and time scale $t = l/u(l)$. Here the distributions of TKE among the eddies of different sizes are determined using spectral analysis. The spectral analysis is capable of transforming a time-domain signal to the frequency-domain in terms of a *power spectral density (PSD) as a function of frequency* or wave number k_w. The PSD is calculated by decomposing the time series of instantaneous velocity into a set of periodical sine and cosine functions with various amplitudes and frequencies, then plotting the amplitude of each periodical function with their frequency on a log scale. In turbulence analysis, the spectral density functions are commonly used as energy spectrum function E and spectrum function S_u of velocity fluctuations. A nondimensional correlation function $c_u(y^+)$ between two stream-wise velocity fluctuations $u'(x)$ and $u'(x+y^+)$ having a stream-wise lag distance y^+ can be defined as follows (Nezu and Nakagawa, 1993):

$$c_u(y^+) = \frac{\overline{u'(x)\cdot u'(x+y^+)}}{\overline{u'u'}} = \int_0^\infty S_{uu}(k_w)\cos(k_w y^+)dk_w \tag{6.259}$$

where k_w is the wave number in the stream-wise direction, $S_{uu}(k_w)$ is a one-dimensional spectrum function of u' and is given by

$$S_{uu}(k_w) = \frac{2}{\pi}\int_0^\infty c_u(y^+)\cos(k_w y^+)dk_w \tag{6.260}$$

The relation between equations (6.259) and (6.260) is called the "cosine transform." For the case of no lag, ($y^+ = 0$), equation (6.259) becomes

$$\int_0^\infty S_{uu}(k_w)dk_w = 1 \tag{6.261}$$

The way in which the turbulent kinetic energy (TKE) is distributed over the range of scales is the fundamental characterization of turbulence. For isotropic turbulence, this can be performed by means of the *energy spectrum function* $E(k_w)$ that represents the turbulent kinetic energy k to contain eddies of size l having wave number $k_w\left(=\frac{2\pi}{l}\right)$. By definition, k is the integral of $E(k_w)$ over the full range of wave number k_w. Thus,

$$k=\frac{1}{2}\overline{u_i'u_i'}=\int_0^\infty E(k_w)dk_w \tag{6.262}$$

In developing $E(k_w)$ within the inertial sub-range, $E(k_w)$ is solely dependent on wave number k_w and dissipation rate ε (by Kolmogorov second similarity hypothesis). From an analogy between spatial turbulence field and spectral wave number, the kinetic energy dissipation rate ε is regarded as a fundamental quantity. Therefore, by dimensional analysis, the possible form for the energy spectrum function can be obtained as

$$E(k_w)=C\varepsilon^{2/3}k_w^{-5/3} \tag{6.263}$$

where C is the universal Kolmogorov constant, which was experimentally determined as 1.5 (Zhou, 1993). Equation (6.262) leads to the Kolmogorov hypothesis of energy spectrum function. The inertial sub-range is a range of time-scales or length-scales of turbulence that is not associated with turbulence production. Instead, this range represents the direct energy transfer from the energy containing large scales to the dissipative small scales. The degree to which a linear range can be identified in the log–log plotted spectra is an important marker for the state of the turbulence at the point under consideration. *Kolmogorov hypothesis* of energy spectrum function states: in the inertial sub-range, the energy spectrum function is proportional to –5/3rd power of wave number which is universal. The hypothesis of energy spectrum function is commonly known as *Kolmogorov's –5/3-rd power law.*

The energy spectrum follows a power law $\propto k^{-5/3}$ over a range of scales (i.e., inverse of wave numbers) extending from the integral scales l_o to the dissipation scale l_d. The integral range l_o of scale is called the energy carrying or production range because most of the turbulent energy is produced (usually by some instability mechanism). The range within $l_o \gg l \gg l_d$ is the inertial sub-range because the dynamics are dominated by the inertial terms in the Navier–Stokes equations (Frisch, 1985), where the direct production and dissipation are negligible. The range $l \le l_d$ is called the *dissipation range* because both inertial and dissipation terms are relevant. It may be summarized that: (1) the most of the turbulent energy is contained in large eddies or small wave numbers, (2) most of the dissipation occurs at small eddies or large wave numbers, and (3) between the small and large wave numbers or between the large and small eddies, there is a range of wave numbers responsible for transferring turbulent energy from small to large wave numbers. The –5/3-rd power law can be verified by using experimentally obtained velocity data over a sufficiently long period of time in a turbulent flow. A large Reynolds number associated with large eddies gives rise to a larger inertial sub-range. It is very difficult to reach sufficiently large Reynolds numbers in a laboratory experiment to produce an adequately broad inertial sub-range. A schematic diagram illustrates the energy spectrum $E(k_w)$ in a log–log scale in Figure (6.36). In the figure, the –5/3 slope of the curve in the inertial sub-range corresponds to the Kolmogorov –5/3 power law. On the other hand, energy-containing range corresponds to the small range of wave number k_w and the dissipation range corresponds to large range of k_w.

For example, a typical velocity data in the stream-wise, vertically upward, transverse directions recorded using Acoustic Doppler Velocimeter (ADV) shows a number of spikes. The majority of spikes

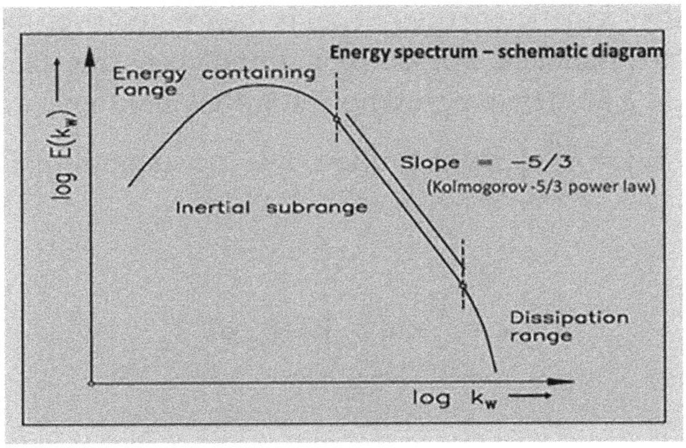

FIGURE 6.36 Schematic diagram of energy spectrum $E(k_w)$ vs. wave number k_w.

are found in the stream-wise component. The spikes in the stream-wise component have a velocity in the correct range. The spike structure was already apparent in velocity samples as shown in Goring and Nikora (2002). Unfortunately, it cannot be determined whether ADV data spikes occur because of the flow disturbance and turbulence caused locally by the sub-merged sensor head or because of data treatment inside ADV. These spikes often occur over several consecutive velocity measurements. It is important to note that each spike in the stream-wise velocity is correlated with an increased amplitude spike in the vertical velocity component.

Spike-removal procedures (Goring and Nikora, 2002) are often applied only to the stream-wise component of the flow vector. Instantaneous Reynolds stress calculations and turbulent kinetic energy (TKE) estimates might be strongly affected if spikes are not removed in both components. After application of a phase-space threshold despiking technique to the raw velocity data, the velocity time-series were again detrended using Butterworth low-pass filter at a Nyquist frequency (f_s) of 20 Hz and lower frequencies such as 10, 5 and 2 Hz, for initial tests before the spectrum analysis. The raw velocity signals were passed through a low-pass filter lower than the Nyquist frequency to obtain smoother signals. To determine the velocity spectra, the total velocity time series data were divided into three ensembles that possessed dynamically similar interaction types between small- and large-scale eddies (Poggi et al., 2004). Each ensemble contained a size of 2,048 data points. Spectral analyses were executed from all the three ensembles and averaged. Thereafter, the stream-wise velocity spectra were plotted, and the data follow the Kolmogorov –5/3 law for 5 Hz, where the inertial sub-range occurs. Thus, the cut-off frequency 5Hz was fixed. The inertial sub-range was defined to include turbulent eddies which were considerably increased compared with the viscous dissipation scales, and considerably reduced compared with the integral length scales of the flow (Katul et al., 2003).

The spectral frequency (f_s) or frequency spectrum is estimated within 0–20 Hz at a 0.0195 Hz interval with 95% confidence limit in the coherence level (Rennie and Hay, 2010), using power spectral density algorithm available in the Matlab software package with Hamming windows (Venditti and Bennett, 2000; Sarkar and Mazumder, 2014). Figure (6.37) shows power spectral density (PSD) of velocity signals against spectral frequency (f_s) in log–log scale for a fully developed flow. The slope of PSD within the inertial sub-range is compared to the Kolmogorov –5/3 law. The power spectra of stream-wise velocity signals suggest approximately a good fit to the slope –5/3 power-law. For example, Figure 6.37 shows the PSD against spectral frequency for $u_m = 40$ cm/s for all three velocity components at $y/h = 0.55$.

The power spectral density $P(f_s)$ is multiplied by the spectral frequency (f_s) i.e., $f_s * P(f_s)\,\text{cm}^2\text{s}^{-2}$ to obtain the velocity spectra and it is plotted against spectral frequency f_s in semi-log scale (Figure 6.38a). The area under the spectral curve represents the total variance and the peaks correspond to energetic

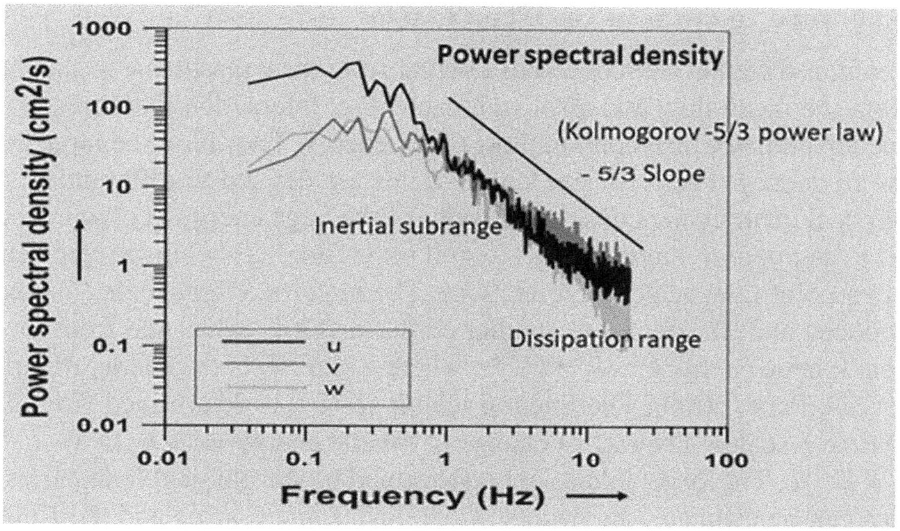

FIGURE 6.37 Power spectral density $P(f_s)$ for all three velocity components. (Taken from ISI laboratory data.)

eddies (Boppe and Neu, 1995). The stream-wise spectra generally obtain peak at frequency ranging ≈ 0.5–1 Hz. The occurrence of peaks at different frequencies differs due to the occurrence of eddies of different shapes and sizes. Velocity co-spectral analyses are performed on the joint time series of stream-wise u and bottom-normal v components of velocity following the similar procedure shown in Figure 6.38b. In a one-dimensional velocity spectrum (e.g., u or w), the variance is proportional to the Reynolds normal stress (i.e., $\overline{u'^2}$ or $\overline{v'^2}$), whereas the covariance between two velocity components is proportional to the Reynolds shear stress $(\overline{u'v'})$. The co-spectral tends to have larger peaks than the corresponding stream-wise spectra (Venditti and Bauer, 2005). The squared-coherency spectra can be thought of as a frequency-specific correlation coefficient, which signifies the cross-spectra at a specific frequency band (Jenkins and Watts, 1968).

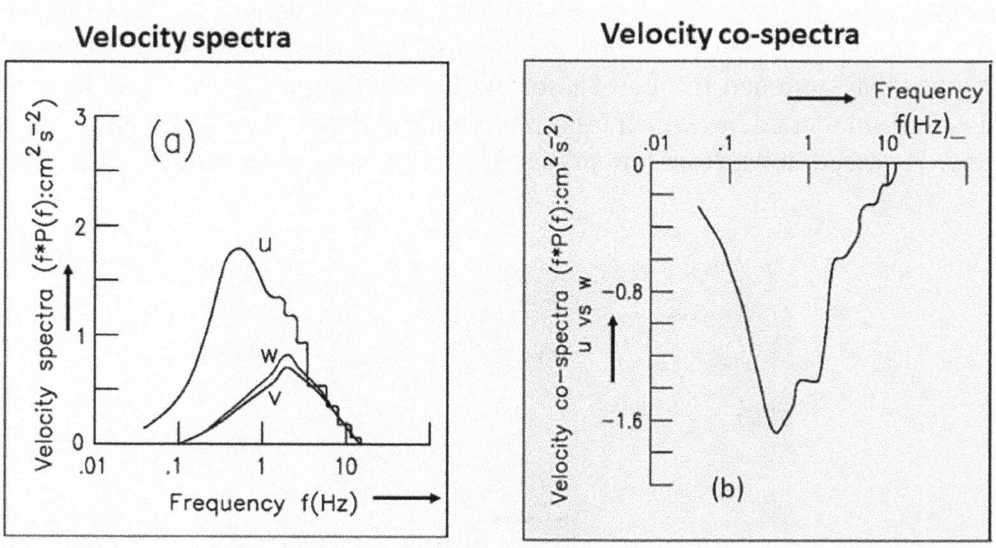

FIGURE 6.38 (a) Velocity Spectra for all three velocity components; (b) Velocity co-spectral plots. (From ISI Laboratory data.)

6.11.2 Turbulent Integral Length Scale and Energy Cascade

Turbulent flow contains different sizes of eddies ranging from the largest scale to smallest. Large eddies obtain energy from the mean flow and also from each other interaction mechanisms. Thus, these are the energy production eddies which contain most of the energy. They have the large flow velocity fluctuations with low frequency. The turbulent length-scales are defined to differentiate the behaviors of eddies, and its physical quantity explains the behavior of the large energy-containing eddies in a turbulent flow referred to as *integral length-scale.* Integral scales are highly anisotropic defined in terms of the normalized two-point flow velocity correlations. The *integral length scale* l_0 is the scale at which the energy is produced and transferred to smaller eddies and it is calculated from the dissipation rate ε and the turbulent kinetic energy k as $l_0 = k^{1.5}/\varepsilon$, where ε (cm^2/s^3) is the energy dissipation rate (Nezu and Nakagawa, 1993; Pope, 2000). The integral length scale l_0 is determined from the characteristic velocity (u_0) and time t_0 scales. The rate of change of kinetic energy $\propto (u_0^2/t_0) = (u_0^3/l_0) = \varepsilon = (k^{3/2}/l_0)$, which implies $l_0 = k^{1.5}/\varepsilon$. The larger eddies are constrained by the physical boundaries of the flow, and the smaller eddies can be estimated by means of the Taylor micro-scale and the Kolmogorov length-scale. The maximum length of these scales is constrained by the characteristic length of the apparatus. For example, the largest integral length scale of pipe flow is equal to the pipe diameter. In the case of atmospheric turbulence, this length can reach up to the order of several hundred kilometres.

Kinetic energy (KE) in turbulent flow is transferred from the mean flow to the turbulence from the large sizes of eddy which is comparable to the length scale of the flow. Larger eddies due to high energy content are unstable and break up into smaller eddies and kinetic energy gets transferred into smaller eddies with dissipation of energy. The transferring process of KE in different steps of eddies with loss of energy is known as *energy cascade* (Figure 6.39). This transfer of energy from larger eddies into smaller eddies continues until the eddy Reynolds number ($Re_l = ul/\nu_f$) is small enough to make stable through viscosity and the kinetic energy is dissipated into heat. The energy cascade occurs from large-scale structures to smaller scale by an inertial and inviscid mechanism. Eventually this process creates structures that are small enough that molecular diffusion becomes important and viscous dissipation of energy finally takes place. The scale at which this happens is the Kolmogorov length scale.

Richardson (1922) suggests a concept of turbulence that a turbulent flow is composed by "eddies" of different sizes. The sizes define a characteristic length scale for eddies, which are also characterized by velocity scales and time scales dependent on the length scale. The large eddies are unstable and eventually break up into smaller eddies, and the kinetic energy of the large eddies is divided into the smaller eddies that stemmed from it. These smaller eddies undergo the same process, giving rise to even smaller and smaller eddies which inherit the energy of their forerunner eddy, and so on. In this way, the energy is passed down from the large scales of the motion to smaller scales until reaching a

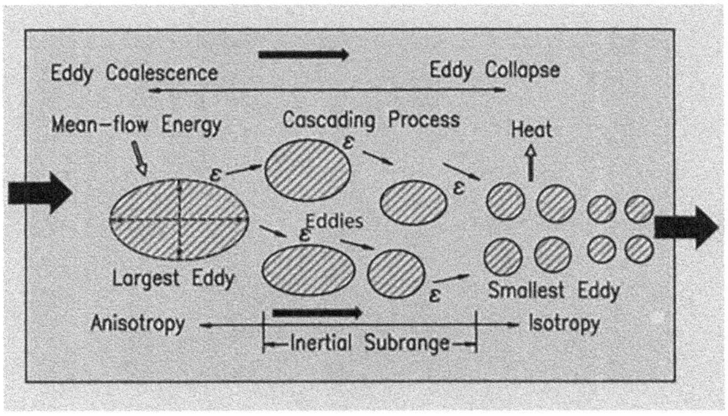

FIGURE 6.39 Steps of energy cascading. (Modified from Nezu, 2005.)

sufficiently small length scale such that the viscosity of the fluid can effectively dissipate the kinetic energy into internal energy.

6.11.3 Taylor's Length Scale

The *Taylor micro-scale* is referred to as an intermediate turbulent length scale at which the viscosity appreciably affects the dynamics of turbulent eddies in the flow. This length scale is conventionally used to turbulent flow which is characterized by a Kolmogorov spectrum of velocity fluctuations. In such flow, the length scales which are larger than the Taylor micro-scale are not affected by the viscosity, and these larger length scales are generally referred as inertial range. The intermediate scales between the largest and the smallest scales which make the inertial sub-range. Taylor micro-scales are not dissipative scale but pass down the energy from the largest to the smallest without dissipation. The length scales which are less than the Taylors micro-scale are subjected to strong viscous forces and then the kinetic energy is dissipated into heat. These shorter length-scales are generally termed as the dissipation range. Taylor micro-scales are often used in describing the turbulence more conveniently as these micro-scales play a dominant role in energy and momentum transfer in the wavenumber space. The Taylor micro-scale λ_t is the measure of typical eddy size in the inertial sub-range and is the relevant length-scale of turbulence, and is given by:

$$\lambda_t = \left(\frac{15\nu_f \overline{u'u'}}{\varepsilon} \right)^{0.5} = \left(\frac{15\nu_f \overline{u'^2}}{\varepsilon} \right)^{0.5} \tag{6.264}$$

where ε is the TKE dissipation rate given by (6.247). For isotropic turbulence, one can write $\overline{u'^2} = 2k/3$, where $k = (\overline{u'^2} + \overline{v'^2} + \overline{w'^2})/2$. Therefore, the Taylor micro-scale λ_t can be written as

$$\lambda_t = \left(\frac{10\nu_f k}{\varepsilon} \right)^{0.5} \tag{6.265}$$

The Taylor micro-scale falls into the large scale of eddies and the small scale of eddy.

6.11.3.1 Kolmogorov Length Scale

It is the smallest scales in the spectrum that form the viscous sub-layer range. In this range, the energy input from nonlinear interactions and the energy drain through viscous dissipation are in exactly balanced. The small scales have high frequency, causing turbulence to be locally isotropic and homogeneous. The scale formed from the dissipation rate and viscosity is called the *Kolmogorov length scale* η_k, which is a measure of smallest eddy present in the flow (Pope, 2000; Li and Katul, 2017), and it is given by:

$$\text{Length scale}: \eta = \left(\frac{\nu_f^{\,3}}{\varepsilon} \right)^{1/4} \tag{6.266a}$$

$$\text{Velocity scale}: u_\eta = (\varepsilon \nu_f)^{1/4} \tag{6.266b}$$

$$\text{Time scale}: \tau_\eta = \left(\frac{\nu_f}{\varepsilon} \right)^{1/2} \tag{6.266c}$$

$$\left(\frac{u_\eta}{\eta} \right) = \frac{1}{\tau_\eta}, \ Re_\eta = \frac{\eta u_\eta}{\nu_f} = 1 \tag{6.267}$$

These scales are the indicative of the smallest eddies present in the flow, the scale at which the energy is dissipated.

6.11.4 Kolmogorov Hypothesis

Kolmogorov (1941) postulated that for very high Reynolds numbers, the small-scale turbulent motions are statistically isotropic (i.e., no preferential spatial direction could be discerned). In general, the large scales of a flow are not isotropic, since they are determined by the particular geometrical features of the boundaries (the size characterizing the large scales will be denoted as *L*). Kolmogorov's idea was that in Richardson's energy cascade, the geometrical and directional information is lost, while the scale of motion is reduced so that the statistics of the small scales has a universal character: they are the same for all turbulent flows when the Reynolds number is sufficiently high. In the process of transfer of turbulent energy from large-scale motion (small wave number) to small scale of motion (large wave number), the influence of initial large scale of motion ceases at a point beyond which viscous dissipation occurs to range of wave number, where the characteristic of turbulence is the same for all flows. That range is called the *universal equilibrium range*. The *universal* means where the character of turbulence is independent of the flow environment, and the equilibrium means the energy input is in balance with the energy dissipation. The Kolmogorov's hypothesis belongs to this range.

Kolmogorov's first hypothesis states that irrespective of large scale of motion, the turbulence in the universal equilibrium range is isotropic. The energy spectrum *E*(*k*) is a function of viscosity ν_f, dissipation ε and wave number *k* (Figure 6.40). Assuming the isotropic turbulence, the energy spectrum function *E*(*k*) in dimensional form, one gets

$$E(k) = u_\eta^2 \eta \theta(\eta k) \tag{6.268}$$

where $u_\eta = (\nu_f \varepsilon)^{1/4}$ is the characteristic velocity (cm/s), and $\eta = \left(\nu_f^{\,3}/\varepsilon\right)^{1/4}$ is the characteristic length (cm), and the θ is a function of η and *k*. The characteristic time τ_η is defined as $\tau_\eta = \eta/u_\eta = (\nu_f/\varepsilon)^{1/2}$. The above parameters are involved in large wave number (small length scale motion), which means, these are the Kolmogorov scales: smallest length and smallest time scales of turbulence because these scales are smaller than those involved in molecular scale in motion. *Kolmogorov's second hypothesis* states that the peak of energy spectrum and that of dissipation spectrum with respect to wave number *k* are sufficiently apart.

The energy spectrum with large eddies (small wave number) contains most of the turbulent energy, while the dissipation spectrum with smallest eddies (large wave number) associates with viscous dissipation. According to second hypothesis, there must be a wave number range between the ranges, where

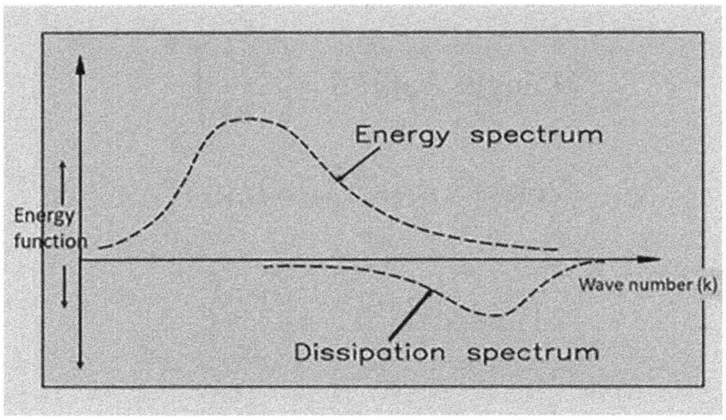

FIGURE 6.40 Energy and dissipation spectra with peaks. (Modified from Sumer, 2002.)

there is no significant energy and dissipation, called universal equilibrium range, where the energy input is in balance with the energy dissipation, and no presence of significant energy (Figure 6.41). In that range, the turbulent energy is transferred from small wave numbers to large wave numbers, and the process is governed by the inertia terms in the Navier–Stokes equation. This range of wave number is called the inertial sub-range. In the inertial sub-range, the energy spectrum is independent of viscosity with no significant dissipation takes place. To get the independent of viscosity in the inertial sub-range, the function $\theta(\eta k)$ is given in the following form as

$$\theta(\eta k) = (k\eta)^{-5/3} \tag{6.269}$$

Inserting all parameter values in equation (6.268), the spectrum function $E(k)$ is given by

$$E(k) = (\nu_f \varepsilon)^{1/2} \left(\frac{\nu_f^{\,3}}{\varepsilon}\right)^{1/4} (k\eta)^{-5/3} \tag{6.270}$$

After putting for η, on simplification, one gets

$$E(k) = \varepsilon^{2/3} k^{-5/3} \tag{6.271}$$

Equation (6.271) is known as the Kolmogorov's –5/3 rd law. Experiments confirm the presence of this sub-range where the spectrum is given by the –5/3 rd law.

Kolmogorov's theory is an asymptotic theory. It has been shown to work well in the limit of very high Reynolds numbers. The exact shape of the normalized spectra may deviate from Kolmogorov's model for intermediate Reynolds numbers. For example, on the order of Reynolds number 10,000, Taylor's Reynolds number $R_T \sim 250$, the exponent of $E(k) \sim k^p$ in the inertial sub-range is often measured to be $p \sim 1.5$ instead of 5/3 (~1.67). Kolmogorov's theory assumes the energy cascade is from large eddies to small eddies; and the transfer of energy from smaller scale to larger scales is known as the backscatter process, although at a much lower rate. The dominant energy transfer is from larger to smaller eddy. The theory assumes that at high Reynolds numbers, the turbulence is highly random, so the large-scale coherent structures may form. The research on the fundamental aspects of turbulence, both experimentally as well as direct numerical simulation (DNS) should be further continued for better solutions.

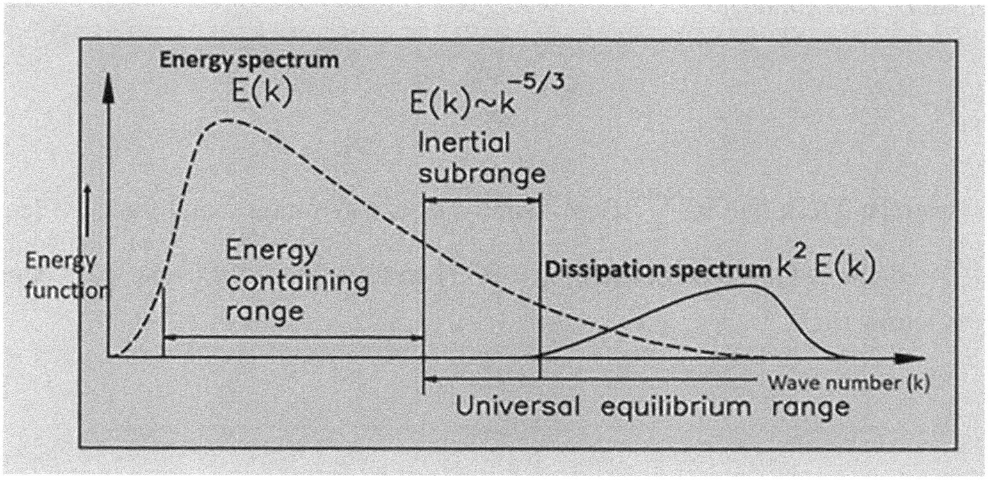

FIGURE 6.41 Representation of energy spectrum. (Modified from Sumer, 2002.)

6.12 ANISOTROPY OF TURBULENCE

In isotropic turbulence, there is no directional preference and a perfect disorderly motion is preserved. Thus, the velocity fluctuations are independent of the axis of reference or invariant to axis of rotation. Thus, the Reynolds normal stresses are identical with respect to its time-averaged value, while in anisotropic turbulence, the velocity fluctuations have directional preference, and thus the Reynolds normal stresses are unequal (Pope, 2000; Jovanovic, 2004). The isotropy is evaluated in statistical sense on the basis of variances and covariance analysis. The dynamics and three-dimensional structures are the basic characteristics of a turbulent flow, however there are some other characteristics such as range of scales, vorticity, dissipativity, etc. The Reynolds stress tensor is the finest device to enumerate such an issue in a turbulent flow, while the correlation coefficient and the Reynolds stress ratio are used for the assessment of the level of turbulence anisotropy (Lumley and Newman, 1977). The symmetric second-order Reynolds stress tensor is given by

$$T = \tau_{ij} = \begin{bmatrix} \overline{u'^2} & \overline{u'v'} & \overline{u'w'} \\ \overline{v'u'} & \overline{v'^2} & \overline{v'w'} \\ \overline{w'u'} & \overline{w'v'} & \overline{w'^2} \end{bmatrix} \tag{6.272}$$

Anisotropic turbulence is defined as the statistical features that have directional preference and mean velocity has a gradient. Any symmetrical matrix could be decomposed into isotropic τ_{ij}^I and anisotropic τ_{ij}^A parts, such as

$$\tau_{ij} = \tau_{ij}^I + \tau_{ij}^A \tag{6.273}$$

The decomposition is performed in the following way:

$$\tau_{ij}^I = \frac{1}{3}\tau_{kk}\delta_{ij}; \quad \tau_{ij}^A = \tau_{ij} - \tau_{ij}^I \tag{6.274}$$

where δ_{ij} is the Kronecker delta function, that is $\delta_{ij}(i \neq j) = 0$ and $\delta_{ij}(i = j) = 1$. Reynolds stress anisotropy tensor b_{ij} is defined as the difference between the ratio of Reynolds stress tensor terms to the turbulent kinetic energy (TKE) and its isotropic equivalent quantity (Lumley and Newman, 1977). The anisotropy tensor b_{ij} is as follows:

$$b_{ij} = \frac{\overline{u_i'u_j'}}{2k} - \frac{\delta_{ij}}{3} \tag{6.275}$$

where k is the average TKE, that is $\frac{\overline{u_i'u_i'}}{2}$. Significantly, b_{ij} is a symmetric and traceless tensor bounded by $\frac{-1}{3} \leq b_{ij} \leq \frac{2}{3}$ and for isotropic turbulence $b_{ij} = 0$. Therefore, the anisotropic part is considered in Reynolds stress tensor as

$$b_{ij} = \frac{\tau_{ij}}{\tau_{kk}} - \frac{\delta_{ij}}{3} \tag{6.276}$$

$$= \begin{bmatrix} \frac{\overline{u'^2}}{\overline{u'^2}+\overline{v'^2}+\overline{w'^2}} - \frac{1}{3} & \frac{\overline{u'v'}}{\overline{u'^2}+\overline{v'^2}+\overline{w'^2}} & \frac{\overline{u'w'}}{\overline{u'^2}+\overline{v'^2}+\overline{w'^2}} \\ \frac{\overline{u'v'}}{\overline{u'^2}+\overline{v'^2}+\overline{w'^2}} & \frac{\overline{v'^2}}{\overline{u'^2}+\overline{v'^2}+\overline{w'^2}} - \frac{1}{3} & \frac{\overline{v'w'}}{\overline{u'^2}+\overline{v'^2}+\overline{w'^2}} \\ \frac{\overline{u'w'}}{\overline{u'^2}+\overline{v'^2}+\overline{w'^2}} & \frac{\overline{v'w'}}{\overline{u'^2}+\overline{v'^2}+\overline{w'^2}} & \frac{\overline{w'^2}}{\overline{u'^2}+\overline{v'^2}+\overline{w'^2}} - \frac{1}{3} \end{bmatrix} \quad (6.277)$$

The nondimensional anisotropic part of the Reynolds stress tensor could be characterized by a set of eigenvalues and the corresponding eigenvectors. The eigen-values of anisotropy tensor b_{ij} are determined by solving the characteristic equation $\det(b_{ij} - \lambda I) = 0$ or $|b_{ij} - \lambda\delta_{ij} = 0|$. Evaluation of this determinant leads to a characteristic polynomial of b_{ij}, which is given by

$$\lambda^3 - I_1\lambda^2 + I_2\lambda - I_3 = 0 \quad (6.278)$$

where the invariants I_1, I_2, and I_3 are defined as follows:

$$I_1 = b_{kk} = 0;\ I_2 = -\frac{b_{ij}b_{ji}}{2};\ I_3 = \frac{b_{ij}b_{jk}b_{ki}}{3} = \det(b_{ij}) \quad (6.279)$$

A cross plot of I_2 against I_3 is termed anisotropic invariant map (AIM). In an AIM, I_2 (positive or zero) represents the degree of anisotropy and I_3 corresponds to the nature of anisotropy. However, the shape of the allowed state area is very narrow and the borders are obviously nonlinear. Thus, to improve the situation, researchers preferred the cross plot of modified invariants ξ and η, defined by

$$\xi^3 = \frac{I_3}{2};\eta^2 = -\frac{I_2}{3} \quad (6.280)$$

The origin of the AIM, i.e., $\xi = \eta = 0$, corresponds to three-dimensional (3D) isotropic turbulence, where the energy ellipsoid has a spherical form. AIMs are plotted in Figure 6.42, using ξ against η at two locations along fully developed flow for $u_m = 23.5$ cm/s. To quantify the 3D isotropy of the turbulence, an invariant function I_F is defined by

$$I_F = 1 + 27I_3 + 9I_2 \quad (6.281)$$

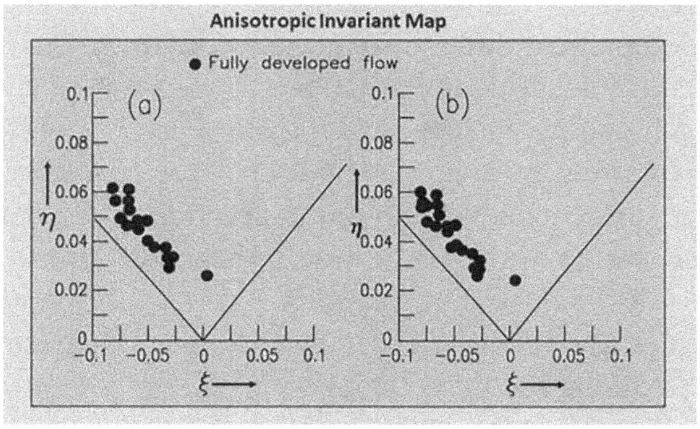

FIGURE 6.42 Typical AIMs by plotting ξ against η at two locations along fully developed flow for u_m=23.5 cm/s. (From ISI laboratory data.)

Thus, when I_F approaches to zero, the turbulence becomes two-dimensional and it becomes unity when turbulence enters a three-dimensional isotropic state.

6.13 TURBULENT FLOW ANALYSIS – APPLICATIONS

The turbulent flow analysis techniques mentioned in the previous sections have wide applications in normal turbulent flow investigations and higher-order turbulent analysis. These techniques are highly useful to understand the turbulence generation and dissipation in experimental works as well as field level applications.

Several experimental studies on turbulent flows over different types of bed roughness have been reported in the flow hydrodynamics with their scope, which are mostly constrained to the friction factors, separation points and wake generation. Large-scale roughness, such as boulders, bluff bodies, dunes and ripples due to the interaction with the turbulent flow are ubiquitous in natural stream or channel, which induce flow separation, circulation and wakes, and hence form resistance. They cause the overall flow resistance, and hence modulation of flow and turbulence characteristics. The study of flows around submerged obstructions in alluvial rivers is very important both in basic sciences and engineering applications. Dunes and ripples-dominated bed load in river and/or laboratory environments are often asymmetric with low-slope in upstream side (stoss) and steep-slope in lee faces at downstream (Guy et al., 1966; Kostaschuk, 2000; Ojha and Mazumder, 2008), while suspension load-dominated environments are often more symmetric with relatively low angle lee faces (Kostaschuk and Villard, 1996; Best and Kostaschuk, 2002; Mazumder et al., 2005a). Several flume and field experiments were performed to estimate the resistance to the flow due to the physical changes of bed forms generated by the turbulent flow (Smith and McLean, 1977; Richards, 1980; Engelund and Fredsoe, 1982; Wiberg and Nelson, 1992; Gabel, 1993; Lyn, 1993; Robert and Uhlman, 2001; Best, 2005; Ojha and Mazumder, 2008; Camussi et al., 2008; Mazumder et al., 2009; Ren and Wu, 2011; and others).

The considerable amount of work as mentioned above has been put-in to study the flow resistance due to structural changes of one or two bedforms in the flow, especially the occurrence of flow separation, recirculation and reattachment, and hence the flow resistance. For example, the turbulence statistics of basic flow and the flow over the submerged obstacle are presented using the above methodologies to understand unknown physics of flow subjected to the frontal collision with the submerged obstacle. Mazumder and Sarkar (2014) investigated experimentally the flow over two artificial upstream-facing waveform structures to identify the spatial changes in turbulence statistics of flow addressing the recirculation eddies in the flow separation regions, the Reynolds stresses, the turbulent bursting events associated with the burst-sweep cycles and drag coefficients. Later Sarkar and Mazumder (2018b) examined the third-order moments, turbulent kinetic energy (TKE) fluxes, turbulent production and dissipation rates, turbulent length scales, and Reynolds stress anisotropy in the wake regions over the upstream-facing waveform structures. Chatterjee et al. (2020) showed experimentally the flow and turbulence characteristics over a forward-facing obstacle with the occurrence of a distinct dipped water surface shape over the obstacle. The obstacle is modeled close to an airfoil shape with a flat top for preventing flow separation (Weinfurtner et al., 2011). An attempt was made to detect dipped water surface shape from the interaction of flow and the obstacle and to adequately address the associated time-averaged turbulence statistics: mean flows, mean-square intensities, Reynolds shear stress, turbulence spectra, eddy viscosity, turbulence production, turbulence kinetic energy, probability-density function (pdf) and the fractional contributions of burst-sweep cycles to the total shear stress along the flow.

6.14 SUMMARY

Some of the important aspects discussed in this chapter are summarized below:

- For channel flow, when the Reynold's number is more than 500, the flow changes from laminar to turbulent flow. Similarly, for pipe flow, when the Reynold's number is more than 2,000, the flow changes from laminar to turbulent flow.
- In turbulent flows, due to violent mixing of fluid particles, transfer of momentum between adjacent fluid layers in a direction normal to the flow causing shear stresses.
- The shearing stress in turbulent flow is given by: $\tau_* = \mu_f \frac{\partial \bar{u}}{\partial y} + \mu_t \frac{\partial \bar{u}}{\partial y}$.
- The governing equations of turbulence are represented by the Reynolds equation, which can be derived from the basic Navier–Stokes equations.
- According to Boussinesq's eddy viscosity hypothesis: $\tau_t = -\rho_f \overline{u'v'} = A_\tau \frac{d\bar{u}}{dy} = \rho_f \epsilon_t \frac{d\bar{u}}{dy}$
- The shear stress according to Prandtl mixing length hypothesis is: $\tau_t = -\rho_f \overline{u'v'} = \rho_f l^2 \left(\frac{d\bar{u}}{dy}\right)^2$ where $l = \kappa_0 y$.
- Turbulent velocity distribution over a flat plate: From Prandtl theory, universal velocity distribution law: $\bar{u} = \frac{u_{*0}}{\kappa_0} \ln y + C$
- Turbulent velocity distribution in smooth pipe: $\frac{\bar{u}}{u_{*0}} = 5.75 \log\left(\frac{yu*}{\nu_f}\right) + 5.5$
- Turbulent velocity distribution in rough pipe: $\frac{\bar{u}}{u_{*0}} = 2.5 \ln\left(\frac{y}{k_s}\right) + 8.5$
- The turbulent kinetic energy k is defined as: $k = \frac{1}{2}\overline{q^2} = \frac{1}{2}\overline{\left(u'^2 + v'^2 + w'^2\right)}$

6.15 EXERCISE PROBLEMS

1. Illustrate the different stages of transition from laminar to turbulent flow in a pipe flow. Explain the principles of stability theory.
2. Derive the Reynolds equations of turbulence from the Navier–Stokes equations by considering the turbulent fluctuations and Reynolds stresses.
3. Illustrate the turbulent boundary layers equations in comparison with laminar boundary layer equations.
4. Illustrate the Prandtl's mixing length hypothesis and its importance in turbulent flow analysis.
5. Derive the expression for turbulent shearing stress using the von-Karman similarity hypothesis.
6. Starting from Prandtl's mixing length theory, derive the Universal velocity distribution law.
7. Illustrate the velocity distribution from the Cole's law wake strength.
8. Differentiate the velocity distributions laws over smooth and rough surfaces.
9. Derive the kinetic energy equation for laminar flows and differentiate it with the turbulent kinetic energy equation for fluctuating flows.
10. Explain the turbulent bursting process and applications of quadrant threshold technique.
11. Define and illustrate turbulent kinetic energy (TKE), TKE flux and turbulent energy budget.
12. Illustrate Isotropic turbulence and derive expression for rate of dissipation.
13. Explain the spectral analysis and elaborate the necessity of it while measuring turbulent flow components using an ADV.

14. Elaborate Taylors' length scale and Kolmogorov length scale with its importance and applications.
15. Illustrate Kolmogorov hypothesis and its importance in turbulent flow analysis. Explain the energy spectrum with the help of Kolmogorov −5/3 law.
16. Define anisotropy of turbulence and explain anisotropic invariant map and applications.
17. On a plain land, a Radio tower to be constructed to a height of 20 m. The wind velocity was measured at 3 m and 9 m, and observed to be 2.6 m/s and 3 m/s. Calculate the wind velocity at 30 m above ground. Assume a fully developed rough turbulent flow, calculate the roughness height of the surface.
18. Calculate the relative roughness for a fully turbulent flow through a rough pipe, given that velocity u_x is 3.5 m/s, for y/a equals 0.25 and velocity u_x is 4.5 m/s for y/a equals 0.5?
19. If the velocities in a 0.4 m pipe carrying oil from the axis are 3.5 m/s and 3 m/s on the centerline and at a radial distance 0.1 m. Calculate the shear velocity and flow rate in the pipe?

7 Turbulent Flow Measurements and Instrumentations

7.1 INTRODUCTION

Due to the complexity of turbulent flows, measurement is very complicated. For the last few decades, starting from the 1920s, scientists and engineers developed various instruments for turbulent flow measurements. Further, in the 20th century, boundary layer theory contributed much to the development of aerodynamics because its knowledge was crucial in the design of airplanes and missiles. The hot-wire anemometry was first used to measure turbulence in airflows in the 1930s. Taylor (1935) initiated to develop the statistical theory of isotropic turbulence which is an "ideal turbulence," but it was important in the turbulence research. The isotropic turbulence was studied to improve the understanding of turbulence and its characteristics with transport phenomena. Particularly, the significance of statistical theory of locally isotropic turbulence was proposed by Kolmogorov (1941). This theory led to well-known Kolmogorov -5/3 law for the spectrum of flow turbulence. Consequently, the theory of locally isotropic turbulence could be applied to the real turbulent flow like the boundary layer flows, channel or pipe flows, jet and wakes, even to geophysical flows, such as atmospheric turbulence and oceanic turbulence. The theory of locally isotropic turbulence can also be applied to the open channel turbulent flow.

The free turbulence such as grid turbulence, and jets and wakes, has been intensively studied theoretically and experimentally by a Cambridge University group, led by Sir G. I. Taylor, G. K. Batchelor, and H. C. H. Townsend. Excellent review of this topic was written by Batchelor (1953), and later by Townsend (1976). Outstanding experimental investigations on boundary layer and duct and pipe flows were conducted by Klebanoff (1954), and Laufer (1951, 1954). These data are still frequently used for understanding basic flow, and these are compared with the data of open channel turbulent flow. Reliable measurements of fluctuating components of velocity had been obtained with the aid of hot-wire anemometers. During 1960, hot-wire anemometer and pressure transducers were used for point measurements of turbulence. The turbulence could be explained theoretically and experimentally with the help of statistical tools, such as covariance, space–time correlations, and spectral analysis; therefore, the turbulence was then considered to be a random and chaotic phenomenon. For many decades, isotropic turbulence was studied in several ways experimentally. The theoretical treatment of isotropic turbulence was associated with more complex flow fields (Townsend, 1976). However, Kline et al. (1967) discovered the bursting phenomena in a turbulent boundary layer.

Then, Brown and Roshko (1974) and Winant and Broward (1974) discovered the vortex-pairing phenomena in the turbulent mixing layer. Summaries and reviews of this literature are available in well-known monographs (Monin and Yaglom, 1971, 1975; Tennekes and Lumley, 1972; Rotta, 1972; Hinze, 1975; Townsend, 1976; Bradshaw, 1976; Schlichting, 1979; Tani, 1980, 1984; Tatsumi, 1986; Landahl and Mollo-Christensen, 1986; Pope, 2000; Schlichting and Gertsen, 2000 and others). In this chapter, various measurement techniques and instrumentation used for turbulent flow measurements and case studies are discussed in detail.

7.2 INSTRUMENTATIONS FOR LABORATORY FLOW MEASUREMENTS

Extensive experimental research has been carried out on the mean flows and turbulence characteristics in open-channel flow using many sophisticated measuring equipment, such as Ott-laboratory current meter, Electromagnetic current meter, Hot-wire anemometers (HWA), laser Doppler anemometers

DOI: 10.1201/9781003000020-7

(LDA), acoustic Doppler velocimeters (ADV), particle image velocimetry (PIV), and various methods of flow visualization such as the hydrogen-bubble and digital imaging technique. All these sophisticated measuring instruments have also shortcomings, which concern with high prices, complicated use, and limited applications. In the early 1970s, the Ott-laboratory propeller-type current meter was used to measure the mean flow only from the revolutions of the propeller per time interval counted electronically without concerning the fluctuations in velocity components. In the later stage, the following electronic measuring gadgets are developed to record the fluid velocity components with fluctuations in a laboratory flume:

1. Hot-wire anemometer (HWA),
2. Laser Doppler anemometer (LDA)
3. Acoustic Doppler velocimeter (ADV)
4. Particle image velocimetry (PIV)
5. Flow visualization using digital imaging.

7.2.1 OTT Laboratory Propeller-Type Current Meter

In early 1970, the OTT C2 small current meter was generally used for flow velocity measurements for shallow waters in small rivers, channels, and laboratories. The measurement range was 0.025–5 m/s, and the accuracy was about ±1%. This current meter provided only the mean velocity of the fluid flow (Figure 7.1). In a broad sense, the mean flow velocity of a small river can be measured using this propeller current meter.

Nowadays velocity measuring instruments are very useful throughout the world to identify the fluid velocity and turbulence structures irrespective of shear turbulence or free turbulence. Brief explanations of more sophisticated fluid velocity measuring equipment are provided in the following sections.

7.2.2 Hot-Wire Anemometer

The hot-wire anemometer (HWA) is a device for measuring the velocity and direction of the fluid flow. This can be done by measuring the heat loss of the wire, which is placed in the fluid stream. The wire is heated by an electrical current. When hot wire placed in the stream of the fluid, in that case, the heat is transferred from wire to fluid, and hence the temperature of the wire reduces. The resistance of wire measures the flow rate of the fluid. The hot-wire anemometer is used as a research device to measure fluid flow. It works from the principle of heat transfer from high temperature to low temperature. The

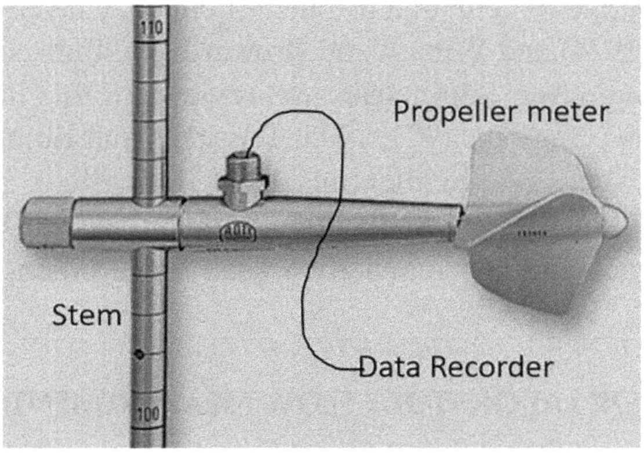

FIGURE 7.1 Photo of OTT laboratory propeller-type current meter taken from Indian Statistical Institute (ISI) fluvial mechanics laboratory at Kolkata, India.

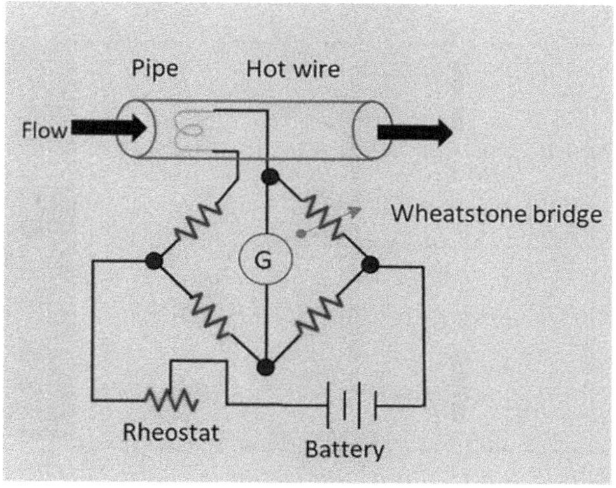

FIGURE 7.2 Typical sketch of hot-wire anemometer.

HWA consists of two main parts: conducting wire and wheatstone bridge. The conducting wire is housed inside the ceramic body. The wires are taken out from the ceramic body and connect to the wheatstone bridge. The wheatstone bridge measures the variation of resistance (Figure 7.2).

7.2.3 Laser Doppler Anemometer (LDA)

Laser Doppler anemometer (LDA), also known as laser Doppler velocimetry (LDV), is a technique of using the Doppler shift in a laser beam to measure the three-dimensional (3D) velocity in transparent or semi-transparent fluid flows, or the linear or vibratory motion of opaque, reflecting, surfaces. The LDA is superior to hot-film anemometer because it can be used to measure multiple components of velocity independently. The measurement with LDA is absolute, linear with velocity, and requires no pre-calibration. The LDA is a widely accepted tool for fluid dynamic investigations in gases and liquids and has been used for more than three decades. It is a well-established technique that gives information about flow velocity. It is a non-intrusive principle and directional sensitivity makes it very suitable for applications with reversing flow, chemically reacting or high-temperature media, and rotating machinery, where the physical sensors are difficult or impossible to use. It requires tracer particles in the flow (Figure 7.3). For example, Ojha et al. (2019) measured three-dimensional velocity using LDA to estimate the turbulence generated by an oscillating grid plate and its impact on sediment transport.

7.2.4 Acoustic Doppler Velocimeter (ADV)

The Micro-Acoustic Doppler Velocimeter (ADV) was introduced in the U.S Army core of Engineers, Waterways Experiment Station (WES) in 1992 to satisfy the need for an accurate current meter that can measure three-dimensional (3D) flows in physical models. ADV is an intrusive device. 16 MHz 3D-Micro ADV is designed to record instantaneous velocity components at a single-point with a relatively high frequency with 100 Hz. This instrument is very popular and easy to use. Measurements are performed by recording the velocity of particles in a remote sampling volume based on the Doppler shift effect. The probe head includes one transmitter and between two and four receivers. The remote sampling volume is located typically 5 cm from the tip of the transmitter, but some studies showed that the distance might change slightly. The sampling volume size is determined by the sampling conditions and manual setup. In a standard configuration, the sampling volume is a water cylinder with a diameter 6 mm and a height 9 mm, although latest laboratory ADVs may have smaller sampling volume

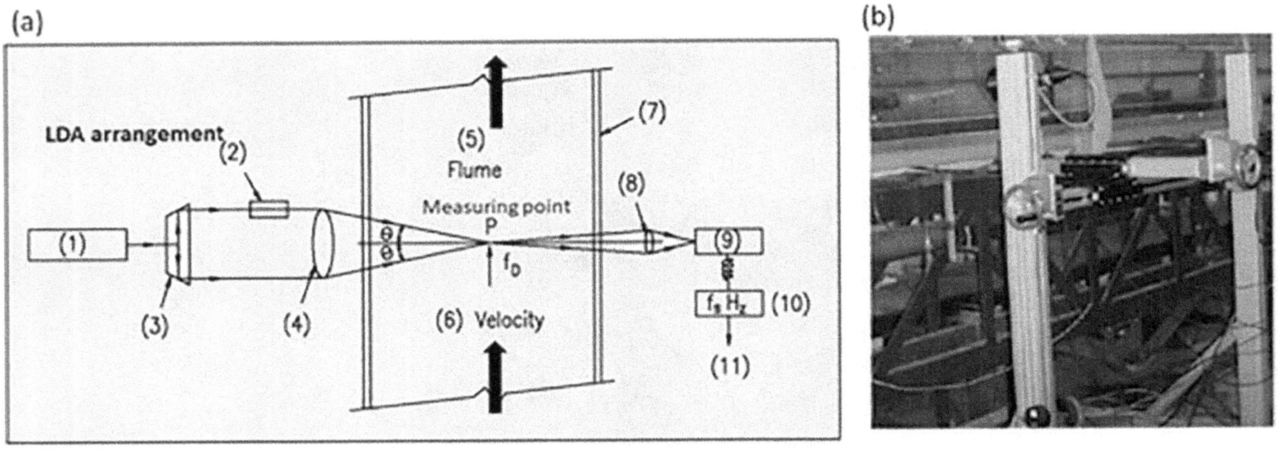

FIGURE 7.3 (a) Typical sketch of laser doppler anemometer (LDA) system consisting of various parts such as (1) laser, (2) brogg cell (40 MHz), (3) beam splitter, (4) front lens, (5) Experimental flume, (6) velocity measurement, (7) glass, (8) lens, (9) photo-multiplier, (10) shifter, and (11) burst signals used in water flume; (b) photo of LDA arrangement (modified from the Technical University of Denmark, fluid mechanics and hydraulic laboratory, Ojha et al., 2019).

(e.g., Sontek micro ADV, and Nortek Vectrino). The ADV is used to measure the instantaneous velocity of water in three dimensions. The principle is: the device sends out a beam of acoustic waves at a fixed frequency from a transmitter probe. These waves bounce off of moving particulate matter in water, and three receiving probes take note of the change in frequency of the returned waves. The ADV then calculates the velocity of the water in the *x*, *y*, and *z* directions. A general schematic of the ADV is shown in Figure 7.4.

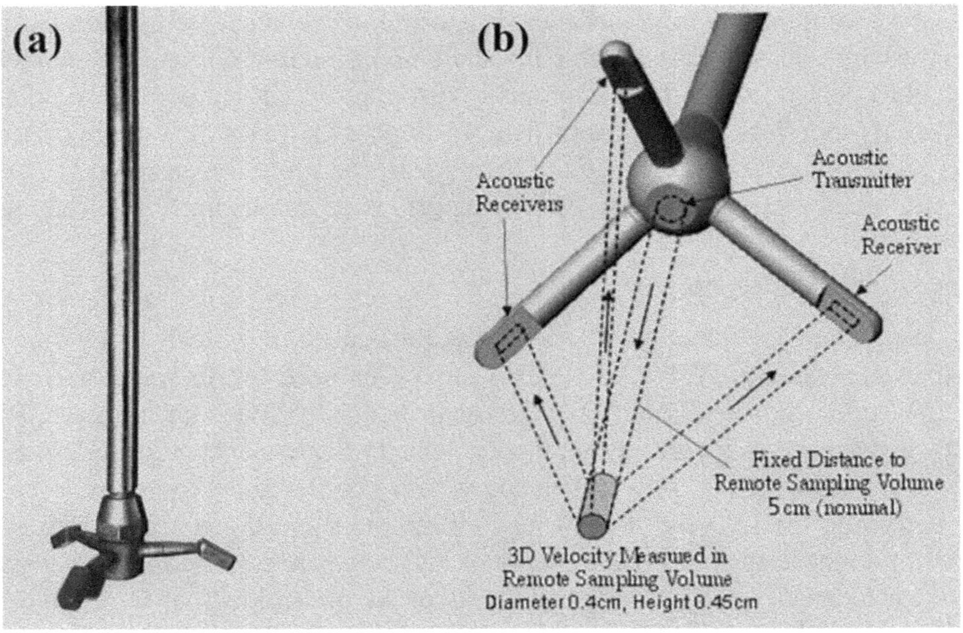

FIGURE 7.4 (a) Down-looking 16 MHz 3D-Micro ADV probe; (b) schematics diagram of the sampling volume (taken from ISI fluvial mechanics laboratory 2002 at Kolkata, India).

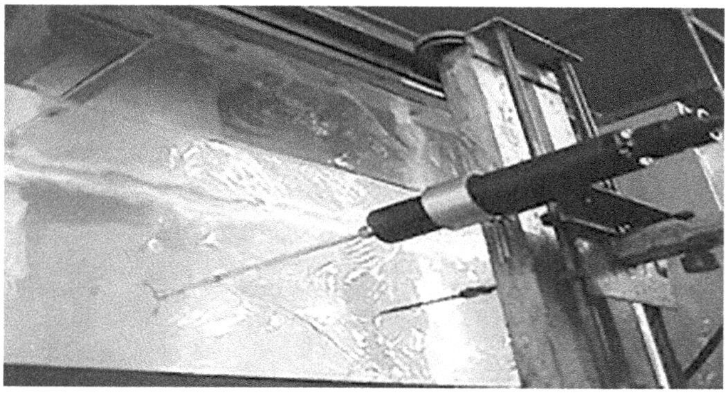

FIGURE 7.5 Data collection using 16 MHz 3D-Micro ADV at the ISI fluvial mechanics laboratory at Kolkata, India.

As shown in Figure 7.5, the instantaneous velocity data are collected using a SonTek 16 MHz down-looking 3-D acoustic Doppler velocimeter (ADV) from the lowest level 0.35 cm to any desired height near the water surface. The sampling volume is located 5 cm below the transmitter probe and the entire probe should be immersed in water. ADV did not collect the velocity data near the free surface. ADV software provided signal quality information in terms of signal-to-noise ratio (SNR) and correlation coefficient (Corr. ADV). SNR is expressed in decibel units, and Corr. ADV is expressed as a percentage. The velocity data are cleaned by removing communication errors, low signal-to-noise ratio (<15 dB), and low correlation samples (<70%) (SonTek Inc., 2001). The ADV sampling volume is $9 \times 10^{-8} m^3$ and is approximately cylindrical oriented along the transmitter beam axis (Lohrmann et al., 1994). Factory calibration of ADV is specified to be ±1.0% of the measured velocity (i.e., an accuracy of ±1 cm/sec is on a measured velocity of 100 cm/sec). The ADV has been validated with several other devices by several investigators and has been used in a variety of applications for turbulence measurements.

7.2.5 Particle Image Velocimetry

Particle image velocimetry (PIV) is an optical method of flow visualization used in education and research (Figure 7.6a). It is mainly used to acquire instantaneous velocity and related properties in fluids. A stereoscopic PIV system from *Dantec Dynamics* (2015) is used to capture the images in a plane for three-dimensional (3D) velocity derivation. The system includes a double-pulsed Nd: YAG laser (1,200 mJ/pulse) with a minimum pulse duration of 4 ns, laser light sheet optic, two Nikon charged coupled device (CCD) cameras, a laser pulse synchronizer, an Intel (R) Xeon (R) CPU E5-1680 V3@3.20 GHzComputer, PIV software (Dantec Dynamics), and a three-dimensional computer-operated traversing system (1,000 mm travel along each axis). Two CCD cameras with resolutions of 2,048 × 2,048 pixels are mounted on a traverse system to capture the images (Figure 7.6b). The PIV system includes the following: (i) a laser pulse synchronizer to synchronize the laser with the cameras and (ii) a calibration target sheet of 450 mm × 450 mm to determine the relation between the pixels. The calibration sheet has black marker on it at equal spaces of 11 mm and a big black dot at the center is considered as the origin for both the cameras. The common field of view for both the cameras is identified with respect to the origin and the number of markers visible in both the cameras in all directions on the calibration sheet. The seeding of the flow is done using silver-coated hollow glass spheres (S-HGS) with a diameter of 10 μm of density of 1.03–1.05 kg/cm³ similar to water (Adrian, 1991; Adrian and Westerweel, 2011; Raffel et al., 2018). The PIV procedure includes seeding of the flow, illumination of the measurement plane using laser and successive capturing of the illuminated plane using the cameras. The whole PIV system is synchronized using a Timer Box. The laser power output is regulated using laser system control. The

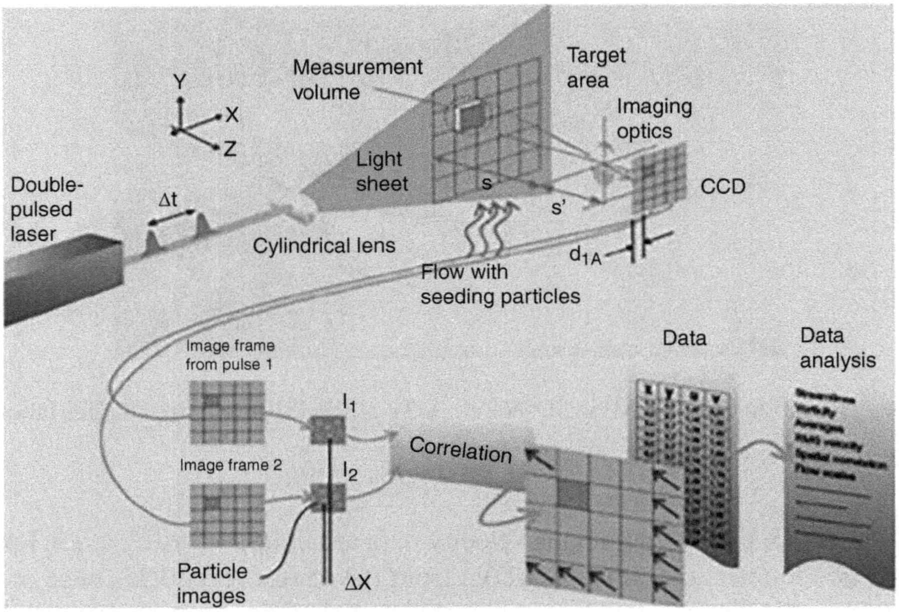

(a)

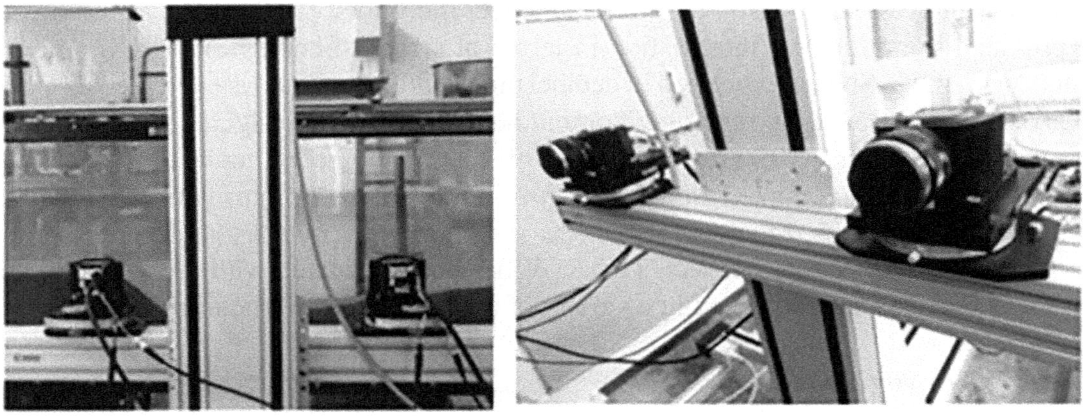

(b)

FIGURE 7.6 (a) Experimental setup for stereoscopic PIV measurements of the flow, showing various parts at Indian Institute of Technology (IIT) Bombay (DANTEC PIV).(b) Photographs of PIV components at hydraulics laboratory, IIT Bombay showing two cameras mounted on traverse: (a) back view; (b) front view.

pulse duration between the lasers is adjusted to optimize the laser intensity and uniform illumination of the flow plane in both frames. PIV images are captured by the two cameras, which continuously transferred the data to the computer at a frame rate of 15 Hz, denoting 15 instantaneous (double frame) vector fields per second. The raw data are then saved to the main database in the PC for post-processing and analysis using *Dantec dynamics* software. The PIV technique is non-intrusive and measures the velocities of micron-sized particles following the flow in a plane.

7.2.6 Flow Visualization Using Digital Imaging

Flow visualization technique using digital imaging in fluid dynamics is used to study the flow patterns in order to get qualitative or quantitative information on the movement of fluid. *Nowadays, flow visualization* is an important *tool* in fluid dynamics research. This technique has been *used* extensively in the fields of engineering, biology, physics, chemistry, etc. This technique is mostly applicable to image acquisition in fluid flow using a High-Speed Digital Imaging System, known as Redlake's MotionScope

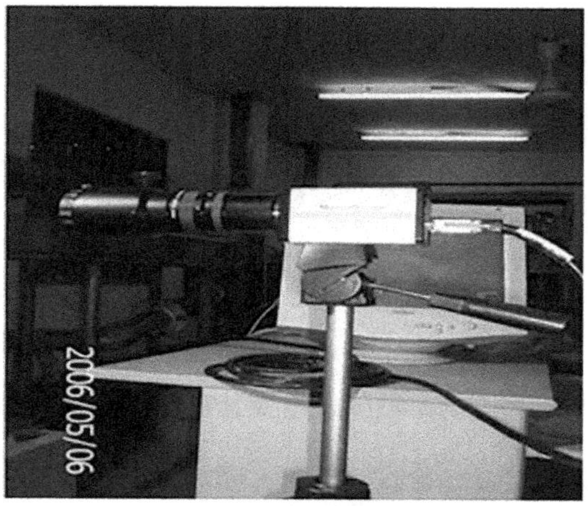

FIGURE 7.7 Redlake motion-scope high-speed camera with stroboscope at ISI fluvial mechanics laboratory (2002) at Kolkata.

or a High-Speed Camera of different frames per second (Figure 7.7). This camera is a complementary metal-oxide-semiconductor (CMOS) digital system that can record a sequence of digital images at a maximum record rate 1,000 frames/sec. The Motion Scope M1 provides 640×512 resolutions at 1,000 fps, whereas the MotionScope M2 offers lightning-fast speeds up to 16,000 fps with reduced vertical resolutions, or maximum resolution of 1,280×1,024 at 500 fps. The system stores the images in the memory of the controller unit. The images can be viewed forward or reverse at a selected frame rate per second, frame by frame or freeze-frame, to analyze the motion during the event. The main advantage of this high-speed motion-scope (HSMS) is that it can record the events at a very high frame rate that is too fast for the human eyes to visualize and also can playback those recorded events in slow motion so that the motion can be analyzed. Dye or magnesium powder is injected into the liquid flow to visualize flow dynamics. The image processing software (Image Pro-Plus) can extract and analyze the images frame by frame from a video captured by HSMS. The main disadvantages of raw images are as follows: non-uniformity of the background, low contrast for small particles, and cluster of particles. This technique is used to study the particle path in suspension as well as in the bottom.

7.2.7 Acoustic Bed Profiler (Ultrasonic Ranging System)

A typical ultrasonic ranging system (URS) is composed of an electronics package and transducers (SeaTek URS 5 MHz Technical note, 2014). The URS is specialized in high-frequency acoustic instrumentation for both laboratory and field applications for the measurement of bedform elevations that occur in the laboratory flume and river flows. The electronic package contains all of the transmitter, receiver, signal processing, and communication electronics. The electronic package streams the range data back to a logging PC, which runs by software to display and record the range data in real time. Different types of transducers are designed; here, the cylindrical transducers are discussed. They are housed in stainless steel housings and are 1 inch in height and 1/2 inch in diameter and are typically bundled four per cable. This routine transducer array was built out of stainless steel and was designed to fit flush with the lid of a pressurized chamber. In general, the transducer array was designed with 32 elements to cover a large section of seabed with high-resolution or in the large section of sand bed in the laboratory. URS is composed of 32 transducers shown in Figure 7.8. The acoustic operating frequency of this system is 5 MHz and transducer diameter of 1 cm. The closest measurement range of this system is 3 cm, whereas the farthest range is 110 cm. The electronic package communicates with a PC through

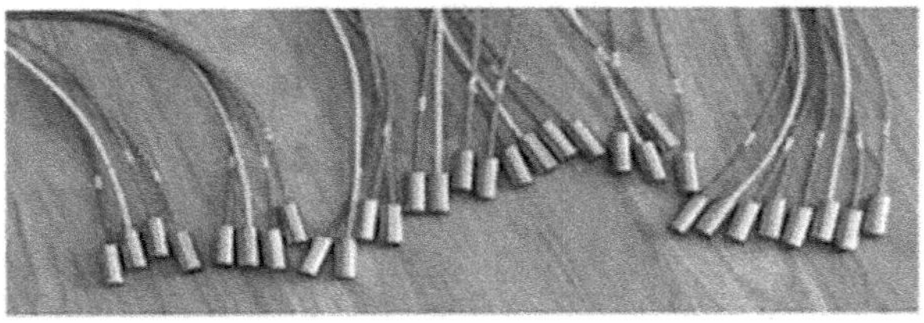

FIGURE 7.8 Thirty-two cylindrical transducers in URS (SeaTek URS 5 MHz technical note, 2014) at ISI Kolkata.

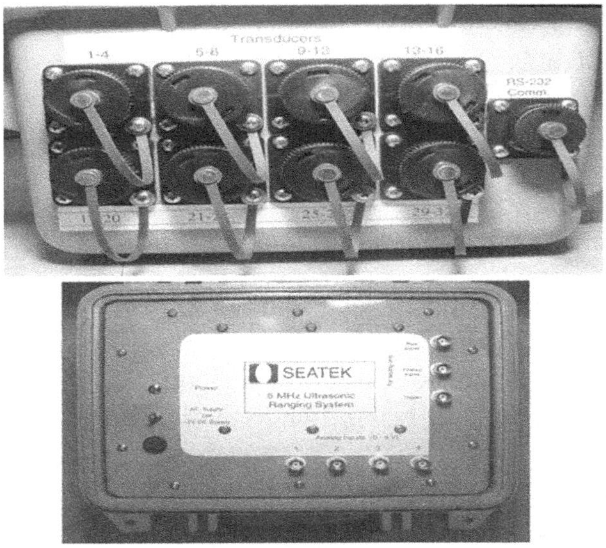

FIGURE 7.9 Data acquisition box with RS-232 port, eight external analog channels, and port for 32 transducers, photograph taken at ISI fluvial mechanics laboratory at Kolkata, India.

a RS232 communication port (Figure 7.9). The electronic package is capable of running up to 32 transducers and sampling up to eight external analog channels.

7.3 FIELD VELOCITY MEASURING INSTRUMENTS

In this section, we have elaborated on the lab-scale flow measuring equipment. Most of the companies simultaneously developed electromagnetic current meters for field experiments, like in rivers, cannels, estuaries, oceans, etc. to measure the mean velocities with fluctuating components. The following current meters are used as:

1. Electromagnetic current meters: Marsh-McBirney (MMB511 and MMB527),
2. Inter – Ocean Current meters S4,
3. Acoustic Doppler Current Profilers (ADCP),

In the following sections, brief details of these equipment are provided.

7.3.1 Electromagnetic Current Meters

The Marsh-McBirney Current Meters (MMB511and MMB527) are manufactured by Marsh-McBirney Inc. (MMB). The water velocity sensor is based on the Faraday principle of electromagnetic induction, where a conductor, such as water moving in a magnetic field, produces a voltage that is proportional to

FIGURE 7.10 Velocity data collection in the McEver's Island site of the Illinois river, River Miles (RM) 50.1: (a) vertical array of three MMB511 current meters (left side), and (b) current meters MMB-527 (right side), (field photograph taken from Bhowmik et al., 1998a, 1998b, 1995a).

the velocity of the water. The basic capabilities of velocity measurements are ±300 cm/s for MMB511 current meter with a probe diameter of 3.9 cm (Figure 7.10a), and ±300 cm/s with direction 0°–360° for MMB527 current meter with a probe diameter of 10.3 cm (Figure 7.10b), whereas their accuracies are ±2% of reading for MMB511 and ±2% of reading ±10° for MMB527 current meter (Bhowmik et al., 1998a, 1998b). Griffith and Grimwood (1981) attempted to measure turbulence in a navigation channel near New Orleans, USA, using an electromagnetic current meter; their sensor consisted of a ball with a diameter of 3.8 cm. West et al. (1986) conducted turbulence measurements in the Great Ouse Estuary, England, using such a meter. It is suitable even for sediment-laden flows.

7.3.2 Inter-Ocean Current Meter (S4)

The Inter-Ocean Current Meter System (1990) known as S4 is a basic electromagnetic current meter with two pairs of internal electrodes and a flux-gate compass with an integral data logger. The S4 current meter is designed to measure with high precision the true magnitude and direction of current using two pairs of titanium electrodes located symmetrically on the equator of the sensor. An internal flux-gate compass provides heading information, used to reference current direction to magnetic North or, for fixed installations; the instrument may be operated in an x–y orthogonal mode, whereby the current vector can be referenced to a landform or structure. S4 can be configured with optional sensors that allow it to operate as a multi-use platform for operations requiring data acquisition of currents, waves, tide, turbidity, and more. S4 can also provide current profiling in conjunction with many of the optional sensors available. The standard S4 includes 64 K bytes of memory expandable to 1 M byte. The features in S4 allows burst mode recording of data over extended periods, previously not possible due to memory limitations with normal burst mode recording. This has particular application for current and wave data recording where users are only interested in recording high-level occurrence of events. The S4 current meters in left and MMB527 meters in right (Figure 7.11) are used for measuring the velocity data from the McEver's Island site in the Illinois River (RM 50.1) to investigate the turbulence in rivers due to the movement of barge traffics (Figure 7.12) and recreational boats. Details are available in the studies of Bhowmik et al. (1992, 1998a, b) and Mazumder et al. (1993).

FIGURE 7.11 Equipment used in the field: two inter-ocean current meters S4 on the left and two MMB527 current meters on the right mounted for deployment in the McEver's Island site on the River Illinois, RM 50.1 for velocity data collection. (By Illinois state water survey, photograph taken from Bhowmik et al., 1998a, 1998b.)

FIGURE 7.12 Typical barge movement in Mississippi river (photograph taken from Bhowmik et al., 1998a, 1998b)

7.3.3 Acoustic Doppler Current Profiler

An acoustic Doppler current profiler (ADCP) is a hydro-acoustic current meter similar to a sonar, used to measure water velocity over a depth range using the Doppler effect of sound waves scattered back from particles within the water column. The term ADCP is a generic term for all acoustic current profilers, introduced by RD Instrument in the 1980s (Model Signature1000, Nortek). The frequency range of ADCP is from 38 KHz to several MHz. The device used in the air for wind speed using sound known as SODAR and works with the same underlying principles. ADCPs contain piezoelectric transducers to transmit and receive sound signals. The traveling time of sound waves gives an estimate of the distance. The frequency shift of the echo is proportional to the water velocity along the acoustic path. To measure 3D velocities, at least three beams are required. In rivers, only the 2D velocity is relevant, and ADCPs typically have two beams. In recent years, more functionality has been added to ADCPs (notably wave and turbulence measurements) and systems can be found with 2–5 or even 9 beams.

Depending on the mounting, one can distinguish between side-looking, downward- and upward-looking ADCPs. A bottom-mounted ADCP can determine the speed and direction of currents at equal intervals all the way to the surface. Mounted sideways on a wall or bridge piling in rivers or canals, it can measure the current profile from bank to bank. In very deep water, they can be lowered on cables from the surface. The primary usage is for oceanography. The instruments can also be used in rivers and canals to continuously measure the discharge. ADCP with pulse-to-pulse coherent processing can estimate the velocity with the precision required to decide small scale motion. As a result, it is possible to estimate turbulent parameters from properly configured ADCPs. A typical approach is used to fit along beam velocity to the Kolmogorov structure configuration and thereby estimate the dissipation rate. The application of ADCPs to turbulence measurement is possible from stationary deployments but can also be done from moving underwater structures like gliders or from sub-surface buoys.

7.4 HYDRAULIC FLUMES

A flume is a man-made channel for water in the form of open-ended structures, whose walls are raised above the surrounding terrain. Flumes are not to be confused with ducts, which are built to transport water, rather than transporting materials using flowing water as a flume. The term flume comes from the Old French word flum, from the Latin flumen, meaning a river. It was formerly used for a stream, and particularly for the tail of a mill race. It is used in America for a very narrow gorge running between precipitous rocks, with a stream at the bottom, but more frequently is applied to an artificial channel of wood or other material for the diversion of a stream of water from a river for purposes of irrigation, for running a sawmill, or for various processes in the hydraulic method of gold-mining. Flumes are specially shaped, engineered structures that are used to measure the flow of water in open channels. Flumes are static in nature – having no moving parts – and develop a relationship between the water depth in the flume and the flow rate by changing the flow of water in a variety of ways. The flume may be divided into several ways like re-circulating flume, open-ended flume, meandering flume, etc.

7.4.1 Re-Circulating Flumes

A recirculating flume is a flume in which the same flow is continuously circulated as shown in Figures 7.13 and 7.14. Other than the function of an open channel and normal studies, recirculating flumes are useful to study the grain-size distributions of sediment in suspension for the simulation of geological processes of sedimentation. Here the details of a recirculating flume built at the Indian Statistical Institute (ISI) Kolkata are illustrated. The sidewalls of the ISI re-circulating experimental flume were made of Perspex windows with a length of 8.5 m, providing a clear view of the flow. Figure 7.13 shows the photograph of the hydraulic channel. The flume consisted of both the experimental and re-circulating channels of same dimensions (10 m long, 0.5 m wide and 0.5 m deep), which looked like oblong in shape. Figure 7.14 shows the schematic diagram of the experimental channel with outlet and inlet pipes (top view (a); and front view (b)). Here in this flume, the commonly used narrower re-circulating pipe located below the experimental channel was avoided. Water was put into the re-circulating flume at a

FIGURE 7.13 Hydraulic flume at the fluvial mechanics laboratory of Indian Statistical Institute (ISI) Kolkata.

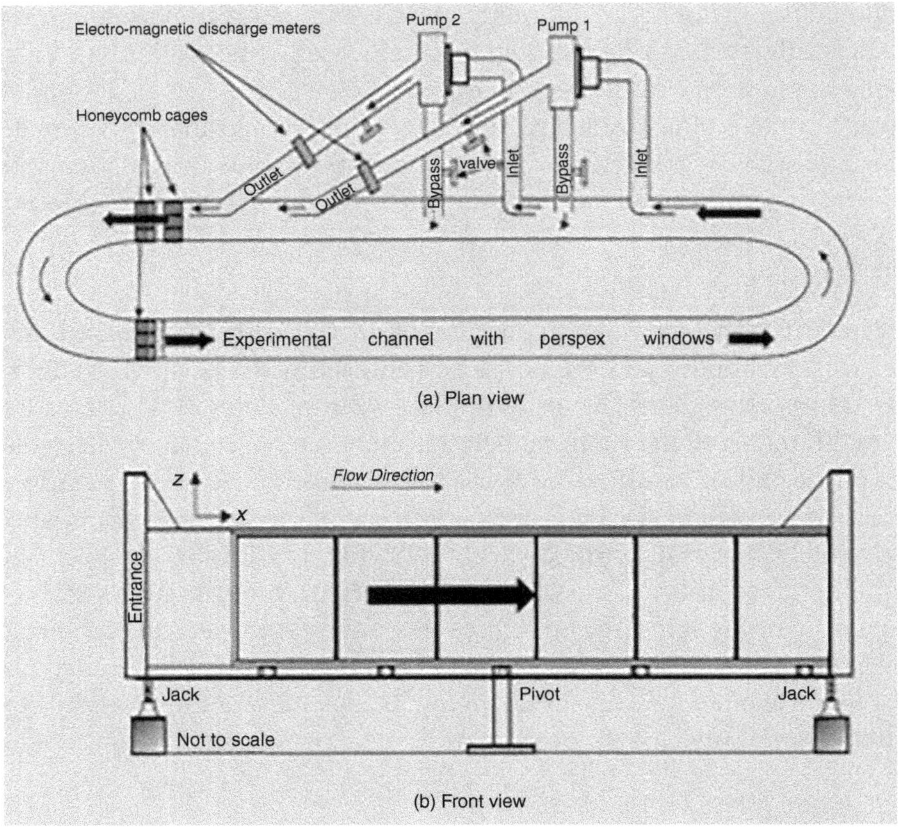

FIGURE 7.14 Schematic diagrams of the experimental channel of fluvial mechanics laboratory (FML) at ISI Kolkata:(a) plan view, and (b) front view.

desired depth for experiment. The main advantage of the present flume was that the whole water body was re-circulated throughout the flume without any disruption of flow due to passing through the narrower cross-sectional pipe, as commonly used. Two non-clogging types of centrifugal pumps for the flow discharge were located outside the main body of the flume. The intake and outlet pipes were freely suspended to allow tilting the flume. The outlet pipes (Pump-1 and 2) were fitted with by-pass pipes and valves, so that the flow discharges were adjusted. Two electromagnetic discharge meters with digital display (Figure 7.15) are fitted with the outlet pipes to facilitate the continuous monitoring of flow. The upstream bend of the channel is divided into three sub-channels of equal width 0.165 m in dimension. Moreover, two honeycomb cages are placed at the back end of the sub-channels in front of the jets of high flow coming out from the outlets, and the third honeycomb cage is placed at the other end of the sub-channels in order to ensure the vortex free and uniform flow of water through the experimental channel. The positions of cages in front of the jets and at the upstream end of the experimental channel substantiate the curvature effect free flow from the measurements.

7.4.1.1 Flume for Digital Imaging

For flow visualization and imaging technique to investigate the movement of sand particle as bed load as well as suspension, a small re-circulating flume of dimension, 3.5 m long, 0.22 m width, and 0.330 m high, is designed at the Fluvial Mechanics Laboratory (FML), ISI Kolkata. Figure 7.16 shows photograph of a Flume of ISI for flow visualization. Figure 7.17 shows the schematic diagram of the experimental setup. The side walls of the flume are made of transparent glass. The test section was located at 2.50 m downstream from the entrance to ensure fully developed turbulence.

FIGURE 7.15 Two electromagnetic discharge meters (left one is low discharge and right one high discharge) installed at ISI fluvial mechanics laboratory at Kolkata.

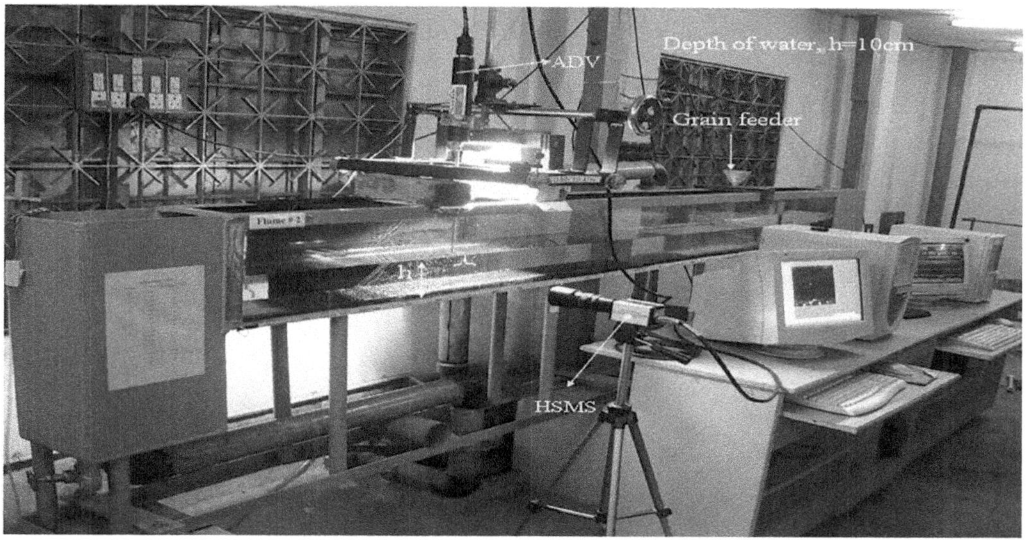

FIGURE 7.16 Flume at ISI fluvial mechanics laboratory for flow visualization and image processing with instruments.

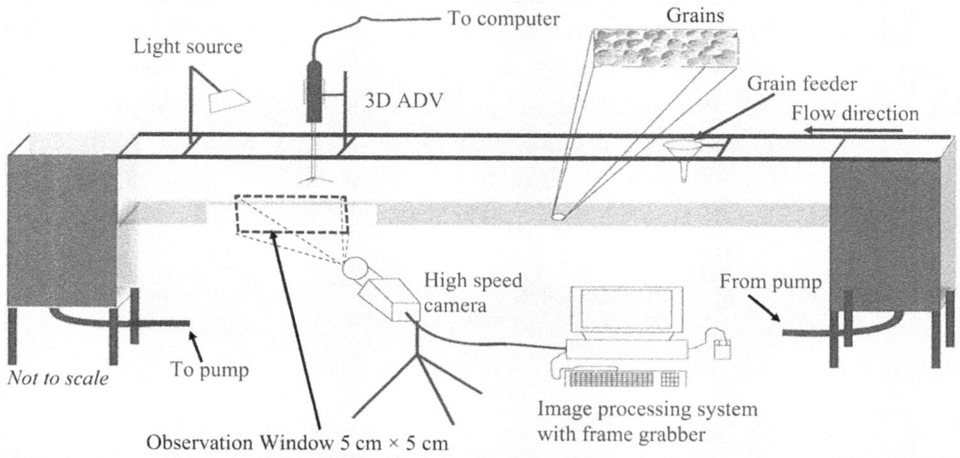

FIGURE 7.17 Schematic diagram of the experimental setup for recording images at ISI laboratory.

Over the past few decades, extensive research works relevant to a variety of scientific disciplines have been carried out in the re-circulating flume under controlled conditions on sediment transport: bed load and grain size distributions in suspension (Ghosh et al., 1981, 1984, 1986; Ghosh and Mazumder, 1981; Ghosh, 1988; Sengupta et al., 1991; Mazumder, 1994; Sengupta et al., 1999; Bhattacharya et al., 2000; Mazumder and Dalal, 2003; Ghoshal, 2004; Mazumder et al., 2005a, b; Ghoshal et al., 2010, 2011, 2013), turbulent flow over a series of artificial dunes (Mazumder and Mazumder, 2006; Mazumder and Ojha, 2007; Ojha and Mazumder, 2008, 2010; Mazumder et al., 2009; Mazumder and Sarkar, 2014; Sarkar and Mazumder, 2014, 2018a, b; Sarkar et al., 2015, 2016), flow characteristics and shear stress over a fluvial obstacle mark (Mazumder et al., 2011; Maity and Mazumder, 2012, 2013, 2014, 2017; Maity et al., 2013), saltation processes of bed materials using digital-imaging (Mazumder et al., 2008; Bhattacharya et al., 2013) and flow characteristics on wave-blocking phenomena (Chatterjee et al., 2018, 2019, 2020). Initially during the 1980–1990s, the Ott-laboratory propeller-type current meter was used to measure only the mean flow. Then later stage around the year 2000 onward with the development of sophisticated measuring techniques and devices, measurements of the turbulent flow in the laboratory, named as FML at ISI-Kolkata, were performed using acoustic Doppler velocimeters (ADV) for velocity recording with fluctuating components, high-speed motion-scope camera (HSMC) for flow visualization, and ultrasonic ranging system (URS) for sediment bed form profiling in the flume. The output of the research work is mostly useful on various problems of interdisciplinary nature, in the areas of turbulence, particle-size distributions, sediment transport (bed load and suspension), combined wave-current flows, flow visualization, and image processing. The methodologies used in the above investigations are related to the applied fluid mechanics, hydraulic engineering, statistics, and geology; and the results of these investigations are of use to hydraulic engineers, sedimentologists, oceanographers and geographers interested in turbulence, sediment transport, and bed form dynamics.

7.4.2 Open Ended Hydraulic Flumes

Three open-ended rectangular hydraulic flumes of different dimensions are setup at the Hydraulics Laboratory, Civil Engineering Department of Indian Institute of Technology Bombay (IIT-B), India for experimental work relevant to study the turbulent flow field and scour around the pier structures. The rectangular flumes are made into three sections with uniform dimension: inlet section, outlet section and test section in the middle. The sidewalls of the flumes are made of transparent Perspex glass to visualize the flow. The water is lifted from the sump to a constant head reservoir (inlet chamber) by a centrifugal pump and re-circulated in the flume. At the entry section of the flumes, honeycomb-type grids and flow straightener in each flume entrance are placed for damping the flow containing large-scale eddies, leading to vortex free uniform flow in the flumes. Adjustable tailgates at the downstream ends of the flume are put to maintain the desirable flow depths in each flume. Variable Frequency Driver (VFD) is used to control the rpm of the pump motor and obtain the desired discharge accurately. The flow rates are monitored using an ultrasonic flow meter (UFM) attached to the feeder pipe of the pump. The dimensions of three similar types of flumes available at the hydraulics laboratory of IIT Bombay are as follows:

1. **Flume-I**: Rectangular flume of dimension 8.5 m long, 0.305 m wide and 0.6 m deep (Figure 7.18),
2. **Flume-II**: Similar flume with dimension 14 m long, 0.5 m wide and 0.9 m deep (Figure 7.19),
3. **Flume-III**: A flume of length 16.5 m long, 1 m wide and 0.85 m deep (Figure 7.20).

Over the past few years, experimental works have been carried out in the hydraulic flumes to investigate the turbulent flow and its characteristics around different shapes of the pier structures relevant to scour formation (e.g., Vijayasree et al., 2018, 2019, 2020; Gautam et al., 2019, 2021, 2022; Misuriya

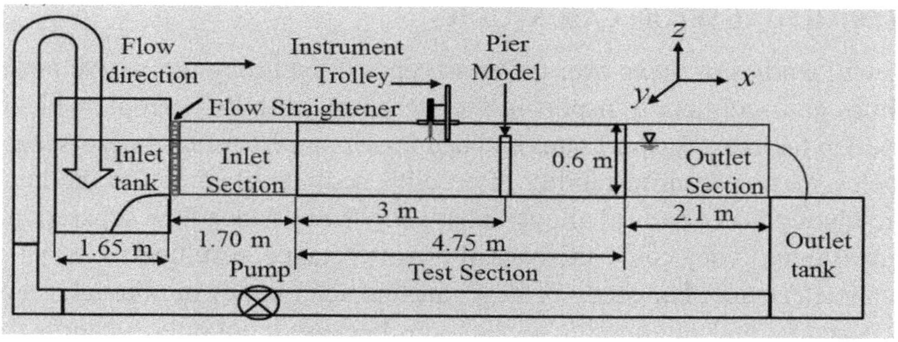

FIGURE 7.18 Flume-I: 8.5 m long, 0.305 m wide and 0.6 m deep at hydraulics laboratory, IIT Bombay.

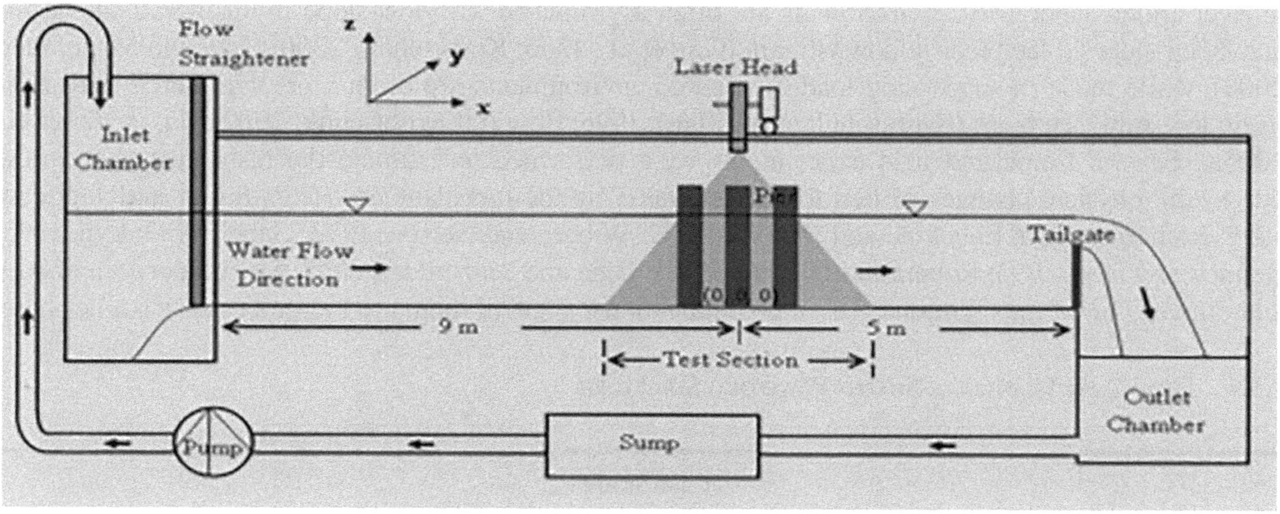

FIGURE 7.19 Flume-II: 14 m long, 0.5 m wide and 0.9 m deep at hydraulics laboratory, IIT Bombay.

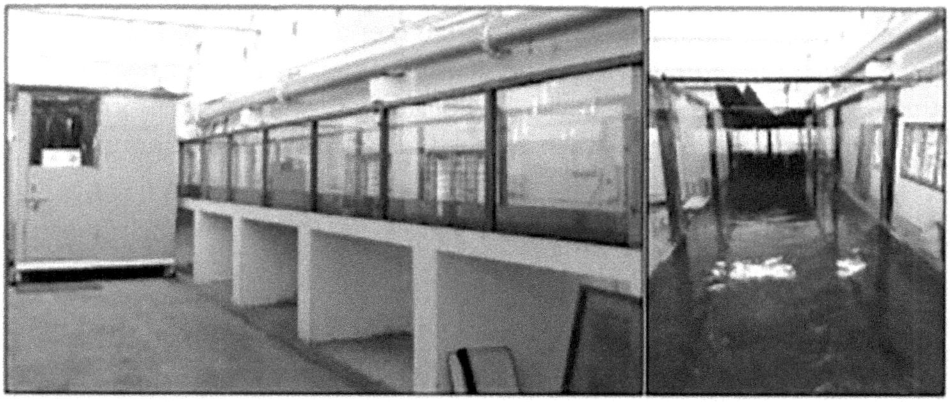

FIGURE 7.20 Flume-III: 16.5 m long, 1 m wide and 0.85 m deep at hydraulics laboratory, IIT Bombay.

et al. 2021, 2023a, 2023b, 2023c; Sahu et al., 2023). The measurements of the turbulent flow with fluctuating velocity components in the laboratory flumes were performed with the aid of acoustic Doppler velocimeters (ADV) and PIV. The instruments ADV and PIV are the intrusive and non-intrusive velocity measuring techniques, respectively. These investigations are carried out for categorization of scour around different shapes of piers, which might provide a clue for suggesting the shape design with possibilities of protection measure against heavy scour.

7.5 FLUME EXPERIMENTAL SETUP: CASE STUDIES

Several experimental studies of flows over different types of bed roughness have been reported in the flow hydrodynamics and sediment transport in the literatures with their scope, which are mostly constrained to the friction factors, transport rates and bed form characteristics. Large-scale roughness, such as dunes and ripples, arising from instability of erodible sediment bed due to the interaction with the turbulent flow are ubiquitous in natural alluvial channel, which induce flow separation and circulation, and hence form resistance. They cause the overall flow resistance, resulting in deviation of mean flow and turbulence characteristics. The study of flows around sand dunes in alluvial rivers is very important both in basic sciences and engineering applications because it not only concerns the mechanism of flow resistance and sediment transport, but also it can help to establish a better mathematical model to simulate sediment-laden flows for flood control, navigation, etc. Dunes and ripples-dominated bedload in river and/or laboratory environments are often asymmetric with low-slope in upstream side (stoss) and steep-slope in lee faces at downstream (Guy et al., 1966; Kostaschuk, 2000; Ojha and Mazumder, 2008), while those in suspension load-dominated environments are often more symmetric with relatively low angle lee faces (Kostaschuk and Villard, 1996; Best and Kostaschuk, 2002; Mazumder et al., 2005a). Several flume and field experiments were performed to estimate the resistance to the flow due to the physical changes of bed forms generated by the turbulent flow (e.g., Smith and McLean, 1977; Richards, 1980; Engelund and Fredsoe, 1982; Wiberg and Nelson, 1992; Gabel, 1993; Lyn, 1993; Bennett and Best, 1995; Robert and Uhlman, 2001; Ojha and Mazumder, 2008; Mazumder et al., 2009; and others). Out of the mentioned research publications, a few of them are focused as case studies here.

7.5.1 Flow Over Triangular-Shaped Waveform Structures

Case Study I

Several investigations had been made under controlled laboratory conditions to estimate the sediment suspension, bed load transportation and the influence of bed roughness on suspension over flat sand beds of heterogeneous composition (e.g., Ghosh et al., 1981, 1986; Ghosh and Mazumder, 1981; Sengupta et al., 1991; Mazumder, 1994; Ghoshal, 2004; Mazumder et al., 2005a, b). Due to the flow over a flat loose sediment bed, when the sediment transport started, different types of bedforms were seemed to be generated, among which single asymmetric, single *nearly* symmetric and series of asymmetric bed forms were observed after a long time, when reached at a nearly equilibrium category. Figure 7.21a

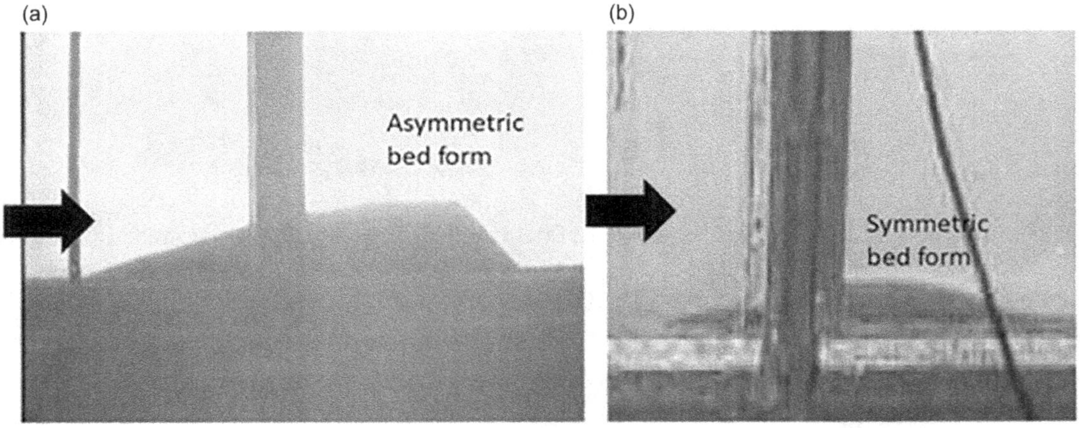

FIGURE 7.21 Photographs of dunes developed during the experiments: (a) asymmetric bed form, and (b) nearly symmetric bed form at the fluvial mechanics laboratory of ISI, Kolkata. (Modified from Mazumder et al., 2009.)

shows the photographs of the waveform of a scalene triangular shape (STS) with asymmetry about the crest, and Figure 7.21b shows *nearly* an isosceles triangular shape (ITS). The dimensions of bed forms, such as amplitudes and wavelengths, depended on the flow velocity and the sand bed materials. The measured velocity data were restricted only to the mean flow for a period of passage of bed forms.

To understand the basic mechanism in the perturbed flow formation, Mazumder et al. (2009) investigated experimentally at ISI flume (Figure 7.13) to explore the influence of artificially made isolated scalene and isosceles triangular shaped bedforms associated with the flow separation, reattachment points, and the perturbed shear layer in the turbulent flow responsible for sediment movement. The approximation of static 2-D bedform was well justified because the speed of the moving bed was small compared to the mean flow. The velocity data were collected using acoustic Doppler velocimeter (ADV) after the equilibrium state. The significant deviations from the mean velocity, intensity, shear stress, eddy viscosity, and the contributions of burst-sweep cycles to the total Reynolds shear stress due to the presence of bed topography were studied. Detailed descriptions of experiments, methodology, results and discussions are available in the studies by Mazumder et al. (2005b, 2009).

Case Study II

To achieve a general output of turbulence characteristics of flow over a series of 2-D asymmetric dunes, which was observed in Mazumder et al. (2005a) over sediment bed, Ojha and Mazumder (2008) addressed the flume experiments at ISI Kolkata (Figure 7.13). Experiments were conducted over 12 asymmetric dunes of mean length 32 cm, crest height 3 cm and the dune width almost as wide as width of the flume, using 3-D micro-ADV (Figure 7.22). The variation of turbulence statistics along the flow affected by the wavy bottom roughness have been investigated in details. Quadrant decomposition of the instantaneous Reynolds shear stress had been adopted to calculate the contribution of ejection and sweeping events to the total Reynolds shear stress generation. The relative dominance of two events is found to contribute in a cyclic manner (spatially) in the near bed region, whereas such phenomenon seems to be disappeared towards the main flow. The detailed descriptions of experiments, methodology, results and discussions are available in Ojha and Mazumder (2008).

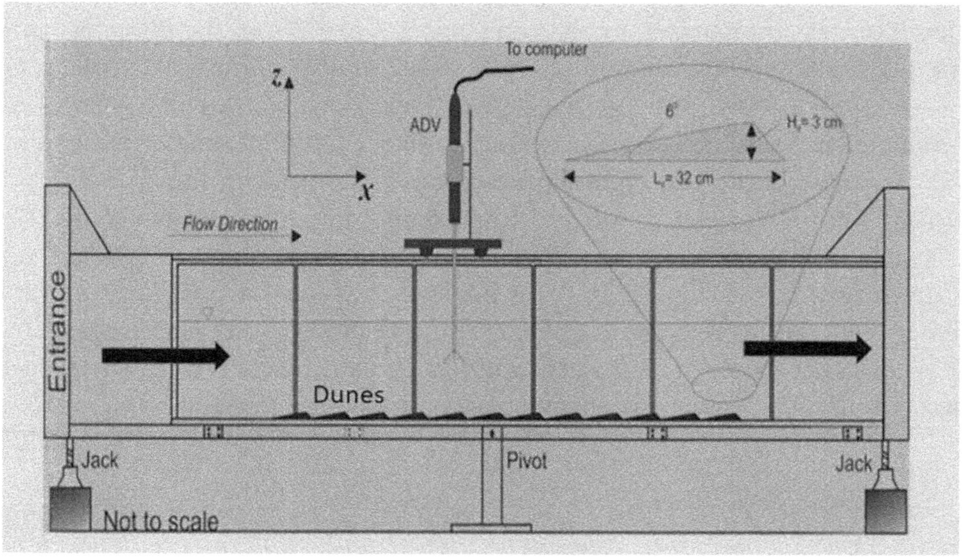

FIGURE 7.22 Flow over a series of dune structures in a laboratory flume at ISI Kolkata. (Modified from Ojha and Mazumder, 2008, 2010.)

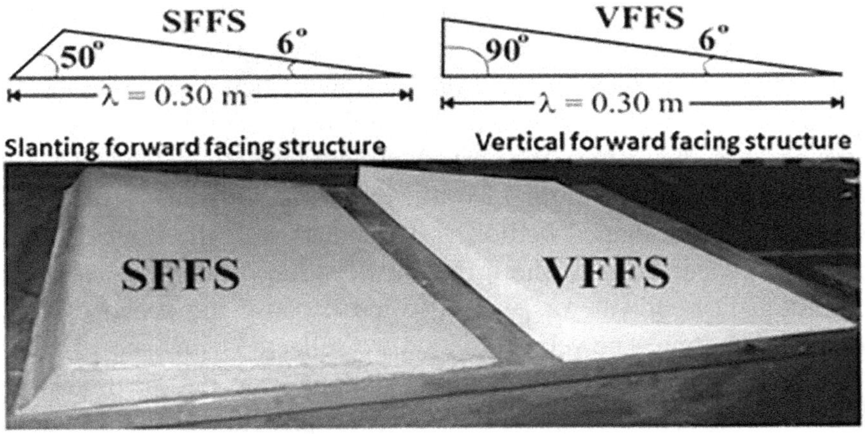

FIGURE 7.23 Schematic diagram of the forward-facing waveform structures. (Photograph taken at ISI laboratory, Kolkata; modified from Sarkar and Mazumder, 2018b.)

Case Study III

The experiments were conducted at the ISI flume (Figure 7.13) to explore the characteristics of turbulent flow and drag over two artificial 2-D forward-facing waveform structures with two different stoss side slopes of 50° and 90° respectively (Figure 7.23). Both structures possessed a common slanted lee side slope of 6°. Here the terms, stoss and lee sides of the structures, are used conventionally according to the flow direction. The velocity data were collected using ADV along the centerline (CL) of the flume. The data were analyzed to identify the spatial changes in turbulent flow addressing the flow separation region with re-circulating eddy, the Reynolds stresses, the turbulent events related to burst-sweep cycles and the drag over two upstream-facing bedforms for a Reynolds number $Re = 1.44 \times 105$. Detailed experiments, data collection, analysis, and discussions are available in the studies by Mazumder and Sarkar (2014).

Subsequently, Sarkar and Mazumder (2014) investigated the turbulent flow characteristics and overall drag over the trough region of a pair of adjacent 2-D forward-facing dune-shaped artificial structures with two different stoss-side slopes and made a comparative study between them. Structures were considered here as equal base length (λ) with a common gentle slope of 6° at the downstream face with similar the stoss-side angles. The velocity data were collected using a 3-D ADV at the flume centerline to analyze the mean flows, Reynolds stresses, spectral analysis with Strouhal number and the overall drag, using the same Reynolds number. Detailed experiments, data collection, analysis and discussions are available in Sarkar and Mazumder (2014). Later, Sarkar and Mazumder (2018b) extended the paper of Mazumder and Sarkar (2014) to investigate the higher-order moments of velocity fluctuations, turbulent kinetic energy (TKE) fluxes, turbulent production and dissipation rates, turbulent length scales (Taylor's and Kolmogorov's scales), and Reynolds anisotropy in the wake regions over two artificial upstream-facing bed forms in tidal environments (Figure 7.23). Such a study has the potential to be useful to researchers who are concerned with the hydrodynamics of flow, mixing process, and length scale characteristics of energy-containing eddies, especially for the characterization of turbulent flow from the anisotropy tensor. Detailed analysis, results and discussions are available in the studies by Sarkar and Mazumder (2018b).

7.5.2 Flow Around Different Shapes of Piers

The interactions between the flow and the bridge pier with sediment bed cause scours which may lead to destruction of bridges. This is a very challenging problem to scientists and river engineers. The hydrodynamics of flow and its consequent effect on the mechanism of scour around the bridge piers are of great interest to the researchers. The flow around a submerged obstacle or a pier generates large-scale eddies like horse-shoe vortices near the obstacle due to the presence of an adverse pressure gradient. The local scour around the piers of different shapes obtained on the basis of field and experimental data were presented by Breusers et al. (1977). They stated that till now no satisfactory results have come out, because the processes involved with the interaction of water and sediment transport associated with the piers of different shapes are too complex, and experimental data are partial and sometimes conflicting. Jones et al. (1993) and Parola et al. (1996) investigated experimentally the geometry of local scour around non-uniform piers mounted on rectangular caissons, and subsequently Melville and Raudkivi (1996) investigated the cylindrical pier placed on cylindrical caissons. The scour-hole structures around submerged piers of different shapes (circular, elliptical, square and triangular) were investigated and the space-time dynamics of scour associated with turbulence were examined by Sarkar et al. (2015). A comprehensive study on the influence of the scour hole on the mean flow and turbulence around a complex pier under equilibrium conditions was explored by Beheshti and Ashtiani (2010).

Based on past research, it is observed that several experiments on local scour around bridge piers of circular or non-circular shape had been carried out over the past years. A few efforts have been directed to study the interactions of flow field with piers (circular or non-circular) experimentally to understand the associated turbulence structures which are accountable for the development of scour geometry. However, a few attempts have been made by Beheshti and Ashtiani (2010) and Gautam et al. (2019) to investigate experimentally the three-dimensional turbulent flow fields around a complex pier with a rectangular pile-cap fixed on a rigid flat surface. The understanding of turbulence properties associated with interactions of fluid flow with different shapes of pier (circular or non-circular) is important for prediction of scour geometry around pier shapes. Some case studies from IIT-Bombay flume experiments on hydrodynamics of flow around different shapes of piers are presented below:

Case Study IV

Vijayasree et al. (2020) investigated the turbulent flow fields and its characteristics along an oblong pier (OP) in a laboratory flume at a constant flow discharge and compared the results with that of circular pier (CP). The novelty of this work was to determine the key parameters of turbulent flow around piers mounted on a rigid flat surface and relate the effective impact of turbulence to the scour structures generated by identical flow conditions. Experiments were conducted in a rectangular flume at the Hydraulic Laboratory of IIT-B flume (Flume-I: Figures 7.18 and 7.24a). Discharge was measured using an ultrasonic flow meter. When the fully developed flow was attained at the flume, experiments were conducted using a single CP and an OP individually mounted on the test section of the flat rigid bed (Figure 7.24b). The CP of diameter 0.03 m was placed in such a way that the locations of both upstream noses of the CP and the OP were the same. The OP of 0.15 m long and 0.03 m wide was mounted vertically on the rigid flat surface and the origin was fixed at the center of the pier. A series of experiments were conducted with the identical flow conditions as that of flat surface without piers. Instantaneous velocity data were recorded using ADV along the CL and three planes parallel to CL at respective distances of 0.025 m, 0.05 m and 0.075 m apart towards right wall due to symmetry (Figure 7.24c) for Reynolds number $Re = 59{,}400$ (Table 7.1). An attempt has been made to address the key parameters of turbulence, such as, mean velocities, Reynolds stresses, turbulent kinetic energy, spectral analysis, and the fractional contributions of burst-sweep cycles to the total Reynolds shear stress along the oblong and circular piers. The influence of flow parameters around the circular pier is considerably higher than that of oblong pier, resulting in more scour around the circular pier, subjected to the identical flow conditions. Detailed experiments, results and discussions are available in the studies by Vijayasree et al. (2020).

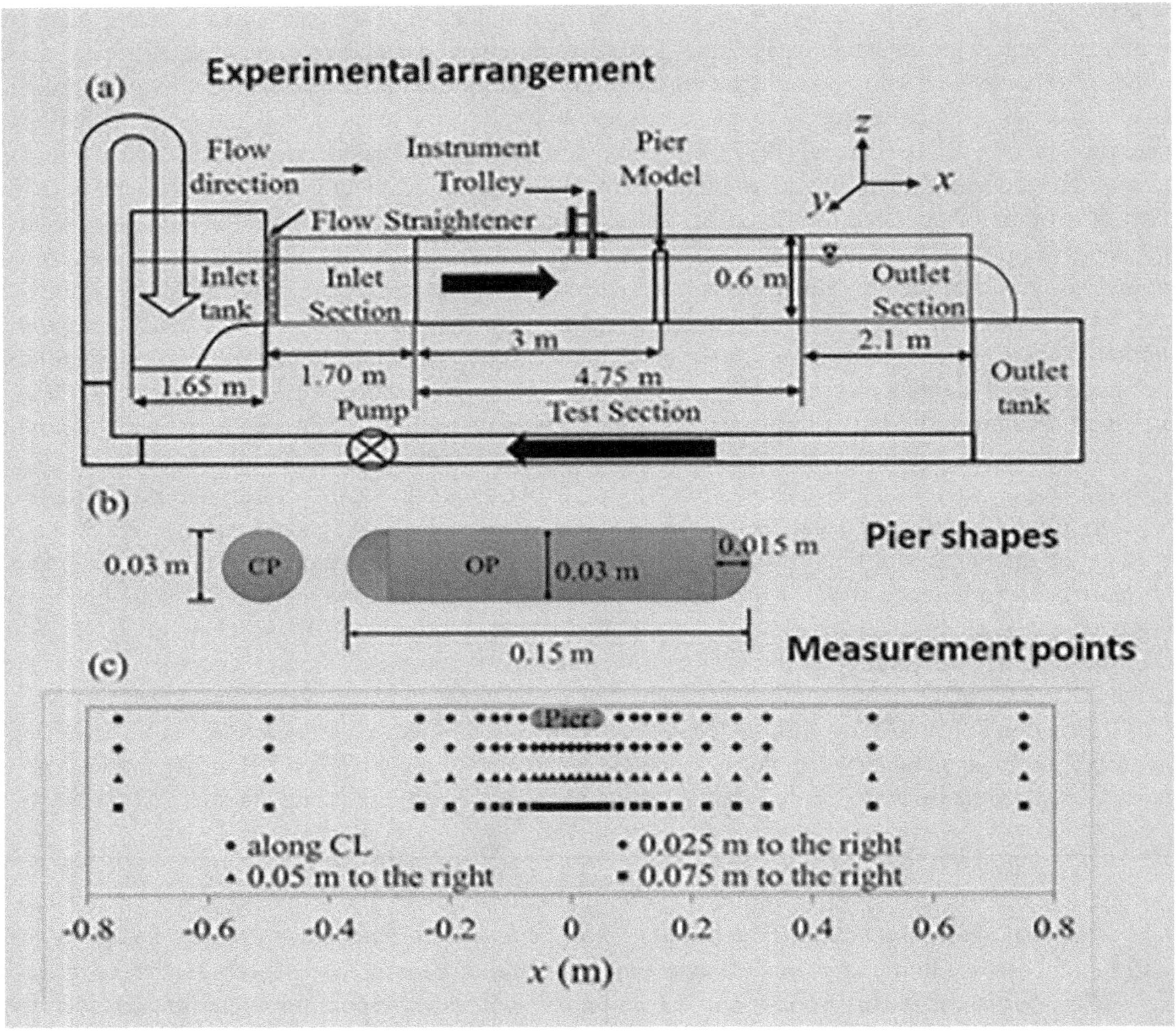

FIGURE 7.24 Schematic diagram of experimental setup (a) Flume-I with experimental setup; (b) Cylindrical pier (CP) and oblong pier (OP); and (c) Velocity data collection on the right side of the pier due to symmetry. (Modified from Vijayasree et al., 2020.)

TABLE 7.1
Hydraulic Parameters

Discharge, Q (m^3/s)	Depth, h (m)	Depth Averaged Velocity, U (m/s)	Froude number, Fr	Reynolds number Re
0.018	0.165	0.36	0.28	59,400
0.015	0.154	0.32	0.26	49,280
0.012	0.143	0.28	0.24	40,040

Case Study V

Gautam et al. (2019) investigated the mean flow and turbulence key parameters around a complex pier, consisting of cylindrical column resting on an elliptical pile-cap over a 2×2 group of piles fixed on a rigid flat surface in a laboratory flume (Figures 7.25 and 7.18) and compared the results with that of a simple pier. The schematic diagrams of simple pier with side, front and top views of complex pier with dimensions (Table 7.2) are provided in Figure 7.26. The velocity data were collected using PIV for three different Reynolds numbers (based on flow depth). More precisely, an attempt has been made to understand the role of an elliptical pile-cap in the flow field and to address the associated turbulence key parameters, such as, mean flow, Reynolds stresses, turbulence kinetic energy, spectral analysis and vorticity. The spectral analysis is performed for both the piers to examine the vortex shedding frequency in the near wake regions. The velocity vector and streamlines in the upstream and downstream in vertical planes of the piers are discussed. Details of experimental setup with PIV, analysis, discussion and conclusions are available in Gautam et al. (2019).

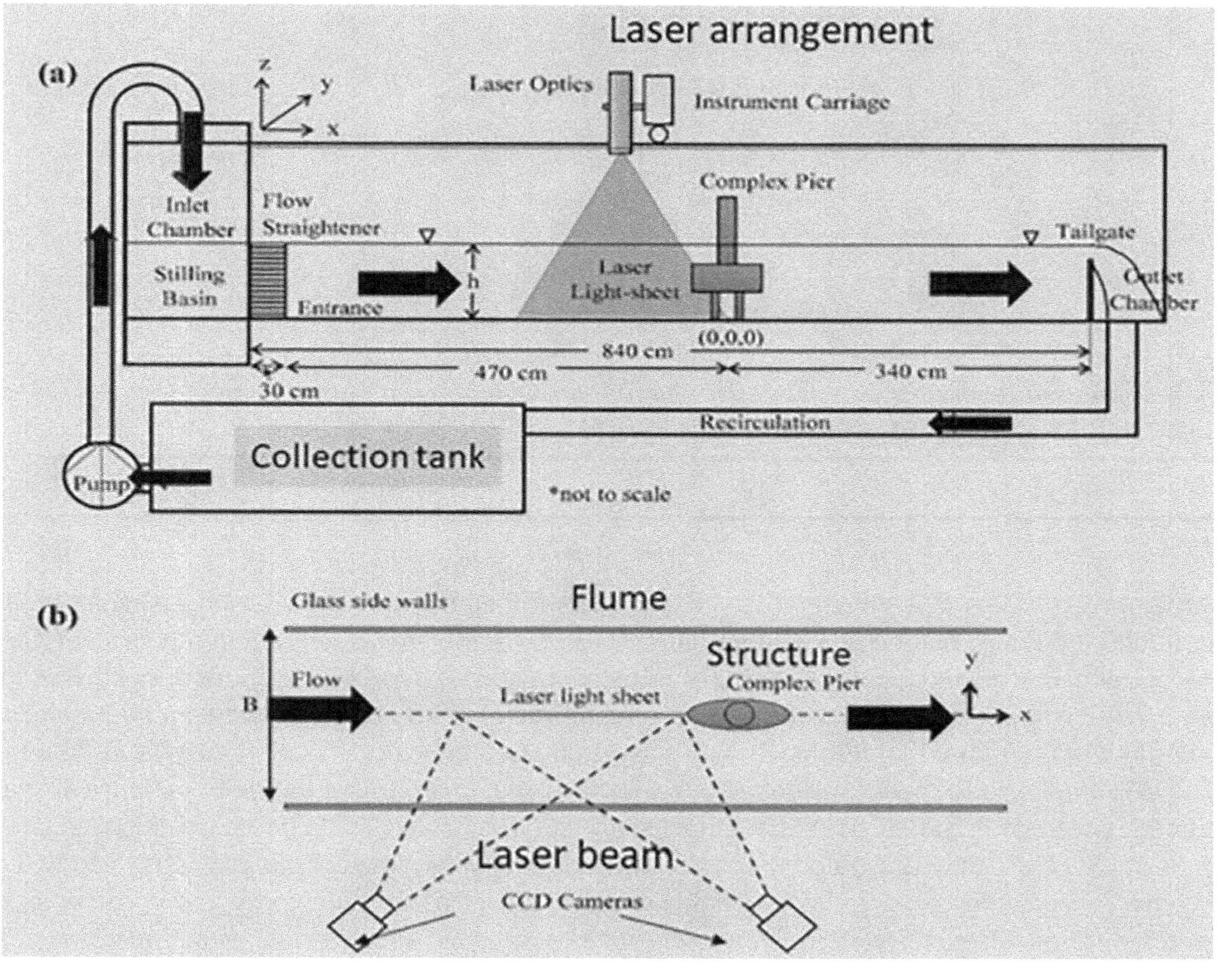

FIGURE 7.25 Schematic diagram of experimental setup (a) side view (xz plane), and (b) top view (xy plane) of the flume developed at IIT Bombay. (Modified from Gautam et al., 2019.)

TABLE 7.2
Dimensions of Complex Pier

Parameter	Definition	Dimension (cm)
L_c	Height of cylindrical column	15
L_{pc}	Height of pile-cap	5
L_p	Height of pile	5
D	Diameter of SP/column of CP	3
D_p	Diameter of piles	1
W_{pc}	Length of pile-cap	12
D_{pc}	Width of pile-cap	4

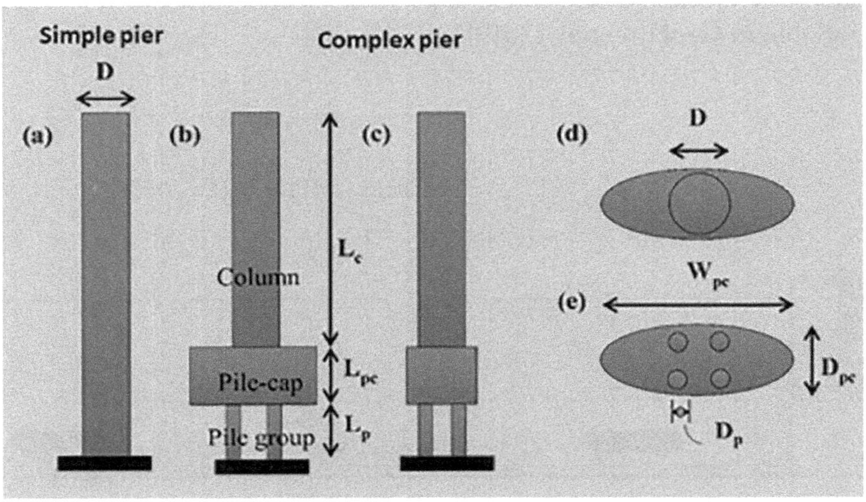

FIGURE 7.26 Schematic diagrams (a) simple pier; (b) side view of complex pier; (c) front view; (d) top view, and (e) bottom view. (Modified from Gautam et al., 2019.)

Case Study VI

Misuriya et al. (2023) made an attempt to examine experimentally the turbulent flow around three in-line circular cylinders, oblong and rectangular shapes of cylinders with a common aspect ratio (length to width ratio is five) mounted individually at the flat rigid surface in a fully developed flow field (Figure 7.27). A series of experiments were performed in a 14 m length, 0.5 m width and 0.9 m deep re-circulating flume (Figure 7.19: Flume 2) with a slope of 0.002 in Hydraulics laboratory of IIT-Bombay. The PIV is employed to record instantaneous velocity components along the three cylindrical shapes mentioned for three Reynolds numbers. Dimensions and the arrangements of the cylinders are shown in Figure 7.28. This study illustrated the various influence flow parameters that are associated with the turbulence, such as mean velocities, Reynolds stresses, turbulent kinetic energy and vorticity around the three different shapes. Detailed descriptions of experiments, analysis, results and discussions are available in Misuriya et al. (2023).

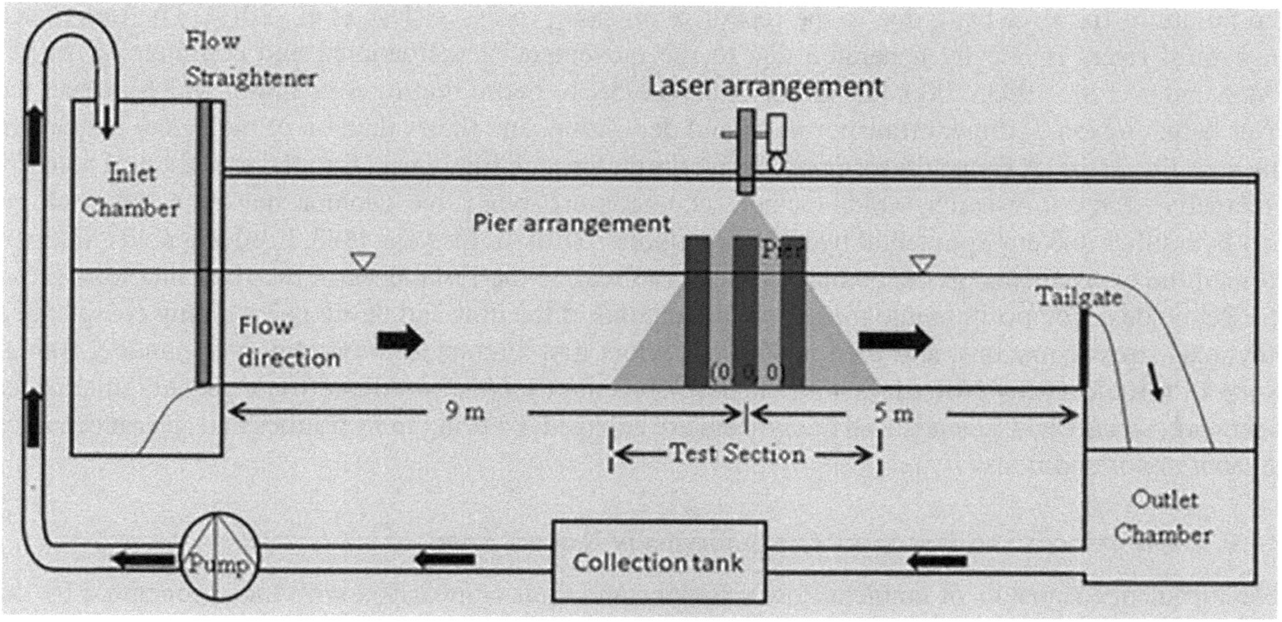

FIGURE 7.27 Schematic diagram of experimental set up at IIT Bombay laboratory. (Modified from Misuriya et al. (2023a).)

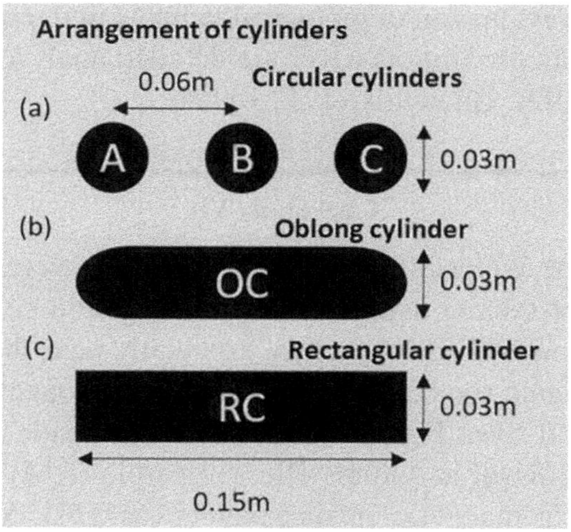

FIGURE 7.28 Arrangements and dimensions of the cylinders (a) three in-line circular cylinders, (b) oblong cylinder, and (c) rectangular cylinder. (Modified from Misuriya et al. (2023a).)

7.6 TURBULENCE IN NATURAL RIVERS

Riverine flows are typically turbulent in nature. Turbulence is ubiquitous and a fundamental engine of transport of water, scattering, mixing and sediment transportation, which has an important role in geophysical disciplines such as the evolution of river morphology, landscapes, channel meandering, etc. River turbulence is responsible for several physical processes such as the transport and mixing of diluted/undiluted substances, river bank erosion, sediment suspension and deposition, and geo-morphological evolution (Schleiss et al., 2014). The interplay between the river turbulence and bank causes the failure of the river bank due to the repetitive erosion processes (Das et al., 2019). The turbulence in natural rivers is heavily generated due to the movement of recreational and commercial traffics (Mazumder et al., 1993, 2006; Bhowmik et al., 1995a, b) pertaining to river flows, which affects the river bank erosion, sediment transportation and deposition, and the evaluation of biological habitats in the sensitive areas. A proper description of river turbulence is fundamental to the evolution of recently emerged ecological research, which includes eco-geomorphology, bio-geomorphology, eco-hydrology, eco-hydraulics and environmental hydraulics (Nikora, 2010). In the year 1883, Lord Reynolds first performed the experiments on the complex problem to realize the turbulence in the river that introduced the Reynolds decomposition into mean and fluctuations of the flow and developed a highly complicated Reynolds stress equations, and then in 1930s, Taylors first attempted to build a mathematical framework for turbulent flow. Nowadays with the development of sophisticated electronic gadgets suitable for fieldwork, researchers, scientists and engineers are engaged, working in hydraulics and river mechanics in laboratory conditions.

7.6.1 Mean Velocity and Turbulence Characteristics in Natural Rivers

The turbulence statistics of ambient flow velocities and their comparison with those generated by the movement of barge traffics within the channel border area in a navigation waterway are presented. The movement of navigation traffic within the restricted waterways, such as Illinois, Mississippi or Ohio Rivers, changes their flow characteristics temporarily in space and time. The understanding of turbulence in natural rivers due to the movement of traffic is important to study the bank erosion, drawdown, sediment re-suspension and the evaluation of biological habitats in the sensitive area. Here, the velocity and sediment data collected from the Illinois River are detailed analyzed and discussed the turbulence parameters (Mazumder et al., 1991; Bhowmik et al., 1995a).

Case Study VII

Scientists from the Illinois State Water Survey (ISWS) of UIUC were engaged in collecting detailed velocity data from a large river system using 2-D electromagnetic current meters. They studied the turbulent velocity structure in natural river systems, especially near the channel border areas. A field study was conducted to investigate the changes in velocity components and turbulent shear stress in the Illinois River, a large natural river. Figure 7.29 shows a schematic diagram of the field setup used for investigation on the Illinois River at Kampsville, river mile (RM) 35.2. The photographs of field data collection sites using electromagnetic current meters (MM511, MM527, and S4) in the Illinois River bank area at the Kampsville, river mile (RM) 35.2 and McEver's Island, river mile (RM) 50.1 are shown in Figures 7.10 and 7.11. Figure 7.30 shows the velocity data collection site near the bank of the Illinois River at Kampsville, river mile (RM) 35.2 with the panoramic view of the field setup in the river. This study aimed to identify the temporal fluctuations of velocity components and turbulent shear stress for natural river flow near the boundary. It illustrated the cross-sectional and vertical variations of velocity components, velocity fluctuating components, turbulent intensities and turbulent shear stress. Details of fieldwork, instruments, analysis, results and conclusions are available in Bhowmik et al. (1995a).

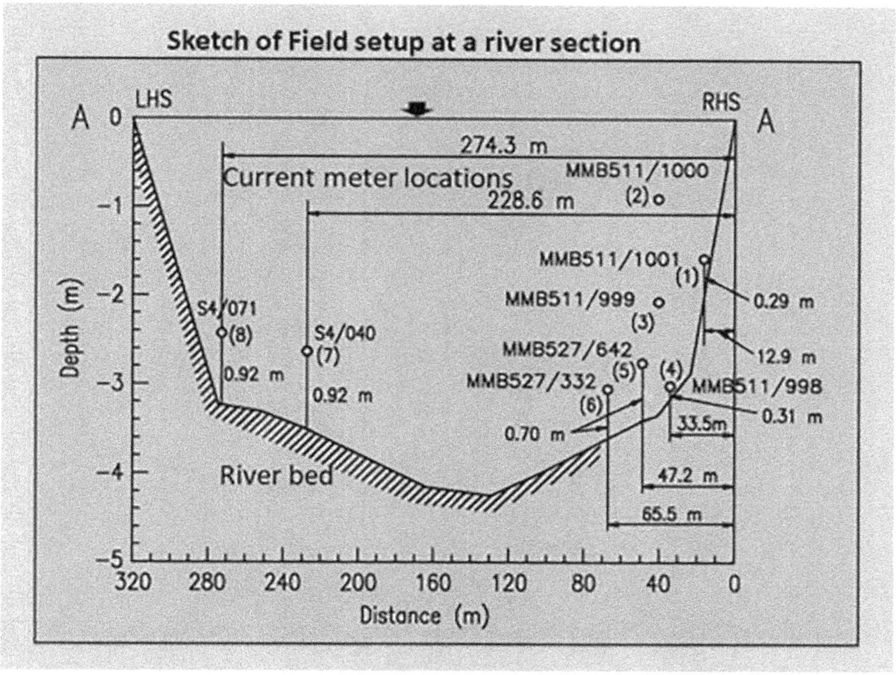

FIGURE 7.29 Schematic diagram showing the data collection setup on the Illinois river at Kampsvillle, river mile (RM) 35.2 deploying the current meters. (Modified from Bhowmik et al., 1995.)

FIGURE 7.30 Data collection site near the bank of Illinois river at Kampsville, river mile (RM) 35.2. (Modified from Bhowmik et al., 1995.)

Case Study VIII

Movement of navigation traffic within restricted inland waterways, such as the Illinois, Mississippi, or Ohio Rivers, changes their flow characteristics temporarily in space and time. The schematic diagram of water motion generated by the barge traffic or moving ship in stagnant water is shown in Figure 7.31 (Blaauw et al., 1984). Physical changes due to the movement of barge traffic were presented in Blaauw et al. (1984), Bhowmik and Mazumder (1990) and Mazumder et al. (1993). The physical changes included the return flow, waves, drawdown, and screw wash. It is well known that due to the movement of the

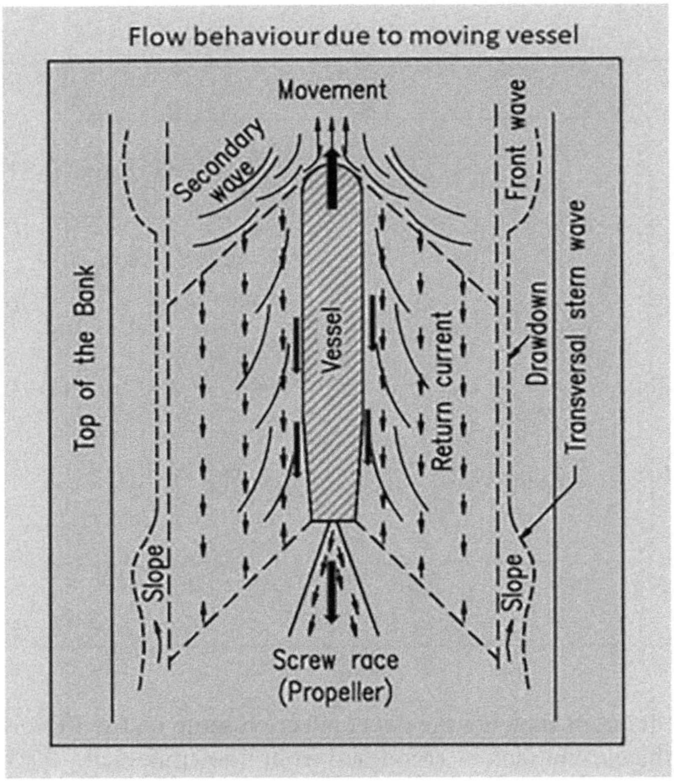

FIGURE 7.31 Schematic design showing the water flow generated due to the moving ship in stagnant water. (Modified from Blaauw et al., 1984, and Mazumder et al., 1993.)

barge traffic, flow velocity beside the barge accelerates in the opposite direction (return current), and the water level drops down within the channel to maintain the continuity of flow. The velocity structure is composed of two parts: barge-induced velocities in the sides and tow-induced propeller jet velocities. The jet flow behind the vessel is generated through the direct action of the propellers, its velocity distribution is normally assumed to be symmetrical around the axis of the propeller, similar to a three-dimensional jet flow. Figure 7.32 shows the plan, side section and cross-sectional views with various physical parameters that describe the characteristics of typical barge-tow convoys on the Upper Mississippi River (UMR). The passage of vessel shows a complex system of primary and secondary effects on the flow pattern.

The following sites were selected, based on the evaluation of plan form characteristics of the UMRS, for this study as: McEver's Island and Kampsville on the Illinois River, and Apple River Island, and Clarks Ferry on the Mississippi River. Here two sites on the Illinois River were selected for field investigation to study the flow velocity, bank erosion and sediment re-suspension due to barge traffic movements. The cross sections of both sites were trapezoidal, but the McEver's Island site was approximately 100 m narrower than the Kampsville site (Mazumder et al., 1993). Figure 7.33 shows the sketches of the coordinate system and cross-sectional profiles in the McEver's Islands and Kampsville sites in the Illinois River. Velocity data were collected using a set of two-dimensional (2-D) electromagnetic current meters installed along a cross section from the shore to the sailing line. The ISWS used two different types of meters MMB511 and MMB527. At both the sites four MMB511s and two MMB527s were used. The accuracy of these meters and their basic capabilities are given in Table 7.3. Further information can be obtained from the user's manual (Inter Ocean Systems, Inc., 1990, *Instruction Manual*). The turbulent characteristics of flows induced by barge traffic in a natural river were analyzed. Parameters

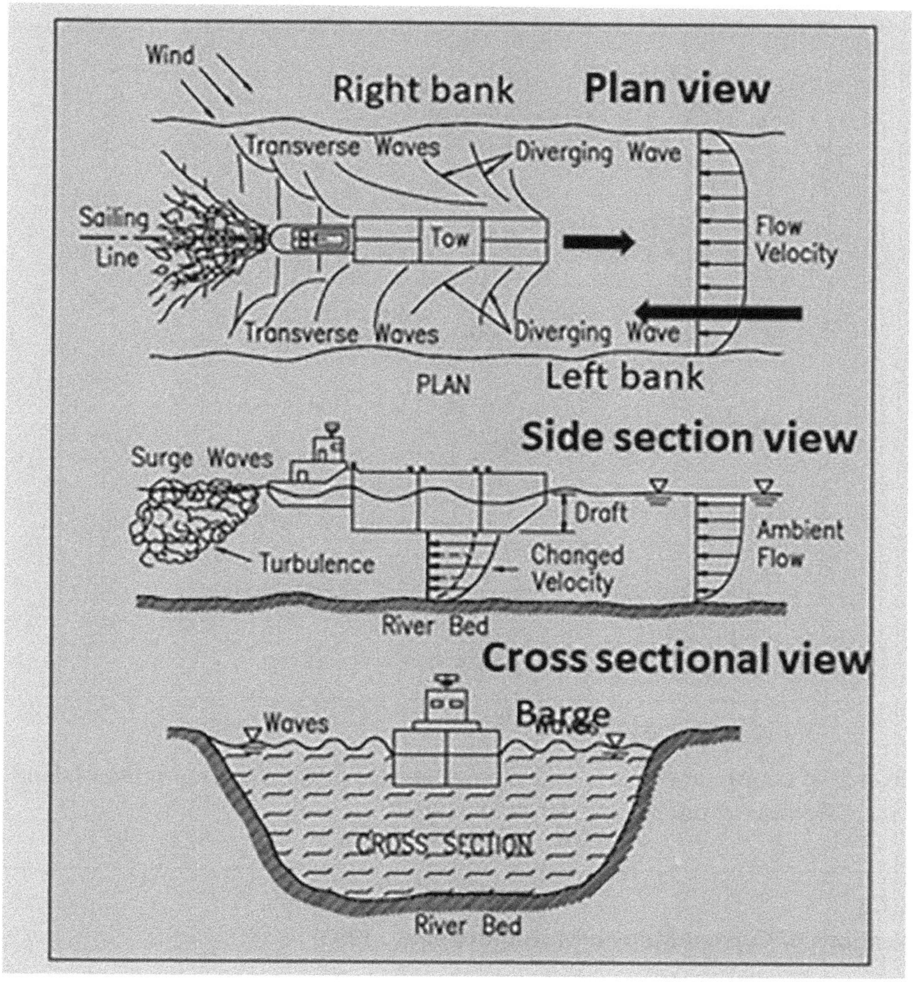

FIGURE 7.32 Flow condition due to movement of barge-tow. (Modified from Bhowmik et al., 1998a.)

included the cross-sectional and vertical distributions of return velocities, Reynolds stresses, and turbulent kinetic energy (TKE). Figure 7.34a and b represent the resultant velocity vectors for the downstream and upstream bound barges in McEver's Island. Results show that significant changes occurred in all of these parameters. Figure 7.35 represents the temporal changes of turbulent fluctuating velocity components and the Reynolds shear stress at a point 10.5 m from the shore when the barge (Illini) moves downstream at the McEver's Island site. It is interesting to note that the largest changes took place in a zone within 10% of the channel width from the shore. For further information, details of fieldwork, instruments, analysis, results and conclusions are available in Mazumder et al. (1993).

Later, Bhowmik et al. (1995a) verified the existing models, using their collected data and found that all the existing models underestimated the average return flow velocity by as much as 30%. The exponential distribution across the width between the barge and the shoreline described by the existing models was not supported by the field data. Therefore, Bhowmik et al. (1995b) suggested that the return flow due to the movement of navigation traffic may follow a parabolic-like distribution. Details are available in the paper of Bhowmik et al. (1995b). Later Mazumder et al. (2006) presented an empirical relation from their data to compute the return flow and drawdown due to the navigation traffic in restricted water ways. Details of the development of empirical relations are available in their paper (Mazumder et al. 2006).

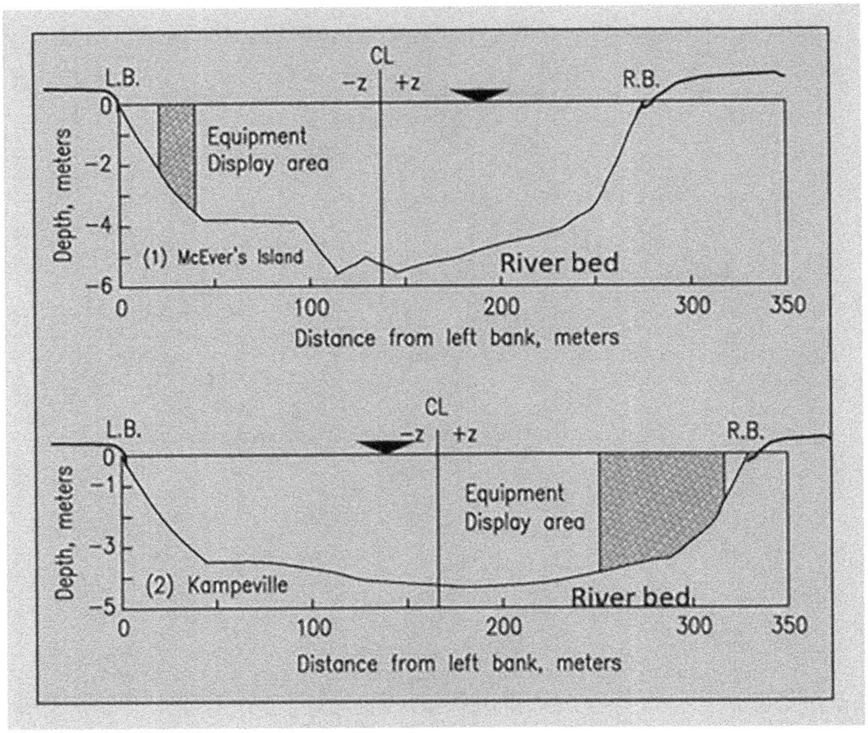

FIGURE 7.33 Sketch of coordinate system and cross-sectional profiles in the McEver Islands and Kampsville sites in Illinois river. (Modified from Mazumder et al., 1993.)

TABLE 7.3
Specifications of Current Meters (Mazumder et al., 1993)

Instrument	Parameter	Range	Accuracy	Probe diameter (cm)
MMB 527	*u*, *v*	±300	±2% of	10.3
	Direction	cm/s	reading	–
MMB 511	*u*, *v*	0°–360°	± 10°	3.9
		300	±2% of	
		cm/s	reading	

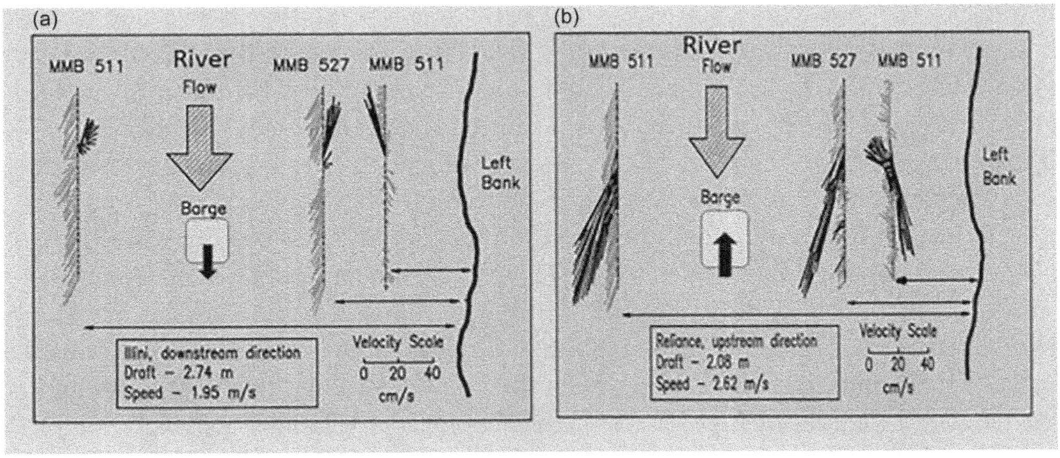

FIGURE 7.34 (a and b) Typical resultant velocity vectors in McEver's Island, (a) downstream-bound barge, and (b) upstream-bound barge. (Modified from Mazumder et al., 1993.)

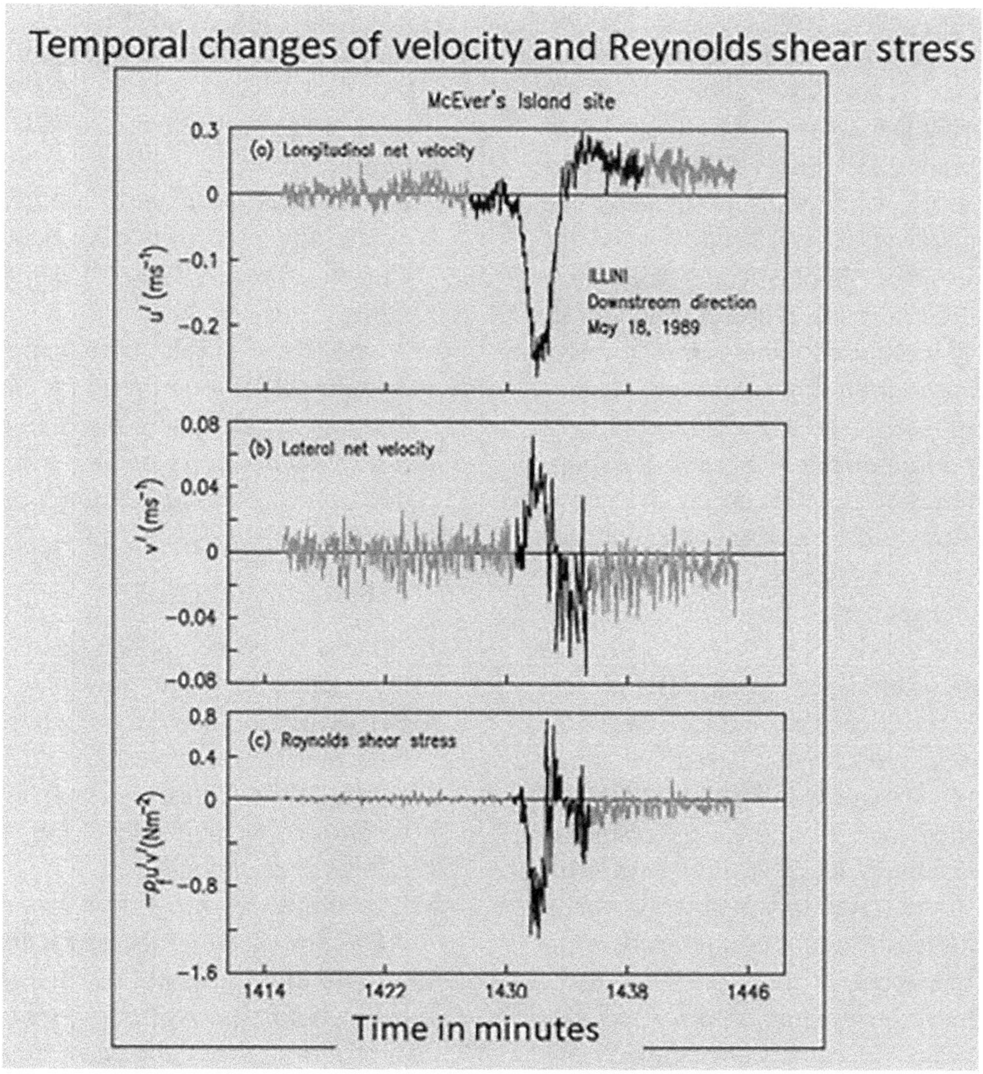

FIGURE 7.35 Typical temporal changes of two velocity components and Reynolds shear stress due to the downstream bound barge movement. (Modified from Mazumder et al., 1993.)

7.7 SUMMARY

Some of the important aspects discussed in this chapter are summarized below:

- Flow visualization technique using high-speed digital imaging is used to study the flow patterns in order to obtain qualitative or quantitative information on the movement of fluid with fine suspension particles.
- Laser Doppler anemometer (LDA), Acoustic Doppler velocimeter (ADV) and PIV are the most commonly used Lab scale turbulent flow measurement equipment. Out of these, ADV is intrusive equipment and LDA and PIV are non-intrusive equipment.
- Extensive experiments are carried out at the FML at ISI-Kolkata using sophisticated equipment (like 3-D Micro acoustic Doppler velocimeters (ADV), High-speed Motion Scope (HSMS) Camera of 1,000 fps with stroboscope light, Electromagnetic discharge meters (EDM), Acoustic Bed Profiler (ABP) with 24 transducers, C. C. D. Cameras, Wave-makers & Wave-absorber, etc), which provided the measurements for experimental fluid mechanics to study the turbulence, sediment suspension, image processing, and wave-blocking phenomenon.

- Three open-ended rectangular hydraulic flumes of different dimensions are set up at the Hydraulics Laboratory, Civil Engineering Department of Indian Institute of Technology Bombay (IIT-B), India for experimental work relevant to study the turbulent flow field and scour around the pier structures. Sophisticated electronic gadgets like acoustic Doppler velocimeter (ADV), PIV are used to measure the velocity structures around the piers.
- Total eight case studies are presented in this chapter on different aspects of experimental study such as: Cases I–III showing the flow and turbulence over the bed form structures; Cases IV–VI for interactions between the flow and the simple and complex bridge piers with sediment bed cause scours; and Cases VII and VIII show the turbulence in natural rivers like Mississippi and Illinois Rivers, to study turbulence statistics of ambient flow and their comparison with those generated by the movement of barge traffics within the channel border area in a navigation waterway. Detailed results are explained in the respective papers.
- The field experimental works have also been carried out using various electronic gadgets like electromagnetic current meters (Marsh-McBirney Current Meters, MMB-511, and MMB-527), and inter-ocean Current meters (S-4). Photographs of field set ups with measuring instruments for collection of velocity data in the McEver's Island and Kempville sites of the Illinois River, RM 50.1 are shown. The inter-ocean current meter (S4) was configured with optional sensors that allowed it to operate as a multi-use platform for operations requiring data acquisition of Currents, Waves, Tide, CTD, turbidity and more.
- A 5 MHz SeaTek Ultrasonic Ranging System (URS) composed of 24 transducers with an electronic package is used for instantaneous bed elevation measurements due to flow over the sediment bed.

7.8 EXERCISE PROBLEMS

1. What are the equipment used for laboratory scale turbulent flow measurements? Briefly discuss each of the equipment with their merits and demerits.
2. Illustrate the working principles of acoustic Doppler velocimeter (ADV) and elaborate its applications.
3. Explain the working principles of PIV and ADV, and discuss their applications.
4. Differentiate the ADV and PIV with respect to its working principles and applications, for lab-scale turbulent flow measurements.
5. What is the important equipment used for field level flow measurements? Briefly discuss each of the equipment with their merits and demerits.

8 Sediment Transport Phenomena

8.1 INTRODUCTION

Sediment is naturally occurring material that is broken down under the processes of weathering, abrasion, and erosion and is subsequently transported by the action of wind, water, or ice flow, or by the force of gravity acting on the particles. For example, sand and silt can be carried out in suspension by the river flow and on reaching the seabed deposited by sedimentation. Slowly they are settled and buried, eventually, they become sandstone and siltstone. The transport process of sediments by water flow is called the *fluvial process* and by wind is called the *aeolian process*.

Mechanism of sediment transport is a branch of basic science in which the processes of erosion, transportation, and deposition of sediment take place under the action of gravity, flowing water, wave, and wind. The sediment material is classified into two categories: *non-cohesive and cohesive sediments*. If the sediment materials are non-cohesive, mechanical forces dominate the behavior of sediment motion in water. The behavior of cohesive sediment is dominated by electrochemical forces that primarily depend on particle size, water chemistry, and mineralogy. For non-cohesive sediment, incipient motion, rolling, sliding, saltation, and suspension of individual sediment particles depend on the physical properties, mineral composition of the particles, characteristics of the flow, and boundary conditions. The sediment motion is the main cause of geo-morphological evolution, fluvial process, and environment change. The processes of erosion of the land surface, transportation of eroded material, and deposition of material in lakes and reservoirs depend on several factors. These factors can be generally classified into the following categories: characteristics of sediment and fluid, flow structure, and channel geometry. In this chapter, the sediment transport phenomenon is discussed with basic theories and solutions.

8.2 PROPERTIES OF INDIVIDUAL SEDIMENT PARTICLES

To understand the mechanism of sediment transport, some fundamental definitions and characteristics of sediment and fluid flow, are discussed here. While dealing with problems of alluvial streams, one is concerned not only with sediment as a collection of several individual particles but also with each particle considered as a separate entity. Properties of sedimentary particles are studied by geologists as well as by engineers. Geologists study the properties in order to trace the origin of sediment and to study the nature of transporting agents. Hydraulic engineers study the properties because of their importance in the phenomena of sediment transport. The most important properties of individual particles are (i) size, (ii) shape, (iii) fall velocity, (iv) mineral composition, (v) surface texture, and (vi) orientation. Among these, in hydraulics, we are interested in size and fall velocity, which are used frequently.

8.2.1 Particle Size

When one deals with a large number of sediment particles, it is very rare that all constituent particles are of uniform size. Of all the properties of sedimentary particles, size is one of the most important and commonly used properties. If all the particles are spherical in size, then specifying the diameter would be enough. However, the sediment particles composing river beds are of numerous shapes from round to flat and to needle-like. These extreme irregularities in their shape confront ordinary classification. For these reasons, the particle size is usually defined by its volume, fall velocity, size of the sieve mesh, or by its intercepts. The particle diameter can be measured by (i) nominal diameter, (ii) fall diameter,

DOI: 10.1201/9781003000020-8

(iii) sedimentation diameter, and (iv) sieve diameter. The following definitions of the size of a particle are used in practice:

Nominal diameter d_n: The nominal diameter d_n of a particle is defined as the diameter of a spherical particle having the same volume and weight as that of a given sediment particle. The nominal diameter provides an idea of the physical size of the particle.

Fall diameter: The fall diameter of a particle is defined as the diameter of a smooth spherical particle of relative density 2.65 and having the same fall velocity like that of a given sediment particle in still water at 24°C

Sedimentation diameter: Sedimentation diameter is the diameter of a spherical particle having the same terminal settling velocity and relative density as the given sediment particle in the same sedimentation fluid under the same condition (Wadell, 1932).

Sieve diameter: Sieve diameter of a particle is arising out of a commonly used method for separating sediment into various size grades, using sieves of decreasing mesh size one below the other. The sieve diameter of a particle is defined as the size of a mesh through which a particle can just pass.

In most cases, a series of sieves are used for the separation of sediment into various size grades. Sieve openings are square in shape. Since a long particle with a small cross-sectional area can pass through a sieve, sieves classify the particles based on the least cross-sectional area of the particle, and thus this classification is not based purely on size. For defining sieve diameters of sediment sizes, Udden (1914)'s geometrical scale of sediment sizes and Wentworth's modification (Wentworth, 1922; Grade and Rangaraju, 2000) of Udden's scale are considered. For sediment sizes (0.02–32 mm) of natural streambeds, the sieve diameter is approximately equal to 0.9 times of nominal diameter d_n (US Interagency Committee, 1957).

Sediment particles are classified into three different categories based on their size: *Silt, Sand, and Gravel.* These classifications are essentially arbitrary but are found in the engineering and geology literature. Table 8.1 shows a grade scale proposed by the subcommittee of hydraulic engineers and geologists of the American Geophysical Union (Lane, 1947). The sizes given in Table 8.1 are the sediment size classification based on the sieve sizes. This size scale is adopted for the sediment population because the sizes are arranged in a geometric series with a ratio of two. Size ranges define the limits

TABLE 8.1
Sediment Particle Size Gradation

Range in mm	Class Name
33–16	Coarse gravel
16–8	Medium gravel
8–4	Fine gravel
4–2	Very fine gravel
2–1	Very coarse sand
1–1/2	Coarse sand
1/2–1/4	Medium sand
1/4–1/8	Fine sand
1/8–1/16	Very fine sand
1/16–1/32	Coarse silt
1/32–1/64	Medium silt
1/64–1/128	Fine silt

of classes that are given names in the Wentworth scale (or Udden–Wentworth scale) used in the United States. One grade scale, known as the phi scale, was first introduced by Krumbein (1934, 1938) as a convenient way of visualizing and statistically analyzing sediment grain size distributions over a wide range of particle sizes. The Krumbein *phi* (ϕ) scale, a modification of the Wentworth scale created by Krumbein (1934), is a logarithmic scale computed by the equation

$$\phi = -\ln_2 d \tag{8.1}$$

where ϕ is the diameter of the particle or grain in mm. For example, for $d = 1$ mm, ϕ is equal to zero; and for $d=4$ mm, ϕ is equal to -2. Equation (8.1) can be rewritten as:

$$\phi = -\ln_2 d = -\frac{\log_{10} d}{\log_{10} 2} \tag{8.2}$$

The main advantage of scale is that it permits the use of the arithmetic scale rather than the logarithmic scale and simplifies the calculation of the various statistical parameters. It is found that for natural sediment, there is a relation between the sieve diameter, d, and the nominal diameter d_n, and is given by $d=0.9\ d_n$.

8.2.2 Particle Shape

The shape of a sediment particle makes the general geometric form apart from the size and material composition. The shape of the particles influences the fluid velocity of the flow, mostly near the bed during the movement of sediment particles. Geologists are interested to study the shape of the coarse particles because the shape focuses the idea on the method of transport and the deposition. Soil scientists are interested to consider the shape as a significant variable to determine the porosity, permeability, and cohesiveness of soils. The shape of the particle depends on the source rock and weathering process, and it is modified by abrasion, corrosion, and breakage. Due to this movement, the sediment particles assume many shapes geometrically, such as cube, sphere, cylinder, cone, and ellipsoid, which are inadequate to describe the shape of the particles.

In sediment analysis, the *sphericity* is defined by Wadell (1932) as the ratio of the surface area of a sphere with the same volume as the particle to the actual surface area of the particle. For the spherical particle, the sphericity is equal to unity, and its value is less than unity for any other shape. In his subsequent paper Wadell (1933), due to some practical difficulties in the concept of sphericity, he redefined it as:

$$\text{Sphericity} = \left(\frac{\text{Volume of the particle}}{\text{Volume of circumscribing sphere}}\right)^{1/3}$$

If d_n is the nominal sphere diameter, and a is a major axis, then the above relationship reduces to the form of sphericity as d_n/a.

Roundness is defined as the average radius of curvature of several corners or edges of a given sediment particle to the radius of curvature of the maximum inscribed sphere (Pettijohn, 1957). Roundness is related to the sharpness of various corners and edges of the sedimentary particle.

The *shape factor* is an important parameter in sediment analysis, and it is defined as:

$$\text{Shape factor} = \frac{c}{\sqrt{ab}} \tag{8.4}$$

where a is the major axis, b is the intermediate axis, and c is the minor axis. The relation between the parameters a, b, and c represents the shape of the particle. Equation (8.4) is the most useful formula for studying the effect of the shape of the particle on the fall velocity. For a sphere, the value of the shape factor is unity and for any other shape, it is less than one. The *flatness ratio* is defined by Wadell (1932) as,

$$\text{Flatness ratio} = \frac{(a+b)}{2c} \tag{8.5}$$

This ratio is equal to or greater than unity. According to Wadell, the value of the flatness ratio varies from 1.05 to 10 for natural sand. There are several parameters to describe the shape of the coarse sedimentary particles. Out of these, sphericity and roundness are the two important parameters considered by geologists. The average sphericity of sediment materials varies from 0.60 to 0.85; while average roundness varies from 0.30 to 0.80. It is important to note that the sphericity and roundness are independent and are not related to the sediment size. For a given stream, it is observed that the roundness increases with sphericity, and both roundness and sphericity increase with the increase in sediment size. The shape factor is the most suitable parameter for fall velocity measurements. For the choice of collective in the mixture, one would consider a coefficient based on average values rather than on the individual particle dimension.

8.2.3 Terminal Fall Velocity of Sediment Particle

The fall velocity of the sediment particle is one of the most important parameters describing the particle in relation to the fluid. The fall velocity of an object in a fluid medium is a particular falling speed at which the drag force and buoyant force exerted by the fluid on the object just equal to the gravity force acting on the object. When a body falls through a fluid, it accelerates until it reaches a constant terminal velocity. This velocity is determined by the density of the fluid, the density of the solid body, the viscosity of the fluid, the shape and orientation of the body and by some length characterizing the size of the body. The velocity of the falling particle may also depend, to some extent, on the size and shape of the vessel, but if the vessel is large enough, its influence may be neglected. The object falls at a constant speed unless it moves in other different mediums. Therefore, the falling under the influence of gravity, the particle will reach a constant velocity, which is called the fall velocity or settling velocity.

We have assumed here all the sediment particles be spherical in shape. Navier–Stokes' equation can be solved for laminar flow around a sphere, neglecting the inertial forces. The solution gives the following expression for viscous resistance:

$$F = 3\pi d \mu_f v_s \tag{8.6}$$

where F is the viscous resistance, d is the diameter of the sphere, μ_f is the dynamic viscosity of fluid and v_s is the terminal fall velocity of the sphere. Under the condition of constant fall velocity, the viscous force F is balanced by the submerged weight of the sphere. Therefore,

$$\frac{\pi d^3(\gamma_s - \gamma_f)}{6} = 3\pi d \mu_f v_s \tag{8.7}$$

Therefore, the terminal velocity is,

$$v_s = \frac{d^2}{18\mu_f}(\gamma_s - \gamma_f) \tag{8.8}$$

in which γ_s and γ_f are the specific weight of the sediment particle and that of the fluid respectively. This equation is generally known as *Stoke's law* for the terminal fall velocity of a sphere and was obtained by Stokes (1851). The following assumptions are made in the derivation of Stoke's law:

1. The inertial forces are neglected from the Navier–Stokes equations. Stoke's law can be used approximately up to a Reynolds number $R_e = \left(\frac{v_s d}{\upsilon_f}\right)$ of unity,
2. Particle shape should be spherical,
3. No-slip condition between the fluid and the boundary of the particle,
4. The particle falls in an infinite and still fluid.

For spherical particle falling with a velocity v_s in a fluid, the following equation may be written as:

$$\frac{\pi d^3(\gamma_s - \gamma_f)}{6} = 3\pi d\mu_f v_s = \frac{C_D A \rho_f {v_s}^2}{2} \tag{8.9}$$

in which C_D is the drag coefficient and A is the projected area equal to $\pi d^2/4$. From equation (8.9), neglecting the inertia forces according to Stokes (1851) one gets the drag coefficient C_D for particle Reynolds number $R_e < 1$ as:

$$C_D = \frac{24\mu_f}{v_s d\rho_f} = \frac{24}{R_e} \tag{8.10}$$

The drag force in the Stokes range is proportional to the first power of the velocity. Equation (8.6) should be applied for the Reynolds number R_e less than 0.5, preferably less than 0.1. Equation (8.10) is another form of Stoke's law. McNown and Lin (1952) suggested the general formula as follows:

$$F = 3K\pi d\mu_f v_s \tag{8.6a}$$

where K is the Stokes number. When viscous force is predominant, the Stoke's number $K = 1$, which gives the Stoke's law (equation 8.6), that means for regular shape. For $K \neq 1$, for irregular shapes, the value of K will be different. For Reynolds numbers greater than 0.1, Stokes law gives results different from those observed in the experiments because of the influence of inertial forces.

Oseen (1927) was the first who assumed some inertia terms in his solution of Navier–Stokes equations. His assumption did not hold at a large distance because acceleration terms were negligible with respect to the frictional terms, which provided the drag coefficient C_D as:

$$C_D = \frac{24}{R_e}(1 + 0.19R_e) \tag{8.11}$$

Goldstein (1929) extended Oseen's equation, which was the more complete solution for Oseen's approximation and the estimated the drag coefficient C_D is as:

$$C_D = \frac{24}{R_e}\left(1 + 0.19R_e - 0.015R_e^2 + 0.0035R_e^3 - \cdots\right) \tag{8.12}$$

which is applicable for $R_e \leq 2$ for experimental data. It seems that the drag coefficient C_D varies gradually from linear to a quadratic relation with respect to particle velocity.

For high Reynolds number $R_e > 2$, the empirical equation for drag coefficient C_D of a sphere is suggested by Schiller and Naumann (1933) experimentally for $R_e < 800$ as

$$C_D = \frac{24}{R_e}\left(1 + 0.150 R_e^{0.687}\right) \tag{8.13}$$

Simultaneously, a quasi-theoretical equation for drag coefficient C_D is suggested by Rubey (1933), combining the Stokes law and the impact formula. He stated that the total resistance to the motion of a particle is the sum of viscous resistance $3\pi d\mu_f v_s$ and impact resistance $\pi d^2 \rho_f v_s^2/4$. Balancing the weight force with the total resistance, and combining with equation (8.9), the drag coefficient C_D is given by:

$$C_D = \frac{24}{R_e} + 2 \tag{8.14}$$

Equation (8.14) shows considerable deviations from the observed value, especially for high Reynolds number R_e. Discussion of this equation is given by Graf and Acaroglu (1967).

Further, Dallavalle (1943) suggested the drag coefficient C_D as:

$$C_D = \frac{24.4}{R_e} + 0.4 \tag{8.15}$$

An equation was suggested by Torobin and Gowin (1959) for $1 < R_e < 10$ and is given as:

$$C_D = \frac{24}{R_e}\left(1 + 0.197 R_e^{0.63} + 0.0026 R_e^{1.88}\right) \tag{8.16}$$

Later, Olson (1961) suggested the drag coefficient C_D can be represented as:

$$C_D = \frac{24}{R_e}(1 + 0.19 R_e)^{1/2} \tag{8.17}$$

Equation (8.17) is valid for the Reynolds number $R_e < 100$. Extensive experiments have been carried out to establish the functional relationship between Reynolds number R_e and drag coefficient C_D. Figure 8.1 shows the variation of C_D with R_e for a sphere over a wide range of Reynolds number. In the same figure, the equations of Stokes, Rubey, and Olson are plotted. A curve for the circular disc is also shown in the figure for comparison.

Fair and Geyer (1963) determined a drag coefficient equation from their experimental data in the range of Reynolds number ($0.5 \le R_e \le 10^4$) and is given by

$$C_D = \frac{24}{R_e} + \frac{3}{\sqrt{R_e}} + 0.34 \tag{8.18}$$

which agrees well within the range of Reynolds number. Swamee and Ojha (1991) developed an equation for drag coefficient C_D in the range ($R_e \le 1.5 \times 10^5$) from the experimental data as:

$$C_D = 0.5\left[16\left\{\left(\frac{24}{R_e}\right)^{1.6} + \left(\frac{130}{R_e}\right)^{0.72}\right\}^{2.5} + \left\{\left(\frac{4000}{R_e}\right)^{2} + 1\right\}^{-0.25}\right]^{\frac{1}{4}} \tag{8.18a}$$

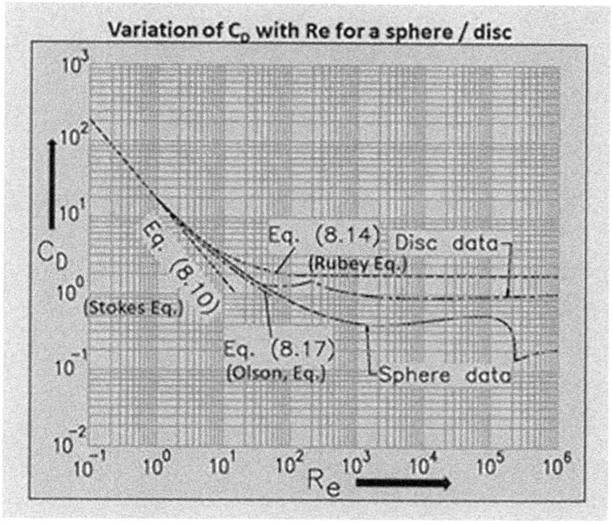

FIGURE 8.1 Variation of C_D with R_e for a sphere and disc. (Modified from Graf, 2010.)

8.3 MOTION WITH LINEAR RESISTANCE

Applying Newton's second law of motion, researchers (i.e., Boussinesq, 1903; Oseen, 1927; Techen, 1947) developed equation of motion for a small spherical particle moving with a variable fluid velocity u under the influence of gravity, and the resulting integro-differential equation is given by

$$\frac{1}{6}\pi d^3\rho_s\dot{u}_s = \underbrace{\frac{1}{6}\pi d^3\rho_f\dot{u}}_{(1)} - \underbrace{\frac{1}{6}\pi d^3\rho_f(\dot{u}_s-\dot{u})}_{(2)} - \underbrace{3\pi d\mu_f(u_s-u)}_{(3)}$$

$$-\underbrace{\frac{3\pi\mu_f d^2}{2\sqrt{\pi\nu_f}}\int_{t_0}^{t}\left(\dot{u}_s(t_1)-\dot{u}(t_1)\right)\frac{dt_1}{\sqrt{t-t_1}}}_{(4)} + \underbrace{\frac{1}{6}\pi d^3\left(\rho_s-\rho_f\right)g}_{(5)} \tag{8.19}$$

where $\nu_f = \mu_f/\rho_f$ is kinematic viscosity, u_s is the velocity of the solid phase, u is the fluid velocity, $\dot{u}_s$ and $\dot{u}$ are the respective accelerations, ρ_s and ρ_f are the respective densities, d is the particle diameter, t_0 is the initial time, and μ_f is the viscosity of fluid. The left-hand side of equation (8.19) represents the force required to accelerate the particle. The term (1) on the right-hand side of equation (8.19) is the fluid acceleration; the term (2) represents the surplus of inertia caused by the pressures, resulting from relative acceleration $(\dot{u}_s - \dot{u})$ or equivalent to the inertia of virtual-mass attached to the solid particle. The term (3) represents the linear resistance or Stoke's term (drag term) for low Reynolds number $R_e < 1$. The term (4) represents the Basset force and gives the force due to the history of the particle, and the last term (5) is the weight force due to gravity. This equation was assumed as a homogeneous fluid velocity field of infinite extent, no interaction between the particles and no particle rotation, and that the linear resistance in relative motion holds true. Equation (8.19) is applicable only for small Reynolds numbers.

8.3.1 Motion with Nonlinear Resistance

If the Reynolds number is very large, the resistance is proportional to the square of the relative velocity as:

$$R_1 \propto (u_s - u)^2 \tag{8.20}$$

No theoretical sound approach for the derivation of equation of motion for high Reynolds number is available, but using some basic assumptions, equation (8.19) of slow motion can be modified for high Reynolds numbers. Equation (8.19) can be modified as: the term (2) of (8.19), which is a virtual-mass term that can be modified by multiplying the displaced volume of fluid, so-called the virtual-mass coefficient k. For a spherical particle, the virtual-mass coefficient $k = 0.5$. Iversen and Balent (1951) pointed out that the coefficient k depends on acceleration parameter and the modified resistance term is written as:

$$R_2 = \frac{1}{6} k\pi d^3 \rho_f (\dot{u}_s - \dot{u}) \tag{8.21}$$

where $k = f\left[\frac{(\dot{u}_s - \dot{u})d}{(u_s - u)^2}\right]$

The modified resistance term (3rd term) becomes

$$R_3 = C_D \pi \frac{d^2}{4} \frac{\rho_f (u_s - u)^2}{2} \tag{8.22}$$

where $C_D = f\left[\frac{d(u_s - u)}{\upsilon_f}\right]$ → steady state drag coefficient, which implies $C_D = f(R_e)$, where $R_e = \frac{d(u_s - u)}{\upsilon_f}$ is the Reynolds number.

The final unsteady equation for a particle moving under the influence of gravity in a fluid flow at a large Reynolds number can be written as:

$$\begin{aligned} \frac{1}{6}\pi d^3 \rho_s \dot{u}_s &= \underbrace{\frac{1}{6}\pi d^3 \rho_f \dot{u}}_{(1)} - \underbrace{\frac{1}{6} k\pi d^3 \rho_f (\dot{u}_s - \dot{u})}_{(2)\ R_2} - \underbrace{C_D \pi \frac{d^2}{4} \frac{\rho_f (u_s - u)^2}{2}}_{(3)\ R_3} \\ &\quad - \underbrace{\frac{3\pi \mu_f d^2}{2\sqrt{\pi \nu}} \int_{t_0}^{t} \frac{\dot{u}_s(t_1) - \dot{u}(t_1)}{\sqrt{t - t_1}} dt_1}_{(4)} + \underbrace{\frac{1}{6}\pi d^3 (\rho_s - \rho_f) g}_{(5)} \end{aligned} \tag{8.23}$$

In steady state, the motion of particle is assumed as

$$0 = 0 - 0 - \underbrace{C_D \pi \frac{d^2}{4} \frac{\rho_f (u_s - u)^2}{2}}_{(3)\ R_3} - 0 + \underbrace{\frac{1}{6}\pi d^3 (\rho_s - \rho_f) g}_{(5)}$$

$$=> C_D \pi \frac{d^2}{4} \frac{\rho_f (u_s - u)^2}{2} = \frac{1}{6}\pi d^3 (\rho_s - \rho_f) g \tag{8.24}$$

which represents the hydrodynamic forces counterbalanced by the gravitational force.

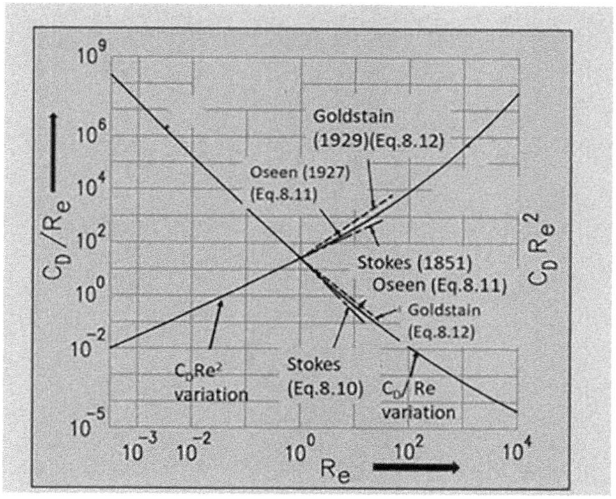

FIGURE 8.2 Curves of $C_D R_e^2$ and C_D/R_e against R_e for spherical particle. Computations of settling velocity and particle diameter are given, using different models. (Modified from Graf, 2010.)

8.3.2 Terminal Velocity

Let us denote the terminal velocity by v_s for velocity of solid particle and by v for velocity of fluid particle. To find the terminal velocity $(v_s - v)$ of a given spherical particle of density ρ_s dropped in a fluid of known properties, Schiller and Neumann (1933) gets

$$C_D R_e^2 = \frac{4}{3}\frac{d^3}{{v_f}^2}\left(\frac{\rho_s - \rho_f}{\rho_f}\right)g \tag{8.25}$$

where $R_e = \dfrac{(v_s - v)d}{\upsilon_f}$ is the Reynolds number. Using equation (8.25), a plot is made for $C_D R_e^2$ vs. R_e in Figure 8.2, and the Reynolds number is obtained and in turn the desired settling velocity is obtained. Equation (8.25) can be written as

$$\frac{C_D}{R_e} = \frac{4}{3}g\left(\frac{\rho_s - \rho_f}{\rho_f}\right)\frac{\upsilon_f}{(v_s - v)^3} \tag{8.26}$$

This equation represents a general approach for the terminal settling velocity of particle. The right-hand side of equation (8.26) is completely known and a plot of $\dfrac{C_D}{R_e}$ vs. R_e is shown in Figure 8.2. Here the particle Reynolds number and in turn the particle diameter are obtained.

From Stoke's law for low Reynolds number $R_e < 1$, the simplification of equation (8.24) gives the terminal settling velocity and is given by

$$v_s - v = \frac{1}{18}d^2 g\frac{\rho_s - \rho}{\mu_f}, \tag{8.27}$$

with density of quartz taken as 2.65 in water at 20°C.

For Stoke's solution $R_e < 1$, the drag for Stoke's law in relative motion is as:

$$R = 3\pi d\mu_f (v_s - v) \tag{8.28}$$

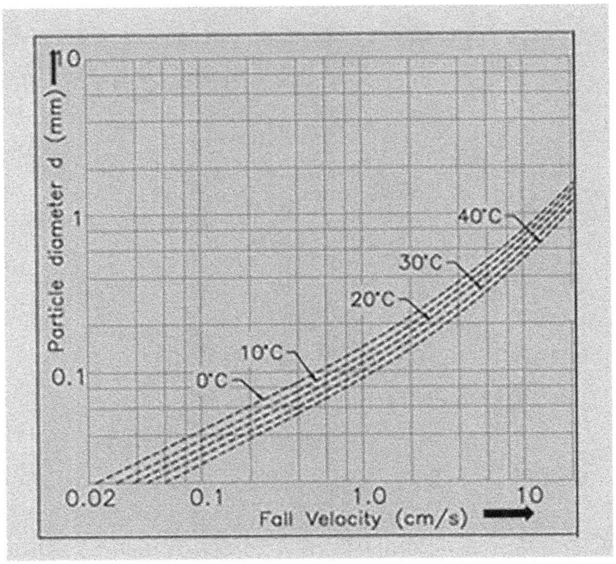

FIGURE 8.3 Fall velocity of spherical particles (relative density is 2.65) in water. (Modified from Garde and Rangaraju, 2000.)

The general drag equation (8.22) is as:

$$R_3 = C_D \pi \frac{d^2}{4} \frac{\rho_f (v_s - v)^2}{2} \tag{8.29}$$

Combining the above two equations (8.28) and (8.29), one gets

$$C_D = \frac{24\mu_f}{d(v_s - v)\rho_f} = \frac{24}{R_e}, \tag{8.30}$$

where $\frac{d(v_s - v)}{\upsilon_f} = R_e$ for $R_e < 0.5$

This equation is another form of Stoke's law. Here one-third of total drag is caused by pressure differences and two-thirds of it is due to frictional force. The drag in the Stokes range is proportional to the first power of the velocity, Reynolds number for which should be smaller than one-half ($R_e < 0.5$), preferably only one-tenth ($R_e < 0.1$).

The fall velocity of spherical quartz particles in water at various temperatures is shown in Figure 8.3. For several hydraulics problems, these fall velocity curves are quite useful because the average relative density of the natural sediment is found to be approximately 2.65. The fall velocity of natural sediment of shape factor 0.70 using the sieve diameter at various temperatures is presented in this figure. The terminal velocity of a particle depends mainly on (i) particle size, (ii) concentration of fluid, (iii) turbulence, and (iv) temperature.

Rubey (1933) suggested a formula for fall velocity of coarse materials which comes beyond Stoke's range. According to Rubey, the total resistance to the motion of a particle is the sum of the viscous resistance and impact resistance as:

$$\frac{4}{3}\pi a^3 g(\rho_s - \rho_f)/6 = \frac{3}{2}\pi a \mu_f v_s + \frac{\pi}{4}\rho_f v_s^2/4 \tag{8.31}$$

The second term of the right-hand side of equation (8.31) is called impact resistance. The fall velocity from the above expression can be written as:

$$v_s = \sqrt{\frac{9\mu_f^2}{\rho_f^2 a^2} + \frac{4}{3} ag \frac{(\rho_s - \rho_f)}{\rho_f}} - \frac{3\mu_f}{\rho_f a} \tag{8.32}$$

This fall velocity formula known as Rubey's formula was used by Einstein in developing the bed load equation. The effect of particle shape on the fall velocity was studied by Albertson (1952), and then by Schulz et al. (1954). They found that the shape of the pebbles and gravels influences the fall velocity appreciable when the Reynolds number is large.

Richardson and Zaki (1954) observed that in the dense sediment suspension, the flow around adjacent settling particles induces a greater drag as compared to that in the clear water, which is called as *hindered settling* effect that means, the settling velocity of particle is fully dependent on the suspended sediment particles. The resulting effect in the settling velocity in sediment-laden flow shows the reduction of settling velocity from that of clear water. According to Richardson and Zaki (1954), the settling velocity or hindered settling of a particle w_s in sediment-laden water with suspension concentration c is given by:

$$w_s = v_s (1-c)^n \tag{8.33}$$

where v_s is the settling velocity in clear water and n is an empirical exponent of reduction of fall velocity, which varies on particle Reynolds number R_e as:

$$\begin{aligned} n &= 4.3\, R_e^{-0.03} \quad \text{for} \quad 0.2 \le R_e \le 1.0; \\ n &= 4.4 R_e^{-0.01} \quad \text{for} \quad 1.0 \le R_e \le 500; \\ n &= 2.39 \qquad\quad \text{for} \quad 500 \le R_e. \end{aligned} \tag{8.34}$$

The exponent n varies from 4.9 to 2.3 for Reynolds number R_e ranging from 0.1 to 10^3. However, the value of n is approximately 4 for the particle size ranging from 0.05 to 0.5 mm. Experiments on fluid sediment mixtures have shown that a substantial reduction of the settling velocity of the particle occurs due to the increased suspension concentration (Woo et al., 1988; Mazumder, 1994). If the density of fluid is increased by suspended sediments, buoyancy force is increased, and hence the substantial reduction of particle fall velocity in suspension occurs. According to Maude and Whitemore (1958), the term hindered settling is used to designate the fall velocity of sediment in suspension resulting from an increase in sediment concentration. They developed a theoretical relationship between the settling velocity of sediment in suspension and the concentration of suspended particles, and the relation is given by

$$w_s = v_s (1-c)^\beta \tag{8.35}$$

where β is a function of particle shape, size distribution and Reynolds number and c is the volume of solid per unit volume of suspension.

Oliver (1961) conducted experiments on settling velocity of particle in sediment-laden water. From his experimental data and those of McNown and Lin (1952), Oliver (1961) developed an equation as:

$$w_s = v_s (1 - 2.15c)\left(1 - 0.75c^{0.33}\right) \tag{8.36}$$

A new formula was proposed by Sha (1965) which was applicable for fine sediment $d_{50} \leq 0.01$ mm as

$$w_s = v_s \left(1 - \frac{c}{2d_{50}^{0.5}}\right)^3 \tag{8.37}$$

Thacker and Lavelle (1977) showed that the reduction of fall velocity may be taken into account by a factor $(1-c)^n$ multiplied by the single fall velocity w_0. Subsequently, Lavelle and Thacker (1978) studied the effect of hindered settling on sediment concentration profiles.

Dietrich (1982) developed an empirical equation for settling velocity, which depends on the size, shape, density and roundness of natural sediment particles. In his equation, there are four dimensionless parameters, such as nominal diameter d_*, settling velocity w_*, Corey shape factor and the roundness index factor, where $d_* = \frac{(\rho_s - \rho_f)gd}{\rho_f w_s^2}$, and $w_* = \frac{\rho_f w_s^3}{(\rho_s - \rho_f)g\upsilon_f}$. He obtained an imperical equation given by:

$$\log w_* = -3.7672 + 1.9294\,(\log d_*) - 0.0981(\log d_*)^2$$
$$-\,0.00575(\log d_*)^3 + 0.00056(\log d_*)^4 \tag{8.38}$$

This equation can be written as:

$$w_s = \frac{\upsilon_f}{d} 10^{-1.2557+0.9765(\log d_*)-0.03273(\log d_*)^2-0.00192(\log d_*)^3+0.0002(\log d_*)^4} \tag{8.39}$$

Dietrich's proposed general formula allows for determining the settling velocity of sediment particles from laminar to turbulent regions, considering the effects of size, density, shape factor, and roundness factor. However, the roundness factor used in his formula is rarely measured in practice, and his formula is relatively difficult to use. Woo et al. (1988) considered the effect of hindered settling of sediment particles to estimate the vertical distribution of large concentration of sediment particle, using log-law and power-law velocity profiles. The derivation of Woo et al.'s model included the effect of viscous shear stress which was a function of suspended concentration.

Cheng (1997) suggested a simplified formula to predict the settling velocity of natural sediment, using the particle Reynolds number and the dimensionless particle diameter and is given by:

$$w_s = \frac{\upsilon_f}{d}\left(\sqrt{25 + 1.2d_*^2} - 5\right)^{3/2} \tag{8.40}$$

where $d_* = d\left(\frac{(\rho_s - \rho_f)g}{\rho_f \upsilon^2}\right)^{1/3}$ is the dimensionless particle diameter. This equation can be used to evaluate the settling velocity of natural sand particles explicitly. This equation is applicable to a wide range of Reynolds numbers from Stokes flow to the turbulent flow regime. Several settling velocity formulas for spherical and non-spherical particles developed by different investigators, such as Sha (1954), Zhang et al. (1989), Van Rijn (1989), Zhu and Chang (1993), are used for comparative study; it shows the highest degree of prediction accuracy and also agrees well with diagram of US Inter Agency Committee (1957).

Jimenez and Madsen (2003) suggested a simple settling velocity formula of sediment particle sizes ranging from 0.063 to 1mm with a modification of Dietrich's formula (1982). They adopted a general expression as:

$$\frac{1}{w_*} = c_1 + c_2\left(\frac{1}{S_*}\right) \tag{8.41}$$

with

$$w_* = \frac{w_s}{\sqrt{\left(\frac{\rho_s}{\rho_f} - 1\right)gd}} \text{ and } S_* = \frac{d}{4\upsilon_f}\sqrt{\left(\frac{\rho_s}{\rho_f} - 1\right)gd} \tag{8.42}$$

where c_1 and c_2 are constants to be determined from the experimental data for a certain range of S_*. values. Baldock et al. (2004) proposed a model of settling velocity of sediment particle at high concentration in suspension. A model for hindered settling of sand grains was also developed by Tomkins et al. (2005). Wu and Wang (2006) derived a general relation for settling velocity, using the extensive data collected from different countries and regions. They claimed that the suggested explicit relation of the settling velocity for given sediment size and shape factor can be easily used. Zhiyao et al. (2008) proposed a simple settling velocity model, using kinematic viscosity as a function of concentration. Pal and Ghoshal (2013) developed an expression to predict the settling velocity of a sediment particle, which is dispersed in a sediment-fluid mixture during a turbulent flow. Recently, Kumbhakar et al. (2017) developed an entropy-based model on the exponent of reduction of settling velocity using the well-known Shannon entropy. The derived expression is tested with a large number of experimental measurements and is also compared with the existing models.

8.4 EFFECT OF PARTICLES ON VISCOSITY

In this section, the determination of viscosity due to suspension is discussed. The basic variables include the nature of the fluid-solid phase, the relative motion between the solid and fluid, and the concentration. For dilute suspensions, Einstein developed a relationship between viscosity and concentration in sediment-laden fluid as

$$\frac{\mu_{\text{susp}}}{\mu_f} = 1 + K_e C \tag{8.43}$$

where μ_{susp} is the viscosity of the sediment-laden fluid (water & sediment mixture), μ_f is the viscosity of liquid, K_e is Einstein's viscosity constant, and C is the volumetric concentration of solid phase, and is defined as

$$C = \frac{\text{Volume of sediment}}{\text{Volume of water} + \text{Volume of sediment}} \tag{8.44}$$

The sediment concentration is usually expressed in mass per unit volume of concentration. The mass density of sediment fluid mixture ρ_m is expressed as

$$\rho_m = \rho_f + \left(\rho_s - \rho_f\right)\text{C} \tag{8.45}$$

The specific weight of sediment fluid mixture γ_m is given by

$$\gamma_m = \gamma + (\gamma_m - \gamma)C = g\rho_m \tag{8.46}$$

The derivation of equation (8.43) is based on the fundamental equations of hydrodynamics. The following are the assumptions:

1. The liquid is Newtonian and incompressible.
2. The inertial terms in the Navier–Stroke's equation vanish or negligibly small. Each particle should be sufficiently far away.
3. The solid phase consists of rigid, smooth and uniform spheres.

Under these conditions, Einstein viscosity constant was found to be $K_e = 2.5$.

Considering the first-order interactive effects, several investigators reached different equations and these are summarized in the form of:

$$\frac{\mu_{\text{susp}}}{\mu_f} = 1 + K_e C + K_2 C^2 + \cdots \tag{8.47}$$

Here the particle–particle interaction and particle–boundary interaction are involved. Ward (1955) suggested a solution for high volume concentration, which is similar to equation (8.43) as:

$$\frac{\mu_{\text{susp}}}{\mu_f} = 1 + 4.5C \tag{8.48}$$

Bagnold (1954) reported the dynamic viscosity of sediment-fluid mixture as:

$$\mu_m = \mu_f \left[1 + \frac{1}{\left(\frac{0.74}{C}\right)^{1/3} - 1}\right]\left[1 + \frac{0.5}{\left(\frac{0.74}{C}\right)^{1/3} - 1}\right] \tag{8.49}$$

where μ_f is the dynamic viscosity of a clear water. Lee (1969) provided an empirical formula for μ_m and is given by:

$$\mu_m = \mu_f (1 - C)^{-\left(2.5 + 1.9C + 7.7C^2\right)} \tag{8.50}$$

For ellipsoidal particles in suspension, Ward (1955) reported the viscosity as:

$$\frac{\mu_{\text{susp}}}{\mu_f} = 1 + K_j C \tag{8.51}$$

where K_j depends on the various axial ratios.

8.4.1 Stable Suspension for Spherical Particles

Several equations have been proposed by researchers to determine the viscosity of suspension. Stable suspension means the equilibrium suspension, where the conservation of mass satisfies. Ward (1955) suggested an equation if the experimental results are expressed in the form as:

$$\frac{\mu_{\text{susp}}}{\mu_f} = 1 + K_e C + K_e^2 C^2 + K_e^3 C^3 + \cdots \tag{8.52}$$

where K_e is an experimentally determined Einstein viscosity constant. For low concentration, it reduced to

$$\frac{\mu_{\text{susp}}}{\mu_f} = 1 + K_e C \tag{8.53}$$

Equation (8.52) predicts the correct viscosity in suspension for volume concentration up to 35%.

From the experimental data, Ward (1955) concluded that the suspension viscosity μ_{susp} is independent of the absolute size of spheres but dependent on the size ratio of the spheres.

8.4.2 Unstable Suspension

In suspension where the sediment particles are either downward or upward settling. The unstable suspensions are more common because of the frequent existence of density difference between the two phases. Experimental work reported by Oliver and Ward (1959) suggested that for low concentrations, equation (8.43) generally explains the data with Einstein's viscosity constant ranging from 3.00 to 3.60 depending on density difference.

For $0.1 < C < 0.3$, the equation may be represented as:

$$\frac{\mu_{\text{susp}}}{\mu_f} = (1 + k_1) + (1 + 2k_1) K_e C + (1 + 3k_1) K_e^2 C^2 + \cdots \tag{8.54}$$

where $k_1 = 0.33 \dfrac{\rho_s - \rho_f}{\mu_f}$, which gives +ve values of k_1 for downward settling suspension, and –ve value for upward settling.

Happel and Brenner (1965) presented an expression as

$$\frac{\mu_{\text{susp}}}{\mu_f} = \frac{(1 - C)^r}{\dfrac{v_{\text{susp}}}{v_s}} \tag{8.55}$$

where $\dfrac{v_{\text{susp}}}{v_s}$ is the ratio of sedimentation velocity v_{susp} of suspension to the individual particles settling velocity v_s. The exponent r varies from 1 to 2.

8.4.3 Suspension of Non-Spherical Particles

The stable suspension of non-spherical shaped particles was studied by Ward and Whitmore (1950) and it is found that the viscosity was dependent on the particle size in contrast to those of spheres. Similar results were also obtained by Moreland (1963). They concluded that the difference between the spherical and non-spherical in suspension was due to layers of immobile fluid in the surface of non-spherical particles. Whitemore (1957) reported this by multiplying the hydrodynamic volume factor k_ω with the concentration C and gave an equation as:

$$C' = k_\omega C \tag{8.56}$$

where C' is the hydrodynamic volume concentration, C is the concentration, k_ω is the hydrodynamic volume factor. It may be considered into one equation by substituting equation (8.49) into equation (8.39), where Einstein's viscosity constant is represented by k_e=2.0. Some detailed discussions are available in Ward (1955).

8.5 BULK PROPERTIES OF SEDIMENTS

To understand the dynamics of a sedimentary particle, one must know the properties of individual particles. However, the next step is to study the properties of a group of particles. These properties are called bulk properties and they are of great use in sediment transport problems. The most important bulk properties of sediments are (i) size distribution of sediments (ii) porosity (iii) specific weight, and (iv) angle of repose. These are discussed in this section.

8.5.1 Grain-Size Distributions of Sediments

When one deals with a large number of sediment particles in nature, it is ascertained that all constituent particles are not of the same size. For this reason, one must know the relative quantity of particles in different size groups. For several practical problems, the size can be taken as a single property of sediment describing the characteristics. The sediments in the range of gravels to sands i.e. 16 mm to 1/16 mm are usually analyzed by sieve analysis. Sieve sizes are based on the geometric scale. American Society of Testing material (ASTM) has specified the United States Standard Sieve series which are based on the geometric scale with a ratio $2^{1/4}$. The sieves to be used for the analysis are once selected and put one over the other with the mesh size decreasing in the downward direction (Figure 8.4). A pan is put at the bottom of the sieve column. The sieve column is usually shaken using an automatically shaking machine. The period of sieving depends on the method of shaking and weight of the sample. The larger the sediment sample in weight, the greater the time required. With automatic shakers, 15–20

FIGURE 8.4 Sieve machine with shaker.

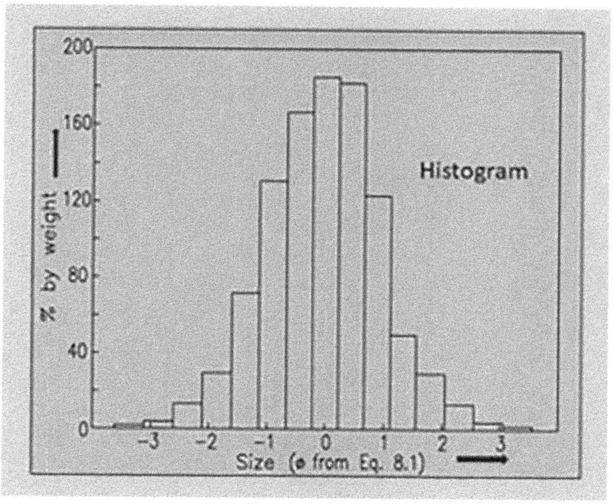

FIGURE 8.5 Histogram of grain size frequency curve (size in ϕ scale).

minutes time is usually adequate for 15 gm of sample. After sieving of particular time duration, the data obtained are in the form of amount of sediment retained over sieves of various sizes. These data are presented in various ways so as to draw definite conclusions and obtain definite information. The statistical representations of data, that are most commonly used, are (i) histogram, (ii) frequency polygons and frequency curve, and (iii) cumulative frequency curve.

8.5.1.1 Characteristics of Frequency Curves

From the size analysis, the total weight of sediment sample and the weights of various samples retained over the individual sieves are obtained. A histogram is a display of statistical information that uses rectangles to show the frequency of data items in successive numerical intervals of equal size. It is graphically displayed of data using bars of different heights, where the ordinate contains the percentage of total sample weight, and the abscissa contains the size. In a histogram, each bar contains the weight of individual size from the range (Figure 8.5). Taller bars show that more amount of sample sediment falls in that range. A histogram displays the shape and spread of continuous sample data. Figure 8.6 shows the frequency curve (line) passing through the central points of the bars with a peak point as the modal size. Once the histograms, frequency curves or cumulative frequency curves are drawn for several samples, it is easy to compare these graphs and draw specific conclusions regarding their similarities and dissimilarities.

Some frequency curves are found to have only one maximum or one mode, while others can have two maxima. The former is called unimodal (Figure 8.7a), while the latter is called bimodal distribution (Figure 8.7b). In exceptional cases, one can have more than two maxima or two modes called polymodal (Figure 8.8). It has been found that, in general, coarse gravels indicate a bimodal distribution, while sands tend to yield a uni-modal distribution. There can be several causes for bimodal or polymodal distribution, one possibility is that the sample is made of two or more populations corresponding to different modes of transport. The other important characteristic is the central tendency which is defined as the tendency of one size grade to be present in larger amounts than the other size grades. This tendency is described in terms of mode, arithmetic mean, geometric mean or median size. The material which has a very small spread is commonly called as flat or uniform (Figure 8.9) while having a large spread is called non-uniform distribution.

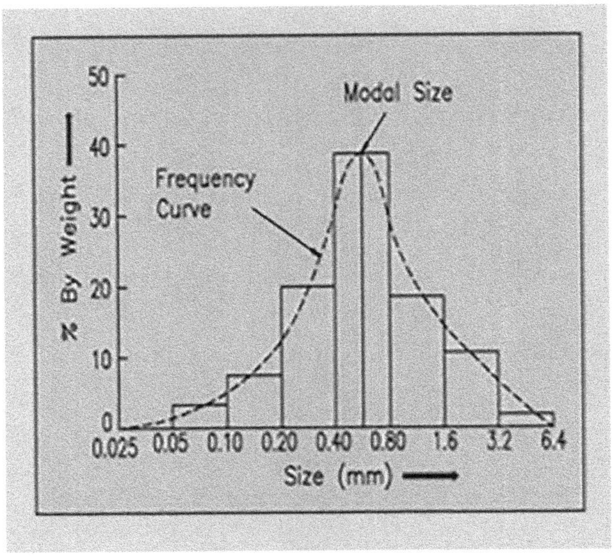

FIGURE 8.6 Frequency curve (line) passing through the central points of the bars with a peak point as the modal size.

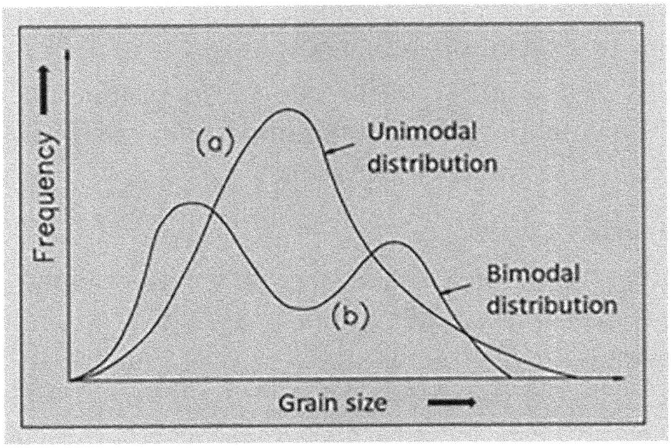

FIGURE 8.7 (a) Unimodal and (b) bimodal size distributions.

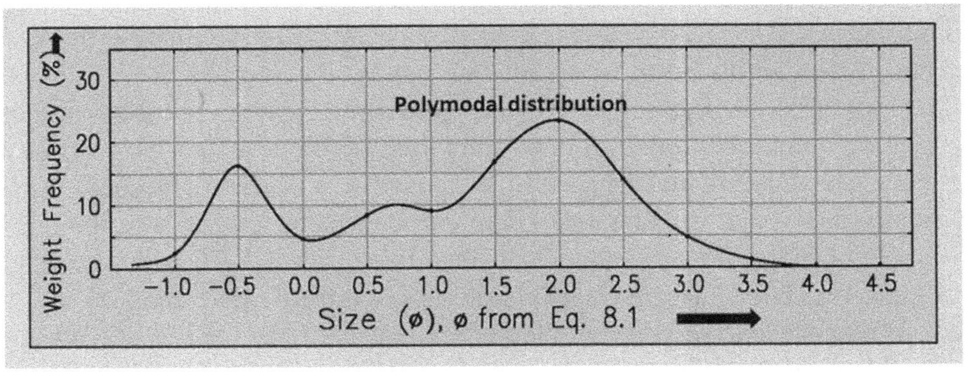

FIGURE 8.8 Polymodal distribution of grain sizes (ϕ scale is given by equation 8.1). (Modified from Sengupta, 1975a.)

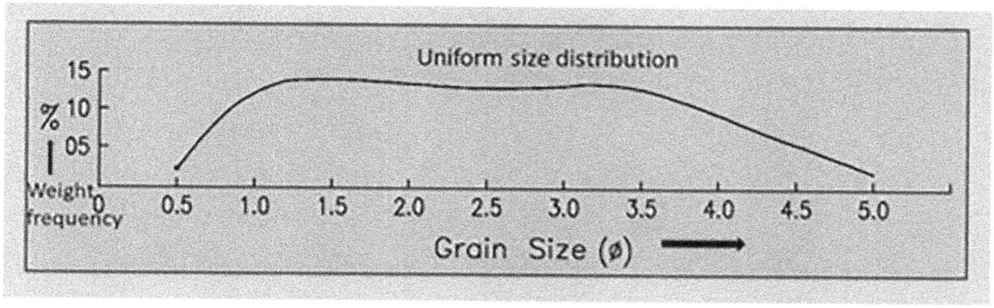

FIGURE 8.9 Flat or uniform size distribution (ϕ scale is given by equation 8.1).

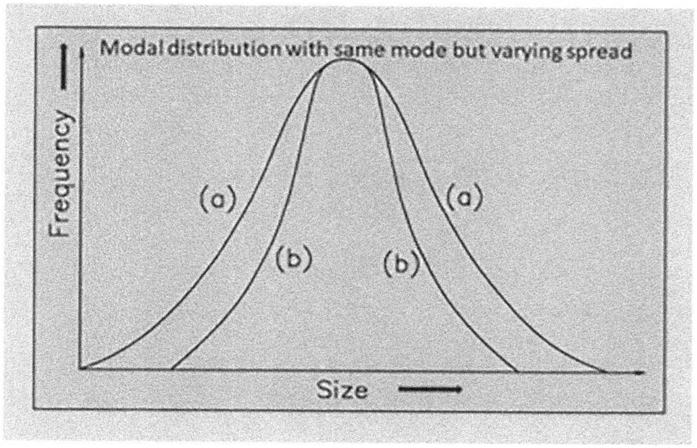

FIGURE 8.10 Frequency curves with the same mode with different spreading (a and b).

Two frequency curves of the same uni-modal can have the same mean size but their spread may be different (Figure 8.10a and b). Here are the frequency curves with the same mode with different spreading. The degree of spread is measured by the standard deviation or quartile deviation.

If the frequency distribution curve is the same about the mean, the distribution is called symmetric distribution (Figure 8.11a). Frequency distribution curves having the same spread can be different in their symmetry. Some curves may be symmetrical while some curves are asymmetrical (Figure 8.11). The measure of symmetry or asymmetry is called skewness of the distribution. It may be either positive or negative depending on whether the skewness is to the left or to the right. Figure 8.11b is called negatively skewed distribution, and (Figure 8.11c) is the positively skewed distribution.

Another property of the frequency curves is the peakedness. Two frequency curves can be alike in their spread, average and their symmetry or asymmetry but can be different in their peakedness. The measure of peakedness is known as kurtosis (Figure 8.12). There are three types of kurtosis: leptokurtic, mesokurtic (normal peak) and platykurtic distributions (Mazumder and Das, 1992).

All these characteristics -unimodel or polymodal distribution, central trendency, spread as measured by the standard deviation, symmetry about the mean as measured by the skewness and peakedness as measured by the kurtosis, help to differentiate between various frequency distributions. The central tendency of the frequency distribution curve can be described by the parameters such as mode, median and mean.

The *mode* is a statistical term that is the most predominant size in the sample and refers to the size corresponding to the maximum ordinate of the frequency curve. *Median size* d_{50} is the sediment size for which 50% of the material by weight is finer.

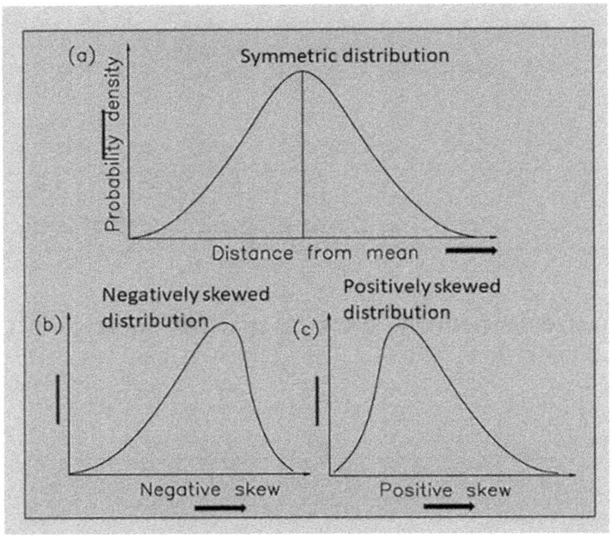

FIGURE 8.11 Skewness of grain size distributions: (a) symmetric, (b) negatively skewed, and (c) positively skewed.

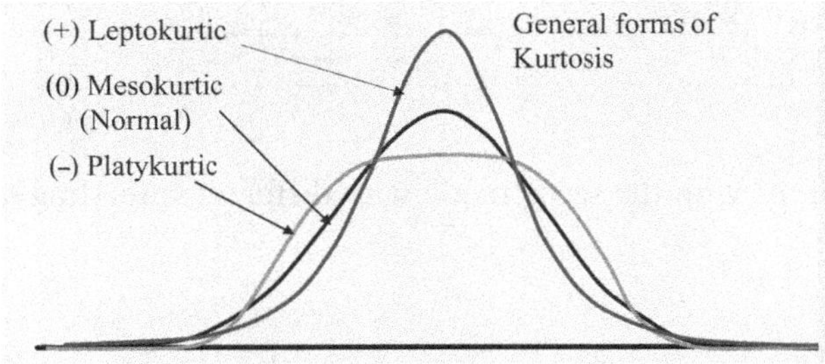

FIGURE 8.12 Kurtosis of the distribution: (+) Leptokurtic, (0) Mesokurtic and (–) Platykurtic. (Modified from Mazumder and Das, 1992.)

The *arithmetic mean* d_a is defined as:

$$d_a = \frac{\sum d_i \Delta p_i}{\sum \Delta p_i} = \frac{\sum d_i \Delta p_i}{100} \tag{8.57}$$

where Δp_i is the percent weight corresponding to the size d_i. For normal distribution, the arithmetic mean, median and mode have the same value.

The *geometric mean* is a mean, which indicates the central tendency of a set of samples by using the product of their values. The geometric mean size d_g is the square root of the product of d_{84} and d_{16} and is given by

$$d_g = (d_{84} d_{16})^{1/2} \tag{8.58}$$

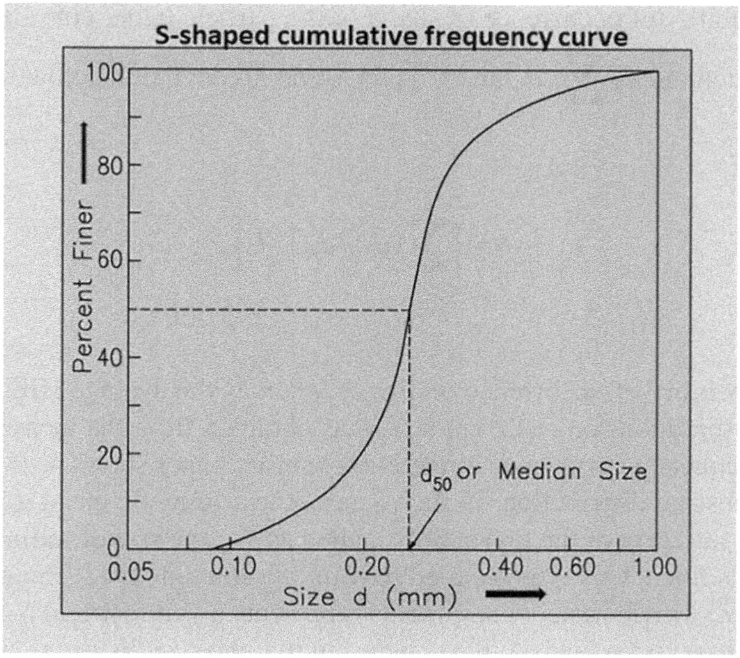

FIGURE 8.13 Typical cumulative frequency curve for sand particles.

where d_{84} and d_{16} represent diameters such that 84% and 16% of the mixture by weight are finer. The geometric standard deviation σ_g is used to determine the non-uniformity of the sediment mixture. The geometric standard deviation σ_g is defined as

$$\sigma_g = \left(\frac{d_{84}}{d_{16}}\right)^{1/2} \tag{8.59}$$

For a given size distribution, if $\sigma_g \leq 1.4$, the sediment size is considered to be uniform (Mazumder et al., 2011; Ojha et al., 2019), otherwise else, sediment is non-uniform. *The sorting coefficient* is defined as $S_0 = \sqrt{\frac{d_{75}}{d_{25}}}$, in which d_{75} and d_{25} are the diameters such that 75% and 25% of the material by weight is finer than these sizes respectively. For completely uniform materials, the sorting coefficient S_0 is unity.

The *cumulative frequency curve* is obtained by plotting the finer or coarser sizes in percent than a given size in the ordinate axis and the size of sediment in abscissa. This frequency curve is plotted on a variety of graph papers such as (i) natural scale for ordinate and abscissa, (ii) linear scale in ordinate and log scale in abscissa, (iii) probability scale in ordinate and linear scale in abscissa, and (iv) probability scale in ordinate and logarithmic scale in abscissa. When the cumulative frequency curve is drawn for size distributions of sediment sizes on ordinary or semi-log papers, one gets an S-shaped curve (Figure 8.13).

A symmetrical *S*-curve in ordinary paper plots as a straight line on the arithmetic probability paper. Such a distribution is known as Gaussian or normal distribution, which follows:

$$f(d) = \frac{1}{\sigma\sqrt{2\pi}} \exp\left\{-\frac{(d-d_m)^2}{2\sigma^2}\right\} \tag{8.60}$$

where $f(d)$ is the probability of occurrence of size d of the particle under consideration, d_m is the arithmetic mean diameter defined as $\sum_{i=1}^{n} f_i d_i$, and σ is the standard deviation defined as

$$\sigma = \left\{ \sum_{i=1}^{n} (d_i - d_m)^2 f_i \right\}^{\frac{1}{2}} \tag{8.61}$$

in which f_i is the probability of occurrence of size d_i taken as the average size in each class, and n is the total number of observations on sediment size d_i as obtained from the measured size distribution.

If the cumulative frequency curve on arithmetic probability paper shows a straight line, the data follows the normal or Gaussian distribution. In such a case, the arithmetic mean d_m can be obtained from the plot by reading the intercept of the line at 50%, called as d_{50}, the size of sediment for which the 50% of material by weight is finer. This size is called *median*. Krumbein (1942) showed that the cumulative frequency curve for size distributions of sediments represents a symmetrical S-curve when plotted on semi-log paper. This observation leads to the conclusion that the size distribution curve of natural sediments shows log-normal. If the data are plotted in log-probability paper, the curve will show the straight line for log-normal distribution.

8.5.2 Grain Size Distributions (Log-Normal, Hyperbolic and Laplace)

The mathematical properties of grain-size distributions have long fascinated to Sedimentologists. Several investigators studied the grain-size distributions in different sedimentary deposits. A discovery of great interest to geologists has been the fact that the sizes of sediment particles in nature tend to follow a log-normal distribution, that is, the logarithm of particle sizes of naturally occurring sand and sandstones is normally distributed. This fact was first pointed out by Krumbein (1934, 1936) and later confirmed by other scientists (Kolmogorov, 1941; Blench, 1952; Harris, 1958; Kennedy and Koh, 1961; Sengupta, 1967, 1979; Ghosh and Mazumder, 1981, Sengupta et al., 1991, 1999; Sengupta, 2007). The probabilistic model was first put forward to explain the log-normality of particle size distributions which was due to Kolmogorov (1941). His model was for a rock being repeatedly struck by hammer. There are several models by different probabilists to explain how the crushing of rocks leads to the log-normal distribution (Figure 8.14a). All these models are the multiplicative form of central limit theorem, and combine the selection and breaking of rocks in different degrees. Details are available in Kondolf and Adhikari (2000).

However, the grain sizes of sand populations in natural sedimentary environments are not always distributed log-normally. Several scientists claim that the size distribution of sediment particles in nature is log-hyperbolic (Barndorff-Nielson, 1977; Bagnold and Barndorff-Nielson, 1980) for desert or aeolien environment (Figure 8.14b) and log-skew-Laplace (Fieller et al., 1984; Feiller and Flenley, 1992) for coastal sedimentary environment (Figure 8.14c). Bagnold and Barndorff-Nielson (1980) provided a precise description of hyperbolic distributions of grain sizes and indicated its wide applicability to sediment transport problems. It may be mentioned here that the log-normal and log-skew-Laplace distributions are the limiting distributions of log-hyperbolic family (Christiansen and Hartmann, 1991). Figure 8.15 shows the fitted densities in logarithmic scale of relative proportion on the ordinate axis: (a) Log-normal, (b) log-hyperbolic and (c) log-Laplace distributions. The name "hyperbolic distribution" derives from the fact that in a plot of the probability density function with a logarithmic scale on the ordinate axis, the curve depicted is a hyperbola, as in the example shown in Figure 8.16.

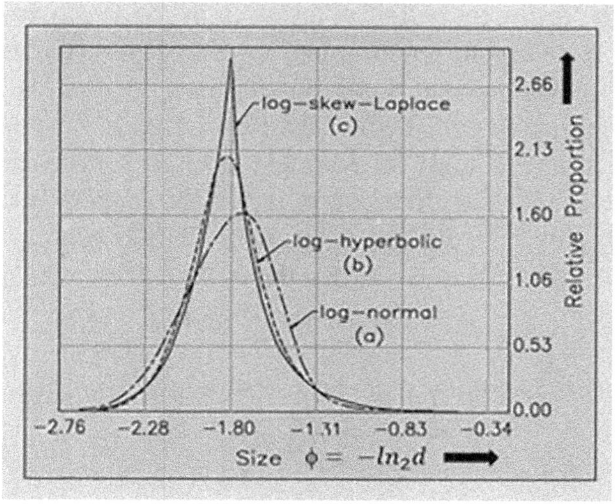

FIGURE 8.14 Fitted densities in natural scale for sand particles. (Modified from Purkait, 2006.)

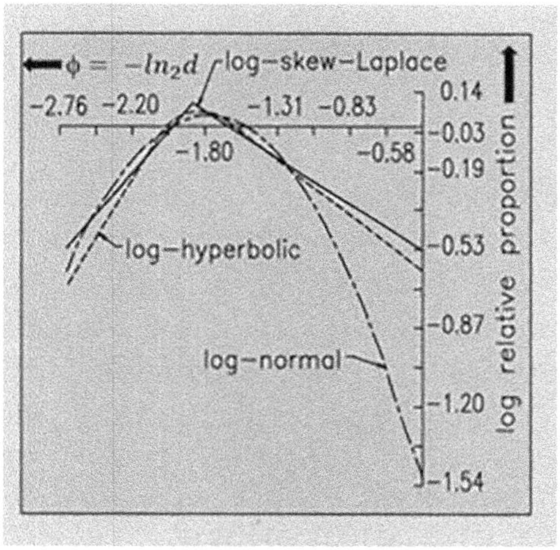

FIGURE 8.15 Fitted densities in logarithmic vertical scale for sand particles. (Modified from Purkait, 2006.)

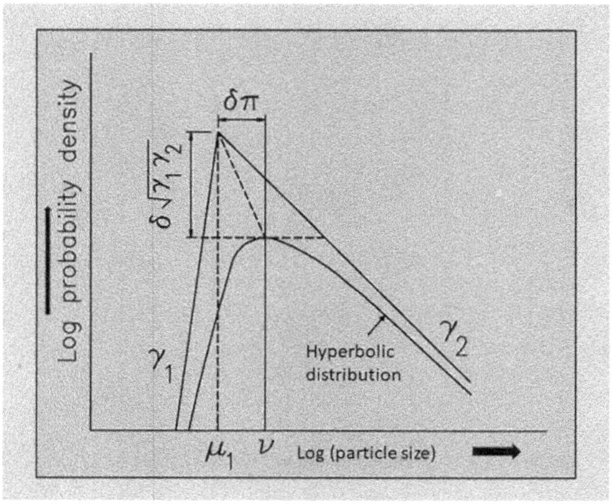

FIGURE 8.16 Sketch of log-probability (density) functions of a hyperbolic distribution. The graph of this function is a hyperbola and the asymptotes (γ_1 and γ_2) of the hyperbola are indicated on the figure. (Modified from Barndorff-Nielson, 1977.)

The size distribution of the sediment bed of many streams, particularly those that contain sand and gravel, departs significantly from log-normal. In many cases, size distributions in streams with sand and gravel beds are distinctly bimodal or multimodal in character. A bimodal size distribution is defined here as one that has two prominent modes with a significant drop in the percentages of sediment in the size grades between them. The field study by Kuhnle (1992, 1993) has identified a stream (Goodwin Creek, Mississippi) with a bimodal size distribution in which the sand fraction of the sediment bed is entrained at lower flow strengths than the gravel fraction. Church et al. (1991) obtained a multimodal size distribution in bed material, while they studied the mobility of the sand and fine gravel fraction of a gravel-bed stream. Kothyari (1995) used the method of power transformation to study the grain-size distribution of a river bed.

The following three density functions that most commonly have been proposed are the normal, hyperbolic and skew Laplace distributions below:

8.5.2.1 Normal Distribution

Among the family of size frequency distributions, the density function of normal (Gaussian) distribution is given by:

$$g(\phi;\mu,\sigma)=\frac{1}{\sigma\sqrt{2\pi}}\exp\left(-\frac{1}{2}\frac{(\phi-\mu)^2}{\sigma^2}\right),\text{ with }\sigma>0 \tag{8.62}$$

where the size $\phi=-\ln_2 d$ (d in mm) is the random variable (Krumbein, 1934) from equation (8.1), μ is the mean of the distribution and $\sigma>0$ is the standard deviation. In the normal distribution, the mean, median and mode values coincide, the skewness=0, and the kurtosis=3.0. Here the normal phi curve may be treated exactly as the Gaussian curve, except that the properties of the curve apply to a logarithm of the diameter instead of the diameter itself. A log-normal distribution (Figure 8.14a) is defined as the logarithm of particle sizes of naturally occurring sand and sandstones normally distributed.

8.5.2.2 Hyperbolic Distribution

Barndorff-Nielsen (1977) first introduced the hyperbolic distribution to describe the mass-size distribution of aeolian sand (Figure 8.14b). The name "hyperbolic distribution" derives from the fact that in a plot of the probability density function of the distribution with a logarithmic scale on the ordinate axis the curve depicted is a hyperbola, as in the example shown in Figure 8.16. Denoting the distributed size variable by ϕ and the probability density function by $p(\phi)$, the four parameter equation of the hyperbolic curve is of the form:

$$p(\phi;\gamma_1,\gamma_2,\mu_1,\delta)=\frac{\sqrt{\gamma_1\gamma_2}}{\delta(\gamma_1+\gamma_2)K_1\left(\delta\sqrt{\gamma_1\gamma_2}\right)}\times\exp\left[-\frac{1}{2}(\gamma_1+\gamma_2)\times\sqrt{\delta^2+(\phi-\mu_1)^2}+\frac{1}{2}(\gamma_1-\gamma_2)(\phi-\mu_1)\right] \tag{8.63}$$

where γ_1 and γ_2 represent the slopes of two linear asymptotes of $\text{Ln}\, p(\phi;\gamma_1,\gamma_2,\mu_1,\delta)$, μ_1 is the abscissa of the intersecting point of two asymptotes γ_1 and γ_2 (Figure 8.16), $K_1()$ is the modified Bessel function of third kind, and scale parameter $\delta(>0)$ can be expressed as $\delta=2(v-\mu_1)\sqrt{(\gamma_1\gamma_2)/(\gamma_1-\gamma_2)}$ with v is the observed mode of the distribution (Figure 8.16). The slopes γ_1 and γ_2 correspond to Bagnold's (1941) fine grain coefficient and coarse grain coefficient respectively. The parameter δ represents the difference in ordinates of the point of intersection of the asymptotes and the maximum of the parabola. When the variable ϕ follows log-hyperbolic distribution (Figure 8.14b), equation (8.63) can be written as:

$$\text{Ln}\, p(\phi;\gamma_1,\gamma_2,\mu_1,\delta) = v^* - \frac{1}{2}(\gamma_1+\gamma_2)\sqrt{\delta^2+(\phi-\mu_1)^2} + \frac{1}{2}(\gamma_1-\gamma_2)(\phi-\mu_1) \quad (8.64)$$

where

$$v^* = \text{Ln}\frac{\sqrt{\gamma_1\gamma_2}}{\delta(\gamma_1+\gamma_2)K_1\left(\delta\sqrt{\gamma_1\gamma_2}\right)} \quad (8.65)$$

is the norming constant, which is adjusted to agree with the observed frequency. For strongly skewed distributions, v may differ considerably from μ_1 (Figure 8.16). Clearly, $v = \mu_1$ if and only if $\gamma_1 = \gamma_2$. In this case the hyperbolic distribution is symmetric around v. Moreover, the normal distribution is the limiting form of the hyperbolic family when the slope of the left asymptote (γ_1) equals the slope of the right asymptote (γ_2) of the hyperbola and $\delta \to \infty$ while $\delta\sqrt{(\gamma_1\gamma_2)} \to \sigma^2$, σ = standard deviation, δ = scale parameter, which is a measure of spread (sorting). Then the log-hyperbolic distribution of grain size data was verified by Bagnold and Barndorff-Nielson (1980), Ghosh and Mazumder (1981), Barndorff-Nielsen et al. (1982), Christiansen et al. (1984), Wyrwoll and Smyth (1985) and Christiansen and Hartmann (1991) and others in different sedimentary environments and found that the grain size distribution is better improved as log-hyperbolic rather than log-normal. The unimodality or log-normality of the weight frequency distribution of grain sizes was recognized as a transitional phenomenon occurring within a critical range of velocity and height above the sand beds of given composition (Sengupta, 1975a, 1975b, 1979). Ghosh and Mazumder (1981) explained the conditions of uni-modality, symmetry and log-normality of grain size distributions in suspension even with a hyperbolic distribution in the bed. The grain size distribution tends to be log-normal or uni-modal for that size range of bed material for which the settling velocity of particles is linearly distributed in log-scale. Sengupta et al. (1991) re-examined the validity of the conventional method of grain-size interpretation based on the log-probability plots proposed by Visher (1965, 1969), and also investigated the relationship between the lognormal and log-hyperbolic distributions. For this purpose, sediment samples were collected from the natural as well as experimental channel.

8.5.2.3 Skew Laplace Distribution

Fieller and Flenley (1992) introduced "log-skew-Laplace" distribution (Figures 8.14c and 8.15), which is essentially described by two straight lines instead of hyperbola. The log-skew-Laplace distribution of particle size is regarded as the limiting form of the hyperbolic family obtained by letting the scale parameter $\delta \to 0$, and can be written as:

$$g(\phi;\alpha,\beta,\mu_2) = (\alpha+\beta)^{-1}\exp\left\{\left(\frac{\phi-\mu_2}{\alpha}\right)\right\} \text{ for } \phi \le \mu_2 \quad (8.66a)$$

$$= (\alpha+\beta)^{-1}\exp\left\{\left(\frac{\phi-\mu_2}{\beta}\right)\right\} \text{ for } \phi > \mu_2, \text{where } \alpha,\beta > 0 \quad (8.66b)$$

where ϕ indicates the observed size variable and α,β,μ_2 are parameters. The three parameters of the distribution $\alpha,\beta,$ and μ_2 are defined respectively as the reciprocals of the arc tangents of the acute angle made by the two lines with the horizontal axis and the abscissa of their point of intersection. Clearly, α is the slope of the left asymptote i.e. coarser fractions, β is the slope of the right asymptote i.e. finer fractions. The values of α and β were estimated from equations (8.66a) and (8.66b).

Purkait and Mazumder (2000) and Purkait (2002) observed that after a considerable transport distance of sediment on the Usri River in India, the conditions of hyperbolic distributions become log-normal. They also studied the grain size distribution pattern within point bars and between the point bars in the light of log-normal, log-hyperbolic and log-skew-Laplace distribution to identify the best fit model. These results are very helpful to palaeo-hydraulic interpretations of ancient fluvial deposits. Purkait (2006) also observed that irrespective of bed form size, the bed load of Usri River sediments followed log-normal distribution. He found from the successive sampling that the bed roughness changes from source to mouth of the stream being pebbly sands at source and sandy downstream at different water discharges. During this investigation, sediment samples were taken all along the thickness of the fore set of three different categories of bed form height, and found that the lognormal distribution pattern is the best-fit statistical model for different bed form sizes.

8.5.3 Porosity and Specific Weight

Porosity and specific weight are two important parameters in sedimentology and porous media problems.

Porosity is defined as the percentage of pore space in the total bulk volume of sediment. The porosity P is given by

$$P = 100 \times \frac{\text{Bulk volume} - \text{grain volume}}{\text{Bulk volume}} \tag{8.67}$$

Moreover, If V_{void} is the volume of void, V_{solid} is the volume of solid, then the porosity is defined as

$$\text{Porosity} = \frac{V_{\text{void}}}{V_{\text{void}} + V_{\text{solid}}} \tag{8.68}$$

An empirical relation for the porosity of unconsolidated saturated sediment is given by Komura and Colby (1963) as:

$$\text{Porosity} = 0.245 + \frac{0.14}{d_{50}^{0.21}} \tag{8.69}$$

where d_{50} is the size in mm. Modification of Komura's relationship was proposed by Wu and Wang (2006) using the observed data from laboratory and field and is given by:

$$\text{Porosity} = 0.245 + \frac{0.214}{(0.002 + d_{50})^{0.21}} \tag{8.70}$$

The volume of pore and hence the porosity are of considerable importance in the problem related to the storage of oil and groundwater. The porosity of sediment is affected by the size distribution, shape of the particles, state and method of deposition and packing. Generally speaking, fine-grained materials will have higher porosity than the coarse-grained materials.

Specific weight or *dry unit weight* is defined as the dry weight per unit volume of sediment in place. The knowledge of correct specific weight is essential to predict the period between the construction of a reservoir and the time when it becomes so nearly full with sediment that it is no longer useful for storage purpose. This period is called the life of the reservoir.

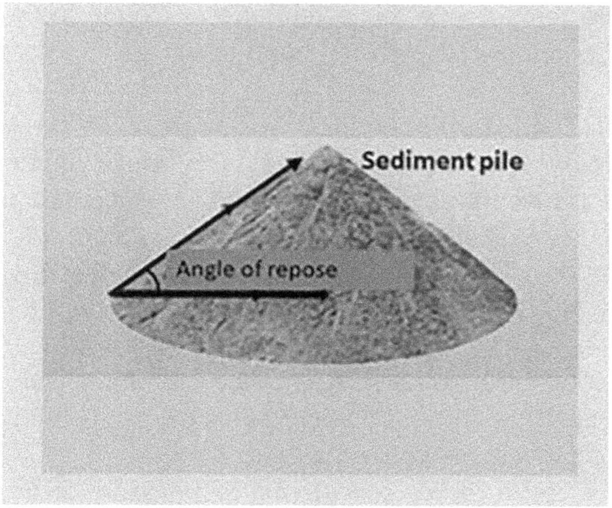

FIGURE 8.17 Angle of repose for a typical sediment pile.

8.5.4 Angle of Repose

The *angle of repose* or *critical angle of repose* of a granular material is the steepest angle of descent relative to the horizontal plane when the particle is on the verge of sliding on the sloping surface of sediment pile. At this angle, the material on the slope face is on the verge of sliding. The angle of repose can range from 0° to 90°. The morphology of the material affects the angle of repose; smooth, rounded sand grains cannot be piled as steeply as can rough, interlocking sands. The angle of repose is equal to the angle of internal friction at the contact of sediment particle. The angle of repose is approximately equal to $\tan^{-1}\mu_f$, where μ_f is the static friction coefficient. The friction coefficient μ_f is described to resist the motion relative to submerge gravity component normal to the sliding. It represents the ratio of tangential resistance force to downward normal force. The angle of repose (Figure 8.17) can be utilized in many applications including the transportation and storage of goods, slope stability, avalanches, barchan dune formation, and many others.

8.6 INCIPIENT MOTION OF SEDIMENT PARTICLES

We consider the case of flow in an open channel of a given slope and with a movable bed made up of uniform non-cohesive sediment material. When experiments are conducted in the channel under steady uniform flow conditions, starting with a very small discharge, it is found that the material comprising the bed is stationary for small discharges, which means, there is no movement of particles. When the discharge is increased to a certain value, it is found that there is a random motion of the individual particles on the bed. In fact, at any given hydraulic condition, some particles move and some do not move. This is owing to the statistical nature of the problem, which implicitly brings out the fact that the flow is turbulent. In other words, the flow condition is such that the sediment particles of given characteristics just start moving. This condition is known as the condition of *critical motion* or the condition of *incipient motion* of sedimentary particles. That depends mostly on the external critical force by which the particle can just initiate dislodge. Vanoni (1964) observed the motion of particles on the bed at low shear stresses, where the particles move in burst frequency condition. A burst frequency may occur between half and one per second at critical condition.

The knowledge of hydraulic conditions at which sediment particles of a given size starts dislodging, is of considerable importance to hydraulic engineers. They are also of great use in studying the mechanism of sediment deposition in reservoirs, surface erosion and in the study of riverbed variation and migration. From a theoretical point of view, the conditions for incipient motion are of great significance because these conditions are associated with the equilibrium of various forces acting on individual particles.

When the sediment bed is coarse and non-cohesive, it is mainly the submerged weight of the sediment particle that resists motion. The coarse particles move as single entities. When the material is fine comprising silts and clays, cohesive forces are predominant, in which, lumps of material move as a unit. It is only in the former case that a satisfactory analysis of the incipient condition exists. Three different approaches are used to establish the condition for incipient motion of sediment particles comprising the bed (Garde and Ranga Raju, 2000). These are discussed below:

1. **Competency**: The size of the bed material is related to either bed velocity or flow velocity, which just causes the particle to move. In other words, the critical fluid velocity equations, considering the impact of fluid velocity on the particle.
2. **Lift concept:** When the upward force considering the pressure differences due to the gradient of velocity is just greater than the submerged weight of the particle, the condition of lift motion is established.
3. **Critical tractive force:** This is based on the idea that the tractive force exerted by the flowing water on the channel bed in the direction of flow is mainly responsible for the motion of the sedimentary particles. In other words, critical shear stress equation; considering the frictional drag of the flow on the particle.

Among these three concepts, the critical tractive force approach is more rational and more sound than others and is now used widely. The competent velocity concept is much simpler, which is still used in the river model works.

8.6.1 Competent Velocity

A competent mean velocity or competent bottom (or bed) velocity is a mean or bed velocity which is just able to dislodge the material of a given size and of a given specific weight. Stream competence refers to the heaviest particles a stream can carry. It depends on both the discharge and the velocity (since the velocity affects the competence and therefore the range of particle sizes that may be transported). As stream velocity and discharge increase so the competence and capacity also increase. The velocity of water is directly affected by channel slope, the greater the slope the greater the flow velocity. In turn, the velocity of water increases the stream competence.

8.6.1.1 Critical Velocity Equations

The condition of incipient movement for non-cohesive, loose solid particles is described in terms of the forces acting on the particle by

$$\tan\varphi = \frac{F_t}{F_n} = \frac{\text{tangential resistive force}}{\text{downward normal force}} \tag{8.71}$$

where F_t and F_n are the forces parallel and normal to the angle of repose φ. Here F_t and F_n are the resultant of the hydrodynamic drag F_D and lift force F_L and the submerged weight W_g. The angle of repose φ is equivalent to the pivoting angle of the superimposed particle resting on the bed particles at a point of contact on which the particle can move. The condition of incipient movement under the action of these three forces becomes from Figure 8.18 as:

$$\tan\varphi = \frac{W_g \sin\alpha + F_D}{W_g \cos\alpha - F_L} \tag{8.72}$$

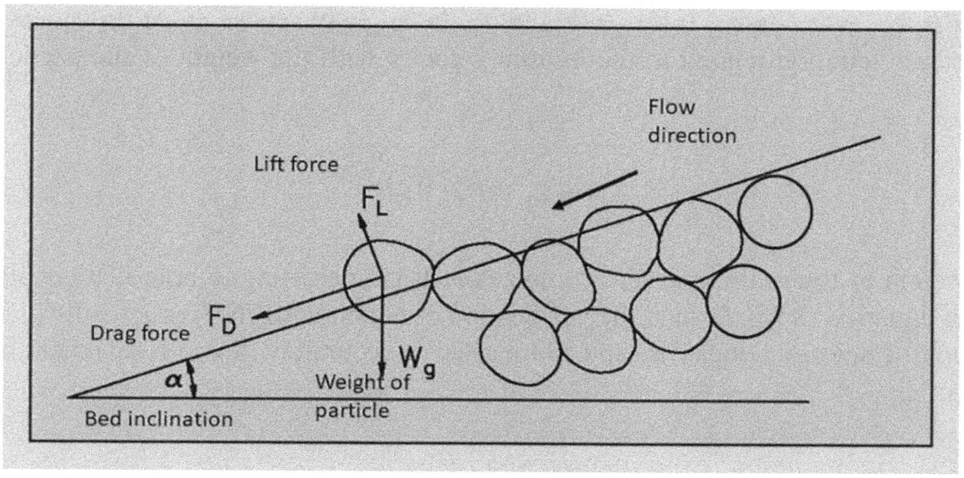

FIGURE 8.18 Force diagram on a particle in loose sediment bed. (Modified from Graf, 2010.)

where α is the angle of inclination of the bed from the horizontal at which incipient sediment movement takes place.

In usual way, the drag and lift forces are respectively expressed as:

$$F_D = C_D K_1 d^2 \frac{\rho_f u_b^2}{2} \tag{8.73}$$

$$F_L = C_L K_2 d^2 \frac{\rho_f u_b^2}{2} \tag{8.74}$$

where u_b is the bottom fluid velocity, C_D and C_L are the drag and lift coefficient respectively, d is the particle diameter, ρ_f is fluid density, and K_1 and K_2 are the particle shape factors. The submerged weight W_g of the particle is:

$$W_g = K_3 \left(\rho_s - \rho_f\right) g d^3 \tag{8.75}$$

where K_3 being the shape factor, ρ_s is the density of solid particle and g is the acceleration due to gravity. Using equations (8.73)–(8.75) in (8.72), we get

$$\frac{\left(u_b^2\right)_{Cr}}{\left(\rho_s/\rho_f - 1\right) g d} = \frac{2K_3\left(\tan\varphi\cos\alpha - \sin\alpha\right)}{C_D K_1 + C_L K_2 \tan\varphi} \tag{8.76}$$

with $\left(u_b\right)_{Cr}$ is the critical bottom velocity at which the incipient motion of the particle takes place. The quantity in the right-hand side (RHS) is referred as the sediment co-efficient A_1 as:

$$A_1 = \frac{2K_3\left(\tan\varphi\cos\alpha - \sin\alpha\right)}{C_D K_1 + C_L K_2 \tan\varphi} \tag{8.77}$$

Here it is observed that the sediment co-efficient A_1 depends on: (1) particle size and shape, size distribution, uniformity, texture; (2) dynamics of flow because this determines the values of C_D and C_L;

(3) channel slope because $\cos\alpha \approx 1$, if $\alpha = 0$; and (4) the angle of repose φ. At this point, Forchheimer (1914) reported an equation related to the bottom velocity with the weight of the particle, and it was expressed as

$$(u_b)_{cr} = K_4 (W_g)^{\frac{1}{6}} \tag{8.78}$$

where W_g is weight of the grain, and K_4 is a constant. It is interesting to note that equation (8.78) is a simple form of equation (8.76). In fact, the verification of equation (8.76) is practically difficult because of the definition of bottom velocity u_b and to measure it accurately. Some researchers have also presented the form as:

$$(u_b)_{cr} = \xi (d)^{\frac{1}{2}} \tag{8.79}$$

which is also similar to equation (8.76) with the value of $\xi \approx 4$ experimentally determined by several researchers.

Mavis et al. (1937) derived an equation (dimension in fps) from about 400 tests, and found the resulting equation as,

$$(u_b)_{Cr} = \frac{1}{2} d^{\frac{4}{9}} \sqrt{\frac{\rho_s}{\rho_f} - 1} \tag{8.80}$$

which is also similar to equation (8.76).

The dependence of the angle of repose φ on sediment particle properties is also shown by Lane (1953), Miller and Byrne (1966) and Zhang et al. (1989). For uniform sediment, angle of repose φ varies between 28° and 32°. In particular, Zhang (1989) proposed a relationship for angle of repose φ for non-cohesive sediment size as:

$$\varphi = 32.5 + 1.27\ d_{50} \tag{8.81}$$

where d_{50} in mm for $0.2 \leq d_{50} \leq 4.4$ mm. For non-uniform sediment, φ varies significantly.

Hjulstrom (1935, 1939) extensively analyzed the data obtained from mono-disperse material on a bed of loose sediment of the same size of particles to understand the erosion, transportation and deposition of sediment materials. He used the average flow velocity because the more correct velocity data were seldom available. He felt that the average velocity is more available than the bottom velocity. He assumed that the average velocity is about 40% greater than the bottom velocity for the flow depth exceeding 100 cm. He conducted his study on a canal with a depth of 100 cm and included grain sizes finer than those considered by Shields (Miedema, 2014). Figure 8.19 shows the limiting region at which incipient motion starts and the line of demarcation between erosion, transportation and deposition (sedimentation). The figure indicates that loose, fine sand is easy to erode, and that the great resistance to erosion in the smallest particle range must depend on cohesion and adhesion forces. His study gives the idea on the subject of erosion, transportation and deposition. The diagram is used extensively by sedimentary geologists (Southard, 2006).

However, its application is limited because a new diagram has to be made for each combination of fluid properties and flow depth. Sundborg (1956) demonstrated the dependence of the critical flow velocity on the water depth, for non-cohesive sediments. Williams (1967) also emphasized the need to

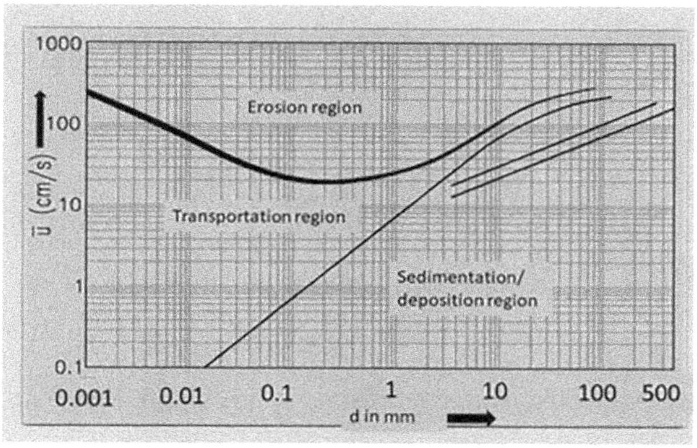

FIGURE 8.19 Hjulstrom diagram: erosion-deposition criteria for uniform particles. (Modified from Hjulstrom, 1935.)

adjust critical conditions for the depth effect. In the Hjulstrom–Sundborg diagram, Sundborg (1956) extended the Hjulstrom diagram by considering separate explicit curves of sediment erosion for four water depths, i.e. 0.01, 0.1, 1 and 10 m.

Barekyan (1963) presented a relationship between the average particle diameter and the mean bottom flow velocity. Jarocki (1963) discussed the formulae (in dimension cm/sec) as:

$$(\bar{u})_{Cr} = 1.4\sqrt{gd}\,\frac{\bar{h}}{7d} \tag{8.82}$$

for relative roughness value $\frac{\bar{h}}{d} > 60$ only, and $\bar{h}$ is the average depth. Carstens (1966) analyzed many published data on incipient motion from a real data set and presented an equation as:

$$\frac{\left(u_b^2\right)_{Cr}}{\left(\frac{\rho_s}{\rho_f}-1\right)gd} \approx 3.61(\tan\varphi\cos\alpha - \sin\alpha) \tag{8.83}$$

Several research papers have been published in the past using the $\bar{u}$ (average fluid velocity) rather than the bottom velocity u_b as the initial motion criteria, only a few are mentioned. Neill (1967, 1968, 1969) presented an equation for scour in the uniform bed materials as:

$$\frac{\bar{u}_{Cr}^2}{\left(\frac{\rho_s}{\rho_f}-1\right)gd} = 2.50\left(\frac{d}{\bar{h}}\right)^{-0.20} \tag{8.84}$$

This equation was established from several experimental data. The criterion to determine the incipient motion of sediment particle using the bottom velocity u_b and average velocity $\bar{u}$ is questionable and criticized by many researchers. Instead of bottom velocity u_b and average velocity $\bar{u}$, many researchers accepted the more satisfactory answers through the bottom shear stress as scour criterion or incipient motion of particle (Rubey, 1938; Lane, 1953). So there exists a certain velocity for each grain, below which it experiences *sedimentation*, while above that certain velocity the particle will erode, which is called *critical scour velocity*. Helley (1969) studied the large particle of diameter $d \approx 33$ cm in the field, and found a good agreement with Hjulstrom's cuve. Garde (1970) developed an equation from a large number of data of incipient condition in hydraulically rough flow regime by competency approach as:

$$(\bar{u})_{Cr} = 1.51\sqrt{\left(\frac{\rho_s}{\rho_f}-1\right)gd} \tag{8.85}$$

Yang (1973) made an excellent review of existing literature, using several experimental data and found some disadvantages of Shields diagram for incipient motion of sediment particles. He developed a new model based on average flow velocity, settling velocity w_s, shear Reynolds number $\left(\frac{u_* d}{\nu_f}\right)$, using the concept of hydrodynamic drag, lift and weight force system on a spherical sediment particle. From the analysis of more than 1,000 data collected from the laboratory flumes and the natural streams published by several authors, he obtained critical velocity equations as:

$$u_{cr}\left(0 < R_{e*} = \frac{u_* d}{\nu_f} < 70\right) = w_s\left(\frac{2.5}{\log R_{e*} - 0.06} + 0.66\right) \tag{8.86a}$$

$$u_{cr}\left(R_{e*} \geq 70\right) = 2.5 w_s \tag{8.86b}$$

In fact, Yang (1973) considered very less data points for the rough bed, i.e., for $R_{e*} = \frac{u_* d}{\nu_f} \geq 70$, the expressions of Neill (equation 8.84) and Garde (equation 8.85) for u_{cr} may be preferred to Yang's equation (equation 8.86b) for critical velocity u_{cr} and settling velocity w_s.

Zanke (1977) suggested an equation for threshold velocity as a function of grain size and kinematic viscosity of water and is given by:

$$(\bar{u})_{Cr} = 2.81\sqrt{\left(\frac{\rho_s}{\rho_f}-1\right)gd} + 14.7 f_1\left(\frac{\upsilon_f}{d}\right) \tag{8.87}$$

where the coefficient f_1 varies from 1 for non-cohesive to 0.1 for cohesive material, and υ_f is the kinematic viscosity of water.

Novak and Nalluri (1975, 1984) investigated experimentally the incipient motion of sediment particles resting on fixed smooth and rough beds. They found individual relationship within the average critical velocity $\bar{u}_{cr}$, the particle size d and the relative density $\left(\frac{\rho_s}{\rho_f}-1\right)$ on smooth fixed beds in circular and rectangular cross-sectional open channel flows over the range of $0.01 < \frac{d}{R} < 1.0$. They also obtained expressions for particle Froude number as a function of shear Reynolds number for both the cross-sectional shapes for incipient motion within the range $10 < R_{e*} < 1000$. Novak and Nalluri (1984) investigated the incipient motion of discrete particles on the beds of various roughness elements, using the same approach as the smooth one; however, data were corrected by Einstein's approach (1950), and the suggested equation for the rough bed is given by

$$\frac{(\bar{u})_{Cr}}{\sqrt{\left(\frac{\rho_s}{\rho_f}-1\right)gd}} = 0.54\left(\frac{d}{R_h}\right)^{-0.38} \tag{8.88}$$

where R_h is the effective hydraulic radius corresponding to the bed. They also showed an interesting relationship between the ratio of incipient motions of particles over smooth and rough beds and the ratio of particle size and bed roughness elements, and the expression is given by

$$\frac{(\bar{u})_{Crs}}{(\bar{u})_{Crr}} = 1 + 1.43\left(\frac{d}{k_s}\right)^{-0.4} \tag{8.89}$$

over a range of $75 > \frac{d}{k_s} > 2$, and $(\bar{u})_{Crs}$, $(\bar{u})_{Crr}$ are respectively the critical velocities of single particles on smooth and rough fixed beds, and k_s is the roughness element of the bed.

Wang and Shen (1985) reviewed the available data and analysis on incipient sediment motion for both unidirectional and oscillating flows according to Shields diagram. They modified the Shields diagram for large and small particle sizes, based on the data collected from China, and the extended Shields diagram was accepted by USACE for incipient motion. Whitehouse and Hardisty (1988) developed a theoretical model for the effect of bed inclination on the threshold of sediment transport for a range of bed slopes, and the model was tested using 672 laboratory experiments. The model suggested a marked decrease in threshold velocities with increase in steeper negative slopes, whereas a general increase for the positive slopes. Diplas et al. (2008) hypothesized that besides the force magnitudes representing the summation of the drag and lift contributions to the particle the duration also should be considered to identify the threshold conditions. The effect of impulse is proposed as the parameter suitable for determining the particle movement.

Gimenez-Curto and Corniero (2009) suggested an explicit dimensionless approach to estimate the incipient motion of granular sediment particles in open channel flow without much complexity. Their model shows the critical dimensionless mean velocity $(\bar{u})_{Cr}$ of a particle size d as a function of bed slope S, angle of repose φ and the specific gravity of the granular sediment, and is given by

$$\frac{(\bar{u})_{Cr}}{\sqrt{\left(\frac{\rho_s}{\rho_f} - 1\right) gd\,\sin(S)}} = 0.78\left[1 + \sqrt{1 + 0.48\left(\frac{\rho_s}{\rho_f} - 1\right)(\cot S - \cot\varphi)}\right]^{3/2} \tag{8.90}$$

One of the advantages of this approach is that it does not require the shear stress and the shear velocity to be calculated. This expression is developed for determining the critical incipient motion condition for granular sediments. Jacobs et al. (2011) suggested a model for erosion threshold of sand–mud mixtures. Miedema (2014) fitted an empirical model to the Hjulstrom diagram. His equation defines the critical velocity for initiation of motion as a function of the grain size in 1 meter water depth:

$$(\bar{u})_{Cr} = 1.5\left(\frac{v_f}{d}\right)^{0.8} + 0.85\left(\frac{v_f}{d}\right)^{0.35} + \frac{9.5(\gamma_s - \gamma_f)d}{\left[1 + 2.5(\gamma_s - \gamma_f)d\right]} \tag{8.91}$$

Rousar et al. (2016) examined the effects of various parameters on the incipient motion of grains, and suggested a theoretical model of lift and drag coefficients to determine the incipient motion of particles, including the effects of velocity fluctuations in addition to the mean velocity. The proposed formulations consider the effects of angle of repose compared well with the experimental data. Woldegiorgis et al. (2018) recommended an explicit analytic expression to determine the incipient motion of consolidated cohesive and non-cohesive sediments, based on the critical erosion idea of the Hjulstrom–Sundborg–Miedema diagram. Generally, the critical stress force needed for the entrainment of consolidated cohesive sediment is much higher than that of non-cohesive sediment. It is generally acknowledged that there is a considerable difference in the tractive force needed to initiate the motion of cohesive and non-cohesive sediments (Fang et al., 2014; Miedema, 2014). As the topic in cohesive sediments is not in the scope of this book, it is not discussed here.

8.6.2 Lift Force Concept

A particle on the bed is lifted when the upward hydraulic force exceeds the submerged weight of the particle (for details, see Chapter 4). When the particle rests on bed, the velocity at the bottom is zero, while the velocity at the top of the particle is positive. As a result, there is a high pressure under the particle and low pressure at the top. This pressure difference creates an upward force and when this upward force exceeds the submerged weight of the particle, the particle is moved upwards. As the particle travels upwards, the velocity difference between the top and bottom of the particle will reduce, thereby reducing the lift. After the particle reaches a certain height, the lift becomes smaller than the submerged weight. The particle then starts falling down during its transport in the forward direction and the particle touches the bed. The same sequence of events repeats and thus the particle travels along a series of curved paths. Jeffreys (1929) showed that a two-dimensional cylinder resting on a bed in an infinite ideal fluid lifted if

$$\left(\frac{1}{3}+\frac{1}{9}\pi^2\right)U^2 > \left(\frac{\rho_s-\rho_f}{\rho_f}\right)ga \tag{8.92}$$

where U is the free stream velocity and a is the radius of the cylinder. For $\dfrac{\rho_s-\rho_f}{\rho_f}=1.65$ and $g = 9.81\ \text{m/s}^2$, the above relation is given by

$$U_{cr}^2 = 11.32\ a \tag{8.93}$$

It is interesting to note that the competent velocity approach and that based on the lift concept give similar results that mean, the square of the competent velocity is proportional to the diameter of the particle. A similar idea was discussed by Reitz (1936) for lift when the circulation and viscosity parameters were included. Lane and Kalinske (1939) stressed the role of turbulence in determining the lift. They assumed the following concepts for lift force: the instantaneous vertical velocity fluctuations near the particles exceed the terminal velocity; the velocity fluctuations follow the Gaussian distribution, and the existence of correlation between the velocity fluctuations and the shear velocity. White (1940) studied the lift of an individual particle and came to a conclusion that the lift is very small as compared to the weight of the particle. However, Einstein and El-Samni (1949) measured the lift force directly as a pressure difference and found that the lift force per unit area of the particle is given by

$$\text{Lift} = \frac{C_L\rho_f\bar{u}^2}{2} \tag{8.94}$$

where $C_L = 0.178$ is the lift coefficient and $\bar{u}$ is the velocity of water at a distance 0.35 d_{35} from the theoretical bed. The value of C_L is valid only for rough bed, and d_{35} is the size for which 35% of the material by weight is finer. The theoretical bed was found to be below the top of the uppermost grains a distance equal to one-fifth the sphere diameter for a bed with hemispheres and 0.20 d_{67} for a gravel bed. They also analyzed the turbulent fluctuations on the lift force. According to Einstein and El-Samni (1949), the dynamic uplift on a single grain subjected to turbulent fluctuations is statistically distributed according to Gaussian distribution. The probabilistic approach to modelling of lift force on sediment transport has been showing a challenge in developing bed load formulas. Einstein (1950) used the probabilistic concept to the lift force to derive the bed load function. The theoretical approach by Iwagaki (1956) showed the similar conclusion as White (1940), that is, the lift is not that important compared to the weight of individual particle. Chepil (1961) pointed out that when the particle starts moving; the lift force tends to decay, and consequently the drag force increases. He showed that the lift and drag ratio is about 0.85 for $47 < ud/\upsilon_f < 5\times10^3$ in wind for hemispherical particle with diameter d. Gessler (1965) showed that the average dynamic uplift is directly proportional to the average bottom shear stress with the aid of logarithmic velocity distribution for open channel flow over a rough boundary, and hence it was also documented that the turbulent fluctuations of the bottom shear stress statistically follow the same distribution, like normal-error law. Coleman (1967) studied the lift forces acting on a plastic and steel spheres individually placed on a theoretical stream bed to determine the variation of lift coefficients with Reynolds number. Apperley (1968) studied a sphere mounted on gravel bed and obtained the lift-to-drag ratio as 0.5 for the roughness Reynolds number $Re_* = \dfrac{k_s u_*}{\upsilon_f} = 70$, where k_s is the roughness parameter, u_* is the shear velocity, and υ_f is the kinematic viscosity. Aksoy (1973) found the lift-to-drag ratio of a sphere of 20 mm diameter is about 0.1 for roughness Reynolds number about 300, while Bagnold (1974) found the ratio is about 0.5 for the Reynolds number value 800 for 16 mm diameter sphere.

Watters and Rao (1971) studied the lift and drag forces on a 95.3 mm diameter sphere placed on different bed configurations and observed negative lift for $20 < Re_* = \dfrac{k_s u_*}{\upsilon_f} < 100$. But Davies and Samad (1978) accounted that the lift force on a sphere adjacent to a boundary becomes negative if both significant underflows occur beneath the sphere and $Re_* < 5$. For $Re_* \geq 5$, however, the lift was found to be positive. Cheng and Chiew (1998) presented a theoretical formulation of the lifting probability for sediment entrainment, which was later modified by Wu and Lin (2002). Both of their works incorporated the probability distribution of instantaneous velocity to explore the relationship between lifting probability and flow condition. The Gaussian and log-normal distributions of instantaneous velocity were adopted in their analyses, respectively. Their optimal choices of a constant lift coefficient were based on the best fitting to the experimental data but not to vary as a function of the flow condition. Moreover, the sediment particle was assumed to lie on a bed of closely packed particles yet such a configuration represented only one of the many possible situations. Wu and Chou (2003) presented the theoretical formulations of the rolling and lifting probabilities of sediment particle in hydraulically smooth and transitional open-channel flows and they found two modes of transportation. This model is an extension of the work of Wu and Lin (2002) considering the derivation of rolling and lifting probabilities in fluctuating velocity and irregularity in grain sizes.

Schmeeckle et al. (2007) measured hydrodynamic lift on a bed sediment particle and found that vertical force correlated poorly with downstream velocity, vertical velocity, and vertical momentum flux whether measured over or ahead of the test particle. Mazumder et al. (2008) obtained a most striking feature that the fluctuation of the angle of orientation of particle from one frame to another also follows the Gaussian distribution, which is closely related to the fluctuating shear stress (Gessler, 1965) because shear stress is differential convection to the fluid layers which causes the rotation to the particle. Therefore, the experiments revealed that the average lift force superimposed with random fluctuations follows the normal-error law. Dwivedi et al. (2011) made an excellent experimental study to estimate the lift force of a spherical sediment particle at different exposures. In their work, Bhattacharyya et al. (2013) extended the Mazumder et al. (2008) model, considering the lifting effect caused by circulation, called the Magnus effect, generated at the rough boundary, which is imperative for better modeling of particle movement over the rough bed. A particle at the bed surface will initiate to move, when the friction force for the particle moving along the bed is exactly balanced by the sum of the drag force and the appropriate component of gravitational force in a sloping bed. In a shear flow, the lift force caused by the fluid velocity gradient is termed as lift due to shear effect F_{Ls}. For a spherical particle in a viscous flow, Saffman (1965, 1968) derived a lift due to shear effect as:

$$F_{Ls} = C_L \rho_f d^2 \left(\bar{u} - u_p\right)\left(\upsilon_f \frac{\partial \bar{u}}{\partial y}\right)^{0.5} \tag{8.95}$$

where C_L is the lift coefficient, and the term $\left(\bar{u} - u_p\right)$ is relative fluid velocity, in which u_p is the mean particle velocity in the x-direction and $\bar{u}$ is the mean fluid velocity. Finally, Bhattacharyya et al. (2013) found analytically the total lift force F_L is a combination of lift due to shear and Magnus effect (Rubinow and Keller, 1961) and is given by:

$$F_L = C_L \rho_f d^2 \left(\bar{u} - u_p\right)\left(\frac{\partial \bar{u}}{\partial y}\right)^{0.5}\left[\upsilon_f^{\,0.5} + 0.5 \operatorname{sgn}(\mathrm{Re}_*) d \left(\frac{\partial \bar{u}}{\partial y}\right)^{0.5}\right] \tag{8.96}$$

where $\operatorname{sgn}(\mathrm{Re}_*) = 1$ for $\mathrm{Re}_* > 1, \operatorname{sgn}(\mathrm{Re}_*) = 0$ for $\mathrm{Re}_* \le 1$ and Re_* is the frictional Reynolds number.

8.6.3 Critical Tractive Force

Out of three approaches of incipient motion of particles, such as competency, lift force and critical tractive force, it is the critical tractive force which has gained most importance to be rational and is discussed here. This approach is based on the consideration of equilibrium of a sediment particle resting on the bed under the action of drag and lift forces caused by the flowing fluid and the submerged weight of the particle. Before starting the critical tractive force, the average shear stress is discussed in detail because the conception of average bottom shear stress is necessary to derive the critical tractive force.

8.6.3.1 Average Shear Stress

Let us consider steady uniform flow in a wide rectangular channel and consider equilibrium of a water prism $ABCD$ under various forces acting on it (Figure 8.20) with smooth water surface. Since there is no acceleration of the fluid, the summation of all the forces acting in the direction of flow must be zero.

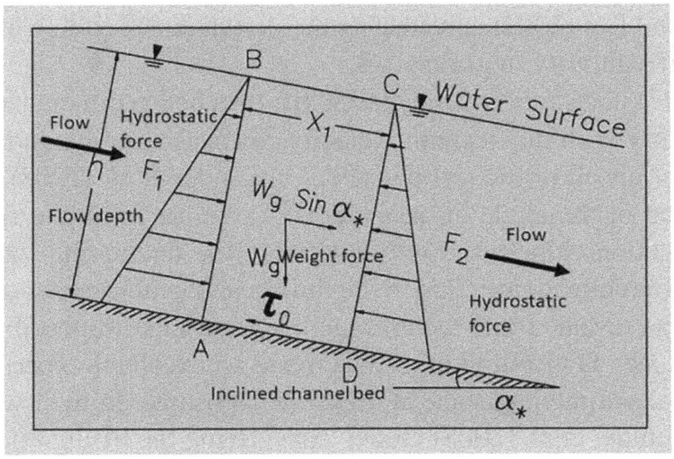

FIGURE 8.20 Forces acting on a water prism.

The forces acting on the prism *ABCD* are: hydrostatic forces, weight force and resistance due to boundary. Resistance between water-air interface is usually very small and can be easily neglected for smooth water surface. Hence, the forces acting on the water prism are as:

$$\Sigma F = F_1 + W_g \sin\alpha_* - F_2 - \tau_0 \times \text{wetted area} = 0 \tag{8.97}$$

where τ_0 is the average shear stress at the boundary, F_1 and F_2 are the hydrostatic forces and w_g is the weight of the water prism. Since the depth of flow is the same at sections *AB* and *CD*, $F_1 = F_2$ and $\tau_0 = W_g \sin\alpha_*$/wetted area. But wetted area is equal to $(b+2h)x_1$, where b is channel width, h is the water depth and x_1 is the length of the prism. Also, weight force $w_g = bhx_1\gamma_f$, where γ_f is the specific weight of the fluid. Therefore

$$\tau_0 = \frac{bhx_1\gamma_f \sin\alpha_*}{(b+2h)x_1} \tag{8.98}$$

Or,

$$\tau_0 = \gamma_f \, R_h \sin\alpha_* \tag{8.99}$$

in which $R_h = \dfrac{bh}{b+2h}$ is the hydraulic radius. For small α_*, $\sin\alpha_* \approx \tan\alpha_* = J$ is the channel slope. Therefore, the average shear stress τ_0 from equation (8.99) is given by:

$$\tau_0 = \gamma_f R_h J \tag{8.100}$$

For very wide channel $R_h \approx h$, and hence $\tau_0 = \gamma_f hJ$ for such channels. The force exerted by water on the channel bed will have the same magnitude but will act on the direction of flow. This shear force can be directly related to the velocity distribution near the boundary and viscosity of the fluid.

8.6.4 Critical Shear Stress

The relationship between the weight component of water column and the friction force at the bottom $\rho_f u_b^2$ was derived by Forchheimer (1914) and can be written as

$$\gamma_f dJ = K_4 u_b^2 \tag{8.101}$$

where h=water depth, J=slope of energy grade line, K_4=a constant (dimension $\dfrac{M}{L^3}$). The expression $\gamma_f hJ$ is the tractive force per unit surface, is written as

$$\tau_0 = \gamma_f hJ \tag{8.102}$$

This equation may be replaced by $\tau_0 = \gamma_f R_h J$, where R_h is hydraulic radius. Using (8.101) and (8.102), the critical velocity equation (8.82) can be written as

$$\frac{\rho_f \left(u_b^2\right)_{Cr}}{\left(\rho_s - \rho_f\right)gd} = \frac{\left(\tau_0\right)_{Cr}}{\left(\gamma_s - \gamma_f\right)d} = A''(\text{say}) \tag{8.103}$$

where A'' is the sediment coefficient, $\tau_0 = K_4 u_b^2$, and $(\tau_0)_{Cr}$ is critical shear stress or critical drag force at incipient motion.

Schoklitsch (1914) proposed an equation based on his experimental data as:

$$(\tau_0)_{Cr} = \sqrt{0.201\gamma_f(\gamma_s - \gamma_f)\lambda' d^3} \ (\text{in Kg/m}^2) \tag{8.104}$$

where d is the mean grain diameter and λ' is the shape coefficient ranging from $\lambda' = 1$ for spheres to $\lambda' = 4.4$ for flat grain. For grain of diameter $d \geq 0.006$ m, an equation was established by Krey (1925) and is given by

$$(\tau_0)_{Cr} = 0.076(\gamma_s - \gamma_f)d \ (\text{in Kg/m}^2) \tag{8.105a}$$

and for $0.0001\,\text{m} < d < 0.003\,\text{m}$

$$(\tau_0)_{Cr} = 0.00285\,(\gamma_s - \gamma_f)d^{\frac{1}{3}} \ (\text{in Kg/m}^2) \tag{8.105b}$$

Kramer (1935) and Tiffany and Bentzel (1935) suggested equations from their own experimental data and is given by:

$$(\tau_0)_{Cr} = 29\sqrt{(\gamma_s - \gamma_f)d/M} \ (\text{in gm/m}^2) \tag{8.106}$$

where M is the uniformity coefficient defined of Kramer (1935). This equation is based on the data of size diameter d ranging from 0.24 mm to 6.52 mm, and the uniformity coefficient M varying from 0.265 to 1.00.

A relation between $(\tau_0)_{Cr}$ and the size d has been developed by Chang (1939) for different ranges of grain sizes, and is given by

$$(\tau_0)_{Cr} = 0.0045\left(\frac{\rho_s - \rho_f}{\rho_f}\frac{d}{M}\right)^{\frac{1}{2}} \quad \text{for} \left(\frac{\rho_s - \rho_f}{\rho_f}\frac{d}{M}\right) > 2.0 \tag{8.107a}$$

$$(\tau_0)_{Cr} = 0.0064\left(\frac{\rho_s - \rho_f}{\rho_f}\frac{d}{M}\right) \quad \text{for} \left(\frac{\rho_s - \rho_f}{\rho_f}\frac{d}{M}\right) \leq 2.0 \tag{8.107b}$$

This Chang equation is based on the data of uniformity coefficient M ranging from 0.23 to 1.0, the relative density changing from 2.05 to 3.89, and the sediment size d varying from 0.134 mm to 8.09 mm. Chang suggested first that the law of critical tractive force varies according to the grain size, and this idea was confirmed by Shields, Kurihara, Iwagaki and others.

Leliavsky (1955) presented the shear stress and grain size relationship as a simple relation as

$$(\tau_0)_{Cr} = 166d \ (\text{g/m}^2) \tag{8.108}$$

where d is the mean grain diameter in mm (Figure 8.21). This equation (8.108) shows justice to all experimental data and it is simple to use. This equation should not be applied if the mean grain diameter d is greater than 3.4 mm.

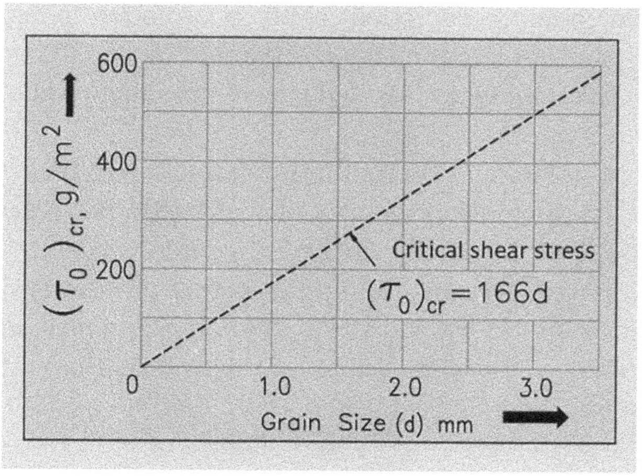

FIGURE 8.21 Critical shear stress as a function of grain diameter. (Modified from Graf, 2010.)

It is seen that the above empirical formulae have some similarities. The critical tractive force depends on the relative density $\frac{\rho_s - \rho_f}{\rho_f}$ and the size of the particle. But these formulae do not take into account the fluid viscosity. Here some formulae are based on the condition of isolated particle movement, and some are based on the general movement of particles. In fact, the above empirical equations can yield the approximate critical tractive stress, but these methods are not directly adequate for estimation of the critical tractive stress. Moreover, these formulae do not through any light on the sediment particle movement mechanism.

Example 8.1

A canal to be designed to carry 50 m^3/sec discharge of clear water without any scour. The canal can be rectangular in cross section with a bed slope of 0.01. The vertical side can be protected with wooden planks. The soil analysis reports show that bed soil is made up of quartz gravel of 50 mm approximate size with a Manning's roughness coefficient of 0.02. Design the canal cross section by considering the critical velocity consideration from Hjulstrom diagram.

Solution

From, Hjulstrom diagram Figure 8.19, for scour criterion, d=50 mm, the critical velocity can be approximately taken as 2.5 m/s. As discharge (Q) is given as 50 m^3/sec, the approximate area of cross section $A = Q/u_{cr}$ = 50/2 5 = 20 m^2.

Now from Manning's equation, $u = \frac{1}{n} R^{\frac{2}{3}} s^{\frac{1}{2}}$; where n is Manning's roughness; R is hydraulic radius and s is bed slope. Here u = 2.5 m/s; n = 0.02 and s = 0.01.

Now solving for R, we will get R = 0.353. Further, $R = A/P$, where P is the perimeter (width B+2×depth (h)).

Therefore, we can get approximate width B as 55.942 m and depth h as 0.357 m.

8.7 SEMI-EMPIRICAL EQUATIONS FOR INCIPIENT MOTION

Several semi-empirical equations have been developed by many investigators (Shields, 1936; White, 1940; Kurihara, 1948; Iwagaki, 1956; Egiazaroff,,1965; and others) to find the incipient motion of sediment particles. All of them were considered the sediment material to be uniform. An accurate determination of the critical shear stress for initiation of motion of the sediment particles in alluvial streams is vital for erosion and sedimentation predictions. This information is necessary for determining the flow strength at which a given size of sediment will begin to move, and is a necessary input to many sediment-transport relations to calculate excess bed shear stress and predict the rate of sediment transport at shear stresses above the critical.

Shields (1936) was the first to study the motion of sediment particles after considering the forces acting on the particle and then applying principles of similarity. His results are very interesting and often quoted and widely used. His pioneering work has inspired several researchers to conduct further studies in many directions. His important findings represent the variation of dimensionless critical shear stress (threshold stress) with the friction/shear Reynolds number corresponding to the sediment threshold. Before this pioneering work by Shield (1936), only empirical relations were there with limited applications.

8.7.1 Shields' Analysis (1936)

For particles of size d and of specific weight γ_s in an ambient fluid of specific weight γ_f, the force required to move the particle is given by

$$F_g = \alpha_1 (\gamma_s - \gamma_f) d^3 \tag{8.109}$$

in which α_1 depends on the sediment characteristics. Similarly, the hydrodynamic force exerted by the fluid on the particle is

$$F_D = C_D \rho_f \frac{u_a^2}{2} \alpha_2 d^2 \tag{8.110}$$

in which u_a is the characteristic velocity, C_D is the drag coefficient of the particle at the Reynolds number corresponding to u_a, and α_2 is a coefficient such that $\alpha_2 d^2$ gives the projected area of the particle. The velocity u_a is considered as the velocity at the top of the particle u_d. Under Karman-Prandtl equation for velocity distribution, the velocity u_d can be expressed as

$$\frac{u_d}{u_*} = f_1\left(\frac{u_* d}{\nu_f}\right) \Rightarrow u_d = u_* f_1\left(\frac{u_* d}{\nu_f}\right) \tag{8.111}$$

in which u_* is the shear velocity $= \sqrt{\tau_0 / \rho_f}$. The shear/friction velocity u_* represents a measure of intensity of turbulent fluctuations. Since $C_D = f_2\left(\frac{u_d d}{\nu_f}\right)$, it follows that

$$C_D = f_2\left(\frac{u_d d}{\nu_f}\right) = f_2\left(\frac{u_d}{u_*}\frac{u_* d}{\nu_f}\right) = f_2\left\{f_1\left(\frac{u_* d}{\nu_f}\right)\frac{u_* d}{\nu_f}\right\} = f_3\left(\frac{u_* d}{\nu_f}\right) = f_3\left(R_e^*\right) \tag{8.112}$$

where $R_e^* = \dfrac{u_* d}{\nu_f}$ is the frictional Reynolds number. Using equation (8.112) for C_D in equation (8.110), the hydrodynamic force F_D is given by

$$F_D = f_3\left(R_e^*\right) \frac{\rho_f}{2} u_*^2 f_1^2\left(R_e^*\right) \alpha_2 d^2 \tag{8.113}$$

Equating F_g and F_D from equations (8.109) and (8.113) for condition of incipient motion and introducing the subscript *cr* for critical, one gets

$$\alpha_1\left(\gamma_s-\gamma_f\right)d^3 = f_3\left(R^*_{e_{cr}}\right)\rho_f \frac{u_{*c}^2}{2} f_1^2\left(R^*_{e_{cr}}\right)\alpha_2 d^2 \tag{8.114}$$

After simplification, equation (8.114) can be written as

$$\frac{\rho_f u_{*c}^2}{\left(\gamma_s-\gamma_f\right)d} = \frac{2\alpha_1}{\alpha_2}\left[\frac{1}{f_3\left(R^*_{e_{cr}}\right)}\cdot\frac{1}{f_1^2\left(R^*_{e_{cr}}\right)}\right] = \frac{2\alpha_1}{\alpha_2} f\left(R^*_{e_{cr}}\right) \tag{8.115}$$

where $R^*_{e_{cr}} = \frac{u_{*c}d}{\nu_f}$ and u_{*c} is the shear velocity at the incipient motion. Thus, for particles of a given shape for critical and incipient condition, one gets the expression for dimensionless critical shear stress θ_c for a particle of given size as:

$$\frac{\tau_{0c}}{\left(\gamma_s-\gamma_f\right)d} = f\left(R^*_{e_{cr}}\right) \tag{8.116}$$

where $R^*_{e_{cr}}$ is the critical Reynolds number with u_{*c} as the critical shear velocity. The variation of $\frac{\tau_{0c}}{\left(\gamma_s-\gamma_f\right)d}$ with shear Reynolds number $R^*_{e_{cr}}$ was obtained by Shields based on experimental data collected in a flume with fully developed turbulent flow (Figure 8.22), using the sediment size ranging from 0.40 mm to 3.4 mm. This is popularly known as *Shields' diagram* and is widely accepted to use in solving the problems of sediment transport.

Shields (1936) suggested a special relation in modern fluid dynamics where the sediment coefficient becomes:

$$\frac{\left(\tau_0\right)_{Cr}}{\left(\gamma_s-\gamma_f\right)d} = f\left(\frac{du_*}{\nu_f}\right) \tag{8.117}$$

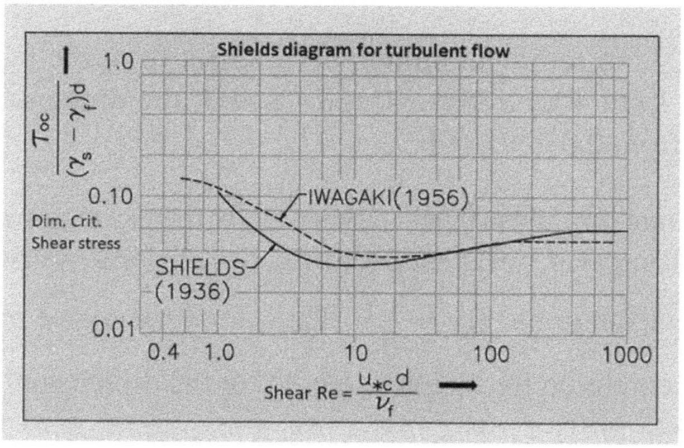

FIGURE 8.22 Shields diagram (condition of incipient motion by Shields and Iwagaki). (Modified from Garde and Rangaraju, 2000.)

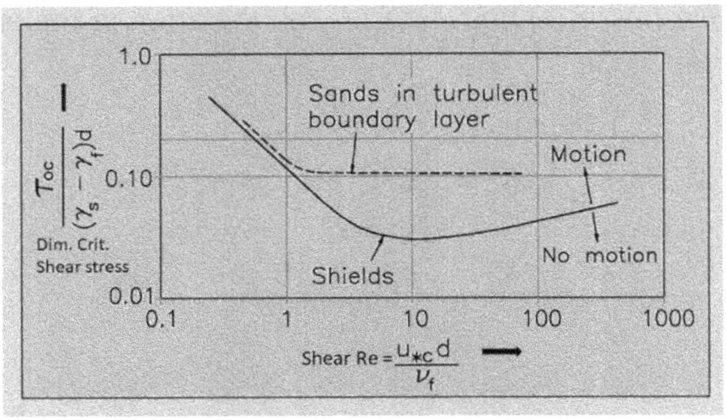

FIGURE 8.23 Shields diagram: dimensionless critical shear stress vs. shear Reynolds number. (Modified from Graf, 2010.)

where $\frac{du_*}{\nu_f}$ is referred to as the shear or friction Reynolds number R_{e*}. To this development, it is necessary to express the characteristic velocity with laws of logarithmic velocity distribution. It might be expressed as the shear Reynolds number R_{e*} in terms of the laminar sub-layer δ_l as:

$$R_{e*} = \frac{u_*d}{\nu_f} = \frac{11.6\ d}{\delta_l} \tag{8.118}$$

From the Shields diagram (Figure 8.23), three distinct regions are noticed:

1. If $d < \delta_l$, upto $R_{e*} \approx 2$, the particles are completely submerged in laminar sub-layer, their movement is under the laminar sub-layer, which is due to viscous action and independent of turbulence. This is similar to smooth boundary layer flow.
2. $d \approx \delta_l$: when the thickness of laminar sub-layer is of the same order as the diameter of the particle, the turbulent eddies disturb the laminar sub-layer, and affect the flow around the particle. At intermediate Reynolds number within $2.5 < R_{e*} < 40$, representing the dip in the curve, there is a zone of transition where the laminar sub-layer partially covers the particles and partially is interrupted. This transition curve has a minimum at $R_{e*} = 10$. Shield's diagram indicates as:

$$\left.\frac{(\tau_0)_{Cr}}{(\gamma_s - \gamma_f)d}\right|_{\text{mini}} \approx 0.03 \tag{8.119}$$

Below this value 0.03, no sediment motion occurs. But from the figure it is seen that within the range of R_{e*} from 40 to 400, that is, $40 \leq R_{e*} \leq 400$, the dimensionless critical (threshold) shear stress increases from 0.038 to 0.06.

3. When $d \gg \delta_l$, the laminar sub-layer δ_l gets destroyed by the existing bed roughness of the grain. The roughness is the source of turbulence and the dimensionless critical shear stress is independent of shear Reynolds number $R_{e*} = \frac{u_*d}{\upsilon_f}$. Thus, for very coarse material at large Reynolds number $R_{e*} \geq 400$, the condition for incipient motion of the dimensionless critical shear stress is as:

$$\frac{(\tau_0)_{Cr}}{(\gamma_s - \gamma_f)d} = 0.06, \tag{8.120}$$

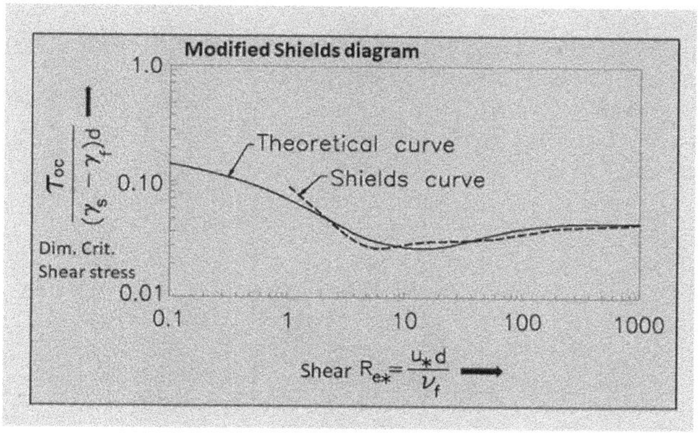

FIGURE 8.24 Shields' diagram with modification. (Modified from Wiberg and Smith, 1985.)

However, Zeller (1963) obtained a constant value to be high as 0.047. Shields diagram is widely accepted because this critical tractive force is verified using lots of experimental data. Since Shields diagram was developed for uniform sand, the non-uniform sand material and sticky or flocculent material give higher critical shear stress $(\tau_0)_{Cr}$ values. In the figure, the grain diameter appears in both ordinate and the abscissa. The diameter d is the representative size, which was taken as the median size, that is, the sieve size for which 50% of the mixture material was finer or coarser. Bogardi (1965) analyzed several data (Russian and Hungarian data) and found a general trend given by the Shields diagram. However, a minimum of dimensionless critical shear stress was shown around 0.015. Some of the limitations of Shields' criterion have been pointed out by Gessler (1965) and Neill (1968) and discussed later. Shields' diagram was subsequently modified by several researchers Miller et al. (1977) and Wiberg and Smith (1985, 1987), Ling (1995) and others. Figure 8.24 shows the modified Shields' curve by Wiberg and Smith (1985).

8.7.2 White's Analysis (1940)

For equilibrium of a single grain resting on a granular bed, an expression for the critical tractive stress was obtained by White (1940). He assumed only the equilibrium of two forces: the submerged weight of the particle and the fluid drag on the particle in the direction of flow, neglecting the lift component of the hydraulic forces. Depending on the shear Reynolds number $\frac{u_* d}{\upsilon_f}$ > or <3.5, he categorized into two sections of analysis: one for Reynolds number $\frac{u_* d}{\upsilon_f} \geq 3.5$, transitional and rough flow regimes; and other one for Reynolds number $\frac{u_* d}{\upsilon_f} < 3.5$, smooth flow regime to analyze the threshold of particle motion.

For hydraulically transitional and rough regimes $\left(\frac{u_* d}{\upsilon_f} \geq 3.5\right)$, the velocity should be large to initiate the larger sediment particle, where the drag action due to skin friction is small compared to the pressure difference. The resultant force due to pressure difference passes through the center of gravity of the spherical particle and is directed to the flow direction (Figure 8.25). Here the packing coefficient η_t is defined as d^2 times the number of particles per unit area, i.e. $\eta_t = Nd^2$, where N is the number of particles per unit area, the shear force per particle is given by $\frac{\tau_0 d^2}{\eta_t} = \frac{\tau_0}{N}$. A sediment particle resting on

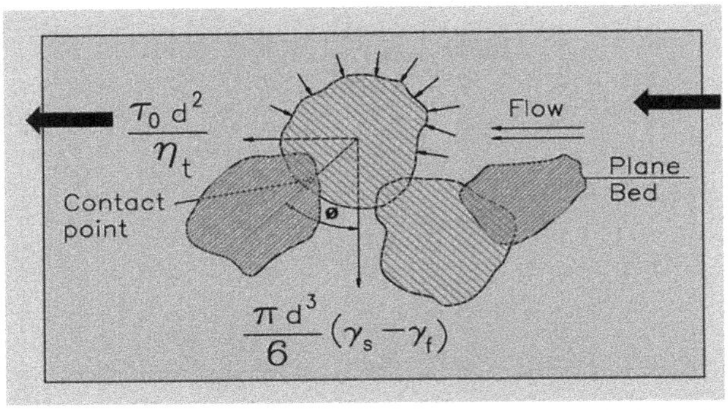

FIGURE 8.25 Force conditions for large velocity case, $\frac{u_* d}{\upsilon_f} \geq 3.5$. (Modified from Garde and Rangaraju, 2000.)

a horizontal bed, when the shear drag is balanced by submerged weight force $\frac{\pi}{6}(\gamma_s - \gamma_f)d^3$, the condition of incipient motion of a particle is given by

$$\frac{\tau_{0c}}{(\gamma_s - \gamma_f)d} = \frac{\pi}{6}\eta_t \tan\phi \tag{8.121}$$

where $\tan\phi$ is the coefficient of friction with ϕ being the angle of repose. White (1940) conducted two individual experiments on the movement of sand particles to verify equation (8.121):- one flow through a converging nozzle and the other one in fully developed flow in parallel walled flume, using the Reynolds number $\frac{u_* d}{\upsilon_f}$ ranging from 33 to 1280. Then he introduced a factor, called the *turbulence factor* T_f from the semi-theoretical analysis of velocity fluctuations in the turbulent flow. The turbulence factor T_f is defined as the ratio of the instantaneous bed shear stress to the mean shear, and hence equation (8.121) can be written as for $\frac{u_* d}{\upsilon_f} \geq 3.5$:

$$\frac{\tau_{0c}}{(\gamma_s - \gamma_f)d} = \frac{\pi}{6}\frac{\eta_t}{T_f}\tan\phi \tag{8.122}$$

where the turbulence factor $T_f = 2.0$ for turbulent boundary layer flow through a nozzle, and $T_f = 4.0$ for fully developed turbulent flow through an open channel with the value of η_t as 0.4. Using $\eta_t = 0.4$, $T_f = 4.0$ and $\tan\phi = 1$ in the above equation for open channel turbulent flow, one gets for $\frac{u_* d}{\upsilon_f} \geq 3.5$ as:

$$\frac{\tau_{0c}}{(\gamma_s - \gamma_f)d} = 0.052 \text{ for } \frac{u_* d}{\upsilon_f} \geq 3.5 \tag{8.123}$$

For hydraulically smooth flow regime ($u_* d/\upsilon_f < 3.5$), the low velocity was required to move the smaller particles, where the pressure forces are very small compared to the viscous force. Since the tangential force on the surface of the particle is exposed, one would expect the force $\frac{\tau_0 d^2}{\eta_t}$ to pass through a point above the center of gravity (CG) of the particle (Figure 8.26). If a coefficient α_t is introduced to consider this effect, the equation of sediment particle threshold for equilibrium condition is given by:

$$\frac{\tau_{0c}}{(\gamma_s - \gamma_f)d} = \frac{\pi}{6}\alpha_t\eta_t \tan\phi \tag{8.124}$$

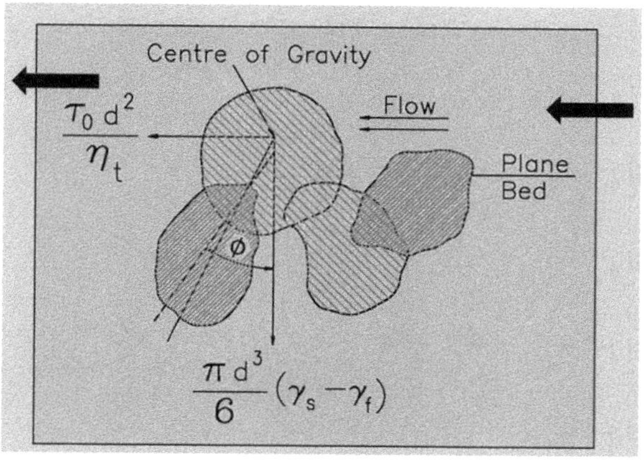

FIGURE 8.26 Force conditions for low velocity $\frac{u_* d}{\upsilon_f} < 3.5$. (Modified from Garde and Rangaraju, 2000.)

Equations (8.121) and (8.124) can be modified, when the equilibrium of a sediment particle is considered in a sloping bed inclined at an angle θ with the horizon, equation can be written as:

$$\frac{\tau_{0c}}{(\gamma_s - \gamma_f)d} = \frac{\pi}{6}\alpha_t \eta_t \cos\theta(\tan\phi - \tan\theta) \tag{8.125}$$

However, White (1940) obtained the average value of $\alpha_t \eta_t$ as 0.34, the expression for critical tractive stress for $\frac{u_* d}{\upsilon_f} < 3.5$ as:

$$\frac{\tau_{0c}}{(\gamma_s - \gamma_f)d} = 0.18 \tan\phi \tag{8.126}$$

Equation (8.126) does not satisfy the Shields criterion for $\frac{u_* d}{\upsilon_f} < 3.5$, but according to Shields the critical tractive stress should be inversely proportional to $\frac{u_* d}{\upsilon_f}$.

8.7.3 Kurihara's Analysis (1948)

Kurihara attempted to extend the work of White and determined the relationship between the turbulence factor T_f and the shear Reynolds number $\frac{u_* d}{\upsilon_f}$. It is noted that according to White, the value of T_f was 4.0 for fully developed turbulent open channel flow, but according to Shields, it was not constant, it would be function of $\frac{u_* d}{\upsilon_f}$. He considered the time-averaged shear stress due to the main flow and the shear stress resulting from the turbulent velocity fluctuations. In addition, considering the fluctuations in pressure, he developed an expression for T_f, which is a function of Reynolds number $\frac{u_* d}{\upsilon_f}$, rms value and probability p of shear stress increment. He showed for different p values, the turbulence factor T_f vs Reynolds number $\frac{u_* d}{\upsilon_f}$ curve showed a similar pattern to Shields curve. He showed several empirical expressions for critical tractive stress and given in Garde and Ranga raju (2000).

Tison (1953) in his experiments of fully developed open channel turbulent flow indicated that the turbulent factor T_f should not be constant; it should vary with Reynolds number and the flow depth. He found the value of T_f varied from 2.46 to 7.00.

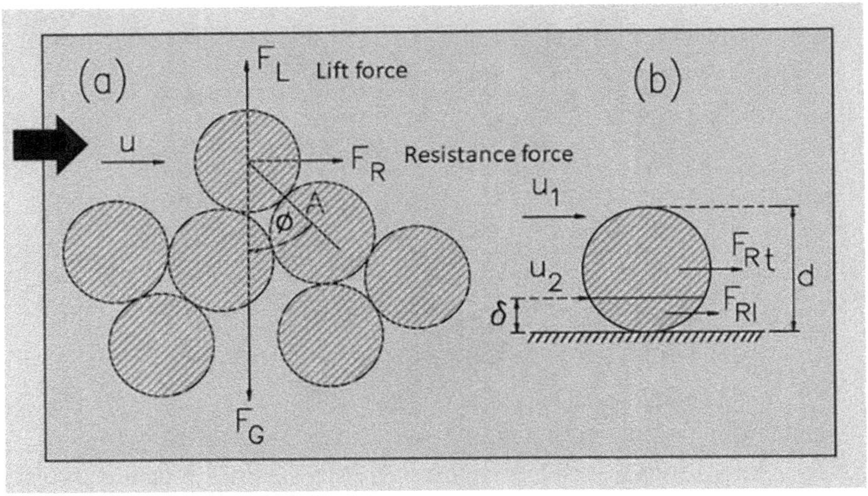

FIGURE 8.27 Force conditions of Iwagaki's analysis: (a) forces on a spherical particle, and (b) fluid forces on a spherical particle. (Modified from Gard and Rangaraju, 2000.)

8.7.4 Iwagaki's Analysis (1956)

The equilibrium of a single particle spherical in shape having a diameter d placed on a rough bed surface was considered by Iwagaki (1956) to determine the necessary conditions for the initiation of the movement of the particle. He considered the forces acting on the particle as lift force F_L due to pressure difference in the vertical direction, resistance force F_R due to turbulent flow and laminar flow and the submerged weight force F_G of the particle (Figure 8.27a and b).

The total resistance F_R can be divided into resistance due to turbulence F_{Rt} and resistance due to laminar flow F_{Rl}, which can be written as:

$$F_R = F_{Rt} + F_{Rl} \tag{8.127}$$

The magnitudes of these two components depend on the viscous sublayer δ in the laminar flow. It is known that when the sublayer $\delta \gg d$, then resistance due to turbulence $F_{Rt} = 0$, and if $\delta \ll d$, then $F_{Rl} = 0$. If $\frac{\pi d^2}{4}\beta_e$ is the area exposed to the turbulent flow and $\frac{\pi d^2}{4}(1-\beta_e)$ be area exposed to the laminar flow, then, F_{Rt} and F_{Rl} are respectively given by

$$F_{Rt} = C_{D1}\frac{\rho_f}{2}u_1^2\beta_e\frac{\pi d^2}{4} - \left(\frac{\partial p}{\partial x}\right)d\beta_e\frac{\pi d^2}{4} \tag{8.128}$$

$$F_{Rl} = C_{D2}\frac{\rho_f}{2}u_2^2(1-\beta_e)\frac{\pi d^2}{4} \tag{8.129}$$

where u_1 and u_2 are respectively velocities in x-direction at a distance $y = d$ and δ, and C_{D1} and C_{D2} are respectively the drag coefficients corresponding to the velocities u_1 and u_2. Here, the pressure gradient $\left(\frac{\partial p}{\partial x}\right)$ can be found considering Euler's equation of motion as follows:

$$-\left(\frac{\partial p}{\partial x}\right) = \rho_f\frac{D\bar{u}}{Dt} \tag{8.130}$$

in which $\frac{D\bar{u}}{Dt}$ is the total acceleration with $\bar{u}$ is the stream-wise velocity component and t is the time. Using equation (8.130) in equation (8.128), the expression of F_{Rt} is given by:

$$F_{Rt} = C_{D1}\frac{\rho_f}{2}u_1^2\beta_e\frac{\pi d^2}{4} + \rho_f\beta_e\frac{\pi d^3}{4}\frac{D\bar{u}}{Dt} \tag{8.131}$$

The expression for lift force F_L due to pressure gradient in the vertical direction is as follows:

$$F_L = -\left(\frac{\partial p}{\partial y}\right)(d-\delta)\frac{\pi d^2}{4} = -\rho_f\frac{\pi d^2}{4}(d-\delta)\frac{Dv}{Dt} \text{ when } d > 2\delta \tag{8.132}$$

$$F_L = -\rho_f\pi\delta(d-\delta)(d-\delta)\frac{Dv}{Dt} \text{ when } d < 2\delta \tag{8.133}$$

where v is the vertical velocity component. Using these general expressions of F_{Rt}, F_{Rl} and F_L, Iwagaki derived three special cases depending on the shear Reynolds number $\frac{u_* d}{\upsilon}$.

Case I: $d \le 2\delta$ or $\frac{u_* d}{\upsilon_f} \le 6.83$

In this case, the sand particle is completely submerged in the viscous sub-layer, which corresponds to smooth flow regime. It means that. $F_{Rt} = 0$, $F_L = 0$, and $\beta_e = 0$. Taking into consideration velocity distribution in smooth flow regime, and the drag coefficient C_{D2} is a function of Reynolds number, i.e., $C_{D2} = C_{D2}\left(\frac{u_* d}{\upsilon_f}\right)$. From the condition of equilibrium, the critical tractive stress is given by (Garde and Rangaraju, 2000):

$$\frac{\tau_{0c}}{(\gamma_s - \gamma_f)d} = C_1 \tan\phi f_1\left(\frac{u_* d}{\upsilon_f}\right) \tag{8.134}$$

where C_1 is an empirical coefficient. This equation holds good for $\frac{u_* d}{\upsilon_f} \le 6.83$.

Case II: $\frac{u_* d}{\upsilon_f} \ge 51.1$

In this case, the laminar sub-layer is vanished due the roughness; the whole particle is exposed to the main flow, the turbulent flow is dominant in the vicinity of the sand particle, so $F_{Rl} = 0$, $\delta = 0$, and $\beta_e = 1$. Then equations (8.131) and (8.132) are written as:

$$F_{Rt} = C_{D1}\frac{\rho_f}{2}u_1^2\frac{\pi d^2}{4} + \rho_f\frac{\pi d^3}{4}\frac{D\bar{u}}{Dt} \tag{8.135}$$

$$F_L = \rho_f\frac{\pi d^3}{4}\frac{D\bar{v}}{Dt} \tag{8.136}$$

The velocity distribution in the rough flow region is assumed as the logarithmic law with zero velocity at the rough surface. Using the concept of Prandtl's mixing length theory, and the minimum scale of eddies, he expressed two differentials $\frac{Du}{Dt}$ and $\frac{Dv}{Dt}$ as functions of known quantities, and for $\frac{u_* d}{\upsilon_f} \geq 51.1$, the critical tractive stress is given by (Garde and Rangaraju, 2000):

$$\frac{\tau_{0c}}{(\gamma_s - \gamma_f)d} = C_1 \tan\phi f_2\left(\frac{u_* d}{\upsilon_f}\right) \tag{8.137}$$

where f_2 is a function of Reynolds number $\frac{u_* d}{\upsilon_f}$.

Case III: $6.83 \leq \frac{u_* d}{\upsilon_f} \leq 51.1$

In this range of Reynolds number, the partial sand particle is immerged in the viscous sub-layer and partial is exposed to the turbulent flow. Therefore, all the forces F_{Rt}, F_{Rl} and F_L exist. Using the idea of Case II, the critical tractive stress is given by:

$$\frac{\tau_{0c}}{(\gamma_s - \gamma_f)d} = C_1 \tan\phi f_3\left(\frac{u_* d}{\upsilon_f}\right) \tag{8.138}$$

In the above equations, f_1, f_2 and f_3 are the known functions of Reynolds number $\frac{u_* d}{\upsilon_f}$. This is developed from the equilibrium condition of a single particle resting on a sand bed. Iwagaki (1956) considered a sheltering effect by introducing an empirical coefficient in the above equations and then all equations (8.134), (8.137) and (8.138) are written in one equation as:

$$\frac{\tau_{0c}}{(\gamma_s - \gamma_f)d} = C_1 \tan\phi f\left(\frac{u_* d}{\upsilon_f}\right) \tag{8.139}$$

where the function f is for different values of $\frac{u_{*c} d}{\upsilon_f}$, and the empirical coefficient $C_1 = 2.5$ for the case of critical condition. The value of $\frac{\tau_{0c}}{(\gamma_s - \gamma_f)d}$ is 0.05 for large values of Reynolds number.

8.7.5 Wiberg and Smith's Analysis (1987)

Wiberg and Smith (1987) derived expression of critical shear stress for non-cohesive sediment particle from the balance of forces acting on individual particle resting at the bed surface. A sediment particle will begin to move when the friction force acting on the particle along the bed surface is exactly balanced by the sum of the drag force and the appropriate component of gravitational force. The forces acting on a sediment particle are submerged weight of the particle W_g, drag force F_D and lift force F_L as follows (Figure 8.28):

$$W_g = \frac{\pi}{6}(\gamma_s - \gamma_f)d^3, \tag{8.140}$$

$$F_D = C_D \frac{1}{2}\rho_f \langle u^2(y)\rangle A_x = C_D \frac{1}{2}\tau_0 \langle f^2(y/y_0)\rangle A_x \tag{8.141}$$

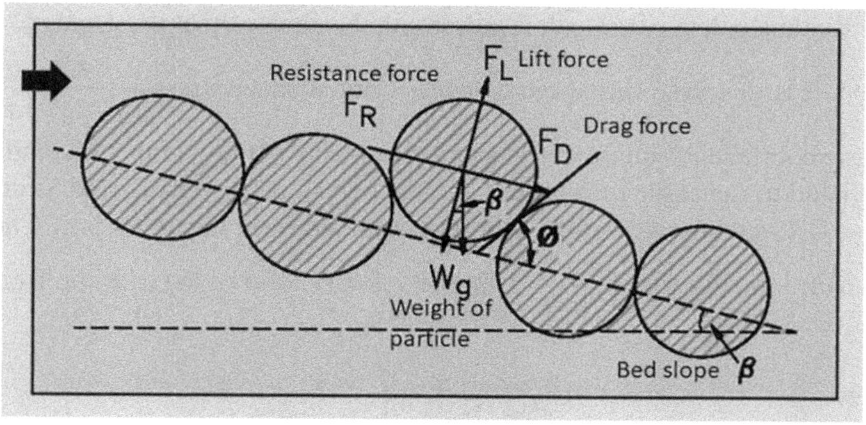

FIGURE 8.28 Balance of forces on a particle at the bed with the angle of repose ϕ, the bed slope β, submerged weight of the particle W_g, drag force F_D, lift force F_L and the balancing resistance force F_R. (Modified from Wiberg and Smith, 1987.)

$$F_L = C_L \frac{1}{2}\rho_f\left(u_t^2 - u_B^2\right)A_x = C_L \frac{1}{2}\tau_0\left[f^2(y_t/y_0) - f^2(y_B/y_0)\right]A_x \tag{8.142}$$

where C_D and C_L are respectively the drag and lift coefficients, $u(y) = u_* f(y/y_0)$ with y_0 is the bottom roughness parameter, $\tau_0 = \rho_f u_*^2$ is the bottom shear stress, A_x is the frontal exposed area, u is the velocity at y level, u_t and u_B are the velocities at the top and bottom of the particle, y_t and y_B are the heights of the top and bottom of the particle from the bed.

The force resisting downstream motion is as:

$$F_R = (W_g \cos\beta - F_L)\tan\phi = F_n \tan\phi \tag{8.143}$$

where F_n is the effective weight of the particle. A grain at the surface will start to move when the resisting downstream force is exactly balanced by the downstream directed drag force and the gravitational force in a slopping bed force. The force balance equation is as follows:

$$(W_g \cos\beta - F_L)\tan\phi = F_D + W_g \sin\beta \tag{8.144}$$

This equation can be rearranged as:

$$\frac{F_D}{W_g} = \frac{\tan\phi\cos\beta - \sin\beta}{1 + \frac{F_L}{F_D}\tan\phi} \tag{8.145}$$

where ϕ is the angle of repose and β is the bed slope.

Using the equations for W_g and F_D from the respective expressions (8.140) and (8.141) with the drag force determined at the critical shear stress $\tau_0 = \tau_{cr}$, equation (8.145) can be written as:

$$\frac{\tau_{cr}}{(\gamma_s - \gamma_f)d} = \frac{2}{\alpha_0 (C_D)_{cr}} \frac{1}{f^2\left(\frac{y}{y_0}\right)} \frac{\tan\phi\cos\beta - \sin\beta}{1 + \left(\frac{F_L}{F_D}\right)_{cr}\tan\phi} \tag{8.146}$$

where $\alpha_0 = A_x/d^2$ dimensionless parameter representing the geometry of the grain $\alpha_0 = 1.5$ for a sphere. From this equation, it is observed that dimensionless critical shear stress $\frac{\tau_{cr}}{(\gamma_s - \gamma_f)d}$ is inversely proportional to the drag coefficient, square of the average velocity $f^2(y/y_0)$ function, and lift-to-drag ratio, but it is directly related to the angle of repose ϕ and the bed slope β. Wiberg and Smith (1985) used the lift coefficient $C_L = 0.2$ in their calculation, as it was obtained from the analysis of Chepil's data (1958). When the flow is hydraulically rough $\frac{u_* d}{\upsilon_f} > 100$, the velocity follows logarithmic from the bed $y > 3k_s$, that is,

$$u(y) = \frac{u_*}{\kappa_0} \ln\left(\frac{y}{y_0}\right) \tag{8.147}$$

where κ_0 is the von-Karman constant 0.407, and $y_0 = k_s/30$ based on the data of Nikuradse (1933). The velocity profile is linear within the viscous sub-layer, and for $\frac{u_* d}{\upsilon_f} < 3$, the velocity profile will be $u(y) = \frac{u_*^2 y}{\upsilon_f}$ in this region. For transitional flow $3 \le \frac{u_* d}{\upsilon_f} \le 100$, Reichardt (1951) derived an equation for velocity of a smooth flow that extends through both the viscous sub-layer and the logarithmic layer and provides a smooth transition between the two regions. Thus Reichartd's velocity profile is given by:

$$\frac{u}{u_*} = \frac{1}{\kappa_0} \ln\left(1 + \kappa_0 y^+\right) - \frac{1}{\kappa_0} \ln\left(\kappa_0 y_0^+\right)\left[1 - e^{\frac{-y^+}{11.6}} - \frac{y^+}{11.6} e^{-0.33 y^+}\right] \tag{8.148}$$

where $y^+ = u_* y/\upsilon_f$, $y_0^+ = u_* y_0/\upsilon_f$, and $y_0 = \upsilon_f/9u_*$ is the height of hydraulically smooth. The coefficient $\frac{1}{\kappa}\ln\left(\kappa_0 y_0^+\right) = -7.78$ is for hydraulically smooth flow. From equation (8.146), the threshold Shields parameter vs Reynolds number is drawn for various roughness parameters (Figure 8.29).

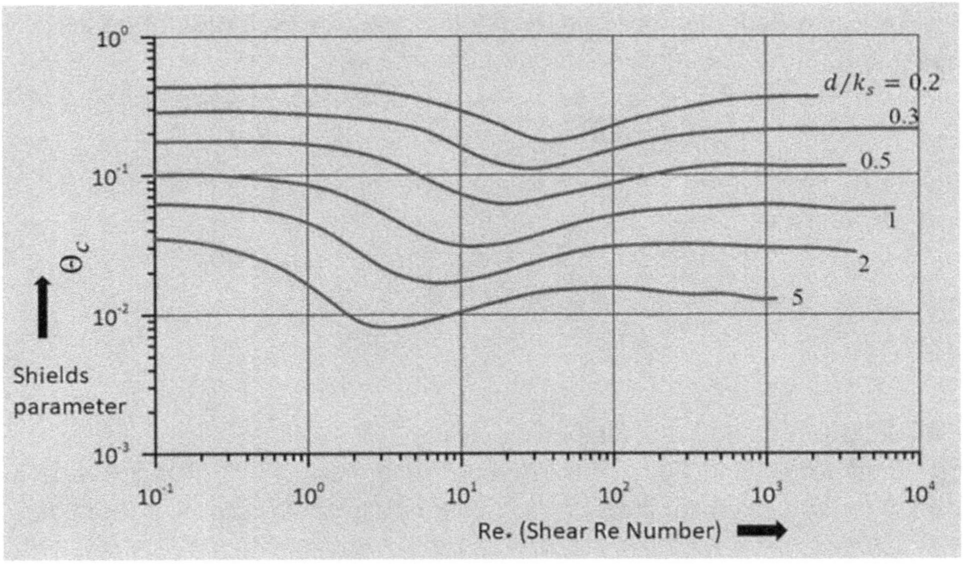

FIGURE 8.29 Shields parameter $\theta_c = \frac{\tau_{cr}}{(\gamma_s - \gamma_f)d}$ against shear Reynolds number $Re_* = \frac{u_* d}{\upsilon_f}$ for different dimensionless particle diameter d/k_s ranging from 0.2 to 5. (Modified from Wiberg and Smith, 1987.)

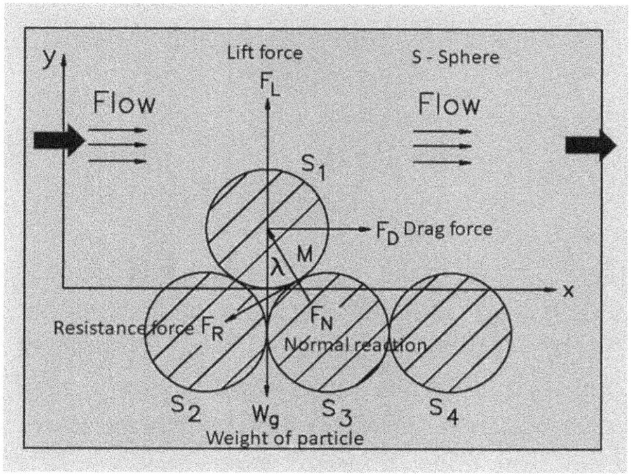

FIGURE 8.30 Forces acting on a spherical particle placed over a bed. (Modified from Ling, 1995.)

8.7.6 Ling's Analysis (1995)

Ling (1995) developed a theoretical model for incipient motion condition of a spherical particle on a horizontal bed in an open-channel flow, based on two fluid dynamic criteria: rolling criterion and lifting criterion. On derivation of his equation, he considered the drag and lift forces as destabilizing agents, and the weight force is the stabilizing agent shown in schematic diagram (Figure 8.30). From the figure, F_D, F_L, W_g, F_N, and F_R are respectively fluid drag on the top sphere, lift force, submerged weight force, normal reaction from the sphere and frictional resistance at the contact. He assumed that the lifting criterion is fulfilled, when the lift force exceeds the submerged weight force of the particle, which means $F_L \geq W_g$. According to van Rijn (1984), the drag force can be expressed as:

$$F_D = C_D \frac{1}{8} \rho_f \pi d^2 u_r^2 \tag{8.149}$$

where C_D is the drag coefficient, and u_r is the relative velocity between the fluid and top solitary sphere. For low Reynolds number, $C_D = 24 \Big/ \dfrac{u_r d}{\upsilon_f}$, the drag force is given by

$$F_D = 3\pi \mu_f \, d u_r \tag{8.150}$$

For turbulent flow at large Reynolds number, the drag coefficient C_D is constant. He considered the total lift force F_L is the sum of shear lift F_{LS} due to Saffman (1965), spin lift or Magnus lift F_{LM} (Rubinow and Keller, 1961) and centrifugal lift force F_{LC}. Thus, the total lift force F_L can be written as:

$$F_L = F_{LS} + F_{LM} + F_{LC} \tag{8.151}$$

in which individual components of equation (8.151) are given by

$$F_{LS} = C_{LS} \rho_f \, d^2 u_r \left(\upsilon_f \frac{\partial u}{\partial y} \right)^{0.5} \tag{8.152}$$

$$F_{LM} = \frac{1}{8}\pi\rho_f\,\omega u_r d^3 \text{ with } \omega \le \frac{1}{2}\frac{\partial \bar{u}}{\partial y} \tag{8.153}$$

$$F_{LC} = \left(\rho_s - \rho_f\right)\frac{\pi}{6}d^3\frac{(d\omega)^2}{4d}\cos\lambda_C \tag{8.154}$$

where C_{LS} is the Saffman lift coefficient, being equal to 1.615, $\bar{u}$ is the mean velocity in the stream-wise direction, ω is the angular velocity, λ_C is the angle between the normal force and gravity force. Saffman (1965) found that the spin lift was less than the shear lift by an order of magnitude, and afterwards many authors ignored the spin lift in their analysis. However, the value of the Saffman lift coefficient was found later to be erroneous. and then Saffman (1968) corrected it from 20.3 to 1.615. This correction puts the spin lift back at the same order of magnitude as the shear lift, therefore both lift forces F_{LS} and F_{LM} should be retained in calculating total lift F_L. The gravitational force on the particle is given by equation (8.140). From equations (8.152) and (8.153), it is observed that the shear lift F_{LS} due to Saffman is proportional to the relative velocity u_r and the Magnus lift F_{LM} is proportional to both relative velocity u_r and angular velocity ω. Ling (1995) assumed the relative velocity u_r as:

$$u_r = u_f - u_p = \alpha u_f \tag{8.155}$$

where u_f is the fluid velocity, u_p is the particle velocity in stream-wise direction and αu_f is the effective fluid velocity, and he assumed the angular velocity ω as:

$$\omega = \frac{2u_p}{\left(d\cos\lambda_C\right)} \tag{8.156}$$

Using equations (8.155) and (8.156) in (8.152) – (8.154), equation can be written as:

$$F_{LS} = \alpha C_{LS}\rho_f\,d^2 u_f\left(\upsilon_f\frac{\partial u}{\partial y}\right)^{0.5} \tag{8.157}$$

$$F_{LM} = \frac{\pi\rho_f}{4\cos\lambda_C}\alpha(1-\alpha)u_f d^2 \tag{8.158}$$

$$F_{LC} = \frac{\pi}{6}\left(\rho_s - \rho_f\right)(1-\alpha)^2 d^2\frac{u_f^2}{\cos\lambda_C} \tag{8.159}$$

Using the above equations in equation (8.151), and following Ling's idea for minimising the parameter α, one gets the expression of total lift force F_L as:

$$F_L = \left[\alpha F_1 + \alpha(1-\alpha)F_2 + (1-\alpha)^2 F_3\right]\tau_0 d^2 \tag{8.160}$$

where $\tau_0 = \rho_f\,u_*^2$ is the bed shear stress, and F_1, F_2, F_3 are respectively the functions of average fluid velocity (Ling, 1995). In order to determine the threshold criterion for particle movement, Ling (1995) assumed three different flow regimes: laminar sub-layer for low shear Reynolds number $\left(\frac{u_* d}{\upsilon_f} < 11.6\right)$, large shear Reynolds number $\left(\frac{u_* d}{\upsilon_f} > 30\right)$ for hydraulically rough regime, and $\left(3 \le \frac{u_* d}{\upsilon_f} \le 30\right)$ for transitional regime.

For low Reynolds number $\frac{u_* d}{\upsilon_f} < 3$, using the linear velocity profile $\frac{u}{u_*} = \frac{y u_*}{\upsilon_f}$, and using equation (8.150) for drag force, and following Ling's calculations, one gets the threshold relation for minimum rolling of incipient motion of uniform particle as:

$$\frac{(\tau_0)_{Cr}}{(\gamma_s - \gamma_f)d} = \frac{\frac{\pi}{6}}{\frac{\alpha}{2}\left(\frac{u_* d}{\upsilon_f}\right) + \frac{6\pi}{\sqrt{2}}} \tag{8.161}$$

for $\frac{u_* d}{\upsilon_f} < 1$, $\frac{(\tau_0)_{Cr}}{(\gamma_s - \gamma_f)d} \approx 0.04$

and the relation for minimum lifting threshold of incipient motion of uniform particle as follows:

$$\frac{(\tau_0)_{Cr}}{(\gamma_s - \gamma_f)d} = \frac{\frac{\pi}{6}}{\alpha F_1 + \alpha(1-\alpha)F_2 + (1-\alpha)^2 F_3} \tag{8.162}$$

which shows that the maximum lift for the linear velocity profile occurs at $\cos\lambda_C = 1$.

For large shear Reynolds number $\left(\frac{u_* d}{\upsilon_f} > 30\right)$, logarithmic velocity distribution is considered, and after detailed analysis following Ling (1995), one gets the rolling threshold criterion as

$$\frac{(\tau_0)_{Cr}}{(\gamma_s - \gamma_f)d} = \frac{\frac{\pi}{6}}{C_{LS} H_1 \left(H_2 \frac{u_* d}{\upsilon_f}\right)^{-0.5} + \frac{\pi}{2\sqrt{2}} C_D H_1^2} \tag{8.163}$$

where C_D is obtained from Schiller and Naumann (1933), and the value $C_{LS} = 1.615$ is taken from Saffman (1965) for all particle Reynolds number up to 500. The lifting threshold for incipient movement of particle in turbulent flow at high Reynolds numbers is similar to equation (8.162).

For transitional regime $3 \le \frac{u_* d}{\upsilon_f} \le 30$, in this case Wiberg and Smith (1987) showed that the formula of Reichardt (1951) agrees well with the measurement of velocity profiles in that region and fits well in both low and high Reynolds numbers. The rolling threshold criterion is given by

$$\frac{(\tau_0)_{Cr}}{(\gamma_s - \gamma_f)d} = \frac{\frac{\pi}{6}}{C_{LS} H_1 \left(H_2 \frac{u_* d}{\upsilon_f}\right)^{0.5} + \frac{\pi}{2\sqrt{2}} C_D H_1^2} \tag{8.164}$$

and lifting threshold criterion is as:

$$\frac{(\tau_0)_{Cr}}{(\gamma_s - \gamma_f)d} = \frac{\frac{\pi}{6}}{\alpha F_1 + \alpha(1-\alpha)F_2 + (1-\alpha)^2 F_3} \tag{8.165}$$

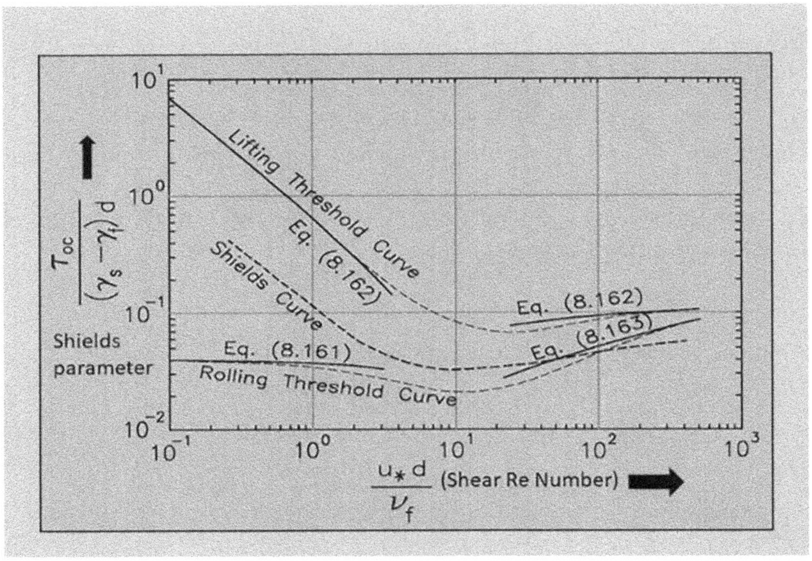

FIGURE 8.31 Threshold Shields parameter $\frac{(\tau_0)_{Cr}}{(\gamma_s - \gamma_f)d}$ as a function of shear Reynolds number $\frac{u_* d}{\upsilon_f}$ for rolling and lifting criteria above and below the Shields curve. (Modified from Ling, 1995.)

For low Reynolds numbers, the curves of equation (8.164) and equation (8.165) coincide with the curves of equation (8.161) and equation (8.162); and for large particle Reynolds numbers, they approach to equation (8.163) and similar to (8.162). So according to Ling (1995), Equations (8.164) and (8.165) are adequate for the use of threshold for all particle Reynolds numbers. The threshold lifting and rolling are far apart at small Reynolds numbers and get closer and closer with the increase of Reynolds number. Perhaps, this explains the unique relationship for critical shear stress that cannot be obtained for low Reynolds number $\frac{u_* d}{\upsilon_f} < 3.5$, (Graf, 1971). Two threshold criteria of incipient motion of rolling and lifting of spherical sediment particle are derived. The rolling threshold provides the minimum dimensionless critical shear stress required to start the particle motion in the form of rolling, and lifting threshold that provides minimum critical shear stress for suspension. The Shields curve along with some representative classical data fall completely between two thresholds lines (Figure 8.31).

8.7.7 Zanke's Analysis (2003)

Zanke (2003) developed a theoretical model for the sediment threshold. He assumed that the threshold bottom shear stress τ_{ocr} for the initial motion in the viscous flow is solely defined by the angle of internal friction ϕ or the angle of repose ϕ_1 of single particle. In turbulent flow, the fluctuations of bed shear stress and the lift force are brought by the velocity fluctuations. Hence, the effective threshold bed shear stress $\tau_0 + \tau_0'$ acting on a particle is greater than the average bed shear stress $\overline{\tau}_0$. On the other hand, the effective weight of the particles is reduced. The threshold of sediment motion can be described solely by the angle of repose of particles and the turbulence parameters. The equation for sediment threshold is given by as:

$$\frac{(\tau_0)_{Cr}}{(\gamma_s - \gamma_f)d} = \frac{0.7\tan\left(\frac{\phi}{1.5}\right)}{\left(1+1.8\frac{u'_{rms,\,cr}}{u_*}\cdot\frac{u_*}{u_{cr}}\right)^2\cdot\left(1+0.4\left(1+1.8\frac{u'_{rms,\,cr}}{u_*}\cdot\frac{u_*}{u_{cr}}\right)^2\tan\left(\frac{\phi}{1.5}\right)\right)} \tag{8.166}$$

where $u'_{rms,\,cr}$ is the threshold turbulence intensity or the root-mean-square of velocity fluctuations at the bed, and u_{cr} is the critical velocity at the bed. According to Nezu and Rodi (1986), the results of the measurements by different authors, the semi-empirical equation for smooth walls is given by:

$$\frac{u'_{rms}\left(y^+\right)}{u_*} = 0.3y^+e^{-0.1y^+} + 2.26e^{-0.881\frac{y}{h}} \cdot \left(1-e^{-0.1y^+}\right) \tag{8.167}$$

where $y^+ = \dfrac{u_* y}{\gamma_f}$ is the dimensionless distance from the wall. The last term $\left(1-e^{-0.1y^+}\right)$ is the modified van Driest damping coefficient. Zanke (2003) used equation (8.167) for $y^+ = k_s^+ = \dfrac{u_* k_s}{\upsilon_f} = \dfrac{u_* d}{\upsilon_f}$ at $y = d$, and the curve may be described to a high degree of accuracy by modifying the factors in the above equation as:

$$\frac{u'_{rms}\left(\dfrac{u_* k_s}{\upsilon_f}\right)}{u_*} = 0.31\frac{u_* k_s}{\upsilon_f}e^{-0.1\frac{u_* k_s}{\upsilon_f}} + 1.8\,e^{-0.881\frac{k_s}{h}} \cdot \left(1-e^{-0.1\frac{u_* k_s}{\upsilon_f}}\right) \tag{8.168}$$

and

$$\frac{u_{cr}}{u_*} = 0.8 + 0.9\frac{u\left(y = k_s\right)}{u_*} \tag{8.169}$$

Using (8.168) and (8.169) in equation (8.166), the dimensionless critical shear stress $\dfrac{(\tau_0)_{Cr}}{(\gamma_s - \gamma_f)d}$ is plotted in Figure 8.32 against shear Reynolds number $Re_{k_s} = \dfrac{u_* k_s}{\upsilon_f}$, and found a good agreement with the existing experimental data from Gilbert (1914), Shields (1936), White (1940), Meyer-Peter and Muller (1948), Yalin and Karahan (1979), but data are not shown in the figure. Other observed data were at the vicinity of the line.

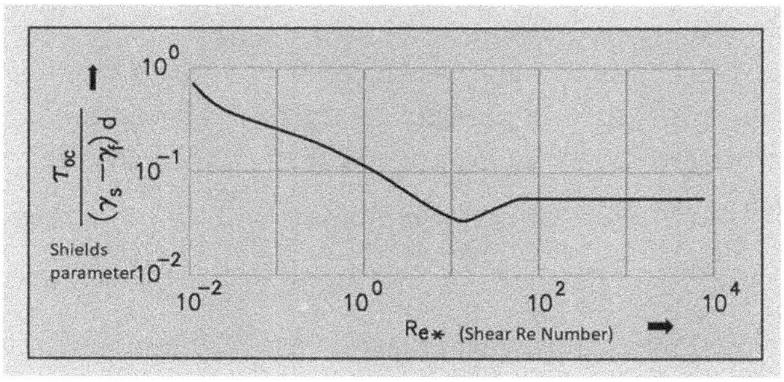

FIGURE 8.32 Threshold Shields parameter $\dfrac{(\tau_0)_{Cr}}{(\gamma_s - \gamma_f)d}$ as a function of shear Reynolds number $Re_{k_s} = u_* k_s / \upsilon_f$. (Modified from Zanke, 2003.)

Example 8.2

For a flow in a wide rectangular alluvial channel, estimate the depth at which sediment particle of 1 mm diameter starts moving. The slope of the channel may be assumed 0.00015. The relative density of sediment is 2.65 and kinematic viscosity of water is $10^{-6} m^2/s$. The dimensionless critical shear stress is given as 0.035 (refer Figure 8.32).

Solution

From the dimensionless critical shear stress, $\frac{(\tau_0)_{Cr}}{(\gamma_s - \gamma_f)d} = 0.035$.
Therefore $(\tau_0)_{Cr} = 0.035\,(\gamma_s - \gamma_f)d = 0.035 \times 9.91 \times 1650 \times 0.001 = 0.565\ \text{N/m}^2$.
The bed material starts moving when $\tau_0 = (\tau_0)_{Cr}$.
For a wide rectangular channel, $\tau_0 = \rho ghS$, where h is the flow depth and S is the bedslope.
Here $0.565 = \rho ghS$. Hence, $9810 \times h \times 0.00015 = 0.565$
Therefore, $h = 0.565/1.4715 = 0.384$ m.
Hence bed material may start to move at a depth of 0.384 m.

Example 8.3

Using the data given in Example 8.2, calculate the critical velocity and shear velocity corresponding to incipient motion using the competency approach.

Solution

From the given data in Example 8.2 and using equation (8.85),

$$(\bar{u})_{Cr} = 1.51\sqrt{\left(\frac{\rho_s}{\rho_f} - 1\right)gd} = 1.51\sqrt{\left(\frac{2.65}{1} - 1\right)9.81 \times 0.384} = 3.765\ \text{m/s}.$$

The shear velocity $u^* = \sqrt{ghS} = \sqrt{9.81 * 0.384 * 0.00015} = 0.0237\ \text{m/s}$.

8.7.8 Further Developments

Cao et al. (2006) developed an explicit expression of the Shields diagram, which furnished the critical Shields parameter determined directly from fluid and sediment characteristics without any trial and error or iteration procedure. They claimed that the traditional relationship between the critical Shields parameter and the shear Reynolds number $Re_* = \frac{u_* d}{\upsilon_f}$, where the shear velocity u_* is yet to be determined. For a specific set of fluid and sediment parameters, one has to way out for some sort of trial and error procedure or iterations to find the critical bed shear stress, which makes its application in river and coastal engineering rather inconvenient. A closer scrutiny of the Shields diagram shows that the critical Shields parameter $\frac{(\tau_0)_{Cr}}{(\gamma_s - \gamma_f)d}$ follows distinct distributions with the shear Reynolds number $Re_* = \frac{u_* d}{\upsilon_f}$ (Graf, 1971; Raudkivi, 1976; Yang, 1996; Chien and Wan, 1999; Yalin and de Silva, 2001). Cao et al. (2006) claimed that the critical Shields parameter $\frac{(\tau_0)_{Cr}}{(\gamma_s - \gamma_f)d}$ did not match with increase in $Re_* = \frac{u_* d}{\upsilon_f}$ following a straight line in the lower region as $Re_* \leq 2$, while $\frac{(\tau_0)_{Cr}}{(\gamma_s - \gamma_f)d}$ is constant for Re_* sufficiently large ($Re_* > 400$), the intermediate region the curve follows a saddle shape (Chien and Wan, 1999).

For the lower and upper regions, calculation of $\frac{(\tau_0)_{Cr}}{(\gamma_s - \gamma_f)d}$ is straight forward with sediment and fluid characteristics, whereas for saddle region it is inconvenient. Guo (2002) observed this phenomenon and he described the logarithmic matching method as follows:

$$\ln\left(\frac{(\tau_0)_{Cr}}{(\gamma_s - \gamma_f)d}\right) = -\ln(\mathrm{Re}_*) + 0.5003\ln\left[1 + (0.1359\mathrm{Re}_*)^{2.5795}\right] - 1.7148 \tag{8.170}$$

for shear Reynolds number $2 \le Re_* \le 60$.

$$\frac{(\tau_0)_{Cr}}{(\gamma_s - \gamma_f)d} = 0.1096(Re_*)^{-0.2607} \quad \text{for } Re_* \le 2 \tag{8.171}$$

$$\frac{(\tau_0)_{Cr}}{(\gamma_s - \gamma_f)d} = 0.045 \quad \text{for } Re_* \ge 60. \tag{8.172}$$

Equations (8.170)–(8.172) are plotted along with that of Yalin and de Silva (2001) in Figure 8.33, and show a good agreement apart from the appreciable discrepancy due to the present modified fit to the lower region. The lower and upper asymptotes for logarithmic matching are also shown in the figure. Cao et al. (2006) claimed that the logarithmic matching formulation equation (8.170) is not yet "well shaped" as the shear Reynolds number Re_* involves the unknown bed shear velocity u_*. Nevertheless, the shear Reynolds number Re_* is well known representation of critical Shields parameter $\frac{(\tau_0)_{Cr}}{(\gamma_s - \gamma_f)d}$, and this Re_* can be defined with the sand particle and fluid characteristics, such as

$$Re_* = \left[d\sqrt{\left(\frac{\rho_s}{\rho_f} - 1\right)gd}\,\frac{}{\upsilon_f}\right]\sqrt{\frac{(\tau_0)_{Cr}}{(\gamma_s - \gamma_f)d}} \tag{8.173}$$

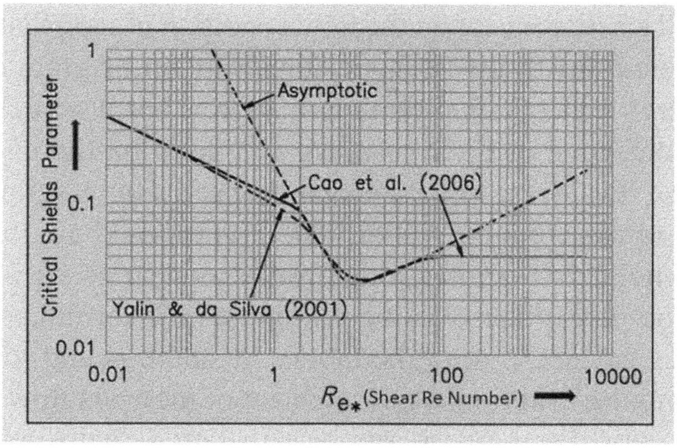

FIGURE 8.33 Formulation of Shields diagram versus shear Reynolds number $Re_* = \frac{u_* d}{\upsilon_f}$ with the relationship of Yalin and da Silva (2001). (Modified from Cao et al., 2006.)

where $d\sqrt{\left(\frac{\rho_s}{\rho_f}-1\right)\frac{gd}{\upsilon_f}}$ can be written as Re for convenient. Now substituting equation (8.173) in right-hand side of equation (8.170), equation (8.170) is converted to a relation between particle Reynolds number Re and the critical Shields parameter $\frac{(\tau_0)_{Cr}}{(\gamma_s-\gamma_f)d}$; and the following expression is given by

$$\ln\left(\frac{(\tau_0)_{Cr}}{(\gamma_s-\gamma_f)d}\right)=-0.6769\ln(Re)+0.3542\ln\left[1+(0.0223Re)^{2.8358}\right]-1.1296 \qquad (8.174)$$

where the shear Reynolds number $6.61 \le Re \le 282.84$.

Using equation (8.173), the lower region equation (8.171) and upper region equation (8.172) respectively can be written as:

$$\frac{(\tau_0)_{Cr}}{(\gamma_s-\gamma_f)d}=0.1414(Re)^{-0.2306} \quad \text{for } Re \le 6.61 \qquad (8.175)$$

$$\frac{(\tau_0)_{Cr}}{(\gamma_s-\gamma_f)d}=0.045 \quad \text{for } Re \ge 282.84. \qquad (8.176)$$

For given values of particle Reynolds number Re, the critical Shields parameter $\frac{(\tau_0)_{Cr}}{(\gamma_s-\gamma_f)d}$ can be explicitly calculated using equations (8.174)–(8.176). The comparative study between the present explicit equations (8.174)–(8.176) with that of Yalin and da Silva (2001) gives good agreement. Therefore, for a specific combination of fluid and sediment parameters, the cririsal Shields parameter $\frac{(\tau_0)_{Cr}}{(\gamma_s-\gamma_f)d}$ can be addressed as a function of particle size d. In accord with the normally used values of $g=9.8\,\text{m/s}^2$, $\frac{\rho_s}{\rho_f}-1=1.65$, and $\upsilon_f=1.0\,E-6\,\text{m}^2/\text{s}$, how the critical Shields parameter $\frac{(\tau_0)_{Cr}}{(\gamma_s-\gamma_f)d}$ varies with the particle size d according to the Shields diagram. Therefore, the explicit formulations should find applications in the general area of sediment transport, which are rendered possible by the log-matching method of Guo (2002).

Valyrakis (2011) proposed a new concept on the flow condition of incipient motion for the entrainment of course grain. He considered mostly two criteria for inception of grain entrainment: (i) the critical impulse and (ii) critical energy concept, and these two approaches were compared. These frameworks were assumed from the perspective of force or energy to describe the phenomenon, considering the momentum transfer from the flow event to the particle respectively. A series of experiments were conducted on mobile particles to examine the validity of the proposed approaches. He claimed that the hydrodynamic forces of sufficiently high magnitude were capable of entraining a particle, only when they lasted long enough. The perception of critical impulse was formulated theoretically for entrainment of both fine and coarse particles, using the modes of saltation and rolling at near-threshold flow conditions. Similar to the impulse criterion, a new concept of incipient flow conditions based on energy concept was proposed. The critical energy criterion was theoretically formulated for the entrainment of spherical particles in saltation or rolling mode (details available in Valyrakis et al., 2011).

Lee and Balachandar (2012) proposed a model of hydrodynamics forces with torque on a particle placed on a bed to investigate the incipient motion and resuspension of particles, using numerical

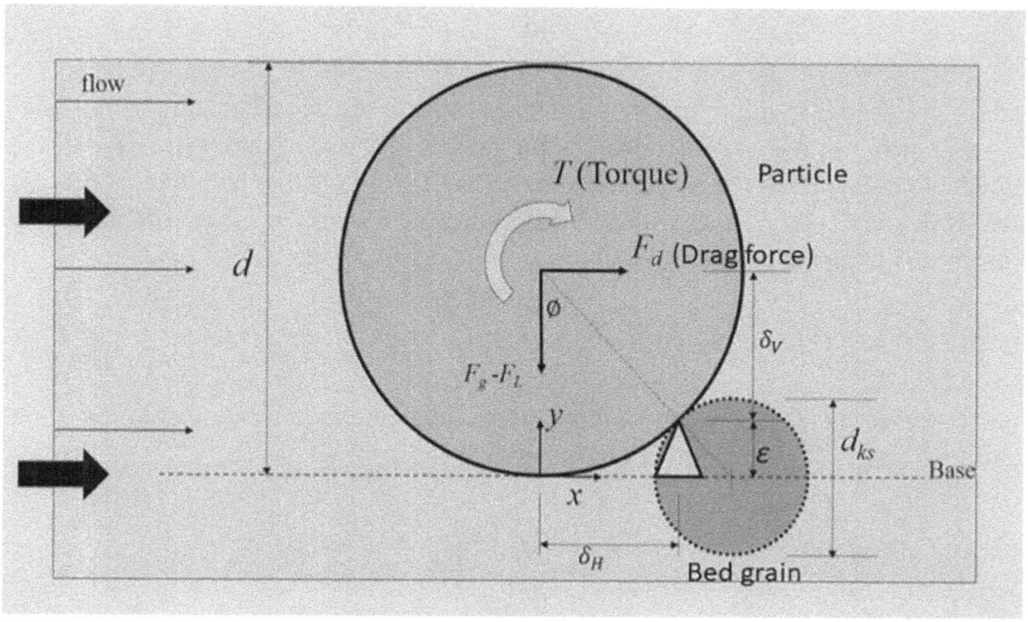

FIGURE 8.34 Schematic diagram of a spherical particle and a bed grain with forces, d is the particle diameter and d_{ks} is the grain roughness. F_D is the drag, F_L the lift, T the torque, F_g is the gravitational force acting on the particle. (Modified from Lee and Balachandar, 2012.)

simulations. They determined the drag, lift and torque coefficients appropriate for a particle placed on a bed and used these models for the incipient rolling motion of the particle.

A spherical particle placed on a bed tends to move (roll and/or slide) by the combined action of hydrodynamic forces, gravity, frictional and other interactive forces on the bed. Here it is assumed the rough bed, comprising spherical sediments, is shown schematically as the dashed circle (Figure 8.34). As the flow passes around the particle, the distributions of pressure (p) and shear stress (τ) around the particle result in a net force on the particle. This hydrodynamic interaction of the surrounding flow with the particle can also be expressed in terms of drag force F_D, lift force F_L and torque T, which are defined as

$$F_D = \int \left(-p\vec{n} + \vec{n} \cdot \tau\right) dS \cdot \vec{e}_x \tag{8.177a}$$

$$F_L = \int \left(-p\vec{n} + \vec{n} \cdot \tau\right) dS \cdot \vec{e}_y \tag{8.177b}$$

$$T = \int (\vec{r} + \tau) dS \cdot \vec{e}_z \tag{8.177c}$$

$$F_g = \frac{\pi}{6} \rho_f \left(\frac{\rho_s}{\rho_f} - 1\right) g d^3 \tag{8.178}$$

where $\vec{n}$ is the unit normal vector to the surface of the sphere and $\vec{e}_x$, $\vec{e}_y$, and $\vec{e}_z$ are the unit vector along streamwise, wall normal and spanwise directions.

An incipient motion of the particle can be determined by the net moment about the pivot point "P" is given by

$$M_P = T + F_D\delta_v - (F_g - F_L)\delta_H \tag{8.179}$$

where δ_v and δ_H are the vertical an horizontal distances from the particle center to the pivot P defined by $\delta_v = d/2 - \varepsilon$ and $\delta_H = \sqrt{\varepsilon(d-\varepsilon)}$, where ε is the height of the pivot from the base of the particle. If $M_P > 0$, the particle will start to roll about the vertex P by lifting off the bed, and if $M_P < 0$, the particle will not move. If $M_P = 0$, the critical condition of incipient motion can be predicted, using δ_v, δ_H and solving the moment balance equation (8.179) as:

$$T + F_D\delta_v - (F_g - F_L)\delta_H = 0 \tag{8.180}$$

Here accurate expressions of F_D, F_L, and T are required to determine the critical threshold condition. The dimensionless drag, lift and moment coefficients are given by:

$$C_D = \frac{F_D}{\frac{\pi}{8}\rho_f u_f^2 d^2},\ C_L = \frac{F_L}{\frac{\pi}{8}\rho_f u_f^2 d^2},\ C_M = \frac{T}{\frac{\pi}{12}\rho_f u_f^2 d^3} \tag{8.181}$$

where u_f is the velocity of the oncoming fluid at the particle center and the area is chosen to be $\pi d^2/4$ for full exposure of a particle. From the sediment-laden flow (Crowe et al., 1998), the coefficients C_D, C_L, and C_M are the functions of only Reynolds number $Re = u_f d/\upsilon_f$, where υ_f is the kinematic viscosity of the fluid. The velocity profile has to be determined from the viscous sub-layer and the logarithmic region. Swamee (1993) proposed an empirical expression for velocity profile, which covers from hydraulically smooth to rough as well as the transitional regions. From the thorough calculations, Lee and Balachandar (2012) determined the hydrodynamic forces (drag F_D, lift F_L, and torque T) on the particle.

The drag coefficient computed from the present simulations is presented in Figure 8.35 as a function of the flow Reynolds number $Re = u_f d/\upsilon_f$. The drag coefficient on a particle nearly sitting on a bed in a fully developed turbulent open channel flow as:

$$C_D = \frac{40.81}{\text{Re}}\left(1 + 0.104Re^{0.753}\right)\left[1 - \text{erf}(0.002Re)\right] + 0.54\,\text{erf}(0.002Re) \tag{8.182}$$

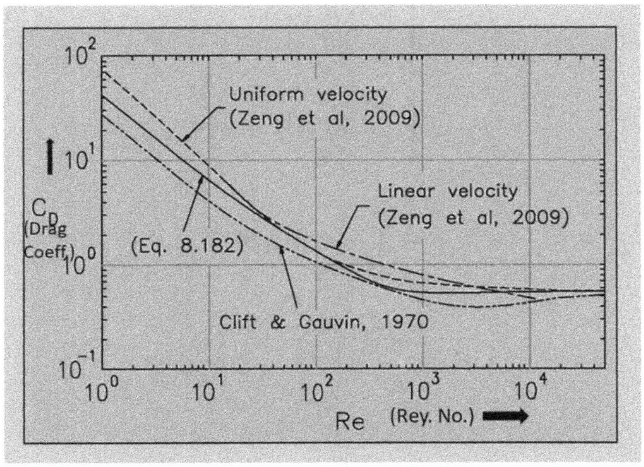

FIGURE 8.35 Comparison of drag coefficients on a spherical particle: Numerical simulations for wall-bounded turbulent flows, solid line represents curve fitting by equation (8.182) for wall-bounded turbulent flow, dash-dotted line for linear velocity, dashed line for uniform velocity. (Modified from Lee and Balachandar, 2012.)

where erf denote the error function. This model assumes that the drag coefficient gradually approaches a constant value of 0.54 at higher Reynolds numbers beyond those considered in the present simulations. In the low Reynolds number regime $Re \ll 10$, C_D from the present simulations with the logarithmic mean velocity are in very good agreement with the corresponding results for the linear mean velocity profile. At large Reynolds number $Re > 10^4$, the drag coefficient for logarithmic mean velocity approaches a constant value of 0.54. In between the regimes ($10 < Re < 10^4$), the drag for the log-law velocity profile is smaller than both the linear and uniform velocity profiles. However, the differences in C_D among the three velocity profiles are quite modest. For elaborate discussions, please follow the paper by Lee and Balachandar (2012).

Similar to drag coefficient, the lift coefficient C_L is also computed and plotted in Figure 8.36. The lift coefficient C_L for a particle on a bed in turbulent open channel flow can be expressed as:

$$C_L = 3.663\left(Re^2 + 0.1173\right)^{-0.22}\left[1 - \operatorname{erf}(0.001Re)\right] + 0.223\operatorname{erf}(0.001Re) \tag{8.183}$$

Using the lift coefficient $C_L = 0.2$ assumed by Wiberg and Smith (1987), they developed their numerical simulation and it was consistent for high Reynolds number.

At finite Reynolds number in a turbulent open channel with log-law velocity profile, we present the curve fit for the moment coefficient (Figure 8.37) as obtained from present numerical simulations as

$$C_M = 15.104(Re)^{-1}\left[1 + 0.0005Re \cdot \operatorname{erf}(0.002Re)\right] \tag{8.184}$$

For $Re \to 0$, the above equation fits the behavior for a small particle embedded in a linear shear flow. With increasing Reynolds number Re, the moment coefficient continues to decrease as inverse power of Re and the trend extends to about $Re \lesssim 10^2$ (Lee and Balachandar, 2012).

Simoes (2014) proposed a method to determine the incipient motion of sediment particles based on the movability number defined by the ratio of shear velocity and the particle settling velocity. His method was verified by a large number of experimental data and was claimed that the approach gives a simple and accurate method of computing the threshold condition for sediment motion. The initiation

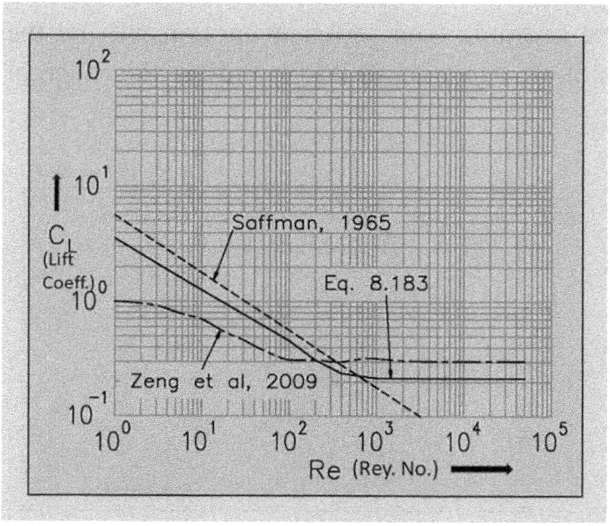

FIGURE 8.36 Comparison of lift coefficients on a spherical particle with existing models. (Modified from Lee and Balachandar, 2012.)

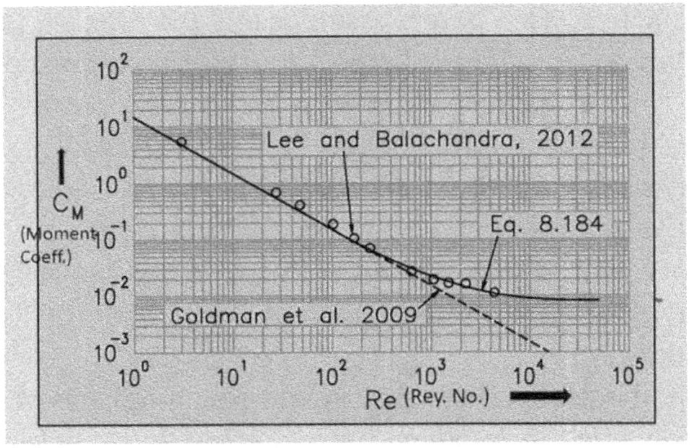

FIGURE 8.37 Comparison of moment coefficients vs Reynolds number with existing experimental data. (Modified from Lee and Balchandar, 2012.)

of ripple formation, Liu (1957, 1958) proposed the movability number of the sediment particle. The relation between the mavability number M_v and the dimensionless shear stress τ_* is obtained as:

$$\tau_* = \frac{4}{3C_d} M_v^2 \tag{8.185}$$

where $\tau_* = \dfrac{\tau}{(\rho_s - \rho_f)gd}$ dimensionless shear stress, and τ is the shear stress. Equation (8.185) represents the movability number M_v is proportional to the square root of dimensionless shear stress τ_*, which results in reduction of data scatter around the curve. Komar and Clemens (1986) showed that the computation of settling velocity w_s for sediment grains is at least as accurate as the determination of the threshold of motion, but they used limited amount of data, so there was a limitation in their results. Paphitis (2001) derived a new relationship using extensive amount of data set to determine the empirical threshold curve as given by:

$$M_{vc} = \frac{0.75}{Re_f} M_v + 14\exp(-2Re) + 0.01\ln(Re) + 0.115 \tag{8.186}$$

where $Re = u_f d/\nu$. Here the hydraulic and sediment motion will be there, if $M_v > M_{vc}$, otherwise there will be no sediment motion. Equation (8.186) is only in the range of $0.10 < Re < 10^5$ as shown in Figure 8.38.

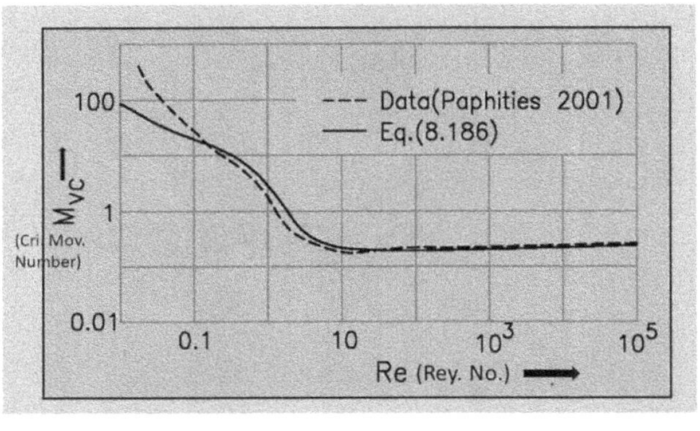

FIGURE 8.38 Threshold of incipient motion of Paphities (2001) and its comparison with observed data of Simoes (2014). (Modified from Simoes, 2014.)

Beheshti and Ashtiani (2008) used the movability number M_v to determine the threshold motion as a suitable parameter, but they found erroneous behavior for large values of dimensionless grain diameter d_* defined as

$$d_* = \left[\frac{(s-1)g}{\upsilon_f^2}\right]^{\frac{1}{3}} d \tag{8.187}$$

where $s = \rho_s/\rho_f$ is the relative density. Simoes (2014) presented his work using more comprehensive data to derive a new empirical relationship, which is more accurate than Beheshti and Ashtiani (2008). Beheshti and Ashtiani (2008) found $M_v (= u_*/w_s)$ to be a more suitable parameter for determining the threshold of motion, tried to eliminate the latter limitation by deriving the following equations:

$$M_{vc} = 9.6674 d_*^{-1.57}, d_* \leq 10 \text{ and } 0.4738 d_*^{-0.226}, d_* > 10 \tag{8.188}$$

where $M_{vc} = u_{*c}/\omega$ is the critical movability number. Equation (8.188) is valid in the range of $0.4 < d_* < 10^3$. Simoes (2014) developed an empirical model based on 517 experimental data set obtained from different sources and physical settings. In his work, he used the settling velocities of particles determined by the procedure given by Dietrich (1982). Using the experimental data set, Simoes (2014) derived a nonlinear equation using the least-square procedure as:

$$M_{vc} = 0.215 + 6.79\, d_*^{-1.70} - 0.075\exp\left(-2.62\times10^{-3} d_*\right) \tag{8.189}$$

Here equation (8.188) developed by Beheshti and Ashtiani (2008) and equation (8.189) developed by Simoes (2014) are plotted in Figure 8.39, which shows a better agreement between the analytical and experimental data. The statistical analysis is presented to show the goodness-of-fit of equation (8.189) to the data. The computation of settling velocity w was made using Dietrich's (1982) method due to its flexibility in incorporating particle shape and roundness. There are other simpler methods of calculating the sediment particle settling velocity, but they have smaller ranges of applicability. He has established

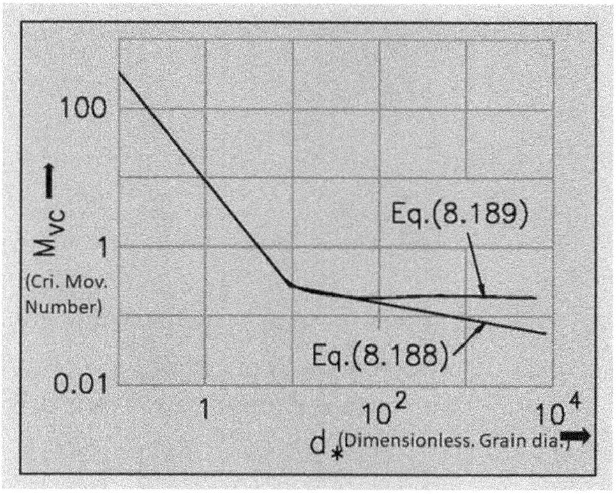

FIGURE 8.39 Critical movability number against dimensionless grain diameter d_*, using equations (8.188) developed by Beheshti and Ashtiani (2008) and (8.189) developed by Simoes (2014). (Modified from Simoes, 2014.)

a new equation, which can be effectively used for the computation of the threshold of incipient sediment motion, providing a simple and more accurate calculation than those based on the traditional Shields parameter.

Roušar et al. (2016) studied the effects of various parameters on the incipient motion of uniform gravels based on the parametric analysis. The main parameters of the flow and the bed materials were the longitudinal bed slope $\sin\alpha$, flow depth h, density of water ρ_f, the kinematic viscosity of water υ_f, the grain diameter d, length of the grain a, width of the grain b, thickness c, the density of the grain ρ_s, and angle of repose $\tan\varphi$ and gravitational acceleration g. The incipient motion of grain is determined by the following functional relationship using dimensional analysis as

$$F_{*c} = F_1\left(\rho_s/\rho_f - 1, Re_*, h/d, \sin\alpha, \tan\varphi\right) \tag{8.190}$$

Introducing the critical Shields parameter for incipient motion on a horizontal surface, the above equation can be written as:

$$\frac{(\tau_0)_{Cr}}{(\gamma_s - \gamma_f)d} = F_2\left(Re_*, h/d, \sin\alpha, \tan\varphi\right) \tag{8.191}$$

Influence of each parameter on the incipient motion of grains is examined from the theoretical and experimental point of view. The dimensional analysis was used to define the main parameters of incipient motion for Shields criteria. Their main contributions were the detailed analysis based on fluctuating components of lift and drag forces, which depend on both velocity components. Here the lift and drag forces are decomposed into the time- and area-averaged component $\overline{F}$ and the fluctuation component F' involving both time fluctuation at a given point and the area fluctuation within an averaged area (Nikora et al., 2001; Yalin, 1972) as:

$$F = \overline{F} + F' \tag{8.192}$$

Similarly, the time- and area-averaged component of lift force $\overline{F_L}$ is given by:

$$\overline{F_L} = 0.5C_L\rho_f A_{xz}\left(\overline{u}_u^2 - \overline{u}_d^2\right) \tag{8.193}$$

and its fluctuating component is

$$F_L' = 0.5C_L\rho_f A_{xz}\left(u_u'^2 - u_d'^2\right) \tag{8.194}$$

where C_L is the lift coefficient, A_{xz} is the projection area in the xz-plane, $\overline{u}$ is the stream-wise flow velocity. Subscripts u and d represent respectively top and bottom of the grain. The time and area averaged component of drag force $\overline{F_D}$ acting in the direction parallel to xz-plane is respectively as

$$\overline{F_D} = 0.5C_D\rho_f A_{xy}\overline{u}_c^2 \tag{8.195}$$

$$F_D' = 0.5C_D\rho_f A_{xy}u_c'^2 \tag{8.196}$$

Here A_{xy} is the projection of the grain surface in xy-plane perpendicular to the flow direction. The ratio of the parallel forces to the normal forces governs the angle of repose specified by $\tan\varphi$ as:

$$\tan\varphi = \frac{F_D}{\frac{1}{6}\pi d^3\left(\gamma_s - \gamma_f\right) - F_L - F_{Dver}} \tag{8.197}$$

where $F_D = \overline{F_D} + F_D'$, $F_L = \overline{F_L} + F_L'$, and $F_{DVer} = \overline{F_{DVer}} + F_{DVer}'$ are average and fluctuating components of drag and lift forces; and the term $(1/6)\pi d^3\left(\gamma_s - \gamma_f\right)$ is the submerged weight force. The drag force in the vertical direction is caused only by the vertical velocity fluctuation w', (where the average velocity is zero, average of its square is not zero, i.e., standard deviation) as:

$$F_{Dver} = \overline{F_{DVer}} + F_{DVer}' = 0 + F_{DVer}' = 0.5 C_{Dver}\rho_f A_{xz} v_c'^2 \tag{8.198}$$

Substituting the drag forces F_D, F_L, and F_{DVer} from respective equations (8.193) – (8.196), and (8.198) in equation (8.197), one gets

$$\tan\varphi = \frac{C_D A_{xy}\left(\overline{u}_c^2 + u_c'^2\right)}{0.33\pi d^3 g\left(\rho_s/\rho_f - 1\right) - C_L A_{xz}\left(\overline{u}_u^2 - \overline{u}_d^2 + u_u'^2 - u_d'^2\right) - C_{Dver} A_{xz} v_c'^2} \tag{8.199}$$

Equation (8.199) can be written as:

$$\frac{C_D A_{xy}\left(\overline{u}_c^2 + u_c'^2\right)}{\tan\varphi} + C_L A_{xz}\left(\overline{u}_u^2 - \overline{u}_d^2 + u_u'^2 - u_d'^2\right) + C_{Dver} A_{xz} v_c'^2 = 0.33\pi d^3 g\left(\rho_s/\rho_f - 1\right) \tag{8.200}$$

Dividing both sides by u_*^2, equation (8.200) can be written as:

$$\frac{1}{0.33\pi d^2 u_*^2}\left[\frac{C_D A_{xy}\left(\overline{u}_c^2 + u_c'^2\right)}{\tan\varphi} + C_L A_{xz}\left(\overline{u}_u^2 - \overline{u}_d^2 + u_u'^2 - u_d'^2\right) + C_{Dver} A_{xz} v_c'^2\right] = \frac{g\left(\rho_s - \rho_f\right)d}{\left(\tau_0\right)_{Cr}} \tag{8.201}$$

Or, this can be written as:

$$\frac{\left(\tau_0\right)_{Cr}}{\left(\gamma_s - \gamma\right)d} = \theta_c = 0.33\pi d^2 u_*^2\left[\frac{C_D A_{xy}\left(\overline{u}_c^2 + u_c'^2\right)}{\tan\varphi} + C_L A_{xz}\left(\overline{u}_u^2 - \overline{u}_d^2 + u_u'^2 - u_d'^2\right) + C_{Dver} A_{xz} v_c'^2\right]^{-1} \tag{8.202}$$

This is the equation for incipient motion of grains. In the case of a spherical grain, $A_{xy} = A_{xz} = d^2$, $C_D = C_{Dver}$, equation (8.202) takes the form as:

$$\frac{\left(\tau_0\right)_{Cr}}{\left(\gamma_s - \gamma_f\right)d} = \theta_c$$

$$= 0.33\pi u_*^2\left[\frac{C_D\left(\overline{u}_c^2 + u_c'^2\right)}{\tan\varphi} + C_L\left(\overline{u}_u^2 - \overline{u}_d^2 + u_u'^2 - u_d'^2\right) + C_{Dver} v_c'^2\right]^{-1} \tag{8.203}$$

Equation (8.203) is solved using the coefficients C_D and C_L; and the velocity measurements. Similarly, for an ellipsoid-sized particle, equation (8.202) can be written as given by Rousar et al. (2016). From their analysis, they found that the effect of angle of repose (tan φ) is very significant and the theoretical developments presented in the paper are in good agreement with their laboratory experiments as well as many others data from the existing literature. Woldegiorgis et al. (2018) developed a unified method that used the dimensionless flow velocity and the flow depth as predictors to determine the incipient motion of particle size for erosion of both the cohesive and the non-cohesive sediments without detailed hydraulic calculations and iterative solutions.

8.8 SHEAR STRESS FOR SLOPING OF RIVER BED AND BANK

Almost all the rivers in nature have slopes, resulting in the flow from the upstream to the downstream. Due to the shear stress at the interface of river bed/bank and flow of water, the river bed/banks traditionally are assumed to become movable, showing the scour and deposition of sediment along the river bed. To design for stable channels, the quantification of critical bed shear stress for sediment entrainment along stream-wise (flow) direction and the selection of appropriate rip-rap size for river bank protection are very important. The initiation of sediment motion is an integral part of understanding of sediment transportation and deposition in rivers. Descriptions for the entrainment of sand particles are involved through the force balance equation of individual grains on the sloping bed. Several semi-empirical equations have been developed by many investigators (Luque and van Beek, 1976; Ikeda, 1982 (for bank slope); Whitehouse and Hardisty, 1988; Chiew and Parker, 1994; Christensen, 1995; Dey, 2003; Armanini and Gregoretti, 2005; Ancey et al., 2008; Lamb et al., 2008; Ferguson, 2012) to find the incipient motion of sediment particles over the channel-bed slope. An accurate determination of the critical shear stress for the initiation of motion of the sediment particles in alluvial streams is vital for erosion and sedimentation predictions.

For example, Luque and van Beek (1976) conducted some experiments in a closed rectangular channel at different slopes of the bed surface, using five different bed materials including sand and gravel. They determined the mean critical bed shear stress from the Shields' grain movement

The rectangular channel was inclined at an angle α with the horizon (Figure 8.18). For the condition of incipient motion grain, the force balance equation (8.72) in the stream-wise direction is considered, where the resultant of the hydrodynamic drag force F_D, lift force F_L and submerged weight force W_g are taken into consideration. Equation (8.72) is written here as:

$$\tan\varphi = \frac{W_g \sin\alpha + F_D}{W_g \cos\alpha - F_L} \tag{8.72}$$

where drag force F_D, lift force F_L and weight force W_g are respectively given by equations (8.73), (8.74) and (8.75); and $\tan\varphi$ is the angle of repose. From their analysis, they obtained for a fixed grain size, the Shields curve decreases with an increase in bed inclination α, and it is obtained from the plot of critical mean bed shear stress τ_{0cr} vs shear Reynolds number Re_* (Figure 8.40), that is,

$$\frac{\tau_{0cr}}{(\rho_s - \rho_f)gd} = f\left(Re_* = \frac{u_* d}{\nu_f}\right) \tag{8.204}$$

Figure 8.40 shows that the measured dimensionless critical shear stress $\frac{\tau_{0cr}}{(\rho_s - \rho_f)gd}$ values fit the Shields curve well for $\beta = 0$, especially for heavy sediment gravel, magnetite and sand of $d = 1.8$ mm.

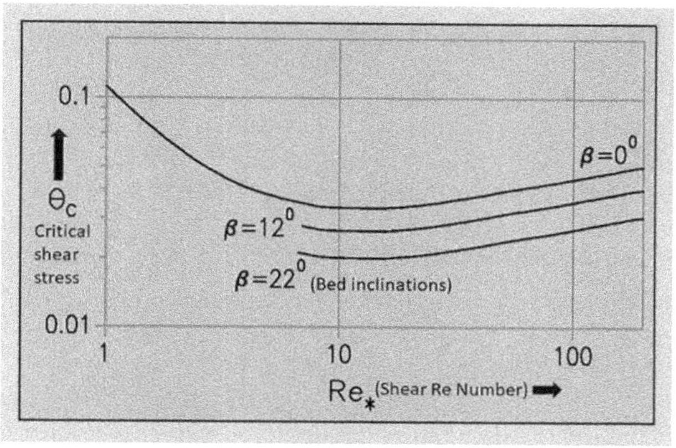

FIGURE 8.40 Normalized critical shear stress $\frac{\tau_{0cr}}{(\rho_s-\rho_f)gd}$ against shear Reynolds number $Re_* = u_* d/\upsilon_f$ for different bed inclinations $\beta = 0°$, $12°$ and $22°$ for critical drag angle $\alpha = 47$. (Modified from Luque and van Beek, 1976.)

This parameter $\frac{\tau_{0cr}}{(\rho_s-\rho_f)gd}$ for these bed materials at downward slopes of $\beta = 12°, 18°$ and $22°$ for $\alpha = 47$ are also plotted in Figure 8.40. They also determined the average particle velocity $\bar{u}_s$ is a function of shear velocity u_* and the critical shear velocity u_{*cr} corresponding to Shields diagram as:

$$\bar{u}_s = 11.5(u_* - 0.7u_{*cr}) \tag{8.205}$$

where the critical shear velocity u_{*cr} is different for different bed materials and different bed inclination α, but in plotting it shows scatter of data points about a single line. According to Chepil (1959), the hydrodynamic drag force F_D acting on grain diameter d at the surface y=0 of horizontal sediment bed ($\alpha = 0$) from (8.72) as:

$$\tan\varphi = \frac{F_D}{W_g - F_L} \tag{8.206}$$

Chepil (1959) found from his experimental data that for loose sediment bed at an angle of repose $\varphi = 24°$, $F_L = 0.85F_D$ and $F_D = 5\pi/4d^2\tau_{0cr}$, where τ_{0cr} is the critical bed shear stress at the threshold of grain movement. Grass (1970) showed that at the Shields' threshold condition, the topmost grain just moved by the time-mean bed shear stress $\bar{\tau}_{ocr}$ at Shields' grain-movement condition. If the bed surface is inclined at an angle α along the flow direction, then the dimensionless critical shear stress $\frac{\tau_{0cr}}{(\rho_s-\rho_f)gd}$ is reduced by a factor $\frac{\sin(\varphi-\alpha)}{\sin\alpha}$ shown by Luque and van Beek (1976). Using equations (8.72) to (8.75), the equations of dimensionless threshold Shields' parameter $\frac{(u_b^2)_{Cr}}{(\rho_s/\rho_f-1)gd}$ on the sloping bed is given by:

$$\frac{(u_b^2)_{Cr\alpha}}{(\rho_s/\rho_f-1)gd} = \frac{2K_3(\tan\varphi\cos\alpha - \sin\alpha)}{C_D K_1 + C_L K_2 \tan\varphi} \tag{8.76}$$

$$\Rightarrow \frac{\left(u_b^2\right)_{Cr\alpha}}{\left(\rho_s/\rho_f - 1\right)gd} = A\left(\tan\varphi\cos\alpha - \sin\alpha\right) \tag{8.76a}$$

where $A = \dfrac{2K_3}{C_D K_1 + C_L K_2 \tan\varphi}$, and $\left(u_b^2\right)_{Cr\alpha}$ is the critical shear velocity at non-zero angle of inclination $(\alpha \neq 0)$ of sediment bed. For horizontal sediment bed ($\alpha = 0$), the equation for incipient Shields parameter equation (8.76a) is given by

$$\frac{\left(u_b^2\right)_{Cr0}}{\left(\rho_s/\rho_f - 1\right)gd} = A\tan\varphi \tag{8.207}$$

If equation (8.76a) is divided by equation (8.207), the ratio of these two equations is given by a factor $\dfrac{\sin(\varphi-\alpha)}{\sin\alpha}$ shown by Luque and van Beek (1976). That is, the ratio of critical at any streamwise bed slope α to the corresponding value at $\alpha = 0$, one gets

$$\frac{\left(u_b^2\right)_{Cr\alpha}}{\left(u_b^2\right)_{Cr0}} = \cos\alpha\left[1 - \frac{\tan\alpha}{\tan\varphi}\right] \tag{8.208}$$

This deviation in critical shear stress is due to the angle of inclination α of sediment bed.

Chiew and Parker (1994) investigated theoretically the threshold condition for the initiation of non-cohesive sediment transport over an inclined horizontal bed surface. They conducted experiments in a closed-conduit flow over the sediment bed and derived force balance equation to evaluate the critical shear stress of sediment particles lying on inclined sloping bed ranging from positive to negative slopes. They also derived a similar force balance equation (8.72) using drag, lift and weight forces with angle of repose over an inclined sediment bed (Figure 8.18) to find the incipient motion of sediment particle and it obtained similar equation as (8.207). It is interesting to note that many researchers such as Stevens et al. (1976), Luque and van Beek (1976), Smart (1984), Dyer (1986) and Whitehouse and Hardisty (1988) have obtained a similar relationship equation (8.208) to determine the critical shear velocity or critical shear stress for sediment particle lying on a sloping bed. In the discussion of Chiew and Parker's approach (1994), Christensen (1995) pointed out that the pressure distribution on a sloping bed is hydrostatic along the normal to the flow and consequently the buoyancy force acts along the normal to the flow rather than along the vertical. Similar results are also obtained by Lau and Engel (1999) using the moment balance equation at limit equilibrium.

Armanini and Gregoretti (2005) studied the influence of the bed slope angle on the incipient motion of sediment grain in both pipe and open-channel flow conditions and obtained substantial different results. They developed a theoretical model based on the force balance equation on the equilibrium of particle lying on an inclined sediment bed, using the relative degree of exposure of the particle to the flow. The relative degree of exposure of the particle to the flow was introduced in the balance equation. In this case, the force balance equation included the lift, drag, buoyancy, and seepage forces together with friction force due to contact with the other particles. The degree of exposure is defined as height above the upstream particle. The degree of exposure e is introduced as two suitable functions $f_1(e)$ and $f_2(e)$ respectively representing as normal projection to the flow and exposed volume of the particle. Using the degree of exposure terms, the force balance equation (8.72) is modified as

$$\tan\varphi = \frac{W_g\sin\alpha + F_D + S_e}{W_g\cos\alpha - F_L} \tag{8.209}$$

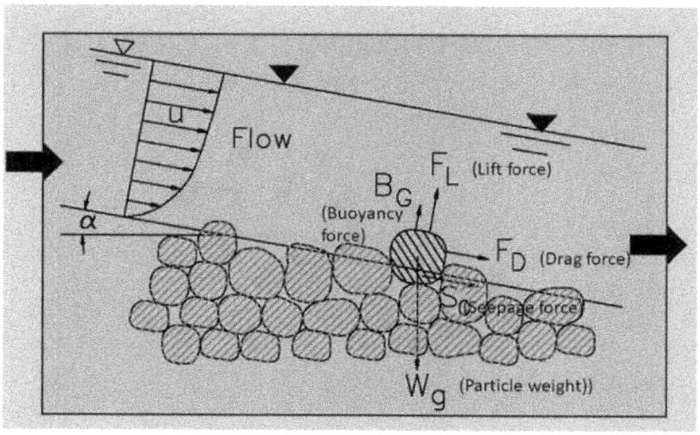

FIGURE 8.41 Forces acting on a single particle with the buoyancy B_G. (Modified from Armanini and Gregoretti, 2005.)

where S_e is the seepage force acting on the lower unexposed surface of the particle (Worman, 1992), and the weight force is not affected by the exposure (Figure 8.41). The forces are given by

$$F_D = C_D K_1 d^2 f_1(e) \frac{\rho_f u_b^2}{2} \tag{8.210}$$

$$F_L = C_L K_1 d^2 f_1(e) \frac{\rho_f u_b^2}{2} \tag{8.211}$$

$$S_e = \rho_f g (1 - f_2(e))(1 - K_2) d^3 \sin\alpha \tag{8.212}$$

where u_b is the fluid velocity at the center level of the particle, C_D and C_L are the drag and lift coefficient respectively, d is the particle diameter, ρ_f is particle density, and K_1, and K_2 are the particle shape factors.

The exposure functions $f_1(e)$ and $f_2(e)$ depend on the particle shape, and for spherical shape, these two functions are given by

$$f_1(e) = \frac{\cos^{-1}(1-2e)}{\pi} + \frac{1}{2\pi} \sin\left[2\cos^{-1}(1-2e)\right] \tag{8.213}$$

$$f_2(e) = e^2 (3 - 2e) \tag{8.214}$$

If $e = 0$, then $f_1 = f_2 = 0$, and for $e = 1$, $f_1 = f_2 = 1$, and for $e = 0.5$, $f_1 = f_2 = 0.5$. Substituting all these above expressions of forces in (8.208) and rearranging, one gets the explicit form with respect to the Shields mobility parameter as (Armanini and Gregoretti, 2005):

$$\frac{\left(u_b^2\right)_{Cr\alpha}}{\left(\rho_s/\rho_f - 1\right) g d} = \frac{K_2 \sin\alpha}{K_1 f_1 \left(\dfrac{u_b}{u_*}\right)^2 0.5\left(C_D + C_L \tan\varphi\right)} \left[\frac{\tan\varphi}{\tan\alpha} - \frac{\rho_s}{\rho_s - \rho_f} - \frac{(1 - f_2(e))(1 - K_2)}{\left(\rho_s/\rho_f - 1\right) K_2}\right] \tag{8.215}$$

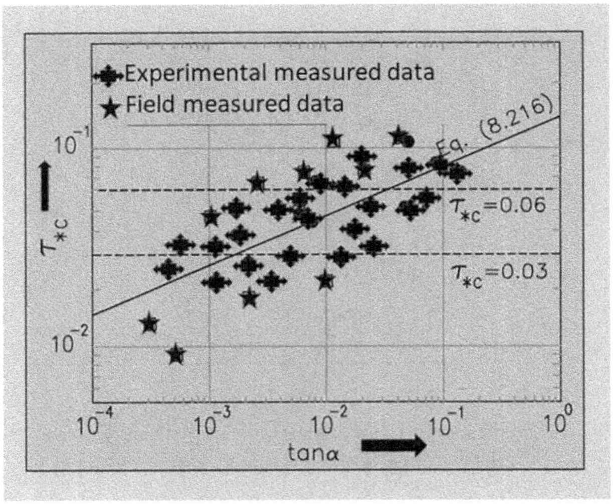

FIGURE 8.42 Plots of channel slope against critical shear stress τ_{*cr} with best fitted line by least squares method. (Modified from Lamb et al., 2008.)

If $e = 1$, that is, for $f_1 = f_2 = 1$, then equation (8.215) reduces to the standard Shields criterion, which is modified according to Christensen (1995) for the sloping effects.

Lamb et al. (2008) studied the critical Shields parameter for incipient sediment motion over steep channel bed. Their finding is a contradictory to the common belief that the increasing slope reduces the bed stability due to the added downstream gravitational force. Data from the flume experiments and natural streams show that the critical shear stress for incipient sediment motion increases with increase in channel slope, indicating more stable in steeper slope for the particles of the same size. They observed that the critical Shields stress for incipient motion is a function of channel slope. This discrepancy is explored by the simple force balance equation, including increased drag from the walls and bed morphology, friction angles, grain emergence, flow aeration, and local flow velocity and fluctuations. It is observed that increased drag due to bed morphology does not appear to be the cause of slope dependency, whereas the grain emergence and changes in local flow velocity and fluctuations seem to be responsible for slope dependency. Figure 8.42 shows the compilation of existing experimental and field data (i.e. Gilbert, 1914; Meyer-Peter and Mueller, 1948; Paintal, 1971; Luque and Beek, 1976; Day, 1980a, 1980b; Wilcock and Southard, 1988; Komar and Carling, 1991; Wilcock, 1993; and others) at slope $\tan\alpha$ versus dimensionless critical Shields stress $\tau_{*cr} = \frac{\left(u_b^2\right)_{Cr0}}{\left(\rho_s/\rho_f - 1\right)gd}$, where α is the bed-slope angle from horizontal. The best fit line of critical shear stress τ_{*cr} in a least square sense with regression coefficient R^2 of 0.41 is given by

$$\tau_{*cr} = 0.15(\tan\alpha)^{0.25} \tag{8.216}$$

From the Figure 8.42 it is also observed that the typical upper $\tau_{*cr} = 0.06$ and lower values $\tau_{*cr} = 0.03$ are assumed for a gravel bed. Here the data have been filtered for $Re_* = \frac{u_* d}{\upsilon_f} > 10^2$. The data of Figure 8.42 represent the regime of τ_{*cr} ranging from 0.03 to 0.06. Yalin and Karahan (1979) and Wilcock (1993) suggested a constant value of 0.047 for mixed gravel, which is widely used. From the figure, it is observed that much of the data does not fall within the range $0.03 < \tau_r < 0.06$. Most of the data show the scattering effect, which may be due to the differences in frictional angles, drag from the channel walls and bed morphology, sediment shapes and size distributions. The trend of increasing the critical shear stress τ_{*cr} with the angle $\tan\alpha$ is significant even though the data have not been corrected to account for these effects. Shvidchenko and Pender (2000a, b) and subsequently Shvidchenko et al. (2001) indicated that the incipient motion of grain is slope dependent even on low slopes ($\tan\alpha < 0.01$) and small particles $\left(Re_* = \dfrac{u_* d}{\upsilon_f} < 10^2\right)$, which suggests that slope-dependent critical shear stress is applicable for lowland rivers and the steep mountain streams. However, the empirical trend toward higher critical shear stress on steeper slopes has been fitted using either linear or power law regressions (Mueller et al., 2005; Recking, 2009; Parker et al., 2011).

Lamb et al. (2008) also used the force balance equation for initiation of particle motion in the coordinate system parallel to sediment bed as

$$\tan\varphi = \frac{(F_g - F_B)\sin\alpha + F_D}{(F_g - F_B)\cos\alpha - F_L} \tag{8.217}$$

where

$$F_D = C_D A_{xs} \frac{\rho_f u_b^2}{2} \tag{8.218a}$$

$$F_L = C_L A_{xs} \frac{\rho_f u_b^2}{2} \tag{8.218b}$$

$$F_B = \rho_f g V_{ps} \tag{8.218c}$$

$$F_g = \rho_s g V_p \tag{8.218d}$$

where V_p is the total volume of particle, V_{ps} is the submerged volume of particle with buoyancy force F_B and weight force F_g are defined by Lamb et al. (2008). Using equations (8.218a–d) in equation (8.217), and rearranging in terms of critical Shields stress as:

$$\frac{\tau_{cr}}{(\gamma_s - \gamma_f)d} = \frac{2}{(C_D)_{cr}} \frac{1}{f^2(y/y_0)} \frac{\tan\phi\cos\beta - \sin\beta}{1 + \left(\dfrac{F_L}{F_D}\right)_{cr} \tan\phi} \left[\frac{V_p \rho_f}{A_{xs} d(\rho_s - \rho_f)} \left(\frac{\rho_s}{\rho_f} - \frac{V_{ps}}{V_p}\right)\right] \tag{8.219}$$

Equation (8.219) is identical to the expression (8.146) derived by Wiberg and Smith (1987) except for the term in the third bracket which accounts for the grain emergence. They have also shown the plots of critical shear stress against channel slope $\tan\alpha$ for different Reynolds numbers, and it is shown that critical shear stress increases linearly with increase in channel slope. The plot of near-bed peak turbulence intensity vs. relative roughness is also shown, using the existing data set. There is a negative correlation between the two parameters. The increasing trend of turbulence intensity leads to decreasing trend of relative roughness, which shows from the wide range of roughness types including boulders and gravels in natural stream (Nikora and Goring, 2000), and gravels, spheres, and wire mesh in laboratory flumes. The importance of individual parameters in equation (8.219) is found in detail in Lamb et al. (2008). From several experimental and field studies, Ferguson (2012) found that the critical Shields stress value is higher in steep shallow flows. This was also explained by the force balance equation on the individual grains. A quantitative model was developed based on the assumptions that the critical Shields stress increased with the inclination of the sediment bed, and the model was verified with the low-gradient gravel bed rivers. Ancey et al. (2008) also investigated experimentally the entrainment and motion of coarse particles in a shallow water stream down a steep slope.

8.9 RIVER BANK EROSION

Erosion of river bank is also one of the main disasters along rivers. Numerous factors interact to cause bank erosion or collapse. To examine the interacting factors, a combination of hydrodynamics forces with bank sediments depending on the bank slope and seepage is an effective methodology for understanding the mechanics of river bank erosion. Several investigations have been made to develop semi-empirical equations for the incipient motion of sediment particles over the sloping river bank (Ikeda, 1982; Xie et al., 2009; Lu et al., 2012; and others).

For example, Ikeda (1982) proposed a theoretical model using the hydrodynamics forces to determine the critical shear stress of non-cohesive sediment in a side-sloping bed with an inclination α with the horizon. Based on the force-balanced analysis, Xie et al. (2009) developed a generalized theoretical expression to describe the critical shear stress for incipient motion of non-cohesive sediments particles on a river bank slope with outflow seepage and the flow direction. It is important to note that the effect of seepage on the stability of riverbanks should be considered into account to develop the theoretical model. If the threshold shear stress of sediment motion decreases, the riverbank gets more easily eroded, that means, scour gets accelerated, and consequently river bank becomes steeper, hence the bank collapses due to seepage force, which exerts on the soil particles. Therefore, the conditions of incipient sediment motion with seepage force in addition to the other hydrodynamics forces are important to investigate the mechanism of river bank collapse. Several investigations of seepage on slope stability are documented by Taylor (1948), Howard and McLane (1988) and Worman (1992), taking into consideration of both seepage and surface flow. Traditionally, the dimensionless critical shear stress $\frac{\tau_{cr}}{(\gamma_s - \gamma_f)d}$ in terms of Shields curve relates with the grain Reynolds number $\left(Re_* = \frac{u_* d}{\upsilon_f}\right)$, where $\tau_{cr} = \rho_f u_{*cr}^2$ is the critical shear stress at the top of the sediment particle. Here τ_{cr} should have a definite relationship to the drag force F_{Dc} acting on sediment particle at the critical state. To derive the equation for critical shear stress τ_{cr} or F_{Dc}, all forces acting on sediment particles are analyzed.

Let us consider a bank $A_1B_1C_1D_1$ with a slope angle β (Figure 8.43). Assuming the water flows parallel to the slope surface and it moves forward at an angle α with the bank line. Sediment particles are

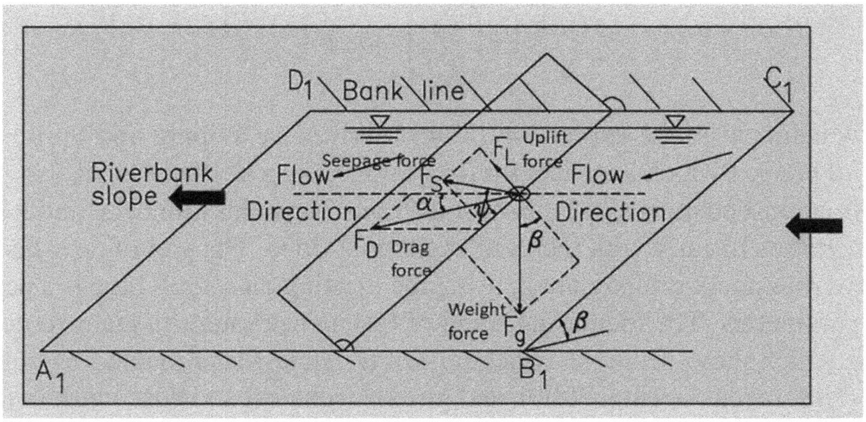

FIGURE 8.43 Sketch of forces acting on a sediment particle on the river bank. (Modified from Xie et al. 2009.)

assumed to be non-cohesive and equal to the mean grain diameter d_{50}. All four forces, i.e., drag force F_D, lift force F_L, weight force F_g, and seepage force F_S, acting on a sediment particle on the slope surface are respectively given by:

$$F_D = \left(\frac{\pi}{8}\right) C_D d^2 \rho_f u_b^2 \tag{8.220a}$$

$$F_L = \left(\frac{\pi}{8}\right) C_L d^2 \rho_f u_b^2 \tag{8.220b}$$

$$F_g = \frac{\pi}{6}(\gamma_s - \gamma_f) d^3 \tag{8.220c}$$

$$F_S = \frac{\pi}{6}\gamma_f h_i d^3 (1+e) \tag{8.220d}$$

where u_b is the approached fluid velocity over the river bank at the sediment particle, h_i is the hydraulic gradient of seepage, e is the void ratio of sediment mass. The approach velocity u_b at the critical state is related to the critical shear velocity u_{*c}, that is, $u_b = u_{*c}/\sqrt{f_*}$, where f_* is the friction factor. The relationship between τ_c and drag force F_{Dc} can be expressed as $\tau_c / F_{Dc} = 8 f_* / \pi C_D d^2$. The seepage factor R_s is defined as:

$$R_s = \frac{h_i \gamma_f (1+e)}{\gamma_s - \gamma_f} \tag{8.221}$$

Using equation (8.221), equation (8.220d) can be written as

$$F_S = \frac{\pi}{6} R_s (\gamma_s - \gamma_f) d^3 = R_s F_g \tag{8.222}$$

If the four forces acting on a sediment particle are projected on the slope surface, two resultant forces: shear force F_R on the slope surface and the normal force F_N perpendicular to the surface are generated and are respectively given by:

$$F_R^2 = (F_D \sin\alpha + F_g \sin\beta + F_S \cos\psi)^2 + (F_D \cos\alpha)^2 \tag{8.223}$$

$$F_N = F_g \cos\beta - F_L - F_S \sin\psi \tag{8.224}$$

where ψ is the angle of seepage force to the slope surface (Figure 8.43). The equilibrium condition for a sediment particle at a critical state is given by

$$\tan\varphi = \frac{F_R}{F_N} = \frac{\sqrt{(F_D \sin\alpha + F_g \sin\beta + F_S \cos\psi)^2 + (F_D \cos\alpha)^2}}{F_g \cos\beta - F_L - F_S \sin\psi} \tag{8.225}$$

where $\tan\varphi$ is the angle of repose. If seepage force is zero ($F_S = 0$), equation (8.225) is given by:

$$\tan\varphi = \frac{F_R}{F_N} = \frac{\sqrt{(F_D \sin\alpha + F_g \sin\beta)^2 + (F_D \cos\alpha)^2}}{F_g \cos\beta - F_L} \tag{8.226}$$

Equation (8.226) represents the equilibrium condition for sediment particles at the critical state at a bank slope angle β and flow forwarding at an angle α with the bank line. If the flow forwarding angle $\alpha = 0$, equation (8.226) is given by

$$\tan\varphi = \frac{F_R}{F_N} = \frac{\sqrt{(F_g \sin\beta)^2 + (F_D)^2}}{F_g \cos\beta - F_L} \tag{8.227}$$

which coincides with the result of Ikeda (1982), who described the results for only bank slope β with horizontal surface.

Squaring both sides of equation (8.226) and after calculations, one gets

$$\tilde{F}^2\left(1 - F' \tan^2\varphi\right)\tau_{*c}^2 + 2\tilde{F}\left(F' \cos\beta \tan^2\varphi + \sin\alpha\cos\beta\right)\tau_{*c} + \sin^2\beta - \cos^2\beta\tan^2\varphi = 0 \tag{8.228}$$

where $F' = \dfrac{F_L}{F_D}$, $\tilde{F} = \dfrac{6F_D}{\pi d^2 \rho_f u_{*c}^2}$, and $\tau_{*c} = \dfrac{\rho_f u_{*c}^2}{(\gamma_s - \gamma_f)d}$

Equation (8.228) is the quadratic equation of τ_{*c}. The positive root of τ_{*c} is the critical shear stress for incipient motion of sediment particle for non-zero angles of β and α. If flow forwarding angle $\alpha = 0$, ,equation (8.228) reduces to the equation of Ikeda (1982) and is given by

$$\tilde{F}^2\left(1 - F' \tan^2\varphi\right)\tau_{*c}^2 + 2\tilde{F}\left(F' \cos\beta \tan^2\varphi\right)\tau_{*c} + \sin^2\beta - \cos^2\beta\tan^2\varphi = 0 \tag{8.229}$$

The positive root of the above equation (8.229) is given by:

$$\tau_{*c} = \frac{-F'\cos\beta\tan^2\varphi + \sqrt{\cos^2\beta\tan^2\varphi + F'^2\sin^2\beta\tan^2\varphi - \sin^2\beta}}{\tilde{F}\left(1 - F'^2\tan^2\varphi\right)} \tag{8.230}$$

If there is no bank slope, i.e., $\beta = 0$, then the critical shear stress is given by:

$$\tau_{*c} = \frac{\tan\varphi}{\tilde{F}\left(1 + F'\tan\varphi\right)} \tag{8.231}$$

Similarly, squaring both sides of equation (8.225) and simplifying for the drag force at the critical state is obtained as (Xie et al., 2009):

$$F_{Dc} = \frac{F_g}{1 - F'^2\tan^2\varphi}\left[\sqrt{A^2\left[(F'\tan\varphi)^2 - \cos^2\alpha\right] + 2ABF'\tan^2\varphi\sin\alpha + B^2\tan^2\varphi}\right.$$
$$\left. - \left(A\sin\alpha + BF'\tan^2\varphi\right)\right] \tag{8.232}$$

where $A = \sin\beta + R_s\cos\psi$, and $B = \cos\beta + R_s\sin\psi$ are the defined parameters.

For bank slope $\beta = 0$, and seepage force $R_s = 0$, one gets $A = 0$, and $B = 1$. Now substituting the values of $A = 0$, and $A = 0$, equation (8.232) can be written as:

$$F_{Dc0} = \frac{F_g \tan\varphi}{1 + F' \tan\varphi} \tag{8.233}$$

From equations (8.232) and (8.233), one gets the critical Shields number ratio as:

$$\frac{\tau_{*c}}{\tau_{*c0}} = \frac{u_{*c}^2}{u_{*c0}^2} = \frac{F_{Dc}}{F_{Dc0}} \tag{8.234}$$

Equation (8.234) represents the ratio of critical bed shear stress on any bank slope with seepage angle to that on horizontal bed. Some particular cases are derived from the above equations: (1) if $R_s = 0$, there is no seepage force through the bank; (2) if $\alpha = 0$, the channel is straight and the flow proceeds parallel to the bank line; (3) if $\alpha = 90°$, the flow follows the direction of the gradient of the slope; (4) if $\beta = 0°$, bank slope is zero, i.e., horizontal slope. The conditions of bank slope $(\beta = 0°)$ and the angle of seepage force $(\psi = 90°)$ coincide with the experiments performed by Cheng and Chiew (1999), and these are verified with their laboratory data (Xie and Yu 2009). It is important to note that for $\alpha = 0$ and $\psi = 90°$, the variation of ratio of critical shear stress $\frac{\tau_{*c}}{\tau_{*c0}}$ with β for various values of hydraulic gradient h_i, demonstrates that $\frac{\tau_{*c}}{\tau_{*c0}}$ decreases with β for a given hydraulic gradient h_i and the decreasing rate is relatively large for $\beta > 10°$. For a given β, the dimensionless critical shear stress $\frac{\tau_{*c}}{\tau_{*c0}}$ decreases sharply with hydraulic gradient h_i. Lu et al. (2012) performed the experiments to study the influence of ground water seepage on the initiation of non-cohesive sediment movement on the river bank slope. Using the force balance equation including the seepage hydraulic gradient, the threshold shear stress condition was derived to study the effect of seepage from hydraulic gradient and the riverbank slope on the sediment motion. The laboratory experiments show that the critical shear stress increases with decreasing riverbank slopes and decreases with increase in seepage hydraulic gradient.

8.10 PROBABILISTIC CONCEPTS OF INCIPIENT SEDIMENT MOTION

Turbulent flow and its corresponding shear stress are random in nature; so it is worthy to examine the incipient motion or entrainment of sand particle, using probabilistic concepts. Numerous sediment transport theories in terms of initiation of sand particles have been developed by using either a deterministic or a probabilistic model. The application of stochastic concept is endorsed by Lane and Kalinske (1939) and Einstein (1942). The turbulence characteristics depend on the nature of flow field near a certain particle under consideration surrounded by different particle sizes, their orientations and compactness. The entrainment or incipient motion depends on the instantaneous velocity fascinated by near-bed velocity fluctuations in addition to the mean flow. The empirical relationship of a critical value for initiation of sediment particle is a difficult proposition because of the randomness of turbulence. So it is suggested that the entrainment mechanism is related to the fluctuations of the velocity rather than the average velocity. The initiations of particle motion have to be expressed with the concept of probability, which relates to instantaneousous shear forces. Several investigations have been made to develop the probabilistic concepts for the incipient motion of sediment particles over the sediment bed (Gessler, 1965, 1970; Grass, 1970; Irvine, 1971; Engelund and Fredsoe, 1976, Fredsoe and Deigaard, 1992; Cheng and Chiew, 1998; Wu and Lin, 2002; Wu and Chou, 2003; and others).

For example, Gessler (1965) made the following assumptions for studying the problems of bed-load transport: (1) the turbulent fluctuations of the bed shear stress are distributed according to normal-error law, and (2) a grain starts to move when the "effective bottom shear stress" on the grain exceeds a critical value which is a function of the grain size and the grain Reynolds number. According to Gaussian distribution, the probability p that in given flow the local shear stress acting on an individual grain being eroded from the bed is given by

$$p = 1 - q\left\{\frac{\tau_0}{(\gamma_s - \gamma_f)d} = \theta < \theta_c\right\}$$

$$\text{or } p = 1 - \frac{1}{\sigma_1\sqrt{2\pi}}\int_{-\infty}^{\frac{\theta_c}{\bar{\theta}}-1} \exp\left(-t^2/2\sigma_1^2\right)dt = 1 - q_{st} \tag{8.235}$$

where $\frac{\theta - \bar{\theta}}{\theta}$, follows a Gaussian distribution with mean zero and variance σ_1^2 with σ_1 is the standard deviation of the shear stress fluctuations, which is approximately 0.57, θ_c is the critical shear stress of the individual grain under consideration, q_{st} is the probability of particles not being eroded determined from the Shields' diagram, t is the dummy variable and the dimensionless average bottom shear stress $\bar{\theta}$ is $\bar{\theta} = \frac{\bar{\tau}_0}{(\gamma_s - \gamma_f)d}$. For a given grain size, the dimensionless critical shear stress $\bar{\theta}_c$ is constant. In this method, grain Reynolds number being known, the values of $\bar{\theta}_c$ could be read directly from the modified Shield's curve provided by Gessler (1965, Figure 8.8). The critical shear stress is obtained from a graph of dimensionless critical shear stress, Shields' parameter, versus grain Reynolds number. The grain will be removed if $\theta > \theta_c$. For coarse material, the standard deviation was determined experimentally to be $\sigma_1 = 0.57\bar{\tau}$, a value which is equivalent to the root mean square of the velocity fluctuation of $u'/u_* \approx 2.4$ at a distance from the rough wall equal to the height of the controlling roughness element. Gessler (1970) suggested this incipient motion model that the self-stabilization process must be of probabilistic nature as well. For maximum $d_{\max}$ and minimum $d_{\min}$ sizes of sediment particles of a bed sample, the retained fraction of sediment having the size less than d is given by:

$$p(d) = \int_{d_{\text{dim}}}^{d} p_f(d)\,dt \tag{8.236a}$$

where $p_f(d)$ is the frequency function of the original distribution. The frequency of retained sediment particles is $p_{\text{Ret}}(d) = k_1 q_{st} p_f(d)$, in which k_1 is a constant to be determined by the condition as: $\int_{d_{\min}}^{d_{\max}} p_{\text{Ret}}\,dt = 1$. The expression for the sediment size distribution of the armor coat or retained particles is as follows:

$$p_{\text{Ret}} = \frac{\int_{d_{\min}}^{d} q_{st} p_f(d)\,dt}{\int_{d_{\min}}^{d_{\max}} q_{st} p_f(d)\,dt} \tag{8.236b}$$

and the grain size distribution of sediment particles being eroded p_{Ero} is given by

$$p_{\text{Ero}} = \frac{\int_{d_{\min}}^{d} (1-q_{st})p_f(d)\,dt}{\int_{d_{\min}}^{d_{\max}} (1-q_{st})p_f(d)\,dt} \tag{8.236c}$$

It is important to note that the dimensionless critical shear stress is assumed to be 0.047 for all grain sizes. In fact, Gessler (1970) studied the grain size distributions for armour coat as well as eroded material for three different shear stresses respectively $\tau_0 = 0.5$ lb per sq ft, 1.0 lb per sq ft, and 1.63 lb per sq ft. It is observed that there is a relatively small variation in the armor coat distribution with bed shear stress. Estimation of the grain size distribution of the armor coat, as well as eroded material, makes possible to calculate the amount of material which will erode prior to the development of a stable condition ensuing from the armoring. Detailed discussions are available in Gessler (1970).

Grass (1970) employed the experimental technique to study the individual bed grain instabilities related to the instantaneous fluid forces acting on the bed, which gave valuable insight into the mechanism involved in the bed instability. To achieve the objective of bed instability process, he considered certain assumptions with regard to the mechanisms. When the fluid velocity flowing over a bed of loose sediment particles is increased sufficiently, individual bed grains start to move in an intermittent and random fashion. This initial bed instability results from the interaction between two statistically distributed random variables: (1) due to random shape, weight and placement of individual grain in the bed, the critical shear stress associated with bed materials will have a probability distribution. If this distribution is measured, then the initial movement characteristics of the bed material can be defined., and (2) due to the turbulence close to the boundary generating the shear stress distribution, the other random variable in the process of initial bed instability results. The boundary turbulence generates random local instantaneous bed shear stress, which has a probability distribution. The flow conditions at the stage of initial bed grain movement can then be defined in terms of the lower and upper limits in these two probability distributions. Grass defined the background flow distribution in terms of the critical bed shear stress distributions associated with the bed material as:

$$\overline{\tau} + n\sigma = \overline{\tau}_c - n\sigma_c \tag{8.237}$$

where n is a multiplicative factor that determines the intersecting point of two distributions, σ is the standard deviation of the distribution of instantaneous shear stress τ values about the mean $\overline{\tau}$, σ_c is the standard deviation in the distribution of instantaneous critical bed shear stress, $\overline{\tau}_c$ is the average values of measured critical instantaneous shear stresses. From the experimental results, he found the approximate relationships as $\sigma_c/\overline{\tau}_c \approx 0.3$ and $\sigma/\overline{\tau} \approx 0.4$. Using the approximate relationships in equation (8.237), one gets

$$\frac{\overline{\tau}}{\overline{\tau}_c} = \frac{(1-0.3n)}{(1+0.4n)} \tag{8.237a}$$

The mean bed shear stresses $\overline{\tau}$ were calculated from the measured $\overline{\tau}_c$ values for $n=0$, 1, 2, which showed straight lines on a logarithmic scale. For $n = 0.625$, the line agrees with that of Shields. Details are available in Grass (1970).

Cheng and Chiew (1998) derived a theoretical model for the initiation of sediment particles, using a pickup probability concept. The pickup probability is considered when the instantaneous lift force acting on the particle exceeds its effective weight force (Einstein, 1950). Hence, the pickup probability for a bed particle can be defined as that of the lift force greater than the effective weight force of the particle:

$$P = P(F_L > F_g) \tag{8.238}$$

where F_L is the instantaneous lift force acting on a particle, and F_g is the submerged weight force of the particle. This pickup probability is also defined as the fraction of the time when the lift force is greater than the effective weight for a given time interval, or the percentage of the number of particles in motion on a fixed area of bed surface (Einstein, 1942, 1950). Now lift force is given by:

$$F_L = \frac{\pi}{8} C_L d^2 \rho_f u_b^2 \tag{8.239}$$

and the weight force is as:

$$F_g = \frac{\pi}{6}(\gamma_s - \gamma_f) d^3$$

Using equation (8.239) and equation (8.240) in equation (8.238), one gets

$$P = P\left(u_b^2 > B^2\right) = P(u_b > B) + P(u_b < -B) \tag{8.241}$$

where $B = \sqrt{\dfrac{4(\gamma_s - \gamma_f)d}{3C_L}}$

Figure 8.44 shows the probability distribution of equation (8.241). If the velocity u_b assumes normal distribution, its density function can be written as

$$f(u_b) = \frac{1}{\sigma\sqrt{2\pi}} \exp\left\{-\frac{(u_b - \bar{u}_b)^2}{2\sigma^2}\right\} \tag{8.242}$$

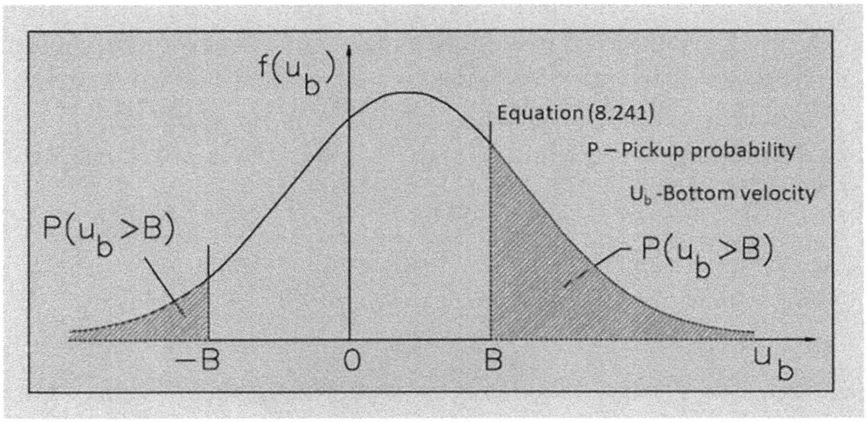

FIGURE 8.44 Pickup probability distribution.

where $\bar{u}_b$ is the time-mean value of the instantaneous velocity u_b and the longitudinal intensity of turbulence is σ. From Equation (8.241), one can write as

$$P = 1 - \int_{-B}^{+B} f(u_b)\,du_b = 1 - \int_{-B}^{+B} \frac{1}{\sqrt{2\pi}} \exp\left\{ -\frac{(u_b - \bar{u}_b)^2}{2\sigma^2} \right\} d\left(\frac{u_b - \bar{u}_b}{\sigma} \right) \tag{8.243}$$

Using the error approximation analysis less than 0.7% from Guo (1998), equation (8.243) is approximated as:

$$P = 1 - 0.5 \frac{B - \bar{u}_b}{|B - \bar{u}_b|} \sqrt{1 - \exp\left[-\frac{2(B - \bar{u}_b)^2}{\pi\sigma^2} \right]} - 0.5 \sqrt{1 - \exp\left[-\frac{2(B + \bar{u}_b)^2}{\pi\sigma^2} \right]} \tag{8.244}$$

Here the approached velocity u_b of equation (8.244) is determined from the universal log-law. Following Cheng, and Chiew (1998), the approached velocity u_b on the bed $y_b = 0.6d$ with $k_s = 2d$, one gets $\bar{u}_b = 5.52u_*$. Einstein and El-Samni (1949) performed an experiment for Reynolds number Re_* ranging from 3300 to 6500 and proposed a lift coefficient of 0.178, and then the approach velocity u_b at the level $0.35d$ above the theoretical bed level for their computation. Li et al. (1983) measured the hydrodynamic forces acting on a single sphere lying on a smooth bed, and they used lift coefficient $C_L = 0.18$.for high Reynolds number Re_* with the velocity at the level $0.5d$ from the bed surface.

Cheng and Chiew (1998) determined the pickup probability from equation (8.244) using approached mean velocity $\bar{u}_b = 5.52u_*$ and standard deviation $\sigma = 2.0u_*$ for hydraulically rough bed and is given by

$$P = 1 - 0.5 \frac{0.21 - \sqrt{\theta C_L}}{\left|0.21 - \sqrt{\theta C_L}\right|} \sqrt{1 - \exp\left[-\frac{\left(0.46 - 2.2\sqrt{\theta C_L}\right)^2}{\theta C_L} \right]}$$

$$-0.5 \sqrt{1 - \exp\left[-\frac{\left(0.46 + 2.2\sqrt{\theta C_L}\right)^2}{\theta C_L} \right]} \tag{8.245}$$

Equation (8.245) represents the pickup probability depending on the dimensionless shear stress for known lift coefficient C_L. Figure 8.45 shows the family of probability p against the dimensionless shear stress $\theta = \dfrac{\rho_f u_*^2}{(\gamma_s - \gamma_f)d}$ for various values of C_L ranging from 0.1 to 0.4. The dimensionless shear stress

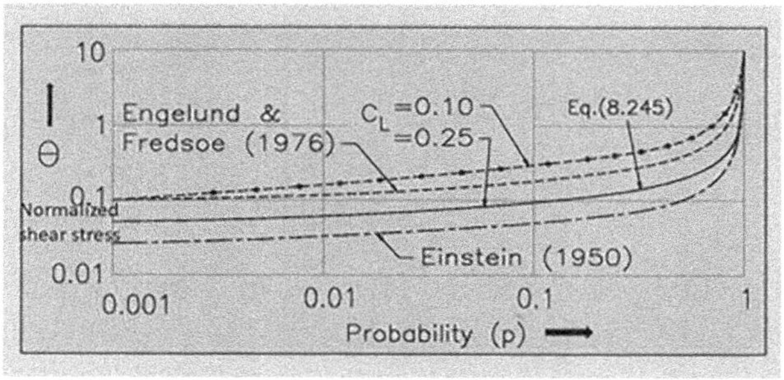

FIGURE 8.45 Pickup probability p against normalized shear stress θ for lift coefficients, using equation (8.245) and verified with the experimental data. (Modified from Cheng and Chiew, 1998.)

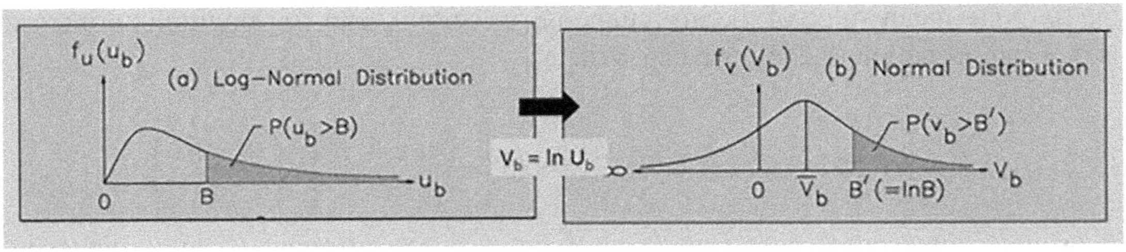

FIGURE 8.46 Plots of (a) log-normal and (b) normal distributions. (Modified from Wu and Lin, 2002.)

$\dfrac{\rho_f u_*^2}{(\gamma_s - \gamma_f)d}$ for high Reynolds number $Re_* > 500$ in the Shields curve has often been cited in the literature with a value of approximately 0.05. Based on the results, the pickup probability p is about 0.6% for $\theta = \dfrac{\rho_f u_*^2}{(\gamma_s - \gamma_f)d} = 0.05$. The critical condition of incipient motion computed by Shields criterion is equivalent to the pickup probability of about 0.6%.

Cheng and Chiew (1998) presented a theoretical model of the lifting probability for sediment entrainment, using the normal distribution for instantaneous velocity, and later the model was modified by Wu and Lin (2002), using the log-normal distribution.

Wu and Lin (2002) presented the formulation of the pickup probability for sediment entrainment under the log-normally distributed instantaneous velocity. Figure 8.46a and b shows log-normal and normal distribution. The results were compared with the published experimental data and the earlier existing pickup probability curve for the normal velocity distribution. If v_b denotes the logarithm of u_b (i.e. $v_b = \ln u_b$), the probability density function v_b can be expressed as:

$$f_v(v_b) = \frac{1}{\sigma_v\sqrt{2\pi}}\exp\left[-\frac{(v_b - \overline{v_b})^2}{2\sigma_v^2}\right] \tag{8.246}$$

where $\overline{v_b}$ and σ_v are respectively the mean and standard deviation of v_b. The pickup probability given in equation (8.241) should be modified as the following:

$$P = P(u_b > B) = P(v_b < \ln B) = 1 - P(-\infty < v_b < \ln B) \tag{8.247}$$

Using this idea of log-normal distribution, in a similar way, equation (8.245) is modified as:

$$P = 0.5 - \frac{\ln B - \overline{v_b}}{2\left|\ln B - \overline{v_b}\right|}\sqrt{1 - \exp\left[-\frac{2(\ln B - \overline{v_b})^2}{\pi\sigma_v^2}\right]}$$

To calculate the pickup probability, one needs the mean $\overline{v_b}$ and standard deviation σ_v with mean velocity $\overline{v_b} = 5.52u_*$ and the turbulence intensity $\sigma_v = 2u_*$ for log-normal distribution. Herein two methods are used for the transformation, namely, the analytical method and the first-order approximation method. For analytical method, the values of $\overline{u_b}$ and σ_u^2 yield $\overline{v_b} = \ln(5.19u_*)$ and $\sigma_v^2 = 0.123$. Using $\overline{v_b}$ and σ_v^2, above equation results in revised form as:

$$P = 0.5 - \frac{\ln\left(\frac{0.049}{\theta C_L}\right)}{2\left|\ln\left(\frac{0.049}{\theta C_L}\right)\right|}\sqrt{1-\exp\left[-\frac{2}{\pi}\left[\frac{\ln\left(\frac{0.049}{\theta C_L}\right)}{0.702}\right]^2\right]} \tag{8.248}$$

For first order approximation method, the equation will be as:

$$P = 0.5 - \frac{\ln\left(\frac{0.044}{\theta C_L}\right)}{2\left|\ln\left(\frac{0.044}{\theta C_L}\right)\right|}\sqrt{1-\exp\left[-\frac{2}{\pi}\left[\frac{\ln\left(\frac{0.044}{\theta C_L}\right)}{0.724}\right]^2\right]} \tag{8.249}$$

Equations (8.248) and (8.249) both indicate that, for a given lift coefficient C_L, the pickup probability is only dependent on the dimensionless shear stress $\theta = \frac{\rho_f u_*^2}{(\gamma_s - \gamma_f)d}$. For comparison, the theoretical pickup probabilities with experimental data are plotted in Figure 8.47. The optimal C_L=0.21 for log-normal, and C_L=0.25 for normal distributions.

In their analysis, Choi and Kwak (2001) presented three modes of incipient motion of sediment particles, namely, rolling, sliding and lifting, using the concept of the probabilistic approach from the deterministic point of view. They also used the forces acting on a spherical particle lying on the bed, such as weight, drag, lift and resisting forces, and are given by

$$F_g = \frac{\pi}{6}(\gamma_s - \gamma_f)d^3 \tag{8.250}$$

$$F_D = \frac{\pi}{8}C_D d^2 \rho_f u_b^2 \tag{8.251}$$

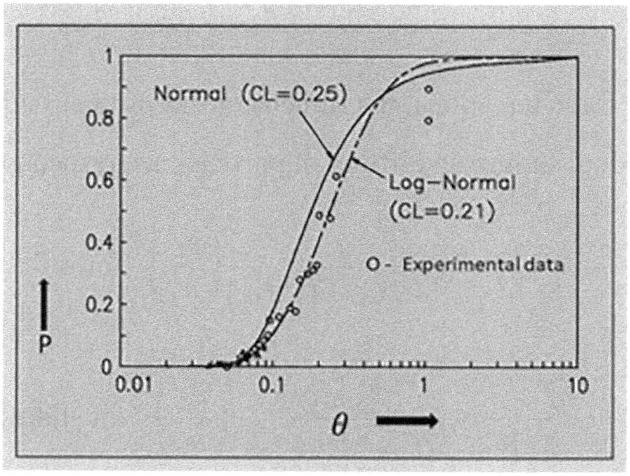

FIGURE 8.47 Comparison of theoretical pickup probabilities P against dimensionless shear stress θ with experimental data. The optimal C_L=0.21 for log-normal, and C_L=0.25 for normal distributions, compared with the experimental data. (Modified from Wu and Lin, 2002.)

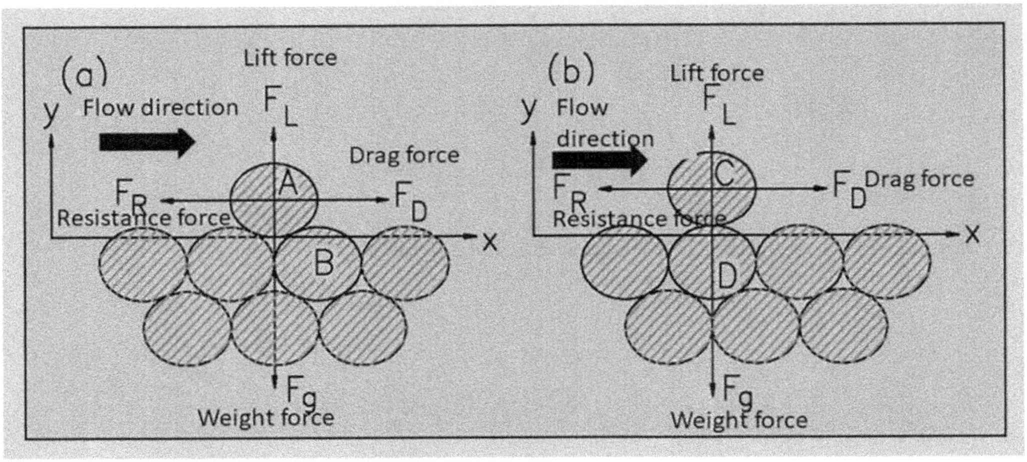

FIGURE 8.48 Schematic diagram: (a) rolling and lifting, and (b) sliding and lifting of particle. (Modified from Choi and Kwak, 2001.)

$$F_L = \frac{\pi}{8} C_L d^2 \rho_f u_b^2 \tag{8.252}$$

$$F_R = \tan\phi\left(F_g - F_L\right) \tag{8.253}$$

where d is the particle diameter and u_b is the effective velocity near the bed. For the closely packed three-dimensional arrangement of spheres shown in Figure 8.48, following the moment analysis of Coleman (1967), the threshold condition for a particle A to roll over a particle B is given by

$$2\sqrt{2}F_D + F_L = F_g \tag{8.254}$$

The sliding motion of particle C occurs, when F_D is the same as F_R, which means

$$F_D = F_R \tag{8.255}$$

From the above-mentioned force equations, the dimensionless critical bed shear stress $\frac{\tau_{0c}}{\left(\gamma_s - \gamma_f\right)d}$ in threshold conditions for rolling, sliding and lifting of particles are respectively given by:

$$\frac{\tau_{0c}}{\left(\gamma_s - \gamma_f\right)d} = \frac{2}{3\left(\sqrt{2}C_D + 0.5C_L\right)}\left(\frac{u_b}{u_*}\right)^{-2} \quad \text{for rolling} \tag{8.256}$$

$$\frac{\tau_{0c}}{\left(\gamma_s - \gamma_f\right)d} = \frac{4\tan\phi}{3\left(C_D + \tan\phi\, C_L\right)}\left(\frac{u_b}{u_*}\right)^{-2} \quad \text{for sliding} \tag{8.257}$$

$$\frac{\tau_{0c}}{\left(\gamma_s - \gamma_f\right)d} = \frac{4}{3C_L}\left(\frac{u_b}{u_*}\right)^{-2} \quad \text{for lifting} \tag{8.258}$$

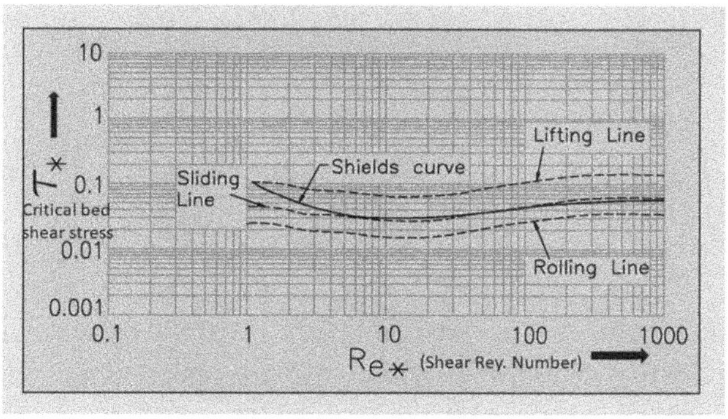

FIGURE 8.49 Threshold conditions for rolling, sliding and lifting with respect to Re_*. (Modified from Choi and Kwak, 2001.)

where the effective velocity u_b is defined at $b = y = d/2$ and has to be determined. Figure 8.49 shows the threshold conditions for lifting, sliding and rolling with respect to shear Reynolds number Re_*. The effective velocity u_b is determined from the three different layers: viscous sub-layer, transitional layer and log-layer for rough boundary for friction Reynolds number $Re_* > 10$. In their analysis, they considered Reichardt's formula (1951) for velocity in a smooth transition between two regions. Choi and Kwak (2001) assumed the effective velocity u_b as normally distributed following Cheng, and Chiew (1998), the probability density function can be written as Equation (8.242) with the standard deviation or turbulence intensity.

Nezu and Nakagawa (1993) proposed a semi-empirical relation of turbulence intensity very close to the wall as:

$$\frac{\sigma}{u_*} = D_u \exp\left(-\frac{y^+}{Re_*}\right)\left(1-\exp\left(-\frac{y^+}{D_b}\right)\right) + D_c y^+\left[1-\left(1-\exp\left(-\frac{y^+}{D_b}\right)\right)\right] \tag{8.259}$$

where D_u, D_b and D_c are the empirical constants with respective values 2.3, 10 and 0.3; $y^+ = (u_* y/\upsilon_f)$ dimensionless distance, and Re_* is the frictional Reynolds number. In the intermediate and outer regions, Nezu and Nakagawa (1993) found an expression of turbulence intensity as:

$$\frac{\sigma}{u_*} = D_* \exp\left(-\frac{y}{h}\right) \tag{8.260}$$

where the value of empirical constant D_* is 2.3 for the intermediate region and 2.26 for the outer region. Choi and Kwak (2001) used the equation (8.259) for turbulence intensity near the wall region with a value of $\frac{\sigma}{u_*} = 2$ for the outer region. The initiation of particle motion by rolling in the flow, the condition (8.254) holds good, and the probability P_r of rolling is given by:

$$P_r = P(2\sqrt{2}F_D + F_L > F_g) \tag{8.261}$$

Using F_D, $F_L$and F_g, one gets

$$P_r = P\left(u_b^2 > R^2\right) = P(u_b > R) + P(u_b < -R) \tag{8.262}$$

where the parameter R is defined as Choi and Kwak (2001). Using equation (8.262), equation (8.243) can be written as:

$$P = 1 - \int_{-B}^{+B} f(u_b)\,du_b = 1 - \int_{-R}^{+R} \frac{1}{\sqrt{2\pi}} \exp\left\{-\frac{(u_b - \bar{u}_b)^2}{2\sigma^2}\right\} d\left(\frac{u_b - \bar{u}_b}{\sigma}\right) \tag{8.263}$$

Integration of the above equation by using error function estimates the rolling probability.

Similarly, the sliding and lifting probabilities $(P_s,\ P_l)$ are respectively defined by

$$P_s = P\left(u_b^2 > S^2\right) = P(u_b > S) + P(u_b < -S) \tag{8.264}$$

$$P_l = P\left(u_b^2 > L^2\right) = P(u_b > L) + P(u_b < -L) \tag{8.265}$$

where the parameters S and L are defined in Choi and Kwak (2001). Similarly, one can find the probabilities from equation (8.263) for sliding and lifting only by replacing the parameter R with the respective parameters. Each probability equations indicate the percentage of the number of particles in motion (rolling, sliding and lifting) on a fixed bed area (Einstein, 1950; Cheng and Chiew, 1998). From the calculation they observed that Shields curve corresponds to a probability of 0.70, 0.45, and 0.04 for rolling, sliding, and lifting, respectively, at large values of Reynolds number Re_*, which means if the sediment particles are exposed to Shields critical stress, the chances will be about 70% for rolling, 45% for sliding, and 4% for lifting, respectively. Details are available in Choi and Kwak (2001).

Wu and Chou (2003) investigated the rolling and lifting probabilities by introducing the probabilistic concepts of turbulent fluctuation and bed grain geometry, in hydraulically smooth-bed and transitional open-channel flows. Two threshold conditions were identified to define the probabilities of entrainment in the rolling and lifting modes. The lifting probability was derived by Einstein (1950) for the bed load function (discussed in the next Chapter), the rolling probability was derived by Sun and Donahue (2000) in the fractional bed load equation and the sliding probability was derived by Paintal (1971) for bed load model. Wu and Chou (2003) considered the fluctuations of velocity and the randomness of bed grain geometry in the derivation of rolling and lifting probabilities, which is an extension of previous work by Wu and Lin (2002). Wu and Lin (2002) studied the pickup probability of sediment using the log-normal velocity distribution.

Consider a spherical particle of size d resting on the bed consisting of identical spheres as shown in Figure 8.50. Here the coordinates x, y, and z axes represent the longitudinal, vertical, and transverse directions, respectively. The virtual bed level ($y = 0$), where the flow velocity is zero, is set at a distance below the top of the bed grains. Here, sphere 1 (S1) is in contact with an upstream at sphere 2 (S2) and a downstream at sphere 3 (S3). The points of contact between the sphere S1 and S3 leveled at C is at a distance h' from the bed level, the bottom of the solitary particle is at $y = \delta$ from the bed level. The lower and the upper limits of δ depending on minimum and maximum compactness of the bed particles are $\delta = -0.75$d and 0.116d, respectively. When $\delta = -0.75$d, $S1$ is at the lowest possible position to protrude into the flow; when $\delta = 0.116$d, $S1$ is resting at the highest possible position to remain stable. The initial position of $S1$ is oriented randomly relative to the bed level, thus δ is treated as a random variable within the two limits, $\delta = -0.75d$ and 0.116d; the probability density function $P(\delta)$ is expressed as:

$$P(\delta) = \frac{1}{(0.116d + 0.75d)} = \frac{1.155}{d} \quad \text{for} -0.75d \le \delta \le 0.116d \tag{8.266}$$

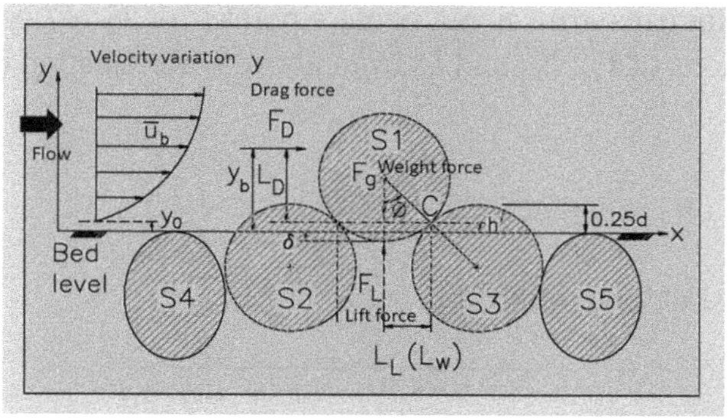

FIGURE 8.50 Schematic of longitudinal section (*xy*-plane) of bed grain geometry, flow velocity and forces acting on S1. (Modified from Wu and Chou, 2003.)

The external forces acting on the submerged $S1$ include weight force F_g, hydrodynamic drag force F_D, and lift force F_L, which are given by (8.250) to (8.252) as:

$$F_g = \frac{\pi}{6}(\gamma_s - \gamma_f)d^3 \tag{8.250}$$

$$F_D = \frac{\pi}{8}C_D d^2 \rho_f u_b^2 \tag{8.251}$$

$$F_L = \frac{\pi}{8}C_L d^2 \rho_f u_b^2 \tag{8.252}$$

Notations are already explained earlier. The area-averaged instantaneous velocity u_b is used to account for the effect of fluctuating fluid forces. The drag coefficient of a sphere C_D depends on the Reynolds number (Schiller and Naumann, 1933) as:

$$C_D = \frac{24}{Re}\left(1 + 0.15Re^{0.687}\right) \tag{8.267}$$

where $Re = \dfrac{\bar{u}_b d'}{\upsilon_f}$, $d' = 0.75d + \delta$, and $\bar{u}_b$ is the spatiotemporal mean velocity. Equation (8.267) is valid for $Re \le 1.754$, while $C_D = 0.36$ for $Re > 1.754$ in smooth bed flow (Graf, 1971). Chepil (1958) provided the average ratio of lift to drag is nearly a constant value at about 0.85 from his data set for $Re = \dfrac{u_* d'}{\upsilon_f} \le 5000$. (For details description, readers are suggested to see the paper by Wu and Chou, 2003). The spatiotemporal mean velocity $\bar{u}_b$ is determined from the universal log-law with $y_0 = \dfrac{k_s}{30}$, $k_s = 2d$ and is given by equation (8.7) of Wu and Chou (2003).

Wu and Chou (2003) determined the threshold conditions for rolling, using the force moments causing the particle motion to exceed those keeping the particle at rest, and the condition can be expressed as:

$$F_D L_D + F_L L_L > F_g L_g \tag{8.268}$$

where L_D, L_L, and L_g are moments arms about contact point C of F_D, F_L, and F_g respectively. Now the combination of force equations of F_D, F_L, and F_g from above with equation (8.268), one gets the following as:

$$u_b^2 > B_R^2 \tag{8.269}$$

where B_R is the rolling threshold parameter as:

$$B_R = \sqrt{\frac{2L_g}{C_D L_D + C_L L_L} \frac{\pi(\gamma_s - \gamma_f)d^3}{6A\rho_f}} \tag{8.270}$$

They also determined the threshold conditions for lifting to occur, using the concept of dynamic lift on a sediment particle exceeds its submerged weight, and the condition is expressed as:

$$F_L > F_g \tag{8.271}$$

Employing the forces in Equation (8.271), one gets as:

$$u_b^2 > B_L^2 \tag{8.272}$$

where B_L is the lifting threshold parameter, and is given by

$$B_L = \sqrt{\frac{2\pi(\gamma_s - \gamma_f)d^3}{6A\rho_f C_L}} \tag{8.273}$$

The ratio of the parameters of rolling and lifting thresholds for $C_L = C_D$ is given by

$$\frac{B_R}{B_L} = \sqrt{\frac{L_L}{L_D + L_L}} < 1 \tag{8.274}$$

equation (8.274) represents the threshold for lifting is higher than that of rolling, which agrees with the results of Ling (1995). Once the thresholds for lifting and rolling are identified with the probability distribution of the instantaneous velocity, the probabilities of entrainment can be defined.

If the instantaneous velocity u_b follows the log-normal distribution, the probability density function (pdf) of $v_b = \ln u_b$ can be represented as:

$$f(v_b) = \frac{1}{\sigma\sqrt{2\pi}} \exp\left\{ -\frac{(v_b - \overline{v_b})^2}{2\sigma^2} \right\} \tag{8.275}$$

where $\overline{v_b}$ and σ are mean and standard deviation of v_b, respectively. Because u_b is a non-negative variable, the threshold conditions demonstrated in equations (8.269) and (8.272) can be modified as $u_b > B_R$ for rolling and $u_b > B_L$ for lifting. If $B_R < u_b < B_L$, the particle entrains in a pure rolling mode, keeping in contact with the bed. However, if $u_b > B_L$, the incipient motion of particle occurs in simultaneously rolling-lifting mode (Figure 8.51). In other words, the particle is lifted off the bed while it starts to roll. Because the lifted particle is no longer in contact with the spheres below it, herein we identify this type

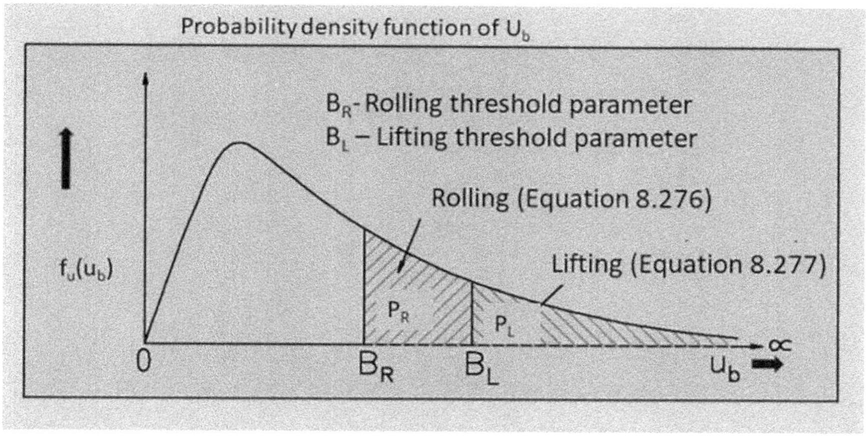

FIGURE 8.51 Probability density function of u_b and definition of rolling and lifting probabilities. (Modified from Wu and Chou, 2003.)

of entrainment for $u_b > B_L$, as the lifting mode. Based on these, it is now possible to define the rolling and lifting probabilities.

The probability of entrainment in the rolling mode is expressed as:

$$P_{\text{rolling}} = P(B_R < u_b < B_L) = P(B_R' < v_b < B_L') = P(v_b < B_L') - P(v_b < B_R') \tag{8.276}$$

where $B_R' = \ln B_R$, and $B_L' = \ln B_L$. Here the velocity v_b is normally distributed; equation (8.276) is approximated and given in Wu and Chou (2003). Similarly, the probability of entrainment in the lifting mode can be expressed as:

$$P_{\text{lifting}} = P(u_b > B_L) = P(v_b > B_L') = 1 - P(-\infty < v_b < B_L') \tag{8.277}$$

Following the same procedure, using the velocity v_b normally distributed; one can get the equation as given in Wu and Chou (2003). Using the method proposed by Wu and Chou (2003), the rolling probabilities corresponding to a range of dimensionless shear stress $\left(\text{Shields parameter} = \dfrac{\rho_f u_*^2}{(\gamma_s - \gamma_f)d} \text{ within 0.001 to 10}\right)$ are evaluated numerically. Similarly, the lifting probabilities corresponding to a range of dimensionless shear stress θ between 0.001 and 10 can be evaluated numerically. Figure 8.52 shows the relationships between the computed results and the dimensionless shear stress $\theta = \dfrac{\rho_f u_*^2}{(\gamma_s - \gamma_f)d}$, where one can see the distinct difference between the lifting and rolling probabilities. The lifting probability increases monotonically with θ, whereas the rolling probability increases with θ in the region of $\theta < 0.15$, then reduces for larger values of θ, and these are verified with the results of Guy et al. (1966), Luque (1974), Jain (1992) and Papanicolaou (1999). Because rolling and lifting are mutually independent modes as illustrated in Figure 8.51, the total probability of incipient motion (P_T) is equal to the sum of rolling and lifting probabilities, i. e., $P_T = P_{\text{rolling}} + P_{\text{lifting}}$, and is shown in Figure 8.52. Details descriptions are available in Wu and Chou (2003). With the approach of pickup probability, the bed load function may be derived.

Several investigators such as Mingmin and Qiwei (1982), Keshavarzy (1997), Papanicolaou (1999), Dancey et al. (2002) Papanicolaou et al. (2001, 2002) and others, developed probabilistic models for predicting sediment entrainment for different flow regimes. Keshavarzy (1997) introduced the probabilistic

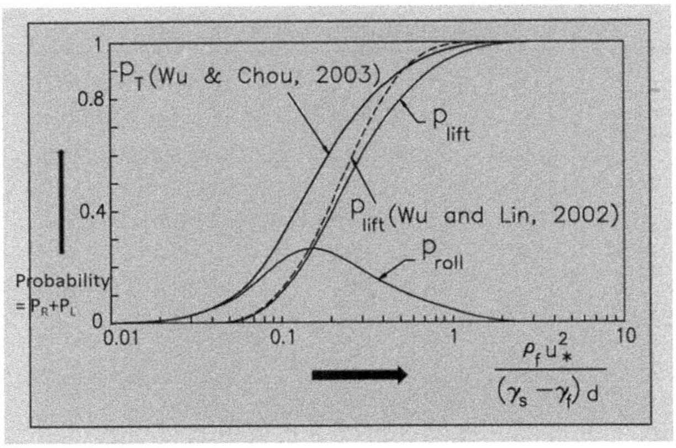

FIGURE 8.52 Entrainment probabilities vs. dimensionless shear stress. (Modified from Wu and Chou, 2003.)

aspects of particle motion through the turbulence characteristics of the flow, where the energy aspects were considered through an instantaneous force balance model. In this manner, a stochastic-deterministic model for particle motion was developed. Dancey et al. (2002) investigated the probability of individual particle, considering the statistical nature of sediment movement in turbulent flow, where the mechanism was strongly dependent on the packing density of sediments.

8.11 THRESHOLD OF INCIPIENT MOTION DUE TO TURBULENT BURSTING

The experimental work has shown that the nature of flow near the wall in the turbulent boundary layer is repetitive, and the flow at the wall is quasi-cyclic process, which is a deterministic sequence of events that occurs randomly in space and time. Kline et al. (1967) first discovered the idea of turbulent bursting to explain the turbulent features based on the transfer of momentum between the turbulent and laminar regions near a boundary. Taking into consideration of quasi-cyclic process of flow pattern and the basic flow features, Sutherland (1967) first investigated the particle entrainment in turbulent flows by making use of turbulence near the wall. His study was considered to include both the initiation of particle motion and the suspension of particles by the flow. The concept of initiation of sediment particles was implemented only after the discovery of turbulent bursting phenomena (explained in Chapter 6). He also argued that sediment entrainment was associated with the near-bed turbulent eddies which help to roll the particles. Offen and Kline (1973, 1974, 1975) proposed models explaining the complete cycle of events. According to their investigations, the flow consists of transverse changes of low-speed and high-speed streak regions, which form locally and momentarily in the boundary. The former and the latter events are called subsequently as ejections and sweeps in these two phases of this sequence. In the ejection phase, low-speed fluid is ejected away from the wall, and in the sweep phase, high-speed fluid penetrates towards the wall, both in the form of three-dimensional disturbance. Major production of turbulence and subsequent dissipation occur in this region, and naturally the size of eddies decreases through diffusion. They also viewed the ejection of low-speed fluid (that is, lift-up of the wall streak) as an upwelling motion of this sub-boundary layer similar to local separation of flow due to a temporary local adverse pressure gradient. Immediately after the wall streak is lifted up, a local convected recirculation cell will develop below the lifted streak (Figure 8.53a).

As the ejection grows, both the lifted fluid and the recirculation cell will move away from the wall and develop in size (Figure 8.53b). The flow along the lowest side of the convected cell will be in the reverse direction with respect to an observer moving with the convection speed of this recirculation cell. The lifted wall streak continued to grow with a passage over the next low-speed streak downstream (Figure 8.53c); and the condition will exist for the occurrence of another lift as this structure continues

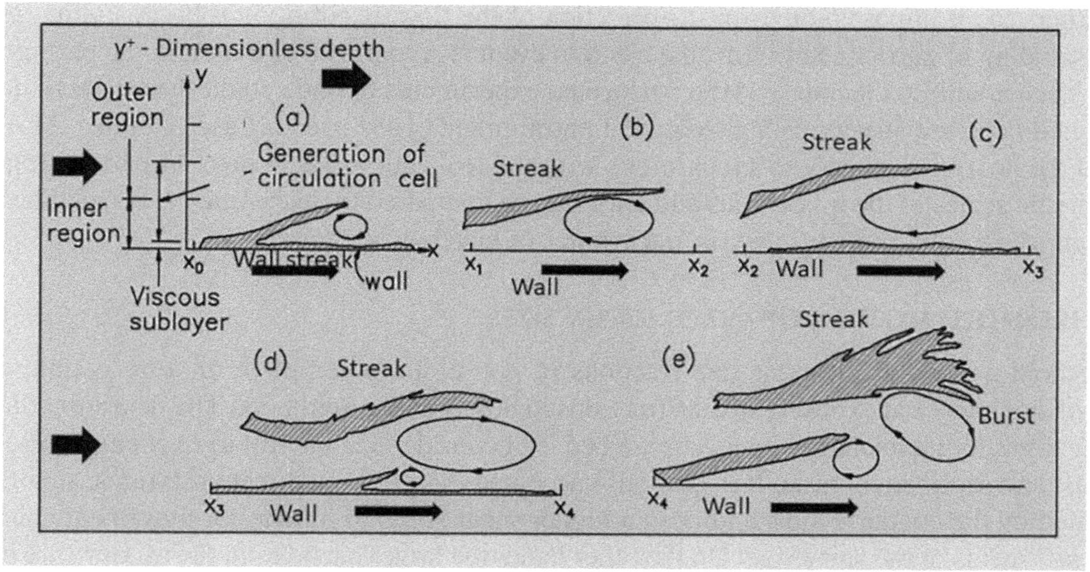

FIGURE 8.53 Instantaneous views of a turbulent burst. (Modified from Offen and Kline, 1975.)

to pass overhead (Figure 8.53d), and then the lifted fluid breaks up as it interacts with the next lifting streak (Figure 8.53e). The whole sequence of events during the lifetime of a wall streak, i.e., appearance and growth of the wall streak leading to lift-up from the wall, and breakup of coherency of the structure, is called bursting. There has been a major development on the relationship between turbulent bursting and the transport of sands in marine and river environments (Jackson, 1976; Sumer and Oguz, 1978; Nelson et al., 1995; Cao, 1997). For example, Jackson (1976) was the first to provide a new insight into the bursting process of turbulent boundary layers to study the mechanics of sedimentation, and genesis of bedforms in natural geophysical flows. The experiments suggest that the turbulent boundary layer can be divided into an inner zone and an outer zone. The inner zone is designated by a viscous sub-layer displaying span-wise alternations of high and low-speed streaks and periodically disrupted by lift-ups of low-speed streaks, and the outer zone is designated by an oscillatory growth and breakup stages of bursting.

The burst cycle exists in turbulent boundary layers of all natural flows. This bursting process promotes more entrainment of sediment particles than the tractive force alone. Jackson pointed out that the high-speed streaks preferentially suffer sweeps, whereas low-speed streaks undergo lift-up, in the following manner: to preserve the continuity, fluid nearest to the bed must be transported laterally from high-speed streaks to low-speed streaks, which would account for grain segregation into the latter. From that idea, the researchers are involved to study the role of turbulence bursting on sediment entrainment. In entrainment of sediment particles, Clifford et al. (1993) and Nelson et al. (1995) suggested to correlate the local sediment movement with the turbulent events because of the change in the near-bed turbulence. They observed that durations of sweeps and outward interactions near the bed strongly influence the sediment transport process. Moreover, they also showed that the magnitude of the outward interactions increased relative to other events. Cao (1997) discussed a theoretical model of sediment entrainment from the flat and loose sediment bed, using the averaged bursting period scaled on the inner variables and spatial scales of burst. The entrainment flux was compared with the available laboratory data set from hydraulically smooth and the transitional bed situations (van Rijn, 1984), and found satisfactory results of the developed entrainment function. In their study, Keshavarzy and Ball (1997) recognized the importance of bursting events in transfer of momentum into the laminar boundary layer. Particularly, the sweep event is the most important bursting event for entrainment of sediment particles because it imposes forces in the direction of the flow, resulting in rolling, sliding, and occasionally saltating of particle. Similarly, the ejection event is recognized to maintain the particles in suspension. Sechet and Le Guennec (1999) performed experiments to understand the interaction between the near-bed coherent structures and sediment entrainment. They used a Laser Doppler Anemometer to record the near-bed velocity to synchronize with the real-time measurement of particle trajectories. Characteristic scales of the trajectories and bursting are compared to determine the correlation between the motion of the solid particles and the occurrence of ejections and/or sweeps.

8.12 THRESHOLD MOTION OF MIXED GRAIN SIZES

The threshold motion of different size fractions of non-uniform sediment mixture is quite complex because of shielding and exposure of size fractions in non-uniform sediment. The determination of critical shear stress for incipient motion in stream bed is of considerable interest to river engineers, geomorphologists, and sedimentologists. It is generally accepted that in the sediment mixtures, smaller grains are shielded by the coarse grains, and need a higher shear stress to initiate for movement compared to uniform sediments of the same size. On the other hand, the larger particles in the mixture are entrained at a lower shear stress than a bed of uniform sediment because of increased exposure and instability. From the observations of the researchers in the laboratory studies (Day, 1980a, 1980b; Dhamotharan et al., 1980; Misri et al., 1984) and in the field (Milhous, 1973; Parker et al., 1982), the critical shear stress for individual size fractions is an integral part of mixed-size sediment transport. It was observed that the critical shear stress of individual size fractions in mixed-size sediments is appreciably different from that of uni-size sediments. The threshold shear stress of a given size fraction depends on the sediment mixture and the surface texture.

Egiazaroff (1965) first proposed an equation to determine the dimensionless critical bed shear stress $\frac{(\tau_0)_{Cr}}{(\gamma_s-\gamma_f)d_i}$ of size fraction d_i of non-uniform sediment mixtures. He assumed the velocity at a distance of $0.63d$ is equal to the fall velocity of the sediment. The expression of critical shear stress for any size d_i is given by

$$\frac{(\tau_0)_{Cr}}{(\gamma_s-\gamma_f)d_i}=\frac{0.1}{\left[\log 19\left(\frac{d_i}{d_{50}}\right)\right]^2} \tag{8.278}$$

where d_{50} is that size for which 50% of the mixture is finer (or median). For fine-graded mixture $d_i < d_{50}$, the resistance towards incipient motion is increased, the opposite is true for coarse-graded mixture for $d_i \gg d_{50}$. Ashida and Michiue (1971) modified the critical shear stress equation (8.278) for $\frac{d_i}{d_{50}} < 0.4$. The equation is given by

$$\frac{(\tau_0)_{Crd_i}}{(\tau_0)_{Crd_{50}}}=\frac{0.85}{\frac{d_i}{d_{50}}} \tag{8.279}$$

where $(\tau_0)_{Crd_{50}}$ is the dimensionless critical shear stress for the median size d_{50}. The following modifications of equation (8.279) were suggested by Hayashi et al. (1980) as:

$$\frac{(\tau_0)_{Crd_i}}{(\tau_0)_{Crd_{50}}}=\frac{1}{\frac{d_i}{d_{50}}}\text{ for }\frac{d_i}{d_{50}} < 1.0 \tag{8.280}$$

$$\frac{(\tau_0)_{Crd_i}}{(\tau_0)_{Crd50}} = \left(\frac{\log 8}{\log 8 + \log\left(\frac{d_i}{d_{50}}\right)} \right)^{-2} \quad \text{for} \ \frac{d_i}{d_{50}} > 1.0 \tag{8.281}$$

Equations (8.280) and (8.281) can be used to find the critical shear stress for any size particle in the mixture provided $(\tau_0)_{Crd50}$ should be known. However, the variation of $(\tau_0)_{Crd50}$ is very limited. The value of critical shear stress $(\tau_0)_{Crd50}$ for the median size d_{50} is not determined satisfactorily. Qin (1980) investigated the incipient motion of non-uniform sediment. Parker and Klingeman (1982) also determined the dimensionless critical shear stress for the median size $(\tau_0)_{Crd50}$ as 0.0876. They proposed the critical shear stress formula as:

$$(\tau_0)_{Crd_i} = \frac{0.0876}{\left(\frac{d_i}{d_{50}}\right)^{0.982}} \tag{8.282}$$

Simultaneously, Parker et al. (1982) showed in the field study that all grain sizes in the bed of Oak Creek, Oregon started movement at nearly the same bed shear stress. Parker et al. (1982) defined a technique to determine the incipient bed shear stress. The dimensionless sediment transport rate is defined as:

$$W_i^* = \frac{(\gamma_s - \gamma_f) q_{si}}{\rho_f u_*^3 f_i} \tag{8.283}$$

where q_{si} is the unit sediment transport rate for ith size fraction, f_i is the fraction of ith size contained in the mixture bed. The reference shear stress or incipient bed shear stress is defined, when $W_i^* = 0.002$, which corresponds to a very low rate of sediment transport. Using the concept of sediment transport rate equation, several researchers (Wilcock and Suthard, 1988) determined transport function of sediment transport rate. Wiberg and Smith (1987) derived expression of critical shear stress for non-uniform sediment particles. Wilcock and Suthard (1988) studied experimentally the incipient motion of mixed size particles. The motivation of their study was to determine the critical shear stress for incipient motion of individual size fraction within the sediment mixtures. They observed that the relative size of a fraction along with its absolute size largely controlled its initial motion, and was described by three variables: size ratio of the fraction (i.e., the ratio of the ith fraction to the mean grain size of the mixture), percentile position of the fraction, and the standard deviation of the mixture. They only used the relative size fraction parameter (d_i/d_{50}) to their model and is given by

$$\frac{(\tau_0)_{Crd_i}}{(\tau_0)_{Crd50}} = \left(\frac{d_i}{d_{50}}\right)^{\beta} \tag{8.284}$$

where the exponent β is determined using the least square fitting method. The variation of $(\tau_0)_{Crd50}$ and β are determined as 0.024 to 0.087 and −0.65 to −1.06 respectively. In their study, they also observed that the index grain size (mean or median size) and the Shields criterion of entire sediment mixture held for a wide range of grain size distributions, including lognormal, rectangular uni-modal, weekly bimodal and skewed distributions. Simultaneously, Wilcock (1987) suggested two methods to estimate the critical shear stress of individual size fractions in mixed sediment size: using largest grain-size displaced, and the other one using the shear stress produces a small value of transport rate for each fraction. The first method shows the critical shear stresses for largest grains vary with the square root of grain size, and the second method shows the little variation of reference transport critical shear stresses with grain size.

Kuhnle (1992) observed that the critical shear stress increased consistently with the grain size distributions from Goodwin Creek, Mississippi. The sediment size distributions of Goodwin Creek are more strongly bimodal than that of Wilcock and Southard (1988). Wilcock (1993) studied the critical shear stress of individual fractions $(\tau_0)_{Crd_i}$ in uni-modal and weakly bimodal sediment mixtures, and found a little variation with grain size and depends on the mean grain size of the mixture. He used an equation as:

$$\frac{(\tau_0)_{Crd_i}}{(\tau_0)_{Crd_{50}}} = \alpha\left(\frac{S_{*i}}{S_{*50}}\right)^{\beta_1} \tag{8.285}$$

where S_{*i} and S_{*50} are the Shields values of d_i and d_{50} respectively, α and β_1 are the empirical constants expected to vary with the mixture of grain size distribution. For uni-modal size distribution, $\beta_1 \sim 0$, so $(\tau_0)_{Crd_i} = \alpha S_{*50}$ for all fractions. For strongly bimodal distributions, $\beta_1 > 0$ up to order of unity, so that $(\tau_0)_{Crd_i} \to S_{*i}$. For any size distribution, $(\tau_0)_{Crd_i} = \alpha S_{*i}$ at $d_i = d_{50}$. Details are explained in Wilcock (1993).

The empirical parameters α and β_1 of equation (8.285) are also obtained by Patel and Ranga raju (1999) to vary over a large range. Bridge and Bennett (1992) proposed a semi-empirical analysis for the critical shear stress of sediment mixture. Their relationship was expressed in terms of critical shear stress of ith size, ratio of d_i/d_{50}, and the angle of repose φ_i for the size i under consideration. Subsequently, Kuhnle (1993) performed a series of experiments in laboratory flume to study the incipient motion of gravel and sand mixtures of 0:100, 10:90, 25:75, 45:55, and 100:0 ratios of gravel and sand respectively (Figure 8.54) for size distribution. The sediment used in the experiments consisted of a gravel mix ($d_{50} =$ 5.579 mm, range 2 – 8 mm) and a sand mix ($d_{50} = 0.444$ mm, ranging 0.177 – 2.000 mm). The three other sediment mixtures used were the same gravel as previously described and a sand nearly identical to that used in the 100% sand experiments, but with a $d_{50} = 0.444$ mm, in the following proportions: 10% gravel, 90% sand; 25% gravel, 75% sand; and 45% gravel, 55% sand (Figure 8.54). He obtained an interesting observation that the sand in each of the five sediment mixtures initiated to move at nearly the same bed shear stress; and in the 100% gravel experiments, all gravel sizes initiated to move nearly at the same bed shear stress, whereas in the sand-gravel mixtures, increase in critical shear stress was observed for gravel with increase in size (Figure 8.55).

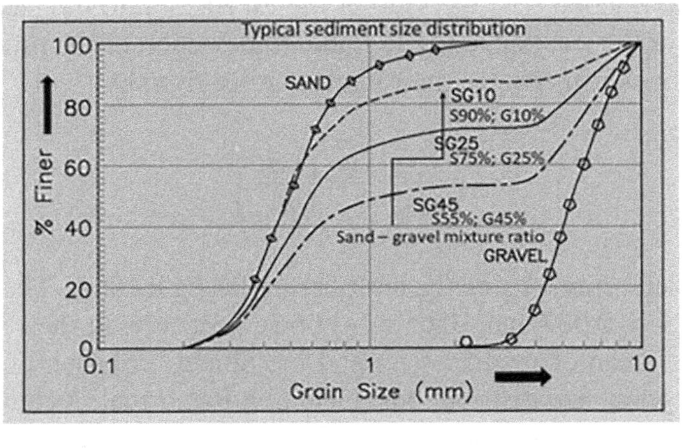

FIGURE 8.54 Typical sediment size distributions used for experiments. (Modified from Kuhnle, 1993.)

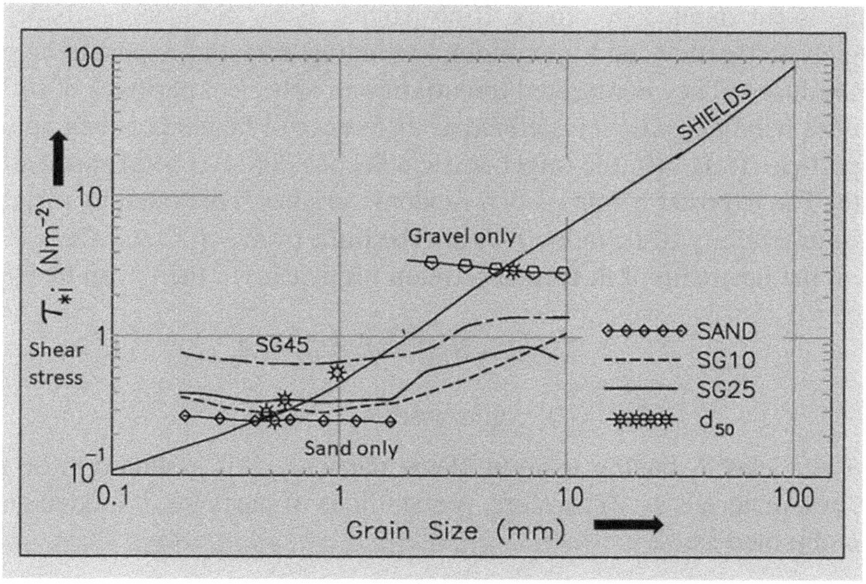

FIGURE 8.55 Plots of bed shear stress against grain size of each fraction for the sand, gravel, SG10, SG25, and SG45 beds. In the 100% sand and the 100% gravel beds, all sizes began to move at nearly the same bed shear stress. (Modified from Kuhnle, 1993.)

It was identified by Kuhnle (1992) from the field study with a bimodal size distribution in which the sand fraction of the bed sediment is entrained at lower flow strengths than the gravel fraction. For the beds composed of sand-gravel mixture all sand sizes showed essentially a constant relation between shear stress and grain-size, whereas for the gravel size fraction critical shear stress increased with increase in size (Church et al., 1991).

Patel and Rangaraju (1999) proposed a relationship for dimensionless critical shear stress τ^*_{Crd50} and geometric standard deviation σ_g for a wide range of field and laboratory data as:

$$\tau^*_{Crd50} = \frac{(\tau)_{Crd50}}{(\gamma_s - \gamma_f)d_{50}} = 0.045\sigma_g^{-0.60} \tag{8.286}$$

where $\sigma_g = (d_{84}/d_{16})^{1/2}$ is the geometric standard deviation of the sediment mixture, in which d_{84} and d_{16} are the respective sizes for which 84% and 16% of the mixture finer by weight. They also proposed critical tractive stress $(\tau_0)_{Cr\sigma}$ is a function of geometric standard deviation σ_g of sediment size, that is, $(\tau_0)_{Cr\sigma} = f(\sigma_g)$ to compute critical tractive stress of a particular size fraction $(\tau_0)_{Crd_i}$ of sediment mixture as:

$$\frac{(\tau_0)_{Crd_i}}{(\tau_0)_{Cr\sigma}} = \left(\frac{d_i}{d_\sigma}\right)^{-0.96} \tag{8.287}$$

where $d_\sigma = d_g\sigma_g$ is the representative size of the sediment mixture with d_g is the geometric mean size of mixture. Equation (8.287) is valid for uni-model and weakly bimodal sediments.

Wu et al. (2000) proposed a method to determine the threshold bed shear stress of a grain size fraction from a mixture of sediment, considering the hiding and exposure mechanism of non-uniform sediment transport. They developed a correction factor which is function of hidden and exposure probabilities, related to the size and gradation of bed materials. The drag and lift forces act on a particle depending on how the particle is surrounded by others. If no particle is on the upstream side, then the particle is exposed completely to the flow, and maximum unwinding area and exposure height, otherwise the exposure height is reduced. They considered the mixture of spherical particles of different sizes. They assumed the exposure height δ_E which is defined as difference of height between apexes of the particle and its upstream particle. If $\delta_E > 0$, the target particle is considered to be exposed state, and if $\delta_E < 0$, it is at hidden state. The exposed height δ_E is a random variable. This random variable δ_E is assumed to follow a uniform probability distribution f within the limit from $-d_j$ to d_i, where d_i is the concerned particle, and d_j is at the upstream of it. If the upstream particle is d_j, then f can be expressed as:

$$f = \begin{cases} 1/\left(d_i + d_j\right), & -d_j \le \delta_E \le d_i \\ 0, & \text{otherwise} \end{cases} \tag{8.288}$$

The probability of particles d_j having in front of the particles d_i is assumed to be the percentage of particles d_j in the bed materials p_{bj}. Therefore, probabilities of particles d_i hidden and exposed by the particles d_j can be obtained as:

$$p_{hi,\,j} = p_{bj}\frac{d_j}{d_i + d_j},\; p_{ei,\,j} = p_{bj}\frac{d_i}{d_i + d_j} \tag{8.289}$$

The total probabilities of particles d_i to be hidden and exposed by the particles d_j can be obtained by summing Equation (8.289) as:

$$p_{hi} = \sum_{j=1}^{N} p_{bj}\frac{d_j}{d_i + d_j},\; p_{ei} = \sum_{j=1}^{N} p_{bj}\frac{d_i}{d_i + d_j} \tag{8.290}$$

where N is the total number of particle size fractions of non-uniform sediment mixture; p_{hi} and p_{ei} are the total hidden and exposed probabilities of particles d_i. Here $p_{hi} + p_{ei} = 1$, in case of uniform sediment mixture. The correction factor exists between p_{hi} and p_{ei}. The hiding and exposure correction factor is defined as: $C_i = \left(\frac{p_{ei}}{p_{hi}}\right)^m$. For uniform sediment mixture, $p_{hi} = p_{ei} = 0.5$ and $C_i = 1$ the hidden and exposed probabilities are equal. In the situation of the non-uniform sediment mixture, $p_{ei} \ge p_{hi}$ for coarse particles, and $p_{ei} \le p_{hi}$ for fine particles. Introducing the hiding and exposure correction parameter C_i, they obtained the formula critical shear stress for incipient motion of non-uniform sediment mixture as:

$$\frac{(\tau)_{Cri}}{\left(\gamma_s - \gamma_f\right)d_i} = \theta_c\left(\frac{p_{ei}}{p_{hi}}\right)^m \tag{8.291}$$

where $(\tau)_{Cri}$ is the critical shear stress for particle d_i in non-uniform sediment mixture, and θ_c represents non-dimensional critical shear stress for corresponding uniform sediment size or mean size. From the laboratory and field data, they obtained the values of $\theta_c = 0.03$ and exponent $m = 0.6$. For the verification of equation (8.291), it is necessary to know the threshold shear stress. Wilcock and Crowe (2003) studied the surface-based transport model of mixed-size sediment. Further, Patel et al. (2010) proposed an equation to predict the critical tractive stress of scaling sediment size in terms of Kramer's uniformity coefficient. The performance of the proposed relationship was tested by comparing observed and predicted results with laboratory and field data, and it shows better for both uni-modal and bimodal grain-size distributions.

8.13 SUMMARY

Some of the important aspects discussed in this chapter are summarized below:

- In sediment transport phenomena, the most important properties of individual sediment particles are (i) size, (ii) shape, (iii) fall velocity, (iv) mineral composition, (v) surface texture and (vi) orientation.
- The fall velocity of the sedimentary particle is one of the most important parameters describing the particle in relation to the fluid. For a particle, falling under the influence of gravity, it will reach a constant velocity, which is called the fall or settling velocity.
- Stoke's law for the terminal fall velocity of a sphere is: $v_s = \frac{d^2}{18\mu_f}(\gamma_s - \gamma_f)$ which is valid upto Reynolds number of unity. The drag coefficient given by Oseen is: $$C_D = \frac{24}{R_e}\left(1 + \frac{3}{16}R_e\right)$$
- The terminal fall velocity of a particle depends mainly on (i) particle size, (ii) concentration of fluid, (iii) turbulence and (iv) temperature.
- For dilute suspensions, Einstein developed a relationship between viscosity and concentration in sediment-laden fluid as: $\frac{\mu_{\text{susp}}}{\mu_f} = 1 + K_e C$.
- The most important bulk properties of sediments are (i) size distribution of sediments, (ii) porosity (iii) specific weight, and (iv) angle of repose.
- The statistical representations of grain-size distribution are most commonly used as: (i) histogram, (ii) frequency polygons and frequency curve, and (iii) cumulative frequency curve.
- The *angle of repose* or *critical angle of repose* of a granular material is the steepest angle of descent relative to the horizontal plane when the particle is on the verge of sliding on the sloping surface of sediment pile. The angle of repose is equal to the angle of internal friction at the contact of sediment particle and is approximately equal to $\tan^{-1}\mu_f$, where μ_f is the static friction coefficient.
- In an open channel flow on a sediment bed, the flow condition is such that the sediment particles of given characteristics just start moving. This condition is known as the condition of *critical motion* or the condition of *incipient motion* of sedimentary particles.
- Three different approaches used to establish the condition for incipient motion of sediment particles comprising the bed are: *Competency, Lift concept and Critical tractive force.*
- There exists a certain velocity for each grain, below which it experiences *sedimentation*, while above that certain velocity the particle will erode, which is called *critical scour velocity.* Garde derived the critical scour velocity for hydraulically rough flow regime as: $(\bar{u})_{Cr} = 1.51\sqrt{\left(\frac{\rho_s}{\rho_f} - 1\right)gd}$
- The critical shear stress can be written as: $\tau_0 = \gamma_f R_h J$, where J = slope of energy grade line, R_h is hydraulic radius.
- Several semi-empirical equations have been developed by many investigators to find the incipient motion of sediment particles.
- Shield's analysis based on the equation $\frac{\tau_{0c}}{(\gamma_s - \gamma_f)d} = f\left(R^*_{e_{cr}}\right)$ and Shields diagram is widely used in solving the problems of sediment transport.
- For river bank erosion, the critical Shields number ratio as: $\frac{\tau_{*c}}{\tau_{*c0}} = \frac{u^2_{*c}}{u^2_{*c0}} = \frac{F_{Dc}}{F_{Dc0}}$. It represents the ratio of critical bed shear stress on any bank slope with seepage angle to that on horizontal bed.
- For sediment transport problems, we can examine the incipient motion or entrainment of sand particle, using probabilistic concepts as turbulent flow and its corresponding shear stress are random in nature.

- The threshold of incipient motion can be explained by turbulent bursting based on the transfer of momentum between the turbulent and laminar regions near a boundary.
- The threshold shear stress of a given size fraction depends on the sediment mixture and the surface texture.

8.14 EXERCISE PROBLEMS

1. In sediment transport, what are the important properties of sediments which influence the movement of sediment particle and illustrate with details?
2. What are the bulk properties of sediments that govern the sediment transport phenomena? Illustrate each property with details.
3. Explain the different approaches used to establish the condition for incipient motion of sediment particles in open channel flows.
4. How the Hjulstrom diagram can be effectively used for the design of an alluvial channel?
5. Using Shields analysis, derive the expression for critical shear stress for incipient motion.
6. What are the effects of sloping of river bed and banks on critical shear stress?
7. Derive the critical Shields number ratio for river bank erosion problems.
8. Illustrate applications of probability concepts for incipient sediment motion. Discuss the pickup probability concept for initiation sediment particle movement.
9. Design a canal cross section by considering the critical velocity consideration from Hjulstrom diagram for the following data. The canal is to be designed to carry 30 m^3/s discharge of clear water without any scour. The canal can be rectangular in cross section with a bed slope of 0.015. The soil analysis reports show that bed soil is made up of quarts gravel of 30 mm approximate size with a Manning's roughness coefficient of 0.025.
10. In a wide river which can be considered as wide rectangular alluvial channel, for the given flow condition, estimate the depth at which sediment particle of 1.5 mm diameter starts moving. The slope of the channel may be assumed 0.0002. The relative density of sediment is 2.6 and kinematic viscosity of water is 10^{-6} m^2/s. The dimensionless critical shear stress is given as 0.032.

9 Bed Load Transport, Suspension, and Total Load

9.1 INTRODUCTION

Bedload transport is one of the important parameters for river engineering, channel hydrodynamics and fluvial hydraulics. Bedload transportation generally occurs due to the interaction of turbulent flow in the channel with the bed materials. When the shear stress acting on the sediment bed exceeds its critical value, there is the initiation of particle motion by various processes of transport over the bed. The physics involved in the initiation of particles movement on the bed can be featured by both deterministic and probabilistic characteristics. The deterministic nature shows from the view that the mean velocity or mean bed shear stress dominates the process, whereas the probabilistic nature stems from turbulence, which is highly irregular. The bed material transported by the flow is the most significant characteristic of the two-phase motion. The movement of particles depends on different aspects, such as flow conditions, nature, size and shape of sediment and the ratio of the densities of sediment and fluids. In this chapter, the basic principles, theories and solution methods of bed load transport including suspension and total load are discussed in detail.

9.2 MODES OF SEDIMENT MOVEMENT

There are three types of modes for the initiation of a sediment particle on the loose sediment bed, i.e., *rolling, sliding, and saltation or jumping or lifting* (Figure 9.1) (Garde and Ranga Raju 2000; Dey 2014). Among them, the first two modes are closely relevant to bedload transport, while the last one acts with both bedload and suspended load transport. In the bed load transport, the sediment movement is usually discontinuous; the particle may roll or slide for some time, halt for a while, and again start rolling or sliding, which means, the particles move intermittently over the bed as a contact process. The movement by saltating or jumping along the bed loses contact with the bed for an instant of time, known as *saltation*. The saltation of a particle is governed by the hydrodynamic drag and lift forces, and bed roughness. Other than the above three modes, one important mode is the state of suspension in which the particles are fully supported by turbulent fluctuations.

The material transported by the flow above the bed is known as the suspended load. Once the particle is picked up by the flow, turbulence strength determines the suspension of particles within the flow. If

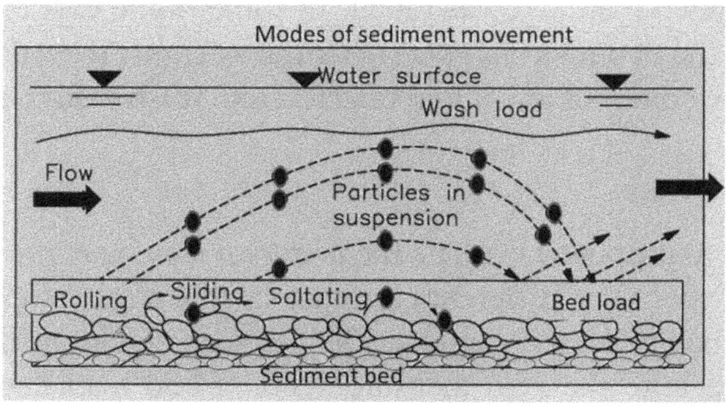

FIGURE 9.1 Typical sketch of rolling, sliding, and saltation or jumping or lifting over the sediment bed. (Modified from Van Rijn 1984; Garde & Ranga Raju 2000; Dey 2014.)

DOI: 10.1201/9781003000020-9

the turbulence level is high enough, then the particle will continue to move as suspension, otherwise, it will slowly fall onto the bed. In natural stream, wash load is also an important fraction of suspended sediment that is carried by the flow and it always remains in suspension without deposition. It consists of very fine particles, such as silt and clay with very small settling velocity stays in suspension due to turbulent flow. The Subcommittee on sediment terminology of American Geophysical Union has defined the various terms as: Contact load, Saltation load and Suspended load.

9.3 BED LOAD TRANSPORT PHENOMENA

9.3.1 Basic Definitions

In bed load transport, the definitions of some of the important terms used are given here.

Contact load includes the sediment particles that move by rolling or by sliding along the bed, i.e., bed material movement dislodging with the bed.

Saltation load is the material bouncing along the bed or moving directly or indirectly by the impact of bounding particles, i.e., for sometimes, the particles lose contact from the bed. If the bed shear stress increases further, the particles move along the bed by a series of short jumps with approximately the same step lengths.

Bed layer is defined as the thickness of the sediment layer in which the particles, such as sand, silt, gravel, etc. can transport by sliding, rolling or sometimes suddenly jumping very near the bed. According to Einstein's definition (1950), the thickness is approximately two times the diameter of the maximum size of the bed materials.

Bed load is defined as the amount of material carried by the flow in the stream-wise direction within the bed layer. This load is a collection of contact load and saltation load together. The bed load transport rate is expressed as the volume or weight of sediment transported per unit width and time. This transport rate can also be defined as the product of particle velocity along stream-wise direction, the concentration of sediment particles transporting and the thickness of the bed layer.

Suspended load is the amount of materials which are transported with the flow or by the flow as a state of suspension due to turbulent fluctuations. The suspended load transport is an advanced stage of bed load transport. The bed load transport can only occur at low shear stresses, whereas at sufficiently high shear stresses both bed load and suspended load can occur.

Wash load is defined as the very fine material transported in suspension having sizes which are not found in appreciable quantity in the material forming the bed and banks. The presence of wash load in suspension tends to decrease the resistance to the flow, which is more significant than the plane bed (Vanoni and Nomicos 1960).

Total load transport is the total transported material for a given flow condition. The total load is the sum of the bed load, suspended load and wash load. In some flume experiments, wash load is absent, and the total load would be the bed material, but in natural streams, the wash load is always present. In that case, the total load is the combination of bed material and wash load (Garde and Ranga Raju 2000).

9.3.2 Transportation of Bed Load

Numerous attempts have been proposed to relate the *bed load transport rate* to the hydraulic conditions and the sediment characteristics. The different concepts such as shear stress, discharge, and statistical aspect are used to develop the bed load equations. The bed load equations have been developed mainly from the laboratory data because the field measurements of bed load rate are very complicated and very few in number. Even in laboratory experiments, it is difficult to measure the bed load transport accurately, especially for fine materials, because some suspended load is always entrapped with the bed load. Moreover, bed load transport at a given section varies reasonably with time. It is also shown that the bed load transport rate also directly varies with the bed elevation, because the bed elevation changes with time with the passage of ripples and dunes over the bed (discussed in Chapter 10), such changes in the bed load transport are partly discussed. Several equations have been formulated for the prediction of bed load transport rate. Some of these equations are entirely empirical in nature, and some are obtained from dimensional considerations and some are purely based on the semi-theoretical approach. Further, the transportation of bed load, suspended load, and the total load with respective methodologies and possible semi-empirical equations will be discussed.

9.4 BED LOAD TRANSPORT EQUATIONS

9.4.1 Shear Stress Concept for Bed Load Equations

Du Boys (1879) was the first to propose a bed load relation using the bed materials that move in a series of superimposed layers parallel to the bed, the velocity of each layer varies from a maximum for the top layer on the bed surface to zero for the lowest layer at some depth (Figure 9.2). Let N be the number of layers moving, and Δh is the thickness of each layer. The lowest layer is assumed to be zero velocity. If δ_u is the velocity of the second layer from the bottom and a linear variation of velocity is assumed, the velocity of the surface layer will be $(N-1)\delta_u$. Hence the bed load transport q_B in weight per unit width and time is as follows:

$$q_B = \gamma_s N\Delta h(N-1)\frac{\delta_u}{2} = \gamma_s \Delta h N\frac{(N-1)\delta_u}{2}, \tag{9.1}$$

where $\gamma_s = g\rho_s$ is the specific weight of a grain, $N\Delta h$ is the thickness of the sediment material, moving with an average velocity of $(N-1)\delta_u/2$.

Since the lowest layer is at rest, the resisting force at the elevation must be equal to the tractive force $\tau_0 = \gamma_f hJ$ offered by the uniform flow on the bed and is given by

$$\tau_0 = \gamma_f hJ = (\gamma_s - \gamma_f)N\Delta h \tan\phi \tag{9.2}$$

where J is the channel slope, $\tan\phi$ is the dynamic friction coefficient. The "critical condition" at which the sediment motion is just begin to be in motion is given by $N = 1$, then

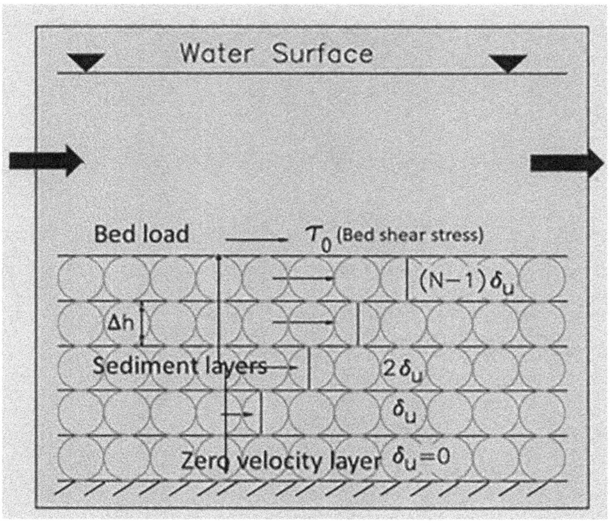

FIGURE 9.2 Schematic of bed load movement as layer by layer. (Modified from Garde & Ranga Raju 2000; Dey 2014.)

$$(\tau_0)_{cr} = (\gamma_s - \gamma_f)\Delta h \tan\phi \tag{9.3}$$

Then, it results from the above relations as:

$$\tau_0 = N\tau_{0cr} \Rightarrow N = \frac{\tau_0}{\tau_{0c}} \tag{9.4}$$

Then using equation (9.4), the bed load transport equation (9.1) can be written as:

$$q_B = \frac{\gamma_s \Delta h \delta_u}{2} \frac{\tau_0}{\tau_{0c}} \left(\frac{\tau_0}{\tau_{0c}} - 1 \right)$$

$$q_B = A(\tau_0 - \tau_{0c})\tau_0 \tag{9.5}$$

$$\text{where } A = \frac{\gamma_s \Delta h \delta_u}{2\tau_{0c}^2} \tag{9.6}$$

Equation (9.5) is known as Du Boy's bed load equation with the characteristic sediment coefficient denoted by A. The total amount of material Q_b is given by

$$Q_b = A\int_{x_1}^{x_2} (\tau_0 - \tau_{0c})\tau_0 \, dx \tag{9.7}$$

where coordinate x denotes the width of the channel. It is noted that the effect of the presence of bed forms has not been considered in the derivative of Du Boy's equation. The oversimplification of bed load equation is a strong disagreement with observations because the bed load moves as sliding layer, as pointed out by Schoklitsch (1914). However, this equation gives good agreement with field and laboratory data making it the widest use of bed load equation.

However, Donate (1929) showed that Du Boy's equation (9.5) can be derived without making any strong assumption, except related to the function of tractive force τ_0, such that

$$q_B = f(\tau_0) \tag{9.8}$$

It was proposed to approximate the function with a power series and is given by

$$q_B = K_1 + K_2\tau_0 + K_3\tau_0^2 + \cdots \tag{9.9}$$

after neglecting the higher order terms, with K_i s are the coefficients to be determined. Two conditions are applied to determine K_1 and K_2. If the shear stress $\tau_0 = 0$, there is no bedload transport $q_B = 0$, so $K_1 = 0$. If the shear stress τ_0 is finite and the critical condition of τ_0 exists, the bed load transport q_B is just about to initiate, but it is still negligible and

$$K_2 = -K_3(\tau_0)_{cr}, \tag{9.10}$$

So, equation (9.9) reduces to

$$q_B = K_3\tau_0(\tau_0 - \tau_{0cr}) \tag{9.11}$$

Comparing these two equations (9.5) and (9.11), it seems that $K_3 = A$ is the characteristic sediment co-efficient, obviously, both equations are identical. From this concept, it may be concluded that the bed load equation depends on excess shear stress $(\tau_0 - \tau_{0cr})$ or τ_0/τ_{0cr} classified as Du Boys-type equation.

For uniform grains of various sand and porcelain, Schoktitsch (1914) determined the characteristic sediment co-efficient A empirically from the laboratory and field data as

$$A = \frac{10.54}{(\rho_s - \rho_f)g}(\text{in metric unit}). \tag{9.12}$$

Equation (9.12) has the deficiency that the data were limited, and the extensive data were reported by Gilbert (1914), and also analyzed by Donate (1929). In a later stage, Straub (1935) determined the average value of A using the mean diameter of the particle d (mm) within the range $(0.125 < d < 4\text{ mm})$ as

$$A = \frac{0.173}{d^{0.75}} \tag{9.13}$$

The model of Straub (1935) has shortcomings for the data collected in small dimensions with a small range of particle sizes, and insufficient of field measurements. Further, Chang (1939) suggested expressing the characteristics sediment coefficient A as a function of Manning's roughness coefficient n (discussed in the next Chapter). Barekyan (1962) proposed bed-load transport equation using average flow velocity. Based on the stream power concept, Dou (1964) established an empirical equation of bed-load transport for sand.

Schoklitsch (1930) considered the basic Du Boys-type equation related to the bed load transport with excess shear stress, based on laboratory experiments. Du Boys suggested the critical shear stress of equation (9.12), using $\tau_0 = \gamma_f Jh$ with water depth h as:

$$(\tau_0)_{cr} = \gamma_f Jh_{cr} \tag{9.14}$$

Then equation (9.11) can be written as:

$$q_B = A(\gamma_f J)^2 h(h - h_{cr}) \tag{9.15}$$

Sediment material moves if and when the water depth h exceeds the critical water depth h_{cr}. Replacing hJ with representative average velocity $\overline{u}$, one gets from Chezy's equation as

$$Jh = \frac{\overline{u}^2}{C^2}. \tag{9.16}$$

Using equation (9.16), equation (9.15) can be written as

$$q_B = A\gamma_f{}^2\left(\frac{\overline{u}^2}{C^2}\right)\left[\frac{\overline{u}^2}{C^2} - \left(\frac{\overline{u}^2}{C^2}\right)_{cr}\right] \tag{9.17}$$

From equation (9.17) it may be indicated that the bed load motion starts at a certain velocity $\overline{u}_{cr} = C\sqrt{(Jh)_{cr}}$ and increases rapidly as the velocity exceeds this *critical condition* value. Equation (9.17) can be written as

$$q_B = A\gamma_f{}^2\frac{1}{C^4}\overline{u}^2\left(\overline{u}^2 - \overline{u}_{cr}^2\right) \tag{9.18}$$

where C is Chezy's roughness value.

Eventually, Schoklitsch (1930) heuristically suggested a bed load transport equation, independent from Du Boys equation, from the laboratory experimental data as:

$$q_B = A''J(q - q_{cr}) \tag{9.19}$$

where A'' is the new characteristic sediment coefficient, q_{cr} is the water discharge at which the material begins to move (Details are available in Graf 1971).

O'Brien and Rindlaub (1934) proposed a generalization of Du Boy's equation, utilizing the dynamic equilibrium that the shear stress should be constant on planes parallel to the bottom. Under a certain analogy of sediment motion with Bingham plastic fluids (non-Newtonian fluid behavior), the relation was obtained as:

$$q_B = A'(\tau_0 - \tau_{0cr})^m \tag{9.20}$$

where the new parameters A' and m are in functional relationship with median diameter. For mixtures ranging from $0.025 < d < 0.560$ mm, the values of m are confined to a range $1.5 < m < 1.8$.

It is well known that Do Boy's equation is a classical bed load transport equation which characterizes the excess bed shear stress. Shields (1936) successfully applied the critical shear stress to Du Boy's transport equation. He proposed the following empirical equation of bed load transport:

$$q_B = \frac{10 q \gamma_f S}{\left(\dfrac{\rho_s}{\rho_f} - 1\right)} \cdot \frac{\tau_0 - \tau_{0c}}{(\gamma_s - \gamma_f)d} \tag{9.21}$$

where $q = Uh$ is the flow rate per unit width with U is the depth-averaged velocity. Equation (9.21) includes the relative density ρ_s/ρ_f within the range of 1.06–4.20 and the sediment size in the range of 1.56–2.47 mm.

Meyer-Peter and Müller (1948, 1949) conducted a series of experiments for a range of characteristic parameters as follows:

$$1\text{ cm} \le h \le 120\text{ cm}, \quad 0.0004 \le J \le 0.02,$$

$$0.4\text{ mm} \le d \le 30\text{ mm}, \quad 0.25 \le (\gamma_s - \gamma_f) \le 3.2 \tag{9.22}$$

where h is the flow depth, J is the slope, d grain size of a mixture, and $(\gamma_s - \gamma_f)$ is specific weight of a grain in fluid. They developed an empirical relation for bed load transport q_B for a steady, uniform two-dimensional flow over a plane bed as follows:

$$\frac{q_B}{\rho_f u_*^3} = 8\left[1 - \frac{0.047(\gamma_s - \gamma_f)d}{\rho_f u_*^2}\right]^{3/2} \tag{9.23}$$

The experiments were carried out for both uniform materials and mixture. In the case of mixture, the size will be assumed as d_{50}. Multiplying both sides of equation (9.23) by $\left(\dfrac{\rho_f u_*^2}{(\gamma_s - \gamma_f)d}\right)^{3/2}$, one gets the dimensionless bed load transport rate Q_s as:

$$Q_s = \frac{q_B \rho_f^{1/2}}{\{(\gamma_s - \gamma_f)d\}^{3/2}} = 8\left[\frac{\rho_f u_*^2}{(\gamma_s - \gamma_f)d} - 0.047\right]^{3/2} \tag{9.24}$$

From equation (9.24), it is noted that the equation does not contain the flow depth h, it implies that the formula is valid only in the vicinity of the bed. In other words, Mayer-Peter and Muller's equation is a typical bed load transport equation. From the analysis, it indicates that there is no limitation mentioned by the authors, but it shows that this equation should not be used for fine sand size $d \le 0.15$ mm, but if it is used the actual transport rate turns out to be 5–10 times different. If the term $\frac{\rho_f u_*^2}{(\gamma_s - \gamma_f)d} = 0.047$ in equation (9.24), then the transport rate $q_B = 0$, which implies that the sediment transport must begin. It clearly indicates that $\left(\frac{\rho_f u_*^2}{(\gamma_s - \gamma_f)d}\right)_{cr} \approx 0.05$. Since the formula is used for the large grains, the criterion for the initiation of sediment movement should coincide with the value ≈ 0.05 corresponding to the Shields curve. In general, the term $\left(\frac{\rho_f u_*^2}{(\gamma_s - \gamma_f)d}\right)_{cr}$ is not a constant quantity, but it is a function of critical Reynolds number $\left(\frac{u_* d}{\nu}\right)_{cr}$ or $\frac{(\gamma_s - \gamma_f)d^3}{\rho_f \nu^2}$. Using this concept, equation (9.24) is generalized into

$$Q_s = \frac{q_B \rho_f^{1/2}}{\{(\gamma_s - \gamma_f)d\}^{3/2}} = 8\left[\frac{\rho_f u_*^2}{(\gamma_s - \gamma_f)d} - \left(\frac{\rho_f u_*^2}{(\gamma_s - \gamma_f)d}\right)_{cr}\right]^{3/2} \tag{9.25}$$

If the material is non-uniform, the constants were different. Further, according to Meyer-Peter and Müller, the total shear stress was not available for sediment transport in the case of undulation bed, a part of shear stress was used up in overcoming the form resistance of the undulation and the bed load transport was a function of shear stress due to grain only. Here the slope J is split up into J_1 and J_2, as $J = J_1 + J_2$, where J_1 is the slope required for resistance to grains, and J_2 is the slope to overcome the resistance of bed undulations. The value of slope J_1 was estimated by Manning-Strickler formula (discussed in Chapter 10). Similarly, if the hydraulic radius h_r is divided into h_{r1} and h_{r2} as $h_r = h_{r1} + h_{r2}$, where h_{r1} correspondingly for grain, and h_{r2} correspondingly for undulation. Einstein (1950) used this concept of hydraulic radius in his bed load transport equation. Frijlink (1952) proposed a formula that can approximate Meyer-Peter and Müller formula, but it is not a Du Boys-type equation. Chien (1954) used this concept of hydraulic radius in the Meyer-Peter and Müller's equation (9.24), replacing $\rho_f u_*^2$ by $\tau_0 = \gamma_f h_{r1} J_1$ as:

$$Q_s = 8\left[\frac{\gamma_f h_{r1} J_1}{(\gamma_s - \gamma_f)d} - 0.047\right]^{3/2} \tag{9.26}$$

This equation (9.26) is similar to the form of Du Boy's equation because the sediment transport rate is related to the effective shear stress. Later, equation (9.25) was verified by different authors such as Bharat Singh (1961) and Misri (1981) in laboratory experiments, and Hansen (1966) and Amin and Murphy (1981) in Skive-Karup River and Fish Creek at Rome respectively. Details are available in Garde and Ranga Raju (2000). Wong and Parker (2006) reanalyzed the bedload relation of Meyer-Peter and Muller (1948) using their database and developed a better fitting formula, which is the correction of bed load relation, and is given by

$$Q_s = \frac{q_B \rho_f^{1/2}}{\{(\gamma_s - \gamma_f)d\}^{3/2}} = 3.97\left[\frac{\rho_f u_*^2}{(\gamma_s - \gamma_f)d} - 0.050\right]^{3/2} \tag{9.24a}$$

Kalinske (1947) proposed a bed load transport equation based on three important aspects: (1) the critical tractive force to initiate the movement of sediment particles; (2) the fluid force acting on a particle fluctuates about a mean value, the fluctuations may be due to pressure fluctuations or turbulent fluctuations; and (3) the bed load is a function of the number, size, and the average speed of the particles in motion. If p_f is the fraction of the bed occupied by the particles, the number of particles on unit area participating in motion is given by $4p_f/\pi d^2$. The average particle velocity $\bar{u}_{av}$ is given by the average difference of the instantaneous fluid velocity u_b at the bed level and the critical fluid velocity u_{bcr} at the grain level. Then the bed load transport per unit width can be written as

$$q_B = \frac{4p_f}{\pi d^2}\left(\frac{\pi}{6}\right)\gamma_s d^3\left(\overline{u_b - u_{bcr}}\right) \tag{9.27}$$

Simplifying equation (9.27), one gets

$$q_B = \frac{2p_f d\gamma_s}{3}\bar{u}_{av} \tag{9.28}$$

Assuming that u_b varies according to normal error law, then

$$\frac{\bar{u}_{av}}{\bar{u}_b} = f\left(u_{bcr}/\bar{u}_b, \sigma/\bar{u}_b\right) \tag{9.29}$$

where $\bar{u}_b$ is the average velocity at the grain level, and $\sigma = \sqrt{\overline{(u_b - \bar{u}_b)^2}}$ is the standard deviation. Equation (9.29) can be written in terms of bed shear stress as:

$$\frac{\bar{u}_{av}}{\bar{u}_b} = f\left(\tau_{0cr}/\tau_0, \sigma/\bar{u}_b\right) \tag{9.30}$$

Now equation (9.28) can be written as:

$$\frac{q_B}{\bar{u}_b d\gamma_s} = \frac{2p_f}{3}\frac{\bar{u}_{av}}{\bar{u}_b} \tag{9.31}$$

Using (9.30), equation (9.31) can be written as:

$$\frac{q_B}{\bar{u}_b d\gamma_s} = \frac{2p_f}{3} f\left(\tau_{0cr}/\tau_0, \sigma/\bar{u}_b\right) \tag{9.32}$$

Equation (9.32) can be expressed in terms of the parameters $\frac{q_B}{u_* d\gamma_s}$ and τ_{0cr}/τ_0, if $\sigma/\bar{u}_b$ is assumed to be 0.25. Figure (9.3) shows the variation between the parameters $\frac{q_B}{u_* d\gamma_s}$ and τ_{0cr}/τ_0, Experimental data of Kalinske (1947) shows good agreement with the theoretical relation.

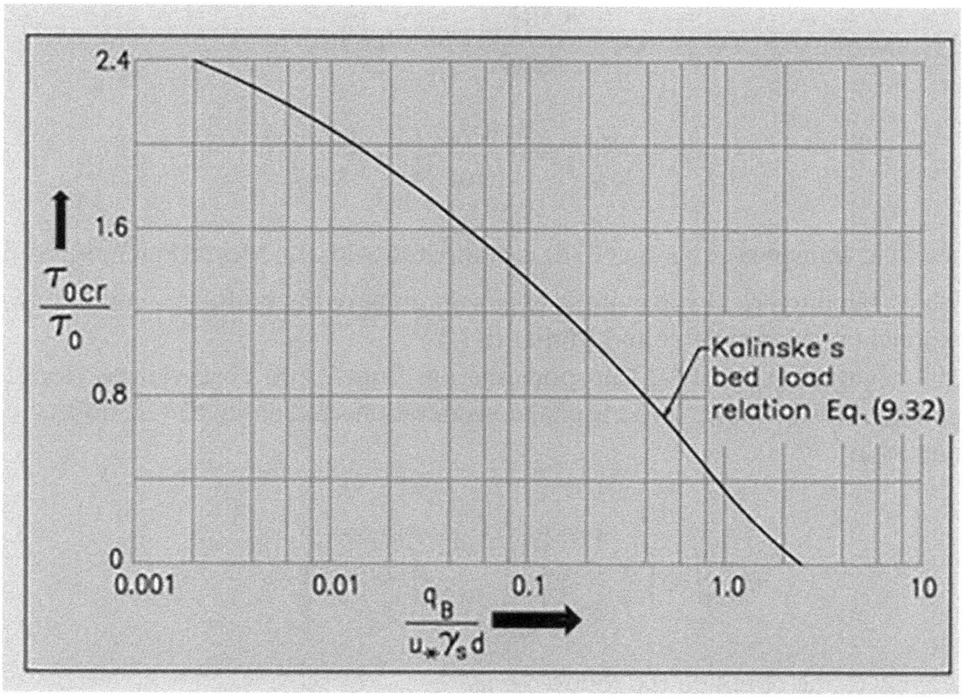

FIGURE 9.3 Sketch of Kalinske's bed load relation. (Modified from Garde & Rangaraju 2000; Kalinske 1947.)

Bagnold (1956) developed a bed load transport formula, using the principles of energy. He proposed a theory for bed-load transport based on the work done by the fluid to transport the sediment. Let W_b be the weight of the submerged particles transporting per unit area in the vicinity of the bed surface, and u_b is the velocity. In unit time, the bed material W_b is transported by the velocity u_b from one section to other. The work done W by the frictional force per unit time is given by:

$$W = q_{sb} \tan\phi \tag{9.33}$$

where $q_{sb} = W_b u_b$ is the weight of the granular materials passing along the vicinity in unit width per unit time and is represented as bed load transport rate. Using the concept of energy, and the bed shear stress, Bagnold assumed the expression as

$$E_P = (\tau_0 - \tau_{0cr})u_d \tag{9.34}$$

where u_d is the fluid velocity in the vicinity of the bed or at $y = d = k_s$. If W is the work done, and E_P is the energy, then the flow itself is a working machine whose efficiency E_{eff} is:

$$E_{\text{eff}} = W/E_P \tag{9.35}$$

Using equations (9.33) and (9.34) in (9.35), one gets the expression for Bagnold's bed load transport q_{sb} as:

$$q_{sb} = \frac{E_{\text{eff}}}{\tan\phi}(\tau_0 - \tau_{0cr})u_d \tag{9.36}$$

To determine the fluid velocity u_d, Bagnold assumed the turbulent flow velocity $u = u_d = 8.50u_*$ at the rough surface $y = d = k_s$. So, equation (9.36) can be written as:

$$q_{sb} = \frac{8.50E_{\text{eff}}}{\tan\phi}(\tau_0 - \tau_{0cr})u_* \tag{9.37}$$

$$\Rightarrow \frac{q_{sb}}{\rho_f u_*^3} = \frac{8.50E_{\text{eff}}}{\tan\phi}\left(1 - \frac{\tau_{0cr}}{\tau_0}\right) \tag{9.38}$$

If Bagnold's bed load transport equation (9.38) is compared with the Meyer-Peter and Müller's equation (9.26), it is solely to the numerical value of the exponent unity of the multiplier containing the ratio $\frac{\tau_{0cr}}{\tau_0}$, whereas the exponent of Meyer-Peter and Müller is 1.5.

Rottner (1959) obtained a bed load transport relation from 2,500 observations from the laboratory flumes. His approach was based on dimensional considerations and described the following four dimensionless parameters as:

$$\frac{q_B}{d\gamma_s\sqrt{gd\left(\frac{\gamma_s}{\gamma_f}-1\right)}}, \frac{h}{d}, \frac{u}{\sqrt{gd\left(\frac{\gamma_s}{\gamma_f}-1\right)}}, \frac{J}{\frac{\gamma_s}{\gamma_f}-1} \tag{9.39}$$

From the data analysis, he obtained the following as:

$$\left(q_B\bigg/\left(d\gamma_s\sqrt{gd\left(\frac{\gamma_s}{\gamma_f}-1\right)}\right)\right)^{1/3} = 0.164 \tag{9.40}$$

for $\left\{u\bigg/\sqrt{gd\left(\frac{\gamma_s}{\gamma_f}-1\right)}\right\} = 1.16$ for all values of h/d.

Finally, he obtained a linear relationship of bed load transport between the parameters $\left(q_B\bigg/\left(d\gamma_s\sqrt{gd\left(\frac{\gamma_s}{\gamma_f}-1\right)}\right)\right)^{1/3}$ and $\left\{u\bigg/\sqrt{gd\left(\frac{\gamma_s}{\gamma_f}-1\right)}\right\}$ for all $\frac{h}{d}$ and he obtained an intersecting point at (0.164, 1.16).

Garde and Albertson (1961) claimed that for uniform material, the bed load transport equation can be expressed as:

$$\frac{q_B}{d\gamma_s u_*} = f(\tau_* - \tau_{*cr}) \tag{9.41}$$

where $\tau_* = \frac{\tau_0}{(\gamma_s - \gamma_f)d}$ is the dimensionless shear stress, and τ_{*cr} is the dimensionless critical shear stress. If τ_* is greater than the critical shear stress τ_{*cr}, then one can write $\frac{q_B}{d\gamma_s u_*} = f(\tau_*)$. During the analysis of their data, they separated the data into the plane bed data and ripple-dune data. On the plane bed condition, the data fell on a single curve in the plot of $\frac{q_B}{d\gamma_s u_*}$ against τ_*, which followed the general trend of Kalinske's relation. On the other hand, for ripple and dune bed data, it showed the considerably scattered plot between $\frac{q_B}{d\gamma_s u_*}$ against τ_*. Further, from the analysis, it is shown that the use of $\frac{u}{u_*}$

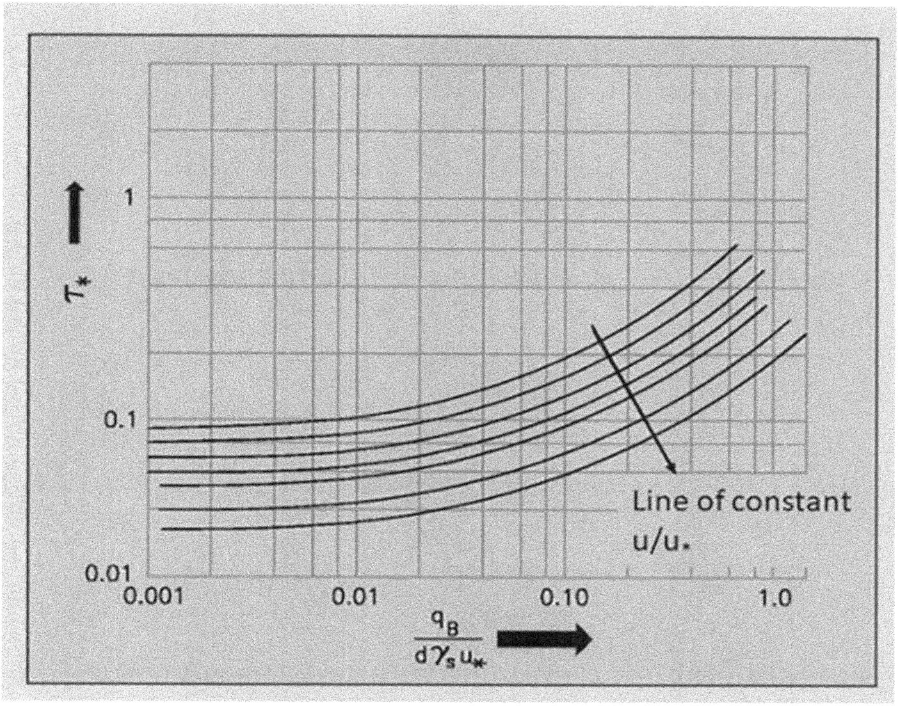

FIGURE 9.4 Variation of $\dfrac{q_B}{d\gamma_s u_*}$ against τ_* for different values of u/u_*. (Modified from Garde and Albertson 1961.)

systematizes the scatter for ripple and dune data and the line of constant $\dfrac{u}{u_*}$ could be drawn on a plot of $\dfrac{q_B}{d\gamma_s u_*}$ against τ_* shown in Figure 9.4.

Yalin (1963) developed an expression for the rate of bed-load transport based on dimensional analysis and the dynamics of the average motion of the grain from the time-averaged characteristics. He expressed the bed load transport rate as:

$$q_b = F_g u_{sg} \tag{9.42}$$

where F_g is the weight of granular material moving over the unit area of the surface of the mobile bed, u_{sg} is the average velocity of the weight. He used the average velocity u_{sg} of the particle as the average jump of the particle. He assumed that the saltation of a particle is analogous to the ballistics of a missile, in the sense that the grain has its maximum level during a saltation owing to its initial velocity when it is lifted from the bed surface, and not to the continuous action of a driving force. Since the transport occurs in the vicinity of the plane rough bed $(k_s \sim d)$, the properties of F_g and u_{sg} behave like two-phase motion, which are the function as follows:

$$\frac{F_g}{\gamma_s d} = f_1(Re_*, \tau_*) \tag{9.43}$$

$$\frac{u_{sg}}{u_*} = f_2\left(Re_*, \tau_*, \frac{\gamma_s}{\gamma_f}\right) \tag{9.44}$$

$$\text{so that } \frac{q_b}{\gamma_s d u_*} = f_1 f_2 \tag{9.45}$$

Under the jumping assumption, Yalin (1963) determined the expression of u_{sg} from a system of differential equations of motion of individual particle. Based on the above concept of jumping condition he arrived at the following expression of the average velocity as:

$$\frac{u_{sg}}{u_*} = C_1\left[1 - 2.307(a_\tau T_\tau)^{-1}\log(1 + a_\tau T_\tau)\right] \tag{9.46}$$

where C_1 is a constant, $a_\tau = \dfrac{2.45\sqrt{\tau_{*c}}}{(\rho_s/\rho_f)^{0.40}}$, and $T_\tau \dfrac{\tau_* - \tau_{*c}}{\tau_{*c}}$. He further assumed that

$$\frac{F_g}{\gamma_s d} = C_2\frac{\tau_* - \tau_{*c}}{\tau_{*c}} \tag{9.47}$$

where C_2 is a constant. Combining equations (9.46) and (9.47), one gets the bed load formula from equation (9.42) as:

$$\frac{q_b}{u_* \gamma_s d} = \text{Const.}\frac{\tau_* - \tau_{*c}}{\tau_{*c}}\left[1 - 2.307(a_\tau T_\tau)^{-1}\log(1 + a_\tau T_\tau)\right] \tag{9.48}$$

This equation (9.48) contains only one constant that must be determined from the experimental data.

Multiplying both sides of equation (9.48) by $\left(\dfrac{\rho_f u_*^2}{(\gamma_s - \gamma_f)d}\right)^{0.5}$, one gets

$$Q_S = \frac{q_b \rho_f^{0.5}}{((\gamma_s - \gamma_f)d)^{1.5}} = 0.635T_\tau\left(\frac{\rho_f u_*^2}{(\gamma_s - \gamma_f)d}\right)^{0.5}\left[1 - 2.307(a_\tau T_\tau)^{-1}\log(1 + a_\tau T_\tau)\right] \tag{9.49}$$

Equation (9.49) is the dimensionless bed load transport formula given by Yalin (1963).

Engelund and Fredsoe (1976) assumed that fluid velocity at the top of a particle is $\alpha_1 u_*$, where α_1 is constant. The drag force acting on the particle is given by

$$F_D = C_D\frac{\pi d^2}{4}\frac{\rho_f(\alpha_1 u_* - u_{sg})^2}{2} \tag{9.50}$$

where u_{sg} is the average velocity of the particle to be determined and C_D is the drag coefficient for the particle suitably modified to include the effect of left. The resisting force is given by

$$F_G = \frac{\pi d^3}{6}(\rho_s - \rho_f)g\beta_1 \tag{9.51}$$

where β_1 is the dynamic friction coefficient. When the particle moves with a constant velocity, equations (9.50) and (9.51) are equal and one gets:

$$\frac{U_{sg}}{u_*} = \alpha_1\left(1 - \sqrt{\frac{A_1}{\tau_*}}\right) \tag{9.52}$$

in which $A_1 = \dfrac{4\beta_1}{3C_D\alpha_1^2}$ is the limiting case of $\tau_{*c} = \dfrac{\tau_{0c}}{(\rho_s - \rho_f)gd}$ the dimensionless critical shear stress for which a particle located on the bed is just immobile. A crude estimate is obtained from the parameter

A_1 by putting the values of $\beta_1 = \tan 27$, $\alpha_1 = 9.0$, $C_D = 0.6$, the value of $A_1 = 0.014$, which is between one-half and one-fourth of the generally accepted values of τ_{*c}. Evaluating the parameter A_1 from the experimental data of Luque and Beek (1976), where A_1 is to be about half of τ_*, so that equation (9.52) can be written as:

$$\frac{U_{sg}}{u_*} = \alpha_1\left(1 - 0.7\sqrt{\tau_{*c}/\tau_*}\right) \tag{9.53}$$

Putting $\alpha_1 = 9.0$, and τ_{*c} is read from the Shields curve, equation (9.53) is used to compute the velocity of sediment particles. Using the particle velocity equation (9.53), Engelund and Fredsoe (1976) obtained an equation for bed load transport. If p is the probability of movement of sediment particles in the surface layer, and $1/d^2$ represents the number of particles per unit area, then the bed load transport rate q_B in weight per unit width is given by:

$$q_B = \frac{\pi d^3}{6}\frac{p}{d^2}u_{sg} \tag{9.54}$$

Using (9.53), equation (9.54) reduces to

$$q_B = 9.3\frac{\pi}{6}dpu_*\left(1 - 0.7\sqrt{\tau_{*c}/\tau_*}\right) \tag{9.55}$$

The dimensionless bed load transport rate Q_S is given from equation (9.55) as:

$$Q_S = \frac{q_B}{\sqrt{\left(\frac{\rho_s}{\rho_f} - 1\right)gd^3}} = 4.8p\left(\sqrt{\tau_*} - 0.7\sqrt{\tau_{*c}}\right) \tag{9.56}$$

Luque and Beek (1976) provided empirical data on Q_S, τ_* and τ_{*c} from their experiments, and found the probability p given by equation (9.56). The results are provided in figure as the probability p versus τ_{*c} for $\tau_{*c} = 0.05$ and 0.06 (Figure 9.5). In this case, the transport rate was so small that all the particles were moving without forming any ripples or dunes. Larger transport rate was no longer applicable, so, necessary information has to be obtained for model to determine p. Guy et al. (1966) proposed a model to determine the probability p for undulation bed with dunes and ripples with sand $d = 0.93$ mm. For dune formation in the bed, it is necessary to divide shear stress into two components: one corresponds to skin friction and other corresponds to form drag. In this case, shear stress due to skin friction is important. At large transport stage, there is suspension. If the bed load is considered as a single layer

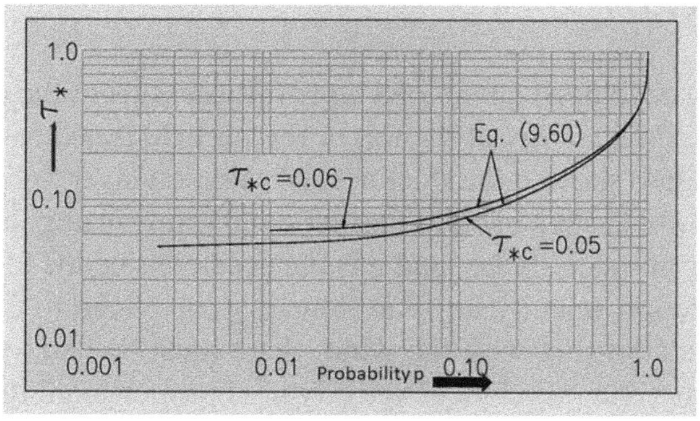

FIGURE 9.5 Plot of probability p against τ_{*c} for $\tau_{*c} = 0.05$, 0.06, $\beta_1 = 0.51$. (Modified from Engelund & Fredsoe 1976.)

of particles, the maximum value of p must be unity corresponding to a simultaneous motion of all particles in the layer.

An expression of p is obtained from the plane bed, where $\tau_0 - \tau_c$ is transmitted as the drag on moving particle. If N be the number of particles moving per unit area, then

$$\tau_0 - \tau_c = NF_G = \frac{\pi d^3 N}{6}(\rho_s - \rho_f) g\beta_1 \tag{9.57}$$

using equation (9.51). Then it can be written as

$$\tau_* - \tau_{*c} = \frac{\pi}{6}\beta_1 N d^2 = \frac{\pi}{6}\beta_1 p, \text{ for } Nd^2 = p \tag{9.58}$$

If $\tau_{*c} = 0.05$, and $\beta_1 = 0.51$, then

$$\tau_* = 0.05 + 0.2668\,p, \tag{9.59}$$

which is given in Figure 9.5 for comparison with Fernandez Luque experiments. For large shear stress, the entire layer of sediment moves as bed load, and the probability approaches to unity for increasing values of τ_*. Engelund and Fredsoe (1976) modified the above equation to get $p = 1$ as $\tau_* \to \infty$ and expressed as (Fredsøe and Deigaard 1992):

$$p = \left[1 + \left(\frac{\frac{\pi}{6}\beta_1}{\tau_* - \tau_{*c}}\right)^4\right]^{1/4} \tag{9.60}$$

which is nearly equal to equation (9.58) for τ_* close to τ_{*c}, and approaches to unity for large values of τ_*. Equation (9.60) is plotted in Figure 9.5 for $\tau_{*c} = 0.05$, and 0.06. If the bed consists of ripples and dunes, the equations should be replaced by shear stress τ_*' due to form drag.

The dimensionless bed load transport rate Q_S is given from equation (9.56) as:

$$Q_S = \frac{q_B}{\sqrt{\left(\frac{\rho_s}{\rho_f} - 1\right) g d^3}} = 4.8\left[1 + \left(\frac{\frac{\pi}{6}\beta_1}{\tau_* - \tau_{*c}}\right)^4\right]^{1/4}\left(\sqrt{\tau_*} - 0.7\sqrt{\tau_{*c}}\right) \tag{9.61}$$

Smith and Mclean (1977) developed a bed load layer thickness from the concept of saltation layer of particles. Balancing the potential energy at the top of the particle trajectory with its maximum kinetic energy on the bed, Owen (1964) showed the thickness of saltation layer to be $\rho_s u_o^2 / 2(\rho_s - \rho_f) g$, where ρ_s and ρ_f are respectively the densities of solid particles and transporting fluid, u_o is the maximum particle velocity at the top layer of the particles, and g is the acceleration due to gravity. Here u_o is assumed to be proportional to the friction velocity for fluid flow. If y_o is related to the thickness of the saltation layer, from the analogous relationship with the wind, there must be bed load transport layer by water flow. Owen (1964) argued that the initial local velocity u_o is proportional to the local shear velocity u_*. Moreover, he assumed that force on the particle, thus the work done per unit volume in lifting the from the bed, is scaled by $\tau_* - \tau_{*c}$, where τ_* is the shear stress, and τ_{*c} is the critical boundary shear stress associated with the initiation of sediment motion. When this approach is considered, the saltation layer or bed load layer thickness ξ_a is given by

$$\xi_a = \frac{\alpha_0 \rho_f \left(u_*^2 - u_{*c}^2\right)}{g\left(\rho_s - \rho_f\right)} = \frac{\alpha_0 \left(\tau_* - \tau_{*c}\right)}{g\left(\rho_s - \rho_f\right)} \text{ for } \tau_* > \tau_{*c} \tag{9.62}$$

where $\alpha_0 = 26.3$ are empirical constants for Columbia River data. The bed concentration defined at the top of the saltation layer ξ_a is given by

$$C_{\xi_a} = \rho_s \gamma_0 \frac{u_*^2 - u_{*c}^2}{u_{*c}^2} \tag{9.63}$$

with $\gamma_0 = 1.56 \times 10^{-3}$ is the empirical constant and $\rho_f u_{*c}^2$ is the critical bed shear stress for bed-load movement of sediment.

Example 9.1

In a wide-open channel flow with uniform flow condition, the flow velocity is 1.5 m/s, flow depth is 3 m, bed slope S_0 is 0.0001, and bed sediment median size d_{50} is 0.6 mm, the coefficient of kinematic viscosity of water is 10^{-6} m^2/s. The critical (threshold) shear stress for the bed is observed to be 0.85 Pa. Calculate the bed material transport rate (in volume rate per unit width) by the DuBoy's formula. Assume that characteristic sediment coefficient (A) as 4.5×10^{-6} kg^{-2} m^4 s^3.

Solution:

The Du Boy's formula is: $q_B = A(\tau_0 - \tau_{0c})\tau_0$. Here flow depth $h = 3$ m
The bottom shear stress is $\tau_0 = \rho_f g h S_0 = 1{,}000 \times 9.81 \times 3 \times 0.0001 = 2.94$ Pa.
The critical (threshold) shear stress τ_{0c} is 0.85 Pa and A = 4.5×10^{-6} kg^{-2} m^4 s^3.
Using the DuBoy's formula, the bed material transport rate = $q_B = A(\tau_0 - \tau_{0c})\tau_0$
Therefore $q_B =$ $4.5 \times 10^{-6} \times (2.94 - 0.85)\, 2.94 = 2.76 \times 10^{-5}$ m^2/s.

Example 9.2

In a wide-open channel flow with uniform flow condition, the flow velocity is 2 m/s, flow depth is 3.5 m, bed slope S_0 is 0.0001, and bed sediment median size d_{50} is 0.6 mm and relative density 2.65, the coefficient of kinematic viscosity of water is 10^{-6} m^2/s. Calculate the dimensionless bed load transport rate by the Meyer-Peter and Muller equation.

Solution:

The dimensionless bed load transport rate by the Meyer-Peter and Muller equation is:

$$Q_s = \frac{q_B \rho_f^{1/2}}{\left\{(\gamma_s - \gamma_f)d\right\}^{3/2}} = 8\left[\frac{\rho_f u_*^2}{(\gamma_s - \gamma_f)d} - 0.047\right]^{3/2}$$

The bottom shear stress is $\tau_0 = \rho_f g h S_0 = 1{,}000 \times 9.81 \times 3.5 \times 0.0001 = 3.43$ Pa

Shear velocity $u_* = \sqrt{\tau_0/\rho_f} = \sqrt{3.43/1{,}000} = 0.0585$ m/s.

Using above equation, $Q_s = 8\left[\frac{1{,}000 * 0.0585^2}{(2{,}650 - 1{,}000)0.6/1{,}000} - 0.047\right]^{3/2} = 50.375$

9.4.2 Bed Load Equations Based on Particle Velocity

Van Rijn (1984a) proposed a bed load expression in which the bed load transport rate is defined as the product of the particle velocity u_p, the saltation height h_s and the bed load concentration C_b. The resulting bed load transport equation is given by

$$q_B = u_p h_s C_b \tag{9.64}$$

The saltation height h_s is related to the dimensionless transport stage parameter $T_* = \dfrac{\tau'_* - \tau_{*c}}{\tau_{*c}}$ and dimensionless particle diameter $d_* = d\left[\left\{\left(\dfrac{\rho_s}{\rho_f} - 1\right)g\upsilon^{-2}\right\}\right]^{1/3}$ as:

$$\frac{h_s}{d} = 0.3 d_*^{0.7} T_*^{0.50} \tag{9.65}$$

in which $\tau'_* = \rho_f u_*'^2$ with $u'_* = \bar{u}\left(\sqrt{g}/C'\right)$ is the bed shear velocity related to grains, C' is Chezy's coefficient with respect to the grain, $\bar{u}$ is the mean velocity, τ_{*c} is the critical shear stress according to Shields curve. van Rijn (1984a) computed the transport stage parameter T_* and the particle diameter d_* from the different sets of hydraulic conditions and determined the saltation height shown in Figure 9.6 with an accuracy of about 10%. The figure clearly shows that for small particles, the dimensionless saltation height is smaller than for the large particles at the same transport stage.

To determine the particle velocity u_p for computation of equation (9.64), van Rijn (1984a) assumed the idea of Bagnold (1973) that the mean stream-wise velocity relative to the flow is that velocity at which mean fluid drag on the particle is in equilibrium with the mean frictional force exerted by the bed surface on the particles. The Bagnold's relation was presented by the following expression as:

$$\frac{u_p}{u_*} = \alpha_1 - \alpha_2\left(\tau_{*c}/\tau_*\right)^{0.5} \tag{9.66}$$

where u_* is the friction velocity, $\tau_{*c} = \dfrac{\tau_{0c}}{\left(\rho_s - \rho_f\right)gd}$ is the critical particle mobility number, τ_* is the particle mobility number, and α_1, α_2 are the coefficients to be determined. As the saltation height is a

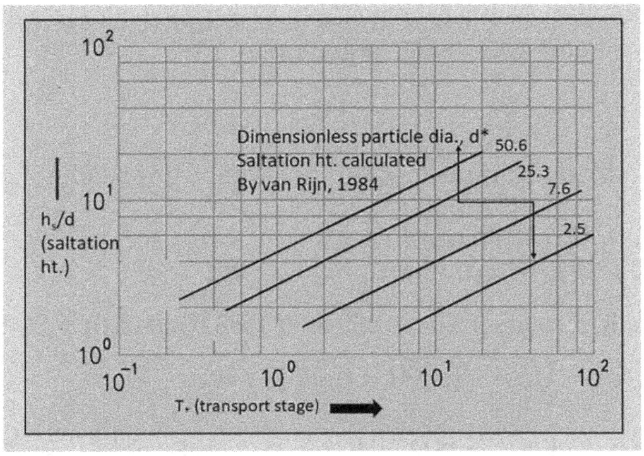

FIGURE 9.6 Saltation height (h_s/d) as a function transport stage T_* for different values of particle parameter d_*. (Modified from Van Rijn 1984a.)

function of dimensionless sediment size d_* and stream-wise velocity along with the particle trajectory, and so the coefficient α_1 will be a function of sediment size d_* and the coefficient $\alpha_2 = 8$. Figure 9.7 shows the computed particle velocity as a function of flow condition and sediment size, and the equation is approximated with about 10% accuracy as:

$$\frac{u_p}{u_*} = 9.0 + 2.6\log d_* - 8\left(\tau_{*c}/\tau_*\right)^{0.5} \tag{9.67}$$

Figure 9.7 was developed from different experimental data (Fernandez Luque 1974; Francis 1973). If the mobility parameter τ_* approaches to the critical value τ_{*c}, the particle velocity $\frac{u_p}{u_*}$ approaches to zero. From the mathematical model, the computational results were approximated about 20% accuracy by the following equation as:

$$\frac{u_p}{\sqrt{\left(\frac{\rho_s}{\rho_f} - 1\right)gd}} = 1.5T_*^{0.60} \tag{9.68}$$

Using the expression similar to equation (9.66) and experimental data fitting, Engelund and Fredsoe (1967) derived the expression as:

$$\frac{u_p}{u_*} = 10 - 7\left(\tau_{*c}/\tau_*\right)^{0.5} \tag{9.66a}$$

The expression (9.66a) is shown in Figure 9.7, which agrees more or less well with Fernandez Luque's data, reported by van Rijn (1984). If both parameters u_p and h_s are known from the measured bed-load transport data, the bed-load concentration is determined. In this section, a general function for the bed-load concentration C_b is derived for both small and large particles. Firstly, however, an effective grain-shear velocity is defined to eliminate the influence of bed forms. When the bed forms are present, the influence of form drag should be eliminated from the model. The measured grain shear stress along a bed form varies from zero in the trough to its maximum value at the top. To compute the average

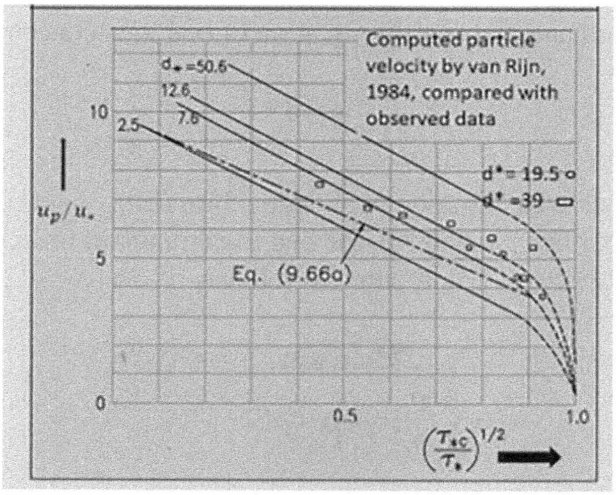

FIGURE 9.7 Particle velocity u_p/u_* as a function of mobility parameter $\left(\tau_{*c}/\tau_*\right)^{0.5}$ for different particle diameter d_*. (Modified from Van Rijn, 1984a.)

bed-load transport, it is essential to estimate the average grain-shear stress at the up-sloping of the bed form. The location where the average grain-shear stress acts, the local flow velocity is about the same at that for plane bed flow with the same mean flow velocity $\bar{u}$ and the particle size. The average grain-shear stress τ_g is given by:

$$\tau_g = g\rho_f\left(\bar{u}^2/C'^2\right) \tag{9.69}$$

$$\Rightarrow \rho_f u'^2_{g*} = g\rho_f\left(\bar{u}^2/C'^2\right)$$

$$\Rightarrow u'_{g*} = \sqrt{g}\left(\bar{u}/C'\right) \tag{9.70}$$

The Chezy-coefficient C' for surface grain roughness of the sediment bed is defined as:

$$C' = 19.43 + 18\,\mathrm{Log}\left(\frac{h_R}{3d_{90}}\right) \tag{9.71}$$

where h_R is the hydraulic radius related to the bed with side-wall correction (Vanoni and Brooks 1957). The average shear stress related to the form drag was determined by averaging the measured pressure distribution over the ripple length resulting in τ'_g. The total average bed shear stress is defined as $\tau_{av} = \tau_g + \tau'_g$, where τ_g is the average grain-shear stress and τ'_g is the average form drag shear-stress (van Rijn 1984).

To compute the bed load transport rates q_B from equation (9.64), the bed load concentration C_b is determined as:

$$C_b = \frac{q_B}{u_p h_s} \tag{9.72}$$

in which the velocity u_p is determined from equation (9.68) and the saltation height or bed load layer h_s is determined from equation (9.65). In the study of the motion of solitary particles, he assumed the particles to be of uniform shape, size and density. It is assumed that the size distribution of the bed material can be represented by the size d_{50} and the geometric standard deviation σ_g. Extensive data taken from Gilbert (1914), Falkner (1935), Tsubaki and Shinohara (1959), Guy et al. (1966), Williams (1970), Luque (1974), and Willis (1979) are analyzed and the equation of bed load concentration can be written as:

$$\frac{C_b}{C_0} = 0.18\frac{T_*}{D_*} \tag{9.73}$$

where C_0 is the maximum bed concentration of about 0.65 for firmly packed grains in absolute volume. Combining equations (9.65), (9.68), and (9.73), the bed load transport (m^2/s) equation (9.64) for particle sizes ranging from 200 to 2,000 μm is computed as follows:

$$\frac{q_B}{d_{50}\sqrt{\left(\frac{\rho_s}{\rho_f}-1\right)g\,d_{50}}} = 0.053\frac{T_*^{2.1}}{d_*^{0.3}} \tag{9.74}$$

To compute the bed load transport equation (9.74), the mean velocity $\bar{u}$, mean flow depth (h), channel width, particle diameters, densities of water and sediment, kinematic viscosity, and acceleration due to gravity are needed.

Bridge and Dominic (1984) developed a theoretical model on dynamics of bed load transport rate from the concept of force balance equation acting on single spherical particle. The bed load transport rate is expressed as:

$$I_B = F_g u_b \tag{9.75}$$

where I_B is the be load transport rate as weight per unit width, F_g is immersed weight of grains moving over unit bed area, and u_b is the mean speed of bed load grains. Here submerged weight F_g and mean velocity of the grain respectively are given by:

$$F_g = (\tau_* - \tau_{*c})\tan\alpha \tag{9.76}$$

$$u_b = a(u_* - u_{*c}) \tag{9.77}$$

where $a = 1/\kappa_0 \ln(y_n/y_0)$ with y_n is the distance of the effective fluid thrust from the boundary related to the saltation height and y_0 is the roughness height. Substituting equations (9.76) and (9.77) in equation (9.75), an expression of bed load transport can be written as

$$I_B = a\tan\alpha(\tau_* - \tau_{*c})(u_* - u_{*c}) \tag{9.78}$$

Equation (9.78) is very similar in general form to those of Meyer-Peter and Muller (1948), Einstein (1950), Bagnold (1956, 1973), Ashida and Michiue (1973), Luque and Van Beek (1976), and Engelund and Fredsoe (1976). The dimensionless form of equation (9.78) is given by:

$$q_B = \frac{I_B\sqrt{\varrho_f}}{\left[(\varrho_s - \rho_f)gd\right]^{3/2}} = a\tan\alpha\left[(\theta_*)^{0.5} - (\theta_{*c})^{0.5}\right]\left[\theta_* - \theta_{*c}\right] \tag{9.79}$$

where q_B is the dimensionless bed load transport rate, $\theta_* = \dfrac{\tau_*}{(\varrho_s - \rho_f)gd}$ is the dimensionless bed shear stress, and θ_{*c} is the critical value of θ_*. The rigorous analysis of forces arising from particle motion is clearly presented in their paper (1984). Later, Bridge and Bennett (1992) developed a model for bed load transport of sediment grains of different sizes, shapes and densities by a unidirectional current. They presented a semi-analytical model for bed load transport of sediment mixtures considering the equilibrium forces. They assumed the fluctuations of bed shear stress to be Gaussian with coefficient of variation being 0.40 and the transport rate was worked out considering the fraction of the time for which a particular shear stress existed.

Smart (1984) conducted a series of experiments in a tilting flume to investigate the sediment transport capacity for steep flow in longitudinal negative slopes from 3% to 20%. He used four different alluvial sediments (coarse sand and gravels) with uniform or non-uniform sizes. Smart parameterized a semi-empirical law for sediment transport suited for both low and steep slopes, and uniform and non-uniform sediments, using the critical Shields parameter. Based on his experimental data and the data of Meyer-Peter and Müller, Smart (1984) developed a bed load transport rate formula in steep channels within the slope range $(0.03 \le J \le 0.2)$ for gravel sizes $(2 \le d \le 10.5)$ mm as:

$$Q_b = \frac{q_B}{d\sqrt{\left(\dfrac{\rho_s}{\rho_f} - 1\right)gd}} = \frac{4C_R}{\sqrt{g}}\left(\frac{d_{90}}{d_{30}}\right)^{0.2} J^{0.6}\left[\theta_* - \theta_{*c}\right](\theta_*)^{0.5} \tag{9.80}$$

Hanes and Bowen (1985) suggested a mathematical model in which grain-grain collisions dominate the dynamics in the granular fluid layer. They determined directly the velocity and the grain transport rate analytically without any fitting parameters. Their model was dependent on the two fluid regions: one was a collision-dominated granular fluid region, and other one was wall-bounded fluid shear region with saltating grains. They used a velocity profile as a function of the stresses applied at the upper surface of the granular fluid. The flux of grain transport in the saltation region is estimated as the product of saltation layer thickness, velocity and concentration. The dimensionless grain transport relation Q_{BS} in the saltation region is given by:

$$Q_{BS} = \frac{\sqrt{\rho_f}\, N_\delta \bar{u} y_m}{d\sqrt{(\varrho_s - \rho_f) g d}} \tag{9.81}$$

where N_δ is the volume concentration of grains within the thickness δ of the granular fluid layer at $y = \delta$, $\bar{u}$ is the horizontal (time-averaged) fluid velocity above granular fluid region, y_m is the vertical maximum height above the granular-fluid region to which grains saltate. Using the functional forms of N_δ, $\bar{u}$ and y_m in equation (9.81), the transport rate Q_{BS} will be an explicit form as a function of familiar non-dimensional Shields parameter $\frac{\tau}{(\varrho_s - \rho_f) g d}$. The immerged weight flux I_G of grains per unit width is given by:

$$I_G = (\varrho_s - \rho_f) g \int_0^\delta Nu \, dy \tag{9.82}$$

Following the analysis of Hanes and Bowen (1985) for N and u, the non-dimensional form of immerged weight grain flux Q_{BG} is given by:

$$Q_{BG} = \frac{I_G \sqrt{\rho_f}}{\left[(\varrho_s - \rho_f) g d\right]^{3/2}} = \frac{18 N_*^3 \sqrt{\rho_f}\, F_2(\lambda_\delta)(\Delta\phi)^2}{(\varrho_s c_1)^{1/2}\left[N_0^2 - N_\delta^2\right]^2} \left[\frac{\tau}{(\varrho_s - \rho_f) g d}\right]^{5/2} \tag{9.83}$$

where $F_2(\lambda_\delta)$ is a function of $\lambda = d/s$, λ is the ratio of the grain diameter d to the separation distance s between the grains at a thickness of the layer δ, $\Delta\phi = (\tan\phi_r)^{-1} - (\tan\phi_\delta)^{-1}$ with ϕ_r is the critical dynamic angle of internal friction, that is, $\tan\phi_r \approx 0.5$, and N_0 is the maximum volume concentration at the bottom ($y = 0$), and other notations are known. The transport rate in the saltation region Q_{BS} and the rate due to immerged weight flux of region Q_{BG} are shown in Figure 9.8 against the dimensionless (normalized) Shields parameter $\theta = \frac{\tau}{(\varrho_s - \rho_f) g d}$.

The theoretical prediction for the total bed load transport is approximated quite well by the *5/2* th degree of polynomial expressed as:

$$Q = Q_{BG} + Q_{BS} = 3.5\left(\frac{\tau}{(\varrho_s - \rho_f) g d}\right)^{5/2} = 3.5\theta^{5/2} \tag{9.84}$$

which is almost the same as the formula of Engelund and Hansen (1967), and equation (9.84) is compared with the bed load transport equation of Meyer-Peter and Muller (1948) in Figure 9.9. The Meyer-Peter and Muller equation was calibrated at low transport rates and it was underestimated the transport

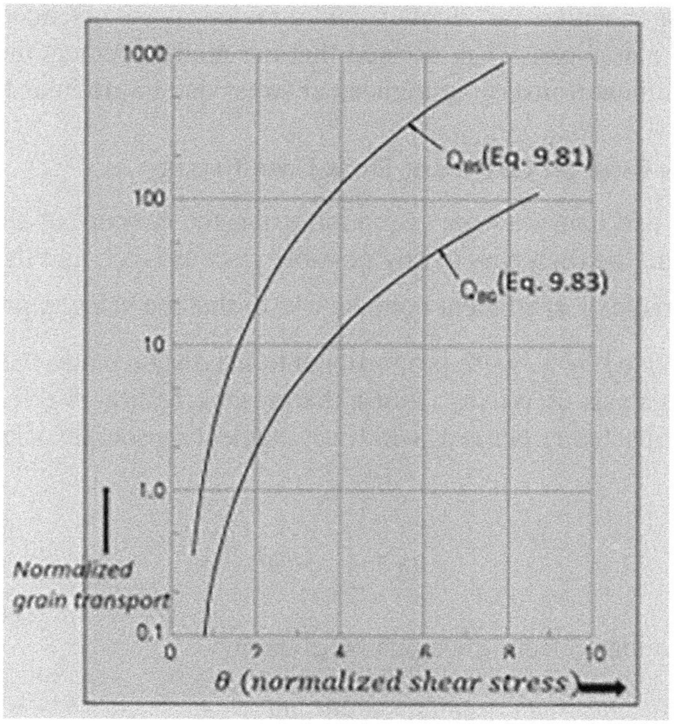

FIGURE 9.8 Normalized grain transport due to grain flow Q_{BG} and saltation Q_{BS} as a function of normalized shear stress θ. (Modified from Hanes and Bowen, 1985.)

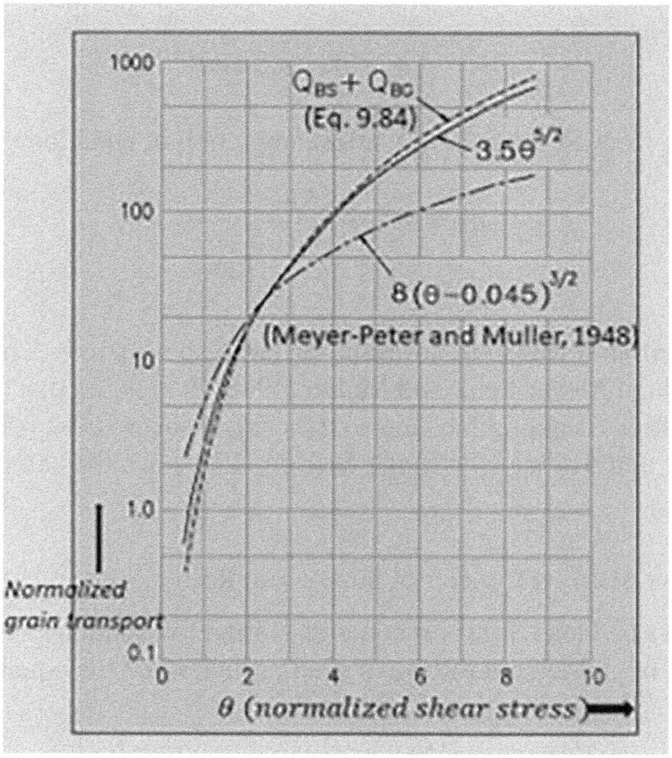

FIGURE 9.9 Bed load transport Q is compared with Meyer-Peter and Muller (1948) with polynomial $Q = 3.5\theta^{5/2}$. (Modified from Hanes and Bowen 1985.)

at slightly higher rates. Later, Hanes and Huntley (1986) reviewed and extended with a discussion of bed resistance and boundary conditions. They showed that the model predicts the reduction of flow resistance and increase in sediment transport at high shear stress due to granular-fluid layer near the bed.

9.4.3 Bed Load Equation Based on Concept of Shear-Layer Thickness

Wilson (1987) derived a bed load transport equation using the concept of sheared-layer thickness h_s, which is much greater than the thickness of any possible viscous layer, and the fluid motion will be too turbulent. It is a characteristic of a turbulent boundary layer that the velocity gradient $\frac{du}{dy}$ can be equated to the friction velocity divided by a length proportional to the distance into the flow. Following the high bed load transport rate analysis of Wilson (1966), the present method is extended to the equivalent of the contact load of solid discharge per unit width q_B. Neglecting the particle slip, bed load discharge can be written as:

$$q_B = \int_0^{h_s} Cu\,dy \tag{9.85}$$

where the velocity u is evaluated from the equation given by:

$$u = \frac{2u_*}{\kappa_0}\sqrt{\eta} \tag{9.86}$$

in which $\eta = \frac{y}{y_s}$ with y_s is the depth extended at which the intergranular resistance equals to applied shear stress. The extended shear layer y_s is given by

$$y_s = \frac{\tau}{\left(\rho_s - \rho_f\right) g C_b \tan\phi} \tag{9.87}$$

Substitution of η using y_s from equation (9.87), after integration of equation (9.85), one gets:

$$q_B = \frac{1.51u_*}{\kappa_0 g\left(\frac{\rho_s}{\rho_f} - 1\right)\tan\phi} \tag{9.88}$$

Equation (9.88) is the bed load transport formula for large shear stress. This equation is compared with the bed-load formula of Meyer-Peter and Muller (1948). It is noted that for the case of high shear stress, the critical shear stress is negligible, and so the critical shear stress is ignored. So, the bed-load formulas of Meyer-Peter and Muller (1948) and Wilson (1987) are the same except some numerical values.

9.4.4 Bed Load Equations Based on Concept of Sliding and Rolling

Madsen (1991) developed a bed load transport equation similar to equation (9.79) of Bridge and Dominic (1984), using the concept of sliding and rolling of sand particles and the equation is given by:

$$q_B = \frac{8}{\tan\phi}\left[(\theta_*)^{0.5} - 0.7(\theta_{*c})^{0.5}\right]\left[\theta_* - \theta_{*c}\right] \tag{9.89}$$

and for saltation of sand particles equation (9.89) is given by

$$q_B = 9.7\left[(\theta_*)^{0.5} - 0.7(\theta_{*c})^{0.5}\right]\left[\theta_* - \theta_{*c}\right] \tag{9.90}$$

Nielsen (1992) developed an equation for the size range ($0.69 \le d \le 28.7$) i.e., for sand and gravel transport is as:

$$q_B = 12[\theta_* - 0.05]\theta_*^{0.5} \tag{9.91}$$

Subsequently, Nino and Garcia (1998) suggested a similar equation for saltating particles with dynamic coefficient of friction $\tan\phi = 0.23$ and is given by

$$q_B = 52.17\left[(\theta_*)^{0.5} - 0.7(\theta_{*c})^{0.5}\right]\left[\theta_* - \theta_{*c}\right] \tag{9.92}$$

Gomez (1991) made an excellent review for complete understanding of bed load transport from the reliable apparatus, sampling techniques and prediction of bed load. Particular emphasis was given to the development of methods of prediction and estimation of bed load discharge, using rigorous sampling and measurement techniques in the field. He also addressed the knowledge of bedload transport dynamics in rivers.

Cheng (2002) derived a simple exponential formula to compute the bed load transport rates for the conditions low to high shear stress. It is known that for large shear stress, the bed load transport rate q_B is proportional to the product of bed shear stress τ and the shear velocity u_*, that is, $q_B \sim \tau u_*$ (Yalin, 1977). The product τu_* represents the energy of the flow near the bed, which is closely related to the motion of the near-bed particles. The dimensionless bed load transport rate is $Q_B = \dfrac{q_B\sqrt{\rho_f}}{d\sqrt{(\varrho_s - \rho_f)gd}}$ and the energy τu_* is normalized as $\Theta = \dfrac{\tau u_*}{\rho_f\left[\left(\dfrac{\varrho_s}{\rho_f} - 1\right)gd\right]^{3/2}}$. Therefore, the proportional relation of $q_B \sim \tau u_*$ can be written as $Q_B = Const.\Theta$, where *Const.* is to be dependent. This linear relationship is consistent with many bed load transport formulas for the case of high shear, even though its physics remains unclear. Cheng (2002) reported one exponential equation for bed load transport rate as:

$$Q_B = 13\Theta^{1.5}\exp\left(-0.05\Theta^{-1.5}\right) \tag{9.93}$$

This above general equation is derived to compute the bed load transport rates for low and high shear stresses, which agrees well with the well-known experimental data set of Gilbert (1914), Meyer-Peter and Mueller (1948), Wilson(1966), and Paintal (1971). For moderate values of shear, the formula is very close to the data set of Einstein (1950) and Meyer-Peter and Mueller (1948); and for the weak transport, it agrees with the empirical relationships of Einstein (1942) and Paintal (1971). Bravo-Espinosa et al. (2003) made an excellent review of several existing bed load transport equations to identify the conditions of bed load transport and performance from the hydraulic and sediment data from 22 alluvial streams in the United States. The observation reveals that the bedload transport conditions in alluvial streams are dependent on individual particle-size fractions rather than the particle sizes represented by one characteristic size. They obtained three conditions of movement in alluvial streams: these are the particle size fraction, the stream power and the transport stage. They compared the several existing models (Parker et al. 1982; Schoklitsch 1962; Meyer-Peter and Muller 1948; Bagnold 1980) with the bed load data collected from the alluvial streams, and from the overall analysis, they found that

Schoklitsch's model had the ability to predict the measured bed load data of eight streams of 22 streams, and the Bagnold equation predicted the measured data for seven streams out of 22 streams. Abrahams (2003) developed a bed load transport equation for sheet flow. When there is no suspended sediment, near the boundary a layer of colliding grains with stationary bed is formed known as sheet flow. An equation for estimating the rate of bed-load transport in sheet flow is developed from an analysis of 55 flume and closed conduit experiments, using the concept of Bagnold's (1966) bed load transport model. This model is based on the principle that a bed load transport is a machine whose performance may be described by the basic energy equation, that is, work rate is the available power multiplied by the efficiency and is given by

$$i_b \tan\alpha = \omega\left(u_s/u\right) \tag{9.94}$$

where $i_b = q_b g\left(\varrho_s - \rho_f\right)$ is the immerged bed load transport rate, q_b is the volumetric bed load transport rate, $\tan\alpha$ is the dynamic friction coefficient, u_s is the mean grain velocity. Equation (9.94) is written as:

$$i_b = \omega e_s/\tan\alpha \tag{9.95}$$

where $e_s = \dfrac{u_s}{u}$ is Bagnold's efficiency of the flow transporting the bedload. Equation (9.95) is Bagnold's basic bedload transport equation. Connecting this equation (9.95) with the equation $i_b = \omega$, equation (9.95) implies

$$e_s = \tan\alpha = u_s/u \tag{9.96}$$

Given that $\tan\alpha = 0.6$ for natural sand, $u_s = 0.6\ u$, and $i_b = 0.6$. The value of 0.60 for e_s is much larger than the value of 0.12 calculated by Bagnold, indicating that sheet flow is a much more efficient mode of bed-load transport than previously thought, which is confirmed from the experimental data. For high bed-load transport rate, Rickenmann (1991) reported that the particles transport like a sheet flow. Subsequently, Abrahams and Gao (2006) developed a bedload transport model over plane bed for rough turbulent open-channel flows.

Mazumder and Dalal (2003) developed a theoretical model to determine the maximum saltation layer thickness of sediment particles in water associated with the migration velocity of the particle in the bed layer. The equation of mean particle velocity at the bed was derived by balancing the horizontal forces acting on the particle in the bed. The formulation explicitly depends on the modified particle velocity, added mass coefficient and take-off angle. The main idea is to determine the average thickness of the moving sediment layer at the near-bed flows. A sediment grain at the bed surface will start to move when the friction velocity for the particle moving along the bed is exactly balanced by the sum of the drag and the appropriate component of gravitational force in a slopping bed (Figure 9.10).

The equation of the horizontal balance of forces can be expressed as:

$$\left(W_g \cos\beta - F_L\right)\tan\alpha = F_D + W_g \sin\beta \tag{9.97}$$

where W_g is the submerged weight force of the particle, F_L and F_D are respectively lift and drag forces on the bed load grains per unit area. The drag force F_D acting on a grain in the direction of the flow is expressed as

$$F_D = \frac{\pi}{8}\varrho_f C_D d^2\left(au_* - u_p\right)^2 \tag{9.98}$$

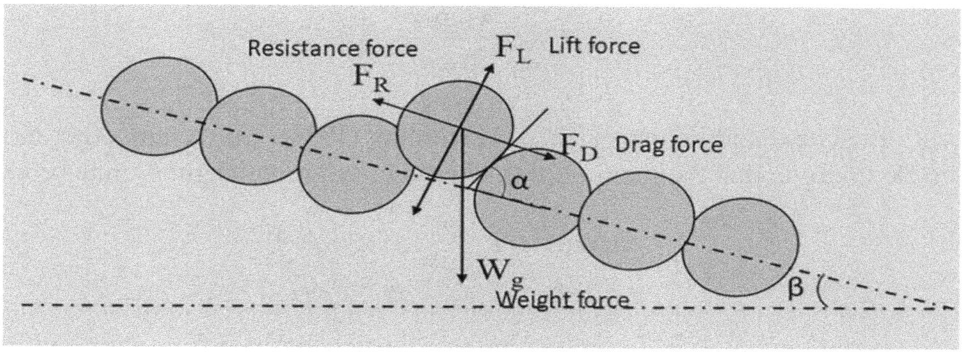

FIGURE 9.10 Balance of Forces on a particle over a bed surface. (Modified from Mazumder and Dalal 2003.)

where the term $(au_* - u_p)$ is the relative velocity, au_* is the fluid velocity at height y_D of the effective drag on the spherical grain of diameter d, and a is the coefficient depending on the height of the effective drag, C_D is the drag coefficient. According to Wiberg and Smith (1985), the velocity at the bottom of the grain is much smaller than the velocity at the top of the grain, and under the assumption, the equation for the mean lift force on the bed load grain is expressed as:

$$F_L = \frac{\pi}{8}\varrho_f C_L d^2 u_T^2 \tag{9.99}$$

The equation of motion for a spherical particle falling vertically in water with its terminal velocity v_s under the influence of gravity is given by:

$$W_g = \frac{\pi}{8}\varrho_f C'_D d^2 v_s^2 \tag{9.100}$$

where C'_D is fluid drag coefficient on a grain at its terminal velocity. Following Francis (1973), $C'_D = C_D$ is assumed for this analysis. Using the above equations (9.98) to (9.100) in equation (9.97), and using some other assumptions (Mazumder and Dalal 2003), the expression of mean velocity u_p of any bed load grain is obtained as:

$$u_p = au_*\left[1-\left(\frac{\tau_{0c}}{\tau_0}\frac{(\cos\beta-\lambda_T)\tan\alpha-\sin\beta}{(\cos\beta-\lambda_{Tc})\tan\alpha_c-\sin\beta}\right)^{0.5}\right] \tag{9.101}$$

where $\tau_{*c} = \dfrac{\tau_c}{(\varrho_s-\rho_f)gd}$ is the dimensionless critical shear stress, and $\lambda_T = \dfrac{\lambda u_T^2}{v_s^2}$, and $\lambda = C_L/C_D$ is the ratio of lift and drag coefficients on a grain. This equation (9.101) serves as a more general expression of mean velocity u_p of the bed load grain than those obtained from the previous studies. This equation is used to determine the average saltation layer thickness. When $\beta \to 0$ and $\lambda_T \ll 1$, equation (9.101) reduces to

$$u_p = au_*\left[1-\left(\frac{\tau_{0c}}{\tau_0}\frac{\tan\alpha}{\tan\alpha_c}\right)^{0.5}\right] \tag{9.102}$$

which is similar to the expressions given by Egelund and Fredsoe (1976), Bridge and Dominic (1984), Mazumder (1994); and corresponding expression for C_D as:

$$C_D = \frac{4\tan\alpha_c}{3\tau_{0c}a^2} \tag{9.103}$$

which agrees with Engelund and Fredsoe (1976), Mazumder (1994). Unlike any other models, it would be more reasonable to argue that the maximum thickness of saltation height h_s may be expressed as:

$$h_s = \frac{u_p^2 \sin^2(\sigma)}{2g_e}, \; g_e = \frac{g(\varrho_s - \rho_f)}{\varrho_s - \rho_f C_m} \tag{9.104}$$

where g_e is the equivalent gravitational acceleration of a particle derived from the resulting equation of dynamic of motion of a particle with respect to the surrounding fluids (Murphy and Hooshiari 1982), $C_m = 0.5$ is the added mass coefficient for spherical particle (Wiberg and Smith 1985) and σ is the take-off angle to be determined from the experimental data.

The theoretical-empirical models of average particle velocity (9.101) and the thickness of saltation layer (9.104) are used to verify the data from Luque and Beek (1976), Ghosh et al. (1981), Wiberg and Smith (1985), Wiberg and Rubin (1989) and Sumer et al. (1996). The models are tested for the range of grain sizes within 0.013–0.33 cm of density 2.64 g/cm^3 and different transport stage $T_* = \tau_0/\tau_{0c}$ varying from 1.24 to 30.00 in which the values of u_* lie between 2.5 and 7.50 cm/s. To determine the migration velocity u_p of any bed load grain from equation (9.101), it is important to know the values of parameters a, τ_0, τ_{0c}, λ, $\tan\alpha$, $\tan\alpha_c$, v_s, The dimensionless critical shear stress τ_{0c} is taken from the theoretical curve of Wiberg and Smith (1985) for different Reynolds number $Re_* = u_* d/\upsilon_f$. Values of average particle velocity (calculated and observed) and the calculated saltation height are shown in Figures 9.11 and 9.12.

A physical mechanism may be assumed which converts the average streamwise velocity to a vertical velocity $u_p \sin\sigma$ through the collisions with particles resting on the bed, where σ is the angle of inclination of particles with respect to the horizontal surface. In the literature (Middleton and Southard 1984), it is suggested that many of the saltating particles lift up almost vertically ($\sigma = 90°$) from the bed surface, which may not be true. Due to collisions with the other particles resting on the bed, the migrating particles might bounce back, and continue to move again with the flow. Following the argument

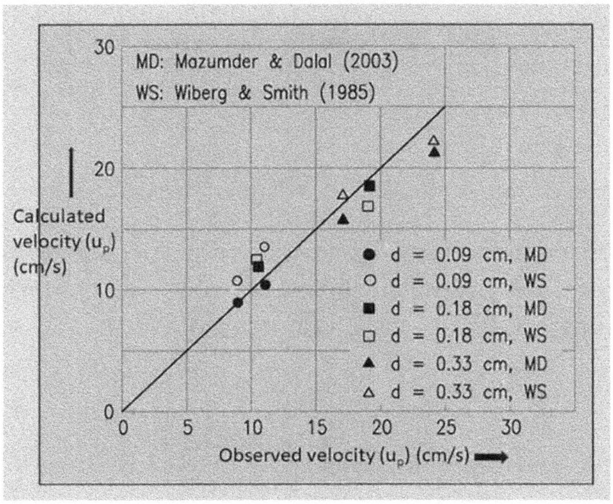

FIGURE 9.11 Plot of calculated average particle velocities against the observed velocities by Mazumder and Dalal (2003) (solid symbols) and the computed velocities by Wiberg and Smith (1985) (open symbols). (Modified from Mazumder and Dalal 2003.)

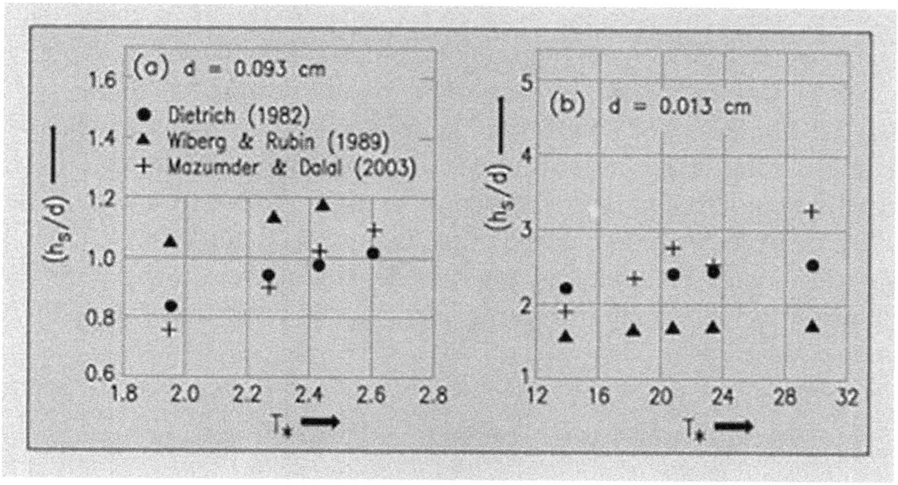

FIGURE 9.12 Plots of normalized heights h_s/d of particles against transport stage T_*; (a) for $d = 0.093$ cm at lower transport stage and (b) for $d = 0.013$ cm at upper transport stage. (Modified from Mazumder and Dalal, 2003).

mentioned above, an important relationship between the transport stage T_* and take off angle σ has been established with higher transport stage T_*, which is available in equation (9.19) of Mazumder and Dalal (2003).

If both parameters u_p and h_s are known, multiplying u_p and h_s by the bed load concentration C_b, the bed load transport rate q_B can be determined. Following van Rijn (1984)'s idea, equation (9.73) is used to determine the bed load concentration C_b as $C_b = 0.18C_0\left(T_*/D_*\right)$, where C_0, T_* and D_* are defined in van Rijn (1984). Combining Eqs. (9.101), (9.104) and (9.73), the bed load transport (m²/s) for particle sizes ranging from 200 to 2,000 μm is computed from equation (9.64), which is similar to the equation as:

$$\frac{q_B}{d_{50}\sqrt{\left(\frac{\rho_s}{\rho_f}-1\right)gd_{50}}} = \frac{u_p h_s C_b}{d_{50}\sqrt{\left(\frac{\rho_s}{\rho_f}-1\right)gd_{50}}} \tag{9.105}$$

Wong and Parker (2006) reviewed the bed load formula developed by Meyer-Peter and Muller from their experimental data and determined two new power laws from the statistical fitting of the experimental results; one is given by:

$$Q_b = \frac{q_B}{d\sqrt{\left(\frac{\rho_s}{\rho_f}-1\right)gd}} = 4.93(\theta_* - 0.0470)^{1.6} \tag{9.106}$$

It keeps the same critical Shields value 0.0470 used by Meyer-Peter and Muller, while a new value 1.60 as an exponent is obtained from a statistical fitting of the experimental results. The other one is given by:

$$Q_b = \frac{q_B}{d\sqrt{\left(\frac{\rho_s}{\rho_f}-1\right)gd}} = 3.97(\theta_* - 0.0495)^{3/2} \tag{9.107}$$

It keeps the same exponent 1.50 as found by Meyer-Peter and Muller, whereas the effective critical Shields value is changed to 0.0495 from the statistical fitting. In either case, the final outcome of the reanalysis is a line of best fit that gives estimates of the bedload transport rate.

Lajeunesse et al. (2010) developed a model on the motion of individual bed load particles entrained by a turbulent flow in a small experimental flume. Their plan was to characterize the motion of individual bed load particles in terms of the macroscopic sediment transport and to analyze the measurements within the frame of the erosion-deposition model proposed by Charru (2006). They performed systematic measurements of the velocity and surface density of the moving grains together with the lengths and durations of the particle flights, using a high-speed video imaging system. They suggested a bed load transport equation as:

$$q_B = 10.6\left[(\theta_*)^{0.5} - (\theta_{*c})^{0.5} + 0.025\right][\theta_* - \theta_{*c}] \tag{9.108}$$

The numerical constant 0.025 represents the nonzero particle velocity at the threshold may be neglected for the practical purpose. Equation (9.108) was tested against direct measurements of the sediment transport rate at the flume outlet, performed with the scale. This equation does not involve any adjustable parameter and the measurements with the scale are independent of all the quantities measured from image acquisition. Subsequently, the model of Lajeunesse et al. (2010) for a sediment bed of uniform grain size was generalized by Houssais and Lajeunesse (2012) to the case of a bimodal one, provided the dependency of the critical Shields parameter with the surface fraction of small grains is taken into account. The investigation was made for a bimodal sediment bed, composed of a mixture of two populations of grains of size $d_1 = 0.7$ and $d_2 = 2.2$ mm respectively. Using high-speed video cameras, the average particle velocity and the surface density of moving particles, defined as the number of moving particles per unit surface of the bed, were recorded. These two quantities are recorded separately for each population of grains as a function of dimensionless Shields parameter and the fraction of the bed surface covered with small grains. Finally, Houssais and Lajeunesse suggested the sediment flux of each size fraction as:

$$q_B = 56.6\frac{\rho_f}{\rho_s}\theta_{*i}\left[(\theta_{*i})^{0.5} - (\theta_{*ic})^{0.5} + 0.022\right][\theta_{*i} - \theta_{*ic}] \tag{9.109}$$

They also showed that the erosion-deposition model developed by Lajeunesse et al. (2010) for a bed of uniform sediment bed is generalized for the case of a bimodal one.

Zhong et al. (2011) developed a bed load entrainment rate function based on the upward flux by integrating the particle velocity distribution defined by the kinetic theory. The upward motion was determined by a sediment particle velocity, which was obtained by solving the Boltzmann equation with the assumption that the sediment-laden flow is dilute. In kinetic theory, the entrainment rate is equivalent to the volumetric flux in the upward direction, which is estimated by integrating the particle velocity over all realizations of the upward motion of the sediment near the riverbed region

$$E_R = \frac{\pi}{6}d^3\int_{-\infty}^{\infty}\int_{0}^{\infty}\int_{-\infty}^{\infty} u_y f\,du_x\,du_y\,du_z \tag{9.110}$$

where E_R is the bed sediment entrainment rate measured in volume, u_i is the stochastic velocity of sediment particle, f is the velocity distribution function of the particles, and d is the diameter of particle. Determination of entrainment rate E_R is possible if the velocity distribution function of sediment particles is known. In kinetic theory, the velocity distribution function f is determined from the differential equation obtained from Chapman and Cowling (1970), satisfying the law of conservation of mass. The velocity distribution function is obtained as (Chapman and Cowling 1970)

$$f(x_i,u_i,t)=\frac{N_0}{(2\pi K)^{1.5}}\exp\left[-\frac{V_p^2}{2K}-\frac{\varphi(x_i)-\varphi(x_{i0})}{K}\right] \tag{9.111}$$

where N_0 is the number of particles of per unit volume, or number density at a reference point x_{i0}, and V_p^2 is the square value of particle peculiar velocity, and K is the granular turbulent temperature, and the function $\varphi(x_i)$ is given by

$$\varphi(x_i)=-\int\frac{\sum F_i}{m}dx_i \tag{9.112}$$

where $m=\rho_s\pi d^3/6$ is mass of the sediment particle of diameter d, $\sum F$ is the resultant of external forces acting on a sediment grain. Interested readers may see the reference in detail (Zhong et al. 2011). Using equation (9.112) in (9.111) and after calculation, the velocity distribution function f can be written as:

$$f=\frac{N_0}{(2\pi K)^{1.5}}\exp\left[-\frac{V_p^2}{2K}-\frac{y-y_0}{K}\left\{(\rho_s-\rho_f)\frac{g}{\rho_s}-\frac{3C_L\rho_f u_*^2}{4d\rho_s}+\frac{3C_L\rho_f|u_y|u_y}{4d\rho_s}\right\}\right] \tag{9.113}$$

where y_0 is the level of reference point where the number density N_0 is defined. Equation (9.113) indicates the normal velocity distribution of sediment particle with respect to the fluctuations of particle velocities. According to the definition of flux in kinetic energy (9.110), the volumetric upward flux of sediment moving in the near wall region is

$$E_R=\frac{C_{y0}\sqrt{K}}{\sqrt{2\pi}\left(1+\frac{3C_D\rho_f y}{2\rho_s d}\right)}\exp\left[-\left(\frac{y}{Kg(\rho_s-\rho_f)\frac{g}{\rho_s}\left(1-\frac{\theta_*}{\theta_{*L}}\right)}\right)\right] \tag{9.114}$$

where $C_{y0}=\frac{\pi d^3 N_0}{6}$ is the reference concentration at $y=0$, and $\theta_*=\frac{\tau}{(\varrho_s-\rho_f)gd}$ is the Shields parameter, and $\theta_{*L}=4/(3C_L)$ is a parameter related to the lift coefficient. Equation (9.114) indicates that the entrainment rate of bed sediment is a decreasing function with respect to y, which also observed by Noguchi and Nezu (2009). When the entrainment flux of sediment particles by a turbulent flow from a loose and erodible channel bed is considered, the reference point for the particles picked up by the flow is naturally from the bed surface, i.e., $y_0=0$. Now $C_{y0}=C_m p$ is the volumetric concentration of movable sediment at $y=0$ rather than C_m, the volumetric concentration of static bed materials, and p is the incipient probability of sediment particles resting on the river bed. The granular turbulent temperature $K=\alpha u_*^2$, where $\alpha=\frac{3C_t^2 D_p}{3\sqrt{c_\alpha}}$, with $c_\alpha=0.09$. Then equation (9.114) leads to

$$E_R=\frac{C_m p\sqrt{\alpha u_*}}{\sqrt{2\pi}\left(1+\frac{3C_D\rho_f y}{2\rho_s d}\right)}\exp\left[-\left(\frac{y}{Kg(\rho_s-\rho_f)\frac{g}{\rho_s}\left(1-\frac{\theta_*}{\theta_{*L}}\right)}\right)\right] \tag{9.115}$$

Finally, dividing both sides of equation (9.115) by $\sqrt{gd\left(\frac{\rho_s}{\rho_f}-1\right)}$, the dimensionless entrainment rate is given by

$$Q_B = \frac{E_R}{\sqrt{gd\left(\frac{\rho_s}{\rho_f}-1\right)}} = \frac{C_m p\sqrt{\alpha\theta_*}}{\sqrt{2\pi}\left(1+\frac{9C_D\rho_f}{8\rho_s}\right)}\exp\left[-\left(\frac{3\rho_f}{4\rho_s(1/\theta_* - 1/\theta_{*L})\frac{1}{\alpha}}\right)\right] \quad (9.116)$$

To calculate the entrainment rate from equation (9.116), the incipient probability p, the volumetric sediment concentration C_m, the drag coefficient C_D, and the parameter α must be determined. For details, readers should consult the paper (Zhong et al. 2011). Haddadchi et al. (2013) analyzed 14 sets of gravel-bed river data, considering the bed load and bed materials; and tested the existing models such as Engelund and Hansen, Van Rijn, Einstein for bed material grain size and Shocklitsch, Meyer-Peter and Mueller, and Frijlink for bed load grain size. They found that the collected bed load and bed material data from the Narmab River, north-eastern Iran fitted well. Kumbhakar et al. (2018) developed an analytical model on bed load layer thickness, using a probabilistic approach based on the Shannon entropy theory. The entropy theory was developed by Shannon (Singh 1998), using the method of Lagrange multipliers for the maximization of entropy function to determine the least biased probability distribution. Using the Lagrange multipliers, the probabilistic model of dimensionless bed-load layer thickness is developed. The formula of bed-load layer thickness is a function of dimensionless shear stress, specific gravity of sediment particle and the particle diameter and is determined from a nonlinear regression analysis. The proposed model was verified, using the wide range of experimental data available in the literature and showed a good agreement with the model.

9.4.5 Bed Load Equations for Non-Uniform Sediment

For non-uniform sediment in alluvial channel, formulations of bed load transport phenomenon have been of great interest. Unlike the transport rate of uniform sediment, the problems related to the non-uniform sediment transport are much more complicated because of the sheltering effects during the interactions of sediments of different sizes. Different semi-empirical equations have been developed by many investigators (Einstein 1950; Ashida and Michiue 1973; Parker et al. 1982; Misri et al. 1984; Patel and Rangaraju 1996; Wu et al. 2000 and others) to determine the fractional bed load transport rate of non-uniform sediment. Einstein (1950)'s model was the pioneer to understand the transport rate of non-uniform sediment. Based on the concept of critical shear stress of incipient motion for non-uniform sediments in terms of the hiding and exposure mechanism, Parker et al. (1982) suggested the following threshold for the incipient motion of non-uniform sediment as given by:

$$W_i^* = \frac{(\gamma_s - \gamma_f)q_{si}}{\rho_f u_*^3 f_i} \quad (9.117)$$

where q_{si} is the volumetric transport rate per unit width for the ith fraction of bed-load; $f_i = p_{bi}$ is the gradation of the i-th fraction of the mixture bed. The reference shear stress or incipient bed shear stress is defined, when $W_i^* = 0.002$, which corresponds to a very low rate of sediment transport (discussed also in Chapter 8). The formula for the uniform bed-load transport rate Q_B or the total transport rate of non-uniform bed-load can be written as:

$$Q_B = \frac{q_b}{d\sqrt{(\varrho_s - \rho_f)gd}} = f_1\left(\frac{\tau_* - \tau_c}{\tau_c}\right) \quad (9.118)$$

Equation (9.118) is extended to establish the relationship for the fractional transport rate of non-uniform bed load. The non-dimensional fractional bed-load transport rate Q_{Bi} is defined as:

$$Q_{Bi} = \frac{q_{bi}}{p_{bi} d_i \sqrt{(\varrho_s - \rho_f) g d_i}} \tag{9.119}$$

where q_{bi} is the transport rate of the ith fraction of bed load per unit width (m/s). The bed shear stress τ_* is usually calculated as $\tau_* = \gamma_f R_h J$ with all usual meaning.

Proffitt and Sutherland (1983) developed a bed load transport method to predict the transportation of non-uniform sediments, using an exposure correction factor which allowed increasing the mobility of coarse particle and decreasing the mobility of fine particle in the mixture. The exposure correction factor was developed using the concept of Paintal (1971), and Ackers and White (1973) transport formulae.

Misri et al. (1984) developed a conceptual model for interaction of particle sizes in the mixtures that means the effect of particular size of sediment on the other size of the mixture in the transport rates. His conception is mostly on the roughness of sediment mixtures, how the particles are sheltered by the other particles in the mixture. They claimed that the sheltered parameter developed by them is different from that of Einstein's sheltering coefficient. Einstein only reported the sheltering parameter of finer particles by the coarser ones, but made no allowance for the increased exposure to the flow of the coarser particles in the case of non-uniform sediments. Misri et al. (1984) introduced herein the sheltered parameter for both effects, through sheltering and increased exposure. Here the composition of the transported material was considered in a way that the fraction of a certain size d_i in the bed material is designated as i_b, and the fraction of the same size in the transported material as i_B. The ratio between the fractions (i_B/i_b) was plotted against dimensionless Shields parameter $\tau_0' = \dfrac{\tau_0}{(\varrho_s - \rho_f) g d}$. It is observed that the ratio is greater than unity for the coarser fractions and smaller than unity for the finer fractions. Moreover, there is a tendency to approach unity at higher shear stresses. They observed that the value of (i_B/i_b) increases with increase in (τ_0'/τ_{0c}') and attained a peak value around $\tau_0'/\tau_{0c}' = 1$, and decreases further increase in (τ_0'/τ_{0c}'). For the mixtures, $i_B/i_b \to 0$, there is no motion at $\tau_0'/\tau_{0c}' = 0.18$ and 12.0. Thus, the following relations may be drawn as:

$$\text{For coarsest grain in motion, } \tau_{0c}' = 5.56\,\tau_0 \tag{9.120a}$$

$$\text{and for finest grain in motion, } \tau_{0c}' = 0.083\tau_0' \tag{9.120b}$$

Therefore, for a known shear stress, one can find the critical shear stress τ_{0c}' from equation (9.120) and from the Shields' criterion, the approximate limiting sizes in transport can be obtained.

Patel and Rangaraju (1996) conducted extensive experiments at IIT-Roorkee on the fractional bed load transport rates of non-uniform sediments, covering the wide range of flow conditions. The semi-empirical model developed by Misri et al. (1984) was tested and modified, incorporating the several parameters of non-uniformity of bed load transport. They expressed fractional bed load transport parameter Q_{Bi} as a function of $\xi_B \tau_{*i}'$ for the flume data and field data of non-uniform sediments. Here, the parameter ξ_B known as hiding-exposure correction factor is introduced to the threshold Shields parameter τ_{*i}' corresponding to the fractional size d_i. They have introduced two parameters C_s and C_m depending respectively on τ_0'/τ_{0c}' and M (Kramer's uniformity coefficient). The correction factor ξ_B is written as:

$$C_m \xi_B = 0.0713 \left[C_s \frac{\tau_0}{(\varrho_s - \rho_f) g d_i} \right]^{-0.75144} \tag{9.121}$$

where the parameter C_m is given by
$C_m = 1$ for $M > 0.38$, and

$$C_m = 0.7092 \log M + 1.293 \text{ for } 0.05 < M < 0.38 \tag{9.122}$$

and the parameter C_s depends on the ratio of τ_0'/τ_{0c}' is given by:

$$C_s = \text{Exp}\left[-0.1957 - 0.9571\left(\log \tau^*_{\text{dim}}\right) - 0.1949\left(\log \tau^*_{\text{dim}}\right)^2 + 0.0644\left(\log \tau^*_{\text{dim}}\right)^3\right] \tag{9.123}$$

with $\tau^*_{\text{dim}} = \tau_0'/\tau_{0c}'$ is the dimensionless shear stress corresponding to grain resistance.

Wu et al. (2000) developed a correction factor which is a function of hidden and exposed probabilities to study the fractional bed load transport rate of non-uniform sediment mixtures. The dimensionless fractional transport rate of bed load Q_{bi} is defined as:

$$Q_{bi} = \frac{q_{bi}}{p_{bi}\sqrt{\left(\gamma_s/\gamma_f - 1\right) g d_i^3}} \tag{9.124}$$

where q_{bi} is the transport rate of the ith fraction of bed load per unit width, and. p_{bi} is the percentage of the ith fraction of bed material. From the several laboratory and field data of non-uniform sediment, they used the least square analysis of curve fitting and found the following formula for the fractional transport rate of non-uniform bed load as:

$$Q_{bi} = 0.0053\left[\left(\frac{n^*}{n}\right)^{1.5} T_{*i}\right]^{2.2} \tag{9.125}$$

where $T_{*i} = \dfrac{\tau_0}{\tau_{0c}} - 1$ is the dimensionless excess shear stress used as an independent parameter in the relationship of Q_{bi}, n is Manning's roughness coefficient for the channel bed, and n^* is the Manning's roughness corresponds to grain roughness, which is calculated from the relation $n^* = (d_{50})^{0.167}/20$. The model (9.125) is tested for a wide range of laboratory and field data and compared with the existing empirical models. The predictions by the proposed formula are considerably good shown in Figure 9.13. They have used the total number of data points as 752. All the collected laboratory and field data for

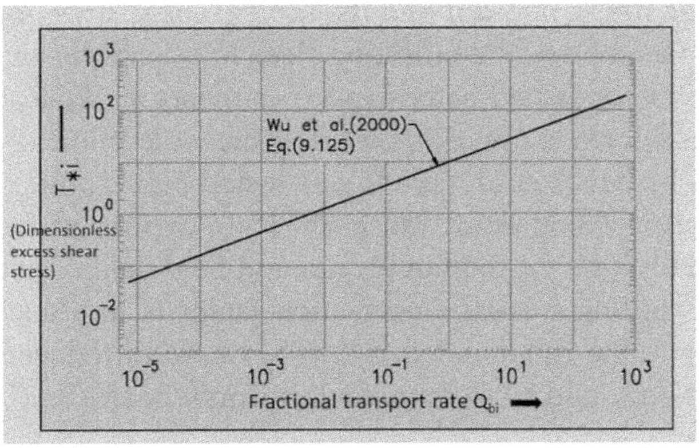

FIGURE 9.13 Normalized fractional transport rates of bed-load parameter Q_{bi} against the normalized excess shear stress T_{*i} for non-uniform sediment mixture of different existing data with Wu et al.'s (2000) equation (9.125).

$T_{*i}>0$ are included. The data distribute along a straight strip, in which Q_{bi} ranges from 10^{-5} to 10^2 and T_{*i} from 10^{-2} to 10^2. The data scattering is attributed to the complexity and stochastic behavior of non-uniform bed-load transport process and the measurement error.

Wu et al. (2004) assumed that the non-uniformity of bed material affects the bed load transport. They extensively studied the effects of non-uniformity of sediment size in bed materials on the bed load transport, using statistical approach, and have claimed that the sediment transport equations based on d_{50} for uniform sediment usually underestimate the transport rates for non-uniform sediments. So, they theoretically derived a size gradation correction factor which is function of the geometric standard deviation of bed material, based on the log-normal size distribution of the sediment mixtures. The correction factor K_d was originally developed for single particle size based on d_{50} for uniform sediments and improved the accuracy of transport calculations for sediment mixtures of non-uniform sizes. A size gradation correction factor K_d is defined by Wu et al. (2004) as ratio of total bed material transport rate by size fractions for log-normal distribution and the size based on d_{50}, that is,

$$K_d = \frac{\text{Bed material transport by size fraction for log-normal distribution}}{\text{Bed material transport based on size } d_{50}}$$

$$= \frac{\int_{-\infty}^{\infty} Q_p f(u)\,du}{Q_{d50}} \tag{9.126}$$

where $Q_p \propto Cd^{-b}$ and $Q_{d50} \propto Cd_{50}^{-b}$ with b is an exponent determined from the experimental data; d_{50} is the median diameter, and $f(u)$ is the density function of log-normal size distribution given by $f(u) = \frac{1}{\sqrt{2\pi}}\exp\left(-\frac{u^2}{2}\right)$ with zero mean and unit standard deviation, which is known as standardized normal distribution. Then equation (9.126) can be written as:

$$K_d = \int_{-\infty}^{\infty} \left(\frac{d}{d_{50}}\right)^{-b} f(u)\,du \tag{9.127}$$

Finally, after simplification, the form of K_d is as follows:

$$K_d = \exp\left[\frac{1}{2}\left(b \ln \sigma_g\right)^2\right] \tag{9.128}$$

where σ_g is the dimensionless geometric standard deviation of bed material, which is equal to $\sqrt{d_{84}/d_{16}}$. Here the correction factor K_d increases with increase in σ_g, with minimum value unity with corresponding to uniform distribution $\sigma_g = 1$, which means that sediment transport equation for uniform sediment based on d_{50} usually underestimates the transport rate for non-uniform mixtures. Therefore, the correction factor K_d can be used to determine the correct prediction for non-uniform sediment mixtures. The improvement on transport rate by K_d factor is significant for data with standard deviation $\sigma_g > 2$ for bed material, while the correction is negligible for data with $\sigma_g < 1.5$.

9.5 PROBABILISTIC CONCEPT FOR BED-LOAD TRANSPORT

9.5.1 Einstein's Bed Load Equations

9.5.1.1 Basic Concepts

Einstein (1942, 1950) was the first to attempt a semi-theoretical solution to the problem of bed load transport. His bed load transport concept appears to be the most popular among all the formulae suggested so far for the prediction of transport rate. His method was based on a few important grounds supported by experimental evidence. His formulation was derived from probabilistic considerations. He opposed the idea of the existence of definable critical condition of sediment movement. According to him, all the particles of uniform sediment should start moving at a shear stress just exceed the critical, if the critical shear stress exists. So, he assumed that the sediment particle will start movement, if the instantaneous hydrodynamic lift force exceeds the submerged weight force of the particle. Once the particle starts motion, there is a probability of the particle being deposited at all points of the bed where the particle may not be dislodged again. According to him, as the instantaneous lift force is random, the particle's motion is related to the random instantaneous lift force, so the movement of particles is also random, and consequently all the parameters like jump, length, height, time of fly and halt, are also random. Therefore, the Einstein's model of bed load transport shows somehow a departure from the Du Boys-type and Schoklitsch-type equations. He also did not consider the effect of bed forms on the bed load transport. There were two basic co-ordinations which broke the past ideas:

1. The definition of a critical value for the initiation of sediment movement is difficult. Thus, such a criterion was avoided. The critical value for initiation of sediment motion was disagreed.
2. It is suggested that the bed load transport is related to the fluctuations of the velocity rather than to the average velocity. The beginning and the end of the particle motion have to be expressed with the concept of probability, which relates the instantaneous hydrodynamic lift forces to the particle's weight.

Experimental evidence shows that there exists an intimate relationship between the bed and bed load:

1. A steady and intensive exchange of particles was observed to exist between the moving bed load and the sediment bed.
2. The bed load moves slowly downstream; the motion of an individual particle is one of quick steps with comparatively long intermediate rest periods.
3. The average step length made by a bed load particle appears to be independent of the flow condition, the transport rate, and the bed composition, and is always the same.
4. Different transport rates can be achieved by a change of the average time between two steps and the thickness of the moving layer.

9.5.1.2 Physical Model

The bed load equation communicates an equilibrium condition of the exchange of bed particles between the bed layer and the sediment bed. This implies that the number of particles deposited per unit time and per unit bed area must be equal to the number of particles eroded per unit time and per unit bed area.

Deposition: Let i_s be the fraction of bed load in a given grain size d, and g_s is the bed load transport rate in weight per unit time and unit width, $g_s i_s$ is the rate at which the given grain size i_s moves through unit width per unit time. If the weight of the particle is $\gamma_s k_2 d^3$, then the number of particles deposited N_d per unit time per unit area is written as

$$N_d = \frac{i_s g_s}{(A_L d)(\gamma_s k_2 d^3)} = \frac{i_s g_s}{A_L d^4 k_2 \gamma_s} \tag{9.129}$$

where $A_L d$ is the individual step length of a given size d, A_L is a constant of the bed load at unit step, k_2 is a constant of particle volume and $\gamma_s = g\rho_s$ is the specific weight of the solid particle.

Erosion: A particle of size d will be eroded depending on the availability of the particle and on the flow conditions, depending on the level of turbulence. If i_b is the fraction of bed material in a given grain size, then the number of particles of diameter d in a unit area of the bed surface $i_s i_b / k_1 d^2$, with k_1 is the constant of grain area. Let p_s represents the probability of this particle being eroded in unit time; the number of particles eroded N_e per unit area per unit time is given as:

$$N_e = \frac{i_b p_s}{k_1 d^2} \tag{9.130}$$

If t_1 is the time necessary to replace a bed particle with a similar one or the time consumed by each exchange, a probability p representing the fraction of total time during which exchange occurs may be defined as

$$p = p_s t_1 \tag{9.131}$$

Since p_s represents the number of exchanges per second, using equations (9.131) in (9.130) and one gets:

$$N_e = \frac{i_b p}{k_1 d^2 t_1} \tag{9.132}$$

As there is no direct technique available to determine the exchange time t_1, Einstein (1942) suggested a dependency with particle settling velocity w_0, that is, the time t_1 is given by:

$$t_1 \propto \frac{d}{w_0} \Rightarrow t_1 = \frac{A_3 d}{w_0} = A_3 \sqrt{\frac{d\rho_f}{g(\rho_s - \rho_f)}} \tag{9.133}$$

where A_3 being a constant of time scale and w_0 is the fall velocity of the particle. Combining equations (9.132) and (9.133), the number of particles eroded N_e in the scour is given by

$$\therefore N_e = \frac{i_b p \sqrt{\gamma_s - \gamma_f}}{k_1 A_3 d^{5/2} \rho_f^{1/2}} \tag{9.134}$$

Equilibrium: Since the deposition rate balances the erosion rate or scour rate, equations (9.129) and (9.134) are equated, which means, $N_d = N_e$ as:

$$\frac{i_s g_s}{A_L d^4 k_2 \gamma_s} = \frac{i_b p \sqrt{\gamma_s - \gamma_f}}{k_1 A_3 d^{5/2} \rho_f^{1/2}}$$

$$\Rightarrow \frac{i_s g_s}{A_L k_2 \gamma_s d^4} = \frac{i_b p}{k_1 A_3 d^2} \sqrt{\frac{\gamma_s - \gamma_f}{\rho_f d}} \tag{9.135}$$

Equation (9.135) is the bed load equation due to Einstein (1942, 1950).

9.5.1.3 Determination of Probability of Erosion

The probability of erosion p may be interpreted as the fraction of bed on which, at any time, the instantaneous lift force exceeds the submerged weight. If the probability of erosion p is small, the deposition is mostly everywhere possible. However, at strong sediment transport, p becomes larger, and deposition is not everywhere possible. The deposition of a particle after a jump length of $A_L d$ is possible, only if p is small. Einstein interprets that p may be used to calculate the distance $A_L d$, which a particle travels between consecutive places of rest. If p is small, the distance of travel is virtually a constant and $A_L d = \lambda_d d$, where λ_d is a single step of bed load and having a value of about 100. Suppose p is large, only $(1-p)$ particles find a change of deposition after travelling the jump length of $100d$, while p particles stay in motion even after travelling $100d$. Out of these p particles, $p(1-p)$ particles are deposited after travelling $2\lambda_d d = 200d$, while p^2 particles are not deposited even within $200d$. Thus total travel distance can be expressed in a series

$$A_L d = \sum_{n=0}^{\alpha}(1-p)p^n(n+1)\lambda_d d$$

$$=\left[(1-p)+2p(1-p)+3p^2(1-p)+4p^3(1-p)+\cdots\right]\lambda_d d$$

$$=(1-p)\left[\frac{1}{(1-p)^2}\right]\lambda_d d = \frac{\lambda_d d}{1-p} \tag{9.136}$$

Using equation (9.136) for $A_L d = \dfrac{\lambda_d d}{1-p}$ in equation (9.135), we get the bed load equation as:

$$\frac{i_s g_s}{A_L k_2 \gamma_s d^4} = \frac{i_b p}{k_1 A_3 d^2}\sqrt{\frac{\gamma_s-\gamma_f}{\rho_f d}}$$

$$\frac{i_s q_s}{\lambda_d d k_2 \gamma_s d^3}\cdot\frac{k_1 A_3 d^2}{i_b}\sqrt{\frac{\rho_f d}{\gamma_s-\gamma_f}} = \frac{p}{1-p}$$

$$\frac{p}{1-p} = \left(\frac{k_1 A_3}{k_2 \lambda_d}\right)\left(\frac{i_s}{i_b}\right)\left(\frac{q_s}{\gamma_s}\right)\sqrt{\frac{\rho_f d}{(\rho_s-\rho_f)gd^4}}$$

$$\frac{p}{1-p} = \left(\frac{k_1 A_3}{k_2 \lambda_d}\right)\left(\frac{i_s}{i_b}\right)\Phi \tag{9.137}$$

where $\Phi = \dfrac{q_s}{\gamma_s}\sqrt{\dfrac{\rho_f}{(\gamma_s-\gamma_f)d^3}}$

The parameter Φ is the dimensionless measure of bed load transport and is called the intensity of bed load transport. Equation (9.137) can be rewritten as

$$\frac{p}{1-p} = A_*\left(\frac{i_s}{i_b}\right)\Phi = A_*\Phi_* \tag{9.138}$$

where $A_* = \dfrac{k_1 A_3}{k_2 \lambda_d}$ and $\Phi_* = \left(\dfrac{i_s}{i_b}\right)\Phi$.

Here, A_* is a constant determined by the experiments and Φ_* is the intensity of transport for an individual grain size, and these are the dimensionless parameters.

The probability of erosion depends on hydrodynamic lift force and particle weight. This can be expressed as

$$p = f\left(\frac{\text{effective weight of particle}}{\text{hydrodynamic lift}}\right)$$

$$\text{or, } p = f\left(\frac{k_2(\rho_s - \rho_f)gd^3}{\frac{1}{2}c_L\rho_f k_1 d^2 u_b^2}\right) \tag{9.139}$$

where c_L is the lift co-efficient is equal to 0.178. The effective velocity u_b may be approximated by the velocity at the edge of the laminar sub-layer if the wall is smooth and is given by

$$u_b \approx 11.6\ u_* \approx 11.6\sqrt{gR'_hJ} \tag{9.140}$$

where R'_h is the hydraulic radius w.r.t. the grain.

Equation (9.139) can be written as:

$$p = f\left[\frac{2k_2(\rho_s - \rho_f)gd^3}{c_L\rho_f k_1 d^2(135R'_hJg)}\right] \tag{9.141}$$

$$p = f\left[\frac{2k_2}{c_L}\left(\frac{\rho_s - \rho_f}{\rho_f}\frac{d}{JR'_h}\right)\cdot\frac{gd^2}{k_1 d^2(135g)}\right]$$

$$= f\left[\left(\frac{2k_2}{135\ c_L \cdot k_1}\right)\left(\frac{\rho_s - \rho_f}{\rho_f}\frac{d}{JR'_h}\right)\right] = f(B_*\psi) \tag{9.142}$$

where $B_* = \dfrac{2k_2}{135k_1c_L}$, $\psi = \dfrac{\rho_s - \rho_f}{\rho_f}\dfrac{d}{JR'_h}$.

Here B_* is a universal constant of scale of ψ to be determined experimentally, with J is the slope.

For weak sediment transport, the simplification of equation (9.138) reduces to

$$\frac{p}{1-p} = \frac{p(1+p)}{(1-p)(1+p)} = \frac{p(1+p)}{1-p^2} = p$$

$$p = A_*\Phi_* \tag{9.143}$$

Combining equation (9.142) with (9.143), a new form of the bed load equation is obtained as:

$$A_*\Phi_* = \left(\frac{i_s}{i_b}\right)\Phi = p = f(B_*\psi) \tag{9.144}$$

The two constants A_* and B_*, and the function are determined empirically. Experimental data of Gilbert (1914) and Meyer-Peter at al. (1934) were used to plot model parameters Φ versus ψ in Figure 9.14.

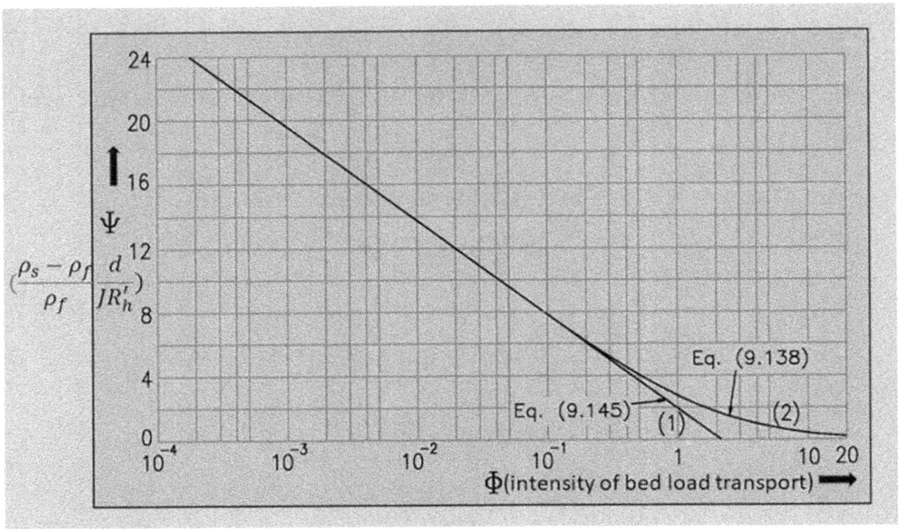

FIGURE 9.14 Trend of model parameters ψ versus Φ for different grain sizes. (Modified from Einstein 1950; Graf 2010.)

From the figure it shows that all the data $\Phi < 4$ can be represented with a single curve (1), the fitted equation is given by

$$0.465\ \Phi = e^{-0.391\psi} \tag{9.145}$$

For strong sediment transport, the curve given by equation (9.145) deviated from the data for large value of Φ, that is, $\Phi > 0.4$. This is obviously to be expected because bed load motion is strong, the approximation given by equation (9.144) fails. The complete equation (9.138) was taken to predict the curve (2), which explains the data well.

Sand mixtures: Einstein determined the effective diameter of sand mixture from some data of U.S. Waterways Experiment station (1935). It was observed that the grain size of the bed materials of which 35%–45% is finer is the most usable value.

Analytical relation for the determination of probability of erosion p may be expressed as the probability of the ratio of effective weight force to instantaneous lift force, which has to be smaller than unity:

$$\frac{2k_2\left(\rho_s - \rho_f\right)gd^3}{c_L\rho_f k_1 d^2 u_b^2\left(1+\eta\right)} < 1 \tag{9.146}$$

$$\Rightarrow \frac{1}{2}c_L\rho_f k_1 d^2 u_b^2\left(1+\eta\right) > k_2\left(\rho_s - \rho_f\right)gd^3 \tag{9.147}$$

Einstein and El-samni (1949) found that $c_L = 0.178$, and the random function η is distributed according to the normal-error law, where $\eta_0 = 1/2$ is the standard deviation of η. The velocity u_b is found to be at a distance of $0.35d_x$ from the theoretical bed, where d_x is the characteristic grain size at the mixture, with

$$\begin{aligned} d_x &= \frac{0.77d_{65}}{x_c},\ \text{if}\ \frac{d_{65}}{\delta' x_c} > 1.80 \\ &= 1.39\ \delta',\ \text{if}\ \frac{d_{65}}{\delta' x_c} < 1.80 \end{aligned} \tag{9.148}$$

where $\delta' = 11.6\ v_f/u_*$, and x_c is the correction factor for viscous effect. The particle size smaller than d_x are sheltered by the bigger particles and experience smaller lift force than that given by the equation of F_L. Therefore, the lift force on these particles must be corrected by division with a parameter ξ which is a function of $\frac{d}{d_x}$. Hiding factor ξ is function of $\frac{d}{d_x}$, which is plotted in Figure 9.15a. Einstein (1950) suggested two correction factors ξ for hiding the smaller particle between larger ones or in laminar sub-layer and ζ as pressure correction for changing the lift coefficient in mixtures with various roughness which is a function of k_s/δ and is plotted in Figure 9.15b. Both the correction factors were determined experimentally by Einstein (1950).

If u_b is expressed with a logarithmic velocity distribution at a distance $0.35d_x$ from the bed, then the velocity is obtained as:

$$u_b = 5.75u_* \log\frac{(30.2)\times(0.35d_x)}{(d_{65}/x_c)} = 5.75\ u_* \log\left(\frac{10.6d_x}{d_{65}/x_c}\right) \tag{9.149}$$

Equation (9.149) is for the rough and smooth boundaries as well as for the transitional $u_* = \sqrt{gJR_h'}$. Now introducing equation (9.149) in equation (9.147), one gets the results as:

$$1 > \frac{1}{1+\eta}\left(\frac{\rho_s-\rho_f}{\rho_f}\frac{d}{JR_h'}\right)\left[\frac{2k_2}{5.885k_1}\right]\left(\left[\log^2\left\{\frac{10.6d_x}{\frac{d_{65}}{x_c}}\right\}\right]\right)^{-1} \tag{9.150}$$

$$\Rightarrow 1 > \frac{1}{1+\eta}\psi\ BB_x^{-2} \tag{9.150a}$$

where ψ is the flow intensity as in equation (9.142), $B = \frac{2k_2}{5.885k_1}$ is a constant of the scale ψ, and B_x^{-2} is the value determined from the experimental data with $B_x = \log\frac{10.6d_x}{d_{65}/x_c}$ from the Figure 9.15c. Furthermore, Einstein suggested two correction factors ξ and ζ, then it is written as:

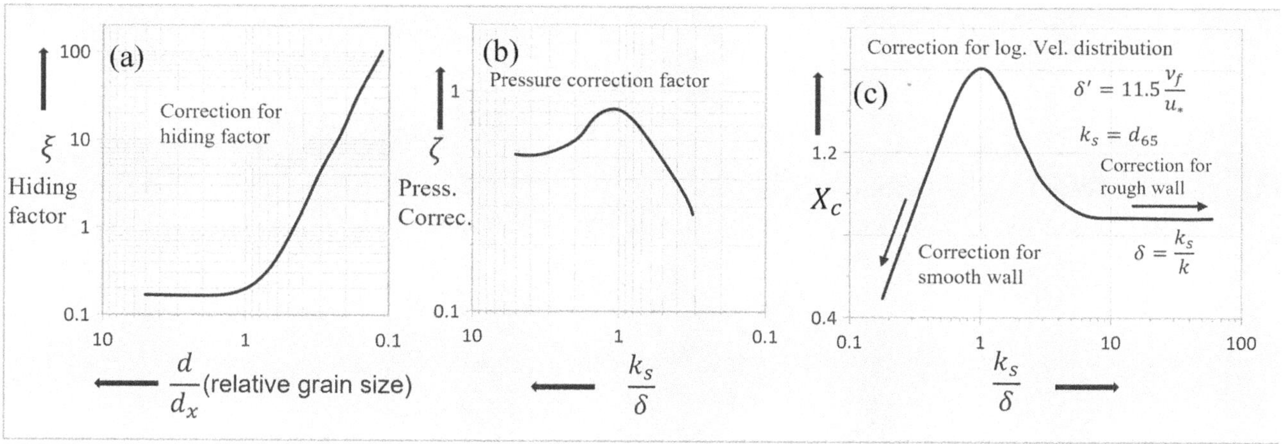

FIGURE 9.15 Correction factors: (a) Hiding factor ξ against d/d_x according to equation (9.151), (b) pressure correction factor ζ against k_s/δ according to equation (9.151), and (c) Correction factor in log-law. (Modified from Einstein 1950.)

$$1 > \frac{1}{1+\eta}\xi\,\zeta\,\frac{B}{\beta^2}\frac{\beta^2}{B_x^{-2}}\psi = \frac{1}{1+\eta}\xi\,\zeta B'\frac{\beta^2}{B_x^{-2}}\psi \tag{9.151}$$

where $B' = \dfrac{B}{\beta^2}$, $\beta = \log 10.6$.

Equation (9.151) can be written as

$$|1+\eta| > \xi\,\zeta B'\frac{\beta^2}{B_x^{-2}}\psi \tag{9.152}$$

The correction factor ξ and ζ are unity for uniform sediment, while $\beta/B_x = 1$ for uniform sediment which $d_x = 1$. Here η may be either positive or negative, the lift is always positive and the absolute-value sign on the left-hand side of equation (9.152) is in place. It is more convenient to substitute $\eta = \eta_0\eta_*$ in (9.152), and square and divide both sides by η_0, equation (9.152) results as:

$$\left(\frac{1}{\eta_0}+\eta_*\right)^2 > \xi^2\zeta^2\psi^2\left(\frac{\beta^2}{B_x^{-2}}\right)^2\left(\frac{B'}{\eta_0}\right)^2$$

$$\text{Or,}\left(\frac{1}{\eta_0}+\eta_*\right)^2 > B_*^2\psi_*^2 \tag{9.153}$$

where $B_* = \dfrac{B'}{\eta_0}$, $\psi_* = \xi\zeta\psi\left(\dfrac{\beta^2}{B_x^{-2}}\right)$.

From equation (9.153), the limiting case of motion may be found, if $\left(\dfrac{1}{\eta_0}+\eta_*\right)^2 = B_*^2\psi_*^2$

$$\Rightarrow (\eta_*)_{\text{lim}} = \pm B_*\psi_* - \frac{1}{\eta_0} \tag{9.154}$$

The probability distribution is according to the normal-error law, such that the probability p for motion becomes

$$p = 1 - \frac{1}{\sqrt{\pi}}\int_{-B_*\psi_*-\frac{1}{\eta_0}}^{+B_*\psi_*-\frac{1}{\eta_0}} e^{-t^2}\,dt \tag{9.155}$$

where t is the variable of integration.

Combining equation (9.138) with (9.155) results in the final bed load equation suggested by Einstein (1950) is

$$1 - \frac{1}{\sqrt{\pi}}\int_{-B_*\psi_*-\frac{1}{\eta_0}}^{+B_*\psi_*-\frac{1}{\eta_0}} e^{-t^2}\,dt = p = \frac{A_*\Phi_*}{1+A_*\Phi_*} \tag{9.156}$$

where A_*, B_* and η_0 are the universal constants to be determined from the experimental data. These constants are obtained from uniform sediments by using the data of Gilbert (1914) and Meyer-Peter et al.

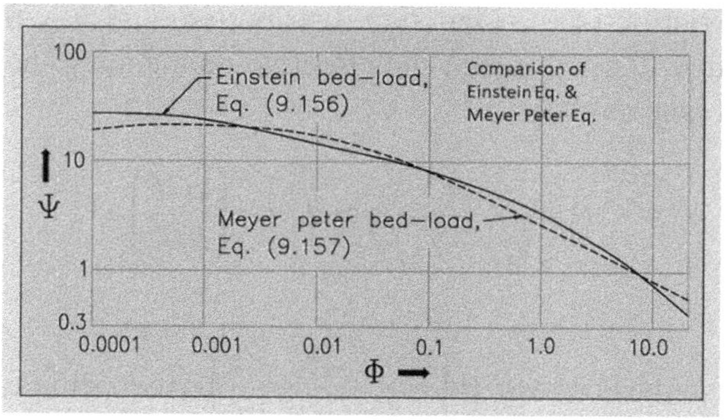

FIGURE 9.16 Plots of comparison of bed load equation of Einstein (1950) with Mayer-Peter and Muller (1948). (Modified from Graf 2010.)

(1934): $A_* = \dfrac{1}{0.023} = 43.5$, $B_* = \dfrac{1}{7} = 0.143$, and $\eta_0 = 1/2$ is the standard deviation of η. The relationship between the parameters ψ_* and Φ_* is plotted in Figure 9.16.

From the knowledge of the final bed load transport equation (9.156), Einstein (1950) continued to develop the relationship between the sediment travelling along the bed and that moving into suspension. His model does not depend on any kind of critical values. Einstein's bed load equation (9.156) and Meye-Peter and Müller (1948)'s bed load equation were considered here to represent the most complete equations. Chien (1954) made the comparative study with these two equations and showed that Meye-Peter et al.'s bed load equation can be written in terms of Φ and ψ as:

$$\Phi = \left(4\psi^{-1} - 0.188\right)^{1.5} \tag{9.157}$$

Equations (9.157) and (9.156) are presented in Figure 9.16 for uniform material, and they showed a good agreement. It is important to note that the sediment mixtures, d_{35} for Einstein relation and d_{50} for Meyer-Peter et al. relation showed a good agreement.

9.5.2 Gessler's Equation (1965)

Gessler (1965) studied the beginning of motion of non-cohesive sediment mixtures of large grain size distributions theoretically and experimentally. He assumed a flat bed of loose grains. Over this flat bed, a current move and a certain shear of stress exerts on the bed. When the shear stress exceeds a critical value, the forces act on an individual grain, so that the grain moves from its position of rest. In his course of work, he made the following assumptions for studying the problems of bed-load transport: (1) the turbulent fluctuations of the bed shear stress are distributed according to normal-error law, and (2) a grain starts to move when the 'effective bottom shear stress' on the grain exceeds a critical value which is a function of the grain size and the grain Reynolds number.

When the dynamic uplift on a single grain is distributed according to Gaussian distribution (Einstein and El Samni 1949), the same is also true for the bottom shear stress. The average dynamic uplift $\bar{a}_L$ is directly proportional to the bottom shear stress as $\bar{a}_L = 17.4\, C_L \bar{\tau}$, where C_L is a lift constant and $\bar{\tau}$ is the average bottom shear stress. If $\tau' (= \tau - \bar{\tau})$ is the absolute fluctuation of bottom shear stress, the probability of τ' being smaller than t can be written as:

$$P(\tau' < t) = \frac{1}{\sigma\sqrt{2\pi}} \int_{-\infty}^{t} e^{-\left(\frac{x^2}{2\sigma^2}\right)} dx \tag{9.158}$$

where σ is the standard deviation of the Gaussian distribution, and x is the dummy variable of integration. The motion of grain will occur when the effective momentary shear stress acting at the grain exceeds a critical value and is given by:

$$\theta = \frac{\tau}{(\gamma_s - \gamma_f)d} > \theta_c = f\left(\frac{u_* d}{\upsilon_f}\right) \tag{9.159}$$

where θ is the dimensionless shear stress $\theta\left(= \frac{\tau}{(\gamma_s - \gamma_f)d}\right)$. According to Gaussian distribution, the probability p of a grain being eroded from the bed is given by

$$p = 1 - q\{\theta < \theta_c\} = 1 - q\{\theta' < \theta_c - \bar{\theta}\}$$

$$\text{Or, } p = 1 - \frac{1}{\sigma_1\sqrt{2\pi}} \int_{-\infty}^{\theta_c/\bar{\theta}-1} \exp\left(-t^2/2\sigma_1^2\right)dt \tag{9.160}$$

where $\frac{\theta'}{\theta} = \frac{\theta - \bar{\theta}}{\theta}$, follows a Gaussian distribution with mean zero and variance σ_1^2, and the dimensionless average bottom shear stress $\bar{\theta}$ is $\bar{\theta} = \frac{\bar{\tau}_0}{(\gamma_s - \gamma_f)d}$. Here, probability $q\{\theta < \theta_c\}$ of critical shear stress of a sediment particle is not being exceeded to move, or the probability of a grain remaining stationary. For a given grain size, the dimensionless critical shear stress $\bar{\theta}_c$ is constant. In this method, grain Reynolds number being known, the values of $\bar{\theta}_c$ could be read directly from the modified Shield's curve provided by Gessler (1965, fig.8). The bed load i.e., the concentration at the bed layer for a sand bed was obtained by multiplying the bed concentration of different sizes by the probability of each grain size. This definition is valid for independent of any grain size distribution, unit grain size as well as sediment mixture. To supplement and verify the theory, natural armoring of channel bottoms consisting of sand-gravel mixtures was investigated in the laboratory and was supplemented by observations in the field.

Cheng and Chiew (1998) explored a relationship between the pick-up probability of sediment particles and the flow condition over the bed, using the concept of Einstein's definition of pick-up probability. The pick-up probability for a bed particle is considered as the fraction of the time when the instantaneous lift force is greater than the effective weight force of the particle for a given time interval (Einstein 1942, 1950). Later, Wu and Lin (2002) developed a mode of pick-up probability for sediment entrainment, using log-normally distributed instantaneous velocity. The results are compared with the experimental data and verified with the previous model of pick-up probability derived from the normal velocity distribution. Detailed discussions for both models are provided in Chapter 8.

9.5.3 Modified Einstein's Approach

The beginning of Einstein's bed-load formula involves several simplified assumptions relating to the step length of a particle, time of exchange, and erosion probability/lifting probability of particle. Later, Wang et al. (2008) suggested a modified version of the Einstein formula without relying on his assumptions, which incorporated a non-uniform sediment model; finally, the modified model was tested using both field and laboratory data. Conceptually, the model was assumed as the step length of a particle increased with lift force exerted by the flow, but decreased with the submerged weight of the particle. As a result, the following relationship was proposed as:

$$\frac{L}{d} = \frac{\lambda \gamma_f J R_h}{(\gamma_s - \gamma_f)d} = \frac{\lambda}{\psi}, \quad \psi = \frac{\rho_s - \rho_f}{\rho_f} \frac{d}{JR'_h}$$

$$\Rightarrow L = \lambda d / \psi \text{ is the step length.} \tag{9.161}$$

where λ is a constant. The rate of particle deposition per unit area r_d can be expressed as:

$$r_d = \frac{r_b}{L_0} = \frac{r_b}{\lambda d} \psi (1 - p) \tag{9.162}$$

where $L_0 = \dfrac{\lambda d}{\psi (1 - p)}$, and p is probability of the lift force being greater than the submerged weight of the particle. The number of particles per unit area is given by $1/(k_2 d^2)$, and weight of a particle is given by $k_1 \gamma_s d^3$. The total weight is $(k_1 \gamma_s d^3)/(k_2 d^2)$. Here the total weight $(k_1 \gamma_s d^3) p/(k_2 d^2)$ is removed from the bed per unit time and area. Einstein assumed that the exchange time, t, for a particle to be removed from the bed is given as $t = k_3 d/\omega$, where k_3 is a constant, and ω is the settling velocity of the particle. The rate of sediment erosion per unit r_e is given by:

$$r_e = \frac{(p k_1 \gamma_s d^3)/(k_2 d^2)}{k_3 d / u_*} = \frac{k_1}{k_2 k_3} \gamma_s p u_* \tag{9.163}$$

The equilibrium occurs, when the rate of deposition (9.162) and erosion (9.163) are equal and is given by:

$$\frac{r_b}{\lambda d} \psi (1 - p) = \frac{k_1}{k_2 k_3} \gamma_s p u_* \tag{9.164}$$

$$\text{Or, } p = \frac{A_* \phi}{\psi^{-3/2} + A_* \phi} \tag{9.165}$$

where $A_* = \dfrac{k_1 k_3}{\lambda k_2}$ is a constant, and $\phi = \dfrac{r_b}{\gamma_s} \sqrt{\dfrac{\rho_f}{(\gamma_s - \gamma_f) d^3}}$.

According to Wang et al. (2008), the particle is moved only if the lift force exceeds the weight force of the particle, then the relation is as $1 + \eta_0 \eta_* > B_1 \psi$, where B_1 is the coefficient, η_0 is the standard deviation of fluctuating lift force, and η_* is the dimensionless parameter characterizing the fluctuating lift force. The probability of erosion of particle is given by:

$$p = \frac{1}{\sqrt{\pi}} \int_{\eta_*}^{\infty} e^{-t^2} dt = \frac{1}{\sqrt{\pi}} \int_{(B_1 \psi - 1)/\eta_0}^{\infty} e^{-t^2} dt = \frac{A_* \phi}{\psi^{-3/2} + A_* \phi} \tag{9.166}$$

Equation (9.166) is computed from measured data of Einstein (1950), using $B_1/\eta_0 = 0.07$, $1/\eta_0 = 2$, and $A_* = 20$. This equation is known as the modified version of Einstein's bed load formula for uniform sediment. Accordingly, Einstein's bed load formula is also modified for the non-uniform sediment size (Wang et al. 2008).

Recently, Zee and Zee (2017) introduced a new formula to modify Einstein's bed load transportation, using a new time factor instead of Einstein's time factor. Einstein's time factor t_E was used for the fall velocity in still water of a given material, it was also applied in the context of bed load transport. Zee and Zee (2017) introduced a new time factor t_Z in Einstein's original formula, replacing the time factor t_E. Einstein's time factor t_E is given by: $t_E \approx d/v_s$ implies as

$$t_E = A_1 d/v_s \tag{9.167}$$

where A_1 is a constant of proportionality, and v_s is the settling velocity of the bed load particle in water. The new time factor t_Z is proposed as:

$$t_Z \approx v_s/g$$

$$t_Z = A_2 v_s/g \tag{9.168}$$

where g is the acceleration due to gravity and A_2 is the constant of proportionality. The new time factor t_Z obeys the rules of Einstein's intension of introducing the time factor t_E. Using the settling velocity v_s expression from Rubey (1933) in equations (9.167) and (9.168), the expression of t_E and t_Z respectively given by:

$$t_E = A_2 d/v_s = \left(A_3/F\right)\sqrt{\frac{d\rho_f}{g\left(\rho_s - \rho_f\right)}} \tag{9.169}$$

$$t_Z = A_1 v_s/g = A_1 F\sqrt{\frac{d\left(\rho_s - \rho_f\right)}{g\rho_f}} \tag{9.170}$$

where $F = \sqrt{\frac{2}{3} + \frac{\left(6\mu_f\right)^2}{gd^3\rho_f\left(\rho_s - \rho_f\right)}} - \sqrt{\frac{\left(6\mu_f\right)^2}{gd^3\rho_f\left(\rho_s - \rho_f\right)}}$.

Using Einstein's time factor t_E from Einstein (1942), the function ϕ_E is defined as:

$$\phi_E = F^{-1}\frac{q_b}{g\left(\rho_s - \rho_f\right)}\sqrt{\frac{\rho_f}{gd^3\left(\rho_s - \rho_f\right)}} \tag{9.171}$$

and correspondingly using the new time factor t_Z, Zee and Zee (2017) found the expression for the function ϕ_Z as:

$$\phi_Z = \frac{Fq_b}{g\left(\rho_s - \rho_f\right)}\sqrt{\frac{\left(\rho_s - \rho_f\right)}{gd^3\rho_f}} \tag{9.172}$$

Combination of equations (9.171) and (9.172) leads to

$$\frac{\phi_Z}{\phi_E} = F^2\frac{\rho_s - \rho_f}{\rho_f} = E$$

$$\Rightarrow \phi_Z = E\phi_E \tag{9.173}$$

Interested readers are suggested to refer for detailed analysis given in Zee et al. (2017) for modified bed load transport rate.

9.6 SALTATION

In the beginning of the chapter, the term 'saltation' is explained. The transport of particles by jumping over the bed is called the saltation, which follows the trajectories like ballistic. Saltation is described as the unsuspended transport of particles over a sediment bed, in the form of consecutive hops in the near-bed region. This process is due to the combination of lift F_L and drag F_D force and submerged weight F_G of the particle in the region of high shear stress near the bed. The material transported in this process is known as the saltation load (Figure 9.17).

When the particle travels by saltation or jumping under hydraulic condition, the equilibrium of a particle in this position is considered. The equilibrium of forces is maintained through the submerged weight force and the resultant hydrodynamics force. If the resultant hydrodynamic force is little increased, the condition is attained when the particle is on the verge of movement. Due to the increase of hydrodynamics force, the successive positions of particle before losing contact with the bed are observed and these are shown in Figure 9.17. When the lift force is greater than the submerged weight, the particle will start to pick-up. As the particle picks up, some area below the particle is exposed to a stagnation pressure leading to an increase in lift force, which efforts upward velocity to the particle. During the passage of particle motion, it follows different positions as: position (a) to position (b) and then position (c) shown in Figure 9.18. The process of movement is a combination of rolling and sliding, and finally taking off with a rotation about a point of contact. A lift force acting on the particle produced by rotation/circulation is called the Magnus effect. In position (a), the particle will start to slide or roll over a

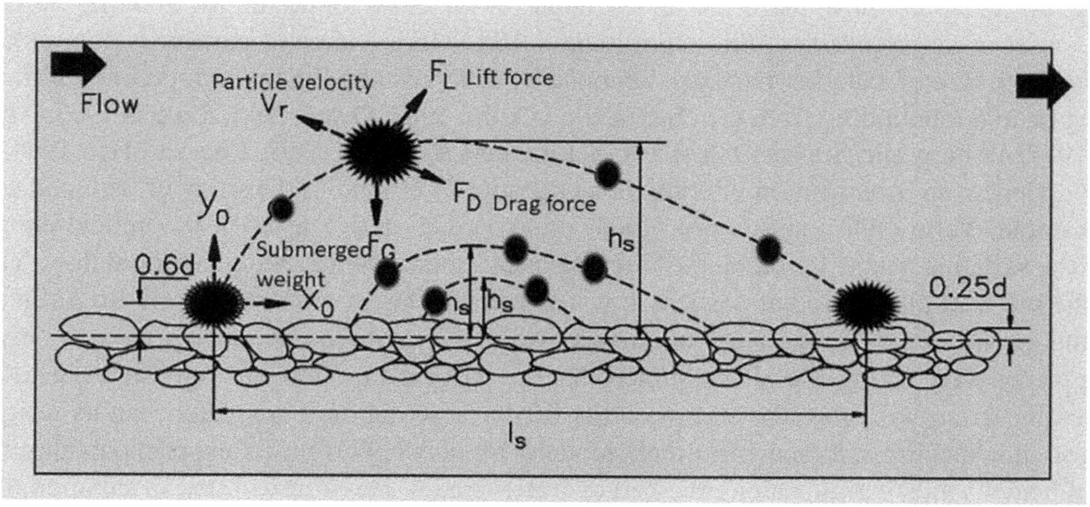

FIGURE 9.17 Sketch of particle saltation. (Modified from Van Rijn 1984a; Mazumder and Dalal 2003, Dey 2014.)

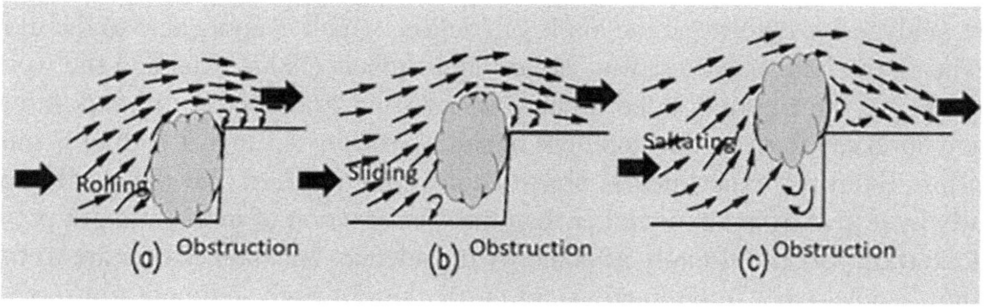

FIGURE 9.18 (a–c) Consecutive positions of saltating particles. (Modified from Garde & Ranga Raju 2000.)

layer of other particles with a moderate velocity (position b), then it may jump over small projection of other particles. When a particle in saltation completes its trajectory and hits the bed, form a parabola, again the particle either springs back or dislodges. In all these processes, the particle spins around itself with motion in the trajectory, and moves with several orientations. Again, a new step of saltation may start with an impact against the bed and the lift force.

Probably, the jumping motion of grains was first described by Gilbert (1914). Later on, Bagnold (1941) in his classic work on grain motion in aeolian conditions described saltation process extensively. He visualized that the grains in saltation moved like ping-pong balls. When the particle is in the air, it receives a supply of energy from the forward fluid pressure. At the impact of almost horizontally moving particle, the same particle and/or possibly other colliding ones are ejected almost vertically upward; and other path of a particle initiates. He stated that due to the impact with other particle an upward force exists, but the model of saltation had not been well accepted. It is more convincing to relate the initial upward motion to the lift force experienced by the particle. Bagnold (1941) observed that the total sand transportation in the air moves about the three-fourths amount in saltation and rests on the surface creeping. Einstein (1942) suggested that during the saltation of particle in the fluid flow, the beginning and ending of the motion of sand particles were expressed by stochastic concept. Due to the influence of turbulent fluctuations, the saltation relates to the ratio of submerged weight to the hydrodynamic lift force induced by the particle. In the subsequent papers, Einstein and El Samni (1949) and Einstein (1950) explained rigorously the lift force as well as the influence of turbulent fluctuations on the lift forces. Einstein conducted a flume study and found that the saltation length is a function of particle size, shape, and flow characteristics. Chepil (1945, 1958) found the lift and drag forces are of the same order of magnitude from the experimental data. The saltation of particle is dominated by the flow conditions, size of saltating particle, components of forces and the random process of the particle impacting on and rebounding from the channel bed (Yalin 1963; Owen 1964; Francis 1973; Abbott and Francis 1977; Wiberg and Smith 1985, 1989; Sekine and Kikkawa 1992; Lee and Hsu 1994; Hu and Hui 1996). Hence, an investigation of continuous saltation is crucial to the study of bed load transport.

For example, Yalin (1963) and Owen (1964) attempted to derive a set of theoretical equations to describe the saltating grains based on the force balances acting on a particle in a fluid flow. Yalin considered the problem of saltation of particle in water, whereas later considered in air. So Yalin assumed the lift and gravity forces for saltation of particle, when it leaves the bed, during the rest only gravity and drag forces were acting. He also assumed that the particle's motion was initially vertical and suggested that the grain will leave the bed when the lift force acting on it is greater than its weight. They did not use any specific saltation from their models. Francis (1973) in his experiments described the motion of heavy solitary grains along the bed of water stream. He observed that in saltation mode, the grains follow low and smooth trajectories with very largely ballistic forces, but in suspension, grains follow much longer and higher paths with wavy pattern due to the irregular turbulence in the stream. He also confirmed that the change from saltation to suspension occurs when the vertical components of turbulent velocity are approximately equal to the settling velocity of grains. The photographs of trajectories are analyzed to determine the angle of friction, which is applicable to the motion of grains retarded by striking the bed. Subsequently, Abbott and Frencis (1977) extended the work reported by Frencis (1973) to describe rigorously the trajectories of solid grains by multi-exposure photographic technique. They observed that the development of saltation from rolling or sliding is much rapid than that of suspension. From the experimental observations, they inferred that the starting of a trajectory depends on fully hydrodynamic forces rather than the conservation of momentum of previous trajectories. The bedload transport is definitely affected by turbulence, but such effects are to be insignificant compared to the particle–particle interactions, in particular, for high sediment transport rates. Francis (1973) demonstrated that the particle saltation could even occur in the absence of turbulence, i.e., in laminar flow. Extensive laboratory observations (Francis 1973; Abbott and Francis 1977) show characteristics of saltating particle.

Wiberg and Smith (1985) developed a theoretical model for paths of a single grain moving as bed load, based on a set of differential equations, which describe the trajectory of a saltating grain as a function of time. The model is primarily with saltation, but rolling is also included as a limiting case, as the boundary shear stress approaches its critical value. This specific model determines the particle trajectory height, length and the particle velocity. They suggested a set of differential equations to examine the problem of grain saltation from the initiation of motion through the entire grain trajectory. The grain trajectories are calculated from the effect of drag, lift, gravity, and virtual mass forces, as is generally true for equations of the motion of a particle in a fluid flow. They derived systematic expressions from the primitive equations for flows with Reynolds number, $Re > 1$. Their model differs from the empirical and semi-empirical relations commonly used to estimate the transport parameters and rates because the applicability of their mode is not limited like empirical models. Initially, they examined their model for single hops, and then they included the collisions at the bed. Subsequently, they tested their model to determine the saltation height h_s, saltation length l_s, and average velocity u_p of grain trajectories and verified these parameters from the data provided by Abbott and Francis (1977), data from photographs of saltating particles by Francis (1973), and the average particle velocity from Luque and van Beek (1976). The plots of saltation height and length against transport stage τ_b/τ_{cr} are shown in Figure 9.19a,b. The calculated and measured averaged particle velocities are shown in Figure 9.19c and the trajectories of particles are shown in Figure 9.19d.

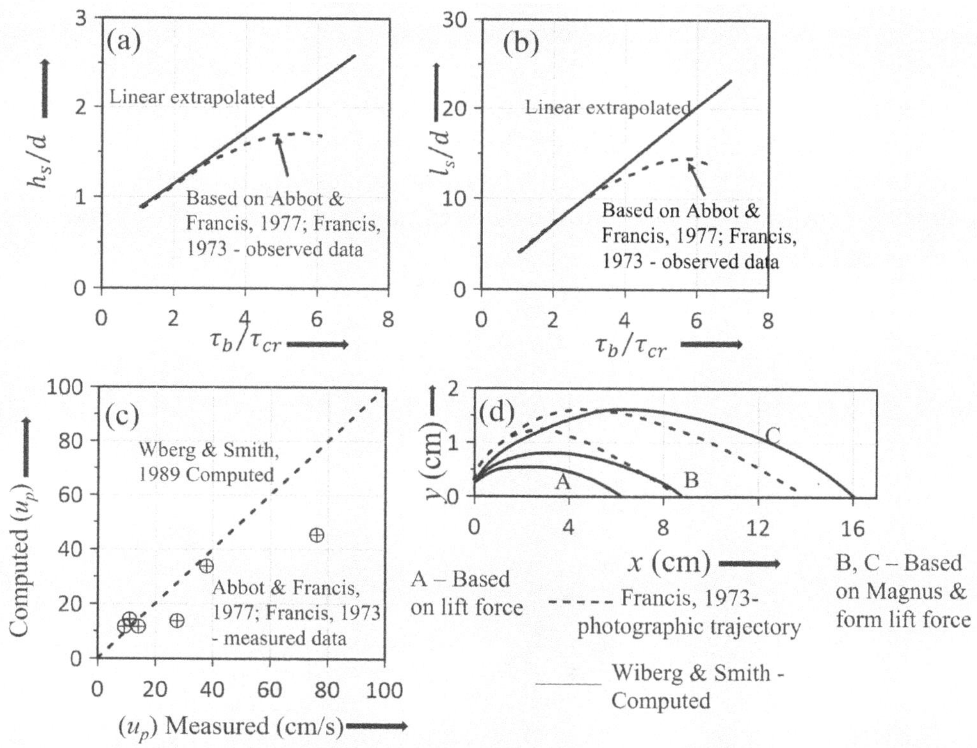

FIGURE 9.19 (a) Normalized calculated saltation height h_s/d as a function of transport stage τ_b/τ_{cr}, (b) Normalized calculated saltation length l_s/d against transport stage τ_b/τ_{cr}, (c) measured and computed average particle velocity u_p cm/s, (d) Predicted trajectory of particle along x-direction; Curve A – computed from lift force, and curves B and C – computed from Magnus and form lift forces. (Modified from Wiberg and Smith 1985.)

Subsequently, Wiberg and Smith (1987, 1989) derived respectively the expressions for the critical shear stress (Shields parameter) as a function of shear Reynolds number for different values of roughness parameter and the bed load sediment discharge from the concentration of the bed load, the particle velocity and the bed load height. The dimensionless critical shear stress is determined from equation (8.119) from Chapter 8, and the volume discharge of bed load sediment per unit width of flow q_b is given by (Wiberg and Smith 1989):

$$q_b = \int_{\eta}^{h_{sm}} C_b u_p \, dy \tag{9.174}$$

where C_b is the concentration of sediment in the bed load layer, y is the height above the bed, u_p is the stream-wise component of particle velocity, η is at the bed level, and h_{sm} is the height of the bed layer or maximum saltation height. Four sediment sizes $d = 0.035$, 0.080, 0.33 and 2.86 cm with sediment density 2.65 g/cm^3 are considered. The calculated volume flux of bed load sediment q_b (cm^3/cm-s) is plotted against the transport stage τ_b/τ_{cr} in Figure 9.20.

Interested readers may look into papers by Wiberg and Smith (1985, 1987, 1989) for particle trajectories, saltation heights, saltation lengths, average particle velocities, bed load concentration, volume flux of bed load, and the bed load transport equations, which are explicitly analyzed. It is interesting to show in Figure 9.21 a comparative study of several bed load transport equations for the size $d = 0.05$ cm with the calculated curve by Wiberg and Smith (1989).

Hui and Hu (1991) used a high-speed photographic technique to measure the particle motions near the channel bed and proposed the dividing boundaries between rolling, saltating, sliding, and suspending motions. Later, Lee and Hsu (1994) applied in their study a real-time flow visualization technique to study the saltation of particle at the near bed region without interference of flow. They used the CCD camera to record the saltating particles passing through the laser light sheet. The images obtained by camera were processed and stored by an image-processing technique. From the analysis of experimental data, they determined the regression equations of dimensionless average saltation length $\overline{l}_s$, height $\overline{h}_s$, and velocity $\overline{u}_p$, are given by:

$$\frac{\overline{l}_s}{d_m} = 196.3(\theta_* - \theta_{*c})^{0.788} \tag{9.175a}$$

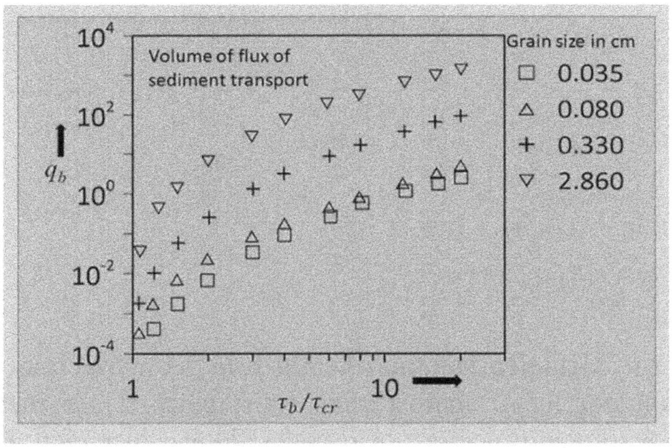

FIGURE 9.20 Volume flux of sediment transport $q_b\left(\text{cm}^3/\text{cm-s}\right)$ against transport stage τ_b/τ_{cr} for different grain sizes $d = 0.035$, 0.080, 0.330 and 2.860 cm. (Modified from Wiberg and Smith 1989.)

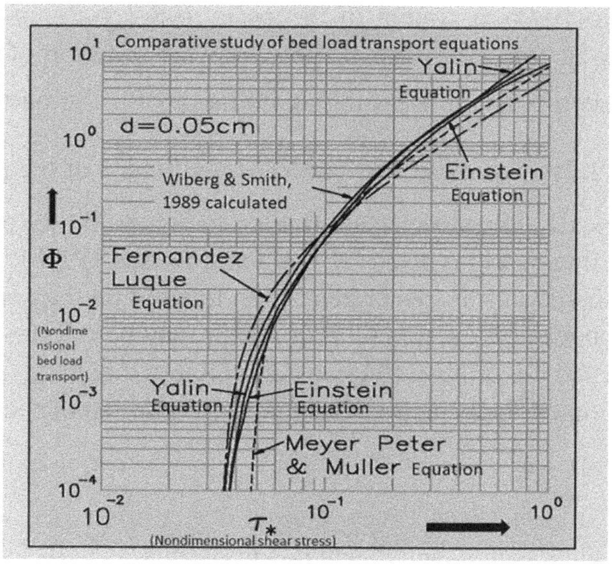

FIGURE 9.21 Plots of comparison of several bed load transport equations with predicted curve for $d = 0.05$ cm. (Modified from Wiberg and Smith 1989).

$$\frac{\bar{h}_s}{d_m} = 14.27(\theta_* - \theta_{*c})^{0.575} \tag{9.175b}$$

$$\frac{\bar{u}_p}{u_*} = 11.53(\theta_* - \theta_{*c})^{0.174} \tag{9.175c}$$

where θ_* and θ_{*c} are the dimensionless shear stress and the critical shear stress respectively, and d_m is the mean particle size in mm. The average saltation length, height, and velocity increase with increase in shear stress. These regression equations are valid for dimensionless shear stress lees that 0.46. They have also developed a theoretical model based on the forces acting on the saltating particles, which include inertia force, submerged weight, lift and drag forces. In addition, the lift force due to Magnus effect was also used. Interesting to note that the maximum saltation height, length, and velocity observed in this study were 9.3 d_m, 106.8 d_m, and 9.9 u_* respectively. The use of Magnus effect is important to estimate the saltation height and it increases by 12%, so the Magnus effect should not be neglected.

In a series of papers, Nino et al. (1994), Nino and Garcia (1994, 1996, and 1998) investigated the particle motion in the near bed region of open channel turbulent flow to understand the bedload transport by saltation and sediment going to suspension. Two experimental studies were performed by Nino et al. (1994) and Nino and Garcia (1996) on saltation of gravel and sand to understand the physics of particle saltation. In particular, they provided a description of the particle collision with the bed, based on the stochastic model for this process and also provided statistics for the geometric and kinematic properties of the saltation trajectories. Moreover, Nino and Garcia (1994) attempted the model for saltation of gravel, where they predicted mean and standard deviations of saltation of gravel-size particles, such as saltation length, height, stream-wise particle velocity, and friction coefficient that are in good agreement with the experimental results of Nino et al. (1994). In fact, the model of saltation of gravel proposed by Nino and Garcia (1994) was applied to that of fine sand and compared the results with experimental data of Nino and Garcia (1996). They also obtained the bed load transport rate, using the saltation model of sand particle. Nino and Garcia (1998) developed the instantaneous equations for particle motion in an unbounded fluid flow proposed by Mei et al. (1991), which were modified by Nino and Garcia (1994) in order to analyze the case of particle saltation in a turbulent boundary layer.

Following Nino and Garcia (1994), a system of dimensionless equations governing particle velocity components in the stream-wise and bed-normal directions for saltation of sand in water is proposed, and the equations are balanced with the Lagrangian equations for the particle trajectory as $u_p = \frac{dx_p}{dt}$, $v_p = \frac{dy_p}{dt}$, where x_p and y_p are stream-wise and vertical coordinates of the particle centroid. The set of equations are solved numerically for particle saltation and compared the results with the experimental observations of saltation of fine sand reported by Nino and Garcia (1996). Finally, the saltation model was applied to estimate the bedload transport rates of sand, based on the concept of Bagnold's formula. For interested readers, the paper of Nino and Garcia (1998) may be referred.

Hu and Hui (1996) suggested the relative thickness of saltation height h_s/d was related to the specific gravity of particles, dimensionless bed shear stress, and the condition of the bed surface. The empirical relations are given by

$$\frac{h_s}{d} = 3.67\left(\frac{\rho_s}{\rho_f}\right)^{1.05} \theta_*^{0.82} \tag{9.176a}$$

for smooth bed surface, and

$$\frac{h_s}{d} = 1.78\left(\frac{\rho_s}{\rho_f}\right)^{0.86} \theta_*^{0.69} \tag{9.176b}$$

for rough bed surface.

Lee et al. (2000) developed a bed load transport model from the particle saltating motion dominated by the forces acting upon the particles. A real-time flow visualization technique was used to measure particle saltating trajectories, corresponding velocities, and impacting and rebounding angles. Based on the experimental data, the regression equations for dimensionless saltating length, height, velocity were developed and compared their results with the numerical simulations. Lee et al. (2000) developed a correlation similar to Hu and Hui (1996) as:

$$\frac{h_s}{d} = 0.088\left(\frac{\theta_*}{\theta_{*c}} - 1\right)^{0.71} D_*^{0.63} \tag{9.177}$$

where D_* is the dimensionless particle diameter. Equation (9.177) has little different coefficient and exponent from van Rijn's simulation (1984).

Cheng (2003) proposed an analytical model to compute the average thickness of bed load layer or saltation height of particles, based on the concept of hydrodynamic diffusion related to particle-particle interactions. He evaluated the thickness h_s of bed load layer, using the diffusivity E and the relative settling velocity ω_r, and defined the bed load layer thickness h_s as:

$$h_s = \frac{1}{\omega_0}\int_0^{c_m} \frac{E}{c\omega_r}\,dc \tag{9.178}$$

where ω_0 is the settling velocity under a very dilute condition, and c_m is maximum concentration of densely packed particles.

Chatanantavet et al. (2013) conducted an experimental study on coarse-grain saltation dynamics in bedrock channel. They recorded the saltation hop height, hop length, and velocity of gravel saltating over a planar bed using 80–160 experiments from high-speed photography and direct measurements.

They obtained that the saltation velocity of bed load is independent of grain density and grain size and is linearly proportional to flow velocity. The saltation height has a nonlinear dependence on grain size. Saltation length increases primarily with flow velocity, and it is inversely proportional to a submerged specific density. Ghoshal and Pal (2014) proposed an expression of bed load layer thickness in an open channel turbulent flow, using the viscous and impact shear stress. Here the viscous shear stress is expressed as a function of effective viscosity of sediment-fluid mixture, velocity gradient, and volumetric concentration of sediment particles.

Due to high sediment concentration near the bed, an impact shear stress is generated and according to Bagnold (1954), the impact shear stress is given by

$$\tau_i = a_i \rho_s \left(\frac{d^2}{G_m} \frac{d\bar{u}}{dy} \right)^2 \tag{9.179}$$

where a_i is the impact coefficient, G_m is the average minimum gap among the adjacent particles, $\bar{u}$ is the stream-wise mean velocity of sediment-fluid mixture, y is the vertical distance. Combining both the shear stresses, and modified settling velocity due to sediment fluid mixture, an expression of bed load layer thickness was developed. He claimed that the model not only estimates the bedload layer thickness for low concentration but it can also estimate the bedload layer for highly concentration or sheet flow. The comparison of the models shows that the proposed model is well applicable for predicting the thickness of the bedload layer or saltation layer for both low and high concentration flow.

9.6.1 Bagnold's Model (1966) due to Energy Concept

In his work, Bagnold (1966) derived an expression for the total load based on the energy concept, which is the combination of bed load and suspended load transport. In the present case of suspension load model, the considerations are no longer restricted to the velocity at the vicinity of the bed, but it is extended by the average velocity u_m to the whole depth h of flow. The total load formula of Bagnold can be derived from the energy concept. Let W_b and W_s be the work rates of the bed load q_{sb} and suspended load q_{ss} respectively. The work rate is given by

$$W_b = q_{sb} \tan\phi \tag{9.180}$$

and the work rate of suspended load W_s by

$$W_s = q_{ss} \frac{v_0}{u_s} \tag{9.181}$$

where $\tan\phi$ is dynamic friction coefficient, v_0 is the terminal velocity of the grains, and u_s is the average transport velocity of the grains in suspension. To derive the expression for suspension, the weight of the grains transported in suspension over a unit area of the bed is G_s, the work done by the gravity acting on these grains per unit time is expressed as:

$$W_s = G_s v_0 = G_s u_s \frac{v_0}{u_s} \tag{9.182}$$

Here $G_s u_s$ is the product of the suspended material travelling in the direction of the flow per unit time and per unit width, which is the transport rate q_{ss} of the suspended load. Now the loss of potential energy E_p of the flow per unit area and time is given by

$$E_p = \gamma J \int_0^h u\,dy = \gamma Jh \frac{1}{h} \int_0^h u\,dy = \gamma Jhu_m = \tau_0 u_m \tag{9.183}$$

where γ is the specific weight, and J is the bed slope. The work rate W_b has the following part of potential energy E_p as:

$$W_b = e_{\text{eff}} E_p \tag{9.184}$$

From the potential energy (available energy), if a part of $e_{\text{eff}} E_p$ is consumed for the work of the bed load, the energy available for the work of the suspended load is obviously as:

$$E_p - e_{\text{eff}} E_p = E_p (1 - e_{\text{eff}}) \tag{9.185}$$

If e_s is the efficiency with respect to the suspended load, then the work rate of suspended load W_s can be written as:

$$W_s = e_s \left[E_p (1 - e_{\text{eff}}) \right] \tag{9.186}$$

Substituting the above equations in W_b and W_s, one gets from equations (9.180) and (9.181):

$$q_{sb} = \frac{e_{\text{eff}}}{\tan\phi} E_p \tag{9.187}$$

$$\text{and } q_{ss} = e_s (1 - e_{\text{eff}}) \frac{u_s}{v_0} E_p \tag{9.188}$$

Adding these two expressions and considering $E_p = \tau_0 u_m$, one gets

$$\frac{q_s}{\tau_0 u_m} = \frac{e_{\text{eff}}}{\tan\phi} + e_s (1 - e_{\text{eff}}) \frac{u_s}{v_0} \tag{9.189}$$

which is the total load formula of Bagnold (1966), where $q_s = q_{sb} + q_{ss}$. It is important to note that Bagnold's derivation of total load transport formula involves neither assumption nor approximation. Other details are available in Yalin (1977).

9.7 IMAGE ANALYSIS ON PARTICLE MOTION

Despite the large number of works addressed to the problems, our understanding of bed load transport remains very low with respect to the reliable predictions of sediment flux, particle movements, even with a good knowledge of free surface water flows. In recent times, the use of image processing techniques for observing the motion of sediment particles has been considered in some research studies to predict the particle trajectory, saltation height, length, resting time of a particle and particle velocity at the near bed region. The analysis of images from a high-resolution digital camera system provides a unique approach to understand the fundamental mechanism of particulate materials. The typical examples include heat exchangers, chemical reactors, sprays, dust clouds, particle jets, sediment transport in rivers, estuaries, and oceans, snow or sand drift, dunes, etc. In addition, the image processing has a wide application area in hydromechanics especially in the determination of sediment grain size distribution, intermittent nature of particle entrainment, and detection of sediment movement. This advanced technique can achieve a more fundamental understanding of physics on the movement of granular materials at the near-bed surface.

During the last few decades, researchers have started to use the *image-processing technique* to record detailed observations and quantitative measurements of sediment motion. For example, several researchers (Drake et al. 1988; Nelson et al. 1995; Kaftari et al. 1995; Pilotti et al. 1997; Nino and Garcia 1998; Keshavarzy and Ball 1999; Ancey et al. 2002, 2003; Frey et al. 2003; Mazumder et al. 2008; Hergault et al. 2010; Bhattacharya et al. 2013; and others) who used the image processing technique as a tool to investigate the intermittent nature of particle entrainment. Drake et al. (1988) described the motion of bed load transport of fine gravel in laboratory, and more rarely in the field, using motion-picture photography. He performed field tests with natural sediments, extracting information from the images. Kaftori et al. (1995) showed that particle motion and entrainment process are controlled by the action of coherent wall structures. Nino and Garcia (1998) employed a high-speed video system to capture the saltation process of particle size of 0.5 mm in an open channel flow over a fixed bed of sand particles of an approximately uniform size of 0.53 mm. They recorded images of the motion of saltating particles of smaller size at a rate of 250 fps. Their observations were on particle resting time, re-entrainment into saltation, particle rotation and transverse motion during saltation, which provided new insights into the physical processes associated with saltation phenomena.

Keshavarzy and Ball (1999) considered the application of image processing to understand the processes of the entrainment of particle from the bed, deposition of particle, speed of particle movement, transport mode and the resting period of particle on the bed. In this study, attention was paid to define the initiation of particle motion over the loose sediment bed, considering the bursting processes from turbulence structure. The image processing technique was utilized to observe the particle motion in a selected area on the bed. They conducted experiments in a straight tilting rectangular flume of 35 m long, 0.61 m width, and 0.60 m height with side glass walls to visualize the flow. The flume bed was made with movable fiber cement sheets. The sand particles of 2 mm nominal diameter were placed over the flume surface. The movement of sand particles was observed and recorded by a high-resolution CCD camera with 25 frames per second, and the images were recorded in video tape. The captured images were digitized using image maker software. To analyze the series of discrete signals spatially and temporally, the statistical tools are applied to these arrays.

To analyze the captured images, two different techniques were used. These are:

1. A probability analysis of the moving particles was determined by counting of number of particles in motion at an instant of time, which was useful to obtain an exceedance probability of particles in motion in time with respect to the exceedance probability of stress fraction of sweep events at the bed. The number of particles entrained into motion was obtained through the difference between the sequential images.
2. Application of cross correlation and Fast Fourier transforms were used to obtain the displacement of particles between the images, and hence the particle velocity.

The difference between two images $f_1(x, y)$ and $f_2(x, y)$ can be expressed as:

$$f_1(x, y) - f_2(x, y) = g(x, y) \tag{9.190}$$

where $g(x, y)$ is new image. Two images are compared by computing the differences between the light intensities at all pairs of corresponding pixels from images $f_1(x, y)$ and $f_2(x, y)$. Subtraction technique was used to obtain the difference between sequences of images. In addition, by applying the convolution technique and a cross correlation tools to the sequences of images, the displacement of particles over the mobile bed was obtained. By counting the number of pixels for displacement, it can be calibrated the distance of movement. If the time between the images is known, the instantaneous velocity of the particle can be obtained. The modification of the Shields diagram was proposed to estimate the particle motion for a desired probability and particle size. Interested readers should follow the paper by Keshavarzy and Ball (1999).

Ancey et al. (2002) considered the basic problem of two-dimensional saltating motion of a spherical particle in a shallow water stream over a steep rough bed to understand the physics of saltation in fluid flow. Subsequently, Ancey et al. (2003) investigated the two-dimensional rolling motion of a single large particle in a similar condition both experimentally and theoretically. They used two theoretical models: the first one is the mean kinetic energy balance to deduce the average particle velocity and the occurrence of rolling regime, and they arrived that the rolling regime is a marginal mode of transport between repose and saltation; the second one is a Markov chain model because of randomness of particle state consisting of resting, rolling and saltating. Finally, the theoretical results are compared with the experimental data and found that the second model provided a correct estimate.

Later on, Ancey et al. (2006) investigated the motion of coarse spherical glass beads entrained by a steady shallow turbulent flow down a steep two-dimensional channel with a mobile bed. Using image-processing technique mechanism, they determined the flow characteristics such as particle trajectories, their state of motion, and the flow depth. In their experiments, they observed that the sediment transport mechanism is an intermittent process. To understand the results, they re-examined Einstein's model of sediment transport which described the statistical properties of variables, like the solid discharge and the number of moving particles. From their analysis, they predicted a negative binomial distribution from the number of moving particles, which fitted best. Subsequently, in the same year, Bohm et al. (2006) explored the experiment for bed load transport to understand in detail the states of motion: resting, rolling and saltating. They developed a particle-tracking method to analyze bed load movement comprising of particle trajectories and motion regimes, rolling or saltation, using image-processing technique. In particular, they observed that the rolling state played an important role in bed load transport, which was neglected (Bagnold 1973). In 2008, Ancey et al. studied in detail the entrainment, deposition and motion of coarse spherical particles in a turbulent shallow water stream down a steep slope. The entrainment, trajectories, and deposition were recorded using a high-speed digital camera. They developed a *birth–death immigration–emigration Markov process model* (Cox and Miller 1966) to describe the particle exchanges between the bed and the water stream. The probability distributions of the solid discharge and deposition frequency were described. Their exhaustive measurements allowed them a thorough statistical analysis of uniform sediment transport based on stochastic Markovian process.

Mazumder et al. (2008) investigated the particle movement under the action of near-bed turbulence over the rough bed surface in open channel flow, using a High-Speed Motion-Scope (HSMS) camera system at the Fluvial Mechanics Laboratory (FML) of Indian Statistical Institute (ISI)-Kolkata. The imaging system was used to record the motion–picture photography of particle movement at 250 fps and the recorded images are analyzed to determine the particle trajectories, saltation heights and lengths of individual particles, angles of orientation, and their impact to the boundary, using Image Pro-Plus (IPP) package. Mazumder and Dalal (2003) suggested that the saltation height of particles at the rough bed surface was influenced by the turbulent bursting events, which occur most violently at the near-bed surface. The frequency distribution of near-bed velocity fluctuating components over the rough bed can be described by Gaussian function, and the distributions of fluctuating shear stress show asymmetry near the rough surface and symmetry away from the bed surface. The frequency distributions of the fluctuating angle of orientation observed frame by frame follow statistically the normal error law for any size particles. The occurrence of normal distribution of angle of orientation might be related to the fluctuating shear stress, which follows normal distribution. The particle motion analysis at the near-bed turbulence using image-processing technique is a new emerging tool to the sediment size-sorting process. Wang et al. (2009) developed a novel method for measuring sand creep along a flat sand bed using a high-speed digital camera to find the velocities and trajectories of creep grains as well as creep fluxes of different sand samples. Lajeunesse et al. (2010) reported an experimental investigation of the motion of bed load particles under steady and spatially uniform turbulent flow above a flat sediment bed of uniform grain size. Using a high-speed video imaging system, the authors studied the trajectories of the moving particles and measured their velocity and the length and duration of their flights, as well as the surface density of the moving particles.

Hergault et al. (2010) studied bed load transport of mixtures of two-size spherical particles (4 and 6 mm) in turbulent supercritical flow at a two-dimensional flume, using image-processing technique and particle-tracking velocimetry algorithms. They compared the experimental results of two sizes, including the size, the position, the trajectory of the particle, the state of motion (rest, rolling, and saltation) and then neighboring configuration of each particle with the earlier one-size experiment of 6 mm size. The sorting of fine particles in the bed formed by larger particles was analyzed taking into account the surrounding configurations. Bombar et al. (2010) performed experiments in unsteady flows through a hydrograph to apply the technique and calculate the bed load transport, using the video recorder. The recorded images are analyzed by image-processing techniques to determine the number and area of active grains moving at any instant as well as the average velocity of the grains. Keshavrzi et al. (2012) studied the initiation and entrainment of sediment transport over the ripples, using the image-processing technique. They analyzed the dominant bursting events and the flow structure over the ripple bed channel. Two types of ripples with sinusoidal and triangular forms were used in their study. From the quadrant decomposition of instantaneous velocity fluctuations close to the bed, the bursting events (outward and inward interactions) were dominant downstream of the second ripple, whereas ejection and sweeping events were dominant upstream of the ripple. Therefore, entrainment would be expected to occur upstream and deposition occurs downstream of the ripple.

Bhattacharyya et al. (2013) studied rigorously the effect of variation of bed roughness and flow intensity on the saltating motion of individual particle of different sizes, using digital imaging technique. Four different fixed rough beds were prepared for experiments including the plane bed surface. For better understanding of saltation characteristics (saltation trajectory, height, length, angle of orientation, etc.) of individual particles under different flow conditions and altered rough beds, a series of flume experiments were conducted. A theoretical model was developed to determine the mean motion of a saltating particle considering the effect of spin near the bed and compared with the observed data. Experiments were conducted in a re-circulating flume (Figure 7.16) of dimensions 3500 mm length × 220 mm width × 330 mm high, designed especially at the Fluvial Mechanics Laboratory (FML), Indian Statistical Institute (ISI) Calcutta. Figure 7.17 shows a schematic diagram of the experimental setup. To obtain a clear view of particle movements, the side walls of the flume are made of transparent glass. The test section was located at 2,500 mm downstream from the entrance to ensure fully developed turbulence. Three individual rough beds $B_{1.0}$, $B_{1.5}$ and $B_{2.0}$ of a monolayer of particles (Figure 9.22) were prepared by gluing sand of respective sieve size of diameters (d) 1.0, 1.5 and 2.0 mm of identical specific gravity 2.65. Each rough bed was painted black for photographic purposes. Halogen light source was used to illuminate the working section appropriately. One 3-D micro acoustic Doppler velocimeter (ADV) with high precision was used to record instantaneous velocity at a rate of 50 Hz; and a high-speed digital imaging system (Redlake, Motion-Scope, PCI) was used to record a sequence of colored digital images of an event at a rate of 60–1,000 fps at the side of the flume.

For all the experiments, the flow depth h was kept constant, 100 mm above the flume base. The hydraulic slope of the flume was of order 0.0001. The flow discharge was kept constant to run the flume, and after a certain time of running the flume, it attained a perfect equilibrium condition over the plane/rough beds. A SonTek 50 mm down-looking 16 MHz 3-D micro ADV was used to record instantaneous

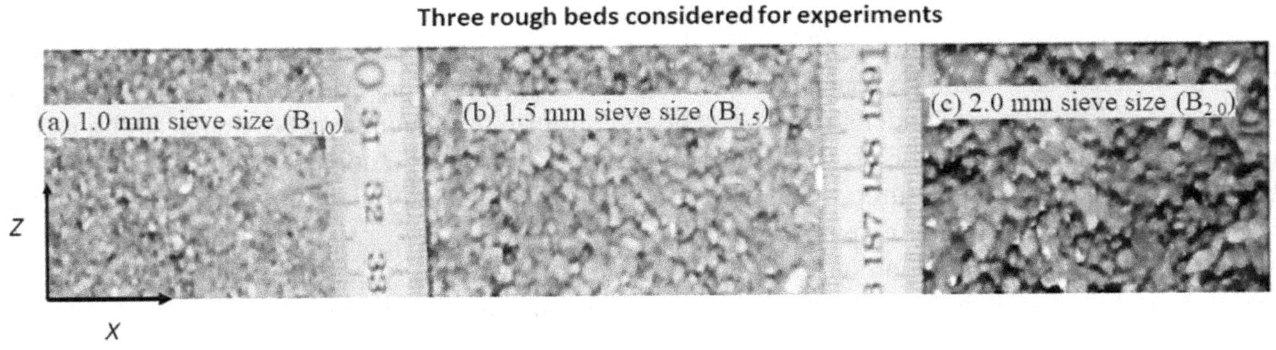

FIGURE 9.22 Three different rough beds in *xz* plane: (a) 1.0 mm sieve size ($B_{1.0}$), (b) 1.5 mm sieve size ($B_{1.5}$) and (c) 2.0 mm sieve size ($B_{2.0}$). (Modified from Bhattacharyya et al. 2013.)

velocity with fluctuations at three different horizontal distances at the flume central line for 300 seconds at a sampling rate of 40 Hz to ensure fully developed flow and two-dimensionality of the flow structure over the rough bed surface (Bhattacharyya et al. 2013). The velocity data were collected using ADV at the mid-section of the channel at about 12 vertical positions, starting from the lowest level 3.5 mm above the bed surface, and the highest level being about 50 mm for each profile. In the present study, the raw ADV data were processed to remove spikes, using the phase space threshold de-spiking technique described by Goring and Nikora (2002).

Since the width–depth ratio was about 2.2, with the occurrence of maximum velocity below the water surface (the dip phenomenon), it is ascertained that flow at the central portion of the flume was two-dimensional (Yang et al. 2004; Absi 2011), and hence the effect of secondary current on saltation was negligible. Therefore, it was ensured that the flow was free from secondary currents at the central portion of the flume. Under the fully developed conditions, three different experiments were conducted, releasing respective sizes of particles independently through the funnel at the centerline of the flume. Particles of identical size were released at a certain time at a distance of 1500 mm upstream from the sampling station, so that particles moving downstream attained steady conditions before coming into view of the camera. Particle images were recorded travelling along the centerline of the flume (± 10 mm) throughout the field of view, ensuring that the particle motion was not influenced by secondary currents. Here the particle velocity was measured in the Lagrangian sense by following the particle in motion under given fluid velocity. A summary of the experimental conditions for three different sets of experiments is presented in Table 9.1. Although the shapes of the particles were irregular, they were assumed to be nearly spherical, with a uniform sieve diameter. This assumption is used in the theoretical model development.

To track the moving particles individually, videos of the moving particles and their collisions with the rough bed were captured at a rate of 250 fps (resolution 480 × 420 pixels square) under the available light source. The recording rate was 8.5 times faster than that of ordinary video recording. The camera was placed 700 mm away from the glass window outside the flume to focus on to a plane at the centerline of the flume covering an area of length of 50 mm and height of 45 mm. These dimensions were adequate to resolve spatially saltating trajectories of particles corresponding to roughly one to two saltations. The motions of individual particles over the plane and rough beds for three different Reynolds numbers $Re = 5.33 \times 10^4$, 6.27×10^4, 7.00×10^4 were recorded. The released particles moved through the centerline of the flume by rolling and saltating over the plane/rough bed surface, inducing a mean slip velocity. Experiments were repeated under identical flow conditions over different rough beds using different sized particles. Videos of the moving particles were recorded. From the series of images, the trajectories and velocities of individual particles were determined. The images extracted

TABLE 9.1
Experimental Conditions

Particles Glued (mm) at the Bed		Reynolds Number (*Re*)		
		$Re_1 = 53{,}300$	$Re_2 = 62{,}700$	$Re_3 = 70{,}000$
1.0	d^a (mm)	1.0	1.0	1.0
		1.5	1.5	1.5
		2.0	2.0	2.0
1.5	d^a (mm)	1.0	1.0	1.0
		1.5	1.5	1.5
		2.0	2.0	2.0
2.0	d^a (mm)	1.0	1.0	1.0
		1.5	1.5	1.5
		2.0	2.0	2.0

Source: Modified from Bhattacharyya et al. (2013).
[a] particle size released in the flow.

from the video recordings were subsequently processed using an algorithm implemented in Image-Pro Plus (IPP) software, and are itemized as follows:

- to enhance the contrast of the images, the background was subtracted;
- the image was converted into a binary image, choosing an appropriate threshold level, and then objects were separated based on proper filtering; and
- the position of the centroid of the particle was then easily obtained with sub-pixel accuracy, and object features such as area, trajectories, angle of orientation of particle in each frame, velocities before and after collision at the bed, etc., were measured.

Using the above algorithm, the images were processed to track the moving particles individually; the processed images of particles are shown in Figure 9.23. As all the data from the videos were in terms of pixel units, we converted those data to metric units. Before each experiment calibration was made using a steel ball of diameter 10 mm painted with the same red color as that of the moving particles. From the calibration, it was obtained that the actual length of 1 pixel equal to 0.07898 mm, with a standard deviation of 0.00371 mm. Therefore, all the features (area, centroid, angle etc.) of sand particles in pixel units were converted to metric units. The collected fluid velocity data were analyzed for different flow conditions. The normalized velocity profiles are computed for all three Reynolds numbers $Re = 5.33 \times 10^4, 6.27 \times 10^4, 7.00 \times 10^4$; and observed that the mean velocity decreased with increase in bed roughness. The log-law, turbulence intensities and shear stress follow the same nature as discussed by Nezu and Rodi (1986), Clifford (1995) and Mazumder et al. (2008) with zero lateral and vertical components of mean velocity shown in Bhattacharyya et al. (2013).

9.7.1 Particle Velocity and Its Characteristics over Rough Beds

The released particles moved downstream over the rough beds mostly in the form of rolling and saltation. A schematic diagram of the saltating particles is shown in Figure 9.24, where h_s and l_s are the saltation height and length respectively. The incipient motion of particles over the bed surface is related to the flow intensity or Shields parameter Θ, defined $\tau_0/(\gamma_s - \gamma_f)d$, where τ_0 is the bed shear stress, γ_s and γ_f are the specific weight of solid and fluid respectively, and d is the particle diameter.

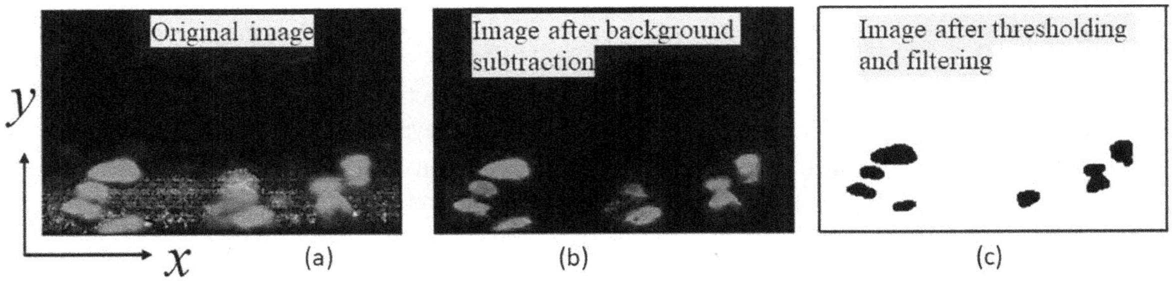

FIGURE 9.23 Experimental observations - (a) Original image, (b) image after background subtraction and (c) image after thresholding and filtering. (Modified from Bhattacharyya et al. 2013.)

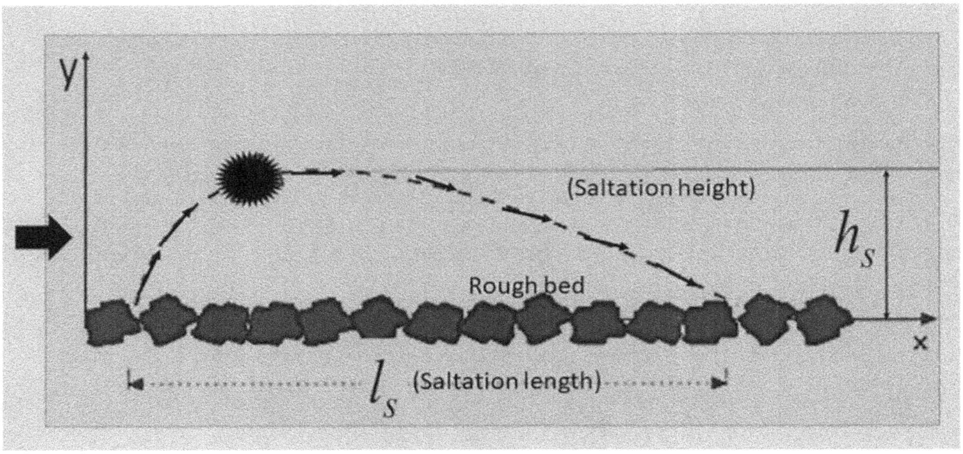

FIGURE 9.24 Sketch of particle saltation. (Modified from Bhattacharyya et al. 2013.)

Typical saltation trajectories over different rough beds are shown in Figure 9.25 for Reynolds number $Re_3 = 7.00 \times 10^4$, where the position of the particle is given by the coordinates (x_p, y_p) as the centroid of the particle image. The height and length of trajectories of different sizes of particles vary under different flow conditions. But, apparently the overall shapes of the trajectories are fairly smooth and of ballistic shape, and interestingly the shape was not changed by the near-bed turbulence. Therefore, normalizing the shape of the trajectories of different sizes of particles over a fixed rough bed $B_{1.0}$, it is likely to fall into a single curve, shown in Figure 9.26. The best-fitted equation of the trajectory of particles predicted from the data with a coefficient of regression $R^2 = 0.96$ is represented by a parabolic function of $y^+ = y_p/h_s$ with respect to $x^+ = x_p/l_s$. The nature of trajectories of saltating particles under different rough bed conditions is almost in agreement with the results of Ancey et al. (2002). The variation of vertical displacement, particle velocity components, the velocity differences between the fluid and the particle along the trajectories are known for Reynolds number $Re_3 = 7.00 \times 10^4$. The variations of stream-wise particle mean velocity against the friction velocity for three different particle sizes (1.0, 1.5, 2.0 mm) over the rough beds are also computed. For three different sizes of particles with three Reynolds numbers, the ranges of saltating height and length are provided in Table 9.2.

It is observed that both saltation height and length decreased with the increase of particle size. During the saltation with bed contacts, the saltation length was about 15–21 times the particle diameter of 1.0 mm, about 8–12 times the 1.5 mm diameter and about 5–11 times the 2.0 mm diameter of particles. The ranges of saltation length were determined for all three Reynolds numbers $Re = 5.33 \times 10^4$, 6.27×10^4, 7.00×10^4. During saltation, particles were also affected by rotation. Saltation height h_s and length l_s of

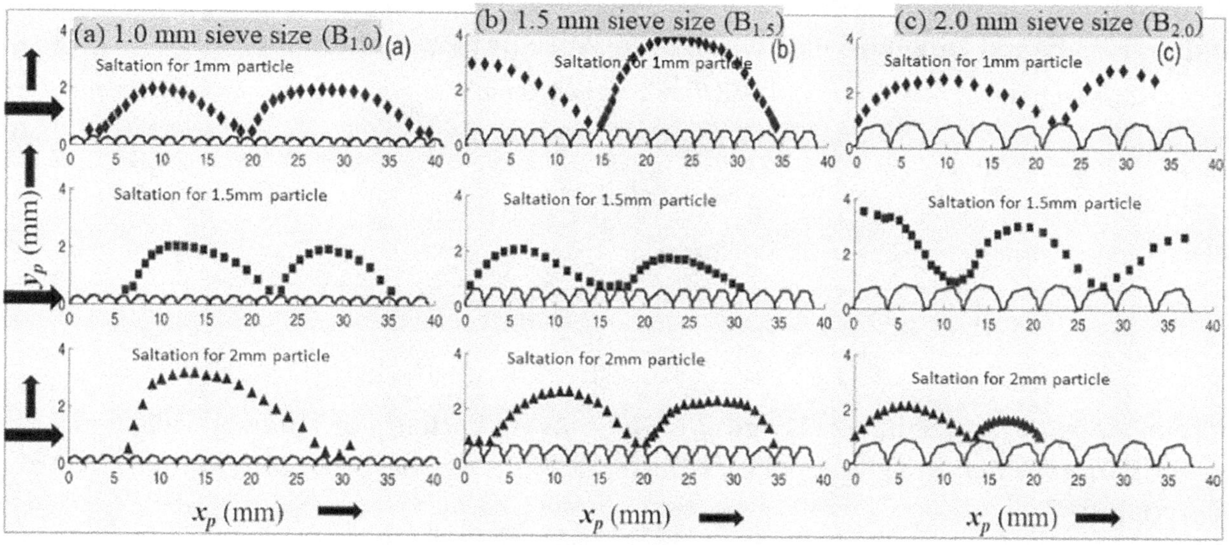

FIGURE 9.25 Plots of saltating particles over three different rough beds: (a) $B_{1.0}$, (b) $B_{1.5}$ and (c) $B_{2.0}$ for flow Reynolds number $Re_3 = 7.00 \times 10^4$. The symbols correspond to the saltating particle of size ♦: 1 mm, ■: 1.5 mm and ▲: 2.0 mm. (Modified from Bhattacharyya et al. 2013.)

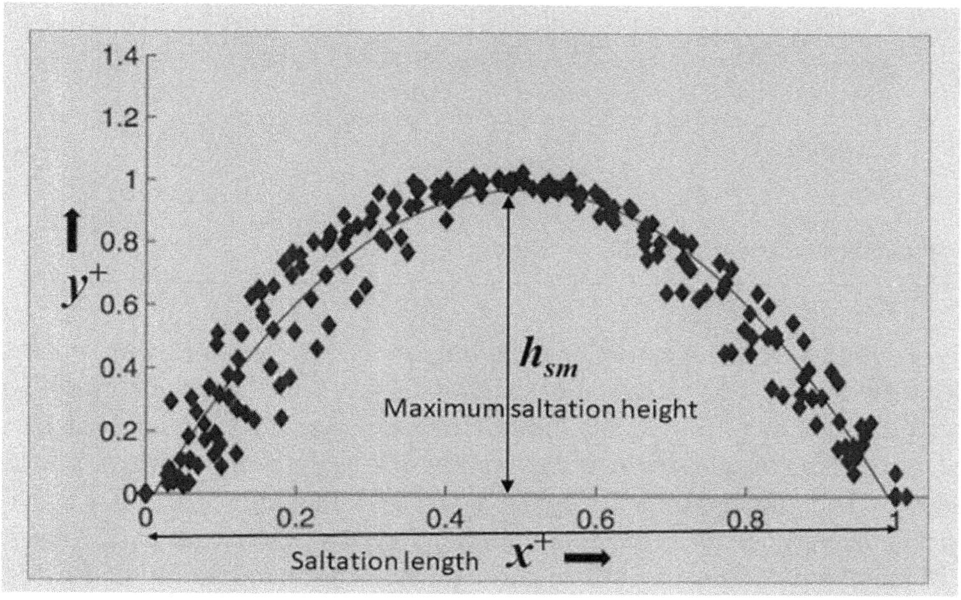

FIGURE 9.26 Trajectories of saltating particles of different sizes moving over the rough bed $B_{1.0}$ for Reynolds number $Re_3 = 7.00 \times 10^4$. The solid line represents the fitted parabola curve. Here dimensionless coordinate system (x^+, y^+) and h_{sm} is the maximum saltation height. (Modified from Bhattacharyya et al. 2013.)

the saltating particles of different size over different k_s are presented in Table 9.3 (Bhattacharyya et al. 2013) under different Reynolds numbers.

Angles of orientation over rough beds are investigated, which represents the frequency distributions of fluctuations of angles of orientation for three sizes of particles from the number of observations $N = 300–330$. Frequency distributions of fluctuating angle of orientation of different saltating particles are shown in Figure 9.27 for (a) $d = 1.0$ mm, (b) $d = 1.5$ mm and (c) $d = 2.0$ mm moving over different

TABLE 9.2
Saltation Height and Length of Particle Sizes for Various Flow Conditions

Size of Particles d (mm)	Saltation Height h_s (mm)	Saltation Length l_s (mm)
1.0	1.5d–2.0d	15.0d–21.0d
1.5	1.0d–1.3d	8.0d–12.5d
2.0	0.7d–0.97d	5.0d–11.0d

Source: Modified from Bhattacharyya et al. (2013).

TABLE 9.3
Saltation Height h_s and Saltation Length l_s of the Moving Particles of Three Sieve Sizes (d) over Different Bed Roughness (k_s) for Reynolds Numbers $Re_1 = 53{,}300$, $Re_2 = 62{,}700$ and $Re_3 = 70{,}000$

		Saltation Height (h_s) (mm)			Saltation Length (l_s) (mm)		
	d (mm)	1.0	1.5	2.0	1.0	1.5	2.0
	k_s (mm)						
$Re_1 = 53{,}300$	0.52	1.49	1.04	0.66	15.51	7.99	5.55
	3.24	1.46	1.04	0.65	19.32	8.73	7.11
	4.53	1.24	1.96	-	17.53	15.02	-
$Re_2 = 62{,}700$	1.21	1.72	1.09	0.92	15.52	9.83	6.63
	4.54	1.64	1.23	0.97	19.42	12.58	7.68
	6.31	-	0.93	1.32	-	12.04	18.32
$Re_3 = 70{,}000$	1.18	2.02	1.14	0.99	20.02	12.18	8.94
	7.79	1.99	1.32	1.50	19.55	13.52	11.37
	9.85	1.87	1.39	1.39	21.52	21.52	11.36

Source: Modified from Bhattacharyya et al. (2013).

rough beds $B_{1.0}$, $B_{1.5}$ and $B_{2.0}$, that means along bed roughness (k_s). Here the histogram shows the variation against bed roughness k_s and the size d of the moving particles. Here the corresponding rows represent the variation in bed roughness k_s and the columns represent the variation in size d of the moving particles. The figures along the horizontal direction (row-wise: (a), (b) and (c)) show how the distribution changes with the gradual increase in bed roughness, when the moving particle size (d) is fixed; and the figures arranged in the vertical direction (column-wise) describe the changes in distribution with different diameters of moving particles, when the bed roughness (k_s) is fixed. From the figures it is observed that the distribution of recorded angles appears almost uni-modal, with a symmetric or asymmetric pattern depending on the particle size (d) and bed roughness (k_s).

The rth moment of the distribution about the mean angle of orientation, denoted by M_r, is given by

$$M_r = \frac{1}{N}\sum_{i=1}^{N}(X_i - \mu_1)^r, \tag{9.191}$$

where μ_1 is the mean of N values of X_i. The statistical parameters are mean μ_1, standard deviation or r.m.s. $\sigma\left(= M_2^{1/2}\right)$, coefficients of skewness $\beta_1\left(= M_3/M_2^{3/2}\right)$ and kurtosis $\beta_2\left(= M_4/M_2^2 - 3\right)$ are computed (Mazumder and Das, 1992) for $Re_3 = 7.00 \times 10^4$ (Table 9.4).

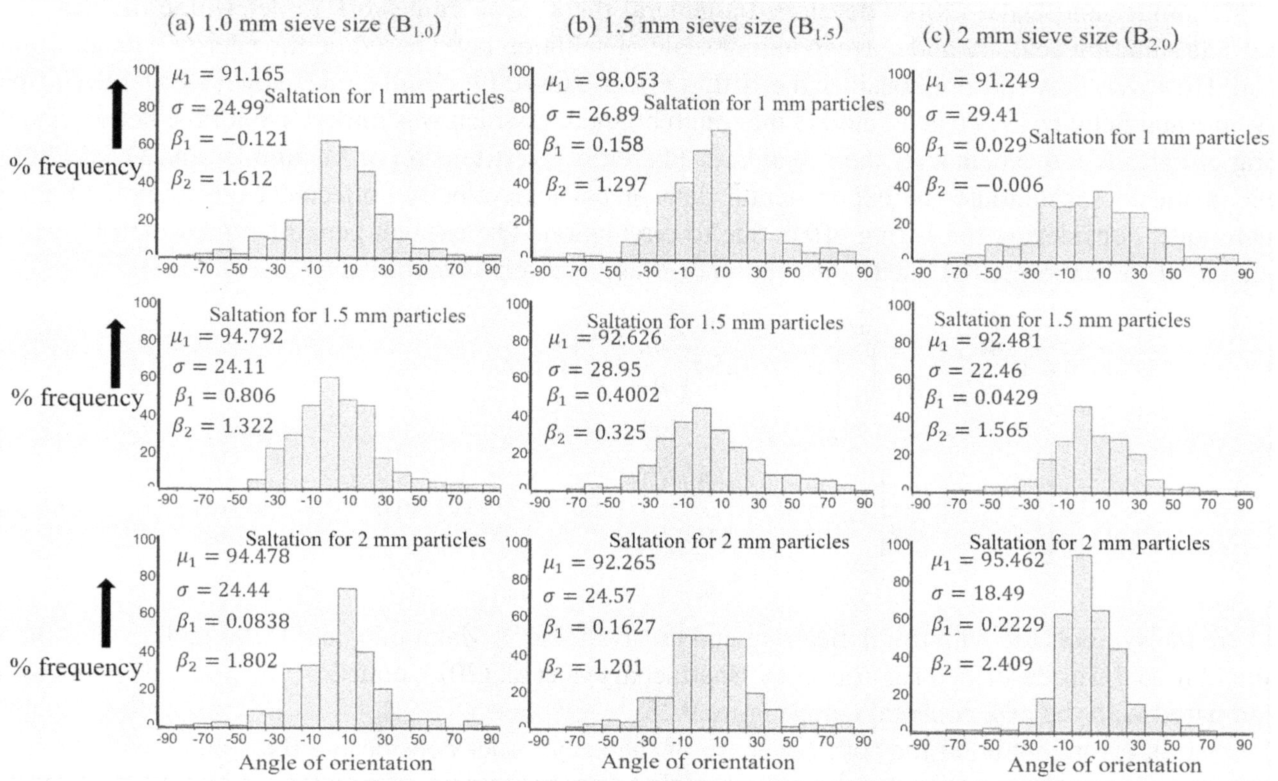

FIGURE 9.27 Frequency distribution of angle of orientation of saltating particles of sieve sizes (a) $d = 1.0$ mm, (b) $d = 1.5$ mm and (c) $d = 2.0$ mm moving over rough beds $B_{1.0}$, $B_{1.5}$ and $B_{2.0}$. (Modified from Bhattacharyya et al. 2013.)

TABLE 9.4

Statistical Parameters of Distributions of Angle of Orientations of Saltating Particles for $Re_3 = 70{,}000$

	k_s (mm) = 1.18	k_s (mm) = 7.79	k_s (mm) = 9.85
d (mm) = 1.0	$\mu_1 = 91.16$	98.05	91.25
	$\sigma = 24.99$	26.88	29.41
	$\beta_1 = -0.12$	0.16	0.03
	$\beta_2 = 1.16$	1.30	−0.01
d (mm) = 1.5	$\mu_1 = 94.79$	92.63	92.48
	$\sigma = 24.11$	28.94	22.46
	$\beta_1 = 0.81$	0.40	0.04
	$\beta_2 = 1.32$	0.32	1.56
d (mm) = 2.0	$\mu_1 = 94.48$	92.26	95.46
	$\sigma = 24.44$	24.57	18.48
	$\beta_1 = 0.08$	0.16	0.22
	$\beta_2 = 1.80$	1.20	2.41

Source: Modified from Bhattacharyya et al. (2013).

Symbols μ_1, σ, β_1 and β_2 represent mean, standard deviation, coefficient of skewness and coefficient of kurtosis respectively.

Mazumder and Dalal (2003) developed a general theoretical framework to determine the near-bed particle saltating velocity and corresponding scale of saltation layer based on the energy balance equation. However, they did not consider the lifting effect caused by circulation, known as Magnus effect (White and Schultz 1977), generated at the rough boundary, which was imperative for the better modelling of particle movement over the rough bed. Therefore, an extension of Mazumder and Dalal (2003) was made to substantiate the experimental data of particle velocity collected using digital imaging technique, considering the lifting effect due to circulation. The explicit general expression of particle velocity u_p (equation 9.101) due to circulation or Magnus effect is modified as given below.

$$u_p = u + \frac{4C_L}{\pi C_D}\left(\frac{\partial u}{\partial y}\right)^{0.5}\left[v^{0.5} + 0.5d\left(\frac{\partial u}{\partial y}\right)^{0.5}\right]\tan\alpha$$

$$-0.5\left[\left(\frac{8C_L}{\pi C_D}\right)^2 \frac{\partial u}{\partial y}\left[v^{0.5} + 0.5d\left(\frac{\partial u}{\partial y}\right)^{0.5}\right]^2 \tan^2\alpha + 4v_s^2(\sin\beta - \cos\beta\tan\alpha)\right]^{0.5} \tag{9.101a}$$

where the symbols are with usual meaning and are available in Mazumder and Dalal (2003). This theoretical model was described rigorously by Bhattacharyya et al. (2013, equation 16), and the results were compared with the experimental data (Figure 9.28).

Overall, errors obtained for both cases are of the same order considering the spin or without spin. Although during the saltation process, the particle lift is due to spin, it is observed from the test results that the consideration of spin (Magnus effect) on the particle motion did not show any improvement in mean particle velocity except for the highest Reynolds number. From the overall observations, it may be concluded that the agreement between the observed and computed mean particle velocity is better when the spin (Magnus effect) is excluded from the model. Detailed investigations are important in formulating a better modelling of bed load transport associated with the rolling and saltation processes from a wide range of parameters, such as particle diameter, bed roughness, flow intensity and flow Reynolds number, which will help the researcher to understand the mechanism of bed load movement over the river bed in the fields of hydraulics, geology and geophysics.

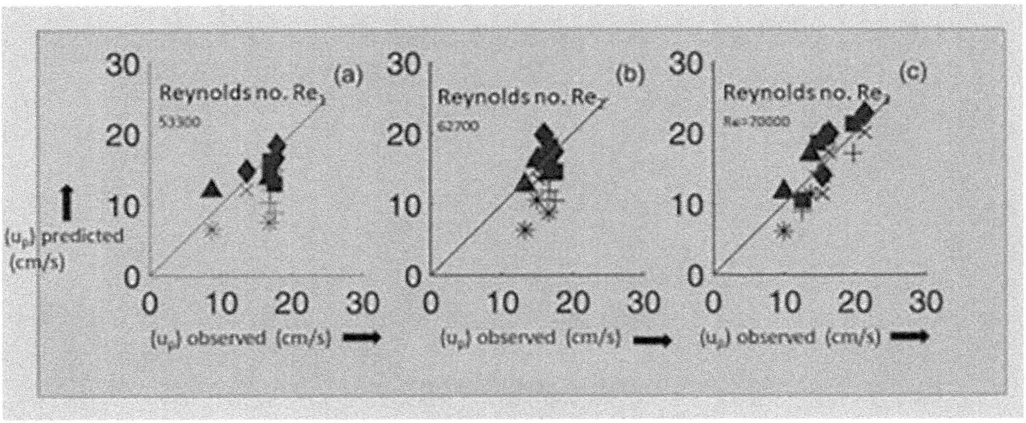

FIGURE 9.28 Plots of predicted vs. observed particle velocity moving over three rough beds for different Reynolds numbers: (a) for Re_1, (b) for Re_2 and (c) for Re_3. Solid symbols represent lift due to shear only and line symbols represent lift due to shear and Magnus force on moving particle of size ♦, ×: $d = 1$ mm, ■, +: $d = 1.5$ mm and ▲, $*$: $d = 2.0$ mm respectively. (Modified from Bhattacharyya et al. 2013.)

9.8 SUSPENDED LOAD

Suspended sediment is conventionally regarded as those sediment particles that are surrounded by the fluid for a period of time, known suspension. The sediment transported by a fluid flow that it is fine enough for turbulent eddies to beat settling of the particles through the fluid. The stronger the flow or the finer the sediment size, the greater amount of sediment that are suspended by turbulence. In the case of suspension, the particles are supported by turbulent flow with fluctuations. The material transported with the flow and by the flow is known as suspended load, whereas the larger solid particles are rolled along the streambed, called the bedload. The mechanism of suspension being transported by the flow is rather complex because of the exchange of both mass and momentum including randomness in the fluid flow. The spreading of sediment particles in the flow by random motion is known as diffusion, while motion due to velocity is known as convection. Due to the weight of the particles, there is a tendency of settling, which is counterbalanced by the random motion of fluid particle, i.e., the fluctuating components of velocity. Particles that move as suspended load or bed load and periodically exchange with the bed are part of the bed-material load. There exists an active interchange between the suspended load and bed load, but also between the bed load and bed itself. The upper boundary of sediment suspension is the free surface, whereas the lower boundary of suspension is on the top of the bed load layer. Therefore, there is a complex boundary at the interface of suspension and bed load layer. Since the sediment particles are held in suspension due to turbulence, the effect of which on the particles may be assumed to be analogue to the diffusion-dispersion process. The diffusion-dispersion model reveals a considerable effect on the suspended sediment concentration, though such a model does not account adequately for all influences. The idea of diffusion-dispersion model of suspension has opened up a new research avenue from the pioneering work of Schmidt (1925) and later Prandtl (1926). The vertical and longitudinal distributions of suspended sediment from the diffusion-dispersion model were developed and found in reasonably good agreement with the observed data from the flume experiments or field measurements.

9.8.1 Theoretical Considerations of Suspension Equation

The suspension equation is based on Fick's law, which states that the mass of a solute, passing a unit area per unit time in a given direction, is proportional to the gradient of solute concentration in that direction, or it defines the flux that moves from high concentration to low concentration at a concentration gradient. Suppose a drop of dye is released in still water, after some time, the dye will be diluted and passed from high to low due to diffusion with a concentration gradient. For one-dimensional diffusion process, it is mathematically written as

$$q = -D_m \frac{\partial C}{\partial x} \tag{9.192}$$

where q is the solute mass flux, C is the mass concentration of diffusing solute, D_m is the proportionality coefficient known as molecular diffusion, assumed constant with respect to concentration C, i.e., $\frac{\partial D_m}{\partial C} = 0$, which truly satisfied in dilute suspension, and the negative sign indicates the solute moves from high to low. If a control volume is considered with a mass $C(x,t)$ per unit volume at a point x at a time t, then the equation of conservation of mass in one-dimensional form can be written as:

$$\frac{\partial C}{\partial t} = -\frac{\partial q}{\partial x} = D_m \frac{\partial^2 C}{\partial x^2} \tag{9.193}$$

This equation is known as diffusion equation and describes how the mass is transferred by Fickian diffusion process. The solution of this equation is a simple normal distribution as:

$$C(x,t)=\frac{M}{\sqrt{4\pi D_m t}}e^{-\frac{x^2}{4D_m t}} \tag{9.194}$$

which describes the spreading by diffusion of an initial solute of mass M introduced at time $t=0$ at plane $x=0$. Here, M is the total mass in the system formed by integrating concentration C along the whole $x-$ axis.

Based on the principle of conservation of mass, equilibrium exists between the rate of change of a property in the control volume and the rate at which the property leaves the control volume, i.e., the rate of change of property due to diffusion. In three dimensions, it can be formulated as

$$\frac{\partial C}{\partial t}=D_m\nabla^2 C \tag{9.195}$$

with $\nabla^2=\frac{\partial^2}{\partial x^2}+\frac{\partial^2}{\partial y^2}+\frac{\partial^2}{\partial z^2}$ is known as Laplace operator. Equation (9.195) is known as three-dimensional diffusion equation in a quiescent medium, similar to the heat conduction equation.

Suppose the influence of an external force exists such that an arbitrary flow of fluid with a velocity vector $\boldsymbol{u}$ is given. The left-hand side of equation (9.195) will be balanced if the rate of change of a mass due to diffusion and convection is considered as

$$\frac{\partial C}{\partial t}=D_m\nabla^2 C-\nabla(C\boldsymbol{u}) \tag{9.196}$$

For incompressible flow, equation (9.196) reduces to

$$\frac{\partial C}{\partial t}=-\boldsymbol{u}\nabla C+D_m\nabla^2 C \quad \because \nabla\boldsymbol{u}=0 \tag{9.197}$$

Equation (9.197) is the standard form of three-dimensional Eulerian diffusion equation in a convective flow field. In Cartesian form, the three-dimensional diffusion equation is

$$\underbrace{\frac{\partial C}{\partial t}}_{\text{I}}+\underbrace{\left(u\frac{\partial C}{\partial x}+v\frac{\partial C}{\partial y}+w\frac{\partial C}{\partial z}\right)}_{\text{II}}=\underbrace{D_m\left(\frac{\partial^2 C}{\partial x^2}+\frac{\partial^2 C}{\partial y^2}+\frac{\partial^2 C}{\partial z^2}\right)}_{\text{III}} \tag{9.198}$$

Since D_m is the molecular diffusion coefficient, equation (9.198) represents a laminar diffusion equation. The term I represents a local change of concentration with time t, the term II within (--) represents the convection of solute, and the term III on the right-hand side represents diffusion with constant molecular diffusivity.

For dispersion in turbulent flow, similar to Reynolds decomposition, the instantaneous concentration and the velocity vector are given by $C=\bar{C}+C'$, $\boldsymbol{u}=\bar{\boldsymbol{u}}+\boldsymbol{u}'$, where $\bar{\boldsymbol{u}}$ and $\bar{C}$ are the time-mean values of velocity vector and concentration at a given position, and C' and $\boldsymbol{u}'$ represent their fluctuations, then equation (9.198) reduces to the turbulent diffusion equation as:

$$\frac{\partial(\bar{C}+C')}{\partial t}+(\bar{u}+u')\frac{\partial(\bar{C}+C')}{\partial x}+(\bar{v}+v')\frac{\partial(\bar{C}+C')}{\partial y}+(\bar{w}+w')\frac{\partial(\bar{C}+C')}{\partial z}$$
$$= D_m\left[\frac{\partial^2(\bar{C}+C')}{\partial x^2}+\frac{\partial^2(\bar{C}+C')}{\partial y^2}+\frac{\partial^2(\bar{C}+C')}{\partial z^2}\right] \tag{9.199}$$

Using similar operators (as explained in Chapter 5), the time-averaging of velocity and concentration, one gets

$$\overline{Cu}=\overline{(\bar{C}+C')(\bar{u}+u')}=\bar{C}\bar{u}+\overline{C'u'} \because \overline{u'}=\overline{C'}=0,\ \overline{\bar{C}u'}=\overline{C'\bar{u}}=0,\ \overline{\bar{C}\bar{u}}=\bar{C}\bar{u} \tag{9.200a}$$

$$\text{Similarly, } \overline{Cv}=\bar{C}\bar{v}+\overline{C'v'},\text{ and } \overline{Cw}=\bar{C}\bar{w}+\overline{C'w'} \tag{9.200b, c}$$

Here the term $\bar{C}\bar{u}$ is known as convective flux and the term $\overline{C'u'}$ is known as diffusive flux due to fluctuations, which are completely nonlinear. Using the averaging techniques using equation (9.200) in equation (9.199) and continuity equation, one gets after simplifications as:

$$\frac{\partial \bar{C}}{\partial t}+\bar{u}\frac{\partial \bar{C}}{\partial x}+\bar{v}\frac{\partial \bar{C}}{\partial y}+\bar{w}\frac{\partial \bar{C}}{\partial z}+\frac{\partial}{\partial x}\left(\overline{C'u'}\right)+\frac{\partial}{\partial y}\left(\overline{C'v'}\right)+\frac{\partial}{\partial z}\left(\overline{C'w'}\right)=D_m\left[\frac{\partial^2 C}{\partial x^2}+\frac{\partial^2 \bar{C}}{\partial y^2}+\frac{\partial^2 \bar{C}}{\partial z^2}\right] \tag{9.201}$$

Elder (1959) found a convenient way to define a coefficient of turbulent diffusion in a closed form solution from the nonlinear terms, similar to Prandtl mixing length theory, such that

$$\epsilon_{sx}\frac{\partial \bar{C}}{\partial x}=-\overline{C'u'},\ \epsilon_{sy}\frac{\partial \bar{C}}{\partial y}=-\overline{C'v'},\text{ and } \epsilon_{sz}\frac{\partial \bar{C}}{\partial z}=-\overline{C'w'} \tag{9.202}$$

where ϵ_{sx}, ϵ_{sy} and ϵ_{sz} are respectively the sediment/mass diffusion coefficients with respect to x, y, and z. Using (9.202) in equation (9.201), and under the assumption that molecular diffusion D_m and mass diffusion coefficients are independent and thus additive, equation (9.201) reduces to

$$\frac{\partial \bar{C}}{\partial t}+\bar{u}\frac{\partial \bar{C}}{\partial x}+\bar{v}\frac{\partial \bar{C}}{\partial y}+\bar{w}\frac{\partial \bar{C}}{\partial z}=\frac{\partial}{\partial x}\left[(\epsilon_{sx}+D_m)\frac{\partial \bar{C}}{\partial x}\right]+\frac{\partial}{\partial y}\left[(\epsilon_{sy}+D_m)\frac{\partial \bar{C}}{\partial y}\right]+\frac{\partial}{\partial z}\left[(\epsilon_{sz}+D_m)\frac{\partial \bar{C}}{\partial z}\right] \tag{9.203}$$

where the eddy diffusivities ϵ_x, ϵ_y and ϵ_z are given as follows:

$$\epsilon_x=\epsilon_{sx}+D_m,\ \epsilon_y=\epsilon_{sy}+D_m,\ \epsilon_z=\epsilon_{sz}+D_m \tag{9.204}$$

In open channel turbulent flow, the turbulent diffusivities are usually considerably larger than the molecular diffusion D_m, using the concept of Reynolds analogy. In this case, the Reynolds analogy is valid if the both sediment/mass and momentum transfers are identical, i.e., in the momentum flux, the kinematic viscosity υ_f is less than the momentum diffusivity or eddy viscosity ϵ_m; and in the sediment/mass flux, the molecular diffusivity D_m is less than the sediment/mass diffusivity or eddy diffusivity ϵ_s.

In laminar flow, the molecular diffusion D_m is dominant, whereas in turbulent flow the momentum or sediment/mass diffusivities are greater than molecular diffusion, i.e., $\epsilon_{m,sx,y,z} \gg D_m$, and hence the molecular diffusion D_m may be neglected. Using this idea, one gets equation (9.203) as follows:

$$\frac{\partial \bar{C}}{\partial t}+\bar{u}\frac{\partial \bar{C}}{\partial x}+\bar{v}\frac{\partial \bar{C}}{\partial y}+\bar{w}\frac{\partial \bar{C}}{\partial z}=\frac{\partial}{\partial x}\left(\epsilon_{sx}\frac{\partial \bar{C}}{\partial x}\right)+\frac{\partial}{\partial y}\left(\epsilon_{sy}\frac{\partial \bar{C}}{\partial y}\right)+\frac{\partial}{\partial z}\left(\epsilon_{sz}\frac{\partial \bar{C}}{\partial z}\right) \tag{9.205}$$

Equation (9.205) is known as the turbulent advection-diffusion equation, which is known as the conservation of mass equation. Here $\bar{u}$, $\bar{v}$, $\bar{w}$ are the mean velocity components or convective velocities tending to move the suspended matter in the respective directions, ϵ_{sx}, ϵ_{sy} and ϵ_{sz} are the corresponding eddy diffusivities in the x, y, z direction respectively. Equation (9.205) is a most general one, applying equally well to solid particles or dispersant in solution form. However, for sediment suspension, the velocity must be considered how the solid particles are moved with corresponding diffusivity of solid particles. In equation (9.205), the over bars denoting the time-averages are no longer needed and therefore dropped. Equation (9.205) may be written as:

$$\frac{\partial C}{\partial t}+u\frac{\partial C}{\partial x}+v\frac{\partial C}{\partial y}+w\frac{\partial C}{\partial z}=\frac{\partial}{\partial x}\left(\epsilon_{sx}\frac{\partial C}{\partial x}\right)+\frac{\partial}{\partial y}\left(\epsilon_{sy}\frac{\partial C}{\partial y}\right)+\frac{\partial}{\partial z}\left(\epsilon_{sz}\frac{\partial C}{\partial z}\right) \tag{9.206}$$

In order to proper understanding of the parameters- momentum diffusivity ϵ_m and the sediment/mass diffusivity ϵ_s, Kalinske and Pien (1943) performed an experiment to study the effects of these two parameters on a mixture of hydrochloric acid and alcohol injected in flowing water. The density of the mixture was that of the flow water. They predicted the concentration distribution using statistical analysis, and compared the results with the measured data, and found good agreement within the expected limit. So, they claimed that there is an indirect proof of mass diffusivity ϵ_s identical to the momentum diffusivity ϵ_m. Jobson and Sayre (1970) also arrived at the same conclusion that the ratio between the diffusivities is approximately unity, i.e., $\epsilon_s \approx \epsilon_m$ in a dilute solution or passive motion of particles with the flow. Nevertheless, the recent development shows that there is a more general relationship, that is

$$\epsilon_s = \beta\epsilon_m \tag{9.207}$$

where β is the proportionality factor which depends on different parameters.

In a steady state condition of concentration $\frac{\partial C}{\partial t}=0$, and no variation in concentration with respect to longitudinal direction x and transverse direction z, that means, $\frac{\partial C}{\partial x}=\frac{\partial C}{\partial z}=0$, and the sediment/mass diffusivity ϵ_{sy} is constant with constant vertical velocity v equated with the settling velocity v_0 of solid particle, then equation (9.206) can be written as:

$$-\frac{\partial v_0 C}{\partial y}=\frac{\partial}{\partial y}\left(\epsilon_{sy}\frac{\partial C}{\partial y}\right) \tag{9.208}$$

$$\Rightarrow 0 = v_0 C + \epsilon_{sy}\frac{\partial C}{\partial y} \tag{9.209}$$

Equation (9.209) represents the state of equilibrium between the vertical rate of sediment motion due to turbulent diffusion and the downward flux of sediment transfer per unit area due to gravity. This equation was first derived by Schmidt (1925) to determine the vertical concentration distribution of fine sand particles in the atmosphere. After integration with respect to y, equation (9.209) reduces to:

$$C = C_a \exp\left[-\frac{v_0(y-a)}{\epsilon_{sy}}\right] \tag{9.210}$$

where C_a is the reference concentration at the level a from the bed. Equation (9.210) shows an exponential curve with the vertical coordinate y for constant values of C_a, ϵ_{sy} and settling velocity v_0, which is a hypothetical phenomenon. The dimensionless concentration C/C_a is a function of sediment/mass diffusivity ϵ_{sy} and the settling velocity v_0. This equation was verified by Hurst (1929) and later by Rouse (1937) from their experimental data. Lane and Kalinske (1941) claimed that sediment/mass diffusivity ϵ_{sy} should not be constant; it should vary with depth h. Considering the sediment diffusivity $\epsilon_{sy} = \kappa_o u_* y(h-y)/h$, and averaging ϵ_{sy} from 0 to h, he obtained $\overline{\varepsilon_{sy}} = u_* h/15$. Lane and Kalinske (1941) suggested equation (9.210) to reduce as:

$$C = C_a \exp\left[-\frac{15 v_0 (y-a)}{u_* h}\right] \tag{9.210a}$$

If the uniform sediments are kept in suspension by turbulent fluctuations, the classical approach, i.e., the Prandtl's mixing length theory, is used to calculate the vertical suspension concentration distribution, similar to steady diffusion equation (9.209). The sediment particles with settling velocity v_0 in the flow are transported from the lower level, where the suspended sediment is $C + l\dfrac{dC}{dy}$, to higher level with suspended sediment $C - l\dfrac{dC}{dy}$ (Figure 9.29), where l is the distance travelled by the particles. The volumes of sediment travelling upward q_u and downward q_d in a unit time through a unit area on a horizontal line $A'A'$ in between the levels are respectively given by:

$$q_u = (v' - v_0)\left(C - l\frac{dC}{dy}\right) \tag{9.211a}$$

$$q_d = (v' + v_0)\left(C + l\frac{dC}{dy}\right) \tag{9.211b}$$

In the steady flow situation, the upward q_u and downward q_d must be balanced to each other, which give as:

$$C v_0 + v' l \frac{dC}{dy} = 0 \tag{9.212}$$

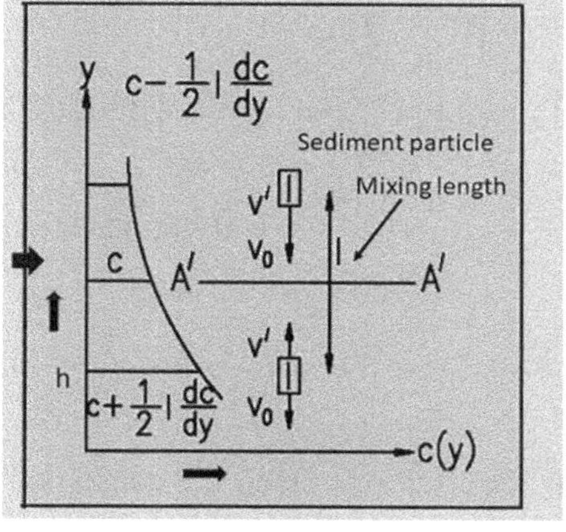

FIGURE 9.29 Sketch of suspended sediment in steady turbulent flow.

Comparing equations (9.209) and (9.212) and using the concept of mixing length theory, the coefficients of $\frac{dC}{dy}$ can be written as $\epsilon_{sy} = v'l = \epsilon_m$, where ϵ_{sy} is the sediment/mass diffusivity and ϵ_m is the momentum viscosity or eddy viscosity. Therefore, equation (9.212) is the same as equation (9.209), if $\epsilon_{sy} = v'l = \epsilon_m$, but in equation (9.207), a factor β is introduced for discrete suspended sediment particles. The β-factor mostly depends on the sediment size, which is ≈ 1 for fine sediment particles in suspension, $\beta < 1$ for coarse sediment particles, and $\beta > 1$ for sand-gravel mixture in suspension over rough beds. The values of β are still under observation. It is observed that the estimated value of β was found 1.8 for moderate velocity 95 cm/s which pick up the fine sediment in suspension and the value β was 2.5 for higher velocity, which pick up coarser sediment in suspension (Mazumder and Ghoshal 2006). The value of β depends on velocity, suspension concentration and the bed conditions, still under experiments.

In unsteady state condition of concentration $\frac{\partial C}{\partial t} \neq 0$ under non-equilibrium in two-dimensional case, the turbulence distribution is uniform and $\frac{\partial^2 C}{\partial x^2} \ll \frac{\partial^2 C}{\partial y^2}$ with constant eddy diffusivity ϵ_{sy}, then equation (9.206) reduces to

$$\frac{\partial C}{\partial t} + u\frac{\partial C}{\partial x} - v_0\frac{\partial C}{\partial y} = \epsilon_{sy}\frac{\partial^2 C}{\partial y^2} \tag{9.213}$$

where v_0 is the settling velocity of the particle. If $u = 0$, equation (9.213) reduces to

$$\frac{\partial C}{\partial t} - v_0\frac{\partial C}{\partial y} = \epsilon_{sy}\frac{\partial^2 C}{\partial y^2} \tag{9.214}$$

and again, for steady state, equation (9.213) reduces to the equation studied by Kalinske (1947) as:

$$u\frac{\partial C}{\partial x} - v_0\frac{\partial C}{\partial y} = \epsilon_{sy}\frac{\partial^2 C}{\partial y^2} \tag{9.215}$$

with the following boundary conditions as:

$$C = 0 \text{ at } x = 0, C = 0 \text{ at } y \to \infty, \text{and } C = C_a \text{ at } y = a = 0 \text{ at the bottom.} \tag{9.216}$$

where C_a is the concentration at the level a near the boundary. The solution of equation (9.215) subject to the boundary conditions (9.216) is given in Graf (1971). Dobbins (1943) solved equation (9.214) subject to the initial and boundary conditions as:

$$C = f(y) \text{ at } t = 0 \text{ for } 0 < y < h \text{ within the region} \tag{9.217a}$$

$$\epsilon_{sy}\frac{\partial C}{\partial y} + v_0 C = 0 \text{ at } y = h \text{ at the water surface} \tag{9.217b}$$

$$\epsilon_{sy}\frac{\partial C}{\partial y} + v_0 C = 0 \text{ at } y = a = 0 \text{ at the bottom.} \tag{9.217c}$$

Here the initial condition (9.217a) represents a given concentration at the vertical depth; the condition (9.217b) represents no net transfer of sediment particles through the water surface, and condition (9.217c) represents the rate of pick-up equals to rate of deposit at the bottom. The solution of equation

(9.214) that subjects the initial and boundary conditions (9.217) with explanations of particular cases is available in Graf (1971).

9.8.2 Velocity and Concentration Distributions in Sediment Mixed Fluid

Studies of mechanics of sediment transport are of fundamental importance to the aspects of many disciplines: geology, hydrology, oceanography, civil and soil engineering, physical geography etc. It is also important for the solution of the practical problems in the field of river sedimentation, erosion and scouring, ancient fluvial deposits, and the field of two-phase flow in particular. Numerous investigations related to sediment-laden turbulent flows in open channels have been undertaken to examine the vertical velocity and sediment concentration profiles. The problems of sediment-laden flows are of direct concern to river engineers and geomorphologists, but also relevant to the field such as coastal sediment transport and transport of solids in pipelines. Several investigations have been reported by different researchers such as Rouse (1937), Vanoni (1946), Einstein and Chien (1955), Vanoni and Nomicos (1960), Thacker and Lavelle (1977), Smith and McLean (1977), Ghosh and Mazumder (1981), Karim and Kennedy (1990), Woo et al. (1988), Ni and Wang (1991), Umeyama and Gerristen (1992), Sengupta et al. (1999), Mazumder et al. (2005a, b), and others.

9.8.3 Rouse Equation (1937) for Sediment Suspension

Assuming the steady state condition $\dfrac{\partial C}{\partial t} = 0$, and no variation of concentration with either x-direction or z-direction, that is, $\dfrac{\partial C}{\partial x} = \dfrac{\partial C}{\partial z} = 0$, then equation (9.206) can be written as

$$v \frac{\partial C}{\partial y} = \frac{\partial}{\partial y}\left(\epsilon_{sy} \frac{\partial C}{\partial y}\right) \tag{9.218}$$

Considering the vertical fluid velocity to be equal to the fall velocity of sediment particle, and replacing ϵ_{sy} by ϵ_s, then equation (9.218) under the equilibrium condition becomes

$$v_0 C + \epsilon_s \frac{\partial C}{\partial y} = 0 \tag{9.219}$$

where ϵ_s is the sediment/mass diffusion coefficient, v_0 is the settling velocity of the particles in still water. Equation (9.219) represents the net upward sediment transport due to diffusion $\epsilon_s \dfrac{\partial C}{\partial y}$ is equal to the net downward sediment flux due to settling $v_0 C$. Equation (9.219) was used by Schmidt (1925) to determine the suspension concentration by the turbulent mixing, and this idea was enhanced by O'Brien (1933). In two-dimensional channel flow, the relative shear stress τ is given by

$$\tau = \tau_0 \left(1 - \frac{y}{h}\right) \tag{9.220}$$

and the logarithmic velocity distribution law is

$$\frac{d\bar{u}}{dy} = \frac{u_*}{\kappa_o y} \tag{9.221}$$

From Reynolds analogy, the shear stress τ_t can be expressed as

$$\epsilon_m = \frac{\tau_t}{\rho_f \dfrac{d\bar{u}}{dy}} \tag{9.222}$$

where ϵ_m is the momentum diffusion coefficient for water. Some researchers have assumed that $\varepsilon_s \approx \varepsilon_m$, while some assumed a relative of the form $\epsilon_s = \beta\epsilon_m$, where β is a constant determined for the experimental data. Combining equations (9.220) to (9.222), the momentum diffusion coefficient ϵ_m can be written as:

$$\epsilon_m = \kappa_o u_* y\left(1 - \frac{y}{h}\right) \tag{9.223}$$

Assuming $\beta = 1$, integrating equation (9.219) and replacing ϵ_s by ϵ_m of equation (9.223), equation (9.219) can be solved as

$$\frac{C}{C_{\xi_a}} = \left(\frac{1-\xi}{\xi}\cdot\frac{\xi_a}{1-\xi_a}\right)^{z_*} \tag{9.224}$$

where $z_* = \dfrac{v_0}{\kappa_o u_*}$ is the Rouse number, $\xi = \dfrac{y}{h}$, and C_{ξ_a} is the concentration of sediment at a reference level $\xi = \xi_a$, with $\xi_a\left(= \dfrac{y_a}{h}\right)$ is a height near the bottom boundary. Equation (9.224) is the suspended load equation introduced by Rouse (1937), popularly known as Rouse Equation for suspension concentration. Equation is used for the calculation of the concentration of a given size with settling velocity v_0 at a dimensionless distance ξ from the bed, if a reference concentration C_{ξ_a} at a reference level ξ_a is known. Equation (9.224) can be written as:

$$\frac{C}{C_{\xi_a}} = \left[\frac{\xi_a}{\xi} - \left(\frac{\xi-\xi_a}{1-\xi_a}\right)\frac{\xi_a}{\xi}\right]^{z_*} = \left[(1-\xi_1)\frac{\xi_a}{\xi}\right]^{z_*} \tag{9.225}$$

where $\xi_1 = \dfrac{\xi-\xi_a}{1-\xi_a}$ is dimensionless variable. Integration of equation (9.225) over the region, where suspended load occurs, say from ξ_a to $\xi = \dfrac{y}{h} = 1$, is obtained as

$$g_s = \int_{\xi_a}^{\xi=1} C\bar{u}\,dy \tag{9.226}$$

where C and $\bar{u}$ are functions of $\xi$and g_s is the suspended load in weight per unit time and width. This equation is limited because each can predict only the relative concentration.

Considering the various values of Rouse number $z_*\left(= \dfrac{v_0}{\kappa_o u_*}\right)$, the dimensionless concentration $\dfrac{C}{C_{\xi_a}}$ from equation (9.225) is plotted against $\xi_1\left(= \dfrac{\xi-\xi_a}{1-\xi_a}\right)$, using the reference concentration $C_{\xi_a} = 0.05$ in Figure 9.30 and found a reasonable agreement with experimental data. Equation (9.225) agrees very well with Vanoni's (1941) measured concentration distribution for $\kappa_o \approx 0.4$. The concentration $\dfrac{C}{C_{\xi_a}}$ from

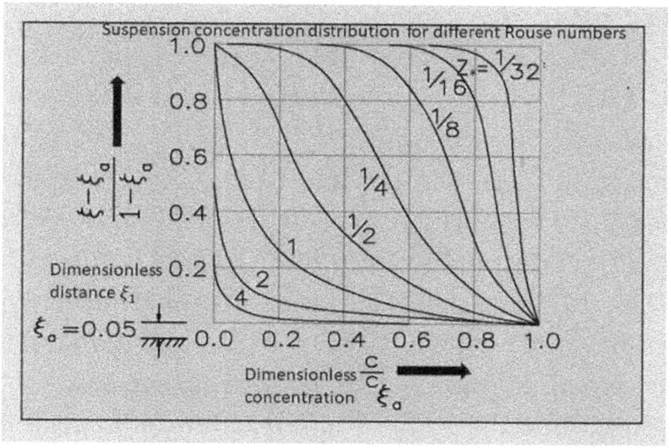

FIGURE 9.30 Suspension concentration distributions for different values of Rouse numbers, using the reference concentration $c_{\xi_a} = 0.05$. (Modified from Rouse 1937, Yalin 1977.)

equation (9.225) is also plotted against ξ for constant values of Rouse number $z_* = 1$ and 2, it observed that the plots show a good agreement with the observed data (Figure 9.31). It is important to note that the concentration C decreases with an increase in vertical distance ξ from the reference level ξ_a near the bed. From the mathematical point of view, the concentration becomes infinity at the bed $\xi = 0$, which is inaccessible. Suspension does not exist too close to the bed, but at the bed particles rather represent part of bed load. Einstein et al. (1940) suggested that the suspension is not possible in the bed layer, which has a thickness of two grain diameters. In fact, at the free surface ($\xi = 1$), the concentration C should be zero, and at the reference level $\xi = \xi_a, C = C_{\xi_a} = 0.05$, as already mentioned above. However, Willis and Coleman (1969) suggested another model for vertically suspended concentration distribution in the form of error function. He also showed the comparative study of his model with the laboratory and field data.

9.8.3.1 Applications

The Rouse equation (9.225) is used to carefully verify the laboratory experimental data on the concentration of suspended material conducted in closed-circuit rectangular channel by Vanoni (1946), indicating the perfect agreement with the observed data. The same equation was used by Zeller (1969) to study the distribution of clay suspension. To check the theoretical equation (9.225) with the observed

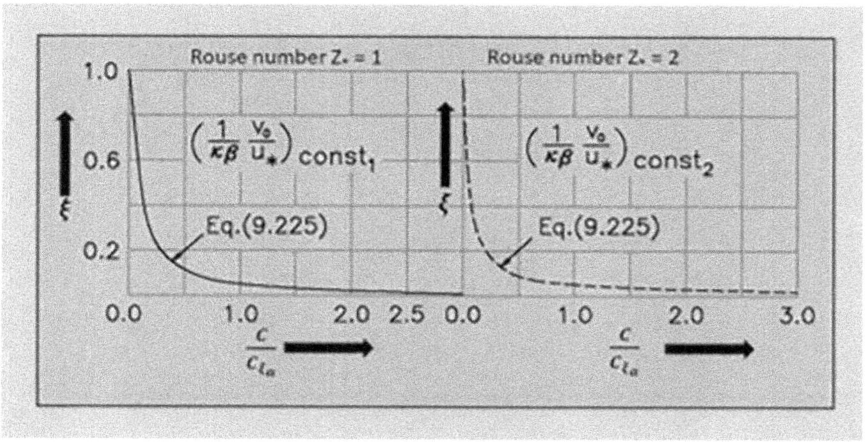

FIGURE 9.31 Plots of concentration c/c_{ξ_a} from equation (9.225) against ξ for constant values of Rouse number $z_* = 1$ and 2. (Modified from Yalin 1977.)

sediment concentration in suspension, Christiansen (1935) analyzed the field sediment suspension data from Colorado River, Imperial canals and River Niles, and in general found good agreement. Straub (1935) also has checked the data from the Missouri River, and Anderson (1942) used the suspended concentration data from Enoree River from a fairly straight section. Further, equation (9.225) was checked to study the concentration distribution from different rivers by Lane and Kalinske (1941) for Mississippi River, by Vanoni (1953) and Harrison (1963) for Missouri River, by Colby and Hembree (1955) for Niobrara River and by Nordin (1963) for Rio Grande; and was found that the equation is of correct form. Sengupta (1975a, 1979) studied the grain-size distributions in suspension above a sand bed of bimodal size distribution ranging from 0.04 to 2.00 mm in a laboratory flume at water velocities from 42 to 160 cm/s. He explained the relative suspension concentration using equation (9.225) to relate the concentration of a suspended particle of a particular diameter to the flow velocity and height of suspension above the bed. Figure 9.32 shows the suspended sediment distribution for two grain sizes (2.50 and 4.75 phi) at three different velocities (43, 104 and 159 cm/s) corresponding to three different experiments. There was only slight change in the relative concentration of coarse material. He obtained log-normality of "weight frequency" distribution of grain sizes was recognized as a transitional feature, occurring through size sorting within a critical range of velocity and height above a sand bed of a given composition (Sengupta 1975). Grain-size frequency distributions of suspended loads at different flow velocities and over sand beds of four different grain-size patterns were studied by Sengupta (1979) in a laboratory flume. Field investigations in an Indian river deposit, followed by flume experiments in laboratory have shown that log-normality of grain-size distribution can occur through a process of size sorting when sediments are taken into suspension from a sand bed of non-lognormal grain-size distribution (Sengupta 1975b).

Although the equation was found to apply the laboratory and field data, the exponent $z_*\left(=\dfrac{v_0}{\kappa_o u_*}\right)$ needed to adjust frequently. Here, the value of z_* is considered using $\epsilon_s \approx \epsilon_m$. If we consider equation (9.207), i.e., $\epsilon_s = \beta\epsilon_m$ in equation (9.225), then we can think about role of β-factor, that means, $z_*\left(=\dfrac{\omega_0}{\kappa_o \beta u_*}\right)$ in equation (9.225). Considering the role of β-factor that the mass diffusivity and momentum diffusivity

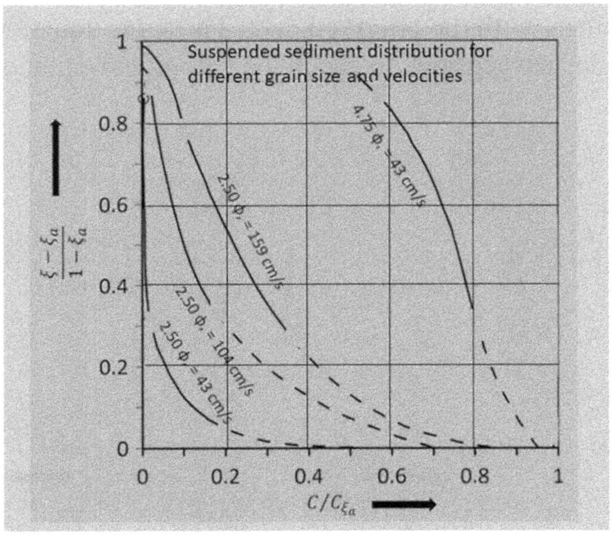

FIGURE 9.32 Plots of suspended sediment distribution in the laboratory flume for two grain sizes at different velocities (2.5 ∅: 43, 104 and 159 cm/s; 4.75 ∅: 43cm/s) corresponding to the three sets of experiments. (Modified from Sengupta 1975a, 1979.)

may not be equal, equation (9.225) should be examined properly for different sediment sizes in suspension, which will provide the β value. Carstens (1952) presented a mathematical expression for β value and concluded that β never exceeds unit but $\beta \le 1$. He showed that for fine particles in suspension, $\beta \approx 1$ or $\epsilon_s \approx \epsilon_m$, and for coarse particles in suspension, $\beta < 1$ or $\epsilon_s < \epsilon_m$. For further discussions about the changes of β-value, readers are suggested to follow the work of Householder et al. (1969) and Jobson and Sayre (1970a, b). For the value of von-Karman universal constant κ_o in open-channel flow without sediment, the value of κ_o in logarithmic velocity law is 0.4. In the presence of sediment concentration in suspension, the values of κ_o were obtained to vary with a tendency to decrease with increase in suspended sediment, verified by Vanoni (1941, 1946) from flume experiments, and by Einstein and Chien (1954) from the field data. A decrease in κ_o-value leads to less effective in mixing, that means, the presence of suspended sediment dampens the turbulence. Vanoni (1963) and Chien (1954) presented a theory based on the rate of frictional energy offered to support the suspended sediment per unit weight of fluid and unit time and made a reasonable relationship with the parameter κ_o-value using the data from the laboratory and the field studies. Barton and Lin (1955) discussed the variation of κ_o from the view of density gradient. Elata and Ippen (1961) obtained interesting results from their experiments that though the κ_o-value decreased with increase in concentration in suspension, the measurements of turbulence showed an increase in intensity with increase in concentration. Hino (1963) developed a general formula for sediment-laden flow, based on the energy equation and the acceleration balance equation of turbulent motion to describe the turbulence characteristics such as the von-Karman constant κ_o, the turbulent intensity, and the diffusion coefficient. The theory and existing experimental data were in good agreement. The theory predicted that the decrease in damping effect or turbulent intensity due to suspended particles was very small, which was contrary to the common accepted belief.

Itakura and Kishi (1980) reported a model of velocity distribution with suspended sediment based on the Monin-Obukhov length, which explicitly depended on von-Karman constant κ_o for the prediction of velocity and concentration profiles, in contrast with the model of Hino (1963) and Willis (1978). Coleman (1981) reported that there is a contradictory nature of increasing or decreasing in von-Karman constant κ_o due to suspended sediment in open channel flow. According to Coleman, the analysis of velocity profiles of sediment-laden flow from the existing experimental data suggests that the von-Karman constant κ_o will be independent of the amount of suspended sediment in an open channel flow, if the wake strength coefficient in flow region is considered, which will be a function of Richardson number. Coleman (1986) conducted flume experiments to study the effect of variation of suspended sediment concentration on velocity profiles, keeping flow depth, slope, and flow discharge essentially constant. He observed from his experiment that the suspended sediment concentration decelerates the flow velocity in the high concentration region near the bed, while to maintain the conservation of mass, there is a compensating acceleration in the upper part of the flow (Figure 9.33).

Coleman (1986) used the sediment-water mixture as a single fluid, where he assumed the kinematic mixture viscosity ν_f and mixture density ρ_f are the function of concentration C, and is given by

$$\nu_f = \frac{\mu_f\left(1+2.5C+6.25C^2+15.62C^3\right)}{\rho_f+\left(\rho_s-\rho_f\right)C} \tag{9.227}$$

If velocities are scaled by the shear velocity u_* and lengths are scaled by ν_f/u_*, an equation of velocity profile over the entire two-dimensional boundary layer flow above the viscous sub-layer is given by:

$$\frac{\bar{u}}{u_*} = \frac{1}{\kappa_o}\ln\frac{u_* y}{\nu_y} + A - \left(\frac{\Delta\bar{u}}{u_*}\right)_R - \left\{\frac{\Delta\bar{u}}{u_*}\right\}_S + \frac{\Pi}{\kappa_o}\omega\left[\frac{u_* y}{\nu_y}\cdot\frac{\nu_\delta}{u_*\delta}\right] \tag{9.228}$$

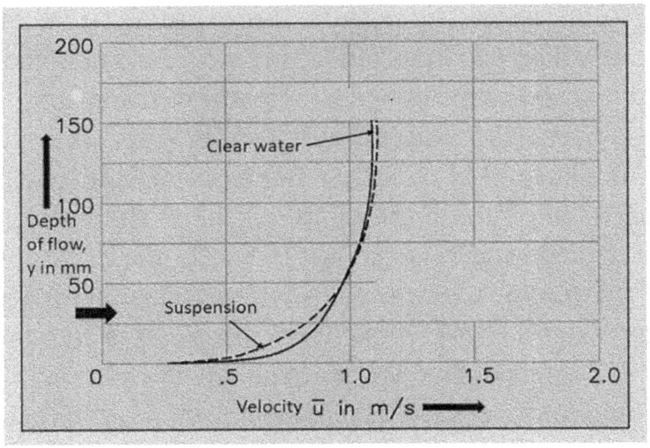

FIGURE 9.33 Profiles of clear water velocity and capacity suspension from the experimental series with sand size 0.105 mm. (Modified from Coleman 1986.)

In equation (9.228), the term A is the integrating constant, the term in the first bracket indicates the velocity reduction factor due to boundary roughness, the term within the second bracket indicates velocity reduction due to suspended sediment, and the last term within the third bracket indicates wake strength function of Cole (1956), which shows the deviation of velocity profile from the near-bed logarithmic profile as the vertical coordinate y increases. Some important observations are noticed due to turbulent flow over the loose sediment beds. Almost all the turbulence parameters such as velocity, von-Karman constant, turbulent intensity, life of eddies, diffusion coefficient, etc. in clear-water flow without sediment are changed due to the complexity of the two-phase flow arising from the fluid-sediment mixture flow above the beds. Chien (1954) pointed out that the logarithmic velocity equation for clear-water fully developed turbulent flow was not adequate in sediment-laden flow due to suspension concentration. The idea was considered by Einstein and Chien (1955) and a new equation was enhanced for the development. Vanoni (1941, 1946) showed from their experiments that the mean velocity for sediment-laden flow was larger than that of clear-water turbulent flow for the same discharge, indicating the apparent reduction roughness coefficient. The increase in velocity does not seem to be an increase in suspended load but does depend mostly on the decrease in depth shown in Figure 9.34 (Vanoni and Nomicos 1960, and Graf 1971).

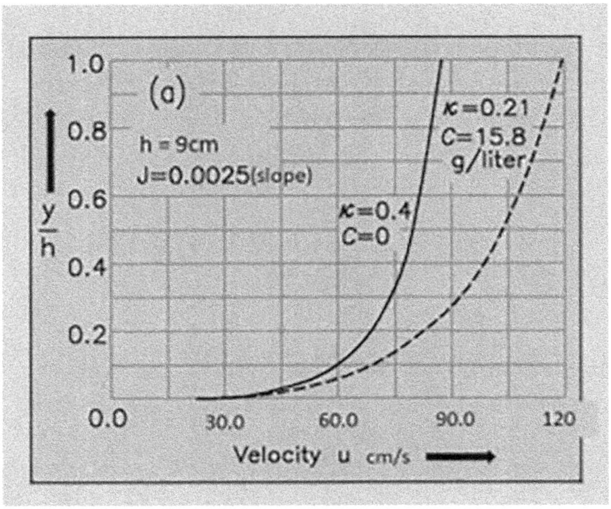

FIGURE 9.34 Velocity profiles for clear water and sediment laden flow. (Modified from Graf 1971.)

Sediment motion over the loose bed is responsible for the development of various bed forms such as ripples, dunes, anti-dunes etc. The change of friction factor due to the bed forms is reported to be much larger than that due to suspended load. They also suggested that the friction factor due to suspension may increase, decrease or no change, depending on the magnitude of these factors. Einstein and Chien (1955) determined the velocity distribution equation in the sediment-laden flow. They considered the sediment-laden flow depth in two regions: one near-bed region where sediments are heavily concentrated and other away from the bed, where the sediment concentrations are relatively small. Here the turbulence is generated at the boundary, where the heavy sediments exist, and it reduces the turbulence level because of spending of energy to suspend sediment particles. When the turbulent flow carries sediments in suspension over the loose sediment bed, the net fluid density ρ_f is related to the volume concentration of suspended sediment C, and is given by $\rho = \rho_f + (\rho_s - \rho_f)C$. In fully developed sediment-laden turbulent flow, the effect of the dynamic coupling between sediment suspension and the turbulent flow is important. Consequently, the turbulent shear stress is modified as:

$$\tau_t = \left(\rho_f + (\rho_s - \rho_f)C\right)\epsilon_t \frac{d\bar{u}}{dy} \tag{9.229}$$

If the concentration is very low, then equation (9.229) reduces to the shear stress due to clear-water flow. Equation (9.229) can be approximated by the bottom shear stress τ_0 as:

$$\tau_0 = \int_0^h \left(\rho_f + (\rho_s - \rho_f)C\right) gJ\, dy \tag{9.230}$$

where g is the acceleration due to gravity, and J is the energy gradient. The modified velocity distribution due to sediment-laden flow is given by

$$\frac{\bar{u}}{u_*} = 5.75 \frac{\sqrt{\rho_f + (\rho_s - \rho_f)\frac{1}{h}\int_0^h C\, dy}}{\sqrt{\rho_f + (\rho_s - \rho_f)C_0}} \log\left(A_c \frac{y}{k_s}\right) \tag{9.231}$$

where C_0 is the sediment concentration at the bed layer, and A_c is the constant to be determined as reference level.

9.8.4 Einstein's Approach (1950)

The Einstein's method is the most advanced one to compute the suspended load, if the velocity and suspension concentration are known. The suspended load g_{ss} in weight per unit time and width is given by:

$$g_{ss} = \int_{y_a}^{h} C\bar{u}\, dy \tag{9.226}$$

where C and $\bar{u}$ are functions of y, and y_a is the lower limit, where the suspended load starts defined by the bed layer thickness. Using the suspension concentration equation (9.224) and the logarithmic velocity distribution in to equation (9.226), one gets

$$g_{ss} = \int_{y_a}^{h} C_{y_a}\left(\frac{h-y}{y}\cdot\frac{y_a}{h-y_a}\right)^{z_*} 5.75 u_*' \log\left(\frac{30.2y}{\delta_c}\right) dy \tag{9.232}$$

where δ_c is the corrected value and u_*' is the shear velocity due to grain only. Putting the reference level $\xi = \xi_a$ with $\xi_a \left(= \dfrac{y_a}{h} \right)$ as a height near the bottom boundary, one gets

$$g_{ss} = \int_{\xi_a}^{1} C_{ya} \left(\frac{\xi_a}{1-\xi_a} \right)^{z*} \left(\frac{1-\xi}{\xi} \right)^{z*} 5.75 u_*' h \log \left(\frac{30.2y}{\dfrac{\delta_c}{h}} \right) dy$$

$$g_{ss} = 5.75 u_*' h C_{ya} \left(\frac{\xi_a}{1-\xi_a} \right)^{z*} \left[\log\left(\frac{30.2h}{\delta_c} \right) \int_{\xi_a}^{1} \left(\frac{1-\xi}{\xi} \right)^{z*} dy + 0.434 \int_{\xi_a}^{1} \left(\frac{1-\xi}{\xi} \right)^{z*} \ln y \, dy \right] \quad (9.233)$$

This equation can be written as:

$$g_{ss} = 11.6 u_*' y_a C_{\xi_a} \left[2.303 \log\left(\frac{30.2h}{\delta_c} \right) I_1 + I_2 \right] \quad (9.234)$$

where the integrals

$$I_1 = 0.216 \frac{\xi_a^{z*-1}}{(1-\xi_a)^{z*}} \int_{\xi_a}^{1} \left(\frac{1-\xi}{\xi} \right)^{z*} dy \quad (9.235a)$$

$$\text{and } I_2 = 0.216 \frac{\xi_a^{z*-1}}{(1-\xi_a)^{z*}} \int_{\xi_a}^{1} \left(\frac{1-\xi}{\xi} \right)^{z*} \ln y \, dy \quad (9.235b)$$

The above integrals I_1 and I_2 are graphically presented in Figure 9.35a,b against the reference level ξ_a. The above equation gives the suspended load rate per unit time and width for a given size fraction location at $\xi = \xi_a$ with reference concentration. It is important to note that at $\xi = 0$ the concentration is infinite, so the concentration equation is not possible to apply the right way at the bed $\xi = 0$. Einstein (1950) assumed the flow layer right way on the top of the bed, where the suspension was not possible, as a bed layer with thickness $2d$ (two times of grain diameter). The materials within the bed layer are the source of the suspended load, and so he determined the lower limit or reference concentration C_{ξ_a}. From the bed load theory of Einstein (1950), if the bed load rate of a given size fraction i_s is $g_s i_s$ with g_s is the bed load in weight per unit time and width, and the velocity of bed load is u_b, the weight of particles of a given size per unit area is $g_s i_s / u_b$. The average concentration is given as:

$$C_{\xi_a} = A_{cc} \frac{g_s i_s}{u_b b_l} \quad (9.236)$$

where A_{cc} is the constant of correction at the bed, $b_l = 2d$ is the bed layer height. Einstein (1950) experimentally determined the above equation as:

$$C_{\xi_a} = 0.086 \frac{g_s i_s}{u_* b_l} \quad (9.237)$$

Now introducing the reference concentration C_{ξ_a} in suspension concentration equation, and assuming $u_*' = u_*$, $y_a = b_l$, one gets

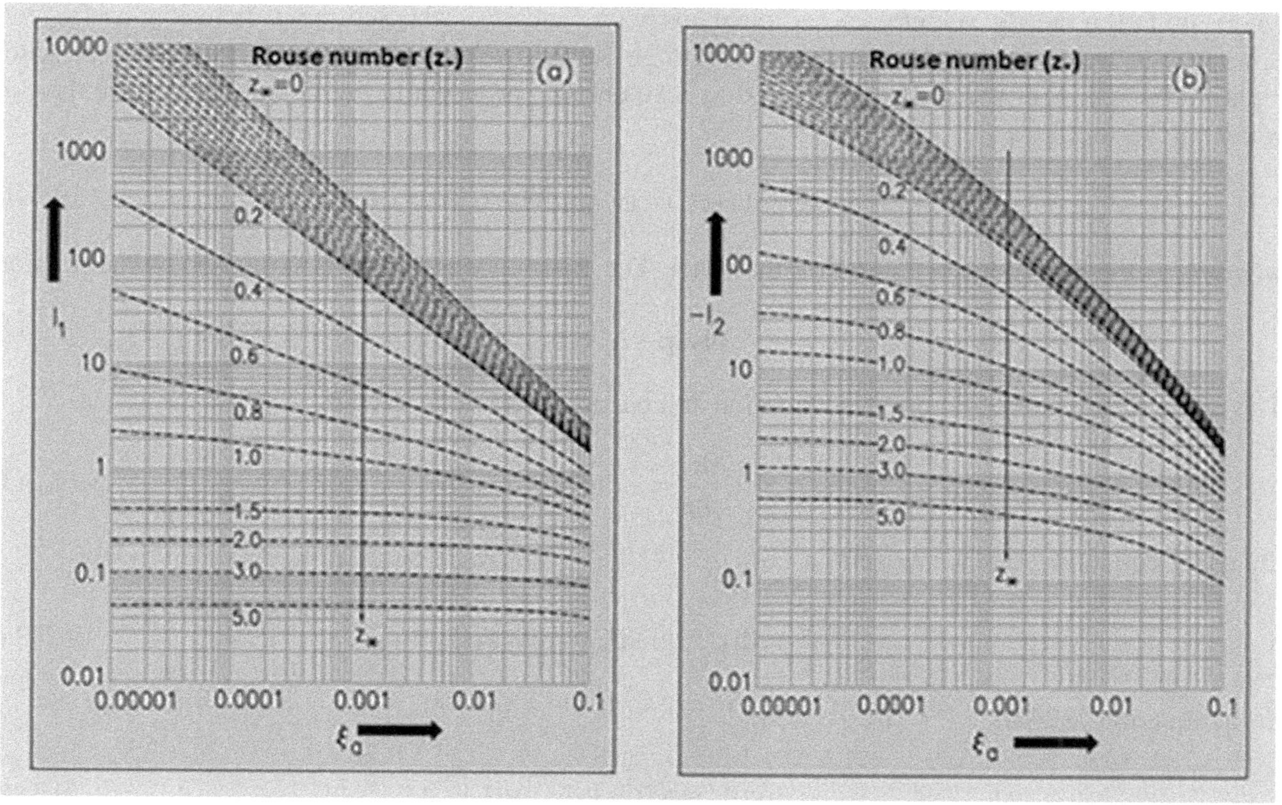

FIGURE 9.35 (a and b) Respective integral equations I_1 and $-I_2$ against ξ_a for different Z_*. (After Einstein 1950.)

$$g_{ss} = g_s i_s \left[2.303 \log\left(\frac{30.2h}{\delta_c}\right) I_1 + I_2 \right] \tag{9.238}$$

The term $2.303\log\left(\frac{30.2h}{\delta_c}\right)$ is known as the transport parameter. So, one obtains a relationship of bed load to suspended load transport for all size fractions for which a bed load function exists, and is given by

$$i_{ss} g_{ss} = g_s i_s \left[2.303 \log\left(\frac{30.2h}{\delta_c}\right) I_1 + I_2 \right] \tag{9.239}$$

This is the dimensionless suspended load transport equation with g_{ss} being the suspended load rate in weight per unit of time and width.

9.8.5 Hunt's Diffusion Equation (1954) for Sediment Suspension

Hunt (1954) derived the diffusion equation for both sediment and water based on the concept of space occupied by the particles suspended in the fluid. If C is the concentration of sediment expressed as a volumetric proportion, $1 - C$ is the concentration for water, where C is expressed as a volumetric proportion. The flux vector denoted by $\mathbf{p}(x, y, z, t)$ for sediment can be written as

$$\mathbf{p} = \mathbf{u}_s C - \varepsilon_s \operatorname{grad} C \tag{9.240}$$

where $\mathbf{u}_s$ is the steady velocity of sediment particles, and ε_s is the sediment diffusion coefficient. Similarly for water, if $\mathbf{q}(x, y, z, t)$ be the flux vector, u the field velocity, $1 - C$ the concentration of water replaced by sediment where C is expressed as a volumetric proportion, the flux vector for the flow of water can be written as

$$\mathbf{q} = \mathbf{u}(1 - C) - \varepsilon_w \operatorname{grad}(1 - C) \tag{9.241}$$

where ε_w is the diffusion coefficient for the water. The continuity condition is expressed by the relation

$$\operatorname{div}(\mathbf{p} + \mathbf{q}) = 0 \tag{9.242}$$

The rate of change of sediment concentration at a particular point is given by

$$\frac{\partial C}{\partial t} = -\operatorname{div}\mathbf{p} \tag{9.243}$$

On using equation (9.240), one gets

$$\frac{\partial C}{\partial t} = \operatorname{div}(\varepsilon_s \operatorname{grad} C) - \operatorname{div}(\mathbf{u}_s C) \tag{9.243a}$$

From equation (9.242),

$$\frac{\partial C}{\partial t} = \operatorname{div}(\varepsilon_\omega \operatorname{grad} C) + \operatorname{div}\{\mathbf{u}(1 - C)\} \tag{9.244}$$

Using the Cartesian coordinates, equations (9.243) and (9.244) may be written as

$$\frac{\partial C}{\partial t} + u_s \frac{\partial C}{\partial x} + v_s \frac{\partial C}{\partial y} + w_s \frac{\partial C}{\partial z} = \frac{\partial}{\partial x}\left(\epsilon_{sx} \frac{\partial C}{\partial x}\right) + \frac{\partial}{\partial y}\left(\epsilon_{sy} \frac{\partial C}{\partial y}\right) + \frac{\partial}{\partial z}\left(\epsilon_{sz} \frac{\partial C}{\partial z}\right) - C\left(\frac{\partial u_s}{\partial x} + \frac{\partial v_s}{\partial y} + \frac{\partial w_s}{\partial z}\right) \tag{9.245}$$

$$\frac{\partial C}{\partial t} + u \frac{\partial C}{\partial x} + v \frac{\partial C}{\partial y} + w \frac{\partial C}{\partial z} = \frac{\partial}{\partial x}\left(\epsilon_{wx} \frac{\partial C}{\partial x}\right) + \frac{\partial}{\partial y}\left(\epsilon_{wy} \frac{\partial C}{\partial y}\right) + \frac{\partial}{\partial z}\left(\epsilon_{wz} \frac{\partial C}{\partial z}\right) + (1 - C)\left(\frac{\partial u}{\partial x} + \frac{\partial v}{\partial y} + \frac{\partial w}{\partial z}\right) \tag{9.246}$$

Equations (9.245) and (9.246) represent the three-dimensional turbulent diffusion equation for sediment and water respectively in the Cartesian coordinate system and they are coupled by the relative vertical velocity of sediment and water with the fall velocity of sediment particle. The Rouse equation and the Hunt equation have been widely used in the field of sediment transport and based on these equations many researchers have developed mathematical models to find the vertical distribution of sediments in a sediment-laden turbulent flow.

In the case of uniform flow where the concentration is constant in time and varies only with the vertical co-ordinate y, the respective equations for sediment and water given by (9.245) and (9.246) reduce to the following:

$$\frac{\partial}{\partial y}\left(\epsilon_{sy} \frac{\partial C}{\partial y}\right) - C \frac{\partial v_s}{\partial y} - v_s \frac{\partial C}{\partial y} = 0 \tag{9.247}$$

$$\frac{\partial}{\partial y}\left(\epsilon_{wy}\frac{\partial C}{\partial y}\right)-(1-C)\frac{\partial v}{\partial y}-v\frac{\partial C}{\partial y}=0 \tag{9.248}$$

The vertical velocity of sediment particle v_s is equated with the sum of the vertical velocity of water v together with settling velocity of particle $-v_0$ in still water, which is as:

$$v_s = v - v_0. \tag{9.249}$$

Using equation (9.249) in (9.247) and (9.248), one gets:

$$\epsilon_{sy}\frac{\partial C}{\partial y}+C\frac{\partial C}{\partial y}\left(\epsilon_{wy}-\epsilon_{sy}\right)+(1-C)Cv_0=0 \tag{9.250}$$

Equation (9.250) is the differential equation for the vertical distribution of suspended sediment with uniform fall velocity v_0 of sediment particle. Except near the bed, the volume of water displaced by the sediment particles could be neglected, which means, the amount of suspended sediment is very small, so equation (9.250) can be written as:

$$\epsilon_{sy}\frac{\partial C}{\partial y}+C\frac{\partial C}{\partial y}\left(\epsilon_{wy}-\epsilon_{sy}\right)+Cv_0=0 \tag{9.251}$$

Here, if $\epsilon_{sy}=\epsilon_{wy}$, equation (9.251) reduces to the Rouse equation (9.219); and equation (9.250) reduces to

$$\epsilon_{sy}\frac{\partial C}{\partial y}+(1-C)Cv_0=0 \tag{9.252}$$

This equation can be integrated as:

$$\ln\frac{C}{1-C}=-v_0\int\frac{dy}{\epsilon_{sy}}+Const. \tag{9.253}$$

In order to solve equation (9.253), we need to find the sediment/mass diffusion coefficient ϵ_{sy}, which is a function of velocity gradient. It is assumed that the turbulent mixing process for the sediment is similar to that for the water. The mean sediment velocity u_s and the mixing length l_s are therefore related as:

$$\tau=\rho_f l_s^2\left(\frac{du_s}{dy}\right)\left|\frac{du_s}{dy}\right| \text{ and } l_s=\kappa_s\frac{du_s}{dy}\bigg/\frac{d^2u_s}{dy^2} \tag{9.254}$$

The sediment mean velocity distribution is given by:

$$\frac{u_m-u_s}{\sqrt{ghJ}}=-\frac{1}{\kappa_s}\left[\sqrt{(1-\xi)}+A_s\,ln\left\{\frac{A_s-\sqrt{(1-\xi)}}{A_s}\right\}\right] \tag{9.255}$$

Since the sediment concentration is found to be zero at the free surface, maximum water velocity and sediment velocity are equal at the free surface. The sediment diffusion coefficient, by analogy with the Boussinesq formula $\tau=\rho_f\epsilon\frac{d\bar{u}}{dy}$, is given by

$$\epsilon_{sy} = \frac{\tau_0(1-\xi)}{\rho_f\left(\frac{du_s}{dy}\right)} \tag{9.256}$$

Using equations (9.256) and (9.255) in equation (9.253) and integrating, the sediment concentration distribution is given by:

$$\left(\frac{C}{1-C}\right)\left(\frac{1-C_a}{C_a}\right) = \left[\frac{\sqrt{1-\xi}}{\sqrt{1-\xi_a}} \cdot \left(\frac{A_s - \sqrt{(1-\xi_a)}}{A_s - \sqrt{(1-\xi)}}\right)\right]^Z \tag{9.257}$$

$$\text{where } Z = \frac{v_0}{\kappa_s A_s \sqrt{ghJ}} \tag{9.258}$$

and C_a is the sediment concentration at reference level a above the bed surface. The concentration is maximum when $C = 1$ at a level given by $\xi = 1 - A_s^2$. When the volumetric concentration of sediment is small, equation (9.257) reduces to the expression as:

$$\left(\frac{C}{C_a}\right) = \left[\frac{\sqrt{1-\xi}}{\sqrt{1-\frac{a}{h}}} \cdot \left(\frac{A_s - \sqrt{(1-\xi_a)}}{A_s - \sqrt{(1-\xi)}}\right)\right]^Z \tag{9.259}$$

The total rate of sediment transportation by weight, per unit width of flow, is given by the integral as:

$$g_s = \int_{\xi_a}^{\xi=1} C\, u_s\, dy \tag{9.260}$$

where C is the concentration expression from (9.257), and u_s is the velocity expression from (9.255).

9.8.5.1 Heterogeneous Sediment Sizes

When the sediment material in suspension is heterogeneous, as in river or canal, equations (9.240) and (9.241) are modified for different fall velocities of different particles. Let the concentration of a fraction of fall velocity v_{0r} be C_r at height y above the bed. Under the equilibrium conditions and assuming the same sediment diffusion coefficient ϵ_{sy} for each size particle and the concentration is a function of y only, the equations are given by:

$$\epsilon_{sy}\frac{\partial c_r}{\partial y} - c_r(v - v_{0r}) = 0, \quad (r = 1, 2, 3, \ldots) \tag{9.261}$$

$$\grave{o}_{wy}\frac{\partial \sum c_r}{\partial y} + \left(1 - \sum c_r\right)v = 0 \tag{9.262}$$

Eliminating v from the above equations, one gets the equation as:

$$\epsilon_{sy}\frac{\partial c_r}{\partial y}\left(1 - \sum c_r\right) + c_r\epsilon_{wy}\frac{\partial \sum c_r}{\partial y} + c_r v_{0r}\left(1 - \sum c_r\right) = 0 \tag{9.263}$$

Now summing individual fraction of concentration as $C_{\text{total}} = \sum c_r$, equation (9.263) reduces to

$$\epsilon_{sy} \frac{\partial C_{\text{total}}}{\partial y} + C_{\text{total}} \frac{\partial C_{\text{total}}}{\partial y} \left(\epsilon_{wy} - \epsilon_{sy} \right) + \left(1 - C_{\text{total}} \right) \sum c_r v_{0r} = 0 \tag{9.264}$$

Comparison of equations (9.250) and (9.264) shows that the total concentration of sediment of all sizes could be obtained from equation (9.250) if a representative fall velocity of the mixture W is so chosen that

$$W = \sum c_r v_{0r} \Big/ \sum c_r \tag{9.265}$$

In general, this relation (9.265) does not satisfy at all depths by a constant value of W. To minimize the error in the diffusion equation, the estimate W can be obtained experimentally from an analysis of the suspended material at various depths and would take the form

$$W = \int_0^h \sum c_r v_{0r} dy \Big/ \int_0^h \sum c_r dy \tag{9.266}$$

Batchelor (1965) found from the analysis of the solution of Rouse equation (9.219) that the usual derivation of the equation contains implicit assumption on the ignorance of particle-particle collisions within the range of low concentration. The solution of this equation near the bed of a channel invariably becomes infinity. Physically, the volumetric concentration should not exceed unity, but there is some inconsistency in the solution of Rouse equation (9.219) near the bed. However, Hunt (1954) suggested a more general diffusion equation whose solutions tend to a finite value at the bed, considering the volume occupied by the sediment particles near the bed. Later, Hunt (1969) re-examined the questions raised by Batchelor (1965) using the more general diffusion equations. He developed more general equations to explain the diffusion of heterogeneous sediments in suspension, from which some results for homogeneous sediment were derived as a special case. He extended the existing method, including the mixture of sizes and sediment volume. Using two size fractions in his work, he observed that the fall velocities of two sizes differ by several orders of magnitude, there was very little overlap in the vertical distributions; the light particles were found in the upper layers of the flow and the heavy particles near the bed. When the fall velocities differ slightly, the distributions overlap to a considerable extent, but only the heaviest material appears in the lowest layer near the bed.

9.9 DIRECT COMPUTATION OF SUSPENSION GRAIN SIZE DISTRIBUTIONS

Several well-known methods for computation of the bed load and suspended load are available (Graf 1971, Garde and Ranga Raju 2000). Computation of the grain-size distribution of the suspended load above a sand bed of given composition is discussed here. The theoretical analyses for sediment suspension are based on a more general diffusion equation whose solutions tend to a finite value at the bed (Hunt 1954). Hunt's diffusion concept is more general in the sense that he suggested two diffusion equations: one for sediment concentration C, and the other one for water $1-C$, so that the total volume is unity. This diffusion concept has so far been the most acceptable one for practical applications as well as verification with experimental data.

Ghosh et al. (1981) developed a method for direct computation of grain size distributions of suspended load from the sand bed materials and flow parameters, using Hunt's diffusion equations. Controlled experiments in laboratory flumes have shown earlier that grain-size distributions of sediments suspended in flowing water bear a definite relationship with the bed material, flow velocity and height of suspension above the bed (Sengupta 1975, 1979). These experimental studies have shown that during the flow a sorting process is initiated immediately above the bed, and the grain-size distribution of the bed layer influences the size distribution of the suspension above (Sengupta 1979). The grain-size distribution of the suspended load above a sand bed must take into consideration a two-stage sorting process: (1) sorting from the bed to the bed layer, and (2) sorting from bed layer to suspension. The Grain-size frequency distributions of the suspended load above three different sand beds (beds 2, 3, and 5) have been computed with the help of the Rouse equation, using the bed layer distributions obtained independently by Einstein (1950) and Gessler (1965) methods as references. Then the suspended load distributions above each of the three sand beds were also computed by the modified Hunt's method. The efficiency of these methods was tested by comparing the computed values with the experimental data of Sengupta (1975, 1979) and Ghosh et al. (1981) collected in close-circuit flumes under known flow conditions.

To develop a mathematical model, the flow was assumed uniform, where the concentration varies only with the vertical coordinate y throughout the depth and the diffusion coefficients of sediment and water are assumed to be the same, the concentration equation (9.252) is considered. In fully developed turbulent flow, considering $\epsilon_{sy} = \epsilon_{wy}$, the diffusion coefficient for sediment ϵ_{sy} can be written as:

$$\epsilon_{sy} = \frac{\tau_0 (1-\xi)}{\rho_f \left(\frac{d\bar{u}}{dy} \right)} \tag{9.267}$$

where $\xi = \frac{y}{h}$. Here, the von-Karman velocity distribution used by Hunt (1954) is used as follows

$$\frac{u_m - \bar{u}}{\sqrt{ghJ}} = -\frac{1}{\kappa_o} \left[\sqrt{(1-\xi)} + A\,\mathrm{Ln} \left\{ \frac{A - \sqrt{(1-\xi)}}{A} \right\} \right] \tag{9.268}$$

which satisfies the boundary condition $u = u_{\max}$ at $\xi = 1$ at the free surface, and A is a constant to be determined. Ghosh et al. (1981) determined the constant A different from that of Hunt (1954). If we put $A = 1$ in equation (9.268), the velocity distribution coincides with that of von-Karman (1930). Ghosh et al. (1981) determined the constant A from the condition $u = 0$ at $\xi_{k_s} = k_s/h$, the dimensionless roughness above the bed (Schlichting 1968). Since roughness k_s is extremely small compared to the depth of flow h, the ratio k_s/h is negligible for higher power of k_s/h. Under this assumption and after making some simplifications (Ghosh et al. 1979), the constant A is obtained as:

$$A = 1 - \frac{1}{2}\frac{k_s}{h} + \exp\left[-1 - \frac{\kappa_o u_m}{u_*} \right] \tag{9.269}$$

Using equation (9.269), the velocity distribution (9.268) is plotted against $\xi = \frac{y}{h}$ in Figure 9.36 (Ghosh et al. 1981). It is seen that the agreement between the observed and expected velocities is very close throughout the vertical height ξ.

Since the velocity distribution is valid only down to $y = 0.25$ cm $= y_1$ (say), a linear velocity distribution is assumed below this height, $y = y_1$ and down to the point $y = k_s$, where the velocity is assumed to be zero (Jobson and Sayre, 1970). The linear velocity profile is of the form

$$u = \frac{u_{y_1}}{y_1 - k_s}(y - k_s),\ y_1 - k_s \neq 0 \tag{9.270}$$

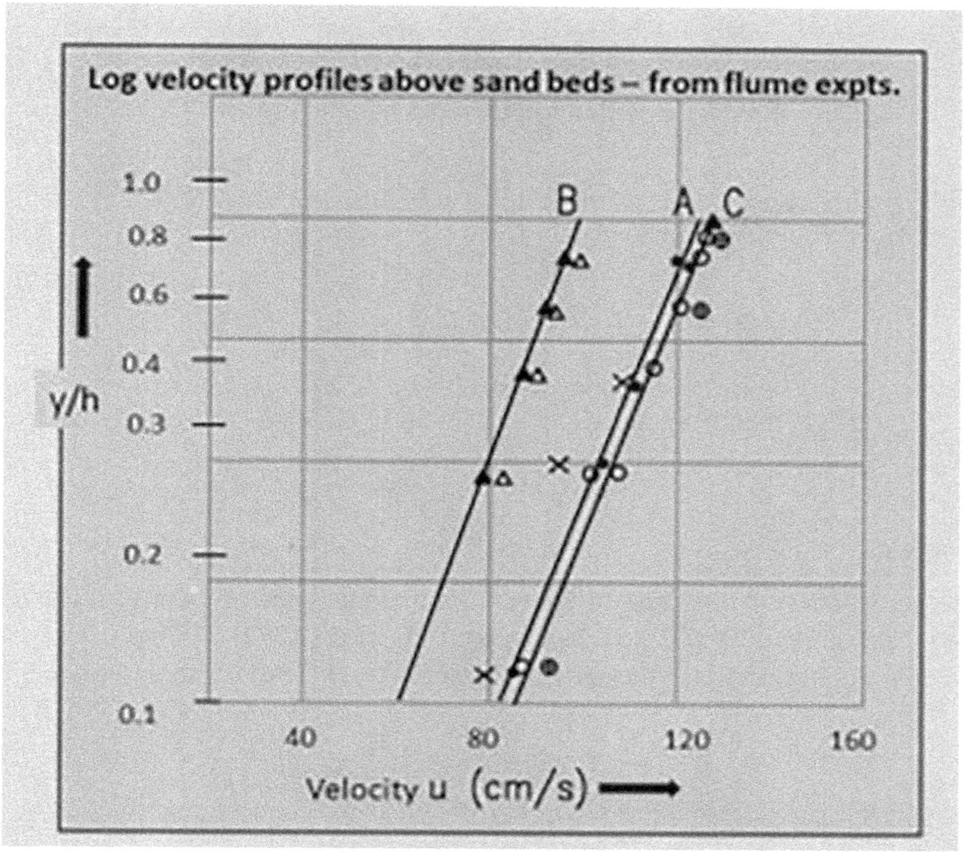

FIGURE 9.36 Profiles of velocity against *y/h* above the sand beds for flume experiments: A for bed-2 (× - observed, * - computed), B for bed-3 (Δ - observed, ▲ - computed), C for bed-5 (⊗ - observed, ⊙ - computed). (Modified from Ghosh et al. 1981.)

where $u_{y=y_1}$ is the extrapolated velocity from equation (9.269) with equation (9.270). Using equations (9.268), (9.269), and (9.270), one gets the expressions for sediment diffusion coefficient:

$$\epsilon_{sy} = 2\kappa_o h u_* (1-\xi)\left\{A - \sqrt{(1-\xi)}\right\} \tag{9.271}$$

and for the linear velocity profile

$$\epsilon_{sy} = \frac{u_*^2}{u_{y_1}}(y_1 - k_s)(1-\xi) \tag{9.272}$$

Integrating equation (9.252) from k_s to ξ using equations (9.271) and (9.272), one obtains:

$$\ln\frac{C_\xi\left(1-C_{\xi_{k_s}}\right)}{C_{\xi_{k_s}}\left(1-C_\xi\right)} = K_1 \ln\frac{h-y_1}{h-k_s} + K_2 \ln\left[\frac{\sqrt{(1-\xi)}}{\sqrt{(1-\xi_1)}}\cdot\frac{A-\sqrt{(1-\xi_1)}}{A-\sqrt{(1-\xi)}}\right] \tag{9.273}$$

where $K_1 = \dfrac{v_0 h u_{y1}}{u_*^2\left(y_1 - k_s\right)}$, and $K_2 = \dfrac{v_0}{A\kappa_o u_*}$.

The first term of the right side of expression (9.273) is found by integrating from k_s to ξ_1 using equation (9.272). The second term is obtained by integrating from ξ_1 to ξ using equation (9.271), which is the same as that of Hunt. In a compact form, equation (9.273) can be written as

$$\frac{C_\xi}{1-C_\xi}=\frac{C_{\xi ks}}{1-C_{\xi ks}}f(\xi,v_0) \tag{9.274}$$

where v_0 is a function size $\phi=-\ln_2 d$ with d in mm from equation (8.1), and

$$f(\xi,v_0)=\left(\frac{1-\xi_1}{1-\xi_{ks}}\right)^{K_1}\left(\frac{\sqrt{(1-\xi)}}{\sqrt{(1-\xi_1)}}\cdot\frac{A-\sqrt{(1-\xi_1)}}{A-\sqrt{(1-\xi)}}\right)^{K_2} \tag{9.275}$$

$$\text{Let } \psi_1=\frac{C_{\xi ks}}{1-C_{\xi ks}}f(\xi,v_0) \text{ and } \psi_1'=\frac{C_{\xi ks}'}{1-C_{\xi ks}'}f(\xi,v_0) \tag{9.276}$$

If we assume $C_{\xi ks}=\alpha C_{\xi ks}'$, then from equations (9.274) and (9.276), C_ξ can be written as

$$C_\xi=\frac{\psi_1}{1+\psi_1}\sim\frac{\alpha\psi_1'}{1+\psi_1'} \tag{9.277}$$

It is reasonable to expect $\alpha\approx 1$ because at the bottom boundary the amount of sediment is relatively large compared to water. Even if α is not close to one, equation (9.277) will hold approximately, if $C_{\xi ks}'$ and the product $C_{\xi ks}'\cdot f$ are small. In the sand beds used, $C_{\xi ks}'$ is small compared to one. Hence,

$$C_\xi'=\frac{C_\xi}{\sum_\phi C_\xi}=\frac{\psi_1'(\phi)}{1+\psi_1'(\phi)}\Big/\sum_\phi\frac{\psi_1'(\phi)}{1+\psi_1'(\phi)} \tag{9.278}$$

Using equation (9.278), it is easy to compute the suspension concentration of a given grain size with a settling velocity v_0 at any height $\xi\geq\xi_1$ above the bed, if the relative concentration C_{k_s}' of the particle at the bed is known. If $\xi=\xi_1$, then equations (9.274) to (9.278) give the average concentration of sediments of different sizes in the bed layer (C_{bl}'). Experimental data suggest that the efficiencies of the different theoretical methods for computation of the bed load and suspended load discussed above have been studied by comparing the computed grain-size frequency distributions with those observed in laboratory flumes under known hydraulic conditions.

Two closed circuit laboratory flumes, one designed at the Uppsala University (Figure 9.42), and the other one at the Indian Statistical Institute, Calcutta (Figures 7.13 and 7.14) were used for this purpose. The equipment used for velocity measurement, sample collection and analyses employed have been described earlier (Sengupta 1979 and Ghosh et al. 1979). Grain-size distributions of suspended loads over six different sand beds were studied during the Uppsala and Calcutta experiments at various heights above the bed (Sengupta 1975, 1979 and Ghosh et al. 1979, 1981). The results of the experiments over three types of sand beds at a height of approximately 20 cm and at flow velocities varying between 98 and 126 cm/s are discussed. The following three sand beds (numbered 2, 3 and 5) were chosen for the present study because of their widely different grain-size distribution patterns (Table 9.5): nearly uniform (bed 2), bimodal (bed 3) and positively skewed with slight bimodality (bed 5). The bed, bed load and suspended load distributions are graphically presented in Figures 9.37–9.39. The observed suspended load distributions above these sand beds are tabulated in Table 9.6. Table 9.6A shows the flow parameters used for experimental study and Table 9.6B shows the grain size distributions in suspension. The experiments were always started with a smooth, flat bed. Bed forms were generated when the competence velocity was exceeded. The dimensions of the bedforms depended not only on the flow velocity but also on the grain-size distribution of the bed material. Wavelengths of the bed forms increased with an increase in flow velocity, but at a velocity exceeding about 120 cm/s, the bed forms were replaced by nearly flat beds in most of the cases. The trends of the observed and computed suspended load

TABLE 9.5
Frequency Distribution of Grain Size of Bed Materials

Sieve Size (d)		Bed 2		Bed 3		Bed 5	
		Weight	Relative Conc.	Weight	Relative Conc.	Weight	Relative Conc.
(ϕ)	(mm)	w_b (kg)	C'_{k_s}	w_b (kg)	C'_{k_s}	w_b (kg)	C'_{k_s}
0.0	0.99	-	-	1.065	0.005	-	-
0.5	0.70	3.969	0.019	14.985	0.050	5.588	0.075
1.0	0.49	30.572	0.148	67.489	0.225	15.024	0.200
1.5	0.35	27.444	0.133	59.300	0.198	7.192	0.096
2.0	0.25	24.035	0.116	15.389	0.051	33.798	0.451
2.5	0.18	31.122	0.150	37.632	0.125	6.380	0.085
3.0	0.12	27.949	0.135	59.874	0.199	5.597	0.075
3.5	0.09	27.036	0.131	37.138	0.124	0.589	0.008
4.0	0.06	19.053	0.092	5.865	0.020	0.765	0.010
4.5	0.04	10.971	0.053	0.651	0.002	0.007	0.0001
> 4.5*	< 0.03	4.839	0.023	0.072	0.0002	0.061	0.0008
Total		206.990		300.000		75.000	

Source: Modified from Ghosh et al. (1981). * – rounded off to 5.0 for the purpose of computation.

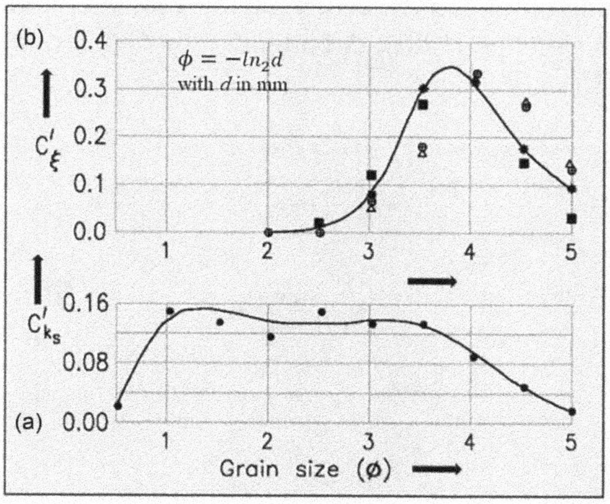

FIGURE 9.37 Relative concentrations for Bed 2, grain size distributions: (a) bed (C'_{k_s}) (b) suspension C'_ξ at $\xi =$ 0.7823.3 cm) * observed, ⊗ computed by the Ghosh et al. (1981), Δ computed by Gessler and Rouse equations, ▪ computed by Einstein and Rouse equations. (Modified from Ghosh et al. 1981.)

distributions, shown in Figures 9.37–9.39, generally agree, but the actual values show marked discrepancies in some cases. To obtain a quantitative idea of these discrepancies, the weighted relative error between the computed and the observed values were computed by the formula (Ghosh et al. 1981). Errors between the computed and observed data are shown in Table 9.7. As seen from this table, none of the methods for computing the suspended load can be regarded as the best for all the beds.

The formula is based on a partial modification of Hunt's (1954) work, changing the velocity profile. While Hunt's equation is valid only for the zone having a logarithmic velocity distribution, Ghosh et al. (1981)'s method uses this distribution down to $y = 0.25$ cm only and then assumes a linear velocity distribution down to the bed to have a more realistic simulation of the actual velocity pattern. Using this modified velocity distribution; two concentration equations for water and sediment have been set

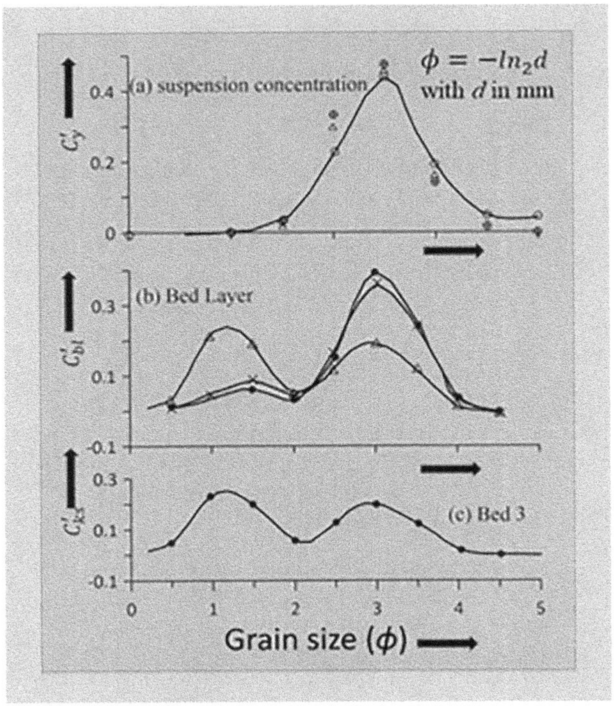

FIGURE 9.38 Relative concentrations for Bed-3, grain size distributions: (a) suspension at y = 17.5 cm ⊙ observed, ⊗ computed by the Ghosh et al. (1981), Δ computed by Gessler and Rouse equations. (b) bed layer ⊙ observed, × computed by the Ghosh et al. (1981), Δ computed by Gessler, (c) grain size distribution of Bed 3. (Modified from Ghosh et al. 1981.)

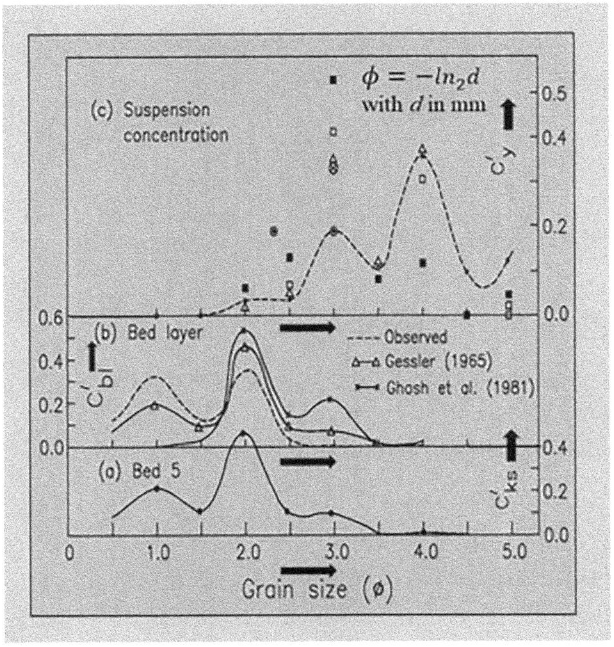

FIGURE 9.39 Relative concentrations for Bed-5, grain size distributions: (a) size distribution of bed 5, (b) bed layer: – – – – observed by Ghosh et al. (1981), × computed by the Ghosh et al. (1981), Δ computed by Gessler (1965), (c) suspension distribution at y = 18.0 cm, * – – – * observed, ⊗ computed by Ghosh et al. (1981), Δ computed by Gessler and Rouse equations, ■ - computed by Einstein equation with hiding correction and Rouse equation, and ⋆ computed by Einstein without hiding correction and Rouse equation. (Modified from Ghosh et al. 1981.)

TABLE 9.6A
Flow Parameters Used in the Experimental Study

	Bed 2	Bed 3	Bed 5
h (cm)	30.0	30.0	30.0
H (cm)	25.0	20.0	20.0
h' (cm)	~ 1.7	~ 2.5	~ 2.0
u_{max} (cm/s)	121.3	97.8	126.0
k_s (cm)	0.0297	0.0451	0.0518
y (cm) = $H - h'$	23.3	17.5	18.0
V (lit.)	5.0	5.0	5.0
J	0.0020	0.0020	0.0022
Temp. (°C)	19.0	19.0	27.0

Source: Modified from Ghosh et al. (1981).

TABLE 9.6B
Observed Grain-Size Distributions of Suspended Loads above the Sand Beds

Sample No.	Average of III2-121-25-25B (Bed 2)		Average of VII-93-20A-B (Bed 3)		Average of 5-124-20A-D(w) (Bed 5)	
d (ϕ)	w_y (g)	C'_y	w_y (g)	C'_y	w_y (g)	C'_y
1.0	-	-	0.004	0.0005	0.005	0.005
1.5	0.032	0.001	0.008	0.001	0.003	0.003
2.0	0.095	0.003	0.033	0.003	0.030	0.035
2.5	0.396	0.013	0.289	0.031	0.033	0.043
3.0	2.438	0.081	2.158	0.228	0.160	0.196
3.5	9.144	0.306	4.197	0.443	0.095	0.112
4.0	9.558	0.320	1.869	0.197	0.306	0.363
4.5	5.320	0.179	0.500	0.053	0.095	0.113
> 4.5	2.903	0.097	0.420	0.044	0.109	0.129

Source: Modified from Ghosh et al. (1981).
d = grain diameter, H = sampling height, h = water depth, h' = bed form height, u_{max} = maximum velocity, k_s = roughness height, y = effective sampling height, V = volume of suspension sample, J = slope, and Temperature.

TABLE 9.7
Estimation of Errors between Computed and Observed Suspended Loads above Beds 2, 3, and 5

Bed No.	Location: Height y (cm)	Estimation Einstein with ξ-Correction	Estimation Einstein without ξ-Correction	Estimation by Gessler Method	Estimation (Ghosh et al. 1981 Method)
2	23.5	0.34	0.27	0.39	0.36
3	17.5	-	-	0.29	0.36
5	18.0	1.18	0.68	0.54	0.53

Source: Modified from Ghosh et al. (1981).
Note: The suspended loads have been computed by application of the Rouse suspension equation (C'_y), using the respective computed bed layer concentration as a reference level distribution.

up as in Hunt (1954). Solving these two equations, the sediment concentration at any height above a bed is obtained in terms of flow parameters and bed materials. For each of the three beds, the trends of the suspended load's grain-size distribution patterns have been computed using the Rouse equation on the bed loads obtained by: (1) Einstein's method (with and without hiding correction), and (2) Gessler's method (1965). Quantitative estimates of the errors between the observed and the computed values indicate that no one method can be claimed to be particularly superior to the others. The possible sources of errors in each of these methods have been discussed by Ghosh et al. (1981). The real advantage of this method is that it affords a direct computation of the suspended load from a bed's grain-size distribution without going through an intermediate stage (bed load) as required by the other existing methods. The applicability of this method for computation of suspended load has been further tested in a natural stream (suspended load data of the Niobrara River Nebraska, Colby and Hembree 1955, p. 134, Table 6). The comparative results for computed and observed suspended load grain size distribution are presented in the paper by Sengupta et al. (1991).

9.9.1 Conditions of Log-Normality in Suspension

Subsequently, Ghosh and Mazumder (1981) developed a theoretical framework to relate the suspended load's grain size distribution, particularly the occurrence of *uni-modality, symmetry and log-normality*, with the flow parameters and grain-size distributions of the bed materials. They found a family of curves relating the height of suspension and shear velocity for uni-modal grain size distribution in suspension with known peaks which are approximately lognormal. Using these curves, it is possible to work out the shear velocities for a given heights when the mode of lognormal grain size distribution in suspension is known.

Controlled experiments conducted by Sengupta (1975, 1979) indicated that under suitable conditions the log-normality can be attained through a process of size-sorting during suspension transportation in water flows, even when the source materials are not log-normal. Log-normality was explained within a critical range of velocity and height above the sand bed of a given composition. It is observed that the grain size distribution has a tendency to be log-normal or uni-modal for that size range of bed material for which the settling velocity of particles is linearly distributed in log-scale. The conditions leading to uni-modality, symmetry and log-normality in suspension for a given hyperbolic bed's grain size distribution have been deduced from the well-known Rouse equation.

Under equilibrium conditions, conventional Rouse equation (9.219) is written, replacing concentration $C = S_y$, $\epsilon_s = \epsilon$ and $v_0 = c(\phi)$ as:

$$c(\phi)S_y + \epsilon\frac{\partial S_y}{\partial y} = 0 \tag{9.279}$$

Integrating equation (9.279) from k_s to y, one gets

$$S_y(\phi) = S_{k_s}(\phi)\left(\frac{d-y}{y}\cdot\frac{k_s}{d-k_s}\right)^{\frac{c(\phi)}{\kappa_o u_*}} \tag{9.280}$$

where $S_{k_s}(\phi)$ is the reference concentration at the k_s level. Equation (9.280) can be written as:

$$S_y(\phi) = S_{k_s}(\phi)\exp(\psi c(\phi)) \tag{9.281}$$

$$\text{where } \psi = \frac{1}{\kappa_o u_*}\ln\left(\frac{y}{d-y}\cdot\frac{d-k_s}{k_s}\right) \tag{9.282}$$

is a parameter characterizing the effect of flow and the expression S'_ψ for relative suspension concentration distribution at any height ψ (varying y) is given by:

$$S'_\psi = \frac{S_\psi(\phi)}{\sum_\phi S_\psi(\phi)} = S_{k_s}(\phi)\exp\{-\psi c(\phi) + g(\psi)\} \tag{9.283}$$

$$\text{where } e^{-g(\psi)} = \sum_\phi S_{k_s}(\phi) e^{-\psi c(\phi)} \tag{9.284}$$

While studying the question of unimodality of $S'_\psi(\phi)$, it is noted that $\ln c(\phi)$ is approximately linear in ϕ; that is, $\ln c(\phi) \approx a + b\phi,\ 1 \le \phi \le 5$, where $a = 3.5758$, and $b = -1.1840$.

Let $C_\psi(\phi) = \ln S'_\psi(\phi)$ and $C_{k_s}(\phi) = \ln S_{k_s}(\phi)$, equation (9.283) can be written as:

$$C_\psi(\phi) = C_{k_s}(\phi) - \psi c(\phi) + g(\psi) \tag{9.285}$$

Now $C_\psi(\phi)$ is expected to be unimodal if one can find a unique solution to the equation for the mode ϕ. Then from equation (9.285),

$$\frac{dC_\psi(\phi)}{d\phi} = \frac{dC_{k_s}(\phi)}{d\phi} - \psi b e^{a+b\phi} = 0 \text{ at } \phi = \hat{\phi} \tag{9.286}$$

Since (I) $b<0$, equation (9.286) can have a solution only when $\frac{dC_{k_s}(\phi)}{d\phi} < 0$ at $\phi = \hat{\phi}$, and (II) solution of equation (9.286) is unique if $\frac{d^2C_\psi(\phi)}{d\phi^2} \le 0$ at $\phi = \hat{\phi}$ for $2 \le \hat{\phi} \le 4$, the range within which the solution is sought. It is clear that a sufficient condition for (II) is $\frac{d^2C_{k_s}(\phi)}{d\phi^2} \le 0$ at $\hat{\phi}$ for $2 \le \hat{\phi} \le 4$.

This condition holds if bed layer concentration $S_{k_s}(\phi)$ is normal or hyperbolic (Barndorff-Nielsen 1977) with a mode to the left of 2. For such $S_{k_s}(\phi)$, the set of values of ψ for which equation (9.286) has a solution will determine the region of uni-modality. For example, Bed 3 in the range $2 \le \hat{\phi} \le 5$ is approximately hyperbolic with mode near 3.0. From Figure 9.40a, it is clear by (I) that equation (9.286) can have a solution only to the right of $\phi = 3.0$ and (II) also holds for $\phi \ge 3.23$. Hence, whenever equation (9.286) has a solution, it is unique so that the corresponding $S'_\psi(\phi)$ is uni-modal. For bed 2, Figure 9.40a, (1) implies that equation (9.286) can have a solution only to the right of $\phi = 3.5$; there is a straight line describes $S_{k_s}(\phi)$ adequately so that (II) holds, i.e., if equation (9.286) has a solution, the solution is unique. From the hyperbolic plot of bed 0 in Figure 9.40b, it follows from condition (I) that equation (9.286) can have a solution only to the right of $\phi = 2$ and condition (II) also holds for $\phi \ge 2.3$. Whenever equation (9.286) has a solution, it is unique, so that the corresponding $S'_\psi(\phi)$ is uni-modal. Thus, the conclusion holds for these beds.

It is also noticed from the normality of suspension distribution above the bed 2, when equation (9.286) has solution and (II) holds at $\phi = \hat{\phi}$. It is checked that

$$c(\phi) \approx c(\hat{\phi}) + \frac{1}{2}c''(\hat{\phi})(\phi - \hat{\phi})^2, |\phi - \hat{\phi}| \le 1 \tag{9.287}$$

and if one assumes

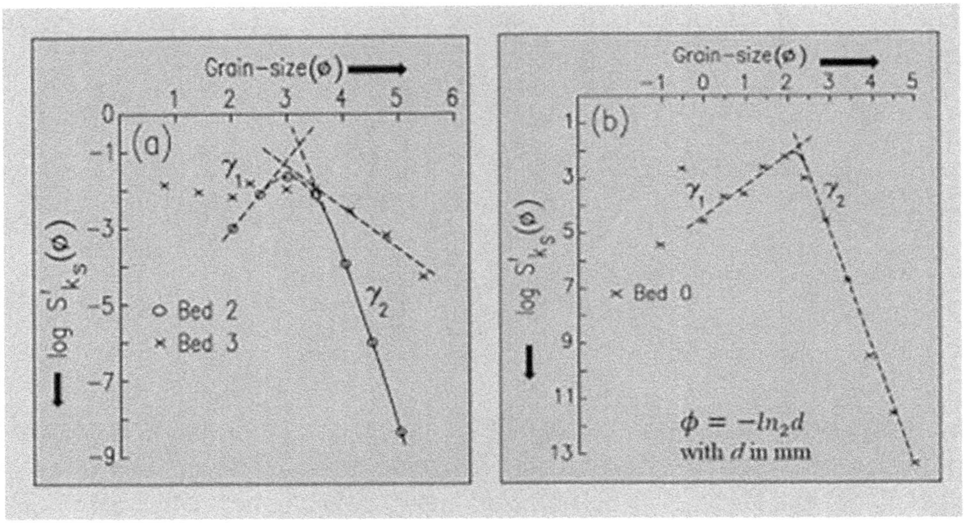

FIGURE 9.40 (a and b) Grain-size distributions of experimental beds 2, 3 and 0 ($\log S'_{k_s}(\phi)$ against ϕ) in log-hyperbolic equation (9.291). (Modified from Ghosh and Mazumder 1981.)

$$S_{k_s}(\phi) \approx S_{k_s}(\hat{\phi}) + \frac{1}{2} S''_{k_s}(\hat{\phi})(\phi - \hat{\phi})^2, |\phi - \hat{\phi}| \le 1, \tag{9.288}$$

Then one gets from equation (9.285)

$$S_\psi(\phi) \approx S_\psi(\hat{\phi}) + \frac{1}{2} S''_\psi(\hat{\phi})(\phi - \hat{\phi})^2, |\phi - \hat{\phi}| \le 1 \tag{9.289}$$

The reason for confining attention to $|\phi - \hat{\phi}| \le 1$ in equation (9.287) is that most of the mass of $S_y(\phi)$ is concentrate here. Clearly, equation (9.288) holds for a lognormal or hyperbolic bed $S_{k_s}(\phi)$ with mode to the left of $\phi = 2$; it holds for bed 2. Then $S_\psi(\phi)$ looks approximately like a normal with mean $\hat{\phi}$ and

$$\text{Variance} = -S''_\psi(\hat{\phi}) = -S''_{k_s}(\hat{\phi}) + \psi b^2 c(\hat{\phi}) \tag{9.290}$$

To study the bed 3 in details, a hyperbolic distribution to $S'_{k_s}(\phi)$ for $2 \le \hat{\phi} \le 5$ is fitted as:

$$\ln S'_{k_s}(\phi) \approx \nu - \frac{1}{2}(\gamma_1 + \gamma_2)\left\{\delta^2 + (\phi - \mu)^2\right\}^{1/2} + \frac{1}{2}(\gamma_1 - \gamma_2)(\phi - \mu) \tag{9.291}$$

where parameters γ_1, γ_2, δ and μ are determined graphically using the geometrical interpretation given by Bagnold and Barndorff-Nielsen (1980). The constant ν has been adjusted to agree with the observed frequency in the range $2 \le \hat{\phi} \le 5$. The estimated values are $\gamma_1 = 1.787, \gamma_2 = 4.294$, $\mu = 3.228$, $\delta = 0.5039$ and $\nu = -0.225$. From Figure 9.40a, it is clear by (I) that equation (9.286) can have a solution only to the right of $\phi = 3$ and (II) also holds for $\phi \ge 3.23$. Hence, whenever equation (9.286) has a solution, it is unique so that the corresponding $S'_\psi(\phi)$ is uni-modal. It is now examined when $S'_\psi(\phi)$ can be symmetrical as well as uni-modal. Expanding $c(\phi)$ around $\hat{\phi}$ up to the quadratic terms, one gets from equation (9.285) as:

$$\ln S'_\psi(\phi) \approx g(\psi) + \nu - \frac{1}{2}(\gamma_1 + \gamma_2)\left\{\delta^2 + (\phi - \mu)^2\right\}^{1/2} + \frac{1}{2}(\gamma_1 - \gamma_2)(\phi - \mu)$$
$$- \psi c(\hat{\phi}) - \psi b(\phi - \hat{\phi})c(\hat{\phi}) - \frac{1}{2}\psi b^2(\phi - \hat{\phi})^2 c(\hat{\phi}) \tag{9.292}$$

In order to get symmetry, $\hat{\phi}$ must be nearly equal to μ and

$$\frac{1}{2}(\gamma_1 - \gamma_2) - \psi bc(\hat{\phi}) \approx 0. \tag{9.293}$$

Note that equation (9.293) follows from (9.286) if $\hat{\phi} \approx \mu$. Then around $\hat{\phi}$, we have:

$$\ln S'_\psi(\phi) \approx g(\psi) + v - \frac{1}{2}(\gamma_1 + \gamma_2)\left\{\delta^2 + (\phi - \mu)^2\right\}^{1/2} - \frac{1}{2}\psi b^2 (\phi - \hat{\phi})^2 c(\hat{\phi}) \tag{9.294}$$

Inspection of (9.294) suggests that the presence of the term $\left\{\delta^2 + (\phi - \mu)^2\right\}^{1/2}$ is likely to lead to less peakedness and hence higher values of the coefficient of kurtosis β_2 than the normal. Confirmation of the expectation is provided in the following numerical calculations.

For the values of $y = 17.5$ cm, $u_* = 6.297$ cm/s, $h = 27.5$ cm, and $k_s = 0.0451$ cm, one gets the value $\psi = 2.7676$. The skewness and kurtosis of the corresponding suspension distribution are respectively 0.343 and 4.152 (Figures 12 and 13 of Sengupta 1979). The above analysis shows how one can study the uni-modality, symmetry and normality of the suspension distribution $S'_\psi(\phi)$ for a given hyperbolic bed distribution $S'_{k_s}(\phi)$.

Now how one can determine the flow parameters y and u_* leading to unimodality for a given bed distribution. Here the relation (9.282) for ψ is used. Let us consider the bed 2. Since by (I) and Figure 9.40a, equation (9.286) can have a solution only for $\hat{\phi} \geq 3.5$ and observed data with $\hat{\phi} > 4.5$ is likely to be scarced, and let us work with $3.5 \leq \hat{\phi} \leq 4.5$. For fixed $\hat{\phi}$ in this range we now solve (9.286) for ψ; for $\hat{\phi} = 3.5$, $\psi = 1.663$ and for $\hat{\phi} = 4.5$, $\psi = 5.441$.

The curves obtained by plotting y against u_* for these two fixed values of ψ in equation (9.282) are shown in Figure 9.41. The zone of these curves gives the values of y and u_* which will give unimodality with peak at some $3.5 \leq \hat{\phi} \leq 4.5$. As noted before these unimodal distributions will be approximately log-normal. A similar analysis was made for bed 3. To achieve symmetry, we kept $\hat{\phi}$ in the range of 3.23 and 3.5. The resulting curves y against u_* are also shown in Figure 9.41. The combination of y and u_* obtained this way agree well with our experimental observations for both beds 2 and 3. The conditions of unimodality, symmetry and log-normality of grain size distributions in suspension are explained in Ghosh and Mazumder (1981), even with a hyperbolic distribution in the bed. This present study suggests that the critical condition of log-hyperbolic distribution leading to log-normality is satisfied at high velocity. The log-normality attained in grain size distribution in suspension is essentially a function of flow velocity. Grain-size data of two bed materials (bed nos. 2 and 3 of Sengupta 1979) are used for the present investigation. Further, for details, readers are suggested to refer the paper by Ghosh and Mazumder (1981).

9.9.2 Computation of Deposition from Suspension

It is important to note that all the grain sizes taken into suspension by flowing water be preserved in the deposits from suspension at reducing flow velocity (Kuenen and Sengupta 1970). The size distribution deposited closely reflects the one present in suspension. This finding is contradictory to the common belief that coarse grains are the first to be discarded at reducing flow velocity. That has been demonstrated through a series of experiments conducted at Uppsala University and Indian Statistical Institute Calcutta. Results of the experiments show that the modal sizes of suspension load at high velocity and that of deposit formed from suspension at a lower velocity are the same. The available theoretical models (Rouse 1938; Hunt 1969; Ghosh et al. 1981) also indicate the depletion of large proportions of coarse grains at reduced flow velocities. A mathematical model for the process of deposition from suspension

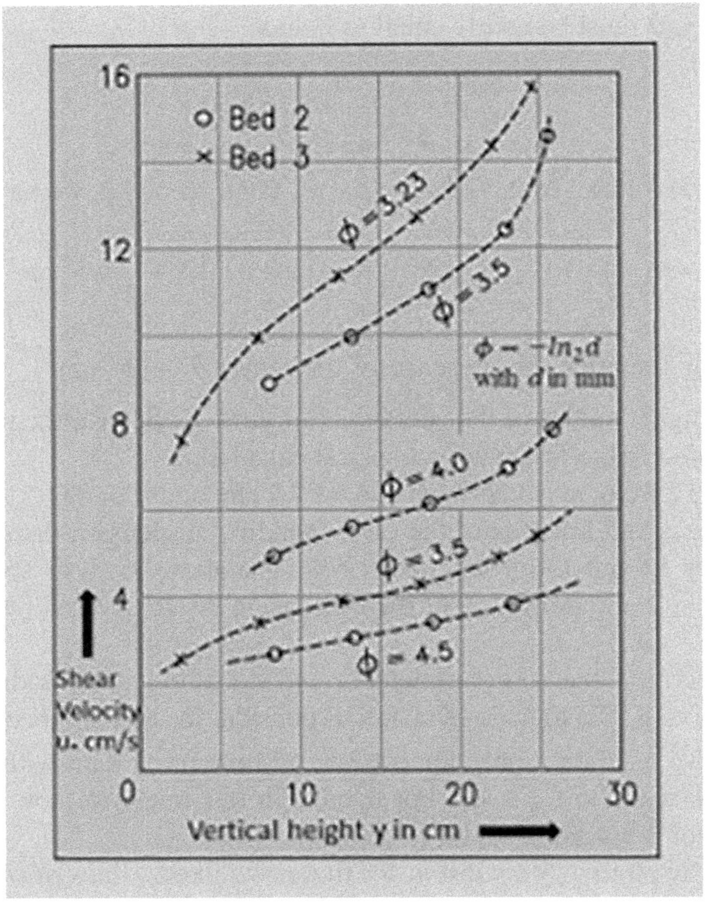

FIGURE 9.41 Vertical height y against shear velocity u_* for uni-modal suspension distributions above beds 2 and 3 with different values of grain sizes. (Modified from Ghosh and Mazumder 1981.)

at a wide range of grain sizes was developed (Ghosh et al. 1986), introducing a dampening factor to the conventional Rouse equation. This model helps to work out for palaeo-flow conditions of suspension current from grain size parameters of ancient deposits.

To resolve this contradiction and to obtain a clearer insight into the process of formation of deposits by unidirectional suspension currents, a series of flume experiments were conducted at the Uppsala Natur Geografiska Institutionen (UNGI), and the Indian Statistical Institute (ISI), Calcutta. This study presents the results of these experiments and also to develop a theoretical model conforming to the experimental findings. The designs of the flumes, experimental procedures, mathematical models and results are briefly described below.

The re-circulating flume used at Uppsala was of a "closed-circuit" type, with a straight working length 10 m, width 60 cm, and height 60 cm (Figure 9.42) (Sengupta 1979). The pumps were provided with devices for continuous speed control, allowing the flume to be operated at any desired flow velocity up to 1.5 m/s at a water depth of about 30 cm. A pack of pipes at the upstream end of the experimental channel ensured vortex-free flow. Velocity recorded by an Ott Laboratory current meter showed that the vertical velocity distributions were logarithmic at a distance of 7 to 8 m from the upstream end of the flume (Figure 9.42). The experiments on deposition were conducted at Uppsala separately for two different sand beds having different types of grain-size frequency distributions (Beds 2 and 3, see Sengupta 1979). The techniques followed for each of these experiments, and the results obtained, are described below.

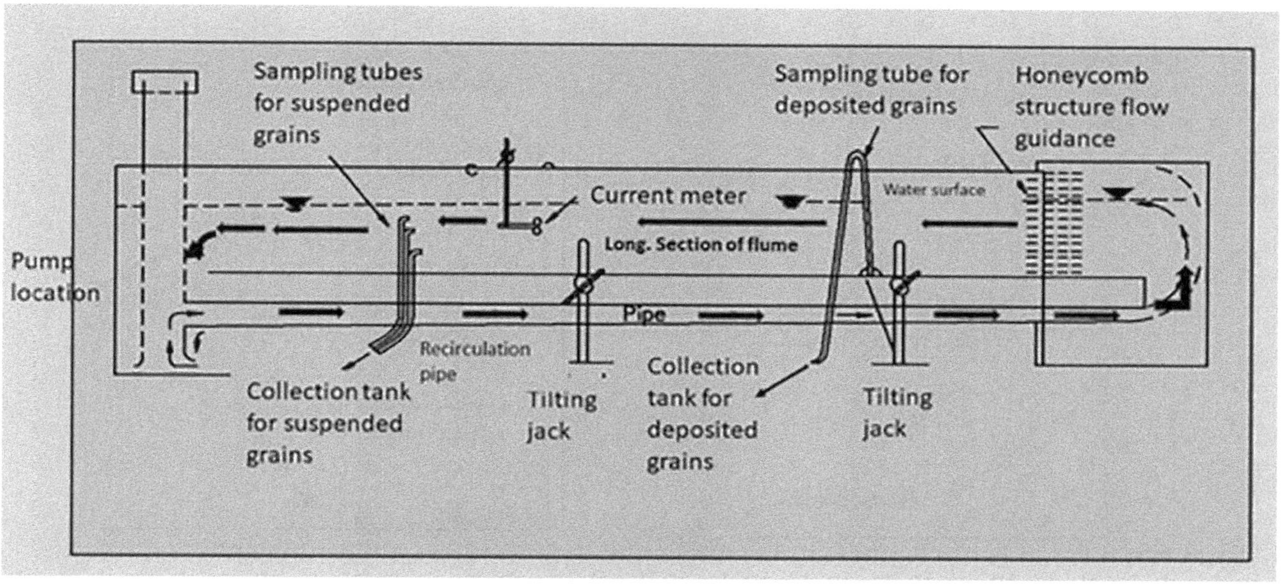

FIGURE 9.42 Experimental set up in the Uppsala University (UNGI) flume. (Modified from Ghosh et al. 1986.)

9.9.3 Uppsala (UNGI) Experiments

Experiments were conducted at Uppsala flume using two different sand beds as follows:

Bed-2. Over the sand bed 2 a high velocity was continued for a considerable time to take into suspension as much sediment as possible. While the pumps were running together with the suspended sediments, the suspended water was transferred into a storage tank with the help of an auxiliary pump. The sand bed 2 was then removed and the flume was thoroughly cleaned. Then the mixture was then pumped back into the clean flume from the storage tank. The total depth of the new water-sediment mixture in the flume became 15.2 cm. A flow velocity of 112 cm/s was generated in the flume to keep all the sediments in suspension again. This new suspension load, which formed the starting material for the experiments on deposition from suspension, was sampled, through siphon tubes located at 5 and 10 cm above the flume base. Grain-size distributions of these suspension load samples are shown in Figure 9.43a. At the next stage of the experiment, the flow velocity was reduced to 96.2 cm/s, and subsequently to 19.2 cm/s. After each stage of velocity reduction, samples of the materials descended from suspension were sucked out of two strips located at a distance of 2.5 and 7.5 m from the upstream end of the flume channel, with the help of a suction pipe of 1 cm diameter inserted from top (Figure 9.42). Grain-size distribution of the materials which remained hidden in the recirculation pipes (Figure 9.43b). The observed deposited materials from suspension on thin layer of (b) at 96 cm/s are shown in Figure 9.43c, and the observed and computed deposited materials from suspension at 19 cm/s are shown in Figure 9.43d. In fact, the grain-size distributions of these samples were studied by sieving after evaporation of water.

Bed-3. The whole experiment was repeated with another sand bed (bed-3) of different grain size distribution. The suspension load was initially sampled at a velocity of 105 cm/s, which was the starting point for the experiment. Grain-size distribution of this suspension load is shown in Figure 9.44a. At the next stage of the experiment, the flow velocity was reduced to 52cm/s and then to 29cm/s. After each stage of velocity reduction, samples of the materials sorted from suspension were sucked out in a similar way. Grain size distributions of these deposited materials (observed) are shown in Figure 9.44b,c. The results of Uppsala experiments demonstrate in a general way the similarity between grain size distribution patterns of the suspension and the deposited loads. The modes in the two sets of curves at

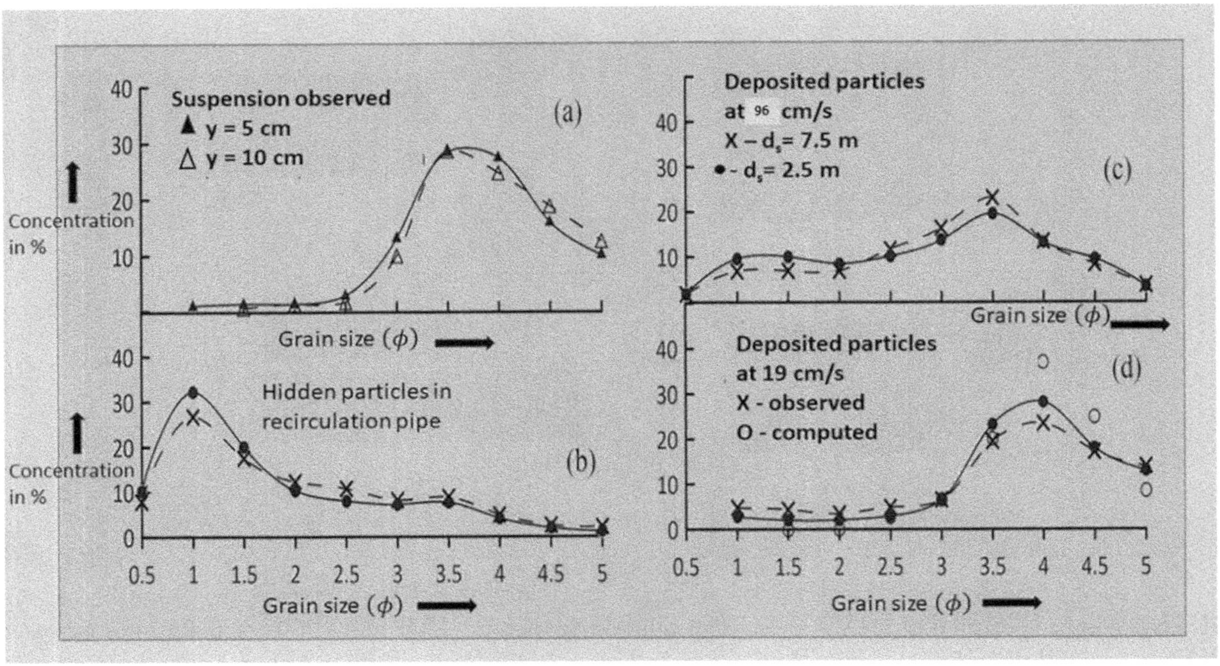

FIGURE 9.43 (a–d) Observed results from Uppsala flume on deposition from suspension load from Bed 2. (a) Grain size distribution of suspension at 112 cm/s at $y = 5$ and 10 cm above the bed. (b) Grain size distribution of the materials hidden in the recirculation pipes (leftover from the previous experiments). (c) Grain size distribution of the deposited materials from suspension on thin layer of (b) at 96 cm/s. (d) Grain size distribution of the deposited materials from suspension on thin layer of (b), at 19 cm/s (observed and computed values). Note that d_s is the distance of sampling from the entry. (Modified from Ghosh et al. 1986.)

around 3.5 phi match closely (Figures 9.43 and 9.44). In each of these curves, however, a second peak can be noticed on the coarser side, at around 1.0 phi.

9.9.4 Calcutta (ISI) Experiments

Experiments were conducted in a closed-circuit laboratory flume especially designed at the Indian Statistical Institute (ISI), Calcutta (Kolkata) to eliminate the difficulties faced at Uppsala flume. Both the experimental and the re-circulating channels of the Calcutta flume were exactly the same dimensions (10 m long × 50 cm wide × 50 cm high). The problem created at Uppsala flume by deposition of a part of the sediment, after being suddenly released into a wider experimental channel from pipes of smaller cross section, was avoided. Generally, a part of sediment was used to remain hidden into the narrower re-circulating pipe located below the experimental channel. The experimental channel had Perspex windows for a length of 6 m. Two non-clogging-type centrifugal pumps providing the flow are located outside the main channels of the flume (Chapter 7, Figure 7.14, ISI Flume for details). The intake and outflow pipes are freely suspended into the re-circulating channel from an overhead structure to allow tilting of the flume. Both the pump outlets are fitted with bypass pipes and valves, so that by manipulating these valves, the flow can be set at any desired speed up to 125 cm/s for a water depth of 30 cm. A honeycomb cage fixed at the upstream bend of the experimental channel ensured smooth, vortex-free flow of water through the experimental channel. The vertical velocity distribution in this flume also closely followed the logarithmic law as in Figure 9.45.

The nature of sediment load to be suspended in this flume at a flow velocity of about 124 cm/s was worked out from the results of the earlier experiments conducted at similar velocities (Ghosh et al.1981, Table 5, sample no. 5-124-20-A-D). A sediment load of 1,235 g having this predetermined grain-size

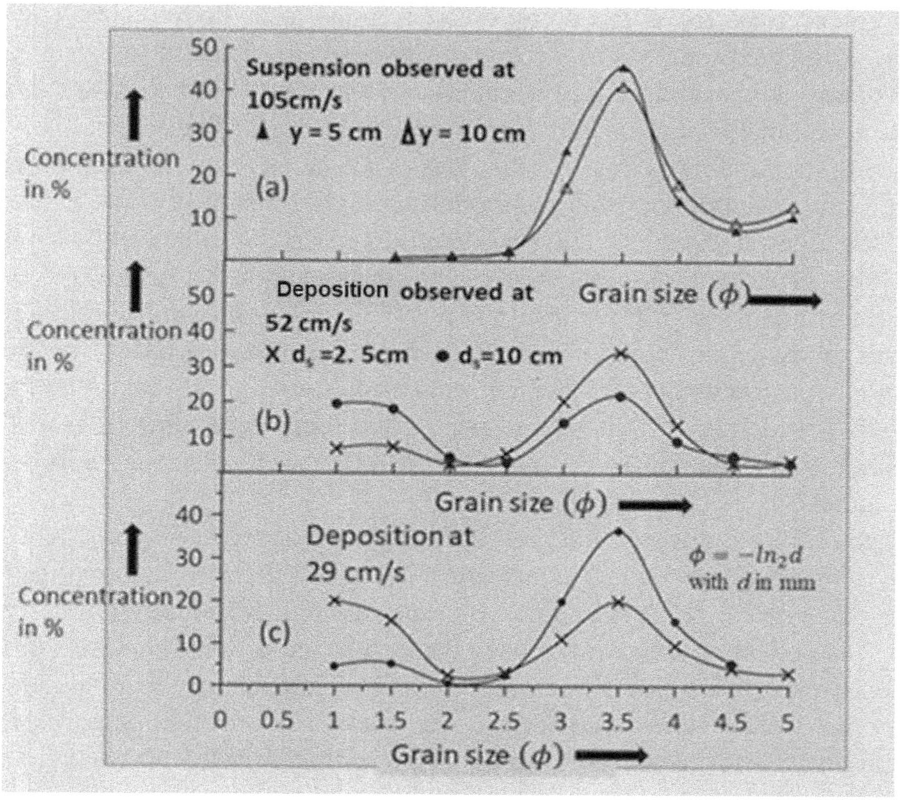

FIGURE 9.44 Observed results from Uppsala flume on deposition from suspension load obtained from Bed 3. (a) Size distribution of suspension load at 105 cm/s at $y = 5$ and 10 cm above the bed. (b) Grain size distribution of the materials deposited from suspension on a layer of 'leftover' at 52 cm/s, (c) Size distribution of the materials deposited from suspension on a layer of 'leftover' at 29 cm/s. Note that d_s indicates the distance of sampling from the entry. (Modified from Ghosh et al. 1986.)

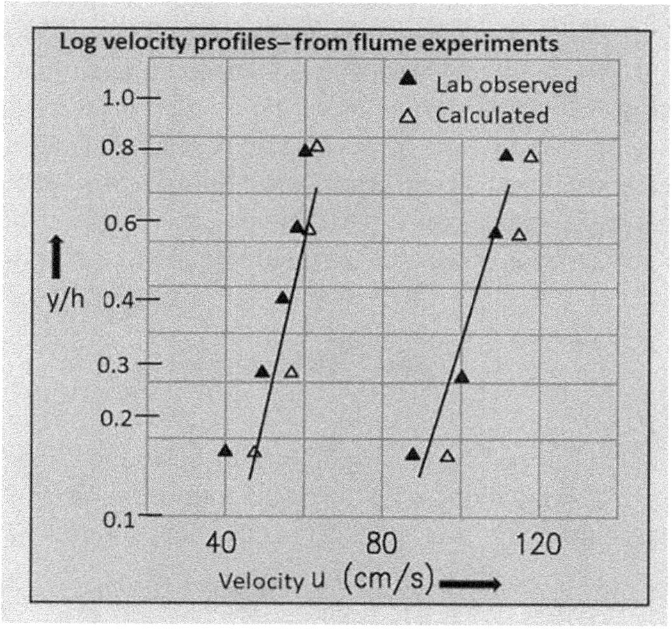

FIGURE 9.45 Logarithmic velocity distributions in the ISI-Calcutta flume.

distribution was quickly inserted at the pump outlet into the clear body of water in the flume (total volume = 4,000 L) while the flume was running at this velocity, to ensure thorough mixing of sediments with the whole volume of water. This load is called *input* in the present work. At this stage, the base of the experimental channel was absolutely free from any sediment load. Since the velocity at which the flow stabilized in the flume was considerably lower (106 cm/s) than that used for calculation of the input, a part of the load was dropped from suspension soon after insertion of the input. Twenty minutes after insertion of the input, when the system reached a steady state as indicated by a homogeneous mixture of water and sediment, the first sample of the materials found on the flume base were sucked out of the flume bed from a distance of 300 cm from the upstream end of the flume with the help of a siphon tube. The tube was run twice across the whole width of the flume to suck out all the sediments dropped on the flume bed, in a narrow strip of about 1 cm, into a container. The width of the suction tube (1 cm) and the angle (20°) at which it was put with respect to the flume base allowed grains as large as 0.30 cm to be easily sucked into the containers placed at the other end of the suction tubes (see Chapter-7, Figure 7.14 ISI Flume).

At the next stage of the experiment, the flow velocity was reduced to 50 cm/s. Allowing 15 minutes to stabilize the system at the new velocity condition, a new set of samples was again collected by siphon tubes from three strips on the flume bed surface located at 300, 500, and 700 cm from the upstream end of the flume. The collections of samples from three different distances were examined to study the nature of variation (if any) in the flow direction. For the development of theoretical model, the average of these three samples were done and provided a good enough sample for comparison. The techniques for processing and analysis of the samples are the same as that described by Sengupta (1975a, b). The results of ISI experiments demonstrate that the grain-size distributions of the input and the corresponding deposits formed from suspension at different velocities are shown in Figure 9.46. The similarity in grain-size-distribution patterns of the suspended loads and the materials deposited from suspension are striking in each case.

9.9.5 Mathematical Model

The essential finding from these experimental data is that the whole range of grain sizes in suspension is dropped to the bed when the flow velocity is reduced. This process cannot be simulated by any available mathematical equation relating suspension and bed loads. The well-known equations, formulated by Rouse (1938) and Hunt (1954, 1969), and modifications thereof, as proposed by Ghosh et al. (1981), indicate rapid exponential decay in concentrations of the relatively coarse-grained materials (coarser than 2.5 φ) in suspension with reducing the flow velocity. A mathematical model is developed which conforms to the experimental findings and can predict the grain-size distribution of the deposited material, given the size distribution in suspension at a higher velocity. The proposed model is as follows:

Suppose we write the concentration S per unit volume for grain size of settling velocity v_s and height y at a given velocity $u^{(1)}$ as

$$S\left(v_s, y, u_*^{(1)}\right) = K^{(1)} f\left(v_s, y, u_*^{(1)}\right) \tag{9.295}$$

where $u_*^{(1)}$ is the shear velocity at the bed corresponding to flow velocity $u^{(1)}$; $K^{(1)}$ is a constant to be determined for each grain size by matching with observed concentration (S_a) at reference level "a"; and f is a function that will be specified. For example, if f is obtained from the Rouse equation, then

$$f = \left(\frac{h-y}{y}\right)^{v_s / \kappa_0 u_*^{(1)}} \tag{9.296}$$

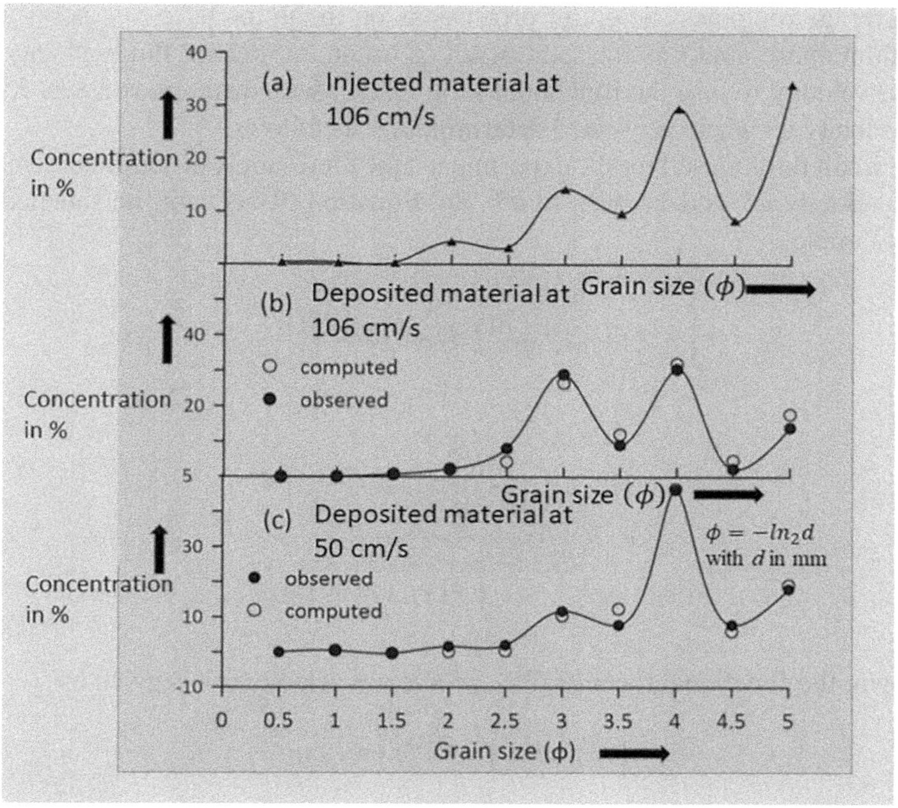

FIGURE 9.46 Plots of the ISI- Calcutta experiments. (a) Grain-size distribution of the input released into suspension at 106 cm/s. (b) Grain size distribution of the materials deposited from suspension at 106 cm/s at a distance 3 m from the upstream end of the flume (observed and computed values), (c) Size distribution of the materials deposited from suspension at 50 cm/s (computed and observed). The plots of the average samples collected from distances 300, 500 and 700 cm from the upstream end of the flume. (Modified from Ghosh et al. 1986.)

with h is the flow depth. Clearly $K^{(1)}$ is determined by

$$K^{(1)} = \frac{S_a}{\left(\frac{h-a}{a}\right)^{v_s/\kappa_O u_*^{(1)}}} \tag{9.297}$$

Substituting the value of $K^{(1)}$ from equation (9.297) and f from equation (9.296) in equation (9.295), we obtain

$$S\left(v_s, y, u_*^{(1)}\right) = \frac{S_a}{\left(\frac{h-a}{a}\right)^{v_s/\kappa_O u_*^{(1)}}} \left(\frac{h-y}{y}\right)^{v_s/\kappa_O u_*^{(1)}} \tag{9.298}$$

Now in order to estimate the total amount of sediment S_t present in the total volume of water in the flume as

$$S_t = A \int_{k_s}^{h} S\left(v_s, y, u_*^{(1)}\right) dy \tag{9.299}$$

where k_s is the average roughness height of protrusions on the flume base (see Schlichting 1967), A is the area of the flume base, and h is the total depth of water. At a lower fluid velocity $u^{(2)}$, which corresponds to shear velocity we use the functional form (9.295) with different constant $K^{(2)}$ for each grain size of settling velocity v_s, where $u_*^{(2)} < u_*^{(1)}$ determine $K^{(2)}$ as follows.

In the closed-circuit flumes used for the experiments, the total supply of sediment remains unchanged, even if the fluid velocity $u^{(1)}$ is decreased to $u^{(2)}$. So, Equation (9.299) will be satisfied for all velocity conditions. For example,

$$S_t = A\int_{k_s}^{h} K^{(1)} f\left(v_s, y, u_*^{(1)}\right)\, dy = AK^{(2)} \int_{k_s}^{h} f\left(v_s, y, u_*^{(2)}\right) dy \tag{9.300}$$

$$\text{Or, } K^{(2)} = \frac{K^{(1)} \int_{k_s}^{h} f\left(v_s, y, u_*^{(1)}\right)\, dy}{\int_{k_s}^{h} f\left(v_s, y, u_*^{(2)}\right)\, dy} \tag{9.301}$$

Once $K^{(2)}$ is known, the functional form (9.295) for a lower velocity $u^{(2)}$ is given by

$$S\left(v_s, y, u_*^{(2)}\right) = K^{(2)} f\left(v_s, y, u_*^{(2)}\right) \tag{9.302}$$

Hence, the deposited sediment concentration S_b at a lower velocity $u^{(2)}$ is determined by the average value between the levels k_s and y_0 as

$$S_b = \frac{1}{y_0 - k_s} \int_{k_s}^{y_0} S\left(v_s, y, u_*^{(2)}\right)\, dy \tag{9.303}$$

where y_0 is the maximum thickness of the bed/bed load sucked by the sampling tube of 1 cm diameter. The value of y_0 is worked out to be 0.3 cm, when the inclination of the siphoning tube is about 20° from the flume base. For the present data, equation (9.302) will be used, but a different choice of f, and hence of $S\left(v_s, y, u_*^{(2)}\right)$ is needed. The modified expression for $S\left(v_s, y, u_*^{(2)}\right)$ given by equation (9.303) is derived in the next section.

In order to determine the concentration function S, the conventional Rouse or the modified Hunt's equation proposed by Ghosh et al. (1981) is used. In practice, however, these two methods are found to work well only when the finer grains ($\phi \geq 3.0$) are estimated at high velocities (≥ 96 cm/s). For other combinations of grain sizes and velocities, the above-mentioned methods overestimate the concentration of the deposits substantially because the value of the exponent increases with increase of settling velocity v_s, for coarser grain or decrease of shear velocity u_*. It has been found that for $v_s / \kappa_o u_* > 1$, these methods lead to a much sharper rate of decay of concentration in suspension than that warranted by observation. To overcome the drawbacks of conventional methods mentioned above, the following model is proposed:

The conventional Rouse equation for suspension may be written as:

$$\epsilon_s \frac{dS}{dy} + v_s S = 0 \tag{9.304}$$

where v_s is the settling velocity of a particle, and ϵ_s, the sediment diffusion coefficient, is given by

$$\epsilon_s = \kappa_o u_* y \tag{9.305}$$

subject to the logarithmic velocity

$$\frac{\bar{u}}{u_*} = \frac{1}{\kappa_o} \log_e \frac{y}{k_s} \tag{9.306}$$

To get rid of dimension, equation (9.304) may be written in non-dimensional form as

$$\frac{1}{S}\frac{dS}{d\eta} = -\frac{v_s}{\epsilon(\eta)} \tag{9.307}$$

$$\text{and } \epsilon(\eta) = \kappa_o \eta (1-\eta) \tag{9.308}$$

where

$$\eta = \frac{y}{h},\ v_s = V_s/u_*,\ \epsilon = \frac{\epsilon_s}{u_* h} \tag{9.309}$$

The terms in equation (9.309) are dimensionless variables.

Equation (9.307) represents the relative change in concentration $\frac{1}{S}\left(\frac{dS}{d\eta}\right)$. The magnitude of exponential decay in concentration indicated by equation (9.307) however, is much larger than that observed in the experiments, particularly near the bed where $\epsilon(\eta)$ is very small. To dampen the effect of exponential decay, therefore, equation (9.307) is modified as

$$\frac{1}{S}\left(\frac{dS}{d\eta}\right) = -\frac{v_s}{\epsilon(\eta)} + \frac{\beta_1}{\epsilon(\eta) S} \tag{9.310}$$

Thus the rate of relative change is reduced by $\frac{\beta_1}{\epsilon(\eta) S}$, where the constant $\beta_1 > 0$ is the dampening factor. To solve equation (9.310), we use the boundary condition at the bed as

$$\epsilon \frac{dS}{d\eta} + (1-\alpha) v_s S = 0 \text{ at } \eta = k_s/h \tag{9.311}$$

where $(1 - \alpha)$ is the probability that a particle which comes in contact with the bed will be lifted. The total sediment concentration S_{tot} supplied in the total amount of water Q in the flume is given by

$$\int_{\frac{k_s}{h}}^{1} S(\eta) d\eta = S_{tot}/Q \tag{9.312}$$

Using the expression of $\epsilon(\eta)$ from equation (9.308), the integration of equation (9.310) with respect to η is

$$S(\eta) = \frac{1}{v_s}\left[\left(\frac{1-\eta}{\eta}\right)^{v_s/\kappa_o} \left(e^{-v_s k_1}\right) + \beta_1\right] \tag{9.313}$$

where the integrating constant k_1 and the dampening factor β_1, determined from the conditions in equations (9.311) and (9.312), are given by

$$\beta_1 = \frac{\alpha}{1-\alpha}\left(\frac{1-k_s/h}{k_s/h}\right)^{v_s/\chi} e^{-v_s k_1} \tag{9.314}$$

with

$$e^{-v_s k_1} = \left(S_{\text{tot}}/Q\right) v_s \Bigg/ \int_{\frac{k_s}{h}}^{1} \left[\left(\frac{1-\eta}{\eta}\right)^{v_s/\kappa_o} + \frac{\alpha}{1-\alpha}\left(\frac{1-k_s/h}{k_s/h}\right)^{v_s/\kappa_o}\right] d\eta \tag{9.315}$$

Therefore, equation (9.313) can be written as

$$S(\eta) = \frac{e^{-v_s k_1}}{v_s}\left[\left(\frac{1-\eta}{\eta}\right)^{v_s/\kappa_o} + \frac{\alpha}{1-\alpha}\left(\frac{1-k_s/h}{k_s/h}\right)^{v_s/\kappa_o}\right] \tag{9.316}$$

The factor $\frac{e^{-v_s k_1}}{v_s}$ in equation (9.316) plays the same role as $K^{(1)}$ in equation (9.295), and the remaining terms as that of f. Hence, equation (9.303) becomes in dimensionless form as:

$$S_b = \frac{1}{\eta_0 - \frac{k_s}{h}} \int_{\frac{k_s}{h}}^{\eta_0} S(\eta)\, d\eta$$

$$= \frac{\frac{\left(S_{\text{tot}}/Q\right)}{\eta_0 - k_s/h} \int_{k_s/h}^{\eta_0} \left\{\left(\frac{1-\eta}{\eta}\right)^{v_s/\kappa_o} + \frac{\alpha}{1-\alpha}\left(\frac{1-k_s/h}{k_s/h}\right)^{v_s/\kappa_o}\right\} d\eta}{\int_{k_s/h}^{1} \left\{\left(\frac{1-\eta}{\eta}\right)^{v_s/\kappa_o} + \frac{\alpha}{1-\alpha}\left(\frac{1-k_s/h}{k_s/h}\right)^{v_s/\kappa_o}\right\} d\eta} \tag{9.317}$$

The integrals $\int_{k_s/h}^{\eta_0}\left(\frac{1-\eta}{\eta}\right)^{v_s/\kappa_o} d\eta$, and $\int_{k_s/h}^{1}\left(\frac{1-\eta}{\eta}\right)^{v_s/\kappa_o} d\eta$ are evaluated by Trapezoidal rule. The probability α is found out from Gessler (1965, equation 8). The technique of computation of the deposited materials, although developed primarily for relatively coarse grains and low velocities, works for fine grains and high velocities as well. In the latter case, the dampening factor β_1 in equation (9.310) turns out to be very small, so that there is not much difference between equations (9.304) and (9.310).

Stochastic interpretations of the processes involved in this model have been presented elsewhere (Ghosh et al. 1984). It is postulated that the flow carries up some materials that come to the bottom and are partially redistributed over the suspension zone. Stochastically, one can think of a particular grain as being displaced according to a Markovian law specified by the diffusion equation. Its behavior at the lower boundary consists of partial reflection and partial jump to a new height. It is believed that equation (9.310), involving the dampening factor β_1, approximates the stationary distribution in this setup. As noted in Ghosh and Mazumder (1981), the Rouse equation itself gives the stationary distribution of the corresponding Markov process when the lower boundary is completely reflecting. Comparison of observed and computed values of suspension and deposits are shown in Figures 9.43 (Bed 2) and 9.44 (Bed 3) for Uppsala (UNGI) experiments, and in Figure 9.46 for Calcutta (ISI) experiments. It is clearly observed from the figures that the nature of the grain-size distributions in both experiments could be closely simulated by the mathematical model developed in this work. The errors between the observed and the computed values (both for absolute values and proportions in the case of the Calcutta experiments) have been computed by the following formula:

$$\text{Error} = \sqrt{\sum \frac{(S_c - S_0)^2}{S_0^2}\left(\frac{S_0}{T}\right)} = \sqrt{\sum \frac{(S_c - S_0)^2}{S_0 T}} \quad (9.318)$$

where S_c = computed values, S_0 = observed values, T = total value of observed. The results are shown in Table 9.8. The errors are within reasonable limits for both experiments. The errors are particularly small when the proportions of deposited materials have been considered.

9.9.5.1 Interpretation of Paleo-Flow Velocities

Controlled experimental studies have shown that the grain-size distribution of suspension under unidirectional flow is related to flow velocity, height of suspension, and nature of bed (source) materials (Sengupta 1975a, 1975b, 1979). Following these clues, a mathematical model was developed by Ghosh and Mazumder (1981) to explain the generation of unimodality, symmetry, and lognormality in suspension loads, even when the size distribution at the bed (source) is not log-normal. An interesting outcome of this study was the formulation of a relationship between the height of suspension and the shear velocity for lognormal distributions with specific mean grain sizes. Thus, the curves provided by Figure 9.41 and Ghosh and Mazumder (1981, Fig. 4) afford an opportunity of computation of paleo-flow velocity (in terms of shear velocity u_*) from the mean grain size of the suspended load, when the height of suspension can be inferred with a reasonable degree of accuracy.

Results of deposition from suspension show that the whole size range present in suspension is deposited, and the mean and the modal sizes of the suspension load may under certain conditions be faithfully represented in the deposits formed from suspension at a reduced flow velocity (Ghosh et al. 1986). Following the arguments given above, it should now be possible to estimate the paleo-flow velocity from the grain-size distribution of the deposits from suspension which are preserved in the geological record. Under suitable conditions, the other unknown parameter can be estimated from the height of sedimentary structures, keeping in mind that dune height bears a well-established relationship with water depth (Allen 1968, p. 171). However, this remains to be tested in the field. In practice, a problem might arise due to variation of magnitude of bed forms due to velocity variations (Harms et al.1982, figs. 2–8).

Sengupta et al. (1991), in their subsequent work, re-examined the validity of the conventional method of grain size interpretation based on log-probability plots and the relationship between the log-normal and log-hyperbolic models. For this purpose, the data collected from the earlier studies from natural stream as well as experimental channels were utilized. Mathematical methods are presented as: (1) for direct computation of suspension grain size distributions from bed's grain size distribution, and (2) for estimation of the range of flow parameters for log-normally distributed suspended loads. This is the

TABLE 9.8
Errors between Computed and Observed Deposits

	UNGI Experiments	ISI Experiments	
Deposits Formed at	**19 cm/s**	**106 cm/s**	**50 cm/s**
E (for absolute values)		0.57	0.23
	-------	------	-------
E (for ratios)	0.42	0.38	0.23

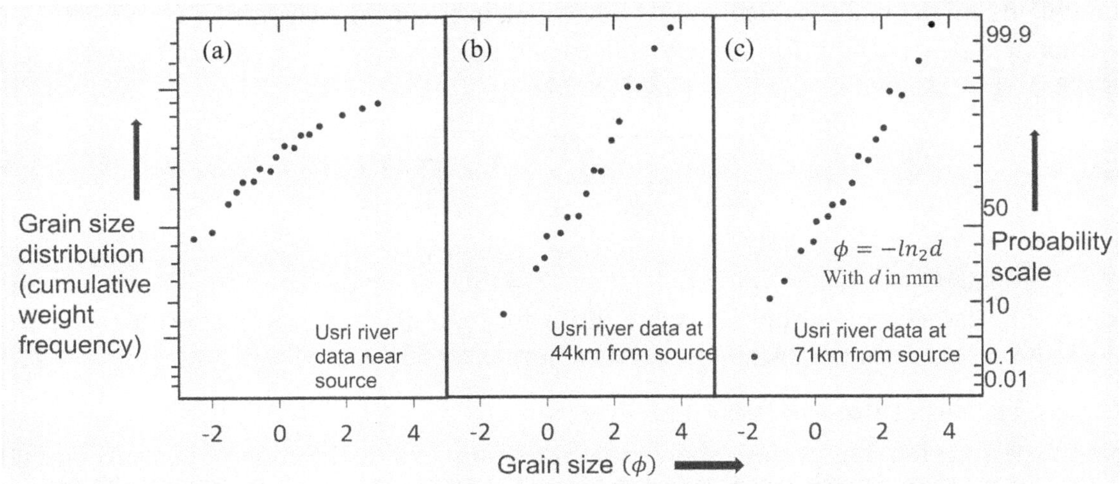

FIGURE 9.47 Grain-size distributions of the Usri River bed samples, showing the development of log-normality with increasing distance from the source. Sample (a) collected from the source, sample (b) from a distance 44 km from the source, sample (c) after 71 km distance from the source. (Modified from Sengupta et al. 1999.)

review work done by the authors with the same grain size problems. A review of the experimental and theoretical studies conducted by other authors is beyond the scope of this book. Grain size distribution of sediment samples collected (USR 43) from the Usri River in Bihar, India were analyzed at different distances from the source. The phi probability plots of the source material were irregular, but with the distance from the source the linearity in plots developed. At about 44 and 71 km downstream, a straight line could be fitted (Sengupta 1975), indicating that the log-normality attained under fluvial transportation, even at non-lognormal at the source (Figure 9.47). Similar feature was also observed in Dwarkeswar River in West Bengal, India with increase in the distance from the source (Sengupta et al. 1999). The mechanism of sorting process from the non-lognormal source to log-normal grain size distribution is investigated with distance from the source. A theory for development of log-normality based on a sorting hypothesis was discussed by Ghosh (1988).

9.9.6 Mazumder's Model (1994)

Mazumder (1994) developed a theoretical model to compute the suspension grain-size distribution based on the Hunt's diffusion equation, taking into account the effect of hindered settling due to the increased suspension concentration. Fluid velocity closest to the bed is estimated using the concept of migration velocities of particles in the bed layer. The fall velocity of sediment v_{sm} varies with concentration C as a result of hindered settling effect (Maude and Whitemore 1958) and is given by

$$v_{sm} = v_0 (1 - c)^{\alpha} \tag{9.319}$$

where v_0 is the fall velocity in clear water and α is the exponent of reduction of fall velocity, which varies from 2 to 5 depending on the Reynolds number, and size of non-cohesive sediment particles. Here the migration velocity u_s of a sediment particle in the bed layer is a function of shear velocity u_* and the critical shear velocity u_{*c} corresponding to the condition of Shield's grain movement of different bed materials. The empirical relation for the migration velocity of a sediment particle in the bed layer is given by Engelund and Fredsoe (1976) as:

$$u_s = \beta u_* \left[1 - 0.7 \sqrt{\frac{u_{*c}}{u_*}} \right] \tag{9.320}$$

where $\beta = 9.0$ for sand particles. As it is difficult to measure the fluid velocity at the top of the bed load layer, it would be more reasonable to assume the migration velocity of the representative size at the level $y = 0.25$ cm (top of the bed layer) to be equal to the fluid velocity at that layer, rather than extrapolation. Therefore, the fluid velocity at the interface will be as: $u = u_s$ at $y = y_1$.

The vertical distribution of sediment concentration above the bed from k_s to y is obtained by combining the settling velocity and modified Hunt's diffusion equation as:

$$\int_{k_s}^{y} \frac{dC}{C(1-C)^{\alpha+1}} = -\int_{k_s}^{y_1} \frac{v_0}{\epsilon_{sb}} dy - \int_{y_1}^{y} \frac{v_0}{\epsilon_{sy}} dy \tag{9.321}$$

where the sediment diffusion coefficient ϵ_{sb} for bed layer zone is as:

$$\varepsilon_{sb} = \frac{u_*^2}{u_s}(y_1 - k_s)(1 - y/h) \tag{9.322}$$

and that for the suspension zone from equation (9.271) as.

$$\varepsilon_{sy} = 2\kappa_o h u_* (1-\xi)\left\{A - \sqrt{(1-\xi)}\right\} \tag{9.271}$$

If $\alpha = 0$ in equation (9.321), the equation is exactly similar to Ghosh et al. (1981) except for the values of u_s and constant A is given by equation (9.269). As α increases to value 5, the particle settling velocity changes due to grain-grain interaction in suspension, but if α is very large, then $v_{sm} \to 0$. To determine the concentration $C(y)$ at any height, numerical integration of equation (9.321) is performed for $\alpha = 0$, 3, 4 and relative bed concentration C_{ks} by trapezoidal rule. The values of α are used from the figure of Maude and Whitemore (1958). Once the concentration is known, the relative concentration C_y' may obtained as

$$C_y' = C_y \Big/ \sum_{\phi} C_y(\phi) \tag{9.323}$$

Equation (9.323) is used to compute the relative suspension concentration C_y' of a given grain size with settling velocity v_0 (in clear water) at any height $y > y_1$ above the bed for a given relative concentration $C_{k_s}(\phi)$. Equation (9.323) corresponding to equation (9.321) gives the average relative bed layer concentration $C_{bl}(\phi)$ of sediment of different sizes. Data used for the verification of the present model are taken from the earlier publications by Sengupta (1975a, 1975b, 1979) and Ghosh et al. (1979, 1981, and 1986). Observed and computed values of suspended grain size distribution above two sand beds have been shown in Figure 9.48. The size distributions of two different beds (2 and 3) are used for computation (Figure 9.48a). Observed and computed values of suspended grain-size distributions above two sand beds are shown in Figure 9.48b,c for several values of α, exponent of reduction of fall velocity. The relative errors between the observed and computed values are shown by the formula (9.318). It is clear that for bed 2, the smallest error is obtained by the present method, whereas for bed 3, the smallest error is obtained by the Gessler's method. On the whole, errors obtained by different methods are the same order, but it seems that the present method for both beds 2 and 3 gives better accuracy than the method developed by Ghosh et al. (1981). The present method can be claimed to be a more realistic one because it accounts the concepts of migration velocity of particles in the bed layer and the hindered settling effect. This present method affords direct computation of suspended load from the bed's grain size distributions without going through an intermediate stage bed load as required by the other methods.

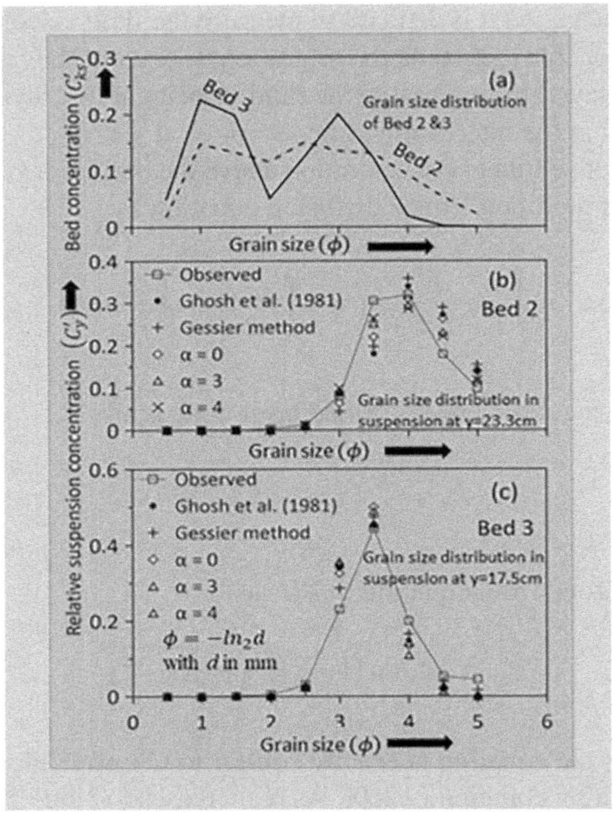

FIGURE 9.48 (a) Grain size distributions of sand beds 2 and 3, (b) Grain size distribution in suspension at $y = 23.3$ cm above sand bed 2, and (c) Grain size distribution in suspension at $y = 17.5$ cm above sand bed 3. (Modified from Mazumder 1994.)

9.9.7 Other Developments

Sengupta et al. (1999) presented an excellent review to summarize the features of grain size sorting process during transportation in the field and in laboratory. Sediments are transported in a stream as traction, saltation and suspension modes. Field data analyses from the Usri, Dwarkeswar Rivers and many other streams showed a general trend of decrease in mean grain size of traction load in the downstream direction, leading to a log-normal grain size distribution. Controlled flume experiments suggest that the process of mechanism of downstream fining is related to the migration of bed forms. The sediment transported over the gentle stoss faces of the bed forms are sorted out by two parts. The coarser materials sliding down the steeper lee faces of the dunes are buried, while the finer fractions are carried forward showing a progressive downstream fining. This mode of grain sorting with a selected range of sizes in the bed is observed to generate the log-normal distribution during transportation. The suspension load over the sand bed was experimentally investigated in different laboratories, using the closed-circuit hydraulic flumes for a desired length of time. The suspension concentration increased with increase in flow velocity and/or decrease of mean grain sizes of the bed materials. An important observation was made that the grain size distribution in suspension was always uni-modal irrespective of the grain size distribution in the bed.

There were two sorting processes: one is from bed-to-bed layer and other one is from bed layer to suspension. The sorting from bed-to-bed layer is computed by one of the well-known bed load equations; and the suspension concentration is computed by Rouse equation or a modified Hunt's equation, which allows the direct computation of suspension grain size distribution from the grain size distributions of bed materials. The necessary conditions for the generation of unimodality, symmetry and log-normality of grain size distributions in suspension were worked out. The log-normality can generate a special case of hyperbolic distribution by a process of grain sorting. Hindered settling effect in dense suspension and similar other processes are likely to be responsible for reducing the settling velocity of the grains. The influence of bed roughness on sediment suspension during transportation was broadly investigated by Sengupta et al. (1999). Their findings showed that increase of bed roughness needed higher velocity for initiation of ripples, and a higher flow velocity was needed for suspending comparable amounts of materials in suspension, when bed roughness was increased. The experimental results on the grain size distribution in suspension as well as deposits are significant because they allow estimating the palaeo-flow conditions from the grain size distribution of the deposits, using the relation developed by Ghosh and Mazumder (1981). Bhattacharya et al. (2000) presented Rouse equation as the steady state distribution of sand grains of different sizes under a Markovian diffusion process governed by the turbulent diffusion. They developed a unified model to estimate the suspension as well as deposition of sand materials.

Mazumder et al. (2005) developed a theoretical model to compute the suspended grain-size distribution based on the coupled diffusion equations of sediment and water, using the computed bed-layer concentration as a reference. The equality in diffusion coefficients in sediment and water is not strictly accurate but is a close approximation for small particles. The effect of suspension concentration into the mean velocity, turbulent and viscous shear stresses owing to the dynamic coupling between the flow and sediments in suspension is investigated. A substantial reduction of particle fall velocity occurred due to the increased suspension concentration. So, the hindered settling due to the increased concentration in suspension was taken into consideration (Richardson and Zaki 1954). According to Mendoza and Zhou (1995), the perturbed mean velocity u_p is considered as the sum of the unperturbed mean velocity u of clear water and the perturbation of mean velocity u_{ps} due to suspension, i.e., $u_p = u + u_{ps}$. The perturbed mean velocity u_{ps} due to suspension is approximated as:

$$\frac{u_{ps}}{u_*} = -\frac{A\delta}{2\kappa_o}\left[D_1(1-\xi)+D_2(1-\xi)^2+\cdots\right]+\left(\frac{1}{\kappa_o}+N_1\right)\ln\xi+N_2 \tag{9.324}$$

where D_1 and D_2 are dimensionless constants representing derivatives of concentration of different order at the upper surface; N_1, N_2 are dimensionless integrating constants; and δ is the reference volume concentration near the bottom of the channel. Here the unperturbed men velocity u is used as usual log-law with the wake-function. Mendoza and Zhou (1995) suggested a unified expression for the perturbed mean velocity u_p as:

$$\frac{u_p}{u_*} = \frac{1}{\kappa_o}\ln\xi + F_1(1-\xi) + F_2(1-\xi)^2 + \frac{1}{\kappa_o}(h/y_0) \tag{9.325}$$

where F_1 and F_2 are dimensionless constants to be estimated from the velocity data. Explicit expressions of F_1 and F_2 are given by:

$$F_1 = -\frac{\delta}{2\kappa_o}AD_1 - N_1, \quad F_2 = -\frac{\Pi}{2\kappa_o}\pi^2 - \frac{\delta}{2\kappa_o}AD_2 - \frac{N_1}{2} \tag{9.326}$$

The concentration gradient $\frac{dC}{d\xi}$ is given in the paper of Mazumder et al. (2005). To solve the velocity and concentration equations, they used the bed layer thickness as reference according to Wiberg and Rubin (1989) and the bed layer concentration as a reference suggested by Smith and McLean (1977). Grain-size frequency distribution of suspended loads at different heights above the four sand beds of different size distributions have been computed using the computed bed load equation as the reference concentration. The validity of the present model has been tested with the experimental data of suspended load samples collected in the laboratory flumes at Indian Statistical Institute (ISI), Calcutta, and Uppsala University under controlled conditions. Comparisons of the results of the present model with observed data show reasonably good agreement.

9.10 SEDIMENT CONCENTRATION IN SUSPENSION

9.10.1 Van Rijn's Equation (1984)

In a steady, uniform flow, the vertical distribution of the sediment concentration profile can be described by:

$$\epsilon_{sy}\frac{\partial C}{\partial y}+(1-C)Cv_{sm}=0 \tag{9.327}$$

in which C = sediment concentration; v_{sm} = particle fall velocity in a fluid-sediment mixture; ϵ_{sy} = sediment diffusion coefficient; and y = vertical coordinate. Experiments with high sediment concentrations in suspension have shown a substantial reduction of the particle fall velocity. For normal flow with suspended particles in the range of 50–500 microns, the reduced fall velocity of particles can be described by Richardson-Zaki type (1954) and Maude and Whitemore (1958) as:

$$v_{sm}=v_0(1-c)^{\alpha} \tag{9.328}$$

In the present analysis, the sediment diffusivity ϵ_{sy} is parabolic distribution in the lower half of the flow depth and a constant value $\epsilon_{s,\max}$ in the upper half of the flow depth (Figure 9.49) are used mainly because it may give a better description of the concentration profile. The parabolic-constant distribution is written as:

$$\epsilon_{s,\max}=0.25\kappa_o u_* h \;\text{ for }\; \xi\geq 0.5 \tag{9.329a}$$

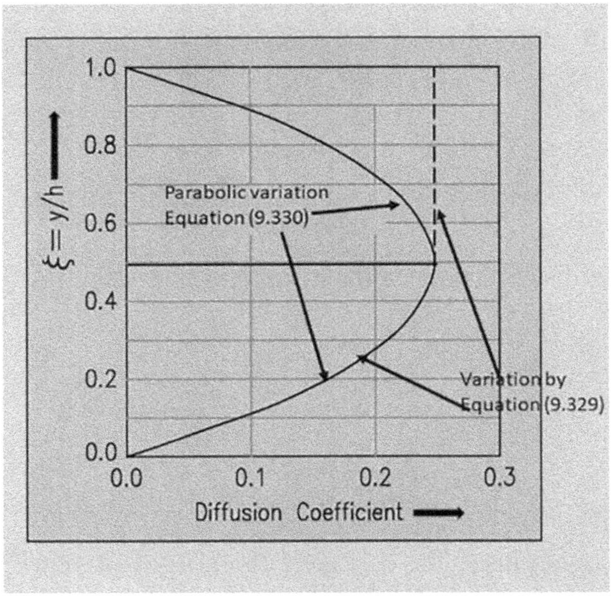

FIGURE 9.49 Profiles of fluid diffusion coefficient. (Modified from Van Rijn 1984b.)

$$\epsilon_{sy} = 4\xi(1-\xi)\epsilon_{s,\max} \quad \text{for} \quad \xi < 0.5 \tag{9.329b}$$

The diffusion of fluid momentum ϵ_m is given by a parabolic distribution over the flow depth h:

$$\epsilon_m = \xi(1-\xi)\kappa_o u_* h \tag{9.330}$$

Equations (9.329) and (9.330) are plotted in Figure 9.49. The diffusion of sediment particles (ϵ_{sy}) is related to the diffusion of fluid momentum ϵ_m by:

$$\epsilon_{sy} = \beta\epsilon_m\varphi \tag{9.331}$$

where β is the proportionality factor which describes the diffusion of sediment particles and diffusion of fluid particles, assumed to be constant over the flow depth and φ indicates the damping factor of the fluid turbulence by the sediment particles and is assumed to be dependent on the local sediment concentration.

Considering $\varphi = 1$ (no damping effect) and using the parabolic-constant distribution according to equations (9.329) and (9.331) with concentration-dependent fall velocity equation (9.328) in (9.327), one gets after integration as:

$$\sum_{n=1}^{4}\left[\frac{(1-C_{\xi_a})^n-(1-C)^n}{n(1-C)^n(1-C_{\xi_a})^n}\right]+\ln\left[\frac{C(1-C_{\xi_a})}{C_{\xi_a}(1-C)}\right]=\ln\left[\frac{\xi_a(1-\xi)}{\xi(1-\xi_a)}\right]^{Z_*} \quad \text{for } \xi < 0.5 \tag{9.332a}$$

$$\sum_{n=1}^{4}\left[\frac{(1-C_{\xi_a})^n-(1-C)^n}{n(1-C)^n(1-C_{\xi_a})^n}\right]+\ln\left[\frac{C(1-C_{\xi_a})}{C_{\xi_a}(1-C)}\right]=-Z_*\left[\ln\left(\frac{\xi_a}{(1-\xi_a)}\right)+4(\xi-0.5)\right] \text{for } \xi \geq 0.5 \tag{9.332b}$$

where $Z_* = \dfrac{v_{sm}}{\beta\kappa_o u_*}$, which implies the influence of the upward turbulent fluid forces and the downward gravitational forces. For small concentrations $\left(C < C_{\xi_a} < 0.001\right)$, equation (9.332) reduces to:

$$\frac{C}{C_{\xi_a}} = \left[\frac{\xi_a(1-\xi)}{\xi(1-\xi_a)}\right]^{Z_*} \quad \text{for } \xi < 0.5 \tag{9.333a}$$

$$\frac{C}{C_{\xi_a}} = \left[\frac{\xi_a}{(1-\xi_a)}\right]^{Z_*}\cdot\exp\left[-4Z_*(\xi-0.5)\right] \text{for } \xi \geq 0.5 \tag{9.333b}$$

The influences of reference concentration C_{ξ_a} and the reference level ξ_a are elaborately explained in van Rijn (1984). He also used the factor- φ as a function of concentration for best agreement with measured concentration as:

$$\varphi = 1+\left(\frac{C}{C_{\max}}\right)^{0.8} - 2\left(\frac{C}{C_{\max}}\right)^{0.4} \tag{9.334}$$

where $C_{\max} = 0.65$ = maximum volumetric bed concentration. In addition, he considered the empirical relation of β -factor described as a function of settling velocity as:

$$\beta = 1+2\left[\frac{v_{sm}}{u_*}\right]^2 \quad \text{for } 0.1 < \frac{v_{sm}}{u_*} < 1 \tag{9.335}$$

According to the present results, β -factor is always larger than unity, indicating the dominating influence of the centrifugal forces.

9.10.2 Woo et al.'s Equation (1988)

In a steady, uniform flow, the vertical distribution of the sediment concentration profile can be described by:

$$\epsilon_{sy}\frac{\partial C}{\partial y}+(1-C)Cv_{sm}=0 \tag{9.327}$$

in which C = sediment concentration; v_{sm} = particle fall velocity in a fluid-sediment mixture; ϵ_{sy} = sediment diffusion coefficient; and y = vertical coordinate. Experiments with high sediment concentrations have shown a substantial reduction of the particle fall velocity due to the presence of sediment particles. The representative fall velocity of the mixture v_{sm} varies with the volumetric concentration C as:

$$v_{sm}=v_0(1-c)^{\alpha} \tag{9.336}$$

They assumed the exponent α varies with the particle Reynolds number and the particle shape and decreases from 4.65 to 2.35 for non-cohesive particles from silts to gravel (see equations (8.33) and (8.34)). The representative fall velocity in clear water v_0 is calculated from the fall velocity v_{0i} of individual size fractions as:

$$v_0=\sum_{i=1}^{N}\frac{v_{0i}\bar{C}_i}{\bar{C}} \tag{9.337}$$

where i denotes each one of the N-size fractions of sands in suspension, whereas the over bar indicates the depth-averaged values of the concentration. At a large concentration in suspension, they considered the total shear stress as a sum of the viscous shear stress and the turbulent shear stress, i.e., $\tau=\tau_t+\tau_v$. The turbulent shear stress τ_t is a function of concentration in the fluid sediment mixture as:

$$\tau_t=\rho_f\epsilon_{sy}\frac{d\bar{u}}{dy}\left(1+\frac{(\rho_s-\rho_f)}{\rho_f}C\right) \tag{9.338}$$

and the viscous shear stress also depends on the concentration C, dynamic viscosity μ_f and the velocity gradient $\frac{d\bar{u}}{dy}$ as:

$$\tau_v=g(C)\mu_f\frac{d\bar{u}}{dy} \tag{9.339}$$

where $g(C)$ is expressed as function concentration, which is given by Thomas (1965) as:

$$g(C)=1+2.5C+10.05\,C^2+0.0027\exp(16.6C) \tag{9.340}$$

which follows Einstein's equation as $g(C)=1+2.5C$ for low concentration. Accordingly, the vertical distribution of shear stress

$$\frac{\tau}{\tau_0}=\frac{1-\xi+\frac{(\rho_s-\rho_f)}{\rho_f}\int_{\xi}^{1}C\,d\xi}{1+\frac{(\rho_s-\rho_f)}{\rho_f}\bar{C}} \tag{9.341}$$

with $\tau_0=\rho_f u_*^2\left(1+\frac{(\rho_s-\rho_f)}{\rho_f}\bar{C}\right)$

Using equations (9.338), (9.339), and (9.340), one gets the sediment diffusion coefficient ϵ_{sy} from the relation $\epsilon_{sy}=\beta\epsilon_m$ as:

$$\epsilon_{sy}=\beta\left[\frac{u_*^2\left(1-\xi+\frac{(\rho_s-\rho_f)}{\rho_f}\int_{\xi}^{1}C\,d\xi\right)}{\frac{d\bar{u}}{dy}\left(1+\frac{(\rho_s-\rho_f)}{\rho_f}C\right)}-\frac{\upsilon_f g(C)}{\left(1+\frac{(\rho_s-\rho_f)}{\rho_f}C\right)}\right] \tag{9.342}$$

where υ_f is the kinematic viscosity of the fluid. To determine ϵ_{sy} from equation (9.342), one needs the find the velocity gradient from the log-law. Substitution of ϵ_{sy} from equation (9.342) and the settling velocity from equation (9.336) in the diffusion equation (9.327), the expression for concentration gradient is given by:

$$\frac{dC}{dy}=\frac{v_{sm}}{\beta}\left[\frac{\frac{d\bar{u}}{dy}F(C)}{\upsilon_f g(C)\frac{d\bar{u}}{dy}-u_*^2\left(1-\xi+\frac{(\rho_s-\rho_f)}{\rho_f}\int_{\xi}^{1}C\,d\xi\right)}\right] \tag{9.343}$$

where $F(C)=C(1+C)^{\alpha+1}\left(1+\frac{(\rho_s-\rho_f)}{\rho_f}C\right)$.

To solve equation (9.343), one needs the velocity gradient from the velocity expression, either from the log-law or from the power law. Considering the power law velocity gradient, equation (9.343) generates an integro-differential equation called "Volterra Integral," which after differentiation with respect to ξ converts to the second-order nonlinear ordinary differential equation. To solve the second-order differential equation, one needs two boundary conditions. The fifth- and sixth-order Runge-Kutta method (Woo 1985) provides a numerical solution of the developed integro-differential equation, using the iterative procedure for which the first guess value is obtained from the Rouse equation. Then the equation is solved.

9.10.3 Umeyama's Equations (1992, 1999)

Umeyama (1992) proposed a theoretical model for velocity distribution in sediment-laden flow by means of a new mixing length concept. For sediment-laden flow, they assumed the mixing length hypothesis as a function of sediment concentration. They confirmed that the mixing length for highly concentrated flow is lower than that for lower concentrated sediment-laden flow. They found from the data analysis that the mixing length hypotheses by Prandtl and von-Karman, utilized to derive the vertical velocity equation, fail to agree with the experimental values for sediment-laden flow in the region outside the near-bed layer. Therefore, in order to improve these hypotheses, the following mixing length was proposed:

$$l = \kappa y (1-\xi)^{0.5\left[1+\alpha\left(C/C_{\xi a}\right)\right]} \tag{9.344}$$

where $\xi = y/h$, α is a constant to be determined by the experimental data, and C_{ξ_a} is the reference concentration at $\xi = \xi_a$. The shear stress for sediment-laden flow may be expressed as:

$$\tau_t = \left[\rho_f + (\rho_s - \rho_f)C\right] l^2 \left|\frac{d\bar{u}}{dy}\right| \frac{d\bar{u}}{dy} \tag{9.345}$$

and shear stress for clear water flow as: $\tau_t = \tau_0 (1-\xi)$

Then equation (9.345) can be written as:

$$\frac{d\bar{u}}{dy} = \frac{u_*}{\kappa y}(1-\xi)^{-0.5\alpha\left(C/C_{\xi a}\right)}\left[1+\left(\frac{\rho_s}{\rho_f}-1\right)C\right]^{-0.5} \tag{9.346}$$

$$\varepsilon_m = \kappa_o y u_* (1-\xi)^{1+0.5\alpha\left(C/C_{\xi a}\right)}\left[1+\left(\frac{\rho_s}{\rho_f}-1\right)C\right]^{0.5} \tag{9.347}$$

From the Rouse equation, one gets the concentration gradient $\dfrac{dC}{dy}$ as:

$$\frac{dC}{d\xi} = -\frac{v_0 C h}{\varepsilon_s} = -\frac{v_0 C h}{\beta \varepsilon_m}$$

$$= -\frac{v_0 C}{\beta \kappa_o \xi u_*}(1-\xi)^{-1-0.5\alpha\left(C/C_{\xi a}\right)}\left[1+\left(\frac{\rho_s}{\rho_f}-1\right)C\right]^{-0.5} \tag{9.348}$$

Integrating equation (9.348) numerically with respect to ξ over the depth, the vertical suspension concentration distribution can be determined using the reference concentration $C = C_{\xi_a}$ at the reference bed level $\xi = \xi_a$. The theoretical concentration distributions have been compared with experimental data based on the laboratory measurements reported by Vanoni (1946) for run 22, and Einstein and Chien (1955) for run 16 and are plotted in Figure 9.50. For a complete discussion of those values, refer to Umeyama and Gerritsen (1992). Here the coefficient β seems to be connected with the magnitude of the sediment load. It is interesting to note that the trend of β with the increase of α is the reverse for the lower and higher concentrated sediment-laden flows. It may be pointed out that Jobson and Sayre (1979) have given experimental evidence that β in an open-channel sediment-laden flow depends on the turbulent characteristics and suggested possibilities of $\beta < 1$ and $\beta > 1$. Umeyama and Gerritsen (1992) studied the vertical distribution of suspended sediment for uniform open channel flow on the same hypothesis and obtained an analytical solution at the fine sediment concentration.

Later, Umeyama (1999) presented a similar model for the vertical distribution of velocity and sediment concentration in a uniform open channel flow carrying higher sediment concentration, using

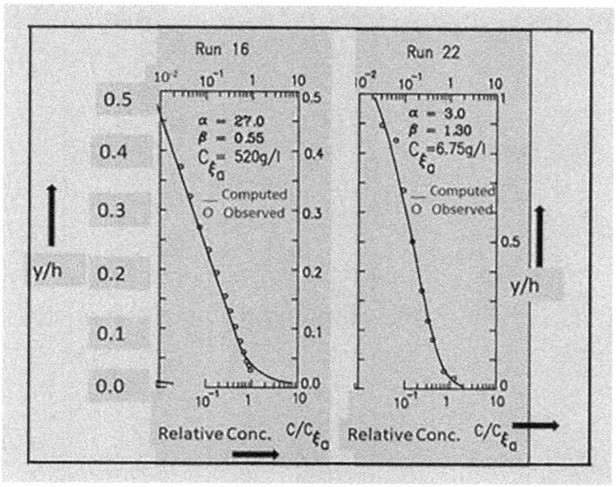

FIGURE 9.50 Computed and measured suspended sediment concentrations using Einstein et al.'s data (1955) for run 16; and Vanoni (1946)'s data for run 22. (Modified from Umeyama 1992.)

hindered settling effect based on Richardson and Zaki (1954)'s formulae (8.33). Here the equation of Umeyama and Gerritsen (1992) is extended using the hindered settling effect to the diffusion equation, and the concentration equation is given by:

$$\text{Log}\frac{C(\xi)}{1-nC(\xi)} = -\frac{v_0}{\beta\kappa_o u_*}\left[1+\left(\frac{\rho_s}{\rho_f}-1\right)C_{\text{bed}}\right]^{-0.5} \cdot$$

$$\cdot\left[\ln\xi+\left(1+\frac{\alpha}{2}\frac{C_{\text{bed}}}{C_{\xi_a}}\right)\xi+\cdots+\frac{1}{k!}\left(1+\frac{\alpha}{2}\frac{C_{\text{bed}}}{C_{\xi_a}}\right)\left(2+\frac{\alpha}{2}\frac{C_{\text{bed}}}{C_{\xi_a}}\right)\cdots\left(k+\frac{\alpha}{2k}\frac{C_{\text{bed}}}{C_{\xi_a}}\xi^k\right)+\cdots\right]\cdots+A' \tag{9.349}$$

where A' = integrating constant. The experimental data obtained by Einstein and Chien (1955) are used to validate the present model. The present study is considered to be the mathematical model on the velocity and concentration distributions for coarse sediment-laden flow. The theoretical findings are compared with the data measured by Einstein and Chien (1955). Details are available in Umeyama (1999). Although the theoretical profile of concentration is similar to the Rouse-type shape, it shows the concentration distribution more accurately in the lower and upper region. The present model explains the velocity and concentration fields with coarse sands quite well due to the increased concentration and compared his results with the experimental data of Einstein and Chien (1955).

Kovacs (1998) proposed a mixing length form based on the concept of damping effect due to suspension in sediment-laden turbulent flow as: $l = y\left(0.4-0.44y/h'\right)\left(1-C^{0.33}\right)$, where h' is effective water depth. The model was verified with several existing experimental data and found that the model is reasonably good.

9.10.4 Cellino and Graf's Equation (2000)

Cellino and Graf (2000) investigated the influence of bed forms in open-channel flow on the suspended sediment concentration distribution given in relation to Rouse diffusion equation. In the Rouse number $Z_* = v_0/\beta\kappa_o u_*$, the sediment diffusion coefficient $\epsilon_{sy} = \beta\epsilon_{my}$. They considered the $\overline{\beta}$-value is the depth-averaged value of β, representing the ratio of the sediment /mass diffusion coefficient ϵ_{sy} and the momentum diffusion coefficient ϵ_{my}, and the vertical distribution is as:

$$\beta(y) = \frac{\epsilon_{sy}(y)}{\epsilon_{my}(y)} = \frac{\overline{C'v'} \Big/ \dfrac{\partial C}{\partial y}}{\overline{u'v'} \Big/ \dfrac{\partial u}{\partial y}} \tag{9.350}$$

where $\overline{C'v'}$ is the sediment flux and $\overline{u'v'}$ is the momentum flux. The depth-averaged β-value is obtained as:

$$\bar{\beta} = \frac{1}{h}\int_a^h \beta(y) \tag{9.351}$$

The experiments were performed in a recirculating tilting flume 16.8 m long and 0.60 m wide. Sediments were added slowly to the uniform flow; the measurements started only after 4 hr of flow circulation when the presence of a sediment layer 2 mm thick on the bed is assured. The measuring section is located 13 m from the entrance of the channel, where the flow is assumed to be established; all measurements were performed at the centerline of the cross section. The three artificial plastic bed forms have been fixed on the flume bed near the measuring section. The ratio of bed-form length and flow depth was about 5, and the ratio of bed form height and flow depth was less than one-sixth. The artificial geometry was slightly modified when sediment deposited downstream of the crest. A set of data from the Rio Grande River, collected by Nordin and Dempster (1963), has been studied by Cellino (1998), who observed that the β-value lies within the range of $0.7 < \beta < 5$. On the other hand, careful experiments by Lyn (1988), Sumer et al. (1996) and Cellino and Graf (1997) observed that β -value is rather small ($\beta < 1$) under saturation condition without the presence of bed forms. To study the dependency of depth-averaged $\bar{\beta}$ on the velocity and concentration distribution on sediment laden flow with and without bed forms, a series of experiments were performed by Cellino and Graf (2000). They studied the velocity, turbulence intensity and Reynolds stress profiles over the bed forms, starting from the crest along the complete trough region. They have also studied the contour plots along the whole region for longitudinal mean velocity, vertical mean velocity, vector plot of mean velocity, longitudinal turbulent intensity, vertical turbulent intensity, Reynolds shear stress, and mean transverse vorticity and streamlines. They have also investigated the dimensionless momentum, ϵ_{my}/hu_* and sediment/mass ϵ_{sy}/hu_* diffusion coefficient profiles. The dimensionless sediment diffusion coefficients ϵ_{sy}/hu_* measured in flows over bed forms are larger in between $0.25 < y/h < 0.70$ than the ones measured in suspension flows over plane bed, indicating that the particles are diffused more efficiently by the high turbulence generated by the bed forms.

Close to the water surface $y/h > 0.70$, the coefficients are similar probably because the effect of the bed forms becomes negligible and close to the bed $y/h < 0.25$, the sediment diffusion coefficients measured in suspension flows over bed forms fall to zero more rapidly becoming smaller than the ones measured in flows over plane bed. This effect is related to the small velocities, turbulence intensities and shear stress measured close to the bed. The dimensionless momentum diffusion coefficients, ϵ_{my}/hu_* measured in flows over bed forms, are smaller than the one measured in flows over plane bed in the upper region of the flow $y/h > 0.36$. Below this level, where the peaks are observed, the coefficients measured over bed forms are slightly larger than those measured in plane-bed flows and close to the bed $y/h < 0.25$ as they become smaller, falling to zero more rapidly than those measured over plane bed. Such a tendency was also reported by Lyn (1993), and Thorne et al. (1996). For the same sediment particles to be considered as fine particles, $v_0/u_* > 0.5$, the value $\beta < 1$ for flow over a plane bed, however in the presence of a bed form, the value $\beta > 1$. This gives the explanation in a way that the field data obtained on the Rio Grande River reported by Nordin and Dempster (1963), where the presence of bed forms is undeniable and consequently the reported β-values are rather large ($\beta > 1$).

9.10.5 Other Models

Cellino and Lemmin (2004) investigated the influence of suspended sediment particles on the burst-cycle dynamics of flow by measuring the instantaneous velocity and suspended concentration profiles. Observations infer that the burst cycle may influence the sediment suspension mechanism through ejection and sweep events. The statistics of ejection and sweep in presence of suspended sediment concentration are analyzed. Experiments were carried out in a re-circulating, tilting open channel, 16.8 m long and 0.6 m wide. The channel bed was rough with a mean equivalent roughness height of 4.8 mm. It consisted of a single layer of gravel glued onto fixed plates (Cellino 1998). Special care was taken to ensure steady and uniform flow conditions. The data were collected on sediment suspension flow using an acoustic particle flux profiler (APFP). The velocity and the suspended concentration profiles were measured in two independent ways from the same acoustical signal: velocity by the Doppler frequency (Rolland and Lemmin 1997), and suspension concentration by ultrasonic echo intensity (Shen and Lemmin 1999). They have analyzed the turbulent events, i.e., outward interaction, ejection, inward interaction and sweep, which represent the quadrant splitting scheme (Lu and Willmarth 1973). The quantitative description of the quadrant splitting indicates that flow structures consisting of ejection and sweep have the two main events composing the burst-sweep cycle throughout the water depth. The analysis confirms that the concept underlying the quadrant splitting technique is valid for both clear-water and suspension flows. The conditional averaging of longitudinal and vertical intensities, and the Reynolds-stress profiles, are of interest because these quantities can be related to the beginning of sediment motion on the bed and the subsequent suspension into the flow (Nezu and Nakagawa 1993). It is important to note that the longitudinal velocity fluctuations decrease close to the bed only for suspension flows, but not for the clear-water flow. The RMS values of the fluctuating vertical velocities associated with the ejection events are larger than those related to the other three quadrants. It was confirmed that the ejection and sweep events are the main phases of the burst cycle. The initial step of the bursting can be represented by the ejection event, whereas the collapse of the coherent structure can be represented by the sweep event. Therefore, there is a strong correlation between the burst cycle and suspended sediment transport. It is shown that the low longitudinal- and vertical-fluctuating velocities occur during an ejection event providing a strong contribution to the instantaneous shear stress. This will contribute to an erosion of the bed and/or to the resuspension of sediment particles. The readers are referred to Cellino and Lemmin (2004) for detailed analysis.

Mazumder and Ghoshal (2002, 2006) developed simple realistic models for velocity from Prandtl's momentum transfer theory and suspension concentration taking into account: (i) the diffusion equation satisfies the continuity of sediment and water using Hunt (1954), (ii) the turbulent shear stress due to Boussinesq's formula and eddy diffusivity are modified in the sediment-laden flow, (iii) the fall velocity of sand particle decreases with increase in suspension concentration, and (iv) modified mixing length hypothesis as a function of concentration. Combining all these physical effects on velocity and suspension concentration equation, one can get the coupled nonlinear equations of velocity and concentration gradients, representing an integro-differential equation (Woo et al. 1988) subject to the boundary conditions. This model is of general applicability in determining the concentration profiles for fine, medium and coarse sands in suspension. To best predict the velocity and sediment data, it is found that there were only two parameters: the mixing length exponent parameter α for the velocity (Umeyama and Gerristen 1992) and the parameter β (ratio of sediment diffusion to momentum diffusion coefficients) for the sediment concentration. The efficiency of the present model has been tested by comparing the computed velocity and sediment concentration with the observed historical data of Vanoni (1946), Einstein and Chien (1955), Coleman (1981), and Lyn (1986). Comparison shows quite a good agreement. It is observed that the estimated value of β was found 1.8 for moderate velocity 95 cm/s which picks up the fine sediment in suspension and the value β was 2.5 for higher velocity 105 cm/s, which picks up coarser sediment in suspension. The value of β depends on velocity, suspension concentration, and bed conditions. Readers are referred to study the papers by Mazumder and Ghoshal (2002, 2006).

Huang et al. (2008) proposed a simple model for the vertical concentration distribution of suspended load, based on sediment diffusion equation in 2D steady turbulent flow. The model avoided the deficiencies as the infinite value of concentration at the bottom and zero value at the surface, existing in the Rouse formula. A theoretical elementary function for suspended load discharge is also derived from the new formula and improves the famous Einstein equation.

Liu et al. (2007) developed a theoretical relationship between sediment concentration and the intensity of vertical fluctuation of particles in sediment laden flow using the idea of two-phase flow and verified their model using the experimental data obtained from the vertical rectangular duct flow. It is shown that the profile of sediment concentration is significantly influenced by particle fluctuation. This leads to a new explanation for the mechanism of two patterns of sediment concentration profiles. They concluded that the vertical fluctuation of particles is one of the most important mechanisms causing sediment particle suspension and maintaining a steady distribution of sediment concentration in the vertical plane.

Dorrell et al. (2018) developed a new mode of threshold movement between the erosional and depositional flow in terms of the volumetric amount of material transported in suspension. In their model, they incorporated the followings: (i) volumetric and particle size limits in the flow ability to transport material in suspension, (ii) particle size distribution effects, and (iii) a particle entrainment function, where erosion is defined in terms of the power used to lift-up sands from the bed. This model offers an order of magnitude, or better, improvement in predicting the erosional and depositional threshold with the existing particle-laden flow models.

9.11 STRATIFICATION EFFECTS BY SUSPENSION CONCENTRATION

Sediments suspension over plane beds influences the flow structure by which sediment particles are carried. For a given distribution of Reynolds stress, a sediment-laden flow typically has a mean velocity profile with larger gradients than the corresponding profile of the clear flow. The influence of suspended sediments on the turbulent flow structure was attempted by many researchers to understand the potential applications in marine and riverine sediment transport (Smith and McLean 1977; Taylor and Dyer 1977; Wiberg and Smith 1983; Glenn and Grant 1987, Villaret and Trowbridge 1991, Ghoshal and Mazumder 2005). The interaction between the turbulent flow of water and sediment suspension is almost based on the approach of the stratified flow analogy (Monin and Yaglom 1971), where the solid particles are continuously distributed in the mixture, and the velocity of sediment particle is assumed to differ from the local fluid velocity with a vertical settling velocity in unbounded stationary fluid flow, which is the same as the terminal velocity in an unbounded, stationary fluid. In most applications, the contribution of particle concentration to the solid-fluid mixture is inhibited to be small, so that the Boussinesq approximation can be used. According to this theoretical framework, the solid particles affect the flow through the turbulent kinetic energy balance, in which a buoyancy term proportional to the turbulent particle flux appears as a sink that extracts energy from the turbulent velocity fluctuations. In effect, the suspended particles stratify the flow and have an influence similar to that of a downward heat flux in the stably stratified atmospheric surface layer (Businger et al. 1971).

Smith and McLean (1977) first used a correction factor as flow stratification effect induced by suspended load transport in order to make the results consistent with the measured spatially averaged shear stress. They performed a series of detailed flow measurements above two-dimensional bed forms of 60–100 m long, 1–3 m high dunes in Columbia River to understand the velocity profiles, local boundary shear stress and sediment transport problems. They determined the spatially averaged velocity profiles and shear stress over one wave length of the bed form. Under such circumstances, the shear velocity and roughness parameter are determined for each stratified layer from the least square fit. In the flow of an erodible sediment bed, the ratio of particle velocity to the local shear velocity is small; the sediment particle will be diffused and transported in suspension, which causes the decrease of sediment concentration with distance into the flow, generating stably stratified boundary layer. The net fluid density ρ_f is related to the concentration of suspended sediment C as: $\rho_f = \rho_f + (\rho_s - \rho_f)C$ so that the density gradient is related to the concentration gradient by:

$$\frac{\partial \rho_f}{\partial y} = (\rho_s - \rho_f)\frac{\partial C}{\partial y} \tag{9.352}$$

and the gradient Richardson number can be written as:

$$R_i = \frac{-g}{\rho_f}\frac{\partial \rho_f}{\partial y}\left(\frac{\partial \bar{u}}{\partial y}\right)^{-2} = \left[\frac{-g(\rho_s - \rho_f)}{\rho_f}\frac{\partial C}{\partial y}\right]\left(\frac{\partial \bar{u}}{\partial y}\right)^{-2} \tag{9.353}$$

The parameter R_i represents the importance of flow stratification in inhibiting the turbulent transfer of momentum and mass, turbulence production being completely eliminated above the critical value of 0.25 or so.

In investigation of the lower part of the atmospheric boundary layer, it is usual to assume a constant stress region in which the non-dimensional shear is

$$\phi_m = \frac{\kappa_o y}{u_*}\left(\frac{\partial \bar{u}}{\partial y}\right) \tag{9.354}$$

and in stably stratified case this is related to the gradient Richardson number by:

$$\phi_m = 1 + \beta(y/L) = (1 - \alpha\beta R_i)^{-1} \tag{9.355}$$

where β is the ratio of the sediment diffusivity ϵ_s to the momentum diffusivity ϵ_m, and α is a constant found to be 4.7 ± 0.5 (Businger et al. 1971); and $L = \dfrac{\rho_f u_*^2}{g\kappa_o \Theta_0}$ is the Obukhov length, and Θ_0 is the density flux in the lowest part of the fully turbulent layer. The non-dimensional constitutive equation by shear velocity yields for a channel flow as:

$$\frac{\epsilon_s}{u_*^2}\left(\frac{\partial \bar{u}}{\partial y}\right) = \frac{\tau}{\tau_0} = 1 - \xi \tag{9.356}$$

For equations (9.354) and (9.355), one gets the analogous expression as:

$$\frac{\kappa_o y u_*}{u_*^2}(1 - \alpha\beta R_i)\left(\frac{\partial \bar{u}}{\partial y}\right) = 1 \tag{9.357}$$

Comparing equations (9.356) and (9.357) with $\xi \ll 1$, one gets

$$\epsilon_s = \kappa_o y u_* (1 - \alpha\beta R_i) \tag{9.358}$$

If $R_i = 0$, $\epsilon_s = \kappa_o y u_*$ or ϵ_s may be written as a general form, so the sediment diffusion coefficient ϵ_s can be written as:

$$\epsilon_s = \kappa_o h f(\xi) u_* (1 - \alpha\beta R_i) \tag{9.359}$$

In steady horizontally uniform flow, conservation of mass equation considering $\epsilon_s = \epsilon_{sy}$ gives

$$\epsilon_{sy}\frac{\partial C}{\partial y} + (1 - C)Cv_{sm} = 0 \tag{9.327}$$

in which C = sediment concentration; v_{sm} = particle fall velocity in a fluid-sediment mixture; ϵ_{sy} = sediment diffusion coefficient; and y = vertical coordinate. When the diffusion coefficients of sediment and momentum are equated in the constant stress layer to give $\beta = 1$ and $\epsilon_{sy} = \kappa_o y u_*$, then equation (9.327) is integrated as:

$$\frac{C}{1 - C} = \frac{C_{\xi_a}}{1 - C_{\xi_a}}\left(\frac{\xi_a}{\xi}\right)^{Z_*} \tag{9.360}$$

where C_{ξ_a} is the concentration at reference level ξ_a and $Z_* = v_{sm}/\kappa_o u_*$. If $Z_* \to 0$, then $C \to C_{\xi_a}$ and $Z_* \to \infty$, then $C \to 0$ except at ξ_a. The boundary conditions are given by:

$$u = 0 \text{ at } \xi = \xi_a \tag{9.361a}$$

$$C = C_{\xi_a} \text{ at } \xi = \xi_a \tag{9.361b}$$

Using equations (9.353), (9.356), (9.359), (9.327) subject to the boundary conditions (9.361), one gets the velocity gradient $\frac{\partial u}{\partial \xi}$ and the concentration gradient $\frac{\partial C}{\partial \xi}$ as:

$$\frac{\partial \bar{u}}{\partial \xi} = \frac{u_*}{\kappa_o}\left(\frac{1 - \xi}{f(\xi)}\right)\left[1 + \alpha_*\left(\frac{\partial C}{\partial \xi}\right)\left(\frac{\partial \bar{u}}{\partial \xi}\right)^{-2}\right]^{-1} \tag{9.362}$$

$$\frac{\partial C}{\partial \xi} = -Z_* C(1 - C)\left[f(\xi)\left\{1 + \alpha_*\left(\frac{\partial C}{\partial \xi}\right)\left(\frac{\partial \bar{u}}{\partial \xi}\right)^{-2}\right\}\right]^{-1} \tag{9.363}$$

where $\alpha_* = \alpha\beta g h\frac{(\rho_s - \rho_f)}{\rho_f}$. Finally, the coupled equations for the velocity gradient (9.362) and the concentration gradient (9.363) are solved numerically for velocity and concentration profiles using the boundary conditions (9.361). The solution gives a complete description of both velocity and concentration of suspended sediment. The theoretical equations are verified with the existing available experimental and field data of velocity and suspended sediment concentration. Detailed explanation is available in Smith and McLean (1977).

Villaret and Trowbridge (1991) developed a simple model for mean velocity and mean sediment concentration based on the existing data collected from the laboratory channel in order to evaluate applicability of the stratified flow to the dilute suspension in turbulent flow. The model was based on the turbulence closure similar to that proposed by Smith and McLean (1977). Following Smith and McLean (1977), they assumed the eddy viscosity ϵ_m and eddy diffusivity ϵ_s as:

$$\epsilon_m = \epsilon_T (1 - \alpha_1 R_F) \tag{9.364}$$

$$\epsilon_s = \frac{\epsilon_T}{\alpha_N}(1 - \alpha_2 R_F) \tag{9.365}$$

where ϵ_T is the effective viscosity that would exist under neutral conditions, α_N is the turbulent Schmidt number for solid particles under neutral conditions, α_1 and α_2 are the empirical constants, and $R_F(y)$ is the flux Richardson number, which may be written as:

$$R_F(y) = -\frac{g\epsilon_s \dfrac{\partial \rho_f}{\partial y}}{\rho_f \epsilon_m \left(\dfrac{\partial \bar{u}}{\partial \xi}\right)^2} = -\frac{g(\rho_s - \rho_f)\epsilon_s \dfrac{\partial C}{\partial y}}{\rho_f u_*^2 (1-\xi)\dfrac{\partial \bar{u}}{\partial y}} \tag{9.366}$$

Using the Richardson number and Cole's wake strength, they obtained the velocity and the concentration profiles, and they presented a comparative study of the theoretical results and the experimental data to test the applicability of stratified flow analogy to the sediment laden flow. The effect of stratification by suspended sediments allows for determining the effect in the measurements, and the constants obtained from the measurements are consistent with previous estimates obtained from measurements in thermally stratified flows. Details are available in their paper.

McLean (1991) used the idea of damping of turbulence to understand the effect of stratification by suspended sediment. As the turbulence is damped due to density stratification by suspended sediments, there is a limiting ability of flow to transport both mass and momentum vertically, resulting the reduction of drag coefficient and the ability of the flow to keep sediment in suspension. This effect appears as a reduction in eddy viscosity ϵ_m and eddy diffusivity ϵ_s due to stratification. The phenomenon was quantified initially for atmospheric flows by Businger et al. (1971) and it was implemented to sediment suspension by McLean (1991):

$$\epsilon_m = \frac{\epsilon_T}{(1 + \alpha_1 \varsigma)} \tag{9.367a}$$

$$\epsilon_s = \frac{\epsilon_T}{(\sigma + \alpha_2 \varsigma)} \tag{9.367b}$$

where ϵ_T is the effective viscosity under neutral conditions, α_1, α_2 and σ are the empirical constants, and ς is the stratification parameter given by:

$$\varsigma = \frac{\rho_f g \epsilon_T}{\tau^2} \overline{\rho_f' v_f'} \tag{9.368}$$

where $\overline{\rho_f{}' v_f'} = (\rho_s - \rho_f)\overline{v' C_n'}$ is the buoyancy flux. The stratification parameter ς is closely related to the ratio of buoyancy production to the shear production in the turbulence kinetic energy equation. As ς increases, the turbulence damping due to stratification becomes larger compared to the production due to shear turbulence. Assuming the $\epsilon_T = \kappa_o h f(\xi) u_*$, the stratification parameter ς can be written as:

$$\varsigma = \frac{\kappa_o hgf(\xi)(\rho_s - \rho_f)}{\rho_f u_*^3 (1-\xi)^2} \sum_n v_n C_n (1 - C_s) \tag{9.369}$$

where v_n = the settling velocity of the nth class; and the total concentration is $C_s = \sum_n C_n$.

The velocity is given by:

$$u = \int_{\xi_0}^{\xi} \frac{u_*^2}{\epsilon_m}(1-\xi)\,d\xi \tag{9.370}$$

where $\xi_0 = y_0/h$ with y_0 is the roughness parameter at which the velocity extrapolates to zero. Assuming the turbulent flux of sediment of the nth class of mixture of different sediment sizes, one can write as:

$$-\overline{v'C_n'} = \epsilon_s \frac{\partial C_n}{\partial y} \tag{9.371}$$

The solution of above equation is given by:

$$\frac{C_n}{1-C_s} = \frac{C_{na}}{1-C_{sa}} \exp\left(\int_a^y \frac{v_n}{\epsilon_s} dy \right) \tag{9.372}$$

Now using equations (9.367) and (9.369), integration of (9.370) and (9.372) is made in an iterative way as long as the parameters u, $f(\xi)$, v_n and ρ_s are known. Once these parameters are known, the net transport of suspended sediment q_s is calculated by integrating the product of velocity and sediment suspension concentration as:

$$q_s = \int_a^h u C_s\, dy = C_a h \bar{u} I \tag{9.373}$$

where h is the flow depth, $\bar{u}$ is the depth-averaged velocity, C_a is the concentration at the reference level a and I is the depth-averaged integration. Smith and McLean (1977) assumed the reference concentration C_a as

$$C_a = \frac{\gamma_0 C_{\text{bed}} \left(\frac{\tau}{\tau_c} - 1 \right)}{1 + \gamma_0 \left(\frac{\tau}{\tau_c} - 1 \right)} \tag{9.374}$$

where $\gamma_0 = 0.004$ is assumed, $C_{\text{bed}} \cong 0.6$ is the concentration at the bed, $\left(\frac{\tau}{\tau_c} - 1 \right)$ is the normalized excess shear stress. The results of suspension concentration including the effects of stratification are shown in Figure 9.51, where the net suspended sediment q_s is plotted against the transport stage $S = \left(\frac{\tau}{\tau_c} - 1 \right)$.

Smith and McLean (1977) assumed that the roughness would be proportional to the excess shear stress, and Wiberg and Rubin (1989) suggested that y_0 should be proportional to the bed-load layer thickness δ_b. Using particle trajectory calculations, dimensionless bed layer thickness δ_b/d is found as:

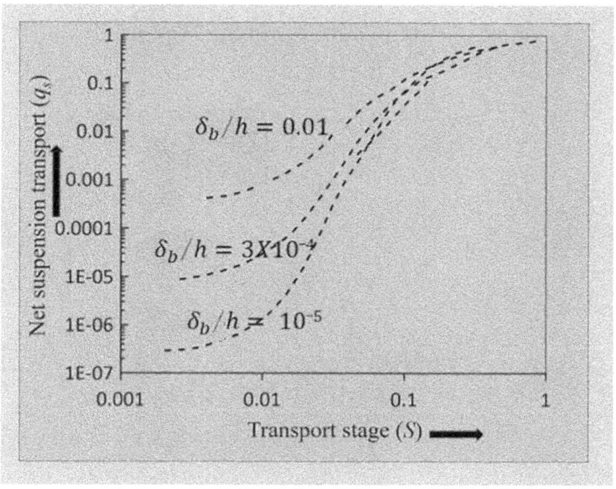

FIGURE 9.51 Net suspension transport (q_s) from equation (9.373) against transport stage $S = \left(\frac{\tau}{\tau_c} - 1\right)$ for different values of δ_b/h, with stratification effect by suspension. (Modified from McLean 1991.)

$$\frac{\delta_b}{d} = \frac{0.68\dfrac{\tau}{\tau_c}}{1 + A_1\dfrac{\tau}{\tau_c}} \tag{9.375}$$

where d = the sediment diameter; and the constant $A_1 = 0.0204(\ln d)^2 + 0.022\ln d + 0.0709$. This yields bed-load layer thicknesses that rarely exceed 3–5d. The roughness is then assumed to be: $y_0 = \alpha_0\delta_b$, where $\alpha_0 = 0.056$. Subsequently, McLean (1992) solved the momentum and suspended sediment concentration equations, including the effects of stratification by suspended sediment and effects due to size distribution and bed forms.

Ghoshal and Mazumder (2005) developed theoretical models for mean velocity and concentration distributions considering the effect of sediment-induced stratification and the modified mixing length due to high suspension together with viscous and turbulent shear stresses, which are the functions of concentration. The models are compared with comprehensive experimental data sets. The comparison reveals that (i) the calculated velocity and concentration profiles agree well with the observed data, (ii) the model constant due to stratification used for verification is consistent with the measurements in thermally stratified flows, and (iii) the higher the sediment suspension, the better the effect of density stratification and the less the impact of mixing length.

The total shear stress τ for turbulent flow derived from the Reynolds stress equation is written as:

$$\tau = \mu_f\frac{du}{dy} + \rho_m\epsilon_m\frac{du}{dy} \tag{9.376}$$

where u is the mean velocity parallel to the wall, μ_f is the coefficient of viscosity, ρ_f is the net density of the fluid–sediment mixture and ϵ_m is the momentum diffusion coefficient. When the turbulent flow carries high sediment in suspension over the erodible sand bed, the density of the mixture may be defined as $\rho_m = \rho_f[1 + AC]$, with $A\left(= \rho_s/\rho_f - 1\right)$ is the relative density, and C is the volume of sediment per volume of the water–sediment mixture. The coefficient of viscosity μ_m of the sediment–fluid mixture is a function volumetric concentration C and is given by Coleman (1981) as:

$$\mu_m = \mu_f\left(1 + 2.5C + 6.25C^2 + 15.62C^3\right) = \mu_f g(C) \tag{9.377}$$

where μ_f is the dynamic viscosity of clear water. Using (9.377) in (9.376), the total shear stress in the sediment-laden turbulent flow is written as

$$\tau = \mu_f g(C)\frac{du}{dy} + \rho_f[1 + AC]\epsilon_m \frac{du}{dy} \tag{9.378}$$

The vertical distribution of total shear stress τ can be written as a function of dimensionless depth $\xi = \frac{y}{h}$ and is given by:

$$\tau = \rho_f u_*^2 \left[1 - \xi + A\int_{\xi}^{1} C\, d\xi\right] \tag{9.379}$$

Combining the above expressions (9.378) and (9.379), the velocity gradient $\frac{du}{dy}$ explicitly yields

$$\frac{\left[(1 + AC)\epsilon_m + v_f g(C)\right]}{hu_*^2\left[1 - \xi + A\int_{\xi}^{1} C\, d\xi\right]}\frac{du}{d\xi} = 1 \tag{9.380}$$

Following Smith and McLean (1977), a relationship between the constant stress layer and a gradient Richardson number R_i has been assumed in the stably stratified boundary layer; and it is extended to a sufficiently large value of y using the Coles' strength parameter Π as:

$$(1 - \gamma\beta_* R_i)\frac{du}{d\xi} = \frac{u_*}{l_*} + \frac{\Pi}{\kappa_o}\pi u_* \sin(\pi\xi) \tag{9.381}$$

where l_* is the mixing length, γ is the ratio of the sediment diffusion coefficient to the momentum diffusion coefficient of water, β_* is a constant found to be 4.7 ± 0.5 by Businger et al. (1971) from the data of the Kansas experiment, and R_i is the gradient Richardson number, defined as the ratio of buoyant production to shear production of turbulent kinetic energy, and is given by equation as:

$$R_i = \frac{-g}{\rho_f}\frac{\partial \rho_f}{\partial y}\left(\frac{\partial u}{\partial y}\right)^{-2} = \left[\frac{-gh(\rho_s - \rho_f)}{\rho_f}\frac{dC}{d\xi}\right]\left(\frac{du}{d\xi}\right)^{-2} \tag{9.382}$$

When $R_i = 0$ the fluid is neutrally stratified, and for $R_i \geq 0.25$, the turbulence production in fluid is completely diminished (Tennekes and Lumley 1980). According to Umeyama and Gerritsen (1992), the mixing length l_* is modified for the entire boundary layer thickness, which is a function of concentration, and is given by:

$$l_* = \kappa_o y(1 - \xi)^{0.5\left[1 + \beta(C/C_{\xi a})\right]} \tag{9.383}$$

where β is a constant to be determined from the experimental data, and $C_{\xi a}$ is the reference concentration at $\xi = \xi_a$ to be determined. Now eliminating the velocity gradient $\frac{du}{d\xi}$ from equations (9.380) and (9.381), one gets the momentum diffusion coefficient ε_m as:

$$\varepsilon_m = \frac{1}{1+AC}\left[\frac{\left(1-\xi+A\int_{\xi}^{1} C\,d\xi\right)hl_*u_*\left(1-\gamma\,\beta_*R_i\right)}{1+\frac{\Pi}{\kappa_o}l_*\pi\sin(\pi\xi)} - v_f g(C)\right] \tag{9.384}$$

Therefore, the velocity gradient $\frac{du}{d\xi}$ can be written as:

$$\frac{du}{d\xi} = \frac{u_*}{l_*}\left\{1+\xi(1-\xi)^{0.5\left[1+\beta(C/C_{\xi a})\right]}\Pi\pi\sin(\pi\xi)\right\}\left[1+\gamma\,\beta_*Agh\frac{dC}{d\xi}\left(\frac{\partial u}{\partial y}\right)^{-2}\right]^{-1} \tag{9.385}$$

Equation (9.385) for the velocity gradient is inaccessible at the free surface $\xi = 1$. The validity of the equation at the free surface is not important, because the available observed data are limited to only $\xi < 1$, so it is not considered the region immediately adjacent to the free surface. The observed maximum flow velocity generally occurs somewhere in the flow below the free surface $\xi < 1$ such a phenomenon may be attributed to secondary circulation or some other effect on the free surface.

In a steady and uniform two-dimensional sediment-laden turbulent flow, where the concentration is constant in time and varies only with vertical co-ordinate y, the vertical distribution of suspended sediment concentration C with particle settling velocity v_0 could be obtained from Hunt's equation (Hunt 1954) as:

$$\epsilon_s\frac{\partial C}{\partial y} + C\frac{\partial C}{\partial y}(\epsilon_m - \epsilon_s) + (1-C)CW = 0 \tag{9.386}$$

where ϵ_s and ϵ_m are the diffusion coefficients of sediment and momentum respectively. If the density of the fluid mixture is increased by sediments, buoyancy force is increased and hence the substantial reduction of particle settling velocity in suspension occurs. The effective settling velocity of the sediment W in sediment-laden flow varies with the volumetric concentration C as a result of hindered settling (Fredsoe and Deigaard 1992), obtained as $W = v_0(1-C)^{\alpha}$, where v_0 is the fall velocity of a grain in a clear fluid and α is the exponent of reduction of fall velocity, which depends on the particle Reynolds number R_g as: $\alpha = 4.35R_g^{-0.03}$ for $0.2 \le R_g \le 1.0$; $\alpha = 4.45R_g^{-0.013}$ for $1.0 \le R_g \le 500$; $\alpha = 2.39$ for $R_g \ge 500$. The momentum diffusion coefficient ϵ_m in a fluid–sediment mixture is related to sediment diffusion coefficient, ϵ_s by $\epsilon_s = \gamma\epsilon_m$. The concentration gradient $\frac{dC}{d\xi}$ can be written as:

$$\frac{dC}{d\xi} = \frac{-hCv_0(1-C)^{\alpha}}{\epsilon_m\left[\gamma+(1-\gamma)C\right]} \tag{9.387}$$

To solve the coupled equations (9.385) and (9.387) by the Runge-Kutta method for velocity and concentration simultaneously, it is necessary to provide the boundary conditions for u and C at the reference level ξ_a as:

$$u = u_{\xi_a} \text{ and } C = C_{\xi_a} \text{ at } \xi = \xi_a. \tag{9.388}$$

The precise location of the reference level is arbitrary. To solve coupled differential equations (9.385) and (9.387), the term $A\int_{\xi}^{1} C\,d\xi$ is omitted, because the inclusion of this integral term makes the problem too complicated (it is encouraged to the readers). At the first step, it is assumed $\beta_* = 0$ (without stratification) in both equations to obtain the first estimates of $\frac{du}{d\xi}$ and $\frac{dC}{d\xi}$ respectively. Next, these two values are used on the right-hand sides of (9.385) and (9.387), for the new estimates of velocity and concentration gradients. The free parameters α and γ are adjusted for a good agreement between theoretical and experimental results.

The theoretical models for the mean velocity and suspension concentration for sediment-laden turbulent flows are compared with the most comprehensive experimental data of Vanoni (1946), Einstein and Chien (1955), Coleman (1981) and Lyn (1986). Here, a comparison with the datasets of Einstein and Chien (1955) is shown only; the detailed comparisons of other data set are available in Ghoshal (2004). These datasets were collected under controlled conditions over plane sediments beds in laboratory channels and these are most frequently used in the literature of sediment transport. To compute the velocity and concentration profiles from (9.385) and (9.387) based on the boundary conditions (9.388), it is important to know the empirical constants κ_o, Π, β_*, β and γ. Here we have used the von-Karman constant $\kappa_o = 0.4$ and the wake strength parameter $\Pi = 0.2$ under fully developed turbulent flow conditions (Nezu and Rodi 1986), and for the stably stratified flow, the empirical constant $\beta_* = 4.7$ is taken from Businger et al. (1971) for computational purpose. The specific gravity of sediments used for all the experiments is 2.65. The quantities β and γ are the free parameters estimated by adjusting to best fit with the observed velocity and concentration data.

Experiments were conducted by Einstein and Chien (1955), in a steel re-circulating flume which was 35.7 cm deep, 30.7 cm wide and 120 m long. The sizes of the mean sediments used in the experiments were the fine sands $(d_{50} = 0.247 \text{ mm})$ for runs S-11 to S-16, medium sands $(d_{50} = 0.94 \text{ mm})$ for runs S-6 to S-10 and coarse sands $(d_{50} = 1.30 \text{ mm})$ for runs S-1 to S-5. The sediment concentration near the bed was in the range of 30–600 g/L. The velocity distribution was measured at 25–31 vertical points between one-third and one-half of the depth to the flume bed. The comparison of computed and observed velocity and suspended sediment concentration profiles for three runs (S-14 to S-16) are made using the reference velocity u_{ξ_a} and reference concentration C_{ξ_a} at the lowest elevation $\xi = \xi_a$ from the bed based on the stratification parameter (Figure 9.52a,b). The fitted concentration profiles are more or less improved by adding stratification for medium and coarse sands (Ghoshal and Mazumder 2005). However, overall, quite good agreement was observed between the computed and observed data for all sand sizes (fine, medium and coarse) including the stratification effect. Moreover, it is assessed that the estimated value of β is 10 for the runs of fine-grained sediments (S-12 to S-16) and in the range of 15–18 for the runs (S-7 to S-10) of medium sands except for the runs S-11 and S-6, where the value of β is 1. The value of β is 8 for coarse-grained sediments (S-1 to S-3). Furthermore, the value of γ increases consistently with increase of grain sizes -fine to coarse-grained sediments (Kaushal et al. 2002). The computed velocity and concentration profiles with and without stratification effects are shown against the observed values of Vanoni (1946), Einstein and Chien (1955), Coleman (1981) and Lyn (1986) in Figure 9.53a,b; and the flux Richardson number R_i against height *y/h* above the bottom is plotted in Figure 9.53c for runs S-11 and S-16 of Einstein and Chien (1955), where the stratification effect is important.

Later, Ghoshal and Mazumder (2006) developed mathematical models to predict the vertical velocity and sediment concentration distributions in a sediment-laden turbulent flow considering a modified mixing length based on density stratification due to suspension. The coupled nonlinear differential

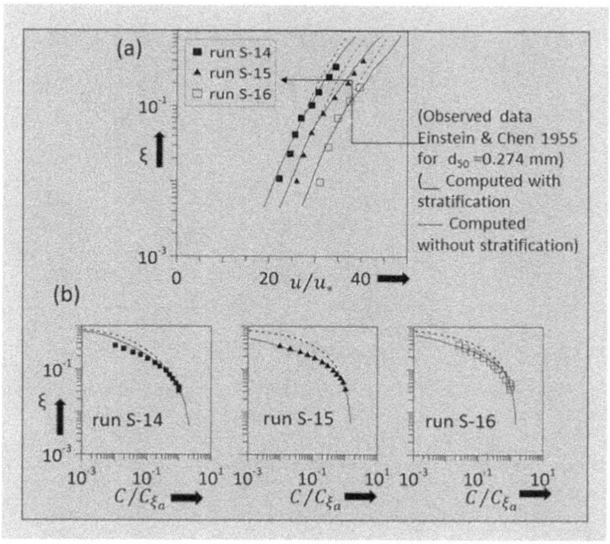

FIGURE 9.52 Comparison of measured and computed velocity (a) and concentration (b) for three observed runs S-14 to S-16 of Einstein and Chen (1955) for fine sands ($d_{50} = 0.274$ mm), continuous line with stratification effect, dashed line without stratification and symbol represents the observed data. (Modified from Ghoshal and Mazumder 2005.)

equations for velocity and concentration arising from the problem have been solved numerically and the results have been compared with the comprehensive existing data. Comparison shows good agreement with the observed data, which proves the validity of the present model.

9.11.1 Estimation of Near-Bed Velocity and Reference Concentration

To solve the coupled differential equations (9.385) and (9.387), we need to determine the reference velocity u_{ξ_a} and the reference concentration C_{ξ_a} at the available reference level $\xi = \xi_a$. The determination of u and C at an arbitrary reference level $\xi = \xi_a$ as a function of flow field is a difficult problem. However, besides using the known velocity and concentration at an arbitrary location, an attempt has been made to estimate the velocity and concentration at a bed layer level. This idea is an alternative way of estimating the bottom velocity and concentration at the bed layer. This seems to be more reasonable from the physical point of view than defining the reference velocity and concentration at an arbitrary location far above the bottom, as already a few diameters away from the bed the sediment particles are kept in suspension by the turbulence of the fluid and should therefore be regarded as sediment in suspension. The velocity of sediment particles in the bed layer is a function of shear velocity u_* and critical shear velocity u_{*c} corresponding to the condition of the Shields grain movement for different grain sizes. The empirical relation for the velocity of a sediment particle in the bed layer $2d$ is given by Engelund and Fredsoe (1976) as

$$u = \delta u_* \left[1 - 0.7 \sqrt{\left(\frac{u_{*c}}{u_*} \right)} \right] \tag{9.389}$$

where the value of δ is 9.0 for sand. Here it is assumed that the migration velocity of sediment size at the top of the bed layer is approximately equal to the fluid velocity at that layer (Mazumder 1994). In this case, it is estimated the value of $\delta = 3.65$ from (9.389) to match the computed velocity u with extrapolated observed velocity of Einstein and Chien (1955) near the bed.

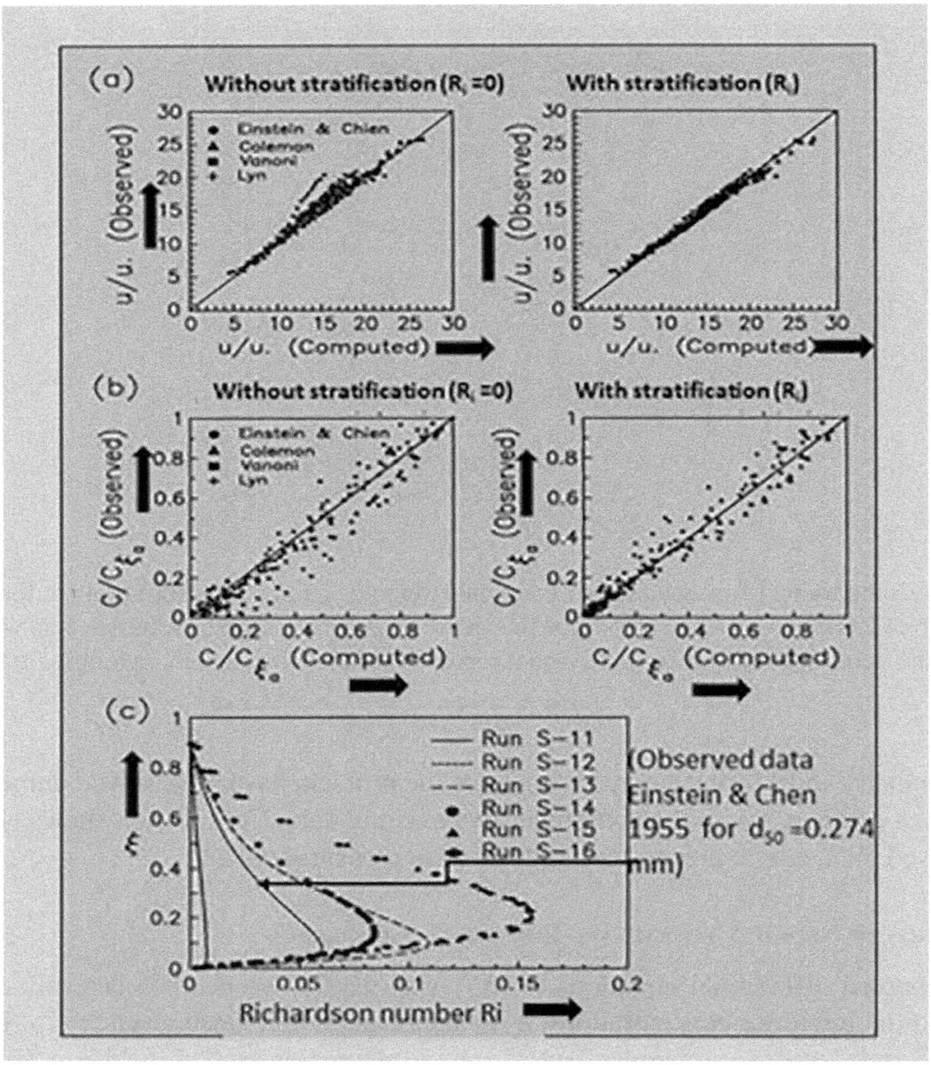

FIGURE 9.53 (a and b) Comparison of calculated and measured velocity and concentration profiles without stratification (left) and with stratification (right), and (c) variation of Richardson number with depth for runs S-11 to S-16 of Einstein and Chien (1955). (Modified from Ghoshal and Mazumder 2005.)

Smith and McLean (1977) assumed the bed layer concentration level C_{ξ_a} at the reference level ξ_a is given by

$$C_{\xi_a} = \gamma_0 \frac{u_*^2 - u_{*c}^2}{u_{*c}^2} \tag{9.390a}$$

$$\text{in which } \xi_a = \frac{\alpha_0 \rho_f \left(u_*^2 - u_{*c}^2\right)}{g\left(\rho_s - \rho_f\right)} \tag{9.390b}$$

The value of the constant γ_0 was taken to be 0.004 by Smith and McLean (1977) and $\alpha_0 = 26.3$ is the empirical constants. We have estimated the value of $\gamma_0 = 0.0055$ as the computation of (9.390) with this value shows approximate agreement between the extrapolated value of the observed concentration of Einstein and Chien (1955) and the computed value of concentration by (9.390) at the level $2d$ except for the run S-11, where the amount of material in suspension is very low. Finally, the expression of reference concentration proposed by Smith and McLean (1977) was used because it was supported by field measurements (Smith and McLean 1977; Dyer 1988).

van Rijn (1984) proposed an equation to estimate the reference concentration at the bed layer level $2d$ and is given by

$$C_{2d} = 0.18 C_{\text{bed}} \frac{S_0}{d_*} \tag{9.391}$$

where C_{bed} is the maximum theoretical value of bed concentration for firmly packed grains, S_0 is the normalized excess shear stress and d_* is the particle diameter, defined as $d_* = d\left(\frac{gA}{v_f^2}\right)^{0.33}$ with d is the median grain size of sediment, A is the relative density of sediment.

Zyserman and Fredsoe (1994) proposed a formula for determining the bed concentration at a reference level $\xi_a = 2d$ as:

$$C_{\xi=2d} = \frac{0.331(\theta - 0.045)^{1.75}}{1 + 0.7196(\theta - 0.045)^{1.75}} \tag{9.392}$$

where θ is the Shields parameter related to skin friction. But the use of this formula gives a highly overestimated concentration in Einstein and Chien's data at the $2d$ level, which was noted by comparing the computed concentration with the extrapolated value of Einstein and Chien's data. So, the estimation of concentration at $2d$ level by (9.392) leads to an erroneous result except in the case of run S-15 and S-16, where the amount of material in suspension is very high.

Cheng (2002) defined the bed concentration at an elevation $2d$ level from the bed, following the ideas of Einstein (1950) as

$$C_{2d} = \frac{C_b}{(1 + 1/\lambda)^3} \tag{9.393}$$

where C_b is the maximum bed concentration for firmly packed grains, λ is the linear concentration given by

$$\lambda = \left(\frac{\theta - \theta_c - \pi p \mu_d}{0.027(A+1)\theta}\right)^{0.5} \tag{9.394}$$

with p is the probability function given by Engelund and Fredsoe (1976) as

$$p = \left[1 + \left(\frac{\frac{\pi}{6}\mu_d}{\theta - \theta_c}\right)^4\right]^{-1/4} \tag{9.395}$$

and μ_d is the dynamic friction coefficient = 1, and A is the relative density.

9.12 BED ROUGHNESS EFFECTS ON SUSPENSION DISTRIBUTIONS

The influence of bed roughness on sediment suspension during transportation was broadly investigated by Sengupta et al. (1999). Their findings showed that an increase of bed roughness needed higher velocity for the initiation of ripples and a higher flow velocity was needed for suspending comparable amount of sediments in suspension, when the bed roughness was increased. The study mentioned above was only concerned with sand-size ranges. To achieve a general output regarding the amount of materials in bed layer and suspension, it is important to investigate the physical mechanism of size sorting, when the bed roughness was gradually increased by adding the coarse grains to the sediment beds. The question was: whether the size distribution during transportation was dependent on bed roughness or merely reflected a part of bed's grain size distributions. McLaren and Bowles (1985) provided a method for estimating the patterns of grain-size distribution of sedimentary deposits in the direction of sediment transport. They found that the size distribution of sediment bed of many streams, particularly the sand-gravel mixture beds depart significantly from uni-modal to log-normal. Kuhnle (1993) conducted a series of experiments to study the shear stress at incipient motion of five sediment beds, comprising the sand-gravel mixture ranging from 0.20 to 10.00 mm. For the bed composed of sand-gravel mixture all sand sizes showed essentially a constant relation between the shear stress and grain size, whereas for gravel size fraction critical shear stress increased with increase in size (Church et al. 1991). They found that the values of the critical shear stress for the sand and gravel fractions in the mixtures were higher when the amount of gravel in the sediment bed was increased. The influence of bed roughness on suspended sediments in turbulent flow was attempted by Mazumder with his group to understand the size sorting process and sediment entrapment in marine and riverine sediment transport (Mazumder et al. 2005; Ghoshal et al. 2010; Ghoshal et al. 2011; Ghoshal et al. 2013, and others).

Mazumder et al. (2001, 2005) investigated the influence of bed roughness on grain size sorting during suspension transportation above the sediment bed in an experimental channel and developed theoretical models for verification with the experimental data. Their analysis allowed to understand whether the size distribution of sediment and amount of sediment in suspension during transportation were significantly dependent on the bed roughness. The theoretical framework based on the concentration and momentum equations mostly depends on two parameters: the bed roughness and the ratio of sediment diffusion coefficient to the momentum diffusion coefficient. The bed roughness is an important parameter because it influences the flow velocity, suspension concentration and sediment pick up process.

The movement of sediment particles in the bed experiences resistance to the flow at the near-bed region due to the collision with the non-moving and moving particles. The series of flume experiments with variety of sediment beds of different bed roughness having same modal grain size have shown that for fixed flow velocity and height, the amount of materials in suspension first increases up to a certain critical value and then decreases with further increase in bed roughness. The bed roughness is computed as d_{65} indicating that the sieve size of which 65% of the mixture by weight is finer (Einstein 1950). A concept of 'critical bed roughness' together with 'sediment entrapment' has been introduced.

A series of experiments were performed in a closed-circuit laboratory flume (Figure 7.14) specially designed at the Indian Statistical Institute (ISI) Kolkata, by placing each of the sediment beds (nos. 10C1–10C5) of known grain size distributions shown in Figure 9.54 with cumulative percentage plots. Grain size distributions of all five bed materials are also shown in Table 9.9. All five different sediment beds were the same modal size with a peak value at 2.0 phi (0.25 mm), but different values of bed roughness with respect to a fixed size distribution (uni-modal or almost normal, bed 10C1). Here three beds 10C1, 10C2, and 10C3 consist of 100% sand ranging from 0.03 to 2.00 mm, whereas the other two beds 10C4 and 10C5 consist of sand-gravel mixture ranging from 0.03 to 8.00 mm in the following proportion: 14% gravel, 86% sand; and 25% gravel, 75% sand. All five sediment beds are shown in Figure 9.55 (Bed 10C1, Bed 10C2, Bed 10C3, Bed 10C4, and Bed 10C5).

The specific gravity of sediment is 2.65. For each experiment, a specific mixture of grain size was laid down on the flume base to produce a uniform bed, generally 2–3 cm thick. The water depth h was kept at a constant of about 35 cm above the flume base for each of the experiments. Bed forms were generated and these are recorded at different flow velocities shown in Figure 9.56a–c with plane bed

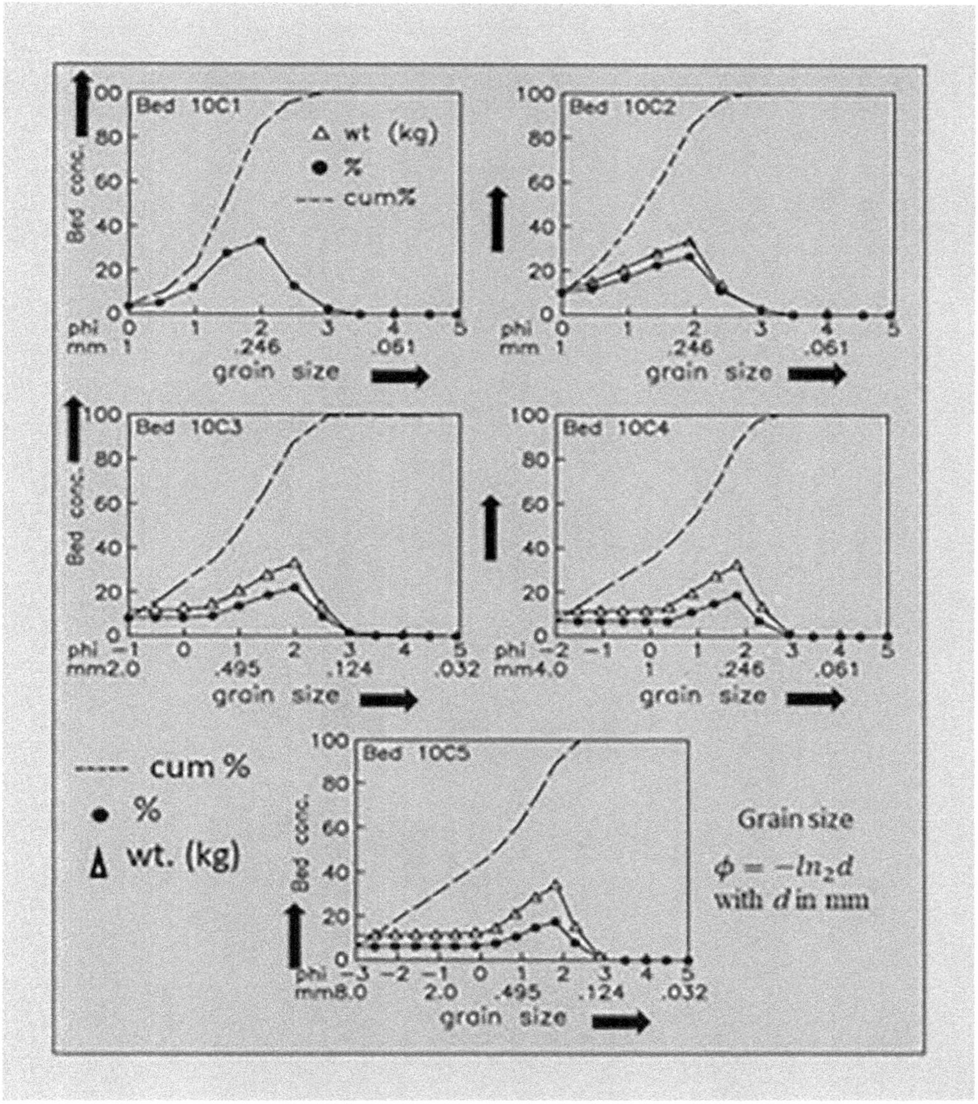

FIGURE 9.54 Size distributions of five sediment beds in weight (kg), percentage and cumulative percentage plots. (Modified from Mazumder et al. 2005.)

Figure 9.56a. The flow velocity was recorded by an Ott Laboratory Current Meter with a propeller of 2.5 cm diameter (Figure 9.57a). The vertical flow velocity profiles collected from all five different sediment beds showed usual log-law-type velocity, even for different values of bed roughness (d_{65}). The data of water surface slope, flow depth, water temperature, bed load and suspension samples, and bed configurations including wavelengths and heights of bed forms were collected and analyzed. Bed load and suspension samples were collected using siphon tubes (Figure 9.57b) mounted on a trolley above the flume at different heights of 5, 10 and 20 cm above the bed. After the evaporation of water from the samples, size analysis of collected samples had been made at ½ phi interval sand weighed the amount of each size class by an electronic digital balance. For illustration of the process of grain sorting during transportation, size distributions of suspended load samples collected at various heights 5, 10, and 20 cm at a maximum velocity (116 cm/s) for all beds, together with size distributions of all five sediment beds and bed loads are displayed in Mazumder et al. (2005). To facilitate comparison, both percentage and actual amount (g/5l) of grain-size frequency distributions of the samples are presented in the figure. Details of velocity, beds, bed loads, and suspended sediment data collected over the beds (numbered

TABLE 9.9

Particle-Size Distribution of Different Sediment Beds for Experiments

Sieve Size		Bed No. 10C1		Bed No. 10C2		Bed No. 10C3		Bed No. 10C4		Bed No. 10C5	
ϕ	mm	Weight (kg)	%	Weight (kg)	%	Weight (kg)	%	Weight (kg)	%	Weight (kg)	%
−3.0	8.000	-	-	-	-	-	-	-	-	12.000	6.061
−2.5	5.657	-	-	-	-	-	-	-	-	12.000	6.061
−2.0	4.000	-	-	-	-	-	-	12.000	6.897	12.000	6.061
−1.5	2.828	-	-	-	-	-	-	12.000	6.897	12.000	6.061
−1.0	2.000	-	-	-	-	12.000	8.000	12.000	6.897	12.000	6.061
−0.5	1.414	-	-	-	-	12.000	8.000	12.000	6.897	12.000	6.061
0.0	1.000	3.193	3.193	11.193	8.955	12.193	8.129	12.193	7.008	12.193	6.158
0.5	0.707	5.189	5.189	14.189	11.351	14.189	9.459	14.189	8.155	14.189	7.166
1.0	0.500	12.346	12.346	20.346	16.277	20.346	13.564	20.346	11.693	20.346	10.276
1.5	0.354	28.341	28.341	28.341	22.673	28.341	18.894	28.341	16.288	28.341	14.314
2.0	0.250	33.480	33.480	33.480	26.784	33.480	22.320	33.480	19.241	33.480	16.909
2.5	0.177	14.151	14.151	14.151	11.321	14.151	9.434	14.151	8.133	14.151	7.147
3.0	0.125	2.674	2.674	2.674	2.139	2.674	1.783	2.674	1.537	2.674	1.351
3.5	0.088	0.232	0.232	0.232	0.186	0.232	0.155	0.232	0.133	0.232	0.177
4.0	0.063	0.269	0.269	0.269	0.216	0.270	0.180	0.269	0.155	0.269	0.136
4.5	0.044	0.078	0.078	0.078	0.063	0.078	0.052	0.078	0.045	0.078	0.039
5.0	0.032	0.044	0.044	0.045	0.036	0.045	0.030	0.045	0.026	0.044	0.023
Total		100.000	100.000	124.999	100.000	149.999	100.000	173.999	100.000	197.999	100.000
Mean grain size (μ)		1.703ϕ (0.307 mm)		1.462ϕ (0.363 mm)		1.099ϕ (0.462 mm)		0.706ϕ (0.613 mm)		0.287ϕ (0.820 mm)	
Bed roughness (d_{65})		1.252ϕ (0.42 mm)		0.951ϕ (0.511 mm)		0.552ϕ (0.682 mm)		0.025ϕ (0.983 mm)		−0.613ϕ (1.529 mm)	
SD (σ)		0.661ϕ (0.632 mm)		0.783ϕ (0.581 mm)		1.089ϕ (0.470 mm)		1.413ϕ (0.376 mm)		1.742ϕ (0.299 mm)	
Coeff. of skewness (β_2)		−0.1713		−0.1088		−0.4490		−0.6550		−0.7726	
Coeff. of kurtosis (β_3)		3.8830		2.7060		2.325		2.1069		1.9616	
Bed conc. (dimless)		0.5700		0.5800		0.6200		0.6500		0.7000	

Source: Modified from Mazumder et al. (2001, 2005).

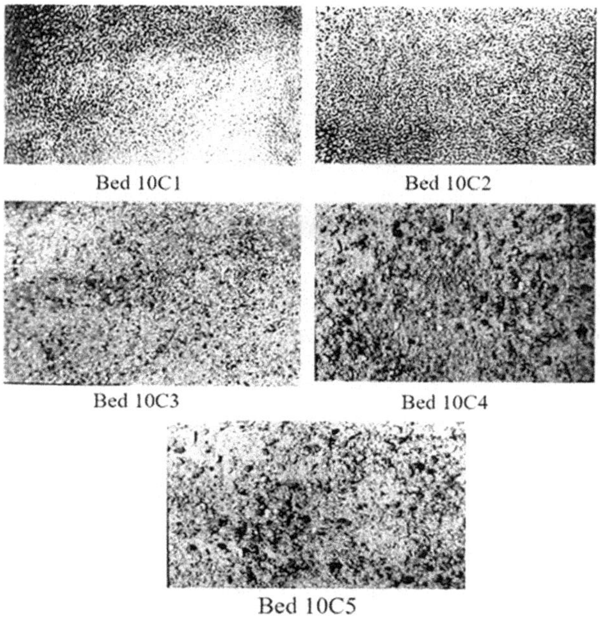

FIGURE 9.55 All five sediment beds of different roughness (sand to sand-gravel mixture) (Photo: Courtesy: ISI Fluvial Mechanics Lab, Prof. B. S. Mazumder.)

FIGURE 9.56 (a–c) Different bed forms developments at the ISI Fluvial Mechanics Laboratory during experiments. (Photo: Courtesy: ISI Fluvial Mechanics Lab, Prof. B. S. Mazumder.)

10C1—10C5) are to be found in Mazumder et al. (2005), and an unpublished report (Mazumder et al. 2001).

In order to develop the theoretical framework, one needs to use the momentum and mass conservation equations to study the velocity and concentration at bed layer and suspension under the influence of bed roughness. Without adjusting the von-Karman and integrating constants, a wake function or divergence function in the log-law is used. The expression of fully developed turbulent flow in the inner and outer regions is assumed as:

$$u = K_1 \ln \xi + K_2 + K_3 \sin^2\left(\frac{\pi}{2}\xi\right) \tag{9.396}$$

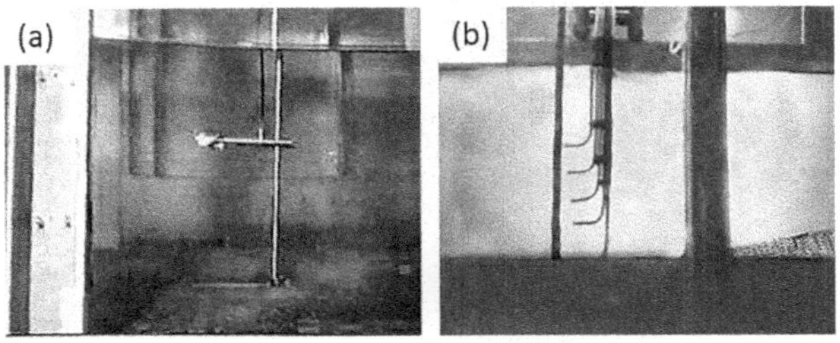

FIGURE 9.57 (a) Flow velocity recorded by Ott Laboratory Current Meter, and (b) bed load and suspension samples collected using siphon tubes above the flume at different heights of 5, 10, 15 and 20 cm above the bed. (Photo: Courtesy: ISI Fluvial Mechanics Lab, Prof. B. S. Mazumder.)

where $\xi = y/h$, $K_1 = \dfrac{u_*}{\kappa_o}$, $K_2 = \dfrac{u_*}{\kappa_o}\ln\left(\dfrac{h}{y_0}\right)$ and $K_3 = \dfrac{u_*}{\kappa_o}2\Pi$ with u_* is the friction velocity, Π is the wake strength parameter.

The coefficients K_1, K_2 and K_3 are estimated from the least square approximation. Figure 9.58 shows the vertical velocity profiles for three different runs $\left(u_{\max} = 68, 102, 116 \text{ cm/s}\right)$ above all five sediment beds.

The suspension equation is approximated as:

$$\epsilon_s \frac{dC}{dy} + C\frac{dC}{dy}(\epsilon_m - \epsilon_s) + C(1-C)v_0 = 0 \tag{9.397}$$

The differential equation for the vertical concentration distribution becomes:

$$\frac{dC}{dy} = \frac{-C(1-C)v_0}{u_*^2\left[\gamma + (1-\gamma)C\right]}\left[\frac{K_1}{\xi(1-\xi)} + \frac{K_3\frac{\pi}{2}\sin(\pi\xi)}{1-\xi}\right] \tag{9.398}$$

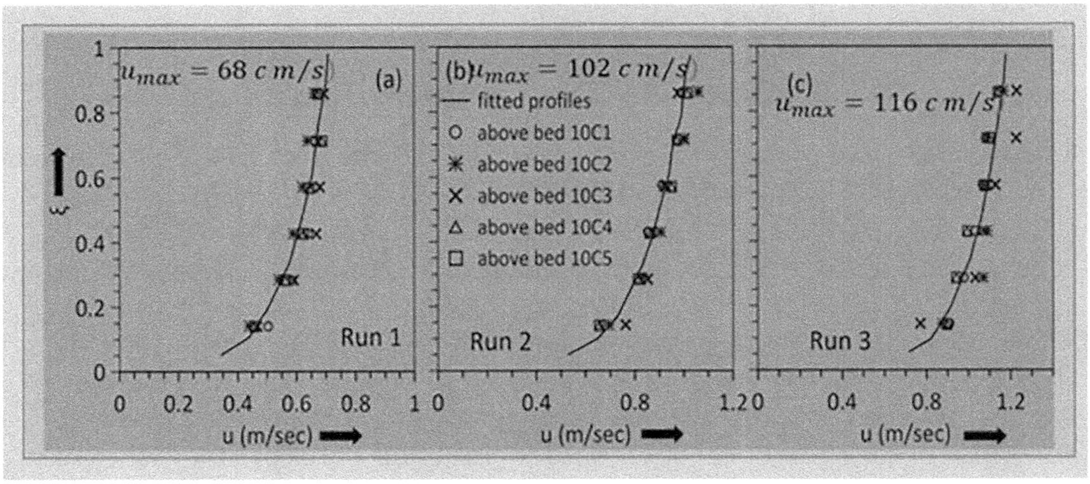

FIGURE 9.58 (a–c) Measured and computed vertical velocity profiles for three different runs $\left(u_{\max} = 68, 102, 116 \text{ cm/s}\right)$ above all five sediment beds (10C1, 10C2, 10C3, 10C4 and 10C5). (Modified from Mazumder et al. 2005.)

with the boundary conditions as: $C = C_{\xi_a}$ at $\xi = \xi_a$, where C_{ξ_a} is the reference concentration at the level ξ_a. To determine the C_{ξ_a} at the level ξ_a, one uses the bed layer thickness according to Wiberg and Rubin (1989) as:

$$\xi_a = \frac{0.68T_* d_{65}}{1 + A_1 T_*} \tag{9.399}$$

where $T_* = \tau_0/\tau_{0c}$ is the transport stage and A_1 is given by $A_1 = 0.0204(\ln d_{65})^2 + 0.022\ln d_{65} + 0.0709$. Following McLean (1991, 1992), the bed layer concentration C_{ξ_a} at the level ξ_a is assumed as:

$$C_{\xi_a} = \frac{\gamma_0 C_b (T_* - 1)}{1 + \gamma_0 (T_* - 1)} \tag{9.400}$$

Here $\gamma_0 = 0.004$, and C_b is the relative bed concentration which varies with the bed, where $C_b = 0.57$, 0.58, 0.62, 0.65, and 0.70 for all respective five beds. The reference bed layer concentration C_{ξ_a} computed from McLean (1991) is compared with extrapolated reference concentration C_{ξ_a} for two different runs (Figure 9.59a), and the variation of C_{ξ_a} from McLean (1991) with respective bed concentrations C_b for two runs shown in Figure 9.59b with the equations $C_{\xi_a} = 0.0003C_b^{-8.76}$ for Run 2 and $C_{\xi_a} = 0.0004C_b^{-8.72}$ for Run 3. Figure 9.60a,b shows the variation of suspension concentration (gm/5l) against bed roughness (d_{65}) of respective beds (10C1–10C5), indicating suspension concentration initially increases, then decreases with the bed roughness d_{65} for both runs 2 and 3. A similar behavior is also observed for γ against d_{65} for respective beds (10C1–10C5) for both runs shown in Figure 9.60c.

For non-homogeneous grain sizes, equation (9.397) is modified to take into account different fall velocities of various particle sizes. According to Hunt (1954) and McLean (1991), equation (9.397) is written as:

$$v_{0n} C_n (1 - C) + \epsilon_s (1 - C)\frac{\partial C_n}{\partial y} + \epsilon_m C_n \frac{\partial C}{\partial y} = 0 \tag{9.401}$$

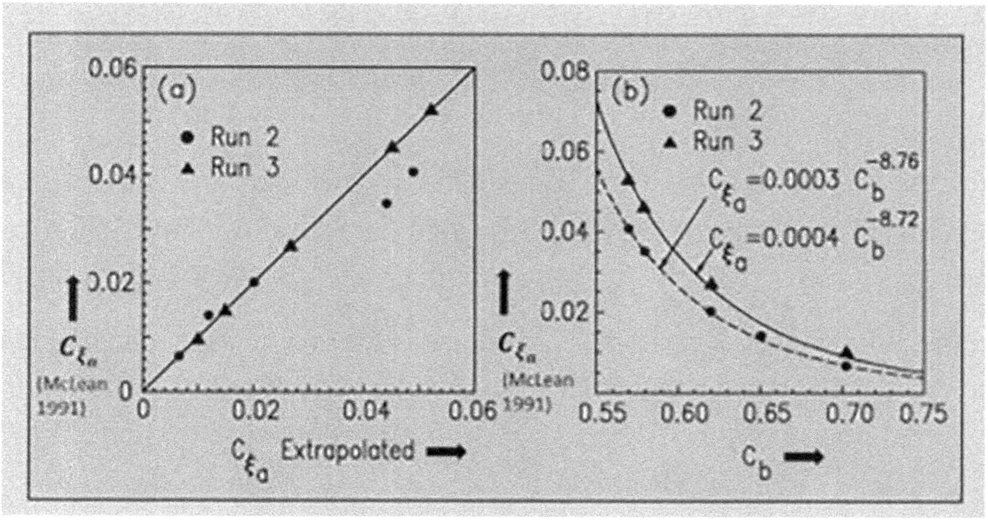

FIGURE 9.59 (a) The reference bed load concentration C_{ξ_a} from McLean (1991) compared with extrapolated reference concentration C_{ξ_a} for two different runs, and (b) variation of C_{ξ_a} from McLean (1991) with respective bed concentrations C_b for two runs showing with the equations. (Modified from Mazumder et al. 2005.)

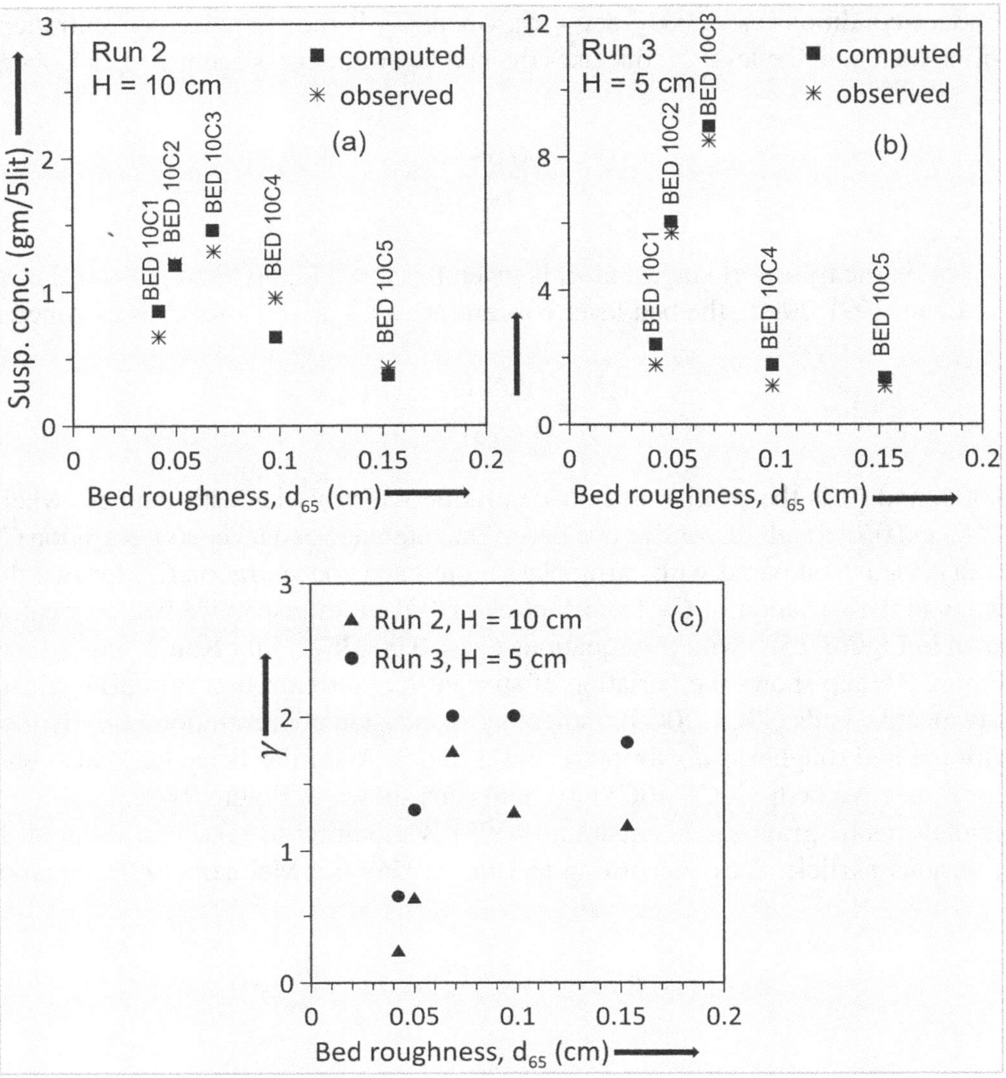

FIGURE 9.60 Variations of suspension concentration (gm/5l) against bed roughness (d_{65}) of respective beds (10C1–10C5): (a) for $H = 10$ cm and $u_{max} = 101$ cm/s (Run-2), (b) for $H = 5$ cm and $u_{max} = 116$ cm/s (Run- 3), and (c) variation of γ against d_{65} for respective beds (10C1–10C5) for both runs. (Modified from Mazumder et al. 2005.)

where C_n is the volume concentration of nth size of sediment and $C = \sum C_n$ is the total concentration, and v_{0n} is the settling velocity of nth size class. Sum over n different sizes on equation (9.401) satisfies equation (9.397) if $v_0 = \sum C_n v_{0n} \big/ \sum C_n$. Using the Reynolds analogy $\epsilon_s = \gamma \epsilon_m$, equation (9.401) is integrated for the concentration of each size class present in the flow, and the solution can be written as

$$C_n = \frac{(1-C)^{1/\gamma}}{\left(1-C_{\xi_a}\right)^{1/\gamma}} C_n(\xi_a) \exp\left(-\int_{\xi_a}^{\xi} \frac{h v_{0n}}{\epsilon_s}\right) \tag{9.402}$$

If $\gamma = 1$, equation (9.402) reduces to that of McLean (1991). To compute equation (9.402), it is necessary to specify the total concentration C at any desired height ξ, the reference concentration C_{ξ_a}, the reference concentration C_n of nth size class at the level ξ_a.

Following McLean (1991), the bed layer concentration as:

$$C_n(\xi_a) = \frac{\gamma_0 C_{bn}(T_{*n} - 1)}{1 + \gamma_0 (T_{*n} - 1)} \tag{9.403}$$

Using computed C, C_{ξ_a} and $C_n(\xi_a)$, equation (9.402) is computed for all grain sizes present in suspension at any particular height above all five sediment beds. The free parameter γ is plotted in Figure 9.61a,b against v_{0n}/u_* by adjusting the bed-fitted data from the experiments within the range $0.01 < v_{0n}/u_* < 1.2$ for all five values of bed roughness d_{65} at two different flow velocities and heights. The empirical equations for free parameter γ for all five different beds are as follows:

For Figure 9.61a,

$$\ln\gamma = -0.181 + 0.08\left(\ln\frac{v_{0n}}{u_*}\right) - 0.175\left(\ln\frac{v_{0n}}{u_*}\right)^2 \tag{9.404}$$

for bed 10C1-10C3, and

$$\ln\gamma = -0.32 + 0.35\left(\ln\frac{v_{0n}}{u_*}\right) - 0.07\left(\ln\frac{v_{0n}}{u_*}\right)^2 \tag{9.405}$$

for beds 10C4 and 10C5.

The R^2 values are 0.965 and 0.960, respectively. Similarly, for Figure 9.61b, the empirical relations are

$$\ln\gamma = 0.234 + 0.39\left(\ln\frac{v_{0n}}{u_*}\right) - 0.09\left(\ln\frac{v_{0n}}{u_*}\right)^2 \tag{9.406}$$

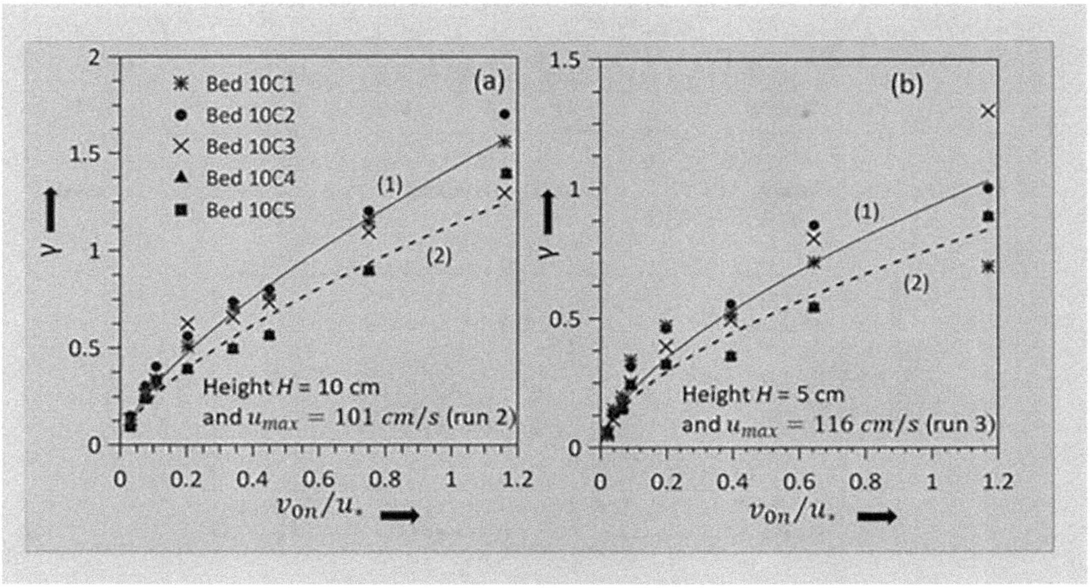

FIGURE 9.61 Plots of variation of γ against settling velocity v_{0n}/u_*. (a) for height H=10 cm and u_{max} = 101 cm/s (run 2), (b) for H=5 cm and u_{max} = 116 cm/s (run 3). Line (1) represents the fitted values for 10C1–10C3, and Line (2) indicates for 10C4–10C5. (Modified from Mazumder et al. 2005.)

for beds 10C1–10C3, and

$$\ln\gamma = 0.096 + 0.655\left(\ln\frac{v_{0n}}{u_*}\right) - 0.012\left(\ln\frac{v_{0n}}{u_*}\right)^2 \tag{9.407}$$

for beds IOC4 and IOC5.

The R^2 values are 0.978 and 0.964, respectively. It appears from the above equations that the γ is a function of settling velocity of particle. Using the above relationships, the value of γ is estimated for each grain size at a fixed velocity and height above all five beds.

The computed grain-size frequency distributions (g/51 by weight) in suspension are plotted together with measured suspension distributions in Figure 9.62 for $H = 10$ cm and $u_{\max} = 101$ cm/s. It is observed from the figure that the observed and computed values of suspension distributions agree well in general; but the actual values show little discrepancies in few of the cases.

Ghoshal et al. (2010) presented the grain-size fractions in the bed load transported over the five heterogeneous sediment beds of different bed roughness (Figure 9.55) studied by Mazumder et al. (2005). The existence of an entrapment factor associated with the sorting process observed from the bed to active layer was modeled based on the modified critical shear stress to estimate the grain-size fractions in the transport layer under given hydraulic conditions. This contribution is modeled in two steps: (i) determination of active layer (near-bed) concentration of each size fraction *transported as bed load* over sand/sand-gravel mixture beds, using modified critical shear stress due to heterogeneity in sediment bed; and (ii) determine the probability density function (pdf) which fits the patterns of grain-size distributions at the active layer (exchange of sediment between bed sub-surface and transport layer) formed from the variations of bed roughness and flow discharge. The question was whether any

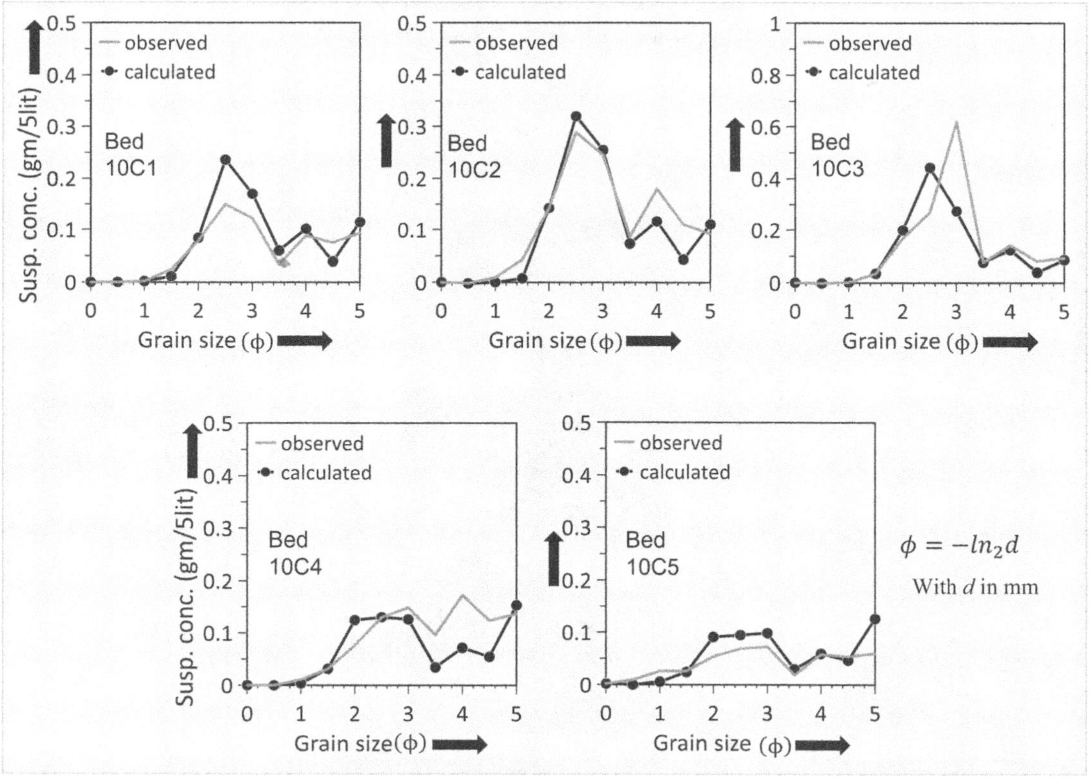

FIGURE 9.62 Profiles of computed and observed grain size distributions of suspension concentration (mg/5l) above all five sediment beds for $H = 10$ cm and $u_{\max} = 101$ cm/s (Run 2). (Modified from Mazumder et al. 2005.)

special grain-size distribution develops in the active layer as an effect of entrapment of finer grains in the interstices of gravel-sized particles, and/or as a result of variations in discharge. A series of experiments were conducted over five different sediment beds (Bed Nos. 10C1–10C5) with known grain-size distributions ranging from 0.032 to 8 mm (Mazumder et al. 2005), with different bed roughness values. Characteristics of all five sediment beds are provided in Table 9.9. Flume experiments showed that the grain-size distributions in both the active layer and the suspension under unidirectional flow were related to the velocity, the characteristics of the bed materials, and the suspension height (Sengupta 1979; Ghosh et al. 1981; Sengupta et al. 1999; Mazumder 1994; Mazumder et al. 2005). These studies showed that the sorting process took place immediately above the bed surface layer leading to a specific size distribution in the active layer, and that the grain sizes of the active layer affected the size distribution in suspension.

The velocity data had been fitted to the log-wake law by the least square method, and the calculated velocities were consistent with the observed data for more than 85%of the flow depth (Mazumder et al. 2005). The shear velocities ($u_* = 4.94$, 6.45 and 7.35 cm/s) were determined by means of the fitted log-wake law for all three maximum velocities. Altogether 15 experiments were conducted in three runs or discharges ($Q = 0.087$, 0.148, and 0.175 m^3/s) for each sediment bed. Details data collections and analysis are available in Mazumder et al. (2005).

Different models such as Gessler (1965), van Rijn (1984), Zyserman and Fredsoe (1994), McLean (1992) were used to estimate grain size fractions of the active layer concentration (near-bed) when the bed materials with different values of bed roughness and flow parameters were known; but none of them was sufficient to describe it properly. The values of critical shear stress were determined by Kuhnle (1993), who obtained the incipient motion following the technique of Parker et al. (1982) for ith grain-size fraction in sand, gravel and sand–gravel mixtures, which were rather different from those obtained from the Shields curve. According to Kuhnle (1993), the transport function depends on grain size and it is strongly non-linear in nature. For the sand–gravel mixtures, the value of modified critical shear stress for each size fraction is higher than that observed for 100% sand beds and lower than that found for 100% gravel beds.

Ghoshal et al. (2010) studied the relative active layer concentrations at three locations (mid-section, inner wall and outer wall) for all five sediment beds and three different maximum velocities ($u_{max} = 68$, 101 and 116 cm/s.). For sand beds (Nos. 10C1–10C3), the computations were made for grain-size fractions up to the size 2.5 phi and for sand–gravel mixtures (Nos. 10C4–10C5) up to 4.0 phi using the modified critical shear stress. The value of the parameter γ_0 was determined as 0.004 by McLean (1992) for a limited set of data contrary to the present experimental data, where parameter γ_0 was a function of grain size. Two relationships between the parameter γ_0 and the size (ϕ) had been established by simple regression for the side-wall boundaries: one for bed roughness (d_{65}) ranging from 0.042 to 0.068 cm (beds 10C1–10C3) and the other for (d_{65}) ranging from 0.099 to 1.53 cm (beds 10C4 and 10C5). Details of experiments are available in Ghoshal et al. (2010).

9.12.1 Patterns of Grain-Size Distributions in the Transport Layer

The patterns of the grain-size distributions in the active layer (near-bed) were studied for various discharges ($Q = 0.087$, 0.148 and 0.175 m^3/s) over all five different sediment beds. It is noticed that the grain size distributions of all five sediment beds follow the log-hyperbolic distribution. The identification of size distribution provided a clue for understanding the transport process in a particular sedimentary environment. The question was whether any special grain-size distribution developed in the active layer as an effect of the entrapment of finer grains in the interstices of gravel-sized particles, and/or as a result of variations in discharge.

The family of size-frequency distributions such as Gaussian distribution, four parameters of hyperbolic distribution, and three parameters of Laplace distributions (discussed in Section 8.5.2) are tested for all five different sediment beds (Bed Nos. 10C1–10C5) including the computed and observed bed loads' grain-size distributions above the five sediment beds for three runs (Run 1, 2, 3) with three different locations such as Inner wall (I. W), outer wall (O.W) and mid-section (M.S).

The weight frequency of the various grain-size fractions of the five sediment beds (Nos. 10C1–10C5) before the experiments has been shown in the probability scale (Ghoshal et al. 2010) as log-hyperbolic distributions. All five sediment beds show a three-segment shape on probability paper. Only in bed 10C1 that the classical three-segment shape is not very clear. All the three models mentioned above are fitted for all five sediment beds, and it is found that the grain-size distributions of all five beds follow a log-hyperbolic form. The estimated parameters and relative errors of the three distribution models fitted for the experimental beds are shown in Tables 9.10a and 9.10b respectively. The formula for calculating the relative error (equation 9.318) is discussed underneath. The grain-size distributions of transport layer samples collected across the width of the channel at three different locations (M. S., I. W. and O. W.) for three values of maximum velocity u_{max} ranging from 68 to 116 cm/s (Runs 1, 2 and 3) are shown in Ghoshal et al (2010) for all five sediment beds (Nos. 10C1–10C5).

A total of 45 samples of bed load collected across the width of the channel for the five sediment beds and three different discharges ($Q = 0.087$, 0.148 and 0.175 m^3/s) are fitted using all three statistical models. For each u_{max}, 15 bed load samples were collected. It was found that (1) at a low velocity, the log-normal distribution fitted best in six cases, (2) at a medium velocity, the log-skew-Laplace distribution fitted best in ten cases, and (3) at a high velocity, log-skew-Laplace distribution fitted very well for seven cases. It was therefore concluded that, at medium and higher velocity, the log-skew-Laplace distribution was the best-fit distribution model among the three models, irrespective of the bed roughness.

TABLE 9.10A

The Parameters Estimated for Three Hypothesized Distributions of Sediment Beds (Modified from Goshal et al. 2010)

Bed	Log-Normal		Log-Hyperbolic				Log-Skew-Laplace		
Number	$\mu(\phi)$	$\sigma(\phi)$	γ_1	γ_2	$\mu_1(\phi)$	$\delta(\phi)$	α	β	$\mu_2(\phi)$
10C1	1.467	0.626	2.140	3.242	1.761	0.746	0.493	0.512	1.664
10C2	1.193	0.799	0.908	3.234	1.819	0.315	0.497	0.863	1.097
10C3	0.807	1.131	0.497	3.232	1.913	0.285	1.117	0.574	1.641
10C4	0.423	1.458	0.315	3.229	1.978	0.258	1.510	0.624	1.591
10C5	0.027	1.776	0.208	3.248	2.057	0.244	1.928	0.668	1.547

TABLE 9.10B

Relative Error (*E*) Computed for the Three Distributions Fitted for Sediment Beds

Bed Number	Log-Normal	Log-Hyperbolic	Log-Skew-Laplace
10C1	0.480	0.351	0.364
10C2	0.394	0.328	0.817
10C3	0.537	0.341	0.466
10C4	0.675	0.342	0.516
10C5	0.889	0.331	0.560

Source: Modified from Goshal et al. (2010).

The grain-size distributions of 45 bed load samples collected across the width of the channel over five sediment beds were compared with the outcomes of three statistical distributions. Increase of the flow velocity led to a change in the grain-size distribution at the sedimentary surface. At medium velocity (101 cm/s), the log-skew-Laplace distribution appears to give the best fit for bed load sediments among the three distributions. The results of the present study may help distinguish between the various sedimentary environments under different hydrodynamic conditions of deposition and bed roughness. The motivation was to determine if the transport process could be deduced from a sedimentary record where grain-size distribution sorting was observed.

Subsequently, Ghoshal et al. (2011) estimated the grain size distributions in suspension at different heights over the same five heterogeneous sediment beds (Bed nos. 10C1 to 10C5) of different values of bed roughness. The analysis had been performed to study the influences of velocity, bed roughness and ripple height on the size sorting, which developed the specific grain-size distribution in suspension. The grain-size distributions of 45 suspension samples collected over five sediment beds are studied. The observed size distributions in suspension were tested using three statistical distributions, such as log-normal, log-hyperbolic and log-skew-Laplace. Tests indicated that log-hyperbolic distribution does not fit in suspension; even the original beds' grain size distributions were log-hyperbolic. From the experimental findings, it is clear that (a) flow velocity influences the overall grain size distribution pattern of suspended sediments leading from log-normal to log-skew-Laplace distribution with velocity, (b) log-normal distribution is the best-fit model irrespective of bed roughness (d_{65}) except in one case (Bed 10C3), which may be identified as critical bed roughness $d_{65} = 0.68$ mm (0.55 phi), and (c) log-normal distribution is the best-fit statistical model irrespective of different ripple heights though the original bed's size distribution in each case was log-hyperbolic. This is similar to the observation in the river Usri where three categories of bed form height were considered for bed load sediments collected from the cross-bedding foresets. From the comparative study, it is interesting to note that the lognormal distribution fits best for 56% and a log-skew-Laplace fits for 44% in suspension distributions for all cases, under unidirectional flow with changes in flow velocity, bed roughness and ripple heights. The pattern of size distribution in suspension may provide a clue for understanding the transport process. The results of this study would be valuable to hydraulic engineers and sedimentologists to predict the nature of bed material from the size distributions of suspended materials.

Rice and Church (2010) presented a model to study the grain-size sorting process within river bars with respect to the downstream fining along the channel. Ghoshal et al. (2013) developed theoretical models for velocity and sediment suspension based on the concept of mixing length that includes the damping effect of turbulence due to sediment suspension in the flow over the sand-gravel mixture bed. It is interesting to note that the non-zero vertical component of velocity generated due to the action of erodible sediment bed follows Beta density function. In their study, they determined the turbulence and its impact on suspension and the progressive downstream fining of grain sizes over the mixture bed. They also proposed a novel statistical clustering/segmentation procedure (Hastie et al., 2001) that splits the stream into three regions namely upstream, downstream and a transitional mid-stream by taking into account the averaged grain-size distribution. Similar principles have been used in turbulence applications (Mazumder 2007). To study the segregation of grain sizes along downstream, the effective bed samples (EBSs) collected from outer, central and inner regions across the channel width were analyzed for two independent experiments carried out for $u_{max} = 105$ and 116 cm/s (details are explained in the paper of Ghoshal et al. 2013). Altogether 50 effective bed samples (EBSs) from upstream to downstream were used for the analysis of segregation. Here the percentage of grain-size distributions of EBS from upstream to downstream are analyzed for outer, central and inner regions of the channel for $u_{max} = 116$ cm/s only. It is interesting to note that the concentration distributions gradually change to almost uni-modal through a process of size-sorting at the bed when sediments are transported from upstream to downstream even from a bed of bimodal distribution of sand-gravel mixture. From the experimental observations, it is noticed that a proportion of sand materials is entrapped in the interstices of the coarse particles and vice versa during transportation. However, for accurate estimation of suspension as well as active-layer concentration, a thorough laboratory study is required to understand the flow behavior and the grain-size fining process over the sand gravel mixture bed. Statistical data analysis of longitudinal fining of grain sizes suggests a selective deposition along the length of the stream into three regions, such as the upstream, the midstream and the downstream of the channel, considering the mean size of distributions along the bed as the feature of interest This research may have a significant contribution to infer the influence of heterogeneity of sand-gravel mixture in suspension as well as longitudinal fining of grain sizes during transportation. The motivation of this study is to understand the bed material character in the river form, sorting process and its role in controlling the sediment flux through landscape.

9.13 TOTAL LOAD

The total amount of sediment material transported per unit time and width through a given section of a river/stream for the given flow and sediment bed conditions is termed total load. The total load of sediment transport can be broadly classified into two categories: bed load rate and suspended load rate. Therefore, the total load is obtained by addition of bed load rate and the suspended load rate. The bed load rate is obtained from the bed load equation derived earlier, and the suspended load is obtained from the suspension load equation. At low transport rate due to low discharge, the most of the sediment move in contact with the bed, so the bed load approximates the total load as well. The total load of sediment practically is made up of solid particles consisting of all grain sizes present in the bed. The bed load and suspended load equations are derived from the supply of particles in the bed materials and it appears as a transported fraction of bed material load. Total load of a stream does not necessarily have to be identical with the bed material load. In the stream, there exist so called wash load which are finer than the bulk of bed material and these are rarely found in the bed. The wash load is very fine material transported in suspension which is not found in bed material forming bed and banks. The wash load depends mainly on the supply of very fine sediment from the watershed. The total load is the sum of the bed load, suspended load and wash load. In the flume experiments, wash load may be absent, and the total load would be the bed material, but in natural streams, the wash load is always present. In that case, the total load is the combination of bed material and wash load (Garge and Ranga Raju, 2000). Einstein (1950) suggested that the limiting sizes of wash load and bed material load may be chosen quite arbitrarily from the mechanical size analyses as the grain diameter d_{10} of which 10% of the bed material is finer. It is important to note that the effect of discharge on the bed material load is clearly defined, but no relationships between the discharge and the wash load is established (Colby 1963; Garde 1970). The total-load of sediment transport rate is estimated as the sum of the bed load and suspended load transport rates, which are computed separately by using the appropriate bed-load and suspended-load formulas. This is an approach of determining the total load of sediment transport by adding two approximate fractional sediment loads. The identification of bed load and suspended load transports is very difficult because of intimate random exchange between two layers. Sometimes total load can be estimated directly from the bed to suspension without going through any intermediate stage like bed load (Ghosh et al. 1981; Mazumder 1994). At high flow discharge, the bed load layer is hardly difficult to separate from the suspended load (Chien and Wan 1999). There are two general methods to estimate the total load: indirect method and direct method. Indirect method is used to determine the total load transport rate as the sum of the bed load and suspended load transport load rates from the bed load and suspended load formulas. In the direct method, total transport load is determined without going through any demarcation of the bed load and suspension load. This method is very useful to rivers or streams to estimate the total load, because the difference between the bed load and suspended load transport is very difficult in the rivers.

9.13.1 Total Load by Einstein (1950)

Einstein (1950) suggested adding the bed load and the suspended-load transport rates for the proper estimation of total load transport. The bed load and suspended load of sediment size fractions are respectively given by $g_s i_s$ and $g_{ss} i_{ss}$, where i_s is the fraction of bed load transport, i_{ss} is the fraction of suspended load transport rate, and g_s is the bed load in weight per unit time and width. The suspended load g_{ss} in weight per unit time and width is given by equation (9.226).

The total load transport rate for a size fraction i_{st} is given by

$$i_{st} g_{st} = g_s i_s + g_{ss} i_{ss} \tag{9.408}$$

where g_{st} is the bed material load rate in weight per unit time and unit width.

Introducing the relation given by equation (9.239) in equation (9.408), one gets

$$i_{st} g_{st} = g_s i_s \left[1 + P_E I_1 + I_2\right] \tag{9.409}$$

where $P_E = 2.303 \log\left(\dfrac{30.2h}{\delta_c}\right)$ is the transport parameter, and I_1 and I_2 are integrals evaluated by Einstein (1950) given in Figure 9.35a,b. The bed load transport rate g_s is obtained from the relation of the parameters ψ_* and Φ_*. It may be mentioned here that the method of Einstein (1950) is complicated and laborious for practical use, but it involves fundamental concepts of physics to study the sediment transport. This equation gives a stream capacity to estimate how much bed material load can transport under uniform and steady flow conditions, excluding any wash load. The total load transport rate is given by (9.409). Einstein's method of computation of bed material load is elegant and it allows only theoretical computation without measuring the suspended or bed load. Colby and Hembree (1955) suggested some modifications of Einstein's procedure for the computation of total load, including the wash load for a given flow (Einstein 1950). In this case, the total load is determined from the knowledge of depth-integrated suspended samples, stream flow measurements and bed sample characteristics (for more details, see Graf 1971). The modified Einstein's procedure suggested by Colby and Hembree (1955) was used by Yang (1996) for applications in the field.

9.13.2 Total Load by Bagnold (1966)

In his work (1966), Bagnold derived a proper expression for the total load based on the energy concept, by summing up both the components: bed load and suspended load transport rates. The expression for bed load transport rate is derived in the bed load section, and this bed load expression is extended to the whole depth of flow for suspended load rate. Let W_b and W_s be the work rates of the bed load q_{sb} and suspended load q_{ss} respectively. Adding these two components, the total load formula of Bagnold (1966) is given by $q_s = q_{sb} + q_{ss}$, in which $q_{sb} = \dfrac{E_{\text{eff}}}{\tan\phi} E_p$ and $q_{ss} = e_s (1 - E_{\text{eff}}) \dfrac{u_s}{v_0} E_p$. Explicitly, the total load rate $\dfrac{q_s}{\tau_0 u_m}$ is given by:

$$\frac{q_s}{\tau_0 u_m} = \frac{E_{\text{eff}}}{\tan\phi} + e_s (1 - E_{\text{eff}}) \frac{u_s}{v_0} \tag{9.410}$$

where E_{eff} is the working efficiency for bed load, e_s is the efficiency with respect to suspended load, τ_0 is the bed shear stress with $\tau_0 u_m$ is the stream power per unit area, v_0 is the terminal velocity of the grains, and u_s is the average transport velocity of the grain in suspension. The total load may be obtained if four parameters E_{eff}, e_s, $\tan\phi$, u_s are known. It is suggested that this equation is equally useful to the turbulent flow as well as laminar flow, but for laminar case second term should not be there.

9.13.3 Total Load by Chang et al. (1967)

Chang et al. (1967) suggested that the total load should be given by the sum of the bed load and suspended load, excluding the wash load, and is given by

$$g_{st} = \int_0^a Cu_s\,dy + \int_a^h Cu_s\,dy = g_{sb} + g_{ss} \tag{9.411}$$

where the first integral term g_{sb} represents bed load moving with be bed layer within the thickness a, and the second integral term g_{ss} represents the suspended load. Chang et al. (1967) used the modified version of Du Boy's expression given by equations (9.5) and (9.6) as:

$$g_{sb} = A_T\bar{u}(\tau_0 - \tau_{0c}) \tag{9.412}$$

where A_T is the bed material discharge coefficient, and $\bar{u}$ is the mean flow velocity. The coefficient A_T is determined experimentally and is functional relation with the bed material, bed configuration, and flow characteristics (Graf 1971). They determined the values of A_T from three different natural streams and they found the values were constant within the range $0.27 < A_T < 1.10$ for different streams. The suspended load was expressed as Einstein (1950) as: $g_{ss} = g_{sb}R_I$, where R_I contains two integrals given by Einstein (1950). Substitution of these expressions, the total bed load is obtained as:

$$g_{st} = g_{sb} + g_{ss} = A_T\bar{u}(\tau_0 - \tau_{0c}) + g_{sb}R_I = A_T\bar{u}(\tau_0 - \tau_{0c})(1 + R_I) \tag{9.413}$$

This equation was tested with 184 flume data set and 57 natural river data and the agreement was found to be satisfactory. This equation was also suggested by Egiazaroff (1965), who considered the bed material transport to be proportional to the excess of mobility.

9.14 SUMMARY

Some of the important aspects discussed in this chapter are summarized below.

- For the estimation of bed load transport, the number of empirical equations was derived based on experimentation and various concepts. DuBoy (1879) derived an equation for bed load transport based on the shear stress concept, $q_B = A(\tau_0 - \tau_{0c})\tau_0$ known as DuBoy's bed load equation with the characteristic sediment coefficient denoted by A.
- Based on a large number of experiments, Meyer-Peter and Müller (1948, 1949) derived equation for dimensionless bed load transport rate.
- Bed load transport equation was derived based on particle velocity. Van Rijn (1984) presented the bedload transport rate equation $q_B = u_p h_s C_b$, where he considered particle velocity u_p, the saltation height h_s and the bed load concentration C_b.
- Bed load transport equation was derived based on the concept of shear layer thickness. Wilson (1987) derived the bed load transport formula for large shear stress.
- Einstein (1942, 1950) was the first to attempt a semi-theoretical solution to the problem of bed load transport using the probabilistic concepts, which appears to be the most popular among all the formulae suggested so far for the prediction of transport rate.

- Wang et al. (2008) suggested a modified version of the Einstein formula without relying on various assumptions, which incorporated a non-uniform sediment model.
- Saltation is an important process in bed load transport. Hu and Hui (1996) suggested the relative thickness of saltation height $\frac{h_s}{d}$ was related to the specific gravity of particles, dimensionless bed shear stress, and the condition of bed surface.
- Recent times, the use of image processing techniques for observing the motion of sediment particles has been considered in some research studies to predict the particle trajectory, saltation height, length, resting time of a particle and particle velocity at the near bed region. The analyses of images from a high-resolution digital camera system provide a unique approach to understand the fundamental mechanism of particulate materials.
- Suspended sediment is regarded as those sediment particles that are surrounded by the fluid for a period of time, known as suspension. The sediment transported by a fluid flow that is fine enough for turbulent eddies to beat the settling of the particles through the fluid. The stronger the flow or the finer the sediment, the greater the amount of sediment can be suspended by turbulence.
- The suspension equation is based on Fick's law, which states that the mass of a solute, passing a unit area per unit time in a given direction, is proportional to the gradient of solute concentration in that direction, or it defines the flux that moves from high concentration to low concentration at a concentration gradient.
- For the estimation of suspended load, the advection-diffusion equation can be used.
- Rouse equation (1937) for sediment suspension is: $\frac{C}{C_{\xi_a}} = \left(\frac{1-\xi}{\xi} \cdot \frac{\xi_a}{1-\xi_a}\right)^{z*}$ where $z_* = \frac{v_0}{\kappa_o u_*}$ is the Rouse number, $\xi = \frac{y}{h}$, and C_{ξ_a} is the concentration of sediment at a reference level $\xi = \xi_a$, with $\xi_a \left(= \frac{y_a}{h}\right)$ is a height near the bottom boundary.
- Einstein's method is the most advanced one to compute the suspended load if the velocity and suspension concentration are known.
- Hunt's diffusion equation for sediment suspension is given as: $\epsilon_{sy} \frac{\partial C}{\partial y} + C \frac{\partial C}{\partial y}\left(\epsilon_{wy} - \epsilon_{sy}\right) + Cv_0 = 0$
- Ghosh and Mazumder (1981) developed a theoretical framework to relate the suspended load's grain size distribution, particularly the occurrence of *uni-modality, symmetry and log-normality*, with the flow parameters and grain-size distributions of the bed materials.
- Mazumder (1994) developed a theoretical model to compute the suspension grain-size distribution based on Hunt's diffusion equation, taking into account the effect of hindered settling due to the increased suspension concentration.
- According to Van Rijin (1984): in a steady, uniform flow, the vertical distribution of the sediment concentration profile can be described by: $\epsilon_{sy} \frac{\partial C}{\partial y} + (1-C)Cv_{sm} = 0$; in which C = sediment concentration; v_{sm} = particle fall velocity in a fluid-sediment mixture; ϵ_{sy} = sediment diffusion coefficient; and y = vertical coordinate.
- Umeyama and Gerristen (1992) proposed a theoretical model for velocity distribution in sediment-laden flow by means of a new mixing length concept. The total amount of sediment material transported per unit time and width through a given section of a river/stream for the given flow and sediment bed conditions is termed total load. The total load is obtained by addition of bed load rate and the suspended load rate.
- According to Einstein (1950), total load transport can be estimated by adding the bed-load and the suspended-load transport rates.

9.15 EXERCISE PROBLEMS

1. Illustrate the bed load transport mechanism with various processes involved. Explain the three types of modes for the initiation of sediment particles on the loose sediment bed.
2. Differentiate between bed load transport and suspended load transport with various mechanisms involved?
3. What is saltation? Illustrate the saltation with a figure and its role in bed load transport.
4. Explain Duboy's bed load equation with its application?
5. Discuss the modification suggested by Schoklitsch (1930) to Duboy's bed load equation with its application?
6. In a wide-open channel flow with uniform flow condition, the flow velocity is 2.5 m/s, flow depth is 3.5 m, bed slope S_o is 0.00015, and bed sediment median size d_{50} is 0.5 mm, the coefficient of kinematic viscosity of water is 10^{-6} m^2/s. The critical (threshold) shear stress for the bed is observed to be 0.7 Pa. Calculate the bed material transport rate (in volume rate per unit width) by the Duboys formula, assuming the characteristic sediment coefficient (A) as 4×10^{-6} kg^{-2} m^4 s^3.
7. Illustrate Mayer-Peter and Muller's equations for bed load transport with its applications and limitations.
8. Elaborate the Bagnold (1956) bed load transport formula using the principles of energy. What is the difference between Bagnold and Mayer-Peter and Muller's equations?
9. Explain the Yalin (1963) bed load equation based on the dimensional analysis and the dynamics of the average motion of the grain.
10. Illustrate van Rijn (1984) proposed a bed load expression based on particle velocity.
11. Elaborate the probabilistic concept for bed-load transport. Explain the assumptions used.
12. Explain the Einstein (1950) bed load equation based on the probabilistic concept?
13. Elaborate the Modified Einstein equation proposed by Wang et al. (2008).
14. Illustrate the image-processing techniques for observing the motion of sediment particles and their application in sediment transport investigations.
15. Illustrate the suspended load transport mechanism with various processes involved. Explain the 3 types of modes for the initiation of sediment particles on the loose sediment bed.
16. Elaborate the turbulent advection-diffusion equation and its use for analysis of suspended load.
17. Write the Rouse equation (1937) for sediment suspension and its application to estimate suspended load.
18. Explain Einstein's (1950) approach to compute the suspended load with its applications.
19. Comment on the direct computation of suspension grain size distribution and its application in bed load and suspended load.
20. Explain Mazumder's theoretical model to compute the suspension grain-size distribution based on Hunt's diffusion equation.
21. Describe the Umeyama equations for velocity distribution in sediment-laden flow by means of a new mixing length concept.
22. For a given river section, explain the total sediment load. Illustrate the total load equation concept by Einstein (1950).

10 Bedform Migration and Scour Structures

10.1 INTRODUCTION

As discussed in previous chapters, when the average shear stress on the loose sand bed materials over an alluvial channel exceeds the critical shear stress, the individual particles initiate to move, which causes them to deform or scour the plane bed. Depending on the flow conditions, the interface of sediment bed and water flow assumes various types of bedforms. Bedforms such as ripples and dunes generated from a flat sand bed interacting with the flow show significant changes in the mean flow, resistance to flow, turbulence, bed load, and suspended load transport. Bedform is defined as the features developed on the loose sediment bed of an alluvial channel due to the flow conditions. The bedform is also called bed undulations or sand waves. In the past, several experimental studies have been conducted to understand the flow resistivity due to changes in turbulent flow generated by the bedforms. The subtle changes in bedform show a strong effect on turbulence and sediment transport. These features of bedform over the bed are termed as *Regimes of flow* introduced by Garde and Albertson (1959) and Garde and Ranga Raju (2000).

For the convenience of description, the bedform features are characterized by the plane bed with/without movement of sediment particles, ripples and dunes, transition, and antidunes (Albertson et al., 1958; Garde and Albertson, 1959; Simons and Richardson, 1962). Best (2005) made an excellent review on the features of the mean flow, turbulence, morphology, and sediment transport associated with ripples and dunes and highlighted the future directions of research. On the other hand, the interactions between the flow and any structure movable or non-movable on the sediment bed cause the sediment erosion around the structure due to vortex generation leading to scour or bed deformation. The hydrodynamics of flow and the consequence effect on the mechanism of scour/erosion around the structure are of great interest to engineers or scientists. Particularly, the scour formation around the bridge piers or other hydraulic structures is the leading cause of the damage of bridges and hydraulic structures across the river or canals (Melville and Raudkivi, 1977). In this chapter, resistance to flow, bedforms, and its migration, and flow and scour around the hydraulic structures or bridge piers are discussed in detail.

10.2 RESISTANCE TO FLOW

In nature, it is observed that there is a strong relationship between the flow regimes and the resistance to flow. Examinations of bed load transport indicate that the bed load transport and resistance to flow are interconnected, and that one depends on the other. Some definite relationships between the mean flow velocity, hydraulic radius, water surface slope, and the characteristics of channel boundary for the case of steady, uniform open channel flow are considered. Such relationships are commonly known as resistance equations. In rigid boundary open channel flow, several resistance equations such as Chezy's equation, Manning's equation, logarithmic equation, etc. are commonly used (Keulegan, 1938, Garde and Raaga Raju, 2000) to study the bed load transport. The constants involved in these resistance equations are well established for subcritical flow in the open channels. A fair amount of research has been performed on the problem of flow resistance in the alluvial channel. The resistance equation is important to design the irrigation channels, river management work, mean velocity distribution, sediment transport studies, etc. In particular, the flow over the sediment bed changes the bed configuration

DOI: 10.1201/9781003000020-10

along with changes in the flow condition. The change in bed condition makes it extremely difficult to describe the resistance to the flow due to bedforms. With the development of bedforms, a part of the sediment load as bed load is transported over the bed, and a part of the materials goes into the main flow of suspension; and eventually, the fluid characteristics changes, which has a significant influence on the mean velocity distributions. For some understanding of the problems associated with the velocity distribution of the resistance to flow in the alluvial channels, a brief overview of the literature concerning the resistance formulae is provided below.

10.3 RESISTANCE FORMULAE

From the basic fluid mechanics theories, we know that the bed shear stress can be written as,

$$\tau_0 = \gamma_f \frac{A}{P} S = \gamma_f RS \tag{10.1}$$

where R is the hydraulic radius, S is the slope of the water surface and $\gamma_f = \rho_f g$. Equation (10.1) plays an important role in developing the flow equation which is common to all shapes of the channel. We can also express the average shear stress τ_0 as,

$$\tau_0 = K\rho_f U^2, \tag{10.2}$$

where K is a coefficient which depends on the nature of the surface and flow parameters and U is the average velocity. Combining equations (10.1) and (10.2), one gets

$$K\rho_f U^2 = \gamma_f RS, \text{ Then, } U = C\sqrt{RS} \tag{10.3}$$

where $C = \sqrt{\dfrac{\gamma_f}{\rho_f}\dfrac{1}{K}}$ is the coefficient which depends on the nature of the surface and the flows. Equation (10.3) is known as Chezy's formula for steady uniform flow. Dimension of C is $\left[L^{1/2}T^{-1}\right]$ and C is known as Chezy's discharge coefficient. It has been found that C varies with the bed and the flow characteristics. The Manning's formula for uniform flow in an open channel is given as:

$$U = \frac{1}{n} R^{\frac{2}{3}} S^{\frac{1}{2}} \tag{10.4}$$

where n is roughness coefficient known as Manning's coefficient. This is essentially a function of the nature of the boundary surface. The values of n for different types of surfaces are available in Chow (1959). This formula is used in uniform flow, which is an acceptable formula. Comparing the Chezy's and Manning's formulae, one gets the Chezy's discharge coefficient as:

$$C = \frac{1}{n} R^{\frac{1}{6}} \tag{10.5}$$

which implies that Chezy's discharge coefficient C is a function of hydraulic radius R and the Manning's roughness coefficient n. From equation (10.5), the following dimensionless relationship can be obtained:

$$\frac{U}{u_*} = \frac{C}{\sqrt{g}} = \frac{1}{\sqrt{g}} \frac{R^{\frac{1}{6}}}{n} \tag{10.6}$$

where u_* is the shear velocity and g is the acceleration due to gravity. Here the coefficient n varies only with the boundary roughness, and to make equation (10.6) dimensionless, the dimension of n will be $TL^{-1/3}$ (Rouse, 1947).

The velocity distributions for turbulent open channel flow past smooth and rough boundaries are respectively given by Prandtl-Karman velocity equations (6.70) and (6.112a) in Chapter 6. Integration of these two equations based on the experimental data, the logarithmic resistance laws are obtained by Keulegan (1938) and are given by:

$$\frac{U}{u_*} = 5.75\log\left(\frac{u_* R}{\nu_f}\right) + 3.25 \text{ for smooth boundaries} \tag{10.7}$$

$$\text{and } \frac{U}{u_*} = 5.75\log\left(\frac{R}{k_s}\right) + 6.25 \text{ for rough boundaries} \tag{10.8}$$

where k_s equivalent sand grain roughness. In the past, several empirical resistance equations have been suggested for channel flow. Among these, the Chezy's and the Manning's equations (10.3) and (10.4) are commonly used. A comparison of equation (10.6) with resistance equations (10.7) and (10.8) shows that Chezy's and Manning's equations (10.3) and (10.4) do not take into account the effect of viscosity or resistance. Hence, these equations are applicable only when viscous effects are negligible or when the boundary is hydro-dynamically rough.

For incompressible fluid, the equation of Darcy-Weisbach friction factor f for pipe flow is given as (Streeter and Wylie, 1975),

$$h_f = f\frac{L}{D}\frac{U^2}{2g} \tag{10.9}$$

where h_f is the head loss due to friction in a pipe of diameter D and length L, and g is the acceleration due to gravity.

For smooth pipe flow, the friction factor f is found to be a function of the Reynolds number, $Re = \frac{UD}{\nu_f}$ only. For rough turbulent flows, f is a function of relative roughness $\frac{k_s}{D}$, independent of the Reynolds number Re. In the transitional region, both roughness and Re play important roles. The roughness magnitude for commercial pipes is expressed as equivalent sand-grain roughness (k_s).

The friction factors f for circular pipe flow in different regions are as follows:

1. For smooth walls with Reynolds number $Re < 10^5$, the friction factor is (Streeter and Wylie, 1975):

$$f = \frac{0.316}{Re^{\frac{1}{4}}} \rightarrow \text{from Blasius formula} \tag{10.10}$$

2. For smooth walls, $Re > 10^5$

$$\frac{1}{\sqrt{f}} = 2.0\,\log Re\sqrt{f} - 0.8 \tag{10.11}$$

which is known as the Karman-Prandtl equation.

3. For rough boundaries, $Re > 10^5$

$$\frac{1}{\sqrt{f}} = -2\log\frac{k_s}{D} + 1.14 \tag{10.12}$$

from the Karman-Prandtl equation.

4. For transitional regions:

$$\frac{1}{\sqrt{f}} + 2\log\left(\frac{k_s}{D}\right) = 1.14 - 2\log\left(1 + 9.15\frac{D/k_s}{Re\sqrt{f}}\right) \tag{10.13}$$

which is from Cole-Brooks-White equation.

The studies on non-circular conduits, such as rectangular, oval and triangular shapes, have shown that by introducing the hydraulic radius R, the formula developed for pipes is also applicable for non-circular ducts. Since for a circular shape $R = D/4$ by replacing D by $4R$, the above equations (10.10) through (10.13) can be used for any duct shape provided the conduit areas are closed enough to the area of a circumscribing circle or semicircle.

For open channel flow, the Reynolds number is $Re = \frac{4RU}{\nu_f}$ and the relative roughness is $k_s/4R$. Then equation (10.9) is modified, putting $D = 4R$ as

$$h_f = f\frac{L}{4R}\frac{U^2}{2g} \tag{10.14}$$

$$\Rightarrow U^2 = \frac{4R \cdot 2gh_f}{fL} = \frac{8gRh_f}{fL}$$

$$\Rightarrow U = \sqrt{\frac{8g}{f}} \cdot \sqrt{R}\sqrt{\frac{h_f}{L}} = \sqrt{\frac{8g}{f}} \cdot \sqrt{RS}, \text{ where } S = \frac{h_f}{L}$$

Note that $\frac{h_f}{L}$ is the slope of energy line$= S_f = S$. Then,

$$U = \sqrt{\frac{8g}{f}} \cdot \sqrt{RS} = C\sqrt{RS} \tag{10.15}$$

which is the Chezy formula, and the Chezy's coefficient C is determined as:

$$C = \sqrt{\frac{8g}{f}} \tag{10.16}$$

Also, the comparison of Chezy's and Manning's formulae gives equation (10.5). Using the Chezy's coefficient C from equations (10.16) in (10.5), one gets the friction coefficient f as:

$$f = \left(\frac{n^2}{R^{1/3}}\right)8g \tag{10.17}$$

Equation (10.17) does not contain the mean velocity U or Reynolds number Re. This equation can be compared with the Prandtl–Karman relationship for rough turbulent flow.

Equating equations (10.6) and (10.8), a useful relationship can be evolved to determine Strickler's Equation as:

$$6.25+5.75\log\left(\frac{R}{k_s}\right)=\frac{1}{\sqrt{g}}\left(\frac{R^{1/6}}{n}\right) \tag{10.18}$$

This relationship shows an approximately linear line if it is plotted $R^{1/6}/n\sqrt{g}$ vs R/k_s in S. I. units, and the relation is given by:

$$\frac{R^{1/6}}{n}=24\left(\frac{R}{k_s}\right)^{1/6} \Rightarrow n=0.0391k_s^{1/6} \tag{10.19}$$

where the roughness k_s is in meter. Strickler in 1923 (Meyer-Peter and Muller, 1948) analyzed the data from several streams of coarse materials without any undulations in Switzerland, and he found the relation as:

$$n=0.0391(d_{50})^{1/6} \tag{10.20}$$

where d_{50} is the sediment size such that 50% of the mixture is finer than this size, which is the median size. This equation (10.20) is commonly known as Strickler's Equation, which is used to determine Manning's coefficient n for a rigid bed where there is no viscous effect. When the channel bottom is made up of roughness with non-uniform grain sizes, which are not moving, Meyer-Peter and Muller (1948) recommended the k_s roughness value as d_{90}, which is the size such that 90% of the material is finer than the size, to be used. This k_s value is slightly different from that of equation (10.20). Einstein (1950) recommended a roughness value k_s in (10.20) as d_{65}, which is the size such that 65% of the mixture is finer than the size. However, Irmay (1949) suggested using the maximum size of the material in the equation. Several authors such as Kamphuis (1974), Hey (1979), Thompson and Campbell (1979) suggested respective values of roughness from their field data in the rigid gravel bed, boulder rivers. It may be mentioned that Hey (1979) found the roughness value k_s equal to 3.5 d_{84} for the gravel bed rivers which agrees well with above findings. The flow over the gravel and boulder rivers is generally characterized by low and partial bed load transport for no significant undulations and relatively low $\frac{R}{k_s}$. Experimental data showed that the resistance characteristics at low $\frac{R}{k_s}$ cannot be described by conventional log-law equation. The resistance law for gravel and boulder rivers is recommended by Thompson and Campbell (1979) for increased resistance at low depth and is given by:

$$\frac{U}{u_*}=5.75\left(1-0.45\frac{d_{50}}{R}\right)\log\left(\frac{2.67R}{d_{50}}\right) \tag{10.21}$$

In general, it is well known that due to flow over the plane alluvial bed, there exists deformation of the bed into ripples, dunes, etc. when the sediment moves. The effects of these bed irregularities on the resistance can be clearly recognized by the sand paper given by Einstein and Banks (1950). In their work, they realized that there are two forms of resistance- one is the grain resistance and the other is the form resistance. The grain resistance mostly comes from the boundary of the channel, which is entirely the resistance offered by the grains. The form resistance comes from the undulations of the bed in addition to the resistance of the sand grains. Therefore, the total resistance in the alluvial channel in general can be considered to be the sum of the grain resistance and form resistance due to bed undulations. The size and shape of the undulations in the river bed and banks change greatly the discharge, hence the large variations in resistance in the rivers. Because of the presence of size, shape, and drag coefficient of bedforms under different flow conditions, the estimation of form resistance is very difficult. Another important factor is the sediment suspension in the alluvial channel, which affects the resistance to flow. It is known that the presence of sediment in suspension damps the turbulence in the flow, and hence the resistance to the flow.

A series of flume experiments relevant to several resistance phenomena have been conducted initially in a laboratory setting under controlled conditions on the diversity of velocity due to roughness protrusions, slopes, and depths (Brooks, 1958; Einstein and Chien, 1955; Vanoni et al., 1961), presence of suspended sediment and bed undulations (Vanoni and Brooks, 1957; Vanoni and Nomicos, 1960; Simons and Albertson, 1963; Ippen, 1973), and ripples-dunes, transitional and anti-dune regimes as river channel roughness (Einstein and Barbarossa, 1952; Garde and Ranga Raju, 1966; Engelund, 1966, 1970; Alam and Kennedy, 1969; Ranga Raju, 1970). Besides these, a vast amount of expressions on resistance to flow and velocity distribution in rigid boundary open channels and in alluvial streams have been developed for many years. In this chapter, as examples of resistance formulae to acquire some knowledge, only a few well-known functional relationships of resistance problems have been discussed. The concepts of the definite relationship between mean velocity, hydraulics radius, slope, channel geometry and bedforms have been widely utilized to determine the equations for resistance to flow.

10.3.1 Lacey's Formula

The pioneering attempt to predict the resistance relationships for alluvial channel flow was made by Lacey (1930). Based on the Indian stable canal data, he obtained the mean velocity U given by

$$U = 10.8R^{0.67}S^{0.33} \tag{10.22}$$

where R is the hydraulic radius, and S is the water surface slope. This equation was tested by several river data and found to be valid at high discharge; however, the equation is not applicable to all stages of the river.

10.3.2 Garde and Rangaraju's Formula

Garde and Rangaraju (1966) developed a formula for mean velocity, using the data collected from flume studies, canals, and natural streams. Using the principles of dimensionless analysis, the functional relationship was written as:

$$\frac{U}{\sqrt{\dfrac{(\rho_s - \rho_f)dg}{\rho_f}}} = K\left(\frac{R}{d}\right)^{0.67}\left(\frac{S}{\dfrac{(\rho_s - \rho_f)}{\rho_f}}\right)^{0.5} \tag{10.23}$$

Initially, the constant K values are respectively determined for plane bed with no motion, ripple-dune, and transitional regime as 7.66, 3.20, and 6.0. The value of K was found as 6.0 for the anti-dune regime. However, on further investigation, the constant was found to be continuous of $U\Big/\sqrt{\dfrac{(\rho_s - \rho_f)Rg}{\rho_f}}$ for ripple, dune and transitional regimes. More details on other resistance formulas are available in Garde and Ranga Raju (2000).

10.4 BEDFORMS

When the average shear stress on the loose sand bed over an alluvial channel exceeds the Shields shear stress, the individual particles just begin to move, which causes to deform the plane bed (Figure 10.1a). In fact, the Shields' relation for initiating the motion of particles is adequate for the development of bedform. After the initiation of motion, the plane bed changes to ripples for sand smaller than 0.5 mm and to dunes for 0.93 mm sand (Simons and Richardson, 1966). Turbulent flow over a plane sand bed causes the bed to deform, which in turn introduces perturbations on the main flow. After a certain time of run, the bedform is developed as regular asymmetrical ripples with steeper slopes downstream faces (Figure 10.1b) over the sand bed surface (Garde and Albertson, 1959; Allen, 1968; Venditti et al., 2005; and others). The effect of the perturbed bed on a turbulent flow is important in determining the formation of bedforms in erodible channels. Bedform is a feature that develops at the interface of fluid and a loose sediment bed, resulting in the bed material being moved by fluid flow. Depending on the flow conditions, the interface of sediment bed and water flow assumes various types of bedforms. Bedforms are also sometimes called irregularities or sand waves. Bedforms such as ripples and dunes generated from a flat sand bed interacting with the flow show significant changes in the mean flow, turbulence characteristics, bed load, and suspended load transport. The ripple dimensions, such as amplitudes and wavelengths, over the bedforms, depend on the flow discharge and the bed characteristics. The ratio of the bed shear stress to the critical shear stress for ripple formation varies from 2 to 10 (Kapdasli and Dyer, 1986). The ripples play an important role in generating flow separation, flow resistance, and controlling the sediment transport over two-dimensional bedforms. Garde and Albertson (1959) introduced the term regimes of flow and defined it as: "As the sediment and the flow characteristics are changed in alluvial channel, the nature of bed surface and water surface changes accordingly. These types of bed and water surface are classified according to their characteristics." These bedforms are often preserved in the rock record as a result of being present in a depositional setting, from the geological point of view. Bedforms are often characteristic of the flow parameters, and may be used to infer flow depth and velocity, and therefore the Froude number.

According to Garde and Albertson (1959), and Simons and Richardson (1962), for convenience, the flow regimes in the alluvial channel are classified into four categories: (i) plane bed, (ii) ripples and dunes, (iii) transition, and (iv) anti-dunes. This classification is based on the magnitude of resistance to flow and on the nature of the bed and water surface configurations. In the lower regime, the bedform is ripples and dunes or both, and the resistance to flow is large, while in the upper regime, the bedform is either plane or has standing waves and anti-dunes; and the resistance to flow is small.

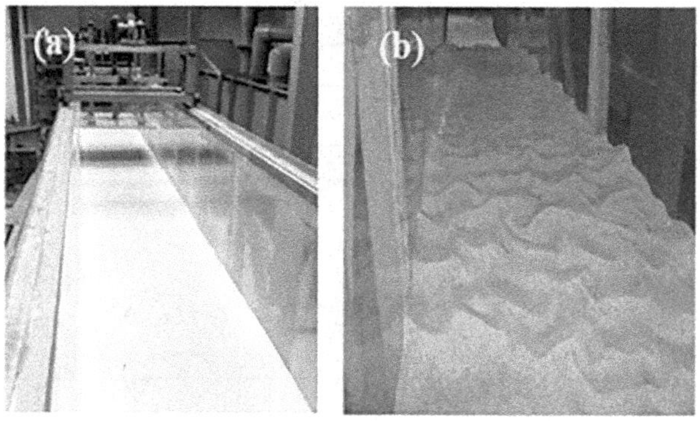

FIGURE 10.1 (a) Flat sand bed (plane bed) of median particle diameter $d_{50}=0.25$ mm, (b) Asymmetrical rippled bedforms developed due to flow over the flat sand bed. (Courtesy: Prof. B. S. Mazumder, Fluvial Mechanics Laboratory, ISI-Kolkata.)

i. **Plane bed:** In this regime, the depth and flow velocity over the bed material of a given size is such that the shear stress on the bed is not large enough to move the sediment particles. That means, the shear stress is less than the critical shear stress. The sediment bed acts like a rigid boundary, and the water surface is fairly smooth with a low Froude number (Figure 10.1a).

ii. **Ripples and dunes:** The ripple and dune type of bedform can appear if the Froude number is less than one ($Fr < 1$), that flow is called tranquil flow (Figure 10.1b). These are the characteristic of subcritical flow in the Froude sense. In fact, ripples and dunes are similar in their shape. Both are triangle-shaped elements having a gentle upstream gradually varying slope and an abrupt downstream surface with a constant slope, which is approximately equal to the tangent of the angle of repose. The mechanism of ripple formation is not yet well understood. The ripples and dunes are asymmetrical in shape (Figure 10.2), they are sharp-crested and travel downstream; and sediments erode from the stoss and redeposit in the lee side. The downstream slope induces the flow separation and migrates in the downstream direction. The ripples and dunes are identified from each other by the differences in their sizes to the flow depth. The size of the ripples is practically independent of the flow depth; while the size of dunes is strongly dependent on the flow depth. In fact, with an increase in discharge, the ripples grow into dunes. Ripples are small triangle-shaped elements, whereas dunes are large triangle-shaped elements similar to ripples, and the ripples disappear at larger shear values. Ripples may be superimposed on the stoss of the dunes; this condition usually prevails when there is a transition between the ripples and dunes. The downstream face of the dune slopes (called lee) is at an angle varying between 30–40 degrees. Ripples are the bedform configurations that are developed in a shear layer of smooth flow at a little excess of bed shear stress, whereas the dunes largely interact with the main flow at a larger excess bed shear stress compared to that of ripples. The length and

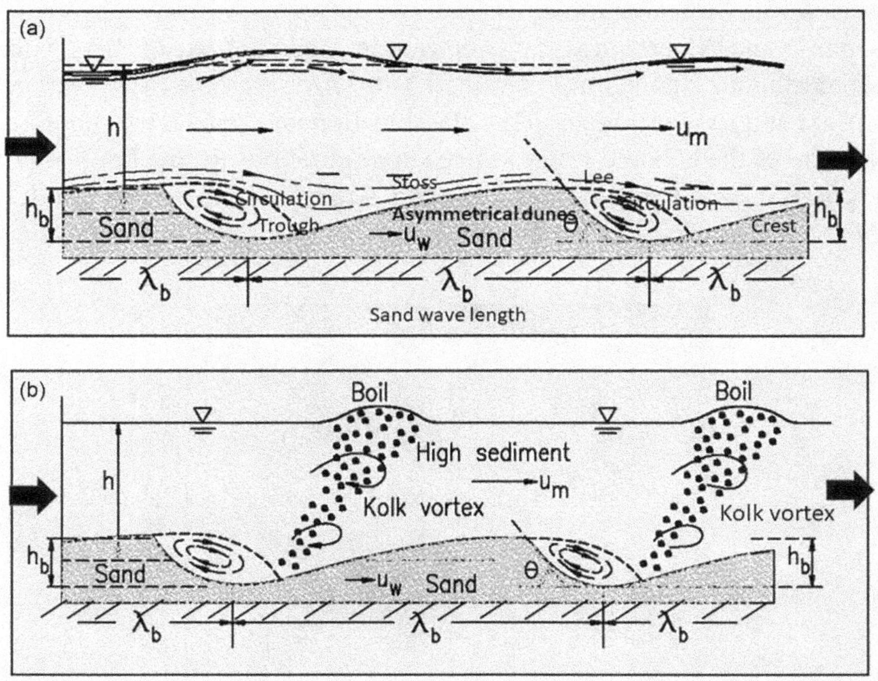

FIGURE 10.2 Schematic diagram of asymmetrical dune shape: (a) with circulation bubbles; (b) with kolks and boils. (Modified from Yalin, 1972; Nezu, 2005.)

height of the sand wave are respectively λ_b and h_b; and the water depth in two-dimensional flow is h. In the case of ripples, the length and height of sand waves are independent of h, whereas, in the case of the dune, they are strongly dependent on water depth h. Ripples and dunes are not fixed, they move in the direction of the flow with a velocity u_w, which is small compared to the flow velocity u_m. Ripples and dunes can occur in both closed conduit and free surfaces, these can occur in river as well as in desert. If the free surface is present, it may or may not be deformed. These are mostly depended on the relative height of the sand waves. If the ratio $\frac{h_b}{h}$ is sufficiently large, the free surface is disturbed, and then the wave shape of the deformed free surface is of out of phase with the wave shape of sand wave surface (Figure 10.2a). Flow separation originates at the dune crest, forms a recirculation bubble like a roller with flow reversal at the lee-side and reattaches to the trough region.

Investigations of dunes have documented the macro-turbulent characteristics of varied flow over the bedforms called "kolks" and "boils" proposed by Matthes (1947). The kolks and boils are the upward tilting vortices of both fluid and sediment originating from downstream of dune crests and at the point of reattachments (Figure 10.2b). In the flow region above the recirculation, a shear layer with high turbulent mixing occurs, where the turbulence production and dissipation take place to a large amount. The study of flows around and above sand dunes in alluvial rivers is very important both in basic science and engineering applications because it is not only concerned with the mechanism of flow resistance and sediment transport but also can help to establish a better mathematical model to simulate the sediment-laden flows for flood control, navigation, etc.

iii. **Transition:** In the transitional regime, the rapid changes in the sand bed and water flow occur with relatively small changes in flow conditions. With an increase in discharge or bed shear stress, the wavelength of the sand waves increases, while the amplitude decreases. In the transitional regime, the bed becomes a plane for relatively fine material under certain flow conditions. Further increase in discharge may produce symmetrical sand waves and water surface waves which are in phase with the sand waves. The water surface shows different forms, like boils, plane surfaces, and standing waves. The transitional regime is unstable compared to the ripple and dune regimes. Consequently, slight changes in slope or discharge change the flow pattern completely. These transitional waves usually form and disappear and do not grow in amplitude.

iv. **Anti-dunes:** If the bed shear stress is further increased and the Froude number is greater than one ($Fr > 1$), the sand waves are referred to as *Anti-dunes*, the shape of which is more symmetrical than the ripples and dunes (Figure 10.3). The development of anti-dunes depends on the flow

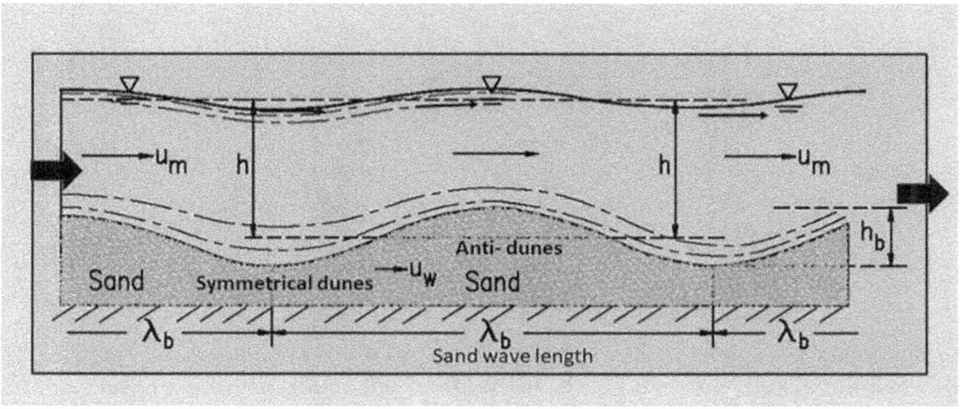

FIGURE 10.3 Shape of anti-dune. (Modified from Yalin, 1972.)

velocity, with an increase in velocity water surface becomes unstable. Due to high velocity, the sand wave changes to a form similar to and in-phase with the surface wave, which may remain stationary or move upstream and downstream, that means, the anti-dunes can move in the positive x-direction, in the negative x-direction or they can be stationary. The velocity of moving anti-dunes u_w is also small compared to the flow velocity u_m. In the case of anti-dunes, the free surface is always deformed; the wave shape of the free surface is in-phase with the wave shape of the sand bed surface (Figure 10.3). For a given size of bed material, the frequency of breaking of anti-dune seems to be a function of flow characteristics and the initial disturbances in the channel (Kennedy 1961; Simons and Richardson, 1962).

Different types of bedforms formed in the alluvial channel with increasing flow velocities corresponding to Froude number Fr are described in Figure 10.4. The following schematic diagrams of bedform are as: (a) Typical ripples form for Froude number $Fr \ll 1$; (b) dunes superimposed by ripples for $Fr \ll 1$ with a weak formation of boils on the water surface; (c) dunes for $Fr < 1$ with boils on the water surface; (d) transitional dunes or washed out dunes for $Fr < 1$; (e) plane bed of grain size $d < 0.4$ mm formed for $Fr < 1$; (f) anti-dunes with standing waves for $Fr > 1$ both bed and water surfaces in same phase; (g) anti-dunes with breaking waves for $Fr > 1$; and (h) anti-dunes for $Fr > 1$ with chutes and pools. The geometry of dune bedforms refers to the representative dune height, length of dunes as a function of the average flow depth, median bed particle diameter and other flow parameters such as mean flow velocity, friction velocity, and grain Reynolds numbers. The foreslope and backslope of the dunes can also be counted as a part of the dune geometry.

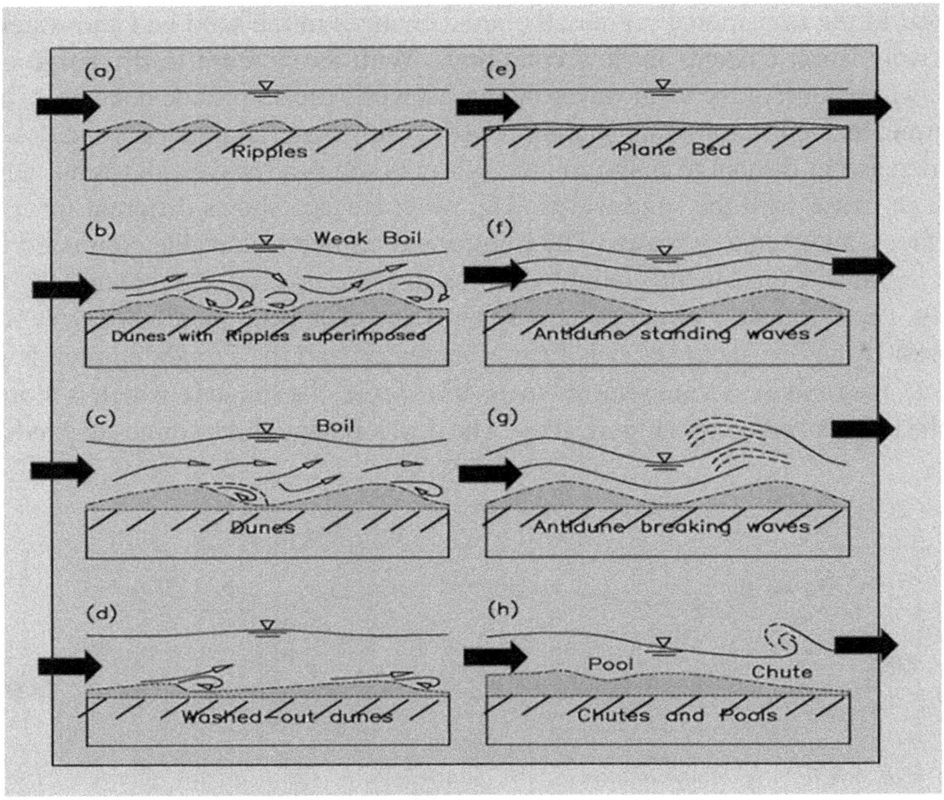

FIGURE 10.4 Schematic diagrams of bedform developed in alluvial channel (a) Typical ripples, (b) Dunes and ripples with weak boils, (c) Dunes with boils, (d) Washed-out and transition, (e) Flat sand bed, (f) Standing waves as antidunes, (g) Antidunes with breaking waves, and (h) Antidunes with chutes and pool. (Modified from Simons and Richardson, 1961.)

10.5 EMPIRICAL RELATIONS FOR RIPPLES AND DUNES

Yalin (1964) reported some geometrical properties of bedforms based on the experimental studies, using the dimensionless variables of height h_b and wavelength λ_b of the bedforms, and he found a dimensionless bedform height as:

$$\frac{h_b}{h} = f\left(\frac{\tau_0}{\tau_{0c}}\right) = f\left(\frac{h}{h_{cr}}\right) \tag{10.24}$$

where h_{cr} is the depth at which incipient motion takes place. He made a series of experiments to establish the functional form (10.24) and obtained the form as:

$$\frac{h_b}{h} = \frac{1}{6}\left(1 - \frac{h_{cr}}{h}\right) \tag{10.25a}$$

Since $h_{cr} < h$, the height h_b cannot exceed one-sixth of the flow depth, so he found as:

$$\frac{h_b}{h} < \frac{1}{6} \tag{10.25b}$$

However, Nordin and Beverage (1965) suggested that the value $\frac{1}{3}$ should be more appropriate than the value $\frac{1}{6}$. After analysis, Yalin again found the following as:

$$\frac{\lambda_b}{d} = f\left(\frac{\bar{h}}{d}\right) \tag{10.26}$$

for large particle Reynolds number $\frac{u_* d}{v_f}$, and $\frac{\lambda_b}{d}$ = constant for small Reynolds numbers. Now the equations can be written as: $\frac{\lambda_b}{h} = 5$ for $d \geq 0.38$ mm, and $\frac{\lambda_b}{d} = 1{,}000$ for $d \leq 0.18$ mm. Yalin reported that the limiting Reynolds number for small and large values is of order 20. He also suggested that the bedform given by this equation $\frac{\lambda_b}{h} = 5$ for $d \geq 0.38$ mm is defined as *dune*, and by this equation $\frac{\lambda_b}{d} = 1{,}000$ for $d \leq 0.18$ mm is defined as a *ripple*. Garde and Albertson (1959) challenged that the ripple dimensions are related to Shields parameter $\tau_* = \frac{\tau_0}{(\rho_s - \rho_f) g d}$ and the grain Reynolds number $\frac{u_* d}{v_f}$. Yalin (1972) argued that the dimensionless ripple length λ_b / d should be a function of grain Reynolds number. Using the flume data, Yalin (1985) showed that $\frac{\lambda_b}{d} = 2250\left(\frac{u_* d}{v_f}\right)^{-1}$ or $\frac{u_* \lambda_b}{v_f} = 2250$. Flemming (1988) proposed a simple empirical relation, based on several measurements of bedform height h_b and length λ_b as $h_b = 0.068 \lambda_b^{0.81}$.

Extensive experimental data by Allen (1968) and Flemming (1988) demonstrate that there is a break in the continuation of bedforms defining two sub-populations of bedforms, which have become almost universally known as ripple and dunes. Allen (1968) indicated that bedform developed in the sand bed has an aspect ratio h_b / λ_b varying between 0.01 and 0.02, although exceptions can occur, and suggested for ripples $0.05 < h_b / \lambda_b < 0.20$, while for dunes $0.01 < h_b / \lambda_b < 0.10$. Bridge (1993) suggested $h_b / \lambda_b < 0.06$ for dunes, and $h_b / \lambda_b < 0.10$ for ripples, but he noted that h_b / λ_b is dependent on the measure of the shear stress.

Ranga Raju and Soni (1976) expressed a formulation in terms of dimensionless variables of height h_b and length λ_b of the bed undulation as:

$$\frac{h_b}{\lambda_b} = 2.16 \times 10^{-5} \left(\frac{h}{d}\right) (\tau'_*)^{-0.33} \tag{10.27}$$

where $\tau'_* = \dfrac{g\rho_f hJ}{(\rho_s - \rho_f) gd}$ is the shear stress. Yalin and Karahan (1979) determined the expression as:

$$\frac{h_b}{\lambda_b} = f\left(\frac{\tau_0}{\tau_{0c}}, \frac{h}{d}\right) \tag{10.28}$$

Bass (1994) suggested the empirical equations for ripple length λ_b and height h_b in equilibrium conditions as follows:

$$\lambda_b = 75.4 \log d + 197 \text{ and } h_b = 18.16\, d^{0.097} \tag{10.29}$$

Garde and Isaac (1993) indicated that the characteristic length of ripples should be ν_f / u_*, further since h_b / h is very small, Froude number is not important in the ripple regime so long as Fr is less than 0.30 or 0.40. Therefore, the following relationship of ripple is written as:

$$\frac{u_* h_b}{\nu_f} \text{ and } \frac{u_* \lambda_b}{\nu_f} = f\left(\frac{\tau_0}{(\rho_s - \rho_f) gd}, \frac{u_* d}{\nu_f}\right) \tag{10.30}$$

The expression (10.30) can be written in terms of h_b, λ_b and water depth h as:

$$\frac{h_b}{h} \text{ and } \frac{\lambda_b}{h} = g\left(\frac{\tau_0}{(\rho_s - \rho_f) gd}, \frac{d}{h} \text{ or } Fr, \frac{u_* d}{\nu_f}\right) \tag{10.31}$$

In the light of the above expressions, the ripple height h_b and the ripple length λ_b can be discussed. From the several existing experimental ripple data (Garde and Isaac, 1993) developed the empirical equations as:

$$\frac{h_b}{d} = 584.45 (\tau_*)^{0.83} \left(\frac{u_* d}{\nu_f}\right)^{-0.66} \tag{10.32a}$$

$$\frac{\lambda_b}{d} = 4115 (\tau_*)^{0.72} \left(\frac{u_* d}{\nu_f}\right)^{-0.32} \tag{10.32b}$$

They found ripple height h_b is predicted more than ±30% accuracy for 42.9% of data and more than ±50% accuracy for 73.8% of data and for the ripple length λ_b for ±30% and ±50% errors are respectively from 52.4% to 90.5% of data (Garde and Ranga Raju, 2000; Table 4.1).

Raudkivi (1997) proposed an empirical relationship between the ripple length λ_b and particle diameter d as

$$\lambda_b = 245 d^{0.35} \tag{10.32c}$$

Fredsoe (1974) developed an equation for dune, using the observed data from Guy et al. (1966) as:

$$\frac{h_b}{\lambda_b} = \frac{1}{8.4}\left(1 - \frac{0.06}{\tau_*} - 0.4\tau_*\right)^2 \tag{10.33}$$

where $\tau_* = \dfrac{\tau_0}{(\rho_s - \rho_f)gd}$ is the Shields parameter. The relation (10.33) is close to the relation given by Yalin (1972), but is deduced from the principle of similarity.

Allen (1978) suggested the following expressions for dimensionless wave height h_b/h as a function of Shields parameter $\tau_* = \dfrac{\tau_0}{(\rho_s - \rho_f)gd}$ as:

$$\frac{h_b}{h} = 0.079865 + 2.238\left(\frac{\tau_*}{3}\right) - 18.12\left(\frac{\tau_*}{3}\right)^2 + 70.9\left(\frac{\tau_*}{3}\right)^3 - 88.3\left(\frac{\tau_*}{3}\right)^4 \tag{10.34}$$

Now considering $\tau_{*c} = 0.045$, equation (10.34) is written as:

$$\frac{h_b}{h} = 0.079865 + 0.0336\left(\frac{\tau_*}{\tau_{*c}}\right) - 0.0041\left(\frac{\tau_*}{\tau_{*c}}\right)^2 + 0.00024\left(\frac{\tau_*}{\tau_{*c}}\right)^3 - 0.000045\left(\frac{\tau_*}{\tau_{*c}}\right)^4 \tag{10.35}$$

van Rijn (1984) developed empirical equations for the computation of dune height h_b

$$\frac{h_b}{h} = 0.11\left(\frac{d_{50}}{h}\right)^{0.3}\left(1 - e^{-0.5T_s}\right)(25 - T_s) \tag{10.36}$$

$$\frac{h_b}{\lambda_b} = 0.015\left(\frac{d_{50}}{h}\right)^{0.3}\left(1 - e^{-0.5T_s}\right)(25 - T_s) \tag{10.37}$$

where $T_s = \dfrac{\tau'_* - \tau_{*c}}{\tau_{*c}}$ is the transport stage, in which $\tau'_* = \rho_f u_*'^2$ with $u'_* = \bar{u}\left(\sqrt{g}/C'\right)$ is the bed shear velocity related to grains, C' is the Chezy's coefficient with respect to the grain, $\bar{u}$ is the mean velocity, τ_{*c} is the critical shear stress according to Shields curve. Here both the functions show maximum values for about $T_s = 5$. From the above two equations (10.36) and (10.37), the bedform length can be derived as: $\lambda_b = 7.3h$. From the verification of field and flume data, there is a relation between the dune length and mean water depth. This result is also reported by Yalin (1964) arrived a relation $\lambda_b = 2\pi h$, which is close to the relation of van Rijn (1984). Yalin (1964) provided experimental data supporting a constant dimensionless dune length. Physically, it indicates that bedform height is reduced for increasing stages of flow, while the bedform length remains essentially unchanged. The formation of ripples, dunes and anti-dunes in the alluvial bed based on the stability analysis was studied by Hayashi (1970), Engelund (1970), Fredsoe (1974, 1986), Parker (1975) and Sumer and Bakioglu (1984). Fredsoe (1986) also reviewed the different physical mechanisms on the formation of bedform geometries using stability analysis. Extensive works on erosion and sedimentation are available in Julien (2010).

Kennedy and Odgaard (1991) proposed bedform height from the laboratory data as:

$$\frac{h_b}{h} = \frac{1}{2}\left[\frac{1.2\lambda\alpha f_0}{8C_D} + \left\{\left(\frac{1.2\lambda\alpha f_0}{8C_D}\right)^2 + \frac{2\pi Fr^2 f_0}{C_D C_1}\left(\frac{f}{f_0} - \frac{1.2\lambda}{2}\right)\right\}^{0.5}\right] \tag{10.38}$$

where f is the Darcey-Weisbach friction factor, f_0 is the rigid flat bed Darcey-Weisbach friction factor, Fr is the Froude number, and $C_1 = 1, C_D = 1, \alpha = 5, \lambda = 1$.

Julien and Klassen (1995) suggested a relation between the bedform height and d_{50} given by:

$$\frac{h_b}{h} = 2.5\left(\frac{d_{50}}{h}\right)^{0.3} \tag{10.39}$$

Karim (1995) proposed a relationship of relative bedform height h_b/h in sand bed flows, depending on the friction velocity u_* and the settling velocity w_{d50} of median bed material size d_{50} as follows:

$$\frac{h_b}{h} = -0.04 + 0.294\left(\frac{u_*}{w_{d50}}\right) + 0.00316\left(\frac{u_*}{w_{d50}}\right)^2 - 0.0319\left(\frac{u_*}{w_{d50}}\right)^3 + 0.0027\left(\frac{u_*}{w_{d50}}\right)^4 \tag{10.40}$$

for $0.15 \le \frac{u_*}{w_{d50}} \le 3.64$ and

$$\frac{h_b}{h} = 0 \tag{10.41}$$

for $\frac{u_*}{w_{d50}} \le 0.15$ and $\frac{u_*}{w_{d50}} \ge 3.64$

Karim (1999) also proposed a method for predicting the relative bedform height h_b/h in sand bed flows, based on the concept of energy loss due to the form drag in open channel flow. This method has a unique feature that can be applied to various bedforms, i.e., ripples, dunes, anti-dunes/standing waves and transitional bed regimes in the alluvial flows. The relative bedform height h_b/h is given as:

$$\frac{h_b}{h} = \left[\left\{J - 0.017\left(\frac{d_{50}}{h}\right)^{0.33} \cdot Fr^2\right\}\left(\frac{\lambda_b}{h}\right)^{1.20}\right]^{0.73} \left(0.47Fr^2\right)^{-0.73} \tag{10.42}$$

for ripples, dunes and transition,

$$\frac{h_b}{h} = \left[\left\{J - 0.017\left(\frac{d_{50}}{h}\right)^{0.33} \cdot Fr^2\right\}\left(\frac{\lambda_b}{h}\right)^{1.20}\right]^{0.73} \left(0.085Fr^2\right)^{-0.73} \tag{10.43}$$

for anti-dunes or standing waves. This equation (10.43) can be solved directly for h_b/h using the appropriate relations given by Karim (1999) for relative bedform length h_b/h depending on the bedform type.

Yang et al. (2005) developed an empirical equation to predict the relative bedform roughness height h_b/h in sand bed flows, using the concept of length of separation region behind the bedform, which depends on bedform height and bedform steepness. They assumed the characteristic length L_s of separation region behind the bedform is proportional to bedform height and it can be expressed as:

$$L_s = \alpha h_b \tag{10.44}$$

where α is the proportionality constant to be determined from the equation as:

$$J = J' + (J'' - J')\frac{\alpha h_b}{\lambda_b} \tag{10.45}$$

where $J = J' + J''$ is the energy slope as a sum of two slopes, with J' is the energy slope due to grain resistance and J'' is the slope due to bedform resistance, and λ_b is the length of the bedform, in which $J' = h'_{df}/L_c$ and $J'' = \frac{h''_{df}}{L_s}$. Here L_c is the characteristic length of direct contact of flow with the bed, L_s is the characteristic length of separation region, h'_{df} corresponds to head loss due to grain resistance, and h''_{df} for head loss due to the bedform resistance. If the drop in the water level over a bedform is h_{df} over

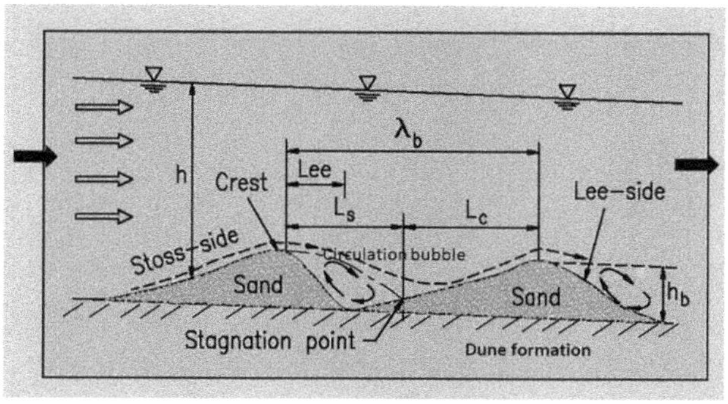

FIGURE 10.5 Dune formation with circulation bubbles and stagnation point. (Modified from Yang et al., 2005.)

the distance λ_b, then we can define the energy slope as $J = h_{df}/\lambda_b$. The total energy loss over a bedform can be expressed as $h_{df} = h'_{df} + h''_{df}$.

The additional energy loss associated with J'' is the result of the sudden expansion of flow at the lee side of the bedform. Engel (1981) found that for the flow over dune, α is practically independent of Froude number, but it is correlated with the relative bedform height $\frac{h_b}{h}$ and it is shown in Figure 10.5. It is clearly observed that the coefficient α decreases systematically with bedform development $\frac{h_b}{h}$ from the lower to upper regimes. It is also clear that the length of bedform λ_b is a function of water depth h and the length of separation is approximately proportional to the length of bedform. The bedform length λ_b is given by the equations as: $\lambda_b = 6.25h$ for dunes given by Julien and Klaassen (1995), $\lambda_b = 2\pi h Fr^2$ for anti-dunes or standing waves given by Kennedy (1963), and for transitional bed regime, Karim (1999) suggested the equation as:

$$\lambda_b = 7.37h\left[0.0014\left(\frac{U}{\sqrt{g(\rho_s/\rho_f - 1)d_{50}^3}}\right)^{2.97}\left(\frac{u_*}{v_0}\right)^{1.47}\right]^{0.295} \tag{10.46}$$

where v_0 is settling velocity of particle. The empirical equation between the coefficient α and the bedform height h_b/h for lower regime as:

$$\alpha = 45\left(1 + 5\frac{h_b}{h}\right)^{-1} \tag{10.47}$$

and for the upper regime,

$$\alpha = 8\left(1 + 5\frac{h_b}{h}\right)^{-1} \tag{10.48}$$

These equations are plotted in Figure 10.6. A large number of flume and field data set is compared to validate the equations.

10.6 SAND BARS

The sand bar is a class of large-scale bedform, the dimensions of which are controlled by the channel width and the depth. The formation and evolution of bars result in distinctive channel patterns. The term bar is used because bed-wave length is proportional to the flow width, and bed-wave height is

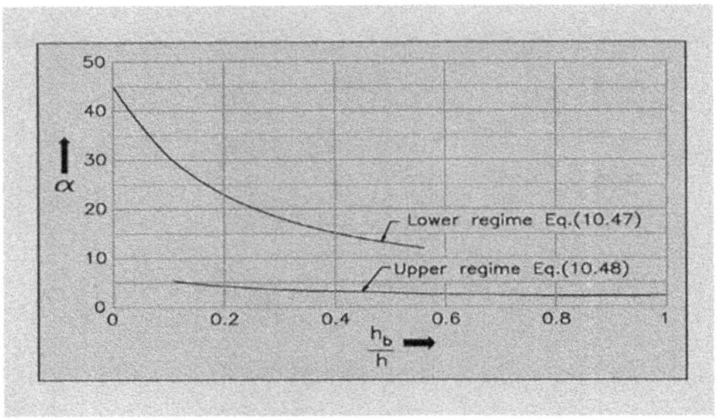

FIGURE 10.6 Plots of coefficient α against bedform height h_b/h for lower and upper regimes. (Modified from Yang et al., 2005.)

comparable to the flow depth. The bed evolves towards a statistically constant geometry composed of single rows or multiple rows of alternate bars that are equilibrium with steady flow and sediment transport conditions. During the evolution, the alternate bars increase in length and height. In plan view, the water flow associated with these bars follows a sinuous path, with a wavelength equivalent to that of the bed waves in a particular row, and with a width equivalent to the bed-wave width. The bar lengths are proportional to their width, and their lengths are comparable to flow depth. Some researchers (Jaeggi, 1984; Ikeda, 1984; Colombini et al., 1987; Nelson, 1990; Garcia and Nino, 1993; Lanzoni, 2000; Knaapen et al., 2001) showed theoretically and experimentally that the ratio of length and width of alternate bars is 3–12. Alternate bars are generally asymmetrical in along-stream cross section, may not have an avalanche face on the downstream side, and generally migrate in the downstream direction. The mathematical expression for the criteria of formation of alternative bars is discussed in Jaeggi (1984). The term alternate bar is used here because the bars occur on the alternating sides of the channel with progressive downstream. Alternate bars have also been referred to as unit bars, linguoid bars, side bars, transverse bars, cross-sectional bars and diagonal bars and riffles (Smith, 1977; Church and Jones, 1982: Bridge, 1993). Some sand bed waves are called transverse bars or linguoid bars, which are actually be dunes (mesoforms) if their lengths are related to the flow depth rather than flow width. Small-scale bedforms such as dunes, ripples, and bed load sheets are commonly superimposed on the alternate bars. In general, four types of bars are distinguished (Figure 10.7): (A) point bars, which occur

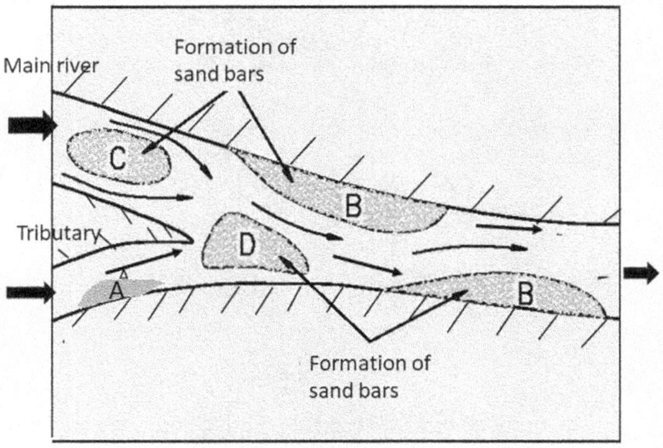

FIGURE 10.7 Schematic of sand bars in a river: (A) point bar, (B) alternate bar, (C) middle bar, and (D) tributary bar.

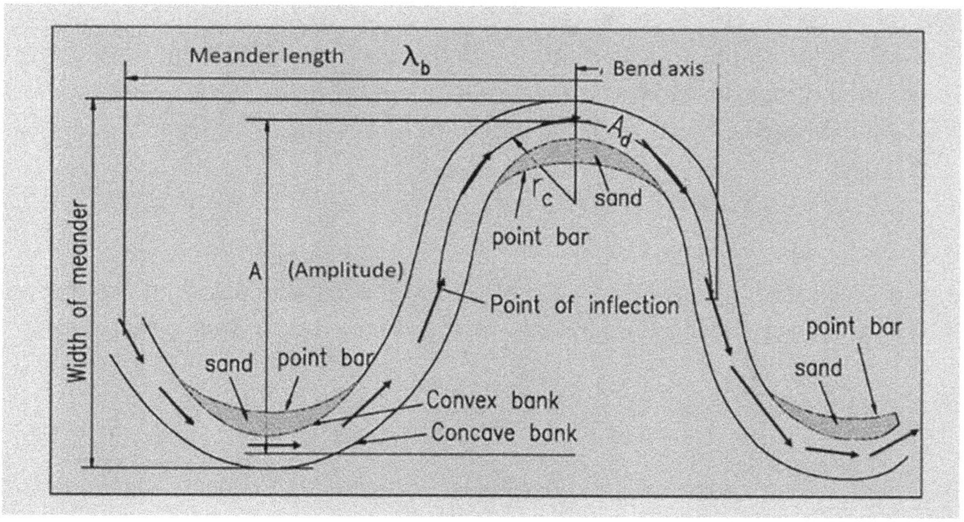

FIGURE 10.8 Point bars in the meandering channel with inflection point, meander length λ_b, amplitude A, mean radius of curvature r_c, arc distance A_d and sinuosity $2A_d/\lambda_b$. (Modified from Leopold et al., 1964.)

on the inner side of the curved channel and vary on the flow conditions, (B) alternating bars generally take place periodically along the straight channel in alternate sides, (C) mid-channel bars which appear at the middle of the straight channel as an isolated bar, which may migrate due to the flow condition, and (D) tributary bars are approximately triangular available at the confluence of the tributaries and main channels. The tributary bars remain almost stationary. Figure 10.8 shows the point bars in the meandering channel which shows inflection point, meander length λ_b, amplitude A, mean radius of curvature r_c, arc distance A_d and sinuosity $2A_d/\lambda_b$.

Yu and Mei (2000) developed a quantitative theory to describe the formation of sand bars under the effect of surface water waves. They derived an approximate evolution equation of bar height under the assumption of gentle slopes of waves and bars and bed load-dominated sediment motion. It was observed that the evolution process of sand bars is a forced diffusion. During formation, bars and waves affect each other through the Bragg scattering mechanism which provides two simultaneous processes: energy transfer between waves propagating in opposite directions and change of their wavelengths. Both effects are developed to be controlled locally by the position of bar crests relative to the wave nodes. Comparative studies are made for the theoretical models with the available experimental data. Also, in their consecutive paper (2008), they reported theoretical and experimental study of sand bar formation under simple harmonic surface waves. Purkait (2002, 2006), in his consecutive papers, studied critically the grain size distribution patterns in a point bar of Usri River, India in the light of log-normal, log-hyperbolic and log-skew-Laplace distributions. The observations were made within the point bars for large and small bedforms; there was a tendency of decreasing mean grain size toward downstream. Between the point bars of large bedforms, there was no consistency in decreasing grain size along the downstream. Sediment samples were collected from the specific size of bedform from the source to the mouth of the river over a distance of 90 km. Data were collected from the four different point bars with hydrodynamic parameters of the river.

10.7 THEORETICAL DEVELOPMENTS

The theoretical study for the generation of bedforms is based on the perturbation analysis referred to as the linear stability analysis. The perturbation analysis involves the linearization of the equations of motion of both fluid and sediment over an infinitesimally small bed. Based on the equation of continuity of sediment, Exner (1925) developed a classical equation to predict the migration velocity of bedforms without considering the frictional effect. If η is the elevation of the sand bed with respect to the horizontal reference or datum and h is the water depth (Figure 10.9), B_w is the channel width, and $\bar{U}$ is the cross-sectional average velocity, then the equation of continuity of flow is given as:

$$(h-\eta)B_w\bar{U} = Q = \text{constant} \Rightarrow \bar{U} = \frac{Q}{(h-\eta)B_w} \tag{10.49}$$

where Q is the flow discharge. According to Exner (1925) the erosion occurs if the discharge increases, and the deposition occurs if the flow decreases in the downstream, and consequently the bed level changes.

Exner's bedform equation is given by:

$$\frac{\partial \eta}{\partial t} + \alpha_E \frac{\partial \bar{U}}{\partial x} = 0 \tag{10.50}$$

where α_E is the Exner erosion coefficient. Combining the above two equations (10.49) and (10.50) with constant width B_w and depth h, one can write as:

$$\frac{\partial \eta}{\partial t} = -\frac{\alpha_E Q}{B_w (h-\eta)^2} \frac{\partial \eta}{\partial x} \tag{10.51}$$

The solution of this differential equation (10.51) subject to the initial condition $t = 0$ is given by:

$$\eta = A_1 + A_2 \cos\frac{2\pi x}{\lambda_b} \tag{10.52}$$

where A_1 is a constant, λ_b is the wavelength of the bedform, and A_2 is the amplitude. Thus, the equation of a bedform for a given time t is

$$\eta = A_1 + A_2 \cos\frac{2\pi}{\lambda_b}[x - c_{BM}t] \tag{10.53}$$

where $c_{BM} = \dfrac{\alpha_E Q}{B_w (h-\eta)^2}$ is known as the bedform migration velocity along downstream. Here c_{BM} is a function of flow discharge Q and Exner erosion co-efficient α_E, that means, c_{BM} increases

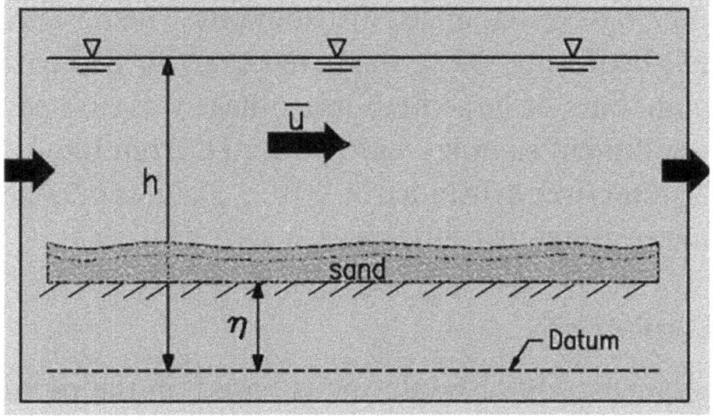

FIGURE 10.9 Sketch of Exner's model. (Modified from Graf, 2010.)

with Q or α_E. The solution of equation (10.53) is shown graphically (Figure 10.10), using $A_1 = A_2 = 1.0, \lambda_b = 20, h = 3$, and $\frac{\alpha_E Q}{B_w} = 1$. The bedform migration velocity c_{BM} shows that the crest of the bedforms moves faster than the trough, and consequently the sinusoidal bed tends to asymmetrical wavy bed with a gentle upstream slope and a steeper one in the downstream side, and an overhanging of the crest is shown from the figure, but practically in this phenomenon, the angle of repose does not occur.

If width B_w is considered to be variable, the resulting equation becomes as:

$$\frac{\partial \eta}{\partial t} = -\frac{\alpha_E Q}{B_w (h-\eta)^2}\frac{\partial \eta}{\partial x} - \frac{\alpha_E Q}{B_w^2 (h-\eta)}\frac{\partial B_w}{\partial x} \tag{10.54}$$

It is noted that equation (10.54) becomes equation (10.51), if the width of the channel is constant. If the width is a function of x as $B(x)$, then the equation will be modified (Graf, 2010). If the frictional effect k_{fr} is considered, the equation of motion with the unsteady flow is written as:

$$\frac{\partial \bar{U}}{\partial t} = -\bar{U}\frac{\partial \bar{U}}{\partial x} - g\frac{\partial h}{\partial x} + g_x - k_{fr}\bar{U} \tag{10.55}$$

Equation (10.55) together with (10.49) and (10.50) represents a set of equations for water and sediment bed movements. For constant width, rearranging and differentiating the above equations, one gets as:

$$\frac{\partial^2 \eta}{\partial t^2} - \left(\frac{gQ}{B_w \bar{U}^2} - \bar{U}\right)\frac{\partial^2 \eta}{\partial t\,\partial x} + k_{fr}\frac{\partial \eta}{\partial t} - \alpha_E g\frac{\partial^2 \eta}{\partial x^2} = 0 \tag{10.56}$$

The solution of equation (10.56) is subject to the initial condition as cosine function given by (10.52) and is given as:

$$\eta = A_1 + A_2 \exp\left[-\left(\frac{k_{fr}}{2} - p\right)t\right]\cos\frac{2\pi}{\lambda_b}\left[x - \frac{1}{2p}\left(\frac{gQ}{B_w \bar{U}^2} - \bar{U}\right)\left(\frac{k_{fr}}{2} - p\right)t\right] \tag{10.57}$$

where p is a function of k_{fr}, λ_b, α_E and $\bar{U}$. If $\left(\frac{k_{fr}}{2} - p\right) > 0$, the amplitude of the bedform decreases with time due to the frictional effect. The longer bedforms migrate with a smaller velocity than the shorter ones, whereas shorter bedforms have faster decreasing amplitude than longer ones. Extension of the

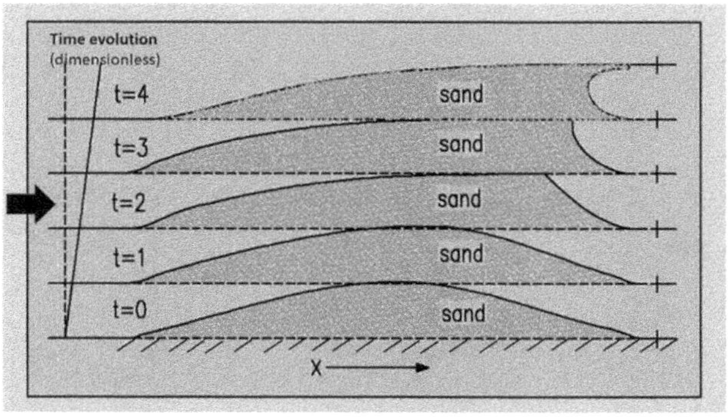

FIGURE 10.10 Time evolution of bedform with constant width channel according to equation 10.53, using $A_1 = A_2 = 1.0, \lambda_b = 20, h = 3$, and $\frac{\alpha_E Q}{B_w} = 1$. (Modified from Graf, 2010.)

foregoing study was performed by Velikanov (1936, 1955) and is discussed in some details by Raudkivi (1967).

Song (1983) suggested a theoretical model to determine the migration velocity of bedforms, assuming the gradually varied steady flow over an erodible bed. Referring the Figure 10.11, the Bernoulli equation and the equation of continuity of water flow are as follows:

$$\frac{\bar{U}^2}{2g}+h+\xi=E \tag{10.58}$$

$$\bar{U}(h+\xi-\eta)=q \tag{10.59}$$

where $\bar{U}$ is the averaged flow velocity, h is the mean depth of water, ξ is the water surface rise, E is the total head, η is the bed elevation with respect to mean bed level, q is the unit discharge.

Assuming the bed load transport q_B is a function of average velocity $\bar{U}$, the following expression can be written as:

$$\frac{\partial q_B}{\partial \eta}=\frac{\partial q_B}{\partial \bar{U}}\cdot\frac{\partial \bar{U}}{\partial \eta} \tag{10.60}$$

Considering the energy E and discharge q as constant, and eliminating ξ from equations (10.58) and (10.59), one gets

$$\frac{\partial \bar{U}}{\partial \eta}=\frac{\bar{U}^2}{q\left(1-Fr^2\right)} \tag{10.61}$$

where $Fr=\dfrac{\bar{U}}{\sqrt{gh}}$ is the Froude number. The wave speed or bedform migration velocity U_b is given by:

$$U_b=\frac{\partial q_B}{\partial \eta}=\frac{\partial q_B}{\partial \bar{U}}\cdot\frac{\partial \bar{U}}{\partial \eta}=\frac{\partial q_B}{\partial \bar{U}}\cdot\frac{\bar{U}^2}{q\left(1-Fr^2\right)} \tag{10.62}$$

Here $\dfrac{\partial q_B}{\partial \bar{U}}>0$, the sign of U_b is determined by the sign of denominator, i.e., $U_b>0$ downstream migration for subcritical flow ($Fr<1$), and $U_b<0$ suggests upstream migration for supercritical flow ($Fr>1$), which agrees well with the known facts.

Milne-Thomson (1960) developed a special problem over a sinuous bottom boundary, using the potential flow concept (Figure 10.12). They considered an irrotational flow of an ideal incompressible

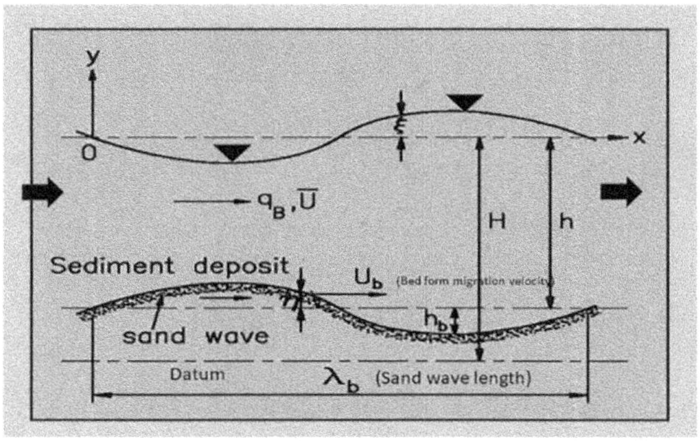

FIGURE 10.11 Schematic of free surface flow over sand waves. (Modified from Song, 1983.)

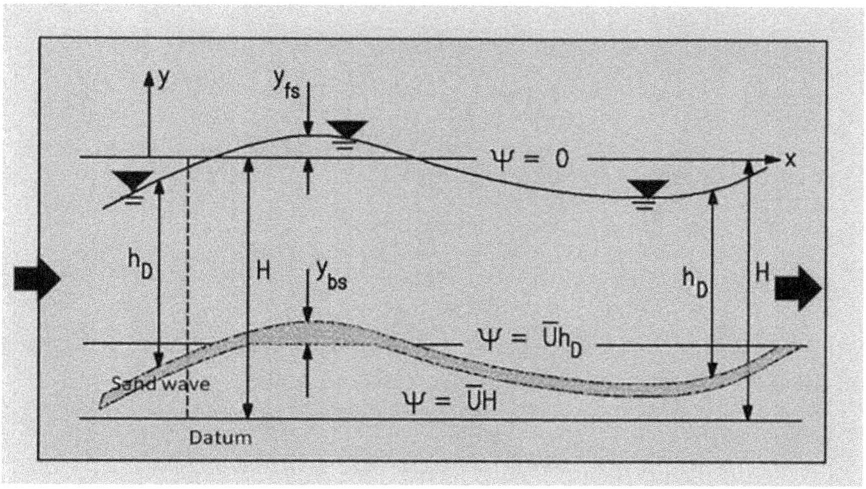

FIGURE 10.12 Steady flow over a sinuous sand bed. (Modified from Graf, 2010.)

fluid subject to the influence of gravity. If $\phi(x, y)$ is the velocity potential, then in two-dimensional flow velocities can be written as:

$$u = \frac{\partial \phi}{\partial x} \text{ and } v = \frac{\partial \phi}{\partial y} \tag{10.63}$$

From continuity equation, one gets Laplace's equation as:

$$\frac{\partial^2 \phi}{\partial x^2} + \frac{\partial^2 \phi}{\partial y^2} = 0 \tag{10.64}$$

subject to the boundary conditions as:

$$\text{for solid boundary}: \frac{\partial \phi}{\partial n} = 0 \tag{10.65a}$$

$$\text{and for the free surface at y} = 0: \frac{\partial \phi}{\partial y} = -\frac{1}{g}\frac{\partial^2 \phi}{\partial t^2} \tag{10.65b}$$

$$\text{In addition, pressure } p \text{ at } y = 0: \frac{p}{\rho_f} = -\frac{\partial \phi}{\partial t} - gy + \frac{p_0}{\rho_f} \tag{10.65c}$$

A flow depth h_D moves with a speed u over a sand bed of sinusoidal wave moving with a velocity c_{BM}, then the profile of bed surface as

$$y_{bs} = a_{bs} \sin \frac{2\pi}{\lambda_b}[x - c_{BM}t] \tag{10.66}$$

If origin is considered at the free surface, the complex potential is given by:

$$w = \bar{U}\left[z + \frac{a_{fs}}{\sinh \frac{2\pi H}{\lambda_b}} \cos \frac{2\pi}{\lambda_b}\{z + iH - c_{BM}t\}\right] \tag{10.67}$$

where $z = x + iy$, and H is determined from

$$\bar{U}^2 = \frac{g\lambda_b}{2\pi}\tanh\frac{2\pi H}{\lambda_b} \tag{10.68}$$

and the profile of the free surface is given by

$$y_{fs} = a_{fs}\sin\frac{2\pi}{\lambda_b}[x - c_{BM}t] \tag{10.69}$$

Now the complex potential equation (10.67) can be separated into real and imaginary parts, and then the velocity potential ϕ and the stream function ψ are found as:

$$\phi = \bar{U}x + a_{fs}\bar{U}\frac{\cosh\frac{2\pi}{\lambda_b}(y+H)}{\sinh\frac{2\pi H}{\lambda_b}}\cos\frac{2\pi}{\lambda_b}(x - c_{BM}t) \tag{10.70}$$

$$\psi = \bar{U}y + a_{fs}\bar{U}\frac{\sinh\frac{2\pi}{\lambda_b}(y+H)}{\sinh\frac{2\pi H}{\lambda_b}}\sin\frac{2\pi}{\lambda_b}(x - c_{BM}t) \tag{10.71}$$

The free surface is designated by the streamline $\psi = 0$, the reference streamline by $\psi = \bar{U}H$, and the streamline on the sand bed by $\psi = \bar{U}h_D$ (Figure 10.12). Substituting these values in the stream function equation (10.71), the following is obtained as:

$$y_{bs} = \frac{a_{fs}}{\sinh\frac{2\pi H}{\lambda_b}}\sin\frac{2\pi}{\lambda_b}(x - c_{BM}t)\sinh\frac{2\pi}{\lambda_b}(H - h_D) \tag{10.72}$$

If equation (10.72) is equated with equation (10.66), one gets

$$a_{bs} = \frac{a_{fs}}{\sinh\frac{2\pi H}{\lambda_b}}\sinh\frac{2\pi}{\lambda_b}(H - h_D) \tag{10.73}$$

Using (10.68) in (10.73), one gets

$$\frac{a_{fs}}{a_{bs}} = \left[\cosh\frac{2\pi h_D}{\lambda_b} - \frac{g\lambda_b}{2\pi\bar{U}^2}\sinh\frac{2\pi}{\lambda_b}h_D\right]^{-1} \tag{10.74a}$$

$$a_{bs} = a_{fs}\left(1 - \frac{g\lambda_b}{2\pi\bar{U}^2}\tanh\frac{2\pi}{\lambda_b}h_D\right)\cosh\frac{2\pi}{\lambda_b}h_D \tag{10.74b}$$

Equation (10.74b) shows the relation of the amplitude of surface waves and that of bedforms. If $\bar{U}^2 > \frac{g\lambda_b}{2\pi}\tanh\frac{2\pi}{\lambda_b}h_D$, the two waves (surface wave and bedform) are in phase, while if $\bar{U}^2 < \frac{g\lambda_b}{2\pi}\tanh\frac{2\pi}{\lambda_b}h_D$, then two waves are out of phase. In fact, the quantity $\left(\frac{g\lambda_b}{2\pi}\tanh\frac{2\pi}{\lambda_b}h_D\right)$ is the speed of propagation of a wave with wave length λ_b, in water depth h_D. If the speed of propagation c^2, then one can write as:

$$c^2 = \frac{g\lambda_b}{2\pi}\tanh\frac{2\pi}{\lambda_b}h_D \tag{10.75}$$

If $\bar{U}^2 = c^2 = \frac{g\lambda_b}{2\pi}\tanh\frac{2\pi}{\lambda_b}h_D$, then the ratio $\frac{a_{fs}}{a_{bs}}$ becomes infinite, and the solution breaks down. Over the special case of flow over fixed sinusoidal sand bed, Anderson (1953) used Exner's erosion equation (10.50) as:

$$\frac{\partial y_{bs}}{\partial t} + \alpha_E \frac{\partial \bar{U}}{\partial x} = 0 \tag{10.50}$$

where y_{bs} is sinusoidal bed height. Anderson (1953) obtained an amplitude ratio of

$$\frac{a_{fs}}{a_{bs}} = \frac{1}{2}\cosh\frac{2\pi}{\lambda_b}h_D \tag{10.76}$$

Combining equations (10.76) and (10.74a), the following equation is obtained:

$$Fr^{-2} = \frac{gh_D}{\bar{U}^2} = \frac{2\pi h_D}{\lambda_b}\left(\tanh\frac{2\pi h_D}{\lambda_b} - \frac{2}{\sinh\frac{4\pi h_D}{\lambda_b}}\right) \tag{10.77}$$

Equation (10.77) suggests a relationship between the Froude number Fr of the flow and the relative wavelength. According to Anderson (1953), the value of $\cosh\frac{2\pi}{\lambda_b}h_D = 2$ divides asymmetrical and symmetrical bedforms. If $\cosh\frac{2\pi}{\lambda_b}h_D > 2$, then the amplitude of the sand wave is less than the one of the surface waves. Due to the continuity, the velocity at the crest is smaller than the one at the trough; therefore, sediment tends to deposit on the crest and is carried to the downstream face. If $\cosh\frac{2\pi}{\lambda_b}h_D < 2$, the opposite is true. Relatively larger velocities at the crest are causing erosion, and the resulting bedforms are symmetrical.

When the bedforms move downstream, the bedform profile is given by:

$$y_{bs} = \frac{2a_{bs}}{\cosh\frac{2\pi}{\lambda_b}h_D}\sin\beta t\cos\left(\frac{2\pi x}{\lambda_b} - \beta t\right) \tag{10.78}$$

and the speed c_{BM} is given as:

$$c_{BM} = \frac{\beta\lambda_b}{2\pi} = \frac{\alpha_E \bar{U}\frac{2\pi}{\lambda_b}\cosh\frac{2\pi}{\lambda_b}h_D}{2\sinh\frac{2\pi}{\lambda_b}h_D} \tag{10.79}$$

Anderson (1953) could not verify his equation (10.79) due to lack of adequate data.

Kennedy (1963) used an Exner-type erosion equation to determine the amplitude a_{bs} of the bed surface wave and the migration velocity c_{BM}. The sediment transport equation is given as

$$\frac{\partial g_{st}}{\partial x} + B_{sw}\frac{\partial \eta}{\partial t} = 0 \tag{10.80}$$

where g_{st} is the local rate of sediment transport per unit width in weight, which is an unknown quantity to be related to the stream velocity and B_{sw} is the bulk-specific weight of the sediment in the bed. Kennedy (1963) suggested that the sediment transport rate g_{st} should be related to the stream velocity and is given as,

$$g_{st}(x,t) = m\left[\frac{\partial \phi}{\partial x}(x-\delta, h_D, t)\right]^n \tag{10.81}$$

where m, n, and δ are constants and are dependent on depth, velocity and properties of fluid and sediment. The quantity δ represents the lag of the fluid and sediment-particle velocity. Substituting (10.70) into (10.81), the following expression is obtained as:

$$g_{st}(x,t) = m\bar{U}^n - \frac{2\pi}{\lambda_b} mn\bar{U}^n a_{fs} \frac{\cosh\frac{2\pi}{\lambda_b}(H-h_D)}{\sinh\frac{2\pi H}{\lambda_b}} \cdot \sin\frac{2\pi}{\lambda_b}(x-\delta-c_{BM}t) + O\left(\bar{U}^n\right) \tag{10.82}$$

Neglecting the higher order terms, the net forward sediment transport rate is given by:

$$g_{st} = m\bar{U}^n \tag{10.83}$$

Combining equations (10.82) and (10.83) together with (10.66) and (10.73), Kennedy (1963) provided a differential equation for a_{fs}, which in turn gives the following relation for bedform migration velocity c_{BM} as:

$$c_{BM} = -\frac{2\pi n g_{st}}{B_{sw}\lambda_b} \coth\frac{2\pi}{\lambda_b}(H-h_D)\cos\frac{2\pi\delta}{\lambda_b} \tag{10.84}$$

and for amplitude a_{bs} of the bedform:

$$a_{bs}(t) = \frac{a_{fs}\sinh\frac{2\pi}{\lambda_b}(H-h_D)}{\sinh\frac{2\pi H}{\lambda_b}} exp\left[\frac{4\pi^2 n g_{st}}{\lambda_b^2 B_{sw}} t \cdot \coth\frac{2\pi}{\lambda_b}(H-h_D)\sin\frac{2\pi\delta}{\lambda_b}\right] \tag{10.85}$$

Equation (10.85) may be interpreted as the small disturbances on the sand bed increasing exponentially if $2\pi/\lambda_b$ and δ make the exponential term in equation (10.85) positive. There is a limitation of maximum height for linearization in the development of the equation. As the amplitude increases, it gives rise to non-linear effects which govern the equilibrium height of fully developed dunes and anti-dunes. Using equations (10.73), (10.84) and (10.85), the various bedforms and the criteria for the occurrence of different bed configurations are shown in Table 10.1 and these may be available in Kennedy (1963) for detailed discussions.

Using the concepts of stability analysis, Kennedy (1963) obtained an expression for the square of Froude number as:

$$Fr^2 = \frac{\bar{U}^2}{gh_D} = \frac{1+\left(\frac{2\pi h_D}{\lambda_b}\right)\tanh\left(\frac{2\pi h_D}{\lambda_b}\right)+\left(\frac{2\pi\delta}{\lambda_b}\right)\cot\left(\frac{2\pi\delta}{\lambda_b}\right)}{\left(\frac{2\pi h_D}{\lambda_b}\right)^2+\left[2+\left(\frac{2\pi\delta}{\lambda_b}\right)\cot\left(\frac{2\pi\delta}{\lambda_b}\right)\right]\left(\frac{2\pi h_D}{\lambda_b}\right)\tanh\left(\frac{2\pi h_D}{\lambda_b}\right)} \tag{10.86}$$

TABLE 10.1
Details of Bed Configurations and Occurrence Conditions

Bed and Surface Profiles		$H-h_D$	$(2\pi/\lambda_b)\delta$	$\sin(2\pi/\lambda_b)\delta$	$\cos(2\pi/\lambda_b)\delta$	**Bed Features Movement**	**Details of Bed Configuration**
1	In phase	Pos.	$0<(2\pi/\lambda_b)\delta<\pi/2$	Pos.	Pos.	Upstream	Anti-dune
2		Pos.	$\pi/2$	Pos.	Zero	None	
3		Pos.	$\pi/2<(2\pi/\lambda_b)\delta<\pi$	Pos.	Neg.	Downstream	
		Neg.	$\pi<(2\pi/\lambda_b)\delta<3\pi/2$	Neg.	Pos.	-	Flat bed
4a	No bed waves	Neg.	$0<(2\pi/\lambda_b)\delta<\pi$	Pos.	-	-	
4b		Pos.	$\pi<(2\pi/\lambda_b)\delta<2\pi$	Neg.	-	-	
4c							
5	Out of phase	Neg.	$3\pi/2<(2\pi/\lambda_b)\delta<2\pi$	Neg.	Pos.	Downstream	Dunes

Source: Modified from Kennedy (1963).

It was suggested that the lag velocity parameter δ can be written as:

$$\delta = jh_D \tag{10.87}$$

For limiting case, $\delta \ll h_D$ or $j \to 0$, equation (10.86) can be written as:

$$\lim_{j\to 0} Fr^2 = \frac{1+\left(\dfrac{2\pi h_D}{\lambda_b}\right)\tanh\left(\dfrac{2\pi h_D}{\lambda_b}\right)}{\left(\dfrac{2\pi h_D}{\lambda_b}\right)^2+\left(\dfrac{6\pi h_D}{\lambda_b}\right)\tanh\left(\dfrac{2\pi h_D}{\lambda_b}\right)} \tag{10.88}$$

Now it is shown that the relation Fr vs. $2\pi h_D/\lambda_b$ for various values of j. A useful relation is obtained from the existing experimental data, which represents a trend line of equation (10.88) plotted in Figure 10.13. The minimum wavelength $\lambda_{b,\,\min}$ for bedforms can be obtained by putting $h \to \infty$ in equation (10.68) and it follows as:

$$\lambda_{b,\,\min} = \frac{2\pi}{g}\bar{U}^2 \tag{10.89}$$

Dividing equation (10.89) by $2\pi h_D$, one gets:

$$Fr_{\max}^2 = \frac{\lambda_{b,\,\min}}{2\pi h_D} = \frac{\bar{U}^2}{gh_D} \tag{10.90}$$

Here $Fr_{\max}$ is the maximum possible Froude number for two-dimensional waves are obtained. For long crested 2-D waves, $2\pi/\lambda_b$ falls within the range $0 \le 2\pi/\lambda_b \le Fr^{-2}$, while for short-crested three-dimensional waves, $2\pi/\lambda_b$ can exceed Fr^{-2} provided for given values of $\bar{U}$ and h_D these waves are shorter than two-dimensional waves. The minimum Froude number for anti-dunes and maximum Froude number for dunes: for a very special case of $H = h_D$, equation (10.68) becomes after dividing both sides by gh_D, it follows as:

$$Fr_a^2 = \frac{\bar{U}^2}{gh_D} = \frac{\tanh\dfrac{2\pi h_D}{\lambda_b}}{\dfrac{2\pi h_D}{\lambda_b}} \tag{10.91}$$

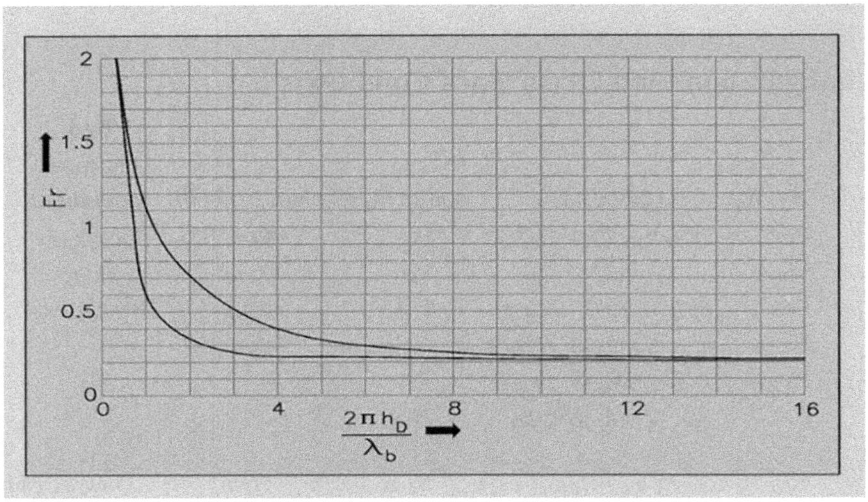

FIGURE 10.13 Froude number Fr vs. $2\pi h_D/\lambda_b$ obtained by Kennedy (1963) trend line from several experimental data of bedform. (Modified from Graf, 2010.)

From the figure, it is observed that the anti-dune data show the best agreement, while practically all dune data show below the curve. Nevertheless, the overall results show rather encouraging. Kennedy (1963) remarked that experimental research should be oriented toward the determination of δ value and its relation to the other parameters. Reynolds (1965) extended Kennedy's idea of two-dimensional to the three-dimensional flow for sand waves on the erodible bed in an open channel flow.

Engelund (1966) considered the problem of bedforms in alluvial channels and suggested some interesting ideas and conclusions. They assumed the equation of motion of real fluid together with equation of continuity for an incompressible fluid. Under certain assumption and after linearization, a differential equation for the unknown bottom shear stress is obtained. They emphasized that the maximum of bed shear stress is located upstream of the minimum depth. The theory was checked with the experimental data and found to be reasonably good agreement. They also obtained another expression for the shear stress at the bed if the characteristics of the loose sand bed and sediment transport are considered. Hayashi (1970) modified the sediment transport rate relation given by equation (10.81) assumed by Kennedy (1963).

Song (1983) determined a solution of flow over a sinusoidal wavy bed to discuss the bedform phenomenon, using the method developed by Milne-Thompson (1960). Song (1983) expressed the complex potential as:

$$w = \bar{U}\left[z + \frac{a_{fs}}{\sinh\dfrac{2\pi H}{\lambda_b}}\cos\frac{2\pi}{\lambda_b}\{z + iH - c_{BM}t\}\right] \tag{10.92}$$

where $z = x + iy$, $\bar{U}$ is the mean flow velocity over the entire flow field. and H is determined from bottom. At $z = -H$, a stream line has a stream function $\psi = -\bar{U}H$. The complex velocity potential is to satisfy the condition of the constant pressure at the free surface. It is therefore required as:

$$\bar{U}^2 = \frac{g\lambda_b}{2\pi}\tanh\frac{2\pi H}{\lambda_b} \tag{10.93}$$

A relation between the amplitude of water waves and the sand waves is given by:

$$\alpha_{bs} = \alpha_{fs}\left[1 - \frac{g}{\frac{2\pi}{\lambda_b}\bar{U}^2}\tanh\frac{2\pi H}{\lambda_b}\right]\cosh\frac{2\pi H}{\lambda_b} \tag{10.94}$$

The speed of small amplitude gravity wave C is given by

$$C^2 = \frac{g\lambda_b}{2\pi}\tanh\frac{2\pi H}{\lambda_b} \tag{10.95}$$

Equation (10.94) indicates that the sand wave and water wave are in phase if the flow is super-critical, but out of phase if the flow is subcritical.

In order to estimate the bedform velocity, the bed load is assumed as a function of the potential flow velocity on the bed. This is justified when there is no flow separation and the boundary layer is thin. In fact, some error is expected for fully grown-up dunes due to separation of flow on the leeside.

Differentiating equation (10.92) with respect to z and by setting $y = -H$, the x-component of the velocity of the bedform is:

$$u_0 = \bar{U} - \alpha_{fs}\bar{U}\frac{2\pi}{\lambda_b}\frac{\cosh\frac{2\pi}{\lambda_b}(H-h)}{\sinh\frac{2\pi H}{\lambda_b}}\sin\frac{2\pi}{\lambda_b}(x - c_{BM}t) \tag{10.96}$$

If the amplitude of the sand wave is not too large, the vertical velocity component is relatively small and the speed is nearly equal to near-bed stream-wise velocity u_0. In this case, the sand wave velocity U_b defined by (10.62) is obtained as:

$$U_b = \frac{\partial q_B}{\partial \eta} = \frac{\partial q_B}{\partial \bar{U}}\cdot\frac{\partial \bar{U}}{\partial \eta} = \frac{\partial q_B}{\partial \bar{U}}\cdot\frac{2\pi}{\lambda_b}\bar{U}U^* \tag{10.97}$$

$$\Rightarrow U^* = \frac{U_b}{\frac{\partial q_B}{\partial \bar{U}}\cdot\frac{2\pi}{\lambda_b}\bar{U}} \tag{10.98}$$

where $U^* = \dfrac{1 - Fr^2\frac{2\pi}{\lambda_b}H\tanh\frac{2\pi H}{\lambda_b}}{\tanh\frac{2\pi H}{\lambda_b} - Fr^2\frac{2\pi}{\lambda_b}H}$ is the non-dimensional parameter. Here the non-dimensional parameter U^* represents the non-dimensional velocity of bedform migration. A linear stability theory for sand dune and anti-dune formations had been re-examined by Colombini (2004) using the rotational two-dimensional flow model. The interaction between two sandy bedforms under low-sediment transport conditions in a laboratory flume was studied by Jerolmack and Mohrig (2005). These bedforms were initially identical and the initial distance between bedform crests was varied in the tests. Even slight changes in bedform troughs were found to have a strong effect on turbulence and sediment transport. On the other hand, based on the Reynolds-averaged Navier–Stokes (RANS) equations, Bose and Dey (2009) developed a theory of turbulent shear flow over bed waves addressing the instability theory for a plane sand bed leading to the formation of bedforms, like dunes and anti-dunes.

10.8 DIFFERENT TYPES OF RIPPLES

After the initiation of motion of particles, the plane bed changes to small ripples of sand sizes smaller than 0.6 mm and to mega-ripples for sand sizes larger than 0.6 mm. After a few runs, sand size of 0.93 mm diameter showed a plane bed with movement at small stream powers before the formation of mega-ripples (Guy et al., 1966). According to Allen (1968) at least for coarse grains, existence of plane bed with movement at low velocities is quite convincing. Southard (1975) pointed out that in sand size finer than 0.1 mm, small ripples directly change to plane bed of upper flow regime, and there is no development of larger bedforms like mega-ripples or sand waves. Reineck and Singh (1973) observed that muddy sediment ripples are erosive, and suggested that sediment with less than 40% sand produces erosive ripples of the longitudinal ripple type. Southard (1975) and Middleton and Southard (1984) considered the term dunes as three-dimensional forms of mega-ripples showing strongly sinuous crests. Two-dimensional forms like straight-crested mega-ripples are mostly described as sand waves. Some concepts of development of bedforms are explored from the physical and mathematical point of view. There exist some physical models of bedforms which are concerned with a particular growth stage of features. On the other hand, the mathematical approach initiated by Exner (1925) has led to theories of bedforms on water-loose bed interface, and the results are compared with the laboratory experiments (Reynolds, 1976) and limited field observations. He realized the complexities in solving the entire problem, and slowly he progressed in developing the complicated problems.

The following transverse bedforms were developed under unidirectional flow conditions: current ripples and dunes formed by unidirectional water flow, ballistic ripples, barchan or crescentic-shaped dunes and transverse dunes formed by the wind, and anti-dunes generated by free surface aqueous flows. The term 'current ripples' is used in the geological context; it is defined as the unidirectional flow ripples which are formed at relatively low flow strengths above the threshold movement on smooth sand bed. Current ripples are transverse ridges, much steeper on the downstream or lee-side than on the upstream or stoss-side with a height of less than 0.04 m and wavelength below 0.6 m (Allen, 1968). Ripples and dunes are the sharp-crested triangular shape and travel downstream, sediment being eroded from the stoss and re-deposited in the lee-side. 'Ballistic ripples' look like current ripples, but commonly somewhat flat. Barchan dunes or crescentic-shaped dunes are formed under unidirectional winds blowing on ground with limited sand supply. They propagate along the main wind direction. The field observations revealed that the shape of barchan dunes is mainly determined by the direction of the wind and sand availability (Bagnold, 1941; Cooke et al., 1993). When the wind is unidirectional and the sand supply is low, flat, crescentic dunes, called barchans, appear and migrate along the wind direction. Crescent-shaped mounds are generally wider than long. The slip-face is on the concave side of the dune (Figure 10.14).

Some types of crescentic dunes move faster over desert surfaces than any other type of dune. There are two types of barchan dunes: Aeolian barchan dunes due to wind flow (Bagnold, 1941; Hersen, 2004) and subaqueous barchan dunes due to water flow (Hersen, 2005). Aeolian barchan dunes are found for decades, whereas the subaqueous barchan dune is recently found which is comparable to aeolian barchans. Experiments in laboratory set-up show that the solitary barchan dunes can be created using periodic and asymmetric motion of a plate underwater to simulate the effect of unidirectional flow (Hersen et al., 2002; Hersen, 2005). However, recently the crescentic scour patterns on the sediment bed are also observed underwater flow conditions (Sengupta et al., 2005; Maity and Mazumder, 2014, 2017), and the subaqueous barchan dunes are also produced in the ISI Laboratory at Kolkata which are comparable with the aeolian barchan dunes. As observed, sand grain can escape from the horns, but not from the main dune body where they are trapped into the slipface. This behavior between the main body and the horns is crucial for understanding the barchan dunes. Figure 10.15 shows the series of barchan dunes developed in the ISI Fluvial Mechanics Laboratory, Kolkata using low amount of sediment in the bed as well as high amount of sand in the bed. Humpback dune and race boat dune are also observed in the ISI Laboratory and are shown respectively in Figure 10.16a and b.

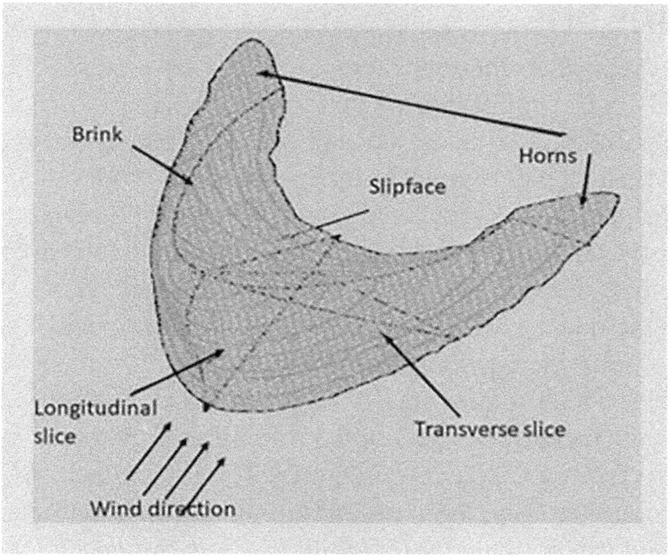

FIGURE 10.14 Typical sketch of Aeolian Barchan dune.

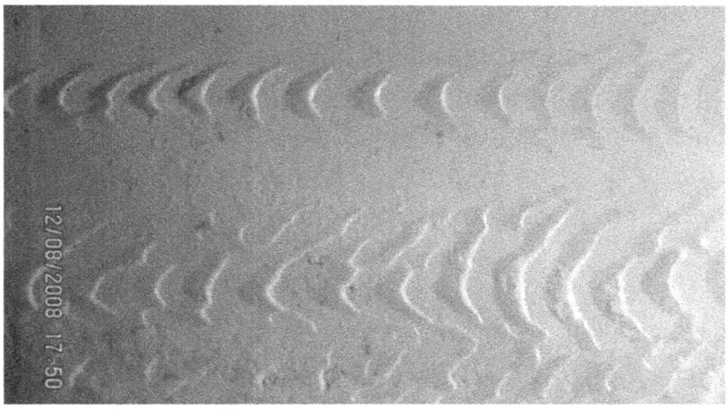

FIGURE 10.15 Subaqueous Barchan dunes developed at Indian Statistical Institute (ISI) Fluvial Mechanics Laboratory, Kolkata. (Courtesy: Fluvial Mechanics Lab., ISI Kolkata.)

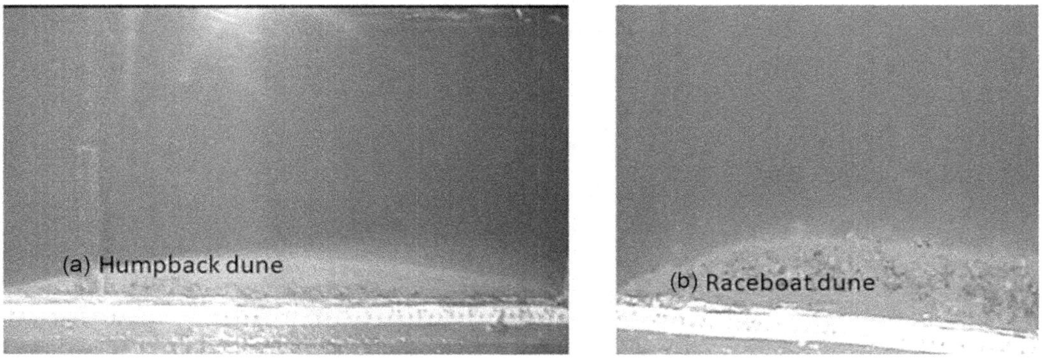

FIGURE 10.16 (a) Humpback dune (b) Race boat dune developed at ISI Laboratory. (Courtesy: Fluvial Mechanics Lab., ISI Kolkata.)

10.9 INITIATION OF BEDFORMS

Gyr and Schmid (1989) investigated the formation of ripples on an erodible sand bed by varying the roughness, the mass density of the uniform grain material and history of the bed evolution. Experiments were conducted with gradually increasing the slope at small increments from a value at which no transport occurred to one at which ripples started forming. From the observations, it is concluded that the ripples originate from transport relics produced by intense sweep. The evolution of the ripples is due to a feedback mechanism between the flow field and the developing bedform. The link between them is the altered sediment transport. Especially if a separation of flow occurs, the wavelength is determined by the separation and reattachment of the flow. They showed three mechanisms of ripple formation in their experiments: (1) gradually increasing the slope with small increments from a value at which no transport occurred to one at which ripples started to occur in isolated rows; (2) slowly increased wall shear stress, which means the increase of shear at the wall takes time to self-stabilize by rearrangement of grains, where the bed surface looks like an "orange peel" (Gyr and Schmid, 1989), whereas this term named as "armoured bed condition" by Williams and Kemp (1972); and (3) rapidly increased wall shear stress, which means the bed can no longer self-stabilize, where sweep events initiate a flow separation are more frequent because sweep of lower intensity can produce relics of the needed size. Raudkivi and Witte (1990) stated that no ripples are formed in laminar flow; the initiation of particle movement must be a function of turbulence. The disturbances arise first from the turbulent bursting process on the sand bed surface. The initial disturbance on the bed surface exerts a deflection on the flow and leads to a subsequent disturbance at the reattachment points at the bed. The development of sand waves is termed as "sand-wavelet" to identify these budding bedforms (Figure 10.17). Ripples or dunes subsequently develop from these sand-wavelets through a process of applied flow with time. The 'sand-wavelet' is defined for the identification of budding bedforms. The process of sand-wavelet generation from a flat bed has been treated as a problem of sand-wave interface instability. The stability theory along this line has been studied by Kennedy (1969), Jain and Kennedy (1974) and Sumer and Bakioglu (1984) using the linearized theory based on the assumption of small sand-wavelet amplitudes.

Baas (1994) performed a series of laboratory experiments for the development of current ripples of very fine sand. An empirical model was developed for the generation of small-scale, unidirectional

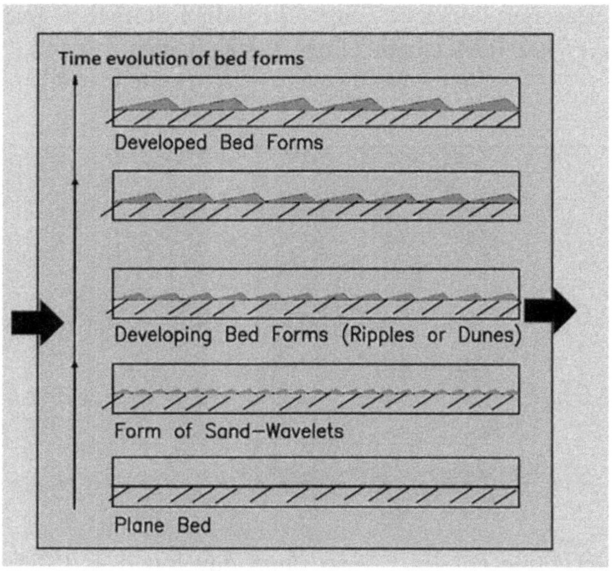

FIGURE 10.17 Development of sand-waves through sand-wavelets (identification of sand-waves from bottom to top with time, and water flow from left to right). (Modified from Coleman and Melville, 1996.)

bedforms in sand of median grain size d_{50} as 0.095 mm from their observed data of steady flow experiments in a flume. Current ripples always attain a linguoid plan morphology with constant average height (13.1 mm) and wavelength (115.7 mm), provided that sufficient time is allowed for their formation. The pattern of these ripples on a flat bed is independent of flow velocity in the ripple stability field. Four major stages of ripple development were distinguished: (1) longitudinal streaks and incipient ripples; (2) straight and sinuous ripples; (3) non-equilibrium linguoid ripples, and (4) equilibrium linguoid ripples. Straight and sinuous ripples are the non-equilibrium bedforms at all flow velocities. Linguoid ripples have lee slope surfaces that are curved generating a lamina similar to catenary and sinuous ripples. Linguoid ripples generate an angle to the flow as well as downstream. Linguoid ripples have a random shape rather than a "W" shape, as described in the catenary description. The characteristics of these stages were studied mainly in the low velocities, in which the rate of morphological change was low. However, the development of stages was similar at higher flow velocities. The time needed to reach equilibrium dimensions is related to the inverse power of flow velocity and ranges from several minutes to more than hundreds of hours. The results of these observed data correspond reasonably well with those of previous studies, provided that various factors, such as experimental methods, sediment characteristics, shallow flow depths and non-equilibrium runs, are considered into account. Details are available in Baas's (1994) paper.

Coleman and Melville (1994) carried out a series of bedform development experiments to study the evolution of bed features from the flat bed conditions. For each experiment, the bed profiles were recorded frequently to study the sand-wave configurations with time due to action of flow velocity. Two sediment sizes with respective geometric mean sizes 0.20–0.82 mm were used for the experiments. The bedforms were generated at a consistent rate from the flat bed conditions and observed that the generation of bedform rate increased with increase in sediment transport rate. They observed from their experimental findings that the rate of bedform generations from the flat bed conditions increases with increasing in flow velocity, decreases in sand size, and flow depth, which agree with those of Raichlen and Kennedy (1965) and Jain and Kennedy (1971). They developed an important relationship between the bedform speed and bedform height. They defined a dimensionless bedform propagation speed c' as

$$c' = \frac{c}{(u_* - u_{*c})(\theta - \theta_{*c})} \tag{10.99}$$

where u_{*c} is the critical shear velocity, θ_{*c} is the Shields parameter with dimensionless bedform height as $h_b' = h_b / d_g$ with $d_g = (d_{84}d_{16})^{0.5}$ is the geometric mean of sediment size. The bedform height and speed relation indicated by the respective pairs of points are approximated as:

$$c'(h_b' - 3.5)^{1.3} = 40 \tag{10.100}$$

Equation (10.100) represents the overall indication of the relation between the bedform speed and bedform height for a given flow system (Figure 10.18). This figure is developed from the trend of experimental data for $d_g = 0.20, 0.82$ mm. For a developing sediment bed, the results indicate that the bedform speed generally decreases with increase in bedform height. For bed developing from the flat bed conditions, bedform speeds therefore generally decrease with time as bedforms grow.

In the subsequent paper, Coleman and Melville (1996) investigated the initiation of bedforms on a flat sand bed, based on the 47-bed development experimental data sets, and compared with existing experimental data. The generation of initial bed waves, i.e., sand-wavelets, on a plane sand bed is found to be initiated by the appearance of random pileup of sediment. When the height of pileup approaches the bed roughness height, further pileups are generated at preferred spacings downstream. The wavelength of sand wavelets is generally insensitive to the applied bed shear stress and is a function of geometric mean size of the sediment, and is described by the empirical formula as:

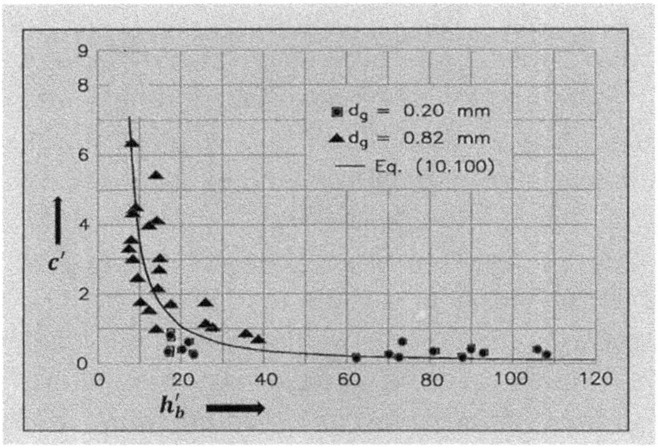

FIGURE 10.18 Plots of dimensionless bedform speed c' vs. bedform height h'_b. (Modified from Coleman and Melville, 1994.)

$$\frac{l}{d_g} = 10^{2.5} R_{*c}^{-0.2} \tag{10.101}$$

where R_{*c} is the critical Reynolds number. The ratio l/d_g is a function of R_{*c}, which shows the sand-wavelet length being relatively insensitive to the applied flow conditions and primarily is a function of sediment. They also showed the variation of the ratio $l/d_g$with relative share stress excess $(\theta - \theta_{*c})/\theta_{*c}$. It is important to note that both ripples and dunes, the characteristics of sand waves, are found to be generated from the sand wavelets.

Venditti and Bennett (2000) studied the turbulent flow over the dune bedforms in a laboratory channel and linked between macro-turbulence and suspended sediment flux. Mainly, they studied the turbulence characteristics and suspended sediment transport over the dunes, the coherent flow structures using spectral and co-spectral techniques, and also determined the link between the turbulent fluctuations in velocity and suspended sediment flux. Turbulent flow parameters demonstrate that the flow separation and perturbed shear layer are the main sources of turbulence production, and that the distribution of suspended sediment is controlled by spatially macro-turbulent flow structures. They observed that the peak spectral energies generally occur at 1–2 Hz for the streamwise velocity and 2–4 Hz for the transverse and bottom-normal velocities. Peak spectral energies for suspended concentration occur near 1 Hz throughout the flow. They also determined the integral time scales for velocity components, and integral length scales for velocity range. For suspended sediment concentration integral time scales and length scales are similar to the streamwise velocity component.

Venditti (2003) addressed the processes responsible for the initiation and subsequent growth of bedforms in alluvial channels. He proposed to determine why a flat sand bed is unstable and how this instability leads to bedform development. In his study, he employed a phenomenological approach to examine the physical processes that transform a flat sand bed to two-dimensional bedforms and then three-dimensional bedforms. He studied in detail: (1) processes of changes from initially flat sand bed to 2D bedforms through laboratory experiments, (2) description of 2D bedform as simple near-bed shear layer instability, (3) process of transition between 2D to 3D bedforms, and (4) any drag reduction from two-dimensional to three-dimensional (2D-3D) bedform transition. In his experiments, $d_{50} = 0.500$ mm was used. He studied the development of dune length λ_b, dune height h_b and migration rate R_m with respect to time t from the results of his experiments. The initial dune height was about 1–4.6 mm depending on the run, and the initial dune length varied between 0.034 m and 0.105 m. The growth of dune bedforms is approximately exponential and can be written as:

$$h_b = a_{h_b}\left[1-\exp\left(-b_{h_b}t\right)\right] \tag{10.102}$$

$$\lambda_b = a_{\lambda_b}\left[1-\exp\left(-b_{\lambda_b}t\right)\right] \tag{10.103}$$

where a_{h_b}, a_{λ_b}, b_{h_b} and b_{λ_b} are the coefficients derived from the least square regression. The bedform migration rates decrease exponentially with time and can be expressed as:

$$R_m = a_{R_m}\exp\left(\frac{b_{R_m}}{x+c_{R_m}}\right) \tag{10.104}$$

where a_{R_m}, b_{R_m} and c_{R_m} are the coefficients derived from the least square regression. All the coefficients are given in Table 2.3 of Venditti (2003). He also studied how the flows used in these experiments compare with conventional models of flow over the sediment bed. Mean velocity components, intensities and Reynolds shear stresses are calculated over the sediment beds. Autocorrelation for the velocity time series was studied to determine the integral time and length scales in the near bed region using the time series data collected over 600 seconds at 5 mm above the bed. The integral length scale is a characteristic length of eddy dimension, and the integral time scale is the time required by an eddy to pass a given point in the flow. He used the Eulerian integral time scale t_E defined as:

$$t_E = \int_0^{t_s} R(t)dt \tag{10.105}$$

where $R(t)$ is the autocorrelation function, t_s is the time step at which $R(t)$ is no longer significantly different from zero (Tennekes and Lumley, 1972). The Eulerian length scale l_E is defined as (Taylor, 1921):

$$l_E = t_E u \tag{10.106}$$

where u is the velocity at that point. From the observed data, he found the mean Eulerian time scale t_E varied within the range from 0.225 s to 0.271 s, while the mean length scale l_E varied between 0.0620 m and 0.0758 m. Thus, he found the average or dominant eddy size is about 0.07 m. Two types of initiation of bedform were observed from the experiment: defect initiation, and instantaneous initiation. The defect bedform initiation was studied by Southard and Dingler (1971) from ripple propagation behind positive defects (mounds) on flat sand beds. Following the work of Southard and Dingler (1971), the defect bedform initiation was examined by Venditti (2003) based on the artificially made defects rather than examining bedform growth from the random features on the bed. Instantaneous initiation bedforms occurred spontaneously over the entire bed surface at different flow strengths. Details are available in his Thesis (Venditti, 2003).

Subsequently, Venditti et al. (2005) examined the initiation of bedform in the unidirectional flow over a flat sand bed composed of homogeneous sand of 0.500 mm diameter. They observed two separate modes of bedform initiation: defect and instantaneous initiation. Defect initiation occurs at the lower flow stage; where sediment transport is sporadic and patchy propagating with flow separation. Instantaneous initiation occurs at the larger flow strengths, where the sediment transport is widespread. They observed that there were negative defects (divots) also in terms of bedform development. Therefore, both positive and negative bed defects were used in their experiments. The defects were generated by either sucking sand into or depositing sand from a large dropper until the desired defect size was attained. The shapes of defects were cone and the diameter was about 30 mm. The negative defects were 8.2–9.4 mm below the mean bed elevation and positive defects were about 8.2–10.4 mm above the bed elevation. However, the conclusion was drawn from the series of runs for defects which varied within the height between 2.5–50 mm. Initially, the bed was covered by lineated striations, oriented along the flow, with spacing approximately equivalent to the expected streak spacing (Best, 1992). These linear streaks did not appear to play any significant role in the development of the bed. Instead, the bed undergoes the following deformations: (1) a cross-hatch pattern is imprinted on the bed, (2) Chevron-shaped scallops develop at the nodes of the cross-hatch, (3) Chevrons begin to migrate and organize into incipient crestlines, (4) crestlines straighten into two-dimensional features, and (5) bedforms grow in height and length. For all details, please refer Venditti (2003) and Venditti et al. (2005). The time required to move through these stages decreased significantly with increasing flow strength. It is important to conclude that the integral length scales derived from the velocity measurements prior to the bedform development are similar to the initial bedform length scales.

10.10 BEDFORM MIGRATION

Free-surface flow over a mobile bed is usually accompanied by entrainment of sediment and formation of bedform; both in turn will influence the flow and its sediment carrying capacity. Several experimental studies of flows over different types of bedforms/roughness have been reported in the hydrodynamics and sediment transport literature with their scopes, which are mostly constrained to the friction factors, transport rates and bedform characteristics. Large-scale roughness, such as dunes and ripples, arising from instability of erodible sediment bed due to the interaction with the turbulent flow are ubiquitous in natural alluvial channel, which induce flow separation, wakes and circulation, and hence form-resistance. The main effect of bedforms in the flow is the separation of flow behind the bedform crest and reattachment point away from the crest; and hence the flow is spatially decelerating there. In the region of recirculation, locally high turbulence is usually observed. Dunes and ripples-dominated bed-load in river and/or laboratory environments are often asymmetric with low slope in upstream side (stoss) and steep-slope in lee-face at downstream side (Guy et al., 1966; Mendoza and Shen, 1990; Kostaschuk, 2000; Venditti, 2003; Ojha and Mazumder, 2008), while those in suspension load-dominated environments are often more symmetric with relatively low angle lee faces (Kostaschuk and Villard, 1996; Best and Kostaschuk, 2002; Mazumder et al., 2005). Several flume and field experiments were performed to estimate the resistance to the flow due to the physical changes of bedforms generated by the turbulent flow (Smith and McLean, 1977; Engelund and Fredsoe, 1982; Wiberg and Nelson, 1992; Gabel, 1993; Lyn, 1993; Robert and Uhlman, 2001; Ojha and Mazumder, 2008; Mazumder et al., 2009; and others). A considerable amount of work has been put into study the flow resistance due to structural changes of one or two bedforms in the flow, especially the occurrence of flow separation and wake, recirculation and reattachment, and hence the flow resistance.

Several detailed experimental and numerical studies of flow over two-dimensional dune type features are found in the literature. Flow over the dunes shows spatial acceleration and deceleration which obliterate the flow field and is classified in to complex flow field. In order to understand the river engineering problems, the lack of adequate knowledge of turbulent flow in alluvial channels have necessitated the experimental information and theory of single-phase flows over rigid boundary. Difficulties present in the study of flow over dunes in the riverine environment are avoided considering the *static features of the dunes* (Figure 10.19) in the laboratory flume. This approximation is justified because the speed of the dunes is very small compared to the stream-wise flow velocity. Some important laboratory work may be mentioned as Vanoni and Hwang (1967), Vittal et al. (1977), Felhman (1985) and others. Of these, Vittal et al. (1977) investigated the velocity distributions over triangular elements simulating ripples and dunes. They presented the form resistance or skin resistance of fixed dune elements and resistance to flow methodology. These are equally important to the detailed field measurements of flow over large dunes at the estuary flow of Columbia River (McLean, 1976) and the data collected on

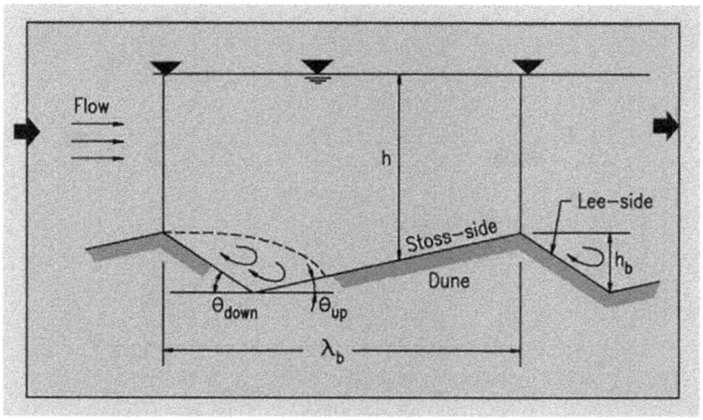

FIGURE 10.19 Sketch of flow domain over static dunes. (Modified from Mendoza and Shen, 1990.)

bedform in the large channels. Turbulent closure models are used for the problem of flow over the bedforms which include mixing length approach (McLean and Smith, 1986), the one-dimensional equation model (Richards and Taylor, 1981), the k-ω turbulence closure model (Puls et al., 1977) and the k-ε model (Mendoza and Shen, 1985a, 1985b). The form and skin resistance over the bedforms were estimated by Mendoza and Shen (1985b) using adhoc modification of the standard k-ε model to balance the solution of the Reynolds-averaged Navier-Stokes equation.

Mendoza and Shen (1990) investigated the turbulent flow field over dunes, utilizing the equations describing the turbulent kinetic energy (TKE), its dissipation rates and the algebraic stress model (second moment turbulence closure) subjected to boundary conditions. A computer code-based finite-difference technique to the PISO algorithm was used to grip the coupling between the continuity and momentum equations to determine the mean velocities, Reynolds stresses, TKE, and pressure distributions over the dune surface. The exact equation governing the transport of turbulent stresses is provided in Mendoza and Shen (1990) as:

$$\text{Convection} = \text{Production} + \text{Dissipation} + \text{Pressure - strain} - \text{Turbulent diffusion} \quad (10.107)$$

The unknown correlations appearing on the right-hand side of equation (10.107) are approximated in terms of the mean velocity components, turbulent stresses, TKE, and the rate of dissipation. The differential equations are solved numerically by finite-difference technique over the small control volumes subject to the boundary conditions. As the transport equation is elliptic in nature, a complete set of boundary conditions is needed at all boundaries of the solution domain, which are bottom boundary, free-surface boundary, and inlet and out boundaries (Mendoza and Shen, 1990). They computed the profiles of mean velocity, shear stress, TKE, skin friction-coefficients and pressure-coefficients, and compared the results with the experimental data of stream-wise velocity, shear stress and TKE from Raudkivi (1967). The computed values of skin-fiction coefficient are shown in Figure 10.20 with measured data of Raudkivi (1963). Downstream from the reattachment point, both computed and measured values show a steadily increasing surface-shear stress. Within the recirculation region, the shear stress is negative with small magnitude, which shows that the magnitude of the shear stress depends on the position along the dune. The maximum shear stress occurs at the crest. Figure 10.21 shows the comparison of computed pressure distribution and the measured distribution reported by Raudkivi (1963). It is observed that the computed values of pressure distribution exhibit a discrepancy when compared with the ones obtained experimentally at the negative pressure zones. Interpretation of the present results combining with visualization experiments conducted by Felhman (1985) on the triangular dunes reveals that the initial rise in pressure over the dunes coincides with the impingement of the shear layer large eddies, which carry appreciable forward momentum upon the bedform surface. Further, the total

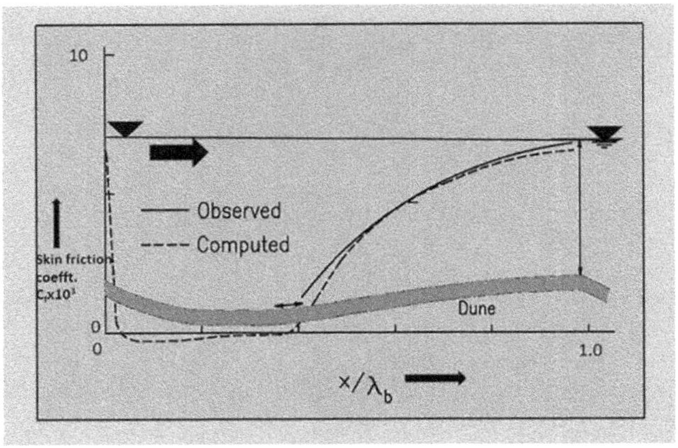

FIGURE 10.20 Measured data of Raudkivi (1963) and computed skin friction coefficients. (Modified from Mendoza and Shen, 1990.)

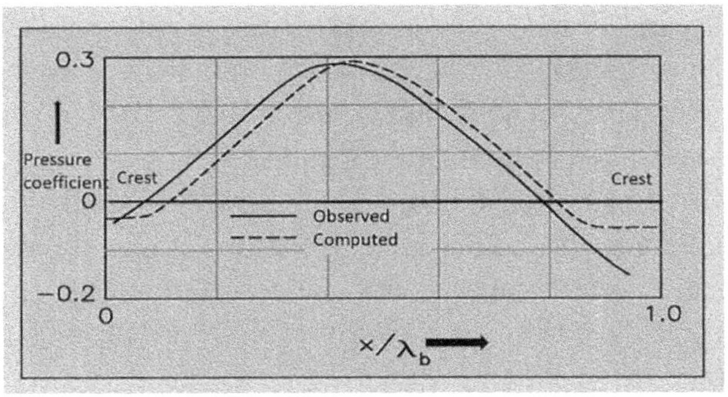

FIGURE 10.21 Measured data of Raudkivi (1963) and computed pressure-coefficient distributions. (Modified from Mendoza and Shen, 1990.)

friction coefficient factor f_t can be estimated according to Einstein (1950) as the sum of form friction coefficient and bed-skin friction coefficient.

10.10.1 Bedform Migration – Model Studies

Lyn (1993) investigated the mean flow and turbulent characteristics in uniform open-channel flows of constant depth over two types of artificial fixed one-dimensional periodic bedforms placed on a flat surface, using two-component laser Doppler velocimeter (LDV). Two types of bedforms, based on triangular elements of the same height (= 20% of the flow depth) and wavelength (~12.5 times the height) but of different shape, were studied. The mean velocity components, turbulence intensities, Reynolds shear stress and eddy viscosity over the two types of bedforms with different Reynolds numbers were examined. The first type of artificial bedform used is shown in Figure 10.22a and consists of a periodic array of 45^0 triangular element of amplitude $h_b \approx 1.2$ cm and wavelength $\lambda_b \approx 15$ cm. Though the dune model is unrealistic, it was studied to examine the effect of bed geometry on turbulence characteristics. It also exhibited features in a clearer form than the dune-bed flow. Measurements closer to the boundary could be made easier since the optical access was maintained. The second type of bedform shown in Figure 10.22b consists of periodic array of triangular elements, and approximate more closely to the geometry of real dune with mildly sloping upstream face, and more sharply sloping downstream face. Since the amplitude and wavelength as well as the 45 angle of the front face are the same as for type (a), the two types may be considered as examples of a class of elements differing only in the upstream of the geometry. Laboratory studies by Vanoni and Hwang (1967) show that dune/ripple steepness h_b/λ_b may

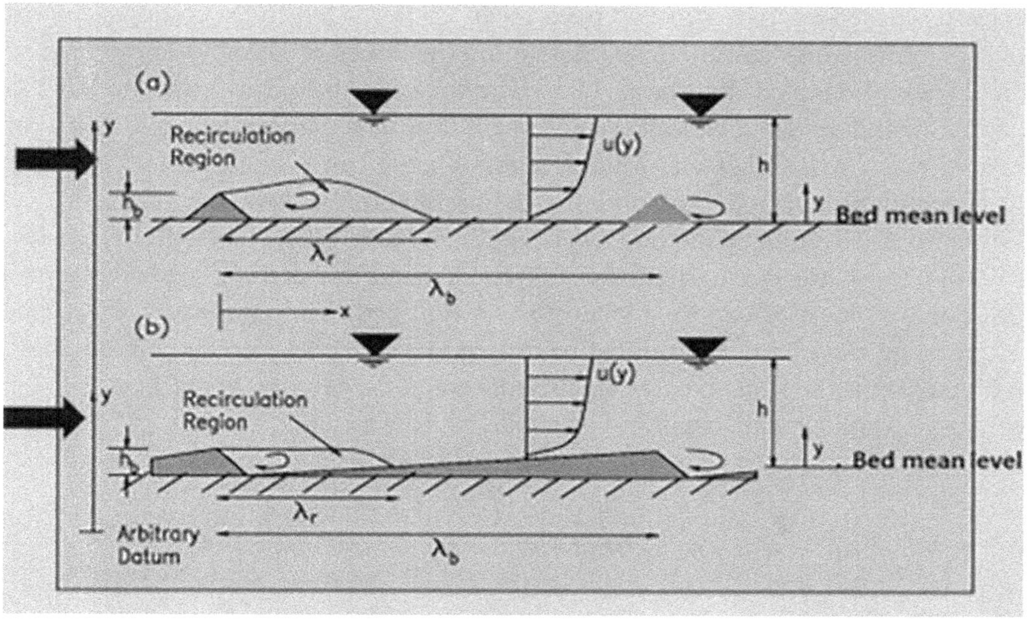

FIGURE 10.22 Schematic of artificial bedforms: (a) First type of bedform with triangular element of amplitude h_b, wavelength λ_b, recirculation length λ_r and (b) Second type of bedform with triangular element of closer to the real dune. (Modified from Lyn, 1993.)

often exceed 0.1, while Fredsoe (1982) supported theoretical predictions that maximum h_b/λ_b would be limited to ≈ 0.06 with experimental evidence from Yalin (1972). The amplitude and wavelength of the artificial bedforms were there chosen to give an intermediate value $h_b/\lambda_b = 0.08$. Flow depth for all three experiments was kept constant at $h = 6.1$ cm with a constant width-to-depth ratio of 4.4. The ratios of flow development length from channel entrance to measurement point to flow depth h and to λ_b were ≈ 150 and ≈ 60 respectively, such that the fully developed channel flow at the measuring location was achieved. Details are available in Lyn (1993).

Nelson et al. (1993) studied the flow measurements over fixed two-dimensional bedforms, using the laser Doppler velocity to investigate the coupling between the mean flow and turbulence, and to examine the effects producing the bedform instability and finite amplitude stability. The coupling between the mean flow and turbulence is explored in both spatially averaged sense by determining the structure of spatially averaged velocity and Reynolds stress profiles, and a local sense through computation of eddy viscosities and length scales. The measurements showed that there is significant interaction between the internal boundary layer and overlying wake turbulence produced by separation at the bedform crest. This study provides a clear picture of the local and spatially averaged flow over the bedforms and has significant implications concerning the interaction between the flow and the erodible bed with respect to the initial bedform instability and finite amplitude stability. Bennett and Best (1995) performed the measurements of flow velocity and turbulent fluctuations over fixed two-dimensional dunes in laboratory flume, using the laser Doppler anemometry. They measured detailed experimental data of the mean flow and turbulence fields over fixed, two-dimensional dunes in order to investigate and document the origins of dune-related micro turbulence, and to use these results to shed light upon the mechanisms of sediment transport and origin of dunes. The flow over the dunes is also assessed by higher-order statistical moments and quadrant analysis. The quadrant-2 i.e., ejection events are concentrated along the shear layer with the occurrence of Kelvin–Helmholtz instabilities and dominate the contribution to the local Reynolds stress; and the quadrant-4 of high magnitude and high frequency occur in the separation zone near reattachment beyond the dune crest. These two events are significant contributors to the local Reynolds stress at each location. From their observations, they claimed that the origin of dune-related macro-turbulence lies in the dynamics of shear layer rather than the turbulence bursting. They also stated that the Kelvin-Helmholtz instabilities control the local flow and the turbulence structure and dictate the modes of sediment transport entrainment and their transport rates. The dune formation, growth and extent of the separation-zone free shear layer and its instabilities may control the differentiation of ripples and dunes in the unidirectional flow.

Karim (1999) proposed a unique method to predict the relative bedform height in alluvial flow, based on the concept of energy loss due to the form drag to the head loss across the sudden expansion in the channel. The proposed method can be applied to different bed configurations such as ripples, dunes, anti-dunes or standing waves and transitional bed regime, which occur in the alluvial flows. A relation for bedform dimension is developed, based on the concept of energy loss h_f produced by form-drag and the head loss across a sudden expansion in open channel flow as:

$$h_f = \frac{K}{2g}\left[\left(\frac{2q}{2h-h_b}\right)^2 - \left(\frac{2q}{2h+(1-2\beta)h_b}\right)^2\right] \tag{10.108}$$

where K is the loss coefficient, g is the acceleration due to gravity, q is the flow per unit width, h is the flow depth, h_b is the bedform height, and β is the parameter defining the point of flow reattachment on the stoss face of a bedform. Derivation of equation (10.108) closely follows the approach of Kennedy and Odgaard (1991) that used a similar concept. Rearrangement of equation (10.108) yields the following:

$$h_f = KFr^2 h_b C_1 \tag{10.109}$$

where Fr is the Froude number, and C_1 is the dimensionless parameter given by:

$$C_1 = (1-\beta)\left(1-\frac{\beta h_b}{2h}\right)\left[1-\frac{h_b^2}{h^2}(0.25-0.5\beta)-\beta\frac{h_b}{h}\right]^{-2} \tag{10.110}$$

Now $h_f = S''\lambda_b$, the energy slope due to form drag S'' can be written as:

$$S'' = KFr^2 C_1\left(\frac{h_b}{\lambda_b}\right) \tag{10.111}$$

Finally, the relative bedform height

$$\frac{h_b}{h} = \frac{1}{KFr^2 C_1}\left(S - f_1\frac{Fr^2}{8}\right)\frac{\lambda_b}{h} \tag{10.112}$$

where f_1 is the Darcy-Weisbach friction factor due to grain roughness only. This equation requires estimating the four quantities: f_1, λ_b, K, C_1, and other parameters are known. The parameter K is known from Shen et al. (1990) as

$$K = 0.44\left(\frac{h_b}{h}\right)^{0.375} \tag{10.113}$$

It may be noted that according to equation (10.113). K varies from 0.18 to 0.30 for h_b/h ranging from 0.10 to 0.35 respectively. The average value of β is equal to 0.22 was suggested by Haque and Mahmood (1985). A modification of K is sought based on the followings: (1) K should a function of h_b/h, bedform steepness h_b/λ_b or h_b/h and λ_b/h; and (2) K should be reduced for antidunes because energy loss due to separation of flow. Therefore, based on these points, the following expressions of K are obtained:

$$K = 0.55\left(\frac{h_b}{h}\right)^{0.375}\left(\frac{\lambda_b}{h}\right)^{-0.020} \tag{10.114a}$$

for ripples, dunes and transition,

$$K = 0.10\left(\frac{h_b}{h}\right)^{0.375}\left(\frac{\lambda_b}{h}\right)^{-0.020} \tag{10.114b}$$

for antidunes or standing waves.

Equation (10.113) was found to work well for ripples and dunes, but it was inadequate for transition regime, when λ_b/h increases rapidly. The dimensionless parameter C_1 varies between 0.8 and 0.9. The parameter f_1 is obtained from the Manning and Strickler relations, which is Van Rijn's relation (1984) as:

$$f_1 = 0.135\left(\frac{d_{50}}{h}\right)^{0.33} \tag{10.115}$$

Introducing the parameter f_1 from equation (10.115), $C_1 = 0.85$ and K from equation (10.114), equation (10.112) is transformed as:

$$\frac{h_b}{h} = \left[\frac{\left\{S - 0.0168Fr^2\left(\frac{d_{50}}{h}\right)^{0.33}\right\}\left(\frac{\lambda_b}{h}\right)^{1.20}}{0.47Fr^2}\right]^{0.73} \tag{10.116a}$$

for ripples, dunes, and transition, and

$$\frac{h_b}{h} = \left[\frac{\left\{S - 0.0168Fr^2\left(\frac{d_{50}}{h}\right)^{0.33}\right\}\left(\frac{\lambda_b}{h}\right)^{1.20}}{0.085Fr^2}\right]^{0.73} \tag{10.116b}$$

for anti-dunes or standing waves.

Equation (10.116) can be solved directly for h_b/h using appropriate relations for λ_b/h depending on the bedform type.

A two-dimensional numerical model, based on the solution of the Reynolds-averaged Navier-Stokes equations and the k-ω turbulence model, was developed by Yoon and Patel (1996) which takes into account sand-grain roughness, over the fixed dune-bed channel. The model predicts the velocity and turbulence characteristics, as well as the pressure and friction distributions along the dune. Details of flow separation and reattachment points are discussed.

Cellino and Graf (2000) studied experimentally the influence of bedforms in open channel flow of suspended sediment concentration, using the sonar instrument for measuring the concentration distribution at one single desired section. Using this sonar instrument, one can determine the instantaneous velocity profiles using back-scattering echo signals (Lhermitte and Lemmin, 1994) and concentration profile at single section (Shen and Lemmin, 1996) as indicative of sediment concentration. They studied the ratio of sediment diffusion coefficient ε_s to the momentum diffusion coefficient ε_m from the measured data, which appeared in the Rouse equation. They observed that due to the presence of bedforms the ratio for suspension flow appears to be greater than the one comparable suspension flow over a plane bed. They also studied the evolution of flow structures along the bedforms, and it is observed that there is a flow separating shear layer and recirculation region behind the bedform crest. They also observed that there are clear peaks in the intensities and shear stress near the boundary of the bedform. The vertical distributions of dimensionless mean and fluctuating concentration are comparable to the ones measured in plane bed suspension flow. The dimensionless mean and fluctuating concentration profiles have their maximum values close to the bed $(y/h < 0.25)$; in the upper part of the flow $(y/h > 0.25)$ the fluctuating concentrations decrease quite rapidly.

From the series of experimental studies over the bedforms in circulating open channel, Mao (2003) investigated the role of turbulent bursting in the movement of suspended sediment particles, when the bedforms are plane surface of different roughness values and of simplified sand waves. He observed that due to the separation of flow at the crest of sand wave, the flow field can be divided into the free turbulent area above the crest and trough turbulent area below the crest. In the free turbulent area, the shear mixing layers are not only present but also the typical turbulent bursting phenomenon is characterized by ejection and sweeping. The sediment particles were pushed over the crest moved further into the suspension. In the trough region, the flow may be separated at the upstream slope and a separation vortex forms near the lee slope. He also observed that there are two flow separation points near the sand bed. The first flow separation point is at the crest and the second separation point is on the next slope. A part of the flow attaches to the next upstream slope forming a separation vortex in the trough, and the other part of the flow attaches behind the second separation point and pushes the sediment particles as bed load towards the next crest. The turbulent coherent structures in the free turbulent area and the trough turbulent area are quite different, and their interaction can result from the so-called Kolk-boil phenomenon. Lots of sediment particles at the trough are ejected by the Kolk-boils to the flow surface. It may be concluded that the interaction of bedforms and bed roughness with turbulent bursting phenomenon plays a key role in the flow and sand movement in alluvial rivers.

Kostaschuk et al. (2005) measured the high-resolution, three-dimensional velocity profiles over the dunes in the Fraser River estuary-Canada, using the acoustic Doppler profiler (ADP) from a moving launch to analyze the flow structures and Reynolds stresses. Quadratic stress models have been recommended for shear stress to estimate the skin friction and form stress, while the law of wall only estimates total stress over dunes. There was limitation of ADP in beam geometry that the ADP beams can encounter the bed at different depth over dunes, so there was a contamination of velocity measurements in some of the lowest bins, especially over steeply sloping dune leesides.

Venditti et al. (2005) investigated the morpho-dynamics of small-scale sand waves superimposed over large-scale migrating dune bedforms. The kinematics and morphology of low amplitude, small-scale sand waves formed over the migrating dunes were studied using the observed data from the laboratory experiments. The natural character of bedforms in alluvial channels has been documented by Jackson (1975). It is known from the existing literature that two, three or more distinct scales of bedforms may occur in the same flow system (Carey and Keller, 1957). In fact, ripples and dunes are frequently superimposed on bars, and ripples are superimposed on dunes (Guy et al., 1966; Jackson, 1976), and also smaller dunes are superimposed on larger dunes (Allen and Collinson, 1974; Jackson, 1976). Even bed load sheets are also found superimposed on bar forms (Bennett and Bridge, 1995a) and dunes. The small-scale superimposed sand waves are referred to as 'sand sheets' because these features could not be classified easily as ripples, dunes or bars. The sand sheet thickness was 10% of the height of the dune on which they were superimposed, they migrated at 8–10 times the dune rate; they had nearly constant length over full range of dune lengths and flow conditions; and they had the aspect ratio of about 0.025. It is important to note that dunes and sand sheets represent distinct scales of sediment transport with different migration rates. In fact, the superimposed sand waves on dunes are often considered simply as an additional roughness element, but such bed waves are the agency responsible for the movement of dune bedforms downstream.

An excellent review on the features of mean flow, turbulence characteristics, morphology and sediment transport associated with river dunes by Best (2005) and he highlighted the future directions of research on predicting the flow resistance, sediment transport and deposition in rivers. The progress of understanding the fluid dynamics associated with alluvial sand dunes has significant importance in determining the principal features of mean and turbulent flows in the field, laboratory and numerical simulation. The review paper is highlighted for five key areas of research in alluvial channels as: (1) influence of dune leeside angle on the flow downstream, (2) the influence of three-dimensionality of dune shape on the generation of turbulence and hence bed shear stress, (3) perturbation of flow field from the bedform superposition, (4) the scale and topology of dune related turbulence and its interactions with sediment transport and flow surface, and (5) influence of suspended sediment on the dune flow and dune morphology.

Flow over the river dunes, which are asymmetric in cross-sectional form, possess an angle-of-repose leeside in a steady, uniform unidirectional flow, can be briefed as five regions: (1) a region of flow separation in the leeside of the dune with occurrence of reattachment point nearly 4–6 dune heights downstream of the crest, (2) a shear layer is generated covering the separation region, which divides the re-circulating flow from the free stream fluid above; and a large-scale turbulence generates in the form of Kelvin- Helmholtz instabilities along this shear layer, and at the same time the free shear layer expands, and creates a wake region which grows and dissipates downstream (Figure 10.23). (3) A region of expanding flow in the consecutive dune leeside. (4) Downstream of the reattachment point a new internal boundary layer develops over the stoss and forms a log-law velocity profile. (5) Maximum horizontal velocity occurs over the dune crest because of reduction of cross section and pressure over the dune crest, and bed shear stresses here are sufficient to generate upper-stage plane bed condition. The differential pressures, generated due to flow separation and acceleration/deceleration associated with the dune-form generate a net force on the dune known as form drag, and this form drag in addition to the grain roughness drag generate dune morphology. The turbulent flow field over dunes is critical in determining instantaneous bed shear stresses, bed load transport rate and sediment suspension, and bed surface and water surface topography.

The quadrant analysis from two horizontal and vertical velocity components at a point shows the four turbulent events such as outward interaction, ejection, inward interaction and sweep in respective four

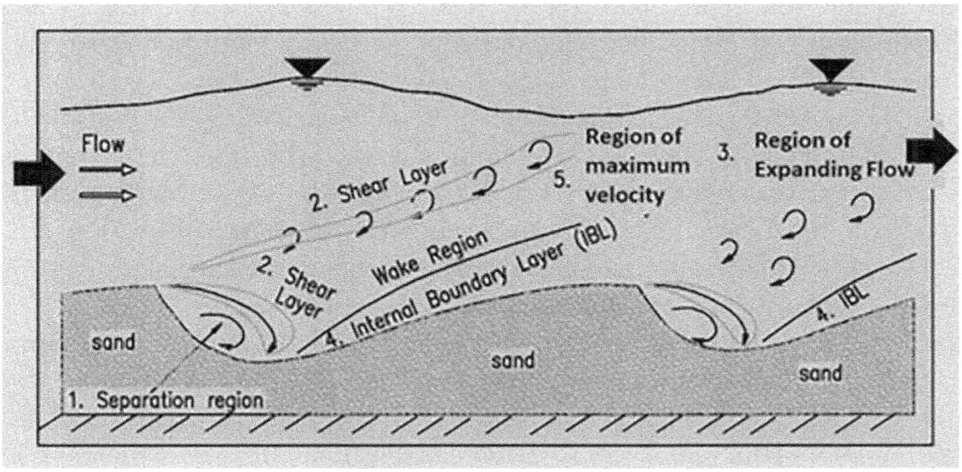

FIGURE 10.23 Sketch of principal regions of flow over asymmetric dunes with circulations. (Modified from Best, 2005.)

quadrants. Positive vertical velocity indicates the flow upward and away from the bed. The quadrant signatures associated with the dunes are described by Nelson et al. (1995), Bennett and Best (1995), Ojha and Mazumder (2008), Maity and Mazumder (2012, 2014, 2017). Large-scale vorticity generated due to Kelvin-Helmholtz instabilities is marked as ejection event (quadrant 2) and arises from the shear layer and reattachment point. These coherent flow structures are advected with the mean flow, often reaching the flow surface and blowing up as 'surface boils' (Jackson, 1975; Best, 2005), which may contain higher amount of concentration of sediment than the surroundings (Venditti and Bennett, 2000). In the issue of three-dimensional dune morphology, Venditti et al. (2005) contented that all the dune bedforms must eventually become three-dimensional due to even minor changes of sand over the bedforms passing from one to another. It is important to note that the crest of the dune can influence the length of separation region and thus influence the leeside Reynolds stresses, drag coefficients and dispersal pattern of sediment. A detailed study of flow over 3D dunes was made by Maddux et al. (2003a, b) in which the dunes were 2D across the flume width, but the height of the crest varied. The work demonstrated that the maximum stream-wise velocities were highest over the nodes rather than maxima in the crest line height, that the turbulence over these 3D dunes was lower than over their 2D counterparts, and that the sinuous crest lines were associated with the presence of secondary currents that were responsible for a large percentage of the momentum flux over the dune. Maddux et al. (2003a, b) also showed that the friction coefficients of 3D dunes were greater than those of 2D counterparts, but the turbulence generated by 3D dunes was weaker than that of 2D case due to the secondary current over the 3D forms. They also argued that form-induced stresses were recognized from the secondary currents and that these augmented the low Reynolds stresses present over the 3D dunes.

The eruption of dune-related turbulence on the flow surface is commonly observed in various natural river channels, which creates large-scale upwellings and boils. Few research studies in laboratory (Nez and Nakagawa, 1993; Best, 2005; Ojha and Mazumder, 2008) and in field (Kostaschuk and Church, 1993; Clifford et al., 1993) have addressed the link between the vortices and dunes to study the coherent flow structures and how they interact with the flow surface. The role of such dune-related large-scale turbulence in the transport of sediment has been addressed by Kostaschuk and Church (1993), Venditti and Bennett (2000), and Cellino and Graf (2000), and applied to several studies of dune formation and stability (Jackson, 1976; Bennett and Best, 1995; Ojha and Mazumder, 2008). It was noticed that the size and intensity of the upwelling was related to the height of dunes and form roughness, and that the generation of turbulence may increase the distinct formation of scour pits. Several studies have shown the plots of boil periodicity versus relative roughness of dunes (Figure 10.24), where relative roughness = dune height/flow depth. Once the relative roughness exceeded 0.2 (i.e., dune height > 0.2 time of flow depth) then boil period became constant, suggesting more frequent boils at this higher relative roughness.

Babakaiff and Hickin (1996) noticed that the chaotic and violently erupting boils which may termed as '*cauliflower*' structure due to upwelling, were only present, when the relative roughness exceeded 0.17. Several studies (Kadota and Nezu, 1999; Le Couturier et al., 2000; Mazumder, 2000) proposed that the large-scale turbulent event takes the form of a loop or horseshoe vortex (Figure 10.25). Kadota and Nezu (1999) suggested that the vortices generated along the separation zone have a slightly different morphology to those that arise at the reattachment region (Figure 10.25).

It is important to note that the morphology of this large-scale turbulence erupts on the water surface (Jackson, 1975; Antonia and Bisset, 1990; Babakaiff and Hickin, 1996), and the recent numerical simulations illustrate the possible internal motions within these rising vortices (Yue et al., 2003, 2005a, b; Hayashi et al., 2003). In addition, recent research has illustrated the key interactions that may occur between vortex rings with a free surface (Rashidi and Banerjee, 1988; Kumar et al., 1999) and how this influences the dynamics of the vorticity. Best (2005) summarized this information and observations of

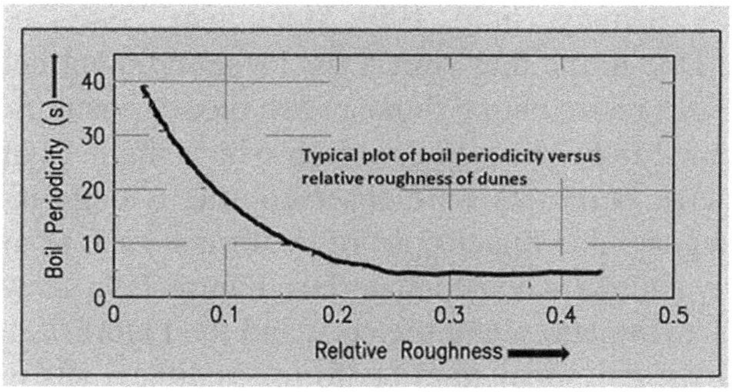

FIGURE 10.24 Schematic of boil periodicity against relative roughness of dunes. (Modified from Babakaiff and Hickin, 1996.)

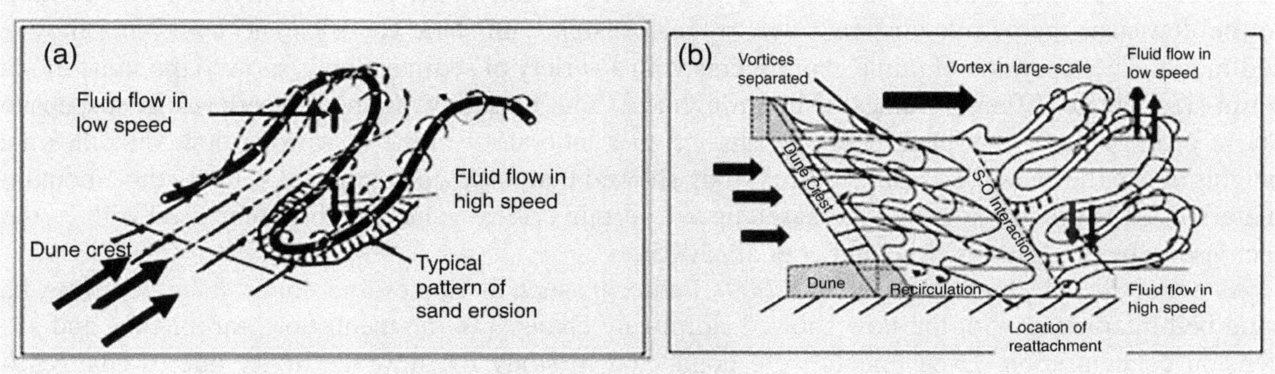

FIGURE 10.25 Sketch of vortex topology associated with dunes. (Modified from Muller and Gyr (1983, 1986), Nezu and Nakagawa (1993), Best (2005).)

the water surface above dune-covered beds to propose a schematic model for the interaction of dune-related turbulence with the flow surface. He also argued that this upwelling and flow surface interaction must induce subsequent downwelling toward the bed in order to satisfy flow continuity.

Using large eddy simulation of the free surface, Yue et al. (2003, 2005a, b) investigated the origins of dune-related turbulence and the vector patterns of flow as vortices approached the water surface. These simulations examined flow over a 0.02 m high dune in flow depths of 0.065 and 0.132 m (relative roughness 0.31 and 0.15, respectively), and with a fixed and deformable free surface, and beautifully illustrate the likely form of vorticity and flow surface interactions. Several key points emerge such as the flow over the dune shows a distinct separation region and the vorticity downstream of the dune is dominated by quadrant 2 (ejection) and 4 (sweep) with Q2 upwelling to the flow surface and inrush of Q4 events toward the bed following Q2 ejection. Near the free surface, the flow is characterized by a series of upwelling and down-welling which are associated with water surface elevation and depressions respectively in the free surface (for details, please see Best, 2005).

10.10.2 Laboratory Flume Studies

Turbulent flow over a plane loose sand bed may cause the bed to deform, which in turn introduces perturbations on the main flow. The effect of the perturbed bed on a turbulent flow is important in determining the formation of bedforms in erodible channels. Several investigations had been addressed under controlled conditions to study the sediment suspension, bed load movement and the influence of bed roughness on suspension over flat sand beds of heterogeneous composition (Ghosh and Mazumder,

1981; Ghosh et al., 1986, 1991; Sengupta et al., 1991; Mazumder, 1994; Ghoshal and Mazumder, 2005; Mazumder et al., 2005a, b). Due to the flow over a flat loose sediment bed (fine sand, coarse sand or mixture of sand and gravel) in a recirculating flume, when the sediment transport started, the different types of bedforms were seemed to be generated, among which single asymmetric, single *nearly* symmetric and series of asymmetric bedforms were observed after a long time of run, when reached at a nearly fully equilibrium category. The dimensions of bedforms, such as amplitudes and wavelengths, depended on the flow velocity and the sand bed materials. Figure 7.21a shows the waveform of a scalene triangular shape (STS) with asymmetry about the crest and the Figure 7.21b shows *nearly* an isosceles triangular shape (ITS). With the development of bedforms, sediment jets were ejected out of the ripple crests and dispersed into the main flow of suspension. This process continued around the whole channel, so that the materials into the suspension came from the whole bed throughout the flume. The bed roughness played an important role because it influenced the flow velocity, suspension concentration and sediment pick-up process. Movement of sediment particles in the bed layer experienced resistance to the flow near the bed downstream due to the collision with static particles and with other moving sediments. The repetition of flume experiments with a variety of sediment beds showed the same modal grain-size, but of different values of bed roughness. The nature of grain-size patterns in suspension above each of these sediment beds was studied in a laboratory flume at various flow velocities and heights above the beds. The results of this study showed that for fixed height and velocity the amount of material (g/L) in suspension first increased up to a certain critical value and then decreased with further increase in bed roughness (Mazumder et al., 2005).

As per studies by Mazumder et al. (2009), bedforms such as ripples and dunes generated from flat sand bed interacting with the flow showed significant changes in the mean flow, turbulence, and suspension concentration. Experiments were conducted to study the flow resistivity due to changes of turbulent flow generated by the bedforms. In fact, due to the complexity of flow structures over moving bedforms in alluvial channels or riverine environments, some of the above-mentioned research considered suspension-free flow over static bedforms (Yoon and Patel, 1996; Venditti and Bennett, 2000; Jerolmack and Mohrig, 2005). The measured velocity data were restricted only to the mean flow for a period of passage of bedforms. In computing, the suspended sediments in relation to the flow parameters and bed material, the precise effect of bedforms associated with the flow separation, wakes, reattachment, and the perturbed shear layer in the turbulent flow responsible for sediment transport is yet unclear. Therefore, the deviations of mean velocity, turbulence, and the contributions of burst-sweep cycles to the total Reynolds shear stress due to the presence of bed topography required a substantial investigation.

Mazumder et al. (2009) investigated the effect of two-dimensional *static* bedform structures on the mean flow, turbulence, and fractional contributions of turbulent bursting events to the Reynolds shear stress. For such fixed bedform structure, no distinction is made between ripples and dunes. The approximation of a static two-dimensional (2D) bedform is well justified because the speed of the moving bed is small when compared with the mean flow. The mobility of bedforms or dunes governs turbulence near the boundary and hence the transition of the bedforms. Although the fixed structure and the surface roughness are not representative of complex bed geometry in the laboratory as well as in nature, the study over the fixed isolated structure/obstacle provides a better picture of turbulence without added difficulties due to a mobile bed. Kostaschuk and Villard (1996) observed two types of dunes, namely symmetric and asymmetric in the Fraser River estuary, Canada, where the velocities on the lee-sides of both the asymmetric and symmetric dunes were directed downstream with no evidence of flow reversal. However, small attention was paid to the turbulence statistics due to these types of bed topographic structures. A laboratory study was performed to understand the flow behavior over the static asymmetric and symmetric waveform structures under identical inflow conditions. This study identified the spatial changes of mean flow and turbulent events associated with the burst-sweep cycle over two types of isolated static waveforms; and facilitated a more detailed comparison of these two types, which are crucial for the process of sediment transport and grain sorting (Mazumder et al., 2009). Two configurations of single waveform geometry (asymmetric or symmetric) were used to compare the turbulence over two different types of structures in Figure 10.26a and b.

The first type (a) was identified as STS (Scalene triangular shape) having a mild slope of 6∞ for the upstream face and a sharply sloping downstream face with lee-side angle of 50°. The second type (b), identified as ITS (Isosceles triangular shape) having equal slopes of 11° for the stoss and lee faces with the same crest height and length as first type, was placed at the same measuring station for the second set of tests in the flume (Figure 7.13). Because of identical dimensions, these two types of structures may be considered as an example of a class of waveforms differing only in geometry due to shifting of crest point from the scalene to the isosceles triangular shape. Both shapes morphologically resemble real dune structure (Kostaschuk and Villard, 1996). The ratio of crest height h_b to wavelength λ_b for both shapes is $h_b/\lambda_b = 0.1$, which is consistent with real dunes (Gabel, 1993; Julien and Klaassen, 1995). The vertical velocity profiles from the upstream to downstream along the flume centerline were measured using a SonTek 5 cm down-looking 3D Micro-acoustic Doppler velocimeter (ADV) for 5 minutes at a sampling rate of 40 Hz to ensure full characterization of the turbulence phenomena. The velocity data were collected with the lowest point in each profile is being 0.35 cm above the flume bed and the highest point is being 24 cm for each profile. For both cases, the experiments were performed at a discharge of $Q = 0.028$ m^{3}s, for a flow Reynolds number $R = u_m h/\nu_f = 1.50 \times 10^5$ or shear Reynolds number $Re_* = u_* h/\nu_f = 7560$ in which $u_m = 0.50$ m/s is the maximum velocity observed at height $y = 0.24$ m above the bed, $u_* = 0.0252$ m/s is the shear velocity over the flat surface obtained from the log-law. Here, interestingly, different flow patterns are observed for each case. Significant flow separation, sudden expansion and flow reversal occur just downstream of the crest of STS and it persists for a longer distance than for ITS. In the recirculation region of STS, the measured velocity profiles are generally represented as

$$\frac{u}{u_{*s}} = \alpha\left(\ln\frac{y}{h} + \beta\right)^{\frac{1}{3}} + \gamma \tag{10.117}$$

where the parameters α, β and γ are determined from the best fit, and u_{*s} is spatially-averaged friction velocity. It may be noted that the local values of shear velocity u_* are estimated by extrapolating the Reynolds shear stress profiles down to the flat/dune surface. The value of u_{*s} is the mean of u_*

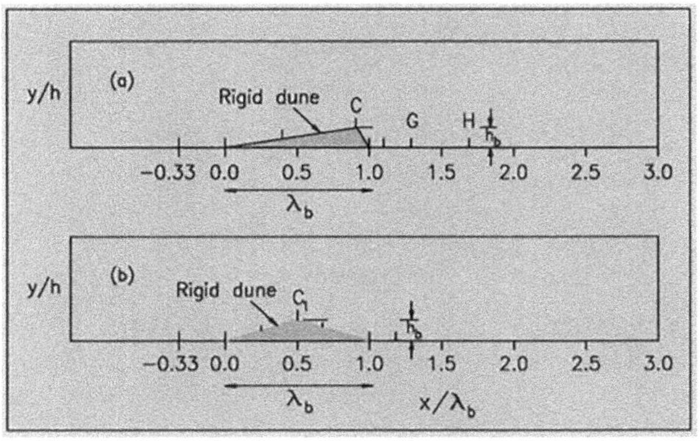

FIGURE 10.26 Two different configurations of single waveform geometry: (a) asymmetric, and (b) symmetric dunes. (Modified from Mazumder et al., 2009.)

values. As the shear stress and the local shear velocity over the obstacles vary along the locations, u_{*s} is used for overall scaling. It is important to note that $u_{*s} = 1.23$ cm/s for STS, and 1.29 cm/s for ITS. It appears further that u increases up to crests (C and C_1) for both cases with the maximum velocity at the dune crests. Interestingly, different flow patterns are observed for each case. Significant flow separation, sudden expansion and flow reversal occurred just downstream of the crest of STS and it persisted for a longer distance than for ITS. In the recirculation region downstream of the crest, three distinct layers are observed: (1) internal layer, $y/h \leq 0.05$, (2) shear layer $0.05 \leq y/h \leq 0.15$, and (3) outer flow region $y/h \geq 0.15$, illustrating the dominant effect of separation. The mean velocity is strongly sheared beyond the crest and has an inflexion point at the crest level. It is evident that the waveform structure with a sharply sloping lee-face has a thicker separation bubble in the recirculation region. It is also observed that the velocity profiles along the flow are almost unaffected above the crest level ($y/h \geq 0.15$). The mean velocity profiles gradually become fully developed farther downstream from the recirculation region, and collapse in the outer flow region, indicating that the mean flow recovers from the perturbation tending to the log-law.

The intensities of turbulence (σ_u, σ_v) in the streamwise and vertical directions were computed for both STS and ITS along stream wise locations. It is observed that all turbulence intensity profiles exhibit the local flow conditions at different positions relative to the structural geometry. The turbulence intensity within the boundary layer develops a strong peak at the onset of the adverse pressure gradient along the shear layer originating at the crest. Note that the zones of high turbulence intensity are characterized by low mean velocity. As it proceeds downstream, the turbulence intensity decreases, which leads to sedimentation. The turbulence intensity within the boundary layer for ITS develops a weaker peak over the mildly-sloping lee-face and decays slowly. The turbulence intensity σ_u/u_{*s} for ITS has a similar trend except at the decelerating region beyond the crest. In this region, two-layer empirical relations of turbulence intensity were proposed as:

$$\frac{\sigma_u}{u_{*s}} = 3.6 \exp\left(7.83\frac{y}{h}\right), \text{ for } 0.014 \leq \frac{y}{h} < 0.12 \tag{10.118a}$$

$$= \frac{2.7\left(\frac{y}{h}\right)}{\frac{y}{h} - 0.06}, \text{ for } 0.12 \leq \frac{y}{h} < 0.80 \tag{10.118b}$$

The relation for σ_v/u_{*s} developed by Nezu and Rodi (1986) does not agree with the measured data even at the flat bed surface. For STS, the revised relation at the flat bed surface is

$$\frac{\sigma_v}{u_{*s}} = 1.20\left[1 - \exp\left(-73.33\frac{y}{h}\right)\right] \tag{10.119}$$

which fitted well with the present data. For the recirculation region, a two-layer empirical relation was proposed as:

$$\frac{\sigma_v}{u_{*s}} = 3.20\exp\left[-\frac{\left(\frac{y}{h} - 0.50\right)^2}{0.10}\right], \text{ for } 0.014 \leq \frac{y}{h} < 0.17 \tag{10.120a}$$

$$= 1.40 \exp\left(-0.27\frac{y}{h}\right), \text{ for } 0.17 \leq \frac{y}{h} < 0.80 \tag{10.120b}$$

Moreover, the two-layer relation for σ_v / u_{*s} seems to agree reasonably well with observations at points *G* and *H*. A comparative study indicates that this component in the recirculation region for STS is much larger than for ITS. Both components of turbulence intensity are dominant in the recirculation region, resulting in erosion and sediment suspension. The dimensionless Reynolds shear stress component τ_{dim} is shown against *y/h* for all streamwise locations in Figure 10.27a for STS and in Figure 10.27b for ITS. At the flat surface, the qualitative behavior of τ_{dim} agrees reasonably well with the data of Nezu and Rodi (1986), but they did not provide any theoretical model to estimate the turbulent shear stress. For the flat surface, Lyn (1993) proposed for the turbulent shear stress in open channel flow as

$$\tau_{\text{dim}} = -\frac{\overline{u'v'}}{u_{*s}^2} = \left(1 - \frac{y}{h}\right) - a_1 \operatorname{erf} c\left[\frac{a_2(y - h')}{h}\right] \tag{10.121}$$

where a_1 and a_2 are empirical constants determined from the test data. For $a_1 \neq 0$, the second term in equation (10.121) likely arises from the secondary circulation. Here, $a_1 > 0$ for downward flow, which reduces the shear stress, and whereas $a_1 < 0$ for upward flow leading to an increased shear stress. Furthermore, Lyn's equation (10.121) does not match with the present data in the recirculation region. Instead, the measured turbulent shear stress τ_{dim} throughout the depth over both STS and ITS for all locations can be represented as

$$\tau_{\text{dim}} = \frac{a}{\sqrt{2\pi}\left(\frac{y}{h}\right)c} \exp\left[-\frac{1}{2}\left(\frac{\ln\left(\frac{y}{h}\right) - b}{c}\right)^2\right] \tag{10.122}$$

where *a*, *b* and *c* are parameters determined from the data along each vertical profile. Equation (10.122) along the stream-wise direction agrees reasonably well for both cases. Note that equation (10.122) upon division by the constant *a*, is a lognormal probability density function. Parameters *b* and *c* can be assumed to be the mean and the standard deviation of the distribution, respectively, and the parameters are directly related to the characteristics of the stress profile. Here *b* and *c* indicate, respectively, the occurrence of maximum and thickness of the region of large stress, whereas the constant *a* represents the magnitude of the maximum stress. The values of the parameters *a*, *b* and *c* are plotted against x/λ_b over both STS and ITS in Figure 10.28a; and the maximum shear stress is shown in Figure 10.28b. It is observed that the values of the constant *a* are almost identical along the flow over both STS and ITS.

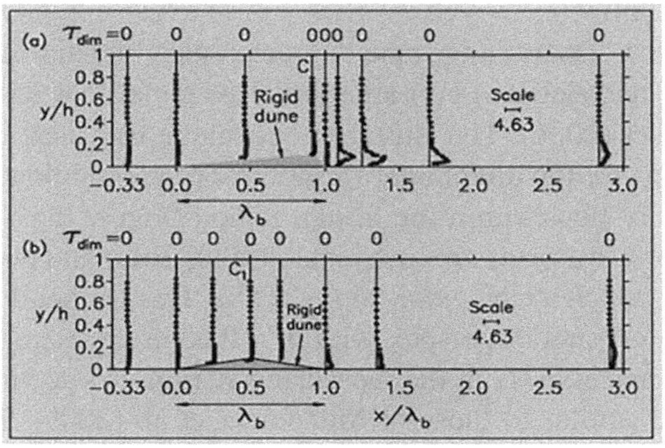

FIGURE 10.27 Plots of τ_{dim} vs. x/λ_b: (a) for STS; (b) for ITS with symbol (.........) represents Lyn's equations (10.121) and (10.122). (Modified from Mazumder et al., 2009.)

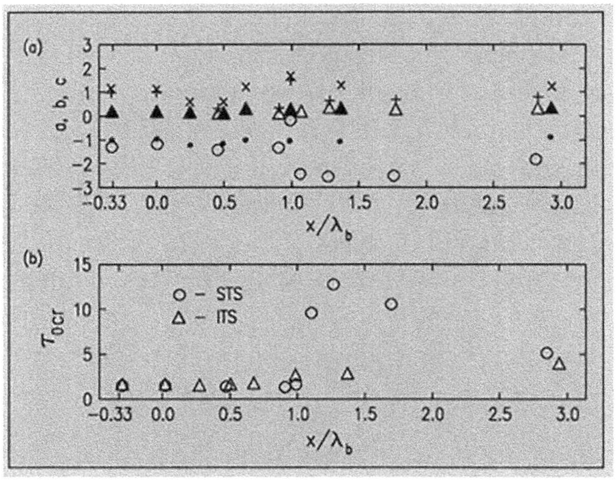

FIGURE 10.28 (a) Profiles of a, b, c (equation 10.122) vs. x/λ_b; Symbols for STS: Δ for a, ∘ for b, + for c; for ITS: solid Δ for a, • for b, × for c. (b) $(\tau_{\mathrm{dim}})_{\max}$ versus $x/\lambda_b a$. (Modified from Mazumder et al., 2009.)

Note that the maximum shear stress for STS is much higher than that of ITS in the recirculation region. The values of b for both cases are almost the same up to $x/\lambda_b = 0.5$, but then b of STS reduces drastically beyond the crest and persists for a longer distance. The values of c show periodicity along the horizontal direction for both STS and ITS. The plotted parameters are related to the production, advection and diffusion of Reynolds shear stress over these two bedforms and also elucidate the similarities and differences in two bedforms.

The conditional statistics of Reynolds stress are also examined for these two types of bedforms: STS and ITS. The Reynolds shear stress $-\rho u'v'$ is sought to examine the turbulence over the structures with respect to the type of turbulent events. Based on qualitative results obtained from the flow visualization on the Reynolds shear stress production, Lu and Willmarth (1973) and Clifford et al. (1993) attempted to determine more quantitative results about the structure of Reynolds shear stress. Keirsbulck et al. (2002) demonstrated that ejection events are predominant in the near-wall region for a smooth wall, which agrees well with the present results. A laboratory study on separated and un-separated flows over the asymmetric and symmetric waveforms are required to examine this result. It was observed that just downstream from the crests, the relative importance of contributions of turbulent events to the Reynolds shear stress production for STS was much higher than that of ITS. The frequency of occurrence of ejection and sweep events beyond the crest of ITS is much higher than that at trough point of STS. This is due to the symmetric waveform with a mild expansion and a low lee-side slope angle, reducing the potential of flow separation. Some characteristics of fluid motion and related turbulent bursting phenomena (sweep and ejection) over an isolated asymmetric waveform structure were studied by Mazumder and Mazumder (2006). The clustering technique was used to characterize the randomness of bivariate distributions for the directional variations of the fluid flow.

The turbulence statistics of flow within the trough region (Figure 10.29) formed by a pair of *scalene triangular-shape* (STS) waveforms are investigated (Pal, 2011) and compared the results with the same formed by a pair of *isosceles triangular-shape* (ITS). Two artificial 2-D asymmetric waveform of similar dimension made of smooth perspex with $h_b = 0.03$ m crest height, $\lambda_b = 0.30$ m length and 0.50 m width, were placed successively at the measuring station 7 m downstream of the channel inlet (source). The structures are similar to those of Mazumder et al. (2009). The velocity data were collected using 3-D acoustic Doppler velocimeter (ADV). The velocity data were analyzed to determine the relative importance of mean flows, Reynolds stresses, turbulent kinetic energy and the contributions of burst-sweep cycles over the trough regions. Quadrant decomposition has also been analyzed for the

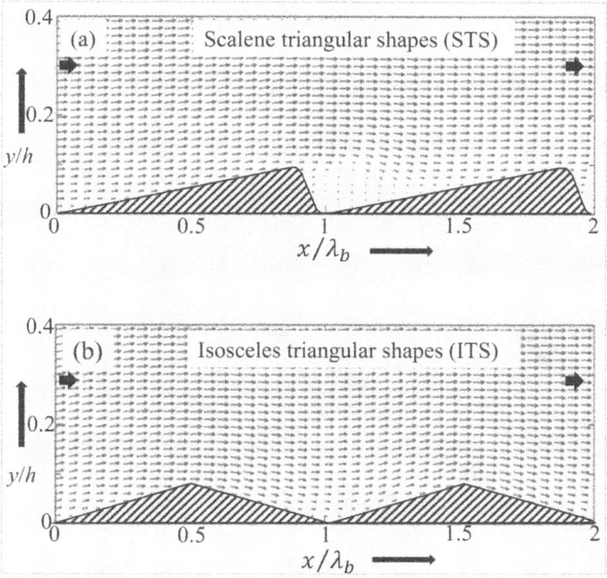

FIGURE 10.29 Mean velocity vector plots over waveform structures: (a) Profiles along the flow over the STS case, (b) Profiles along the flow over ITS case. Flow direction from left to right. x/λ_b is the dimensionless distance along the flow and *y*/*h* is the dimensionless vertical height.

relative importance of bursting phenomena in the near-bed region over the trough regions. The relative importance of contributions of turbulent events to the Reynolds shear stress production for double STS was much higher than that of double ITS just after the crests. The different circulation patterns characterizing fluid flow in the trough regions between two adjacent dunes of different shapes were observed. This research attempts to visualize the fluctuations and re-circulations over the trough regions using the velocity vector (Figure 10.29). The statistical properties of turbulent flow field in trough region were investigated by Mazumder (2007) based on the clustering geometry and interaction of turbulent bursting process, and subsequently, Mazumder (2008) used the statistical clustering based on the information and scale space theory for splitting the trough region into different segments. He also estimated the spatial turbulent kinetic energy (TKE) function in the trough region (Figure 10.30), which revealed interesting features corresponding to the region of high kinetic energy. Understanding of the spatial variation of TKE throws light on the behavior of the energy budget in turbulent motion.

Ojha and Mazumder (2008) investigated the development of a flow region associated with turbulence and shear stress characteristics over a series of 2-D asymmetric dunes placed successively at the flume surface (Figure 7.22). Experiments were conducted over 12 asymmetric dunes of mean length $\lambda_b = 0.30$ m, crest height $h_b = 0.03$ m and the dune width almost as wide as width of the flume (Figure 7.13), using 3-D Micro-ADV at the Indian Statistical Institute, Calcutta, resulting in the steepness, $h_b/\lambda_b = 0.094$ which is consistent with the real dunes in field. The variations of turbulence statistics along the flow affected by the wavy bottom roughness have been studied, after confirming the fully developed flow from the 7th dune (Figure 7.22). The stream-wise and vertical mean velocity components $(\bar{u}, \bar{v})$ at the trough and crest points along the longitudinal direction show the characteristic features of turbulent flow. Two loci of the points of intersections between the velocity profiles on the flat surface and the dune-covered bed led to focus the flow evolution over the dune-covered bed. The line clearly provides the idea of developing flow from the leading dune up to the seventh dune and then it reaches quasi-steady state. It appears that the existence of flow separation, flow reversal and sudden expansion significantly occurs.

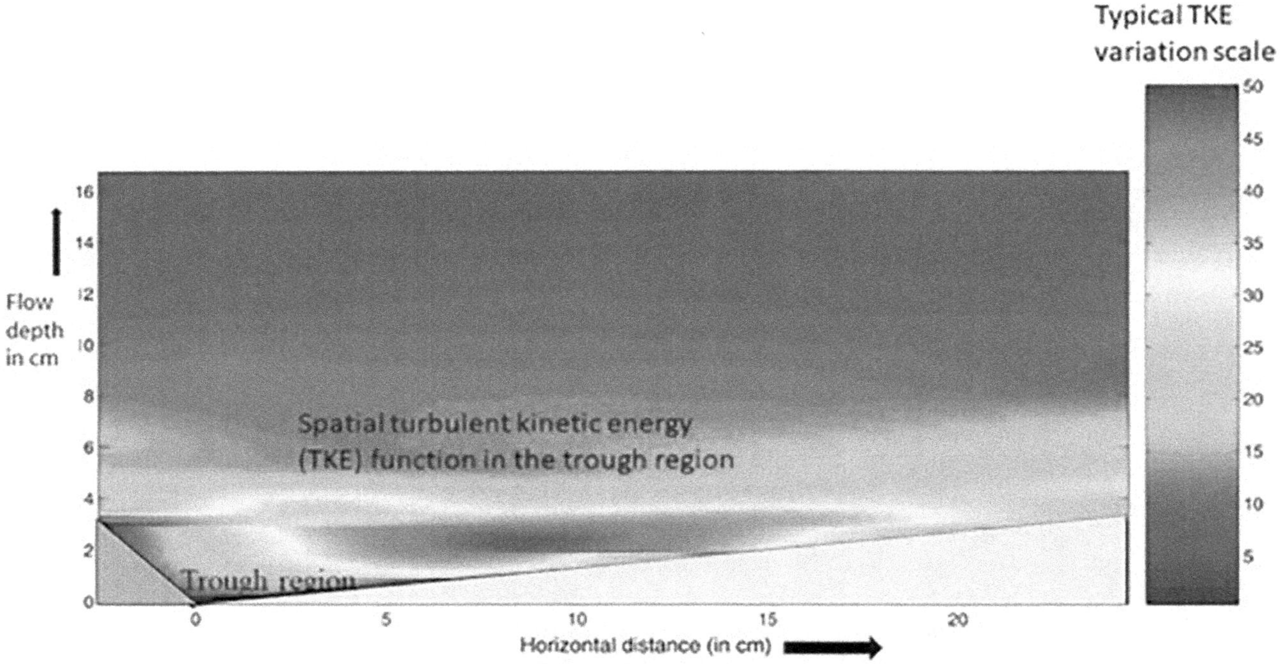

FIGURE 10.30 Spatial changes of TKE in the trough region. (Modified from Mazumder, 2008.)

Quadrant decomposition of the instantaneous Reynolds shear stress has been adopted to calculate the contribution of ejection and sweeping events in shear stress generation. The occurrence of the greatest upward velocity at the crest is likely to be responsible for the reduction of stream-wise flow over the crest. Just after the crest position strong downward flow occurs which leads to decrease the stream-wise velocity. The center of the circulation bubble appears to be located at a distance of about $2.67h_b$ downstream and $0.75h_b$ below the point of separation. The distance between the point of separation and the point of reattachment is about $5.8h_b$, which agrees well with Engel (1981).

Convex streamline curvature serves to stabilize vertical components of turbulence and concave curvature destabilizes flow by conveying turbulent structures toward the surface (Figure 10.31a). The streamline at the crest level changes its curvature from convex to concave. This inflexion in the streamline indicates that the flow separation occurs twice: one at the crest and the other one at a distance about h_b downstream from the crest. Two separation points are marked as circles in Figure 10.31a. The close-up views of vector plots given in Figure 10.31b and c provide clear picture of the flow singularities at the trough and crest positions, respectively. The near-bottom flow region over the dune-covered surface, composed of the separation cell and the internal boundary layer, is influenced by the wakes and the shear layer. These two flow structures are important for understanding sediment movement, the stability of bedforms and dune/ripple migration. Large-scale vorticity is demonstrated as ejection events and arises both along the shear layer and at flow reattachment. These coherent flow structures are advected with the mean flow, often reaching the free surface and erupting as surface "boils." The origin of kolks and boils in the fluvial systems has been vigorously accredited to the boundary-layer bursting process and may contain higher concentrations of sediment in suspension than the surrounding flow. The pressure difference over the dune, generated by flow separation and flow acceleration/deceleration associated with the dune form, generate a net force on the dune. Sweeping and ejections provide an extraction of energy from the complex flow field and these two events are considered to be more important than the interaction events in sediment entrainment. This work elucidates the impression of wavy bed roughness on the main flow, which is responsible for sediment transport process in rivers and estuaries.

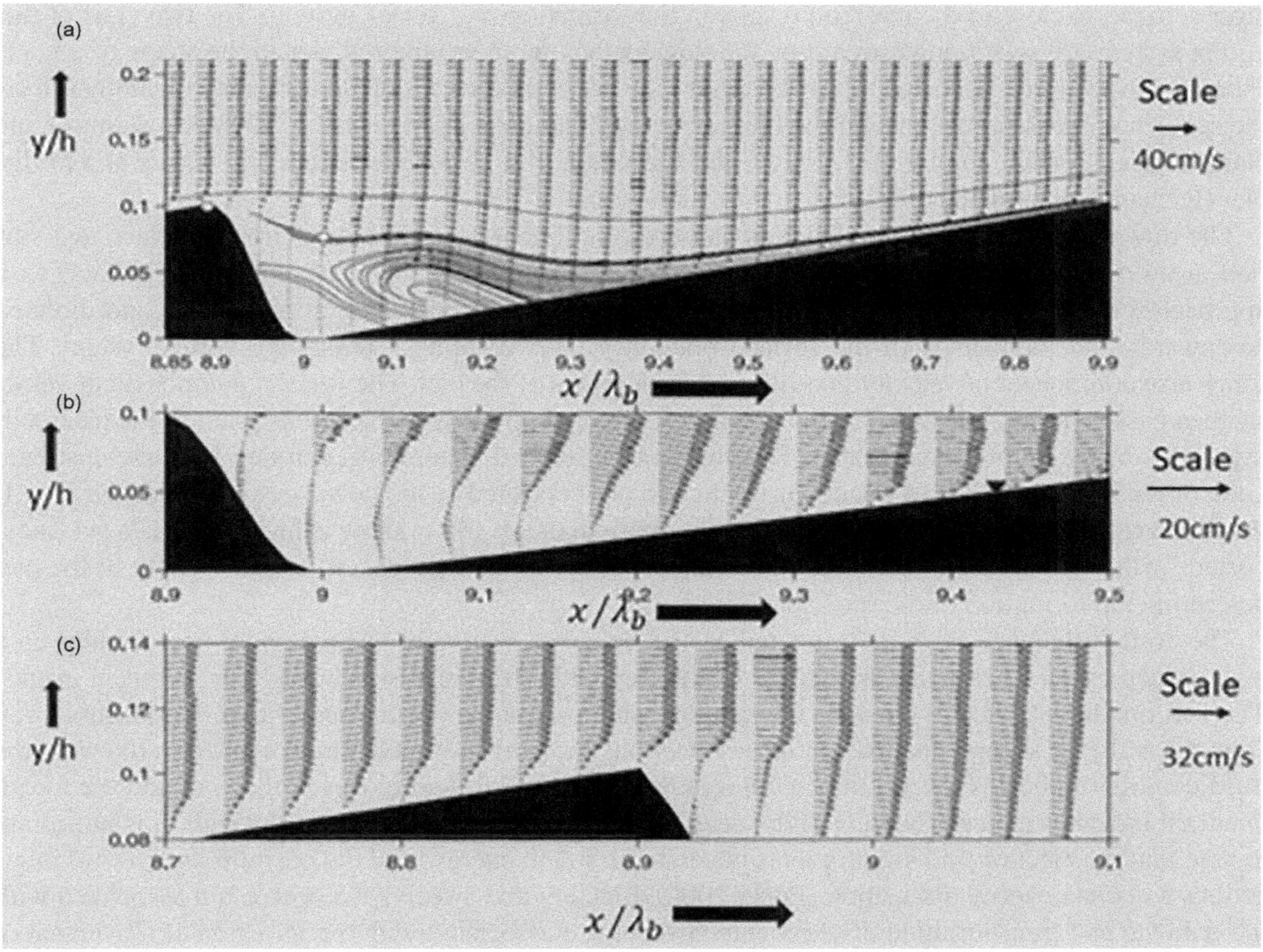

FIGURE 10.31 Plots of velocity vectors: (a) Between 9th and 10th dunes in fully developed region with two points of separation indicated by circles. (b) Close up view of (a) near trough region, and (c) Close-up view over the crest. (Modified from Ojha and Mazumder, 2008.)

10.11 SCOURING PROCESS

The flowing water exerts a shear stress on the sediment bed of the channel or stream, and this shear stress is responsible for the movement of sediment over the bed. If some obstruction is introduced in the channel, the shear stress on the bed around the obstruction increases. That augmented shear stress is responsible for the removal of sand material from the vicinity of the obstruction and this phenomenon is known as *scour*. For instance, when the flow of water passes around immovable objects or obstacles like bridge piers, abutments, spurs, shells, pebbles or wood fragments over a sandy bed, scour marks around the obstacle usually result from the interaction of the local flow field with the sediment bed. Consequently, the object/obstacle restricts the flow in the stream channel developing the excavation of sand on the upstream side of the object accompanied by a series of vortices downstream.

For example, the presence of bridge piers in the natural channel affects the local flow field and causes the formation of local scour, which can damage the bridge structure and ultimately failure of the whole bridge. A bridge pier in a riverbed induces an adverse pressure gradient in the upstream flow resulting in three-dimensional boundary layer separation in front of the pier. A down flow is also developed due to the downward negative stagnation pressure gradient of the non-uniform approach flow adjacent to the upstream face of the pier. The interaction between this downward flow and the horizontal boundary layer separation close to the riverbed results in the formation of a vortex system. The two ends of this vortex system are swept downstream by the flow as they wrap around the pier in the shape of a horse shoe in plan view and hence popularly known as the horse shoe vortex. This vortex significantly affects the scouring process at the base of the bridge piers (Richardson and Panchang, 1998; Muzzammil and Gangadharaih, 2003). Figure 10.32 shows the formation of a horseshoe vortex and wakes at a bridge pier (Richardson and Davis, 1995).

The down flow and horseshoe vortices cause the sediment to get removed from its place and carried away toward the downstream of the pier. The stream-wise component of velocity reduces as it approaches the pier, whereas the vertical component of velocity increases in magnitude and directed downwards and interacts with the surface boundary layer creating a horseshoe vortex system. The transverse component of velocity gets bifurcated sideways of the pier. The vertical component of velocity then passes through the sides of the pier and is directed upwards at the downstream of the pier leading to the formation of wake vortices. Due to this fact, the bed materials get eroded from the upstream node of the pier and travel to the downstream and get deposited at the downstream side of the pier. If the fluid velocity is low, the deposited particles remain there. If the velocity is high, materials get transported further downstream. The turbulent characteristics of the flow pattern in the vicinity of the pier determine the formation of the scour pattern.

The wall turbulence is directly related to the coherent structures, which show sporadically as a burst-sweep cycle. The quadrant analysis of the Reynolds shear stress provides four turbulent events. The first quadrant indicates outward interactions which show the occurrence of high-speed fluid away from the wall; the second quadrant indicates ejections showing low-speed fluid away from the wall, the third quadrant indicates inward interactions showing low-speed fluid moves to the wall and the fourth quadrant indicates sweeps showing high-speed fluid moving to the direction of the wall. It is important to note that the ejection and sweep events generate turbulent energy; and the outward and inward interactions associate energy dissipation (Pope, 2000). Ejection and sweep have been often associated with the removal and transportation of sediments. Sweeps are associated with the initiation of movement of bedload, whereas ejections are responsible for lifting and transporting the bedload.

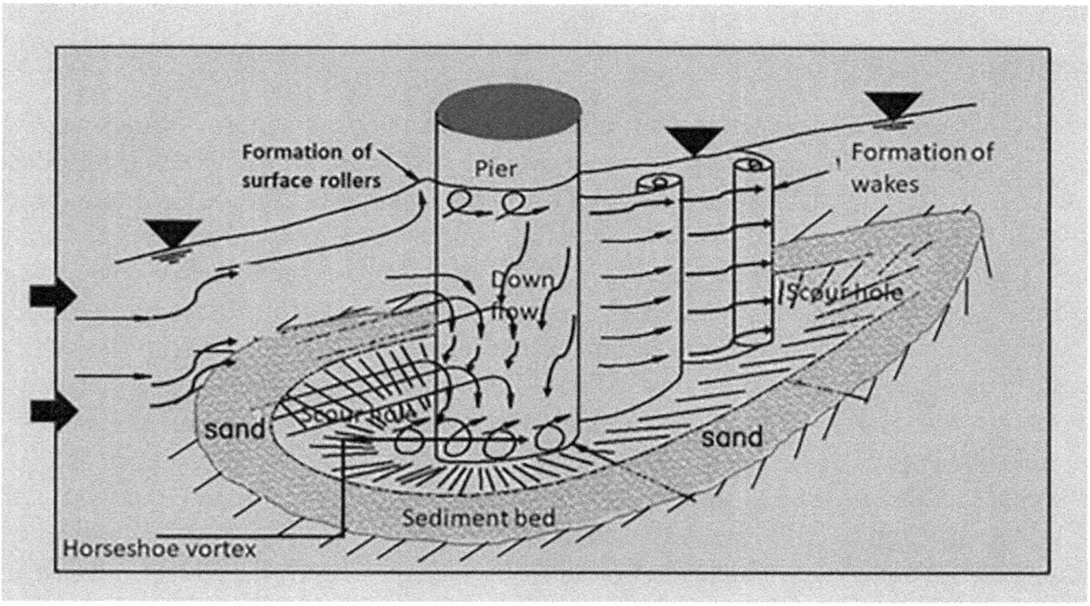

FIGURE 10.32 Flow around a bridge pier. (Modified from Richardson and Davis, 1995.)

10.11.1 Components of Scouring Process

Scouring is a particular form of bed erosion, caused by the impact of hydrodynamic forces, leading to the river-bed lowering. It occurs both during the speat as well as in normal flow conditions. Scour intensity grows with the increase of the flow velocity. A scour hole is a hollow, brought about by scouring. The scouring process consists of such components as:

- aggradation (rise in bed level that comes from the eroded river-bed),
- degradation (lowering of the bed level, which results from the deficiency of sediment material coming from the river basin of the upper part),
- general scour (stream bed lowering along the entire water course bed profile),
- local scour: (local drop of bed level in the vicinity of the structures as a result of erosion),
- clear-water scour: (when there is no debris movement from the areas located),
- live-bed scour: (when the rubble is entrained from higher areas to the scour hole beneath).

In an alluvial stream, there is a continuous transport of sediment as a geometrical process. However, if there is an additional natural or man-made cause to disturb the sediment supply and removal in the reach, such as construction of a bridge, barrage or an off-shore turbine, the stream will have long-term changes in the bed elevation. If there is a progressive build-up of stream bed in a reach due to sediment deposition, it is called aggradation. Conversely, if there is a progressive long-term lowering of the channel bed due to erosion as a result of deficit sediment supply to the reach, it is called degradation.

Extensive investigations relevant to a variety of scour structures for different disciplines have been made in laboratory settings on scour around vertical cylinders or piers embedded in sediment beds (Melville, 1975; Melville and Raudkivi, 1977; Raudkivi and Ettema, 1983; Darghai, 1989; Johnson and McCuen, 1991; Melville and Chiew, 1999; Graf and Istiarto, 2002; Sheppard et al., 2004; Ettema et al., 2006; Unger and Hager, 2007; Debnath et al., 2012; Vijayasree et al., 2019), underwater pipelines (Sumer and Fredsoe, 1990, 2002a, 2002b), around abutments (Kwan and Melville, 1994; Ahmed and Rajaratnam, 2000; Dey and Sen, 2005) and crescentic type scour (Sengupta et al., 2005; Mazumder et al., 2011; Maity and Mazumder, 2012, 2013, 2014, 2017). Local scour around short/finite horizontal cylinders placed in non-cohesive sediment beds has also been investigated (Voropayev et al., 1999, 2003; Sengupta et al., 2005; Catařno-Lopera and Garcia, 2006, 2007; Catařno-Lopera et al., 2007; Mazumder et al., 2011; Maity et al., 2013; Maity and Mazumder, 2014, 2017). In the case of bridges, the estimation of the correct depth of scour below the stream bed is very important since that determines the depth of the foundation. River bridges crossing alluvial beds are highly vulnerable to scouring at piers and abutments. The occurrence of scour around the bridge piers is the main concern about the stability of bridge foundations.

As a consequence, a vast amount of models of local scour around bridge piers have been developed by several researchers for last half a century. There is no scope to go through all these models on scours around bridge piers in this chapter, as most of these models are somehow similar in concepts. But even the scour models which are developed in recent years or decades or so, that are also hardly attempted to consider them here. Therefore, in the present chapter, as examples of scouring models, some models are selected for different scour types.

10.11.2 Scour Below Pipe Lines

In marine environment, the structures are usually exposed to currents, waves and combined waves and currents. The scour processes in the marine environment are more complex than the steady current such as river flow. Pipelines are set up mostly in the marine environments for the transportation of gas and crude oil from offshore platforms. These pipelines are also used for the disposal of industrial and municipal wastewater into the sea. Typically, the pipe size may be from 20–30 cm to more than 1 m in diameter, and length may be from 100 to 1,000 m. The water depth may be from 10 to 100 m. The pipelines are laid on the bed surface, they may be buried or they may be trenched. When the pipelines are exposed to the direct flow on the seabed, and if the seabed is erodible, the scour may occur around the pipe due to the flow action, which may lead to suspension of the pipeline (Figure 10.33). Sometimes the pipeline along the length of suspension may be sagged in the generated scour hole. In the case of sagging, pipelines may reach the bottom of the scour hole, which is eventually buried by the sand materials. There are several highly complex processes in the pipeline-seabed interactions. The scour beneath the pipeline is basically related to the seepage flow in the sand bed, which is driven by the pressure difference between the upstream and downstream sides of the pipe (Figure 10.33).

The critical conditions for the onset of scour around the pipelines have been investigated by several researchers like Mao (1986), Chiew (1990), Sumer and Fredsoe (1991), Sumer and Fredsoe (2002a, 2002b) and others. Mao (1986) performed laboratory experiments to record the time scale; size and shape of scour marks in a steady current and also under wavy conditions. He investigated the role of vortices in front and at the rear of the pipeline; and also discussed the seepage flow underneath the pipe, related to scour (Chiew, 1990). He proposed the existence of two small vortices: one at the upstream and other one at the downstream of the cylinder, to remove sediments from the area around the pipe and the pressure difference between the upstream and downstream sides may cause erosion underneath the pipe. This onset of scour due to the combined action of vortices leads to the formation of small opening under the pipe. Efforts have been devoted to the two-dimensional scour around fixed pipelines in currents (Sumer et al., 1988; Chiew, 1990). Sumer and Fredsoe (1991) performed a series of experiments to determine the critical conditions in the case of waves; and expressed in terms of two parameters, namely, Keulegan-Carpenter number and embedment to diameter ratio. Later, Klompt et al. (1995) extended the Sumer and Fredsoe's (1991) study to the case of combined wave-current flows (Figure 10.34). Subsequently, Sumer et al. (1991) investigated the onset of underneath scour in both currents and waves. The scour depth develops towards the equilibrium stage through a transitional period, for a pipe placed on the bed surface with initially a zero gap. The depth corresponding to the fully developed stage is called the equilibrium scour depth. The scour depth in the case of steady current has been extensively studied by several researchers like Chao and Hennesy (1972), Kjeldsen et al. (1973), Herbich (1981), Mao (1986), Kristiansen and Torum (1989) and others.

For example, Kjeldsen et al. (1973) were the first to develop the empirical relation between the equilibrium scour depth d_s, the pipe diameter p_d and the flow velocity u as follows:

$$d_s = 0.972\left(\frac{u^2}{2g}\right)^{0.2} p_d^{0.8} \tag{10.123}$$

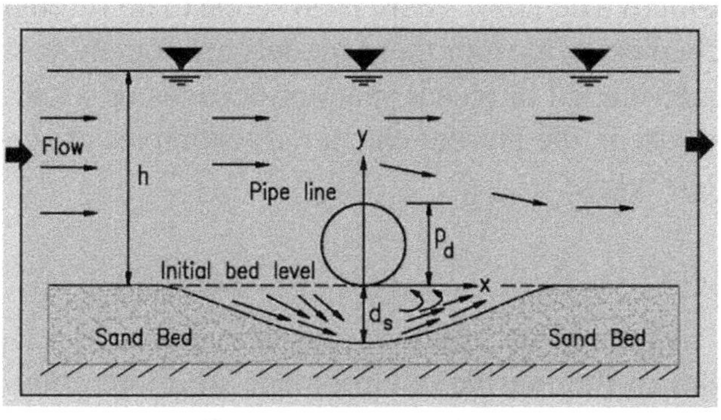

FIGURE 10.33 Sktech of scour hole below a pipe line.

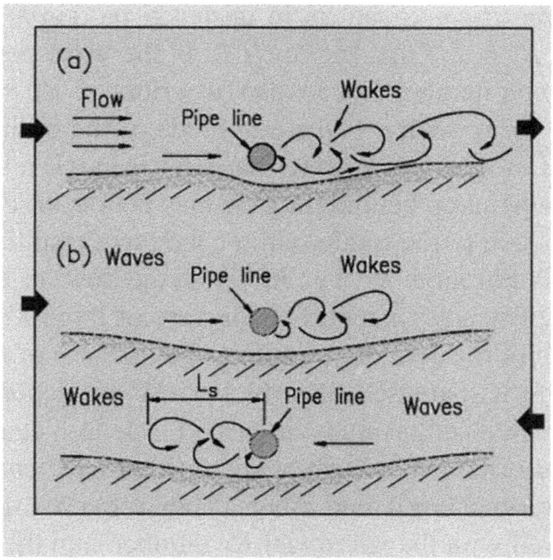

FIGURE 10.34 Sketch of Lee-wake effects: (a) due to current, and (b) due to waves, wake system occurs both sides. (Modified from Sumer et al., 1988.)

The non-dimensional scour depth d_s/p_d is proportional to the Shields parameter θ as:

$$\frac{d_s}{p_d} \propto \theta^{0.2} \tag{10.124}$$

where $\theta = \dfrac{u_*^2}{g\left(\dfrac{\rho_s}{\rho_f} - 1\right)d}$ is the dimensionless Shields parameter. It is noticed that the scour in Kjeldsen et al.'s study was for the live-bed situation ($\theta > \theta_{cr}$) in which θ_{cr} is the critical value of Shields parameter corresponding to the initiation of motion at the bed. The flow actually created by the presence of pipe depends on the following quantities as: the pipe diameter, flow velocity, kinematic viscosity, pipe roughness, and the grain diameter of the bed material. From the dimensional analysis, scour depth d_s/p_d can be found to depend on the following parameters:

$$\frac{d_s}{p_d} = f(k_{s*}, Re, \theta) \tag{10.125}$$

where $k_{s*} = k_s/p_d$ is the relative roughness, $Re = up_d/v_f$ is the Reynolds number and θ is the Shields parameter. From the three parameters equation (10.125), the influence of k_{s*} and Re appears through their effects on the downstream flow of the pipe. If the pipe is hydraulically rough, the wake flow is almost unaffected by Re, while for a hydraulically smooth, some influence of Re is expected in the downstream vortex shedding. In transitional region, the vortex shedding becomes less affected leading to smaller lee-wake erosion, and hence less scour depth.

The scour depth is also induced by waves and tidal flows. When there is a flow condition like tidal flow, there is a scour around the pipeline with wakes downstream of the pipe (Figure 10.34a). The flow attacks the pipe from both sides due to the near-bed flow induced by waves shown in Figure 10.34b. The difference between the case of waves and tidal current and the steady current is that the downstream-wake system occurs on both sides of the pipeline (Figure 10.34b.), i.e., due to waves wake system occurs on both sides.

The formation and extension of wake pattern in unsteady motion are governed by the Keulegan-Carpenter number Kc, where $Kc = u_m T_w / p_d$ in which T_w is the wave period, and u_m is the maximum velocity. The Keulegan-Carpenter number Kc can also be written as $Kc = 2\pi a / p_d$, if the orbital velocity is assumed to vary sinusoidal form, with a is the amplitude of the orbital motion. It is also seen that for small Kc value, the orbital motion of the water particles is small relative to the total width of the pipe. When Kc is very small, separation behind the pipe may not occur. For large Kc the water particles travel quite large distances relative to the total width of the pipe, resulting in separation and probably vortex shedding (Figure 10.35). For large Kc, i.e., $Kc \to \infty$, the flow for each half period of the motion resembles that experienced in a steady current. For the case of large Kc numbers ($Kc \geq 6$), a vortex street is formed on the lee-side of the cylinder, resulting the increase of scour width S_w. The extension of this street, L_s (Figure 10.35) is increased linearly with Kc corresponding to $L_s / p_d = 0.3Kc$. This equation is based on the flow visualization study described in Jensen et al. (1989).

The scour depth in the case of combined waves and current has been studied by Lucassen (1984), Hansen (1992), Sumer and Fredsoe (1996), and others. They observed that the non-dimensional scour depth correlates remarkably well with the parameter Kc number, and the scour depth is represented by the following equation as:

$$\frac{d_s}{p_d} = 0.1\sqrt{Kc} \tag{10.126}$$

for the case of live bed situation ($\theta > \theta_{cr}$). This above relation (equation 10.126) has been confirmed by Gokce and Gunbak's (1991) from their experimental data. In Sumer and Fredsoe's work (1996), experimental data have been obtained covering the wide range of Kc from 5 to about 50 and the full range of $u_c/(u_c + u_m)$ from 0 to 1 in which u_c is the undisturbed current velocity at the center of the pipe. Sumer and Fredsoe (1996) indicated that the scour depth may increase or decrease when a current is superimposed on waves depending on the values of Kc and $u_c/(u_c + u_m)$. Several models of local scour depth about the pipeline have been developed by many authors since few decades shown in Sumer and Fredsoe (2002). Sumer and Fredsoe (1996) showed experimentally the influence of irregular waves on

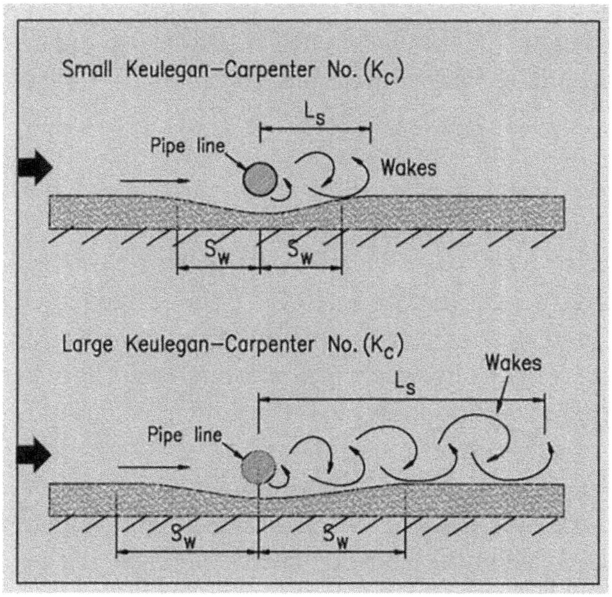

FIGURE 10.35 Stream-wise scour due to lee wakes corresponding to different Kc numbers. (Modified from Sumer et al., 1988.)

scour depth. The effect of pipe position from the bed is also an important factor. The pipe is seldom in contact with the bed, but at some places the pipe is placed above the bed irregularities, and at some locations, the pipe is pressed down the bed. The pipe may also slag down into the scour hole during the self-burial process. The clearance or contactness between the pipe and the bed is defined by the quantity e as a parameter. The smaller the value of clearance e, the larger the influence of the pipe on the equilibrium scour depth. It is interesting to note that no scour occurs if the pipe is placed above $e/p_d = 1$ for moderate Kc number. For increase in Kc values, depth of scour may increase even at larger values of $e/p_d \sim 3$, which may be explained from the lee-wake at larger Kc values. Finally, Hansen et al. (1986) from their experimental data have presented a formula for scour depth be approximated as:

$$\frac{d_s}{p_d} = 0.625 \exp\left(\frac{-0.6\ e}{p_d}\right) \tag{10.127}$$

for $-0.25 \le e/p_d \le 1.2$ for the case of steady currents, and

$$\frac{d_s}{p_d} = 0.1\ \sqrt{Kc}\ \exp\left(\frac{-0.6\ e}{p_d}\right) \tag{10.128}$$

for $0 \le e/p_d \le 2$ for the case of waves, and these equations are valid for the case of live bed condition ($\theta > \theta_{cr}$). Another way of scour depth is due to the effect of vibrations. When a pipeline is laid down on the sea bed, suspended distance of pipeline will develop along the length of the pipeline. This suspended distance undergoes flow-induced vibrations (Sumer and Fredsoe, 1997).

Three equilibrium scour depth structures were reported by Sumer and Fredsoe (1990) for a broad range of the Keulegan-Carpenter number ($10 \le Kc \le 1{,}000$). The width of the scour increased with increase in Kc. The variation of the scour width S_w with Kc (Figure 10.36) be approximated by the following relation as:

$$\frac{S_w}{p_d} = 0.35Kc^{0.65} \tag{10.129}$$

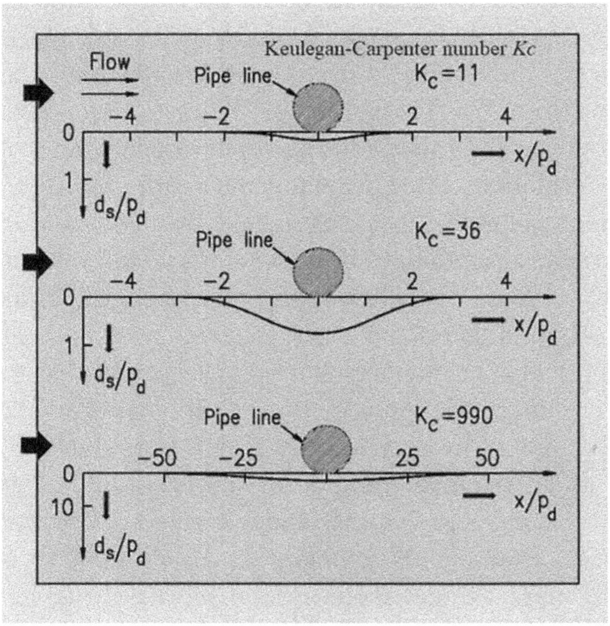

FIGURE 10.36 Schematic diagrams of equilibrium scour depth. (Modified from Sumer and Fredsoe, 1997.)

where S_w is the width measured from the center of the pipe to the end of the scour hole (Figure 10.36). Sumer and Fredsoe (1996) focused their study to determine the influence of combined wave-current environments to the pipeline scour width. Maza (1987) proposed a solution for estimating scour depth that is related to the initial gap-pipe diameter ratio and the flow Froude number. Maza's solution showed that the scour depth d_s increases with increasing Froude numbers. Moncada and Aguirre (1999) extended the approach for estimating both the scour depth and the scour length. Liang and Cheng (2005) developed a vertical 2-D numerical model for local scour beneath pipeline exposed to waves, which are modeled as sinusoidal oscillatory flows. Dey and Singh (2008) conducted research on clear-water scour beneath the pipelines under steady flow. From a large number of experimental data for equilibrium scour depth, a simple cubic polynomial equation describing the equilibrium scour depth was derived. In this book, there is no scope to go through all these models on scour depth about the pipeline due to the steady flow, unsteady flow, combined waves and current flows in this chapter. Interested readers may refer to the book of Sumer and Fredsoe (2002a, b) for their study. The scour depth around the offshore pipelines in the South China Sea was predicted by Cao et al. (2015). Simultaneously, Zhang et al. (2015) made an excellent review on scour propagation beneath the submarine pipeline.

10.11.3 Crescentic Scour

Local scour around any obstruction placed in the sediment bed is a result of the interaction of the local flow field with the sediment particles. When water passes around immovable objects or obstacles like shells, pebbles or wood fragments placed on a sandy bed, scour marks around the obstacle usually result from the interaction of the local flow with the sediment bed. The obstacle restricts the flow in the stream channel developing the excavation of sand on the upstream side of the object accompanied by a series of vortices downstream. A crescentic type scour trough commonly develops around the obstacle with its deepest part on the upstream side and the tail pointing downstream, and consequently the eroded materials from the upstream side are deposited in the downstream side as a sand bar. The larger obstacle, the more turbulence is generated and the more energy is absorbed causing retardation of the flow. Investigations have been made to quantify the scour depths, and the mean flow across the scour marks generated by different types of bluff bodies/obstacles placed on non-cohesive sand beds (Kirkil et al., 2008; Kirkil and Constantinscu, 2015; Mazumder et al., 2011; Euler and Herget, 2011, 2012). In depositional sedimentary structures, scour marks may be termed obstacle marks or comet marks and used as an indicator of palaeo-current direction (Dzulynski and Walton, 1965; Sengupta, 1966, 2007; Karcz, 1968; Melville, 1975; Johansson, 1976; Werner et al., 1980; Allen, 1982; Collinson and Thompson, 1982; Bhattacharya, 1993). Figure 10.37 shows the current crescents in the bedding plane of the ancient sedimentary structure. The formations of crescentic scour marks are clear in cross-bedded sediments depending on the orientation and plunge of the long axes of the pebbles/bluff bodies. A series of current crescents around pebble barriers is shown in Figure 10.38 in a recent stream bed at Subarnarekha River, India. Most of the pebbles are covered up by leaves and other plant materials carried by the stream.

For example, in order to comprehend the basic mechanisms of formations of crescentic scour, Sengupta (1966) carried out field experiments placing pebbles on the loose sand bed in a recent stream to form an imitation of scour mark (Sengupta, 1966; Collinson and Thompson, 1982), to estimate ancient flow conditions in the light of modern analog. Kjeldsen et al. (1973) developed scour holes in a laboratory setup. Werner et al. (1980) made field observations and flume experiments on the nature of comet marks. Field observations suggest that the comet marks are observed typically in shallow seas and these are obstacle induced. Flume experiments were also studied to visualize the flow characteristics in the wake of obstacle marks. Local scour around short or finite horizontal cylinders placed in non-cohesive sediment beds has been investigated by several researchers (Voropayev et al., 1999, 2003; Sengupta et al., 2005; Catǎno-Lopera and Garcia, 2006, 2007; Catǎno-Lopera et al., 2007; Mazumder

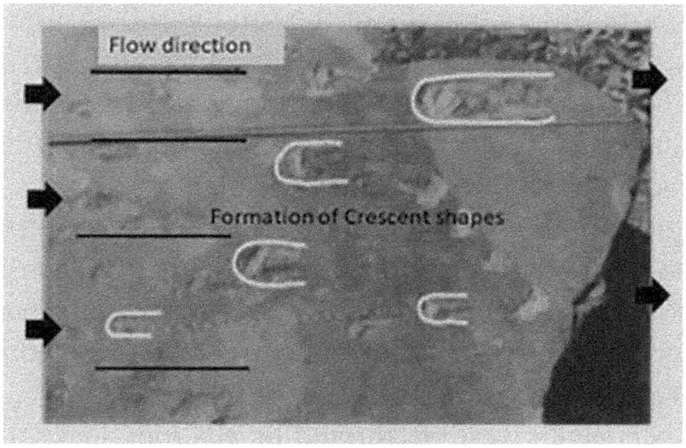

FIGURE 10.37 Current crescents formations in a bedding plane. Proterozoic Kaimur formation, Maihar, MP, India. (Modified from Maity and Mazumder, 2014.)

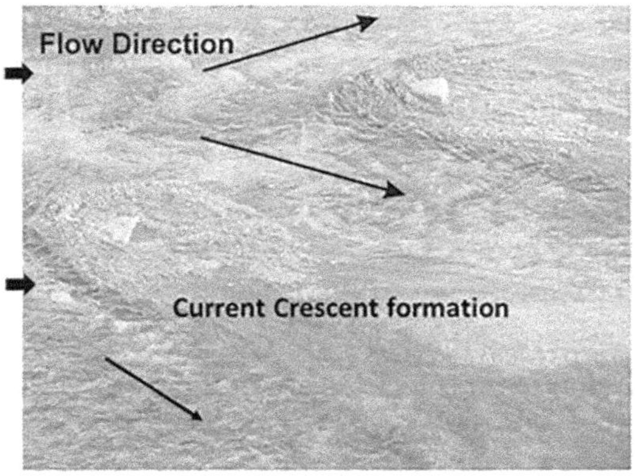

FIGURE 10.38 A series of "Current Crescents" around pebble barriers in a recent stream bed at Subarnarekha River, India. Most of the pebbles are covered up by leaves and other plant materials carried by the stream. (Photograph taken by Mazumder from Subarnarekha, India.)

et al., 2011, Maity et al., 2013). The scour around with the subsequent burial of short horizontal cylinders placed on a sandy bed under progressive waves was investigated by Voropayev et al. (2003). Their motivation was to observe the scour and burial effects of cylindrical objects in coastal environments, where non-linear progressive waves played a significant role. Sengupta et al. (2005) carried out flume experiments to generate the crescentic-shaped scour marks using horizontal short cylinders placed in a sandy bed for possible verification of field observations in a recent stream (Sengupta, 1966). They obtained a positive correlation between the flow velocity and the scour-width irrespective of cylinder diameter and sand size. The local scour and deposition induced by an obstacle in the fluvial environment was studied by Euler and Herget (2012). They obtained a symmetric obstacle mark resulting from flash flood developed around a boulder in the Anapodaris River gorge in Crete in Greece. If the obstacle is strongly inclined towards the downstream direction, maximum scour depth shifts from the frontal to the lateral zones around the obstacle (Nakayama et al., 2002).

Mazumder et al. (2011) carried out a series of experiments in a laboratory flume (Figure 7.13) to study the scour geometry developed around the static short circular cylinders placed on the sediment bed transverse to the flow, and turbulence characteristics of the equilibrium scour geometry. The evolution of the scour mark around the cylinder was recorded using a digital camera. To perform the tests, a sand bed of thickness h' (= 4 cm) and 5 m long covering the entire width (50 cm) of the flume was laid at the bottom. The median particle diameter d_{50} of the sand was 0.25 mm and the standard geometric deviation $\sigma_g = 0.685$. The specific gravity of sediments used for the experiments was 2.65. A series of experiments were conducted over the sediment bed of known grain-size distribution using eight different circular cylinders of diameters $p_d = 2.0$ cm, 2.6 cm, 3.2 cm, 4.2 cm, 5.0 cm, 5.8 cm, 6.0 cm and 7.0 cm of fixed length $L = 10$ cm placed at the center line of the flume.

For each experiment, single cylinder was placed at the centerline over the sand bed transverse to the flow at the measuring station 6 m downstream of the channel inlet (Figure 10.39). Flow depth was kept constant at $h = 0.30$ cm. The turbulent flow field within the scour mark was measured in a laboratory flume, using ADV. The scour marks named as current crescents preserved in geological record are traditionally used as indicators of palaeocurrent direction.

A total of thirty-eight experiments were performed on the identical set-up with different flow discharges ranging from 0.011 to 0.022 m^3/s and different cylinder diameters (mentioned above) to obtain the width of scour mark under the equilibrium condition for each case. The ranges of flow parameters used for the experiments are provided in Table 10.2. The collected velocity data are analyzed to calculate the mean flow and turbulence characteristics at each point. In order to visualize the time-dependent processes of scour mark around a short cylinder of diameter 3.2 cm under the low flow discharge $Q = 0.01472$ m^3/s, a series of photographs was taken about 15–20 minutes interval during the experiment starting from the initial stage of flat bed condition to the equilibrium. The discharge was chosen in such a way that the local flow velocity was less than the critical velocity to initiate the particle movement at the bed. From the photographs, it is clearly observed that for the fixed flow discharge, there is no scour at the initial stage, but as time goes on, the typical scour and the depositional patterns around the cylinder are observed due to the blockage in the flow creating the sufficient shear stress. The initial scour takes place mostly at the upstream boundary and at the end points of the cylinder; consequently, the deposition takes place at the downstream of the cylinder.

Once the scour mark starts at the upstream side parallel to the axis of the cylinder due to the vortex, with increasing time the size and shape of the scour mark vary; and at a time $t_e = 178$ minutes it practically reaches an equilibrium state in size and shape. It is observed from photographs that during the process of evolution with time the scour mark in the upstream side parallel to the cylinder did not show symmetric, which may be due to the nonlinearity of turbulence and wakes; but eventually with

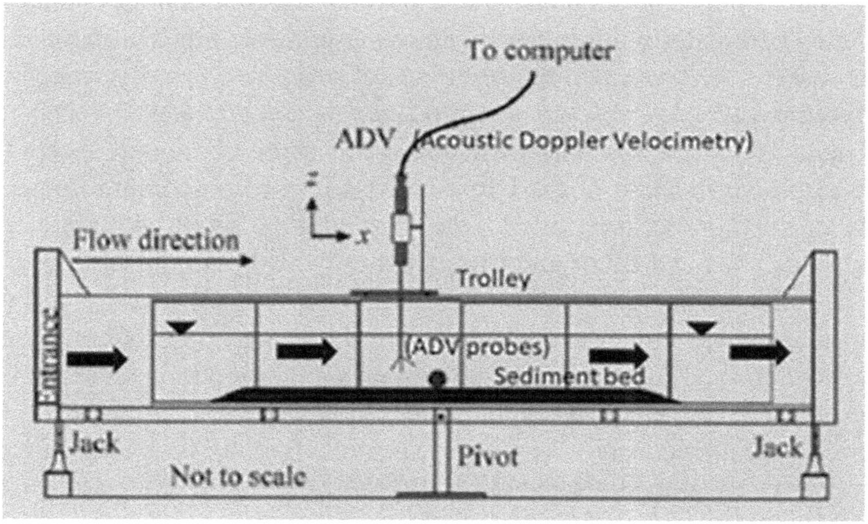

FIGURE 10.39 Schematic diagram of the experimental set-up at the Fluvial Mechanics Laboratory of ISI Kolkata.

TABLE 10.2
Experimental Parameters for Studying the Crescentic Scour

p_d (cm)	a_r (p_d/L)	$Q\times10^{-2}$ (m³/s)	u_m (cm/s)	$Fr\left(=u_m/\sqrt{gh}\right)$	$F_s\left(=u_m/\sqrt{(\gamma_s-1)gd_{50}}\right)$	w_s (cm)
2.0	0.2	1.083	14.64	0.107	2.30	1.0
2.0	0.2	1.250	20.37	0.149	3.20	1.7
2.0	0.2	1.639	27.56	0.202	4.33	2.5
2.0	0.2	1.806	32.21	0.243	5.22	3.0
2.0	0.2	2.056	36.98	0.271	5.81	3.5
2.6	0.26	1.222	19.82	0.145	3.12	1.5
2.6	0.26	1.472	26.46	0.194	4.16	3.0
2.6	0.26	1.500	26.87	0.197	4.22	3.1
2.6	0.26	1.778	31.66	0.232	4.98	4.0
2.6	0.26	2.056	36.70	0.269	5.77	5.0
3.2	0.32	1.139	19.00	0.139	2.99	1.2
3.2	0.32	1.444	24.84	0.182	3.90	2.5
3.2	0.32	1.472	26.63	0.195	4.19	3.5
3.2	0.32	1.778	31.82	0.233	5.00	4.0
3.2	0.32	2.083	36.64	0.268	5.76	5.0
4.2	0.42	1.417	23.28	0.171	3.66	2.5
4.2	0.42	1.472	26.20	0.192	4.12	4.5
4.2	0.42	1.639	27.30	0.200	4.29	4.8
4.2	0.42	1.861	33.62	0.246	5.28	6.0
4.2	0.42	2.083	37.62	0.276	5.91	8.0
5.0	0.50	1.083	14.23	0.104	2.24	1.0
5.0	0.50	1.472	26.34	0.193	4.14	4.0
5.0	0.50	1.806	33.41	0.245	5.25	5.5
5.0	0.50	2.000	36.22	0.265	5.69	6.0
5.8	0.58	1.083	14.58	0.107	2.29	1.0
5.8	0.58	1.361	22.09	0.162	3.47	2.0
5.8	0.58	1.694	28.30	0.207	4.45	5.0
5.8	0.58	1.972	35.34	0.259	5.56	7.0
5.8	0.58	2.222	39.21	0.287	6.16	8.0
6.0	0.60	1.111	18.37	0.135	2.89	2.5
6.0	0.60	1.472	26.45	0.194	4.16	6.5
6.0	0.60	1.556	26.61	0.195	4.18	6.6
6.0	0.60	1.806	33.01	0.242	5.19	7.1
6.0	0.60	2.056	36.55	0.268	5.74	8.0
7.0	0.70	1.278	21.88	0.160	3.44	2.6
7.0	0.70	1.528	26.32	0.193	4.14	5.0
7.0	0.70	1.778	31.39	0.230	4.93	7.0
7.0	0.70	2.083	37.82	0.277	5.94	9.0

Source: Modified from Mazumder et al. (2011).
Note: u_m is the maximum fluid velocity, $a_r\,(=p_d/L)$ is the cylinder aspect ratio, $Fr\left(=u_m/\sqrt{gh}\right)$ is the Froude number, $F_s\left(=u_m/\sqrt{(\gamma_s-1)gd_{50}}\right)$ is the sediment Froude number, and w_s is the scour width.

time (after 2 hours) the scour mark tends to become symmetric in size and shape with deepening about the centerline of the cylinder, which shows the transitional phenomena of the process. Sand bar called ridge also developed in the downstream side at the middle of the cylinder, and with time the ridge grew in length perpendicular to the cylinder with a symmetric scouring in both sides of the ridge. At time

t_e = 178 minutes of equilibrium scour mark, the ridge grew vertically as well little above the cylinder partially covering the downstream side and its length r_e became 20 cm.

In a similar way, two more experiments using two different circular short cylinders of diameters p_d = 4.2 and 6.0 cm of identical length L = 10 cm with the same flow Reynolds number or discharge were performed; and the photographs of equilibrium scour marks developed in the upstream sides of all three cylinders are shown in Figure 10.40. The diameter of cylinder leads to increase the width of scour mark w_s. The equilibrium conditions also occur at different times (t_e = 178, 195 and 300 minutes) for different diameters of the cylinders. The time (t_e) to reach equilibrium increases with increase of diameters. Observed values of depth and width of scour marks, length of the ridge, width of left-end and right-end scour marks generated due to three different cylinders are tabulated in Table 10.3.

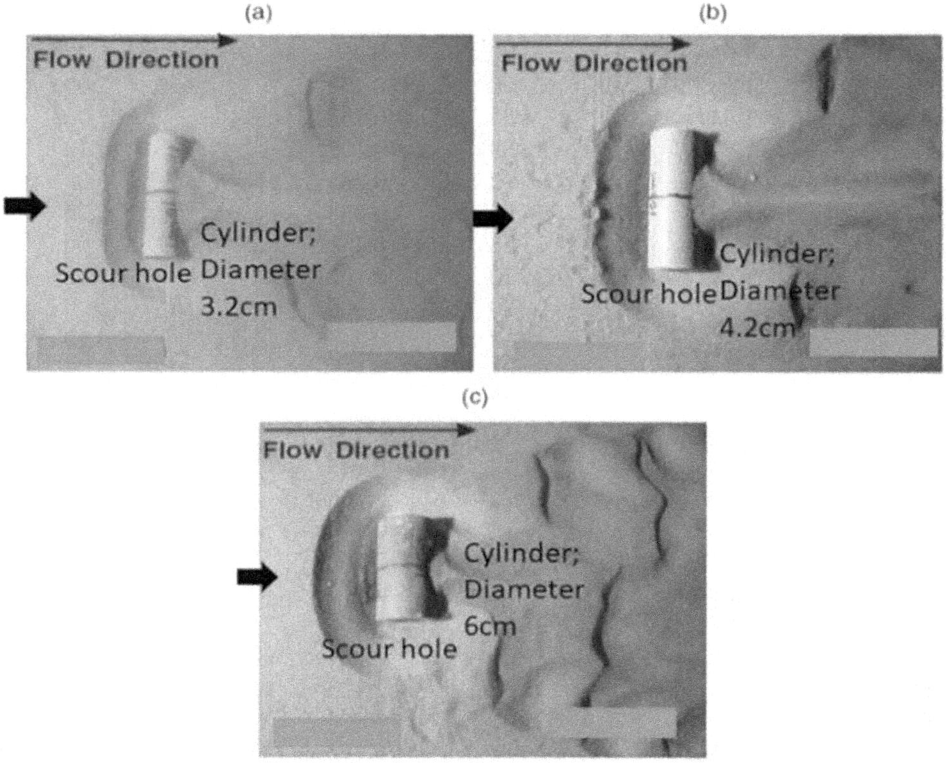

FIGURE 10.40 Development of equilibrium scour holes at the upstream of three cylinder diameters with different times for flow rate of 0.015 m^3/s. (Modified from Maity and Mazumder, 2014.)

TABLE 10.3
Values of Observed Parameters for Reynolds Number Re = 676,00

p_d (cm)	w_s (cm)	t_e (min)	S_{eh} (cm)	r_e (cm)	L_c (cm)	R_c (cm)
3.2	3.5	178	2.56	20.0	3.5	3.5
4.2	4.5	195	2.76	16.6	4.0	4.0
6.0	6.5	300	2.86	7.5	5.0	5.0

Source: Modified from Mazumder et al. (2011).
where t_e is the time of equilibrium scour width, S_{eh} is the equilibrium scour depth, r_e equilibrium ridge length; L_c is the left-end scour width, and R_c is the right-end scour width.

The turbulent shear stress along the flow at the bottom of the scour hole for all three cylinders (p_d=3.2, 4.2 and 6.0 cm) and turbulent shear stress at the bed level (zero level) is plotted against longitudinal distance (Figure 10.41). It is noticed that the shear stress at the bed level (zero level) is always higher than the bottom most shear stress at the scoured region. It is interesting to note that the shear stress along the flow behaves oscillatory in nature with positive and negative values in the scoured region for all the cylinder diameters. It is also noticed that the overall trend of shear stress leads to the negative at the bottom surface of the scoured region, indicating outward flux of momentum and to the positive at the bed level (zero level) indicating inward flux of momentum.

10.11.3.1 General Analysis of Scour Geometry

The equilibrium scour width (w_s) in a steady current is mainly dependent on the following parameters of flow, cylinder and sand bed, i.e.

$$w_s = \left(p_d, L, \rho_f, k_c, d_{50}, \rho_s, u_m, g, v_f\right) \tag{10.130}$$

where p_d is the diameter of cylinder; L is the length of the cylinder, ρ_f is the density of fluid, k_c is the surface mean roughness of the cylinder, ρ_s is the density of solid, u_m is the maximum fluid velocity near the surface, g acceleration due to gravity, v_f is the kinematic viscosity of the fluid.

Combining the parameters from equation (10.130), the following dimensionless form can be deduced

$$w_s = f\left(a_r, R_{cyl}, F_s\right) \tag{10.131}$$

where a_r is the cylinder aspect ratio, R_{cyl} is the cylinder Reynolds number, F_s is the sediment Froude number. From the experimental observations, the dimensionless scour width w_s/L against the sediment Froude number F_s is plotted in Figure 10.42 for different flow discharges ranging from 0.011 to 0.022 m³/s and different cylinder diameters p_d = 2.0 cm, 2.6 cm, 3.2 cm, 4.2 cm, 5.0 cm, 5.8 cm, 6.0 cm and 7.0 cm. From the figure, it shows that for a fixed sediment Froude number F_s ranging from 3.5 to 6, the dimensionless scour-width w_s/L increases with increase of cylinder diameter p_d. Based on the experimental observation, the following empirical relation for the dimensionless scour-width w_s/L as a function of cylinder diameter p_d and F_s has been proposed with coefficient of regression ranging from R^2 = 0.77 to 0.98:

$$\frac{w_s}{L} = aF_s^b \tag{10.132}$$

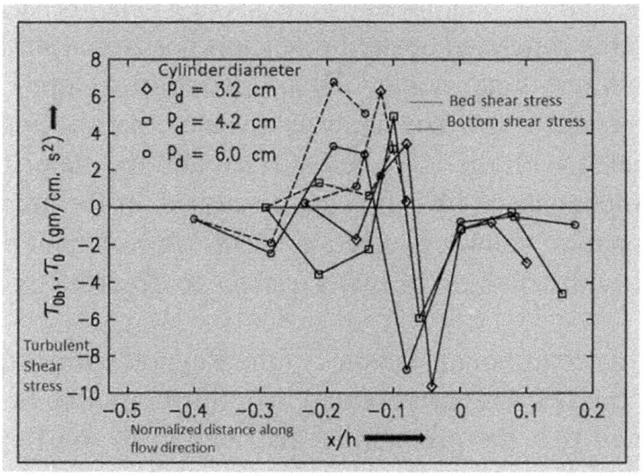

FIGURE 10.41 For Reynolds number Re = 676,00 shear stress along the flow at the bottom in the scoured region and at bed level. (Modified from Mazumder et al., 2011.)

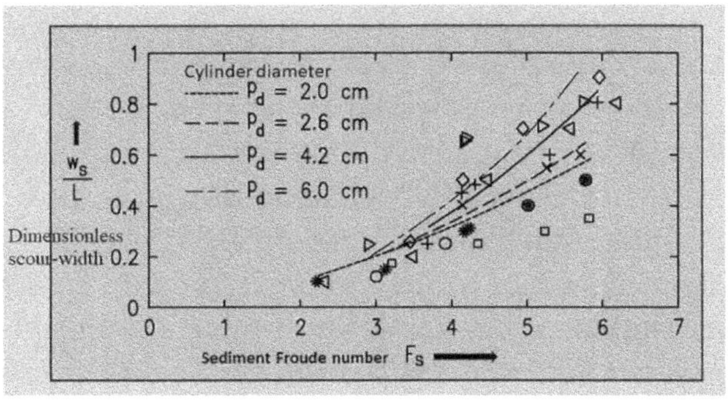

FIGURE 10.42 For different cylinder diameters, profiles of dimensionless scour-width vs. sediment Froude number F_s ($p_d = 2.0\,\text{cm}$ (□), 2.6 cm (*), 3.2 cm (o), 4.2 cm (+), 5.0 cm (×), 5.8 cm (<), 6.0 cm (>) and 7.0 cm (◊)). (Modified from Mazumder et al., 2011.)

with $a = \exp\left[-0.576\ln\left(p_d/L\right) - 4.325\right], b = 0.584\ln\left(p_d/L\right) + 2.563$

Here the coefficients a and b are the function of p_d. In general, the empirical relation (10.132) follows a power function of sediment Froude number F_s for any cylinder diameter (p_d). It is also observed that there is a linear relationship between the scour width (w_s) and the cylinder Reynolds number (R_{cyl}), which agrees well with the relationship obtained by Sengupta et al. (2005).

Maity and Mazumder (2012) studied distributions of turbulent kinetic energy (TKE), turbulent diffusion in both longitudinal and vertical directions at different locations of scour marks, and the contributions of burst-sweep cycles to the Reynolds shear stress over and within the scour marks around the objects. For all the experiments, irrespective of cylinder diameter the magnitude of time fraction and probability of outward interactions (inward interactions) are exponentially related with ejections (sweeps) throughout the depth. All the turbulent events: ejections, sweeps, inward and outward interactions are equally important in the near scour mark region. Maity et al. (2013) studied the growth of scour-width up to a state of equilibrium and developed model of scour-width growth curve using nonparametric regression and smoothing spline techniques. Scour geometry is related to velocity of past water flow and its direction, preservation of fossils in ancient riverbed, etc. With an application of a robust nonparametric method proposed by Dasgupta (2013), the first and higher-order derivatives of the growth curve are estimated and discussed with their applications. Subsequently, Maity and Mazumder (2014) examined the spatial distributions of third-order moments of velocity fluctuations, the turbulent kinetic energy (TKE) fluxes, and the conditional statistics of Reynolds shear stress across the equilibrium crescentic scour structures generated upstream of short horizontal static cylinders. Detailed velocity data were collected using three-dimensional (3D) micro-acoustic Doppler velocimeter (ADV) across and within the equilibrium scour marks. The analysis revealed that the positive and negative values of third-order moments associated with the level bed surface and the scour holes are directly related to coherent structures. The components of TKE flux are discussed for the near-bed region of the level bed surface and scour holes in relation to sweep–ejection events. A cumulant discard method is applied to the Gram-Charlier probability distribution of two variables to describe the statistical properties of the term $u'w'$.

In the quadrant analysis, the conditional statistics of the Reynolds stress along the stream-wise direction across the scour holes could be well represented by a third-order cumulant discard Gram-Charlier distribution. It was confirmed that the probability distribution of the Reynolds stress described the experimental results fairly well. Therefore, the experimental data are used to analyze the turbulent statistics and to predict the trend of coherent phenomena of flow over and within the scour holes generated by the objects/cylinders. The distribution of the joint probability density function in the near-bed region changes cyclically along the scour hole depending on the bottom fluid velocity, which implies a change from upward to downward flux of momentum and vice versa. Both the ejection and sweep events at near-bed points on the level surface are more important than within the scour region; in contrast, both events are stronger for the scour marks than the level bed surface at the outer layer. Sweeps dominate over ejections for the scour hole induced by smaller diameter and ejections dominate for larger diameter. Details are available in Maity and Mazumder (2014). Later Maity and Mazumder (2017) examined the nature of bursting processes using quadrant threshold technique to describe the statistical properties of all three covariance terms $u'v'$, $u'w'$, and $v'w'$ along $(x\ y)$, $(x\ z)$, and $(y\ z)$-planes, respectively, over nearly equilibrium scour hole structure generated by a short cylinder on sediment bed; and verified their results with the existing theory formulated by Nakagawa and Nezu (1977) in $(x\ z)$-plane. Detailed analysis is available in the paper of Maity and Mazumder (2017).

10.11.4 Scour Around Submerged Cylinders of Different Shapes

The flow past a circular cylinder is of significance for many engineering applications, with regards to the flow around e.g., towers, cables and bridge piers. Scouring is an important phenomenon in open channel flow case. It normally takes place in alluvial channels condition, and around human-made structures, like bridge piers. Flow of water interacts with the upstream edge of the obstruction causing an increase in velocity around the sides accompanied by the development of various kinds of vortices like horseshoe vortex, wake vortex for emerged pier-like structures and trailing vortex in addition developed at the top of obstructions for submerged case (Ettema et al., 2006; Tsutsui, 2008; Dey et al., 2008). Scour occurs when fluid flow induces shearing stress on the sand bed surface exceed the shearing strength i.e., greater than the critical bed shear stress. The construction of bridges in alluvial channels causes a tightening in the water flow at the bridge site and hence the considerable scour at the site. Local scour around such engineering constructions changes the river morphology and sea bed, and affects aquatic ecosystems and fish habitats quite sensitively. Local scour around bridge piers was studied by Shen et al. (1969), while Breusers et al. (1977) gave a state-of-the-art review on local scour around piers of different shapes on the basis of field and experimental data. Current research areas include understanding the scour processes, temporal development of scour, predicting scour in sediment beds, parametric studies of local scour, and prediction of scour depth at various types of hydraulic structures. For example, Ansari et al. (2002) studied the influence of cohesion on scour around bridge piers. Ahmed and Rajaratnam (1998) investigated the flow around bridge piers in their laboratory study on flow past cylindrical piers placed on smooth, rough and mobile beds. It was observed that the depth of scour was highly dependent on time. The extent of scour observed downstream of the pier increased as time increased. The flow of fluids around a cylinder of square cross-section is considered to be important problems in the computational fluid dynamics. This is because many of the practical examples involve a similar flow situation. It is clear that extensive work on the flow around different cylindrical shapes viz. circular, square or side by side and two in-line cylinders had been performed over the years, which clearly showed the significant effect of the shape of the cylinder on the flow field and pointed out the importance of study of the flow around different cylindrical shapes (Fael et al., 2014; Sarkar et al., 2015). The effects of pier shape and alignment on the equilibrium scour depth around a single pier has been studied recently by Fael et al. (2014) under clear water flow condition close to the threshold for initiation of sediment motion. They considered five different pier shapes, such as circular, rectangular, square and round-nosed, oblong and zero-spacing pile groups, and made a comparative study with different dimensions of shape. The flow field and scour structures around different shapes of elongated piers, like rectangular, oblong, trapezoidal-nosed, triangular-nosed, and lenticular with a common aspect ratio in identical flow conditions are studied by Vijayasree et al. (2019), using acoustic Doppler velocimeter (ADV). They suggested that the scour depth was dependent on the geometry of the shape.

For example, Sarkar et al. (2015) investigated the high-resolution spatiotemporal local bedform characteristics and coherent structures of turbulent flow around submerged pier-like obstructions of different shapes embedded vertically in the sand bed at a constant flow discharge. They investigated the different deformed bed structures, the statistical characteristics of bed migration, and wavelet power spectral density of temporal bed elevation due to submerged cylinders of different shapes. Experiments were conducted in a laboratory flume (Figure 7.13) using different shapes of cylinders such as circular, elliptical, square and triangular embedded vertically on the sediment bed. For each experiment, a single cylinder was placed at a distance of 6 m from the channel inlet. All the cylindrical structures have a common submergence ratio, i.e., the height of the cylinder to the flow depth is equal to 0.6. For all four experiments, the flow discharge ($Q = 0.015\,\mathrm{m^3/s}$) was chosen in such a way that the bed shear stress away from the cylinder was less than the critical bed shear stress so that the sand particle could not initiate motion, i.e., when there was no sediment motion at the bed except near the cylinder. However, in the vicinity of the cylinder the bed shear stress was high enough to cause sediment motion. In the experiments, the maximum longitudinal velocity occurred at the flow depth of about 20 cm, i.e., the channel width/flow depth was about 2.5, which led to a two-dimensional flow condition at the channel center declining the possibilities of secondary flow (Maity and Mazumder, 2014). Instantaneous sand bed elevations were continuously recorded using SeaTek 5 MHz Ultrasonic ranging system (URS) composed of 24 transducers (Sarkar et al., 2015, Figures 1 and 2). In this experiment, 24 transducers were used out of 32 transducers (Figure 7.8) to collect elevation data in the flume. Each transducer possessed a high precision of ±0.01 cm and was capable of sample collection at 4 Hz rate. The bed evolution due to local scour and deposition at different locations around the cylinder were recorded continuously by URS up to 200 minutes for each of the three experiments. During the experimental time of 200 minutes, i.e., up to the stable bedform, continuous video recording of bedform evolution through the glass side of the flume was made. The video recording was stopped nearly at the time of stable bed condition. The motivation was to study the spatiotemporal evolution of scouring and the bed deformation pattern due to cylindrical obstacles of different sizes. The shape of the scouring zone and the rate of upstream scour depth became negligible after about 200 minutes of run for all the cases. This was confirmed by test runs by critically measuring the scoured bedform patterns continuously for a longer period of time (about 300 minutes) when less than 0.15 cm scour was recorded per hour at the upstream of the cylinder. In this study, this condition was chosen to be the stable bed condition. When the stable bed condition attained, the set-up of URS was removed and the near-bed velocity data were collected at each location of transducer around the cylinder using a SonTek 0.05 m down-looking 3-D Micro acoustic Doppler velocimeter (ADV) for 3 minutes (180 seconds) at a sampling rate of 40 Hz from each position. Experimental set is explained clearly in Sarkar et al. (2015).

The patterns of commencement of bed deformation around different cylinders were quite different. The shapes of submerged cylinders significantly changed the bed surface area locally at the downstream of the respective cylinders. Figure 10.43a–d shows the final stable deformed areas around the submerged cylinders of four different shapes of same size, i.e., circular, elliptical, square and triangular respectively. From the figures it is observed that the scoured radii along and transverse to the flow with respect to the cylinder center, the position of major deposition regions at the downstream, and the overall shape of the distorted area occurred due to the impact of the shape of the cylinder. Two basic forms of scouring took place around the cylinders: one at the upstream of the cylinders due to the formation of horse-shoe vortices causing frontal scour, and the other occurred at the downstream side of cylinders, followed by turbulent wakes causing the deformed bed surface. Equilibrium scour mark attained after a couple of hours with a development of bedform structures at upstream and downstream of cylinders. The scour at the upstream and deposition at the downstream for the elliptical cylinder along the centerline occurred faster than that of the circular cylinder. A symmetric scoured and deposited deformed area about the circular cylinder was observed, but the secondary bedforms created by turbulent wakes

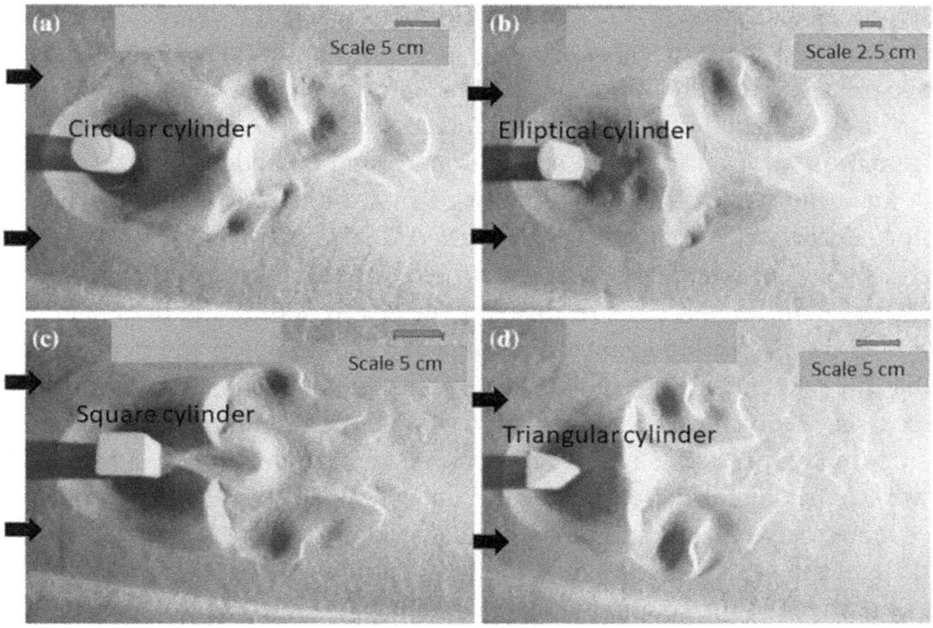

FIGURE 10.43 For cylindrical piers: (a) circular, (b) elliptical, (c) square and (d) triangular shape photos of deformed stable bed surface after 200 minutes of experimental time. (Modified from Sarkar et al., 2015.)

at the downstream were not symmetric. Similar bedform structures were also observed for the case of elliptical cylinder, which implied that circular and elliptical shapes belonged to the same family. No sediment ridge was observed downstream of the circular cylinder, while a clear long sediment ridge along the centerline was identified just at the downstream of elliptical cylinder at the major deposition area. The shapes of the deformed areas (both upstream and downstream) due to square and triangular-shaped cylinders about the flume centerline occurred to be fully symmetric, which could be considered to be in the same class.

For the square cylinder, larger accumulations of sediments were at two sides of the cylinder centerline and for the triangular cylinder, deposition took place around the downstream nose which was to be largest accumulation among the four cases during the whole experimental time. It is noticed that for the square cylinder the sediments were largely accumulated at the two sides about the centerline along downstream, followed by a structure like "water drop" just behind the square cylinder, which did not occur for the triangular cylinder case, but a ridge transverse to the flow behind the triangular cylinder was observed. At some distance downstream of the cylinder center, lunate type of bedforms was created along two sides of centerline with deposition region with a scoured zone at the end of the stoss side in the upstream facing direction for all the cases except the square cylinder. Going to further downstream of this area, the linguoid type of bedforms was observed especially in the case of circular and elliptical cylinders. Details are available in Sarkar et al. (2015).

10.12 SCOUR AROUND CYLINDRICAL BRIDGE PIERS

Local scour due to turbulent flow around a circular cylinder is a long-standing problem in terms of the scour around bridge piers. It is the outcome of the erosive action of the flowing water, which erodes sediment materials from the river bed, banks of natural streams and near obstacles like piers or abutments of a bridge. Immovable obstructions like bridge piers resist the flow leading to an increase in velocity around the sides of the obstacles accompanied by various kinds of vortices like horse-shoe vortices, wake vortices, etc., which generate three-dimensional (3D) character of complex flow field. Hence, the prediction and mechanism of scouring process and the hydrodynamics of flow structures around the bridge piers in rivers are of great importance to civil and mechanical engineers and researchers in the field of environmental fluid mechanics and ecology.

Extensive amount of work on flow and scour around sand-embedded bridge piers were studied by several authors since the work of Tison (1961), who first initiated to study the local scour in rivers through laboratory experimentations, using different models of piers like rectangular, triangular, aerodynamic and lens-shaped structures. Breusers et al. (1977) made a 'state of the art' report to summarize the features of local scour around piers of different shapes, based on the field and experimental data. They also suggested the protection against the scour and the design of the pier structures. It was also clearly stated that till now no entirely satisfactory theoretical and experimental results have come out. Because the processes involved with the interaction of water and sediment transport associated with the pier structures are too complicated, and experimental data are incomplete and sometimes conflicting. The research work on scour formation due to turbulent flow around structures of different shapes or models is rather more recent than that around circular bridge piers.

Melville and Raudkivi (1977) investigated the flow patterns, turbulence intensity distributions and the boundary shear stress in the scour zone of a circular pier under clear water scour conditions. The experiments were confined to the clear water scour condition and the development of one scour hole from the initial flat bed through the equilibrium scour hole condition. A flume of 19 m long, 456 mm wide and 440 mm deep was used for the experiments. The working section was 15 m from the upstream end and a cylinder of 50.8 mm diameter was used as the pier. The flume had a fixed flat bed coated with the sand used, except for a 1.8 m long and 0.13 m deep trough filled with sand at the working section. The sand used in the experiments had a uniform grading curve with d_{35} = 0.300 mm, d_{50} = 0.385 mm, d_{65} = 0.500 mm and a specific gravity of 2.65. A flow of 17.12 L/s was used and it yielded on the bed slope of J = 0.0001 a uniform flow depth y_0 = 0.15 m and a mean approach velocity $\bar{u}$ = 0.25 m/s. The flat bed of the above sand-grain roughness was hydraulically smooth for the threshold of sediment transport condition. The shear velocity for this uniform flow was $u_* = (gy_0J)^{1/2} = 0.0121$ m/s. The study showed that approach flow velocity decelerates as the flow approach toward the pier and reduced to zero near the pier surface. It was also observed that near the pier turbulent intensity is very high in the scour hole.

Kothyari et al. (1992) performed the experiments in a laboratory flume to study the temporal variation of scour depth around circular bridge piers during steady and unsteady clear-water flows in uniform, non-uniform, and stratified sediments. The experiments were conducted in a flume 30.0 m long, 1.0 m wide, and 0.60 m deep, located in the Hydraulics Laboratory of the University of Roorkee, India. Two longitudinal slopes, namely 6.61×10^{-4} and 1.2×10^{-3}, were used. The working section, 2.0 m long, 1.0 m wide, and 1.2 m deep where the piers were located, was 12.0 m downstream from the entrance of the flume. This reach was filled with sediment to a depth of 0.60 m below the bed level of the flume. The first series of experiments was carried out to obtain some idea of the size of the horseshoe vortex under different flow conditions. The horse-shoe vortex is quasi-periodical and is a multiple-vortex system. They observed that the primary vortex in front of the pier is the prime agent causing scour. Details are explained in Kothyari et al. (1992). The experimental data on the temporal variation of scour depth in all these cases for clear-water flow were collected. An attempt has also been made to develop a relationship for the estimation of the maximum scour depth in clear-water flows.

Melville and Chew (1999) developed the clear-water local scour depth around cylindrical ridge piers in uniform sand beds. The influence of flow duration on the depth of local scour was studied. The effect of time on the development of depth of scour at cylindrical piers under clear-water conditions was conducted by a series of experiments, in which scour depth was monitored. The experiments were run for a long period of time to ensure that equilibrium was reached. An expression for the equilibrium depth of local scour d_{es} at a cylindrical pier of diameter D in a uniform sediment can be written as follows:

$$\frac{d_{es}}{D} = f\left(\frac{u}{u_{*c}}, \frac{y}{p_d}, \frac{d_{50}}{p_d}, \frac{t}{t_e}\right) \tag{10.133}$$

where u is the mean approached velocity, u_{*c} is the critical mean approach flow velocity for entrainment of bed sediment, y is the mean approach flow depth, d_{50} is the median size of the sediment, t is the time, t_e is the time for equilibrium depth of scour, and p_d is the cylindrical pier diameter. Here the first three parameters on the right-hand side represent respectively the stage of sediment transport on the approach flow bed, termed the flow intensity $\frac{u}{u_{*c}}$; the depth of flow relative to the width of the pier, termed the flow shallowness $\frac{y}{p_d}$; and the sediment median size relative to the pier width, termed the sediment coarseness $\frac{d_{50}}{D}$. Figure 10.44 shows schematic diagrams indicating the effects of these three parameters such as: flow intensity, shallowness, and coarse sediment on local scour depth. Discussions are given in detail in Melville (1997) and Melville and Chew (1999). It is interesting to note that under clear-water conditions, the local scour depth in uniform sediment bed increases almost linearly with flow intensity to a maximum at the threshold velocity. This maximum scour depth d_{esMax} is called the threshold peak. If the velocity exceeds the threshold velocity, the local scour depth in uniform sediment first decreases and then increases again to a second peak, which is relatively small, but the threshold peak is not exceeded providing the sediment is uniform.

The shallowness parameter influences local scour depth when the horse-shoe vortex, which is the principal cause of scour, is affected by the formation of the surface roller that forms at the leading edge of the pier. Two vortices which are formed have the opposite direction in rotation. The surface roller becomes more dominant and provides the base vortex less capable of entrainment of sediment with reducing the flow depth. Thus, the local scour depth is reduced for shallower flows. In very shallow flows, the local scour becomes independent of flow depth (Melville and Chew, 1999). These trends are evident in the laboratory data of many researchers, including Breusers et al. (1977), Jain and Fischer (1980), Ettema (1980), Chee (1982), and Chiew (1984).

The experiments were conducted by Melville and Chiew (1999) in four different flumes, three at the University of Auckland and one at the Nanyang Technological University. Uniform sands were used at each flume, with sediment coarseness (dimensionless) ranging from 20 to 222. For most of the data, sediment coarseness exceeds 50. Flow intensity ranges from 0.46 to 0.957, while flow shallowness ranges from 0.6 to 12.5. The collected data of equilibrium scour depth are well represented by the following equation:

$$\frac{d_s}{d_{es}} = \exp\left[-0.03\left|\frac{u_{*c}}{u}\ln\left(\frac{t}{t_e}\right)\right|^{1.6}\right] \tag{10.134}$$

where the modulus sign is used to ensure the argument of the negative exponential function. The equilibrium time scale ($t^* = ut_e/p_d$) for the development of a clear-water local scour hole at bridge pier is a function of flow intensity (u/u_{*c}), flow shallowness (y/p_d), and sediment coarseness (p_d/d_{50}). They have also developed an expression for the time taken for equilibrium scour depth based on pier diameter, sediment size and flow velocity.

Balachandar et al. (2000) illustrated the research to understand the effect of tailwater depth on scour hole development, to examine some of the parameters that influence the local scour cycle, and

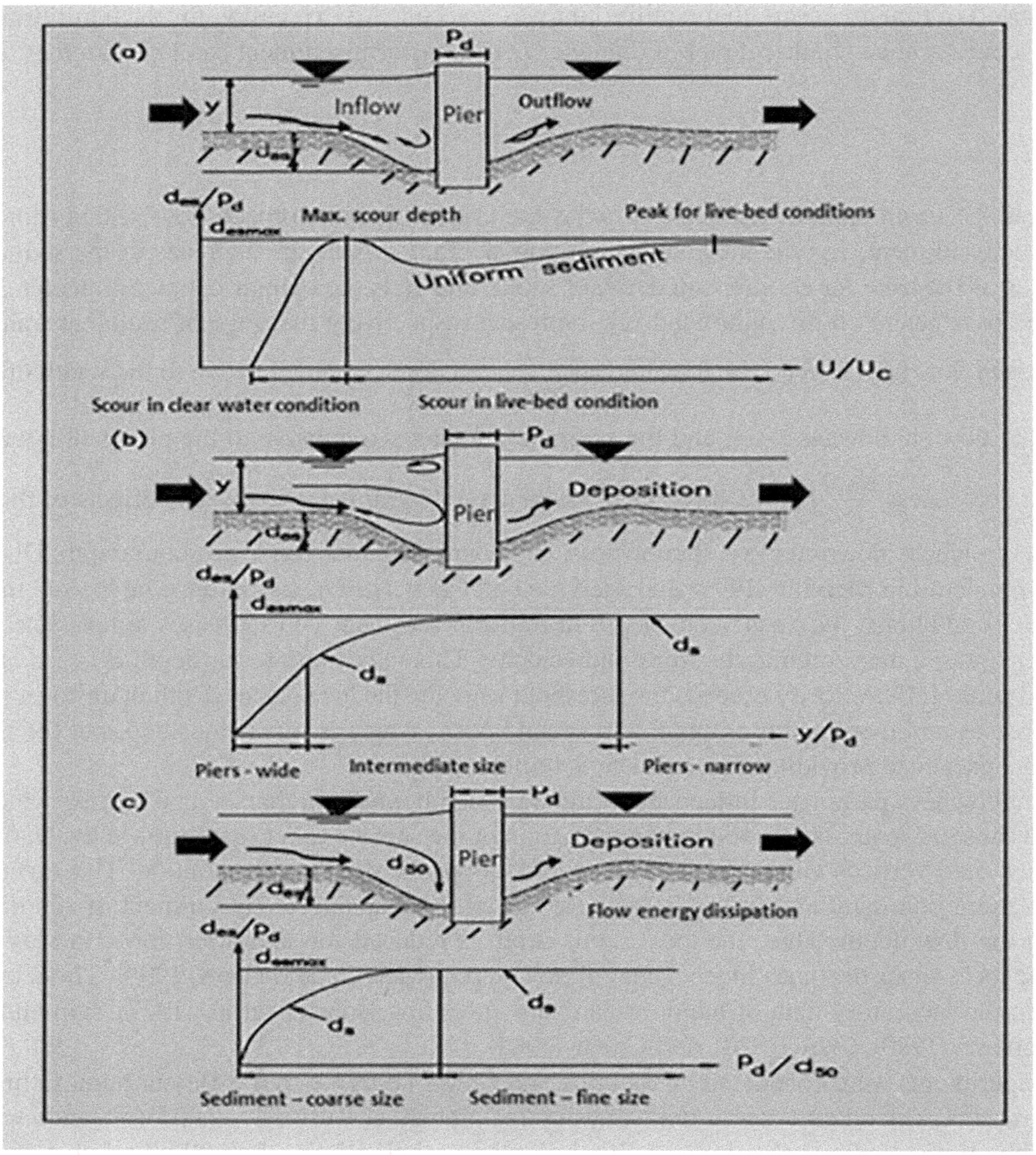

FIGURE 10.44 Schematic of local scour structures due to the influence of (a) flow intensity, (b) shallowness of flow, and (c) sediment coarseness. (Modified from Melville and Chiew, 1999.)

to develop suitable scaling laws. They carried out the experiments using a video imaging technique to obtain the scour profiles, and supplemented by laser Doppler anemometer (LDA) for velocity measurements within the scour region. They proposed scaling laws for predicting the tail-water effect on the scour profile. The streamwise and transverse velocity components indicate that the flow is directed toward the bed during the digging phase and the flow is directed toward the water surface during the refilling phase.

Muzzammil and Gangadhararaih (2003) investigated experimentally the mean characteristics of primary horse-shoe vortex developed in front of a cylindrical pier and believed to be the prime agent responsible of scour during the entire scouring process. A simple and effective method was developed to determine the time-averaged characteristics of the vortex in terms of parameters, like flow velocity, pier and the channel bed. They have also developed an expression for the maximum equilibrium scour depth from the vortex velocity distribution inside the scour hole. The resulting scour prediction equation is compared with those of existing well-known equations and was found to be satisfactory. Experiments were conducted in a glass-walled flume of length 5.0 m and width 0.5 m with fine sediment of median size 0.16 mm and coarse sediment of median size 0.60 mm. The circular hollow glass cylinders were modeled as piers with diameters varying from 31.0 to 78.5 mm. Mudflow visualization technique was used to visualize the horseshoe vortex in the plane of symmetry in front of pier. A detailed description of the mud flow visualization technique, the vortex probe and the vortex gauge, the method of computation of the vortex characteristics, and the measurement techniques are briefly given in Muzzammil and Gangadhararaih (2003). The presence of complex vortex systems has been confirmed by many researchers in the past by investigating the flow patterns using different flow visualization techniques (Dargahi, 1989; Kirkil and Constantinescu, 2015). These studies clearly indicated that the horseshoe vortex system grew near the upstream edge of the cylinder and continued to stretch on the downstream side. The dependency of vortex characteristics on the Reynolds number can also be understood from these studies.

Vijayasree et al. (2019) carried out the experiments in a laboratory flume to the flow field and scour characteristics around different shapes of elongated piers, like rectangular, oblong, trapezoidal-nosed, triangular-nosed, and lenticular on a sediment bed, having a common aspect ratio under identical flow conditions, using acoustic Doppler velocimeter (ADV) measurements. A comparative study was made among the five different shapes of scouring due to each shape under identical flow conditions. They identified the mean velocity components and turbulence characteristics associated with scour feature around the pier of different shapes; and how these are related to the geomorphic properties. They reported that the occurrence of maximum scour depth depends on the geometry of the shape, which perturbs the flow.

Experiments were conducted in a flume at the Hydraulics Laboratory of Indian Institute of Technology (IIT), Bombay, India (Figure 7.18). The flume of 8.5 m long, 0.305 m wide and 0.6 m deep with a constant slope of 0.0003 was used for the experiments. The walls of the flume were made of Plexiglas. The test section about 4.75 m long was filled with sand of depth 0.2 m. The bed material was alluvial quartz with $d_{50} = 0.8$ mm with specific gravity of 2.66, where d_{50} is the mean grain size in the bed. The coefficients of uniformity, curvature and standard deviation of the sediment sample were 1.98, 0.93 and 1.426, respectively. The coefficient of uniformity C_u is defined as ratio $C_u = d_{60}/d_{10}$, where d_{60} and d_{10} represent respectively the sizes for which 60% and 10% of the mixture are finer. The experiments were carried out using respective flow discharges of 0.006, 0.009, 0.012, 0.015 and 0.018 m^3/s with corresponding water depths of 0.111, 0.126, 0.143, 0.153 and 0.165 m above the flat sand bed (Table 1, Vijayasree et al., 2019). Here, the flow intensity ratio was $U_{max}/U_c < 1$ for all five discharges. To identify the turbulence and the scour mechanism with respect to the change in pier shape, flume experiments were performed using five different pier shapes, like rectangular (P1), oblong (P2), Trapezoidal-nosed (P3), Triangular-nosed (P4), and Lenticular (P5), with a common aspect ratio 5 (aspect ratio is pier length/width) under identical flow conditions (Figure 10.45a) with eight schematic measuring locations shown in Figure 10.45b. The blockage ratio was defined as the ratio between the flume width and pier width, which was about 10 times of the pier width for the clear water flow (Shen et al., 1969). The velocity data were used to calculate the following statistical quantities: mean velocities, turbulence intensities as the standard deviations of the measurements of the instantaneous velocities, Reynolds shear stresses and TKE around the pier shapes, and the detailed results are presented in terms of figures in Vijayasree et al. (2019). The patterns of commencement of scour formation in front and around different shapes of bridge piers were quite different. The shapes of piers significantly changed the scour structure locally around the respective piers (Figure 10.46). The depth of scour was highly dependent on the flow velocity and the variation of turbulence intensity in front of the pier.

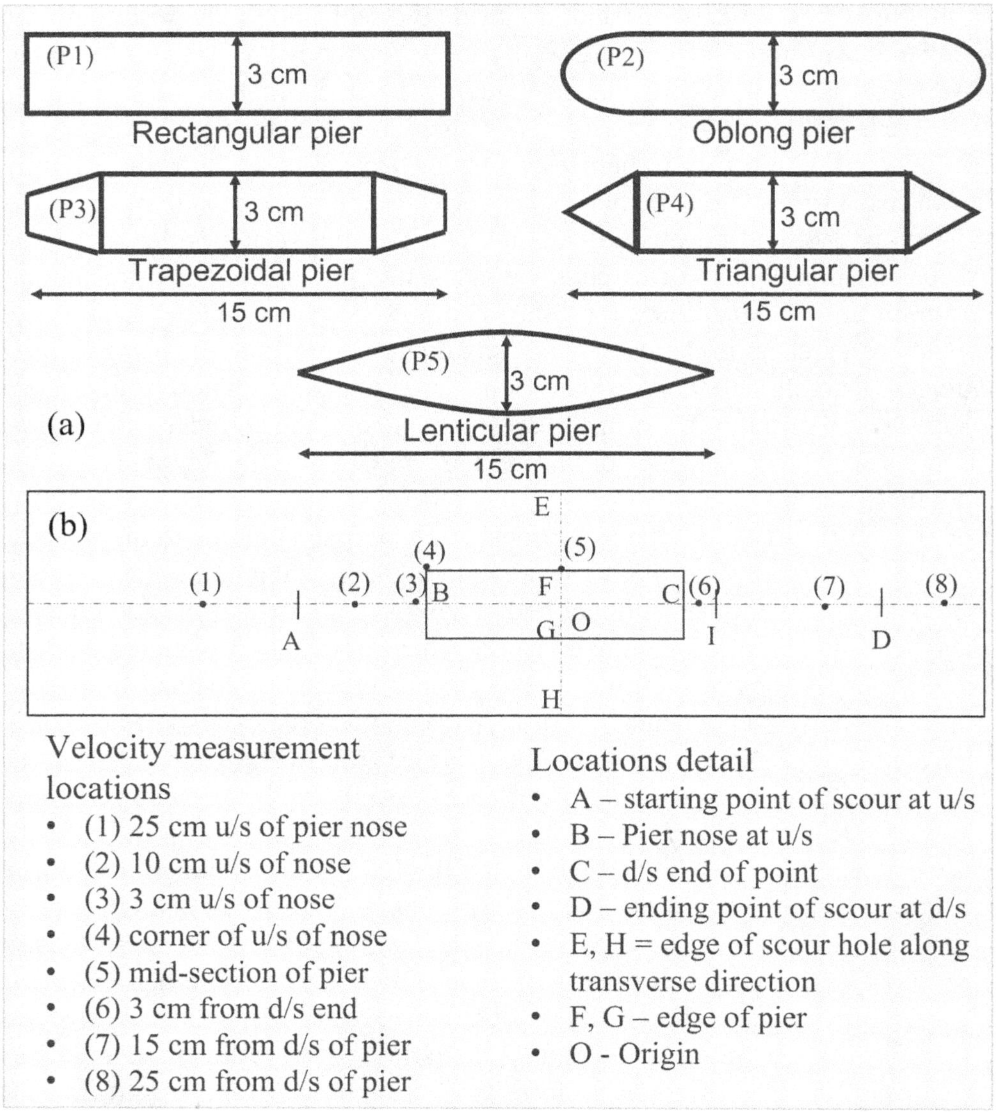

FIGURE 10.45 (a) Various shapes of piers used in the experimental study. (b) Eight measuring locations. (Modified from Vijayasree et al., 2019.)

The maximum scour depth and the depth at the nose increased consistently with increase in discharge for all the pier shapes. The stream-wise velocity decreased for all the pier shapes away from the boundary compared to that of flat surface, but the reduction was prominent near the bed for rectangular, oblong and trapezoidal piers because of the generation of frontal scour. Negative shear stress within and above the scour hole at different locations indicates the outward flux of momentum at these locations, whereas positive shear stress leads to inward flux of momentum. Details of other important results are available in Vijayasree et al. (2019). Subsequently, Vijayasree et al. (2020) studied experimentally the turbulent flow field along an oblong pier mounted vertically over the sediment bed and compared the results with that of a simple cylindrical pier. An attempt has been made to address the various flow parameters that are associated with the turbulence, such as mean velocities, turbulence intensities, Reynolds shear stress, TKE, spectral analysis, and the fractional contributions of burst-sweep cycles to the total shear stress along the oblong and the simple circular pier. The spectral analysis of instantaneous velocity data is executed for both the piers to focus the spectral energy distributions and the vortex shedding frequency with the corresponding Strouhal number. Detailed analyses of the

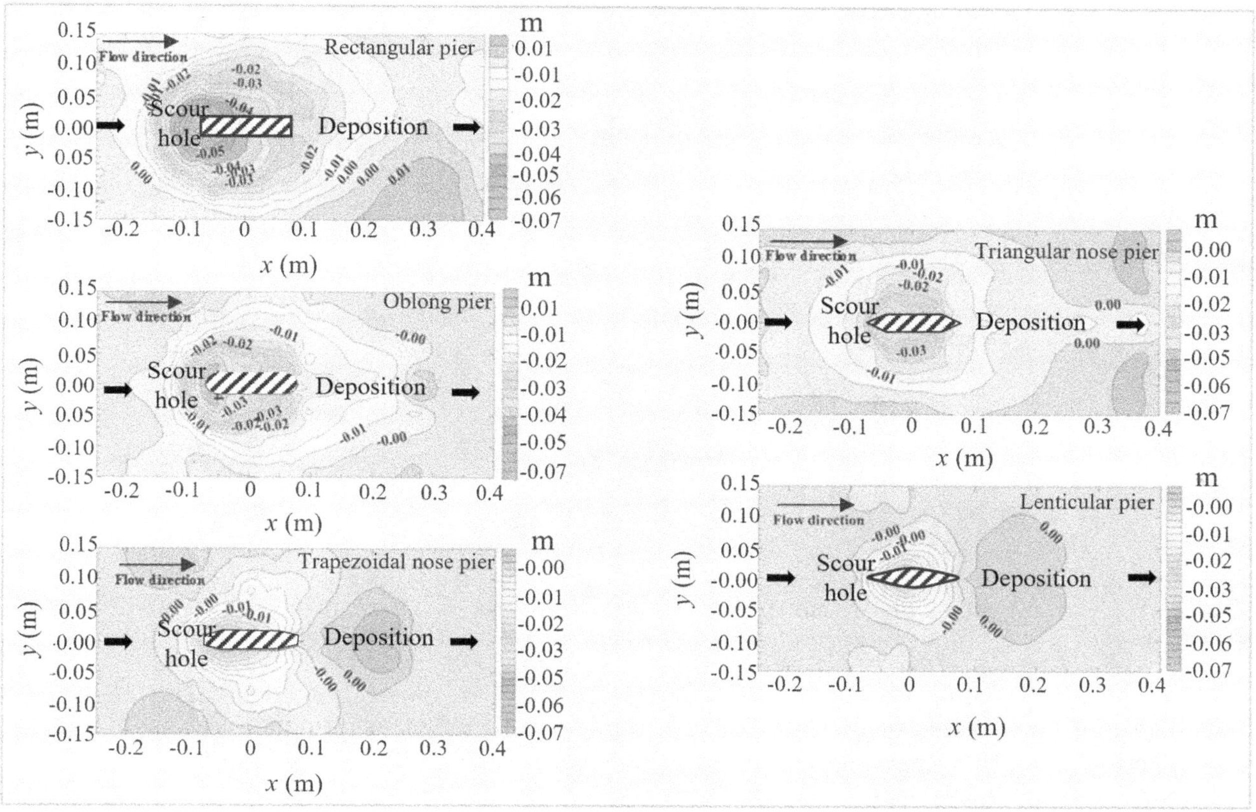

FIGURE 10.46 Scour contours at equilibrium stages for highest discharge. (Modified from Vijayasree et al., 2019.)

hydrodynamics and basic turbulence phenomena around the piers placed on the rigid surface under the interacting flow are shown in Vijayasree et al., (2020).

Misuriya et al. (2021) investigated experimentally the turbulent flow and its higher order statistics around circular cylinders of different diameters mounted vertically on a flat rigid surface, using Particle Image Velocimetry (PIV). They illustrated the time-averaged turbulence parameters, such as mean velocity, intermittency factor, TKE flux, the third-order moment of velocity fluctuations, TKE budget, turbulent production and dissipation rates, turbulent length scales and contributions of burst-sweep cycles to the Reynolds shear stress for three different cylinder Reynolds numbers. On the downstream, the production and dissipation rates increased with cylinder Reynolds number, whereas the opposite trend was observed for dissipation only on the upstream. Increase of Re_d led to increase in Kolmogorov length-scale much higher on the upstream than the downstream, whereas it showed depleting effect on Taylor's length scale on the upstream and increasing effect on the downstream.

Experiments were carried out in a 14 m long re-circulating flume available in the Hydraulics Laboratory of the Civil Engineering Department, Indian Institute of Technology (IIT), Bombay, India. The width and depth of the flume are respectively 50 cm and 90 cm. A schematic diagram of the flume with its accessories for experimental set up is shown in Figure 7.19. A Stereoscopic Particle Image Velocimetry (S-PIV) from Dantec Dynamic, Denmark was used to collect the velocity data at all three directions. Two Charge-Coupled Device (CCD) cameras with resolutions of 2048×2048 pixels were mounted on a traversing system for the precise movement of the camera to capture the images from different sections of the flume (Figure 10.47). This figure shows the arrangement of camera and other PIV details with respect to measuring plane. According to researchers, like Adrian (1991), Adrian and Westerweel (2011), the flow was seeded using silver-coated hollow spherical particle of diameter

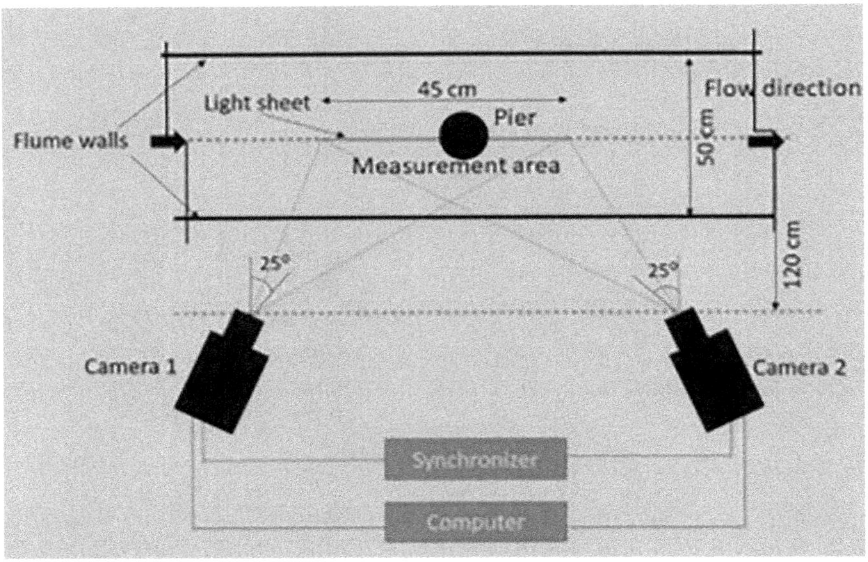

FIGURE 10.47 PIV set up with two cameras: schematic diagram of plan view. (Modified from Misuriya et al., 2021.)

10 μm, made of borosilicate of density 1.03–1.05 kg/cm^3 for PIV measurements. The calibration and data collection using PIV are explained in detail at Misuriya et al. (2021). The results show the effects of different cylinder Reynolds number (Re_d = 12600, 16800 and 21000) on the mean flow and higher-order turbulence parameters. Misuriya et al., (2023a) investigated experimentally the turbulent flow characteristics around different cylindrical objects such as three inline circular cylinders, oblong, and rectangular cylinders using PIV in Flume -II (Figure 7.19).

The above-mentioned literature presents the studies of the turbulent flow mostly on the interaction of flow around a simple bridge pier; however, in practice, many bridges are wide and comprise of many in-line piers with different spacing between the piers. Alam and Zhou (2007) investigated the flow structure around two side-by-side circular cylinders, focusing the ratio between the gap spacing of cylinders and the cylinder diameter. They observed a significant influence of the gap spacing between the cylinders on the time-averaged lift, and two distinct flow structures depending on the ratio. Beg (2010) and Beg and Beg (2015) have studied scour hole characteristics of two piers placed side by side and around two unequal-sized bridge piers in tandem arrangement respectively for various spacing of piers. The local flow field around side-by-side bridge piers with and without scour hole and flow characteristics around single and tandem piers are investigated by Ashtiani and Kordkandi (2012, 2013). Das and Mazumdar (2015) carried out an experimental study to investigate the horse shoe vortex and flow characteristics in a local equilibrium scour hole around two identical cylindrical piers placed along the flow with an eccentricity and concluded that the eccentric arrangement of piers play an important role in the formation of greater scour depth at the eccentric rear pier. Keshavarzi et al. (2017) investigated the turbulent flow interaction experimentally around two in-line circular piers with different spacing and showed the significant influence of spacing between the two in-line piers. The estimation of maximum scour depths due to the spacing between two in-line circular piers aligned in the flow direction was investigated very recently by Keshavarzi et al. (2018) under clear water conditions.

10.13 FLOW AND SCOUR AROUND COMPLEX BRIDGE PIERS

As discussed in previous sections, extensive research work on the scour and the flow field around simple bridge piers have been carried out over the last few decades. Researchers are fully involved to test the various techniques to suppress the vorticity around simple piers. Most studies are performed for bridge piers with a uniform cross-section or circular cross-section. Due to geotechnical and economic reasons, multiple-pile bridge piers and complex piers have become popular nowadays in bridge design (Breusers et al., 1977; Melville and Coleman, 2000; Coleman, 2005; Ashtiani et al., 2010; Beheshti and Ashtiani, 2016). Nowadays, complex piers (CP) are preferred over simple piers to be constructed in rivers system. Complex pier consists of three different structural elements, viz, a cylindrical column resting on a pile cap supported by a number of piles, also known as pile foundation (Figure 7.26). Over its lifetime, the different elements of CP get exposed to the flow because of the varying amount of sand aggradation or degradation around the pier, which poses a challenge to the engineers in predicting the maximum scour depth around such piers.

The Federal Highway Administration (FHWA) recommended a method of superposition for complex piers in its *Hydraulic Engineering Circular No. 18* (HEC-18) to estimate the scour depth d_s around bridge piers being obtained from the sum of contributions to the scour depth caused by each component (Richardson and Davis, 2001) as

$$d_s = d_{scol} + d_{spc} + d_{spg} \tag{10.135}$$

where d_{scol} is the scour component for the pier stem or column, d_{spc} is the scour due to pile cap, and d_{spg} is the scour depth of pile exposed to the flow. Each component in the equation is computed as:

$$\frac{d_s}{h} = 2\ k_1 k_2 k_3 k_4 \left(\frac{p_d}{h}\right)^{0.65} Fr^{0.43} \tag{10.136}$$

for given flow intensity (u/u_{*c}), and flow shallowness (y/p_d).

Here h is the approached flow depth, p_d is the pier diameter, k_1 is the shape factor, k_2 is the angle of attack factor, k_3 is the dune factor, k_4 is the correction factor for the size of sediment bed material, Fr is the Froude number defined by U depth-averaged velocity. Sheppard and Glasser (2009) proposed an alternative methodology for estimating design scour depth using these structures. The interaction of flow with three different structural elements of complex pier generates new vortex systems, as observed by Beheshti and Ashtiani (2010). Over the last few years, quite number of research work on scour patterns around complex piers of varying shapes with different flow conditions and elevations of pile cap are investigated by several researchers (Beheshtil and Ashtiani, 2010; Ashtiani et al., 2010; Moreno et al., 2015; Yang et al., 2018; Gautam et al., 2019, 2022). Some of these studies are focused on examining the maximum scour depths and the application of scour equations around complex pier (CP) without investigating the key parameters of turbulence that are responsible for scour. As there is no scope to discuss all these researches on flow field and scour around the complex bridge piers, only a few important research works, which are basic and important are discussed here.

Coleman (2005) presented a new methodology to predict local scour depth at a complex pier that combined the existing expressions for scouring at uniform piers, caisson-founded piers, pile groups with debris rafts, and pile groups alone. The method recognized the relative scouring potentials of the components of complex piers and the transition of scouring processes occurring for varying pile cap elevation. Sheppard and Glasser (2009) conducted clear-water and live-bed experiments at several laboratories in the USA and New Zealand and developed predictive equations for both simple and complex piers. They discussed predictive equations for local scour around complex structures founded in non-cohesive sediment. They first discussed the equations for simple piers, followed by the equations for the complex pier. Ashtiani et al. (2010) first investigated experimentally the three-dimensional turbulent flow field around a complex bridge pier placed on a rough fixed bed under steady clear-water condition, using the acoustic Doppler velocimeter (ADV) to measure all three velocity components at different locations. All the three elements of complex pier were exposed to the approaching flow. It consisted of a column, a pile cap and a 2×4 pile group found in the sediment bed. A total of 70 experiments were carried out. Three sets of experiments were performed over the entire range of possible pile cap elevations for complex piers with different geometrical characteristics. A range of configurations, including the different pile cap elevations, pile sizes, and shapes of complex piers were considered. Some of the available methodologies to estimate the maximum local scour depth around such complex piers are evaluated. From their data analysis, they estimated the pile cap elevation, thickness of scour depth, time development of scour depth around the complex pier. The maximum scour depth occurs when the pile cap was undercut. Therefore, to estimate the scour depth accurately, the pile cap elevation at which the cap is undercut must be determined. This is highly sensitive to the pile cap elevation when the top of the pile cap is near the bed surface and this sensitivity is more intense for smaller values of the ratio of the column width to the pile cap width. For the cases where the pile cap is located at the bed, the scour process becomes more complicated. The horizontal extensions of the pile cap guard the bed from the scour process, and due to this, it takes a long time to reach equilibrium condition in these cases. With increasing pile cap elevation, the depth of scour may increase or decrease depending on the variation of the equivalent width of the complex pier subjected to the flow.

Simultaneously, Beheshti and Ashtiani (2010) investigated the three-dimensional turbulent flow field around a complex pier with a rectangular pile cap placed on a fixed rough bed. The column and pile cap had a rectangular cross section, while the piles had a circular cross section. The experiments were conducted in a 15 m long, 1.26 m wide, and 0.9 m deep channel. For the experimental runs, a false floor was made on the bottom of the channel. At a distance of 10 m from the upstream end of the channel, there is a 2 m long and 1.26 m wide recess. The piles were rigidly mounted by pressing them tightly in precision-bore holes on a plate located in the flume bed and in holes under the pile cap. In a similar manner, the column was mounted on the pile cap. An acoustic Doppler velocimeter (ADV) is used to record the instantaneous velocity components at different horizontal and vertical planes. The distributions of time-averaged components of velocity field, turbulent intensity components, and Reynolds stresses at different vertical planes are presented. The vector plots of the measured velocity data at different vertical and horizontal planes were used to represent the flow field. The velocity distributions at different vertical sections are compared with log-law velocity profile in open channel flows. Beheshti and Ashtiani (2016) studied experimentally the turbulent flow field around a complex bridge pier with a developed scour hole, using ADV. The complex pier consists of a column, a pile cap and a 2 × 4 pile group. The time-averaged velocity components, turbulence intensities, and Reynolds stresses are presented at different horizontal and vertical planes. Streamlines shows the complexity of the flow around the complex pier. The experiments were conducted in a rectangular channel of 15 m long, 1.26 m wide, and 0.9 m deep. A complex pier was placed in a sediment bed 2 m long, 1.26 m wide, and 0.35 m deep, which was filled with uniform sand of median diameter d_{50} of 0.71 mm, specific gravity of 2.45 and geometric standard deviation σ_g of 1.20. The sediment recess was located at a distance 10 m from the upstream end of the channel.

In order to understand the flow characteristic around the complex piers, Gautam et al. (2019) investigated experimentally the mean flow and turbulence parameters around an elliptical shape of pile cap using PIV and compare the results with that of a simple pier for different flow Reynolds numbers based on the flow depth. Experiments were conducted in a flume of 8.5 m length, 0.305 m width and 0.6 m depth, with a constant slope of 0.0003 (Figure 7.18, Flume 1) at the Hydraulics Laboratory of the Department of Civil Engineering at Indian Institute of Technology (IIT) Bombay, India. The origin was considered at the pier center, where x-axis was along the flow, y-axis vertical to the flow and z-axis was transverse to the flow. The complex pier model was specifically designed as a cylindrical column resting on an elliptical pile cap supported by a 2×2 array of pile group (Figure 7.26).

Prior to starting the experiments, calibration was done using the calibration target of size 450 × 450 mm placed fully under water in the flume to account for refraction due to multiple media (air-glass-water). The two CCD cameras were mounted on the traverse to maintain uniform viewing and capturing of the flow field. Once the fully developed flow was ascertained, experiments were performed with simple pier and complex pier for different Reynolds numbers, viz, 19,320, 29,025 and 38,715. Velocity data collection using PIV and analyses are explained clearly in Gautam et al. (2019). The analyses were performed for all three Reynolds numbers to compare the mean velocity contours between SP (simple pier) and CP (complex pier). The effect of Reynolds number on the mean velocity along upstream and downstream of both SP as well as CP (on the symmetry plane) is discussed. The velocity, intensities and Reynolds shear stresses were normalized using the depth-averaged velocity (U).

The streamline plots describing the role of pile cap along the centerline xy-vertical plane on the upstream and downstream of the piers are shown in Figure 10.48a for Re = 19320, in Figure 10.48b for Re = 29,025, and in Figure 10.48c for Re = 38,715. Interesting phenomena were observed from the figures that the nature of 3-D flow with converging and diverging streamlines along the piers, especially for Re = 19,320 along the simple pier at just downstream. The comparison of TKE contours suggests the influence of pile cap and the group of piles in suppressing the TKE around the pier. Vorticity and power spectra were also performed to investigate the shear layer generation and vortex shedding frequencies. Later Gautam et al. (2022) investigated the turbulent flow characteristics around a complex pier with an elliptical pile cap for understanding the mechanics of flow responsible for scour using an

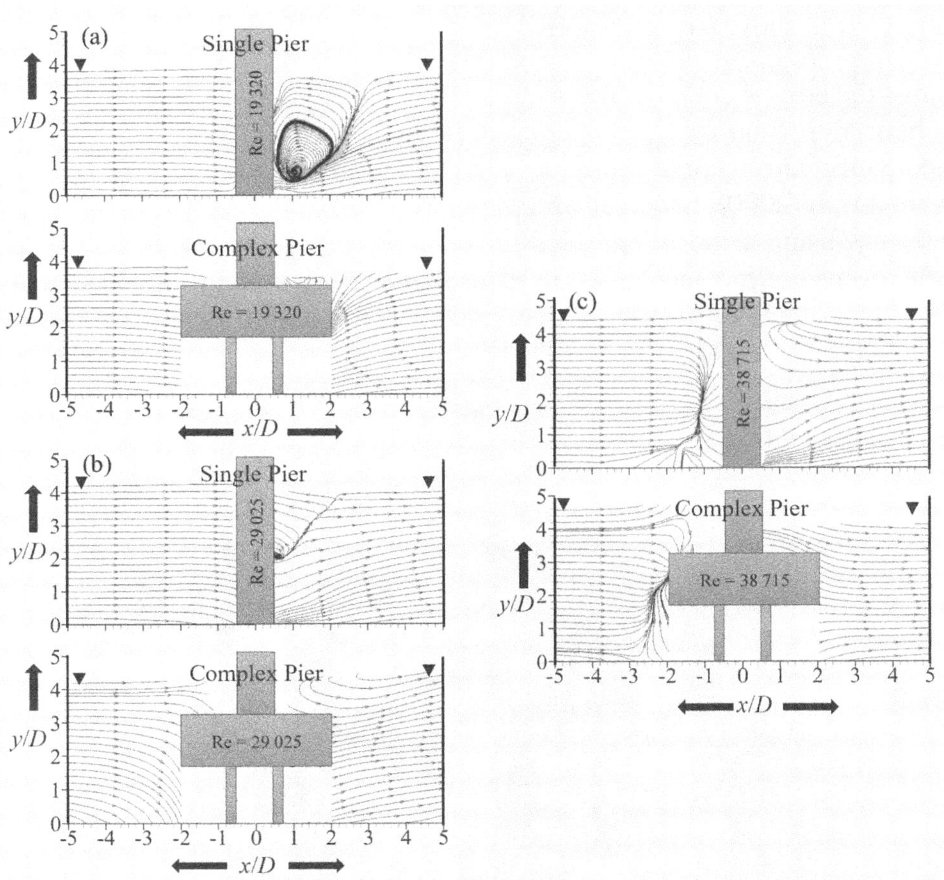

FIGURE 10.48 For simple and complex piers: streamlines along the symmetry plane for different Reynolds numbers (Re). (Modified from Gautam et al., 2019.)

acoustic Doppler velocimeter (ADV) for a Reynolds number of 67,745. They examined mean velocities in horizontal and vertical planes, Reynolds stresses, TKE, and spectral analysis around the CP, which are crucial to estimate the scour patterns. The stream-wise spectra with vortex-shedding frequencies and corresponding Strouhal numbers are focused on three distinct regions generated by CP. This study highlights the flow field around complex pier on a fixed bed, and the presence of a developing or an equilibrium scour hole may alter the flow field.

Further Gautam et al. (2021) carried out scour studies in a larger flume (Figure 7.20) for various pile cap shapes and hydrodynamics features were captured using a PIV. Figure 10.49 shows a typical scour hole developed for rectangular pier cap for fully exposed case. More details on the experimental study using PIV are shown in Gautam et al. (2021).

10.14 SCOUR PROTECTION MEASURES

River bridges over the sediment bed are highly susceptible to scour at piers and abutments. These are important to design for preventing from the severe scouring action safely. They should be monitored periodically to assess scour, which is evaluated according to the level of bed lowering with respect to the footing elevation of the bridge. Assessment of scour vulnerability of a bridge leads to design of appropriate protective measures. The major protective techniques which are employed for preventing or minimizing the local scour at bridge piers can be classified into three categories (Tang et.al. 2009):

1. Bed armoring protective measures,
2. Flow-altering protective measures,
3. Combination of the above two measures.

In order to combat the erosive action of the scour-inducing mechanisms using hard engineering materials or physical barriers such as rock riprap is used. In the latter case, it either inhibits the formation of the scour-inducing mechanisms or to cause the scour to be shifted away from the immediate vicinity of the pier. Chiew and Lim (2000), Lauchlan and Melville (2001), and Dey and Rajkumar (2007) focused on the first category, using armoring devices for reducing local scour at bridge piers. Further flow-altering measures were used to reduce scour by using submerged vanes (Odgaard and Wang, 1987), a delta-wing-like fin in front of the pier (Gupta and Gangadharaiah, 1992), and slot through the pier (Chiew, 1992; Kumar et al., 1999).

FIGURE 10.49 Scour hole developed for a complex pier with rectangular pier cap for fully exposed condition. (Modified from Gautam et al., 2021.)

10.14.1 Bed Armoring Protective Measures

Followings are commonly used for bed armoring protective measures against scour.

10.14.1.1 Riprap

Riprap is the most extensively used protective measure for scour around bridge piers (Figure 10.50). There are many parameters affecting stable riprap size. Placement of riprap as armor layer around the pier at streambed level is a very common method for protecting the scour; the term called riprap stability number (N_c). Most of the equations for riprap design can be rewritten in terms of riprap stability number (Tabarestani and Zarrati, 2012). This parameter indicates the relationship between the flow condition and riprap stone characteristics, and can be written as:

$$N_c = \frac{\rho_f U^2}{g(\rho_s - \rho_f)d_{50}} \tag{10.137}$$

where U is the undisturbed upstream depth-averaged flow velocity, d_{50} is the mean size of stable riprap stones, ρ_f is the fluid density, and ρ_s is riprap stone density. The important factors affecting stability are: relative flow depth, relative stone size and effective pier width. Another observation of riprap layer, the failure varies with variation in stone sizes and pier skew angles. Failure of riprap occurred due to high shear forces at the upstream face of the pier or under combination of forces due to wake vortices and shear stresses downstream of the pier. (Chiew, 1995; Chiew and Lim, 2000; Lauchlan and Melville, 2001; Zarrati et al., 2010). Four modes of failures observed for riprap layer are: (1) Riprap shear failure; (2) Winnowing failure (removal of bed material); (3) Edge failure, and (4) Bedform undermining.

10.14.1.2 Partially Grouted Riprap

Partially grouted riprap is the one method for preventing scour, totally grouted ripraps were used earlier, but it is not recommended these days as this leads to the formation of riprap into a single block and reduce its permeability, and failure due to uplift, or undermining. Partially grouted riprap alleviates

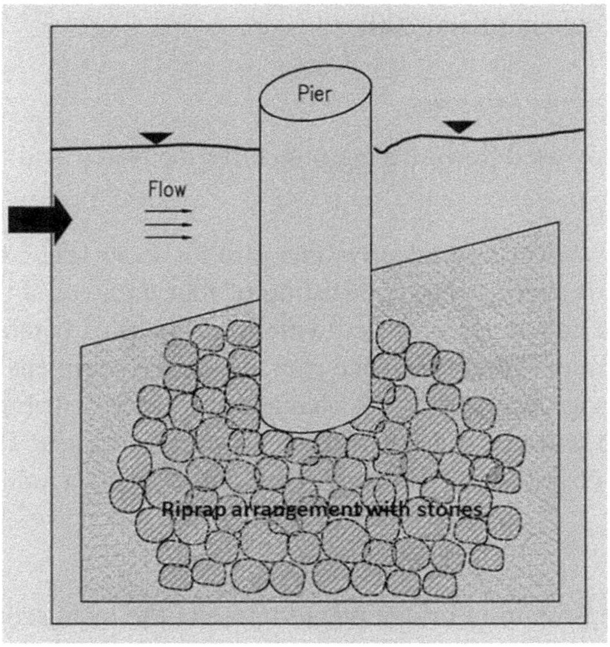

FIGURE 10.50 Typical riprap arrangement with large stones.

the concern and problems associated with the totally grouted one, increase the stability of the riprap unit without sacrificing the flexibility and allows for the use of smaller rocks and thinner riprap layers, thereby reducing the cost of material.

10.14.1.3 Articulating Concrete Blocks (ACB)

Articulating Concrete Blocks can be used to provide a flexible armor layer as a pier scour protective measure. Preformed units which interlock are held together by steel rods or cables are bonded to a geotextile or filter fabric or abut together to form a continuous blanket or mat. Cable tied concrete mats are used often.

10.14.1.4 Gabion Mattresses

Gabion Mattresses are containers constructed of wire mesh filled with loose stones or other similar material. However, rivers carrying coarse bed load can spoil the wire comprising of the gabions. Also, the mattress may pull away from the pier face. Scour can be prevented, by anchoring the gabions to the bed.

10.14.1.5 Grout-Filled Mats

Grout-filled mats are a single continuous layer of strong synthetic fabric sewn into a series of bags or compartments which are internally connected by ducts. The compartments are then filled with a concrete grout that, when set, forms a mat comprised of a grid of pillow-shaped units. Filter points allow for pressure relief through the mat. Grout-filled mats can range from smooth, uniform surface conditions approaching cast in place with concrete in terms of surface roughness, to extremely irregular surfaces exhibiting substantial projections into the flow, resulting in boundary roughness approaching that of moderate size of riprap.

10.14.1.6 Sack Gabions

The sack gabions are increasing in use as an alternative to ripraps as scour protection devices around bridge piers because ripraps are not available or expensive. The gabions have many advantages over other devices such as flexibility, strength, permeability, durability, economy and ecology. It is especially economical because of easy installation, less maintenance, ease of finding suitable fill materials on-site or from nearby quarries. (Yoon and Kim, 2001).

10.14.2 Flow Altering Protective Measures

Followings are the commonly used flow-altering protective measures against scour.

10.14.2.1 Tetrahedral Frames

Tetrahedral frames are a flow-altering protective measure for scour around the pier. These tetrahedral frames are essentially hollow space frames consisting of four equilateral triangles. They are made of reinforced concrete or plain concrete, combined with large sticks of bamboo (Tang et al., 2009). It is observed that at certain locations, their presence can change the sediment transport pattern from erosion to deposition. This change was due to the change in velocity distribution along the width of the river and velocity reduction within the zone of influence of the frames. The tetrahedral frames were found to be a very effective flow-altering protective measure, offering a reduction in scour depth of 50% or more.

10.14.2.2 Ring Columns

Ring Columns are placed upstream of bridge piers to protect the local scour. The protective structure comprises columns of interlocking rings placed upstream of the bridge pier. The columns stabilize rings and allow rings to slide along rails when the scour occurs to achieve a sustained reduction in flow energy. Figure 10.51 shows the arrangements of ring columns and bridge pier (Wang et al., 2011).

Columns of rings may be arranged upstream of the bridge pier to reduce the direct flow impacting the bridge pier and shelter the front of the bridge pier foundation, thereby reducing local scouring depth. The rings are interlaced, generating irregular shaped protective structures. The primary protective function of ring columns is to provide shelter, divert water flow, and dissipate energy to prevent bridge pier scouring. Therefore, parameters like height, diameter, and thickness of the ring columns and distance between the ring columns and pier influence the pier scouring. Scour can be reduced through dispersed flow energy, in turn generating sediment deposition in front of the bridge pier, effectively reducing the depth of bridge pier scour (Wang et al., 2011).

10.14.2.3 Sheath

The sheath faces the problem of local scour attacking its origin itself, modifying conveniently the flow near the submerged structure or object, in order to reduce the intensity and the effect of the mechanisms of local scour (Gris, 2010).

The sheath should be placed around the exterior surface of structures. Those structures can have any shape and can be made of any material.

The sheath has the following basic features:

1. A specially designed artificial *directional* rugosity in the front of the attack.
2. A specially designed *non-directional* rugosity in the rest of the external surface, which gives the sheath a "golf ball" effect; that is, transforms the laminar boundary layer to turbulent relocating the flow separation line downstream of the position corresponding to a smooth pier. This way, the wake reduces its strength, and so do the wake vortices.
3. Optionally, an efficient hydrodynamic, lenticular profile that reduces the area of the front of the attack and thereby the secondary flow and the horse-shoe vortex moment produce a minimum turbulent wake.
4. Also, optionally, the lenticular ensemble could be so made as to rotate freely around the structure to align itself automatically with a variable direction of flow.

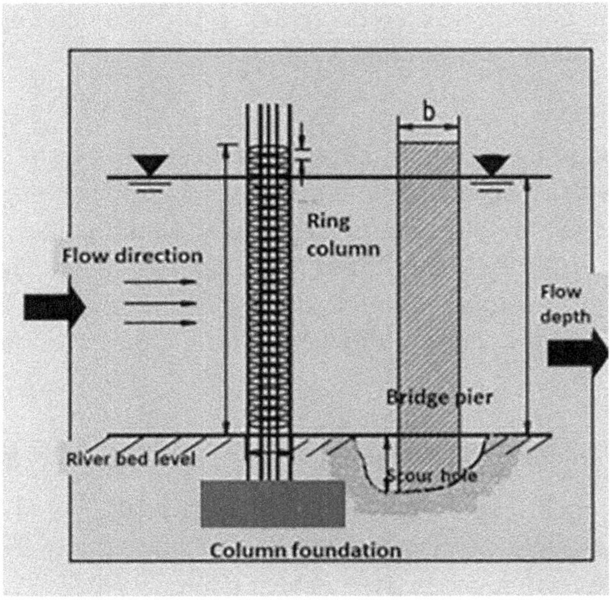

FIGURE 10.51 Schematics of ring column and bridge pier arrangements. (Modified from Wang et al. 2011.)

10.14.2.4 Collar

Collars are widely used as flow-altering protective measure (Heidarpour et al., 2010). A collar at any level above the bed divides the flow into two regions above and below the collar. For the region above the collar, the protective measure acts as an obstacle against the downflow due to which the downflow loses its strength on impingement at the bed, whereas for region below the collar, the downflow and the horseshoe vortex strength are reduced. However, the efficiency of a collar depends on its size and the location on the pier with respect to the bed. The efficiency increases at lower elevations since less flow can penetrate below it. In particular, when a collar is installed below the bed level, although the scour depth in front of pier may reduce with respect to the case of a collar set at the original level, the extension of the scour hole at the pier and the scour depth downstream of it may increase (Kumar et al., 1999).

10.14.2.5 Cable with Collar

Cables are wrapped spirally forming threads on a pier. The scour reduction increases as the cable diameter and the number of threads increases and also when the thread angle decreases (Izadinia and Heidarpour, 2012). Since scour cannot be completely prevented by collar alone, a combination of cable and collar is used. The use of cable reduces the strength of wake vortex and also weakens the downflow and the horse-shoe vortex. Therefore, the simultaneous use of a cable with a collar reduces the scour depth considerably and improves the utility of collar. In many cases, reduction of scour depth in upstream of the pier is about 53% (Izadinia and Heidarpour, 2012).

10.14.3 Combination Methods

For scour protection, instead of using a single protective measure for a particular pier, it is better to have integrated protective measures. This can be a combination of bed armoring and flow altering measures or integrated foundation such as raft foundation with apron. Such a combination would be more efficient and effective in terms of scour protection and foundation stability, though may not be cost-effective, in many cases.

10.14.3.1 Riprap and Collar

When collar and riprap are used together, the collar reduces the riprap extent in the front and sides of the pier. According to Zarrati et al. (2010), a reduction in scour volume of up to 57% of the riprap material can be obtained, by combining a collar and a riprap.

10.14.3.2 Raft with Flexible Apron

The raft has been confirmed to be a good economical method of constructing bridge foundation. The adoption of cut-off, along with an apron is found to be adequate to prevent scour [IRC 78 (2000); IRC 89 (1997)]. Figure 10.52 shows the typical structure of pier with a collar, and threaded pier with collar (Izadinia and Heidarpour, 2012). If the raft is provided with an extension beyond the pier, in the upstream and downstream, the scour may be reduced by 70% (Vijayasree et al., 2018). Provision of a flexible apron helps in reducing the scour in the vicinity of the pier, reducing the danger of undermining the structure.

10.15 EXPERIMENTAL AND FIELD STUDIES ON SCOUR PROTECTION SYSTEMS

Several research works are undertaken across different parts of the world to determine suitable scour protection system. Flume and field experiments are performed to study the protection of scour, using different protection measure techniques by researchers (Chiew, 1992, 1995; Borge and Jangde, 2005; Dey and Sen, 2005; Tang et al., 2009; Wang et al., 2011, 2016a, b; Hamidifar et al., 2017; Vijayasree

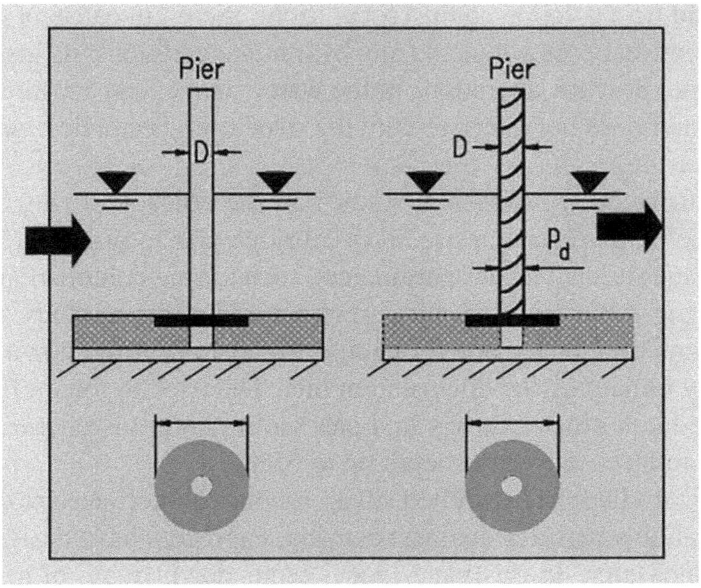

FIGURE 10.52 Pier with a collar, and threaded pier with collar. (Modified from Izadinia and Heidarpour, 2012.)

et al., 2018). For example, Chiew (1992) discussed different methods for preventing scour around bridge piers. It was suggested that the placement of riprap around pier provided the protection of scour. To avoid leaching problems, an inverted filter is normally placed beneath the riprap. Studies showed that placing a one-fourth diameter wide slot near the water surface or the bed level, it is possible to reduce the clear-water scour depth by as much as 20%. A one-half-diameter-wide slot placed near the water surface can reduce the clear-water scour depth by as much as 30%. When a slot is combined with a collar, the study showed that the combination is capable of eliminating scouring altogether. When designed and applied with care, a combination of collar and slot can be an appropriate alternative solution to riprap protection in tackling the scour problems at bridge piers. Later, Chiew (1995) described three failure mechanisms for riprap under clear water conditions, namely, shear failure, winnowing failure and edge failure. Chiew and Lim (2000) identified another type of failure under live-bed conditions, that is, the bedform destabilization of the riprap layer which is caused by the fluctuation of the bed level due to the migration of ripples and dunes.

Dey and Sen (2005) have done extensive studies to determine the scour depth in boulder-bed Rivers under high stream velocities. They did their studies for three different cases namely, for general bed, within channel contractions and at bridge piers. Their studies revealed that riprap pitching helps in reducing the scour depth by about 30%. Reinaldo (2006) studied about the effect of a rip-rap apron around a spill-through abutment. They concluded that the addition of a riprap apron around a spill-through abutment alters the basic form of the abutment, and thereby alters the scour process. As a consequence, the location and form of the resulting scour hole are altered. Increasing the apron width acts to reduce scour depth by several means. Tang et al. (2009) reported that tetrahedral frames are very effective in reducing the scour depth. Tetrahedrons are employed as a flow-altering countermeasure, offering a reduction of scour depth of almost 50% or more. Robie Bonilla (2010) recommended sheath as a protection measure against local scour. The sheath has an especially designed rugosity which reduces the effect of the secondary flow that occurs in the border of attack of the structure. Additionally, the sheath can include complementary elements to give the ensemble a lenticular profile, enhancing the anti-scour performance. Also, it could be configured to rotate freely to align with a variable direction of flow, if needed. The sheath could be a solution against local scour, the main cause of collapse of bridges. The sheath could be cost-effective; could withstand any hydraulic conditions; its installation would not need heavy equipment; does not provoke alterations in the water course; can substitute environment-altering solutions; being underwater does not interfere with the structure's aesthetics; could be used as a preventive or as a remedial measure.

Wang et al. (2011) conducted experimental studies on the efficacy of ring columns as a pier scour countermeasure. Ring columns placed upstream of a bridge pier to protect it from local scouring are typical scouring countermeasures. Various parameters, such as ring column diameter, thickness, height, and distances between ring column and the pier, were considered in the study. Their findings indicated that pores in ring columns provided space for energy dissipation of the down flow, hence, effectively reducing the flow energy impacting the downstream pier. The friction force of ring columns caused by their geometric shape reduces kinetic energy and pier scour depth. In general, it is found that the best pattern of ring columns reduces pier scour depth up to 65%.

Hamidifar et al. (2017) studied the use of bed sill as a scour countermeasure downstream of an apron. In order to protect hydraulic structures against scouring, engineers have designed and used many different types of countermeasures. In this paper, local scour downstream of a rigid apron using a bed sill is studied. They investigated the efficiency of scour reduction by means of a single bed sill located downstream of the apron as countermeasure, and to evaluate its effectiveness at various distances from the end of the apron. It was found that the maximum scour downstream of the apron reduces up to 95%. Furthermore, variations of the characteristic lengths of the scour hole were investigated. Also, it was observed that completely buried sills may not be useful. Finally, an empirical equation was proposed to predict the shape of the scour hole with the sill.

10.15.1 Use of Raft Foundation as Scour Protection Measure

According to IRC78 (2000) guidelines, the mean scour depth (d_{ms}) below the highest flood level (HFL) for natural channels was evaluated from the equation as:

$$d_{ms} = 1.34\left(\frac{D_b^2}{K_{sf}}\right)^{\frac{1}{3}} \tag{10.138}$$

where D_b is the design flow discharge per unit width of effective waterway and K_{sf} is the factor for a representative sample of bed materials obtained up to the level of anticipated deepest scour. The silt factor K_{sf} for a representative sample of bed materials is determined from $1.76(d_m)^{\frac{1}{2}}$, where d_m is the weighted mean diameter in mm.

As per the IRC89 (1997) specifications, where adoption of shallow foundations for bridges becomes economical by restricting the scour, floor protection has to be provided. The floor protection will comprise of rigid flooring with curtain walls and flexible apron so as to check scour, washing away or disturbance by piping action.

Borge and Jangde (2005) have concluded that Raft foundation is the most economical solution in case of weak soils where open foundation is not possible. Raft foundations can be most effectively used on small streams and rivers. Eldho et al. (2010, 2011, 2012) performed the physical model studies on extended raft foundation of bridges as a scour protection measure. Subsequently, Vijayasree et al. (2013) studied the physical model on bridge piers with raft foundation and aprons as a scour protection measure.

Vijayasree et al. (2018) conducted the physical models of scouring behavior around the simple bridge pier resting on raft foundation and its variations with stone aprons in both sides of the pier. The study was

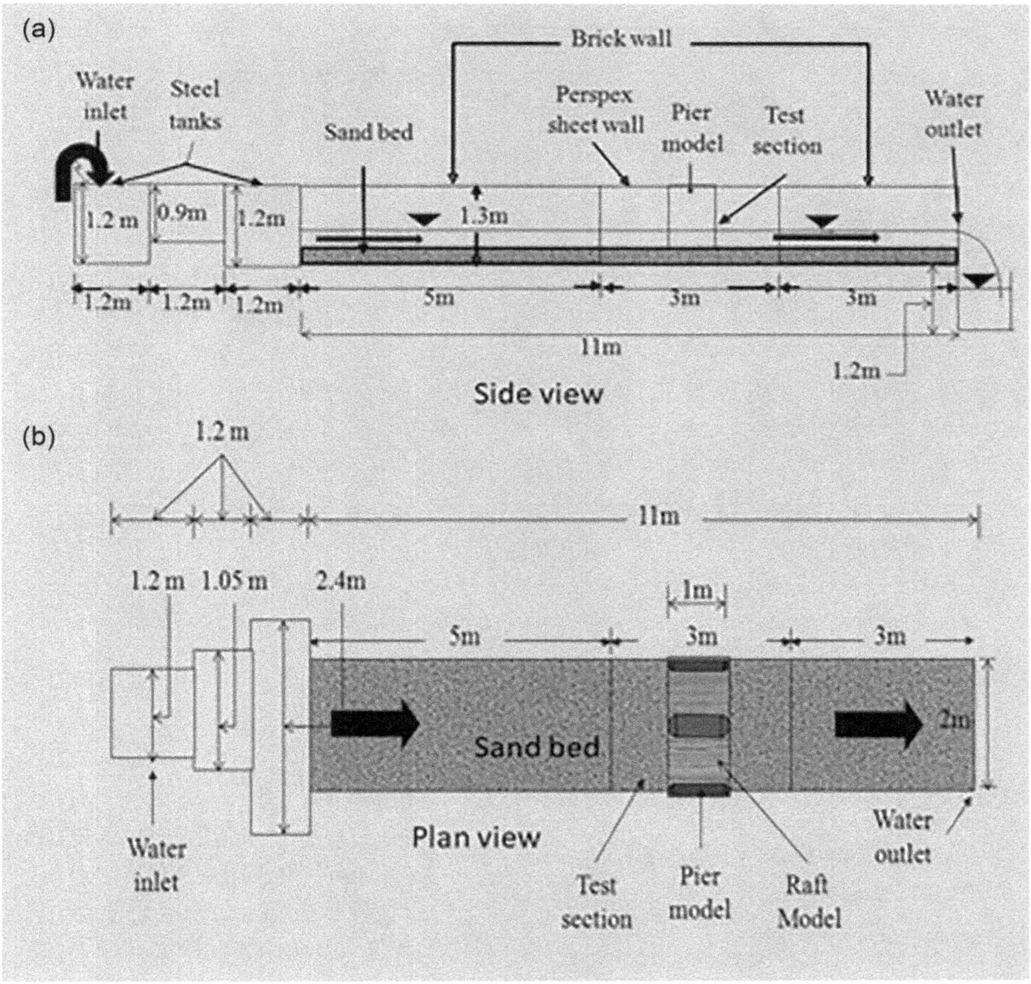

FIGURE 10.53 Sketch of experimental flume at IIT Bombay with sand bed: (a) side view; (b) plan view. (Modified from Vijayasree et al., 2018.)

carried out to identify the scour depth around the piers with raft foundations along with aprons, which provided a clue for suggesting the model design with possibilities of protection against heavy scour. In order to understand the physical models experimentally, the following five cases were examined as: (i) Simple pier, (ii) Pier with raft foundations, (iii) Raft extended to both upstream and downstream by a distance equal to the width of pier, (iv) Raft with stone aprons up to a distance of 0.3 m upstream and 0.5 m downstream, and (v) Raft extended to upstream by a width of the pier and downstream by three times of the pier width with stone aprons in both sides by a distance twice the pier width.

Experiments were conducted in a horizontal masonry flume constructed at the Hydraulics Laboratory of Indian Institute of Technology (IIT), Bombay, India. The flume was 11 m long, 2 m wide and 1.3 m deep, having a plain cement concrete (PCC) floor. The experimental details are given in Figure 10.53a and b. Detailed experimental conditions and results are available in Vijayasree et al. (2013, 2018).

The experimental observations indicated that the strength of the horseshoe vortex reduced considerably due to the presence of rigid raft extension and stone aprons for the case (v), which reduces significantly the scour depth (Figure 10.54). It is concluded that the arrangement can be made as the best way to protect the pier from damages due to heavy scour.

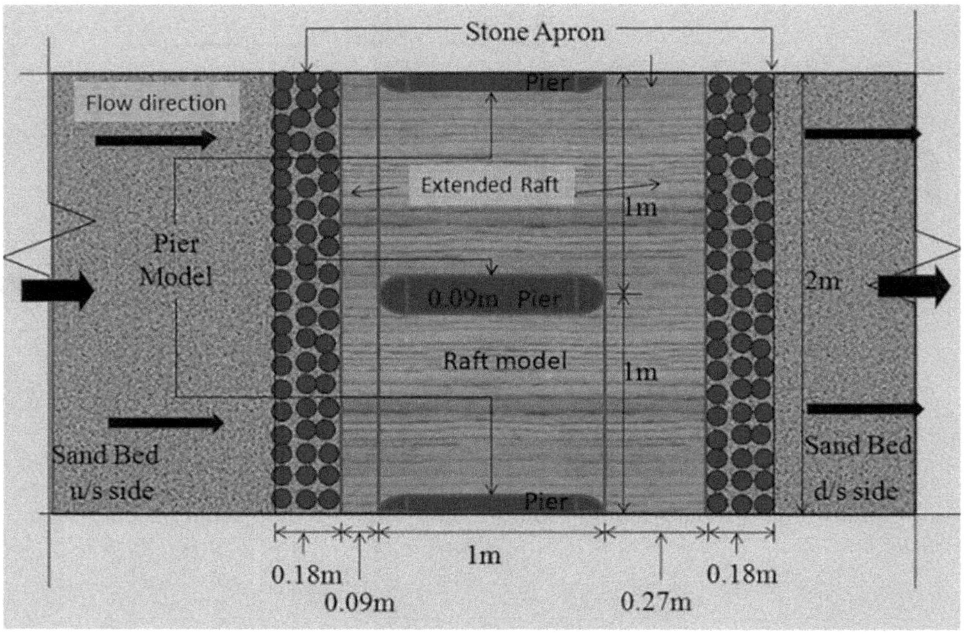

FIGURE 10.54 Photo of rigid raft extension with stone aprons on both sides. (Modified from Vijayasree et al., 2018.)

10.16 SUMMARY

Some of the important aspects discussed in this chapter are summarized below.

- On the assumption of steady uniform flow conditions in open channel flow, the mean flow velocity can be computed using resistance formula such as Chezy's equation $\left(U = C\sqrt{RS}\right)$; Manning's equation $\left(U = \frac{1}{n}R^{\frac{2}{3}}S^{\frac{1}{2}}\right)$; Lacey's formula $\left(U = 10.8R^{\frac{2}{3}}S^{\frac{1}{3}}\right)$; Garde and Rangaraju formula

$$\frac{U}{\sqrt{\frac{(\rho_s - \rho_f)dg}{\rho_f}}} = K\left(\frac{R}{d}\right)^{0.67}\left(\frac{S}{\frac{(\rho_s - \rho_f)}{\rho_f}}\right)^{0.5}$$

- Based on the magnitude of resistance to flow and on the nature of the bed and water surface configurations, the flow regimes in alluvial channel are classified into four categories: (i) plane bed, (ii) ripples and dunes, (iii) transition, and (iv) anti-dunes.
- For plane bed, the sediment bed acts like a rigid boundary, and the water surface is fairly smooth with low Froude number. The ripples are the bedform configurations that are developed in a shear layer of smooth flow at a little excess of bed shear stress, whereas the dunes largely interact with the main flow at a larger excess bed shear stress compared to that of ripples. In the transitional regime, the rapid changes in sand bed and water flow occur with relatively small changes in flow conditions. If the bed shear stress is higher and the Froude number is greater than one ($Fr > 1$), the sand waves are referred to as *Anti-dunes*.
- Number of empirical relations were derived for ripples and dunes based on various experimental observations. For example, Kennedy and Odgaard (1990) proposed bedform height developed from the laboratory data as:

$$\frac{h_b}{h}=\frac{1}{2}\left[\frac{1.2\lambda\alpha f_0}{8C_D}+\left\{\left(\frac{1.2\lambda\alpha f_0}{8C_D}\right)^2+\frac{2\pi Fr^2 f_0}{C_D C_1}\left(\frac{f}{f_0}-\frac{1.2\lambda}{2}\right)\right\}^{0.5}\right]$$

where f is the Darcey-Weisbach friction factor, f_0 is the rigid flat bed Darcey-Weisbach friction factor, Fr is the Froude number, and $C_1 = 1, C_D = 1, \alpha = 5, \lambda = 1$.

- The sand bar is a class of large-scale bedform, the dimensions of which are controlled by the channel width and the depth. The formation and evolution of bars results in distinctive channel patterns.
- Based on the equation of continuity of sediment, Exner (1925) developed a classical equation to predict the migration velocity of bedforms without considering the frictional effect as, $\frac{\partial \eta}{\partial t}+\alpha_E\frac{\partial \bar{U}}{\partial x}=0$, where $\bar{U}=\frac{Q}{(h-\eta)B_w}$, Q is the flow discharge, η is the elevation of the sand bed with respect to the horizontal reference or datum and h is the water depth, B_w is the channel width, and $\bar{U}$ is the cross-sectional average velocity, α_E is the Exner erosion coefficient.
- Large-scale roughness, such as dunes and ripples, arising from the instability of erodible sediment bed due to the interaction with the turbulent flow are ubiquitous in natural alluvial channel, which induce flow separation, wakes and circulation, and hence form-resistance. The main effect of bedforms in the flow is the separation of flow behind the bedform crest and reattachment point away from the crest; and hence the flow is spatially decelerating there.
- Flow over the river dunes, which are asymmetric in cross-sectional form, possesses an angle-of-repose leeside in a steady, uniform unidirectional flow. The turbulent flow field over dunes is critical in determining instantaneous bed shear stresses, bed load transport rate and sediment suspension, and bed surface and water surface topography.
- Turbulent flow over a plane loose sand bed may cause the bed to deform, which in turn introduces perturbations on the main flow. The effect of the perturbed bed on a turbulent flow is important in determining the formation of bedforms in erodible channels.
- Scour is a major cause of failure of many hydraulic structures such as bridge piers. Scour is defined as the removal of river or channel bed material due to the erosive action of flowing water in the stream. In most of the alluvial channels, there is a continuous transport of sediment as a geomorphological process. However, if there are additional natural or man-induced causes such as bridge piers to upset the sediment supply and removal in a reach the stream will have long-term changes in the channel bed elevation leading to major scouring problem and failure of the structure.
- The interaction between the downward flow and the horizontal boundary layer separation close to the riverbed results in the formation of a vortex system. The two ends of this vortex system are swept downstream by the flow as they wrap around the pier in the shape of a horseshoe in plan view and hence popularly known as the horseshoe vortex. This vortex significantly affects the scouring process at the base of the bridge piers.
- The scouring process consists of such components as aggradation, degradation, general scour, local scour, clear-water scour and live bed scour.
- In marine environment, the structures are usually exposed to currents, waves and

combined waves and currents leading major scour around marine structures.

- Crescentic scour: Local scour around any obstruction placed in the sediment bed is a result of the interaction of the local flow field with the sediment particles. When water passes around immovable objects or obstacles like shells, pebbles or wood fragments placed on a sandy bed, scour marks around the obstacle usually result from the interaction of the local flow with the sediment bed.
- The equilibrium scour width (w_s) in a steady current can be represented as a function of dimensionless parameters: $w_s = f\left(a_r, R_{cyl}, F_s\right)$ where a_r is the cylinder aspect ratio, R_{cyl} is the cylinder Reynolds number, F_s is the sediment Froude number.
- On bridge pier scour, large number experimental and field observation studies are available in literature especially, for circular piers and piers with different cross-sectional shapes. Out of the available shapes, the most efficient cross-sectional shape against scour is lenticular shape, though oblong shape is the most commonly used shape.
- An expression for the equilibrium depth of local scour d_{es} at a cylindrical pier of diameter D in a uniform sediment can be written as follows: $\frac{d_{es}}{D} = f\left(\frac{u}{u_{*c}}, \frac{y}{D}, \frac{d_{50}}{D}, \frac{t}{t_e}\right)$ where u is the mean approached velocity, u_{*c} is the critical mean approach flow velocity for entrainment of bed sediment, y is the mean approach flow depth, d_{50} is the median size of the sediment, t is the time, t_e is the time for equilibrium depth of scour, and D is the cylindrical pier diameter.
- Nowadays, complex piers (CP) are preferred over simple piers to be constructed in river system. Complex pier consists of three different structural elements, viz, a cylindrical column resting on a pile cap supported by a number of piles, also known as pile foundation. The scouring problem for CP depends on the exposure to various components of CP and shape of pile cap. Appropriate scour protection measures can be designed to protect the bridge pier and foundation against scour. Commonly three types of protection measures are provided against scour viz. bed armoring protective measures, flow-altering protective measures and combination of the above two measures.

10.17 EXERCISE PROBLEMS

1. Illustrate important resistance formulae used to calculate uniform mean velocity distribution in open channels.
2. Explain the flow regimes categories in alluvial channel according to Garde and Albertson (1959).
3. Elaborate the following empirical relation for Ripples and dunes? Yalin (1964), Ranga Raju and Soni (1976), Yalin and Karahan (1979), van Renjin (1984), Kennedy and Odgaard (1990), Karim (1999), Yang et al. (2005).
4. What is sand bar? Illustrate details with figures.
5. Explain the Exner's bedform equation with its application.
6. Discuss the different types of ripples and its formation.
7. Elaborate the bedform migration phenomena with important parameters.
8. Illustrate the scouring phenomena with important parameters.
9. What are the important components of scoring process? Illustrate with examples.
10. Illustrate the scour below pipelines, the important parameters and elaborate important empirical relations developed.
11. What is crescentic scour? How it is formed? Illustrate with example figures?
12. What are the important parameters of scour formation and equilibrium scour for width and depth?

13. What is a complex bridge pier? What are the components? Elaborate the scoring around complex bridge piers.
14. What are the important techniques used for scour protection?
15. Elaborate the bed armoring protective measures with details.
16. What is the important flow-altering mechanism used for scour protection measure? Illustrate each technique with figures.
17. Illustrate the combination methods for scour protection with details.
18. Explain the raft foundation mechanism and its effect as a protective mechanism against scour.

References

Abbott J. and Francis J. (1977). Saltation and suspension trajectories of solid grains in a water stream. *Philos. Trans. R. Soc. London A.* 284, 225–254.
Aberle J. and Nikora V. (2006). Statistical properties of armoured gravel bed surfaces. *Water Resour. Res.* 42, W11414.
Abrahams A. D. (2003). Bed-load transport equation for sheet flow. *J. Hydraul. Eng. ASCE.* 129 (2), 159–163.
Abrahams A. D. and Gao P. (2006). A bed-load transport model for rough turbulent open-channel flows on plane beds. *Earth Surf. Process. Landforms.* 31 (7), 910–928.
Absi R. (2009). *An Ordinary Differential Equation for Velocity Distribution and Dip Phenomenon in Open Channel Flows.* EBI, Inst. Polytech, Cergy-Pontoise.
Absi R. (2011). An ordinary differential equation for velocity distribution and dip-phenomenon in open channel flows. *J. Hydraul. Res.* 49 (2011), 82–89.
Ackers P. and White W.R. (1973). Sediment transport: New approach and analysis. *J. Hyd. Div. Proc. ASCE.* 99 (11), 2041–2060.
Adrian R. (1991). Particle-imaging techniques for experimental fluid-mechanics. *Annu. Rev. Fluid Mech.* 23 (1), 261–304.
Adrian, R. J., and Westerweel, J., (2011). *Particle Image Velocimetry.* Cambridge University Press, New York.
Agelinchaab M. and Tachie M. F. (2006). Open channel turbulent flow over hemispherical ribs. *J. Heat Fluid Flow.* 27 (2006) 1010–1027.
Ahmed F. and Rajaratnam N. (1998). Flow around bridge piers. *J. Hyd. Engg.* 124 (3), 288–300.
Ahmed F. and Rajaratnam N. (2000). Observations on flow around bridge abutment. *J. Engg. Mech.* 126, 51–59.
Aksoy S. (1973). Fluid forces acting on a sphere near a solid boundary. *Proceedings of the Fifteenth Congress of International Association for Hydraulic Research* (IAHR, Istanbul, Turkey), 1, 217–224.
Alam A. M. Z. and Kennedy J. F. (1969). Friction factors for flow in sand bed channels. J. Hydr. Div. *ASCE.* 95, 1973–1992.
Alam M. M. and Zhou Y. (2007). Flow around two side-by-side closely spaced circular cylinders. *J. Fluids Struc.* 23 (5), 799–805.
Albertson M. I. (1952). Effect of shape of the fall velocity of gravel particles. *Proceedings of the 5th Hydraulic Conference.* Studies in Engineering, University of Iowa, Iowa City, Iowa.
Albertson M. L., Simons D. B. and Richardson E. V. (1958). Discussion of mechanics of ripple formation.., *J. Hydr. Div. ASCE.* 84 (1), 23–32.
Allen J. R. L. (1968). *Current Ripples.* North Holland Publishing Company, Amsterdam.
Allen J. R. L. (1978). L. van Bendegom: a neglected innovator in meander studies. In A. D. Miall (Ed.), *Fluvial sedimentology, Memoir No. 5* (pp. 199–209). Canadian Society of Petroleum Geologists, Calgary.
Allen J. R. L. (1982). Sedimentary Structures their Character and Physical Basis. Elsevier, Amsterdam,, 593.
Allen J. R. L. and Collinson J. D. (1974). The superimposition and classification of dunes formed by unidirectional aqueous flows Sedimentary. *Geology.* 12 (3) 169–178.
Amin M. I. and Murphy P. J. (1981). Two bed-load formulas: An evaluation. Hydraul. Div. ASCE. 107 (HY8), 961.
Ancey C. F., Bigillon F., Frey P., Lanier J. and Ducret R. (2002). Saltating motion of a bead in a rapid water stream. *Phys. Rev. E.* 66 (3), 1–16.
Ancey C. F., Bigillon P. F. and Ducret R. (2003). Rolling motion of a bead in a rapid water stream. *Phys. Rev. E.* 67, 0113–0311.
Ancey C. T., Böhm M. J. and Frey P. (2006). Statistical description of sediment transport experiments. *Phys. Rev. E.* 74, 0113–0302.
Ancey C., Davison A. C., Bohm T., Jodeau M. and Frey P. (2008). Entrainment and motion of coarse particles in a shallow water stream down a steep slope. *J. Fluid Mech.* 595, 83–114.

Anderson A. G. (1942). Distribution of suspended sediment in a natural stream. *Trans. Am. Geophys. Union.* 23, 678–683.

Anderson A. G. (1953). "The characteristics of sediment waves formed by flow in open channels." *Proceedings Third Midwest Conference in Fluid Mechanics,* University of Minnesota, Minneapolis, 379–395.

Ansari A., Kothyari C. and Rangaraju G. (2002).Influence of cohesion on scour around bridge piers. *J. Hydraul. Res.* 40, 717–729.

Antonia R. and Bisset D. (1990). Spanwise structure in the near-wall region of a turbulent boundary layer. *J. Fluid Mech.* 210. 437–458.

Anwar H. O. and Atkins R. (1980). Turbulence measurements in simulated tidal flow. *J. Hyd. Div. ASCE.* 106 (8), 1273–1289.

Apperley L. W. (1968) Effect of turbulence on sediment entrainment. PhD thesis, University of Auckland, Auckland.

Aris R. (1990). Vectors, tensors and the basic equations of fluid mechanics. Dover Books on Mathematics, Paperback, 1990.

Armanini A. and Gregoretti C. (2005). Incipient sediment motion at high slopes in uniform flow condition. *Water Resour. Res.* 41, W12431.

Ashida K. and Michiue M. (1971). An investigation of river bed degradation downstream of a dam. *Proceedings of the 14th Congress International Association Hydraulic Research (IAHR)*, Paris, 247–256.

Ashida K. and Michiue M. (1973). Study on bed load transport rate in open channel flows, *Paper Presented at International Symposium on River Mechanics*, International Association for Hydraulic Research, Bangkok.

Ashtiani B. and Aslani-Kordkandi A. (2012). Flow field around side-by-side piers with and without a scour hole. *Euro. J. Mech. Fluids.* 36, 152–166.

Ashtiani B. and Aslani-Kordkandi A. (2013). Flow field around single and tandem piers. *Flow Turbul, Combustion.* 90 (3), 472–490.

Ashtiani B., Baratian-Ghorghi Z. and Beheshti A. (2010). Experimental investigation of clear-water local scour of compound piers. *J. Hydraul. Eng.* 136 (6), 343–351.

Babakaiff C. S. and Hickin E. J. (1996) Coherent flow structures in Squamish River Estuary, British Columbia, Canada. In P. J. Ashworth, S. J. Bennett, J. L. Best and S. J. McLelland (Eds.), *Coherent Flow Structures in Open Channels* (321–342). John Wiley & Sons, Chichester.

Bagnold R. A. (1941). The Physics of Blown Sands and Desert Dunes. Chapman and Hall, Methuen Halsted.

Bagnold R. A. (1954). Experiments on a gravity-free dispersion of large solid spheres in a Newtonian fluid under shear. *Proc. R.Soc. Lond. Ser. A Math. Phys. Sci.* 225, 49–63.

Bagnold R. A. (1956). The flow of cohesionless grains in fluids Philos. *Trans. R. Soc. London A.* 249 (1956), 235–297.

Bagnold R. A. (1966). An approach to the sediment transport problem from general physics. *US Geological Survey Professional Paper 422-I,* 1–37. https://pubs.er.usgs.gov/publication/pp422I

Bagnold R. A. (1973). The nature of saltation and bed load transport in water. *Proc. Royal Soc. London A.* 332, 473–504.

Bagnold R. A. (1974). Fluid forces on a body in shear flow; experimental use of stationary flow. *Proc Royal Soc. London A.* 340 (1621), 147–171.

Bagnold R. A. (1980). An empirical correlation of bed load transport rates in flumes and natural rivers. *Proc Royal Soc. London A.* 372, 453–473.

Bagnold R. A. and Barndorff-Nielsen O. (1980). The pattern of natural size distributions. *Sedimentology.* 27, 199–207.

Balachandar R. and Bhuiyan F. (2007) Higher-order moments of velocity fluctuations in an open-channel flow with large bottom roughness. *J. Hydraul. Eng.* 133, 77–87.

Balachandar R., Kells J. and Thiessen R. (2000). The effect of tail water depth on the dynamics of local scour. *Can J Civil Eng.* 27, 138– 150.

Baldock T., Tomkins M., Nielsen P. and Hughes M. (2004). Settling velocity of sediments at high concentrations. *Coastal Eng.* 51, 91–100.

Barekyan A. S. (1962). Discharge of channel forming sediments and elements of sand waves. *Trans. Am. Geophys. Union.* 2, 128–130.

Barekyan A. S. (1963). Method of determining permissible velocities. *Soviet Hydrol. Am. Geophys. Union.* 3.

Barman K., Debnath K. and Mazumder, B. S. (2016) Turbulence between two inline hemispherical obstacles under wave-current interactions. *Adv. Water Resour.* 88, 32–52.

Barndorff-Nielsen O. (1977). Exponentially decreasing distributions for the logarithm of particle size. *Proc. Roy. Soc. A* 353 (1674), 401–419.

Barndorff-Nielsen O., Dalsgaard K., Halgreen C., Kuhlman H., Moller J. T. and Shou G. (1982), Variations in particle size distribution over a small dune. *Sedimentology*. 29, 53–65.

Barnes H. T. and Coker E. G. (1905). The flow of water through pipes. *Proc. Roy. Soc. London.* 74, 341.

Barton J. K. and Lin P. N. (1955). *A Study of Sediment Transport in Alluvial Channel*. Colorado State University, Fort Collins.

Bass J. H. (1994). A flume study on the development and equilibrium morphology of current ripples in very fine sand. *Sedimentology*. 41, 185–209.

Batchelor G. K. (1953). *The Theory of Homogeneous Turbulence*. Cambridge University Press.

Batchelor G. K. (1965). The motion of small particles in turbulent flow. *Proceedings of the Second Australasian Conference on Hydraulics and Fluid Mechanics*. University of Auckland, Auckland, 019–041.

Batchelor G. K. (1993). *An Introduction to Fluid Dynamics.* Cambridge University Press, New Delhi, 1993

Batchelor G. K. and Townsend A. A. (1949). The nature of turbulent motion at large wave-numbers. *Proc. Roy. Soc. London A.* 238–255.

Bayazit, M. (1983). Flow structure and sediment transport mechanics in steep channels. In: A. A. Balkema (Ed.), *Mechanics of Sediment Transport*. Euromech, Istanbul.

Beg, M. (2010). Characteristics of developing scour holes around two piers placed in transverse arrangement. *Int. Conf. Scour Eros*, 32, 491–500.

Beg M. and Beg S. (2015). Scour hole characteristics of two unequal size bridge piers in tandem arrangement. *ISH J Hydraul. Eng*. 21(1), 85–96.

Beheshti A. A. and Ashtiani B. (2008). Analysis of threshold and incipient conditions for sediment movement. *Coastal Engg.* 55, 423–430.

Beheshti A. and Ashtiani B. (2010). Experimental study of three-dimensional flow field around a complex bridge pier. *J. Eng. Mech.* 136.

Beheshti AA, and Ashtiani, B. (2016). Scour hole influence on turbulent flow field around complex bridge piers, flow. *Turbul. Combust*. 97, 451–474.

Bendat J. and Piersol A. (2000). *Random Data.* Wiley, New York.

Bennett S. J. and Best J. L. (1995). Particle size and velocity discrimination in a sediment-laden turbulent flow using phase Doppler anemometry. *J Fluids Engg*. 117, 505–511.

Bennett S. J. and Bridge J. S. (1995a). An experimental study of flow, bedload transport and topography under conditions of erosion and deposition and comparison with theoretical models. *Sedimentology*. 42, 117–146.

Bennett S. J. and Bridge J. S. (1995b) The geometry and dynamics of low-relief bed forms in heterogeneous sediment in a laboratory channel, and their relationship to water flow and sediment transport. *J. Sed. Res.* A65, 29–39.

Best J. (2005). The fluid dynamics of river dunes: a review and some future research directions. *J. Geophys. Res.* 110, F04S02.

Best J. and Kostaschuk R. A. (2002). An experimental study of turbulent flow over a low-angle dune. *J. Geophys. Res.* 107(C9), 181–189.

Bharat S. (1961). Bed load transport in channels. *J. Water Energy Int.* 18 (5), 411–430.

Bhattacharya A. (1993). Backwash-and-swash-oriented current crescents: indicators of beach slope, current direction and environment. *Sedimentary Geol.* 84, 139–148.

Bhattacharya R. N., Dalal D. C., Ghosh J. K. and Mazumder, B. S. (2000). Comparison of diffusion based approaches to sediment transport with a stochastic interpretation. In: H. Wang and B. Hu (Eds.), Stochastic Hydraulics 2000 (255–261). Balkema, Rotterdam.

Bhattacharyya A., Ojha S. P. and Mazumder B. S. (2013). Evaluation of the saltation process of bed materials by video-imaging under altered bed roughness. *Earth Surf. Proc. Landforms.* 38 (12), 1339–1353.

Bhowmik, N. G. and Mazumder, B. S. (1990). Physical forces generated by barge-tow traffic within a navigable waterway, Proceedings of the National Conference. San Diego, 604–609.

Bhowmik N. G., Soong D., Adams J. R., Mazumder B. S. and Xia, R. (1998a). Physical changes associated with navigation traffic on the Illinois and Upper Mississippi Rivers. *E*nvironmental Management Technical Center. Onalaska, 205.

Bhowmik N. G., Soong D., Adams J. R., Mazumder B. S. and Xia, R. (1998b). Physical changes associated with navigation traffic on the Illinois and Upper Mississippi Rivers (Appendices). *E*nvironmental Management Technical Center. Onalaska.

Bhowmik N. G., Soong T. W. and Mazumder B. S. (1992). Return flows in Large Rivers associated with navigation traffic. Proceedings of National Conference. Baltimore, 760–765.

Bhowmik N. G., Xia R., Mazumder B. S. and Soong T. W. (1995a). Distribution of turbulent velocity fluctuations in a natural river, *J. Hydraul. Res.* 33, 649–661.

Bhowmik N. G., Xia R., Mazumder B. S. and Soong T. W. (1995b). Return flows in rivers due to navigation traffic. *J. Hydr. Engg.* 121 (12), 914–918.

Blaauw H. G., van der Knaap F. C. M., de Groot M. T. and Pilarcyk K. W. (1984). *Design of Bank Protection of Inland Navigation Fairways.* Delft Hydraulics Lab, The Netherlands.

Blasius H. (1908) Grenzschichten in FlussigkeitenmitkleinerReibung. *Z. Math. Physik Bd.* 56, 1–37.

Blasius H. (1910). FunctionentheoretischeMethoden in der Hydrodynamik, *Zeit. Math. Phs.* 58, 90.

Blench T. (1952). Normal size distribution found in samples of river bed sand. *Civil Engg.* 22, 47.

Bogardi J. L. (1965). European concepts of sediment transportation. *Proc. ASCE.* 91, 1.

Böhm T., Frey P., Ducottet C., Ancey C., Jodeau M. and Reboud J.-L. (2006). Two-dimensional motion of a set of particles in a free surface flow with image processing. *Exp. Fluids.* 41 (1), 1–11.

Bombar G., Guney M. S. and Altınakar M. S. (2010). Application of image processing in unsteady flows to investigate bed load transport. *Mathematical Comput, Appl.* 15 (3), 420–427.

Bonakdari H., Larrarte F., Lassabatere, L. and Joannis C. (2008). Turbulent velocity profile in fully-developed open channel flows. *Environ Fluid Mech.* **8**, 1–17.

Boppe R. S. and Neu W. L. (1995). Quasi-coherent structures in the marine atmospheric boundary layer, *J. Geophys. Res.* 100, 20635–20648.

Borge V.B. and Jangde K.S. (2005). Innovative bridge foundations in weak soils—experiments and practice in Maharashtra. *J. Indian Roads Cong.* 518, 455–484.

Bose S. K. and Dey S. (2009). Suspended load in flows on erodible bed. *Int. J. Sed. Res.* 24 (3), 315–324.

Boussinesq J. (1877). Theorie de I'ecoulementtourbillant. *Mem. Pres. Acade. Sci.* 13, 46.

Boussinesq J. (1896). ComptesRendus de I'Academie des. *Sciences.* 72, 1290–1295.

Boussinesq J. (1903). TheorieAnalytique de la Chaleur. *Sciences.* 2, Hachette Livre – BNF Pub. London, 665 pages.

Bradshaw P. (1976). *Topics in Applied Physics.* Springer-Verlag, Amsterdam.

Bravo-Espinosa M., Osterkamp W. R. and Lopes V. L. (2003). Bedload transport in alluvial channels. *J. Hyd. Engg.* 129 (10), 783–795.

Breusers H. N. C., Nicollet G. and Shen H. W. (1977). Local scour around cylindrical piers. *J Hyd. Res.* 15 (3), 211–252.

Bridge J. S. (1993). The interaction between channel geometry, water flow, sediment transport and deposition in braided rivers. *Geol. Soc.* 75 (1), 13.

Bridge J. S. and Bennett S. J. (1992). A model for the entrainment and transport of sediment grains of mixed sizes, shapes and densities. *Water Resources Res.* 28, 337–363.

Bridge J. S. and Dominic D. F. (1984). Bed load grain velocities and sediment transport rates. *Water Resources Res.* 20, 476–490.

Brooks N. H., (1958). Mechanics of streams with movable beds of fine sand. *Am. Soc. Civil Engineers Trans.* 123, 526–594.

Brown G. L. and Roshko A. (1974). On density effects and large structure in the turbulent mixing layers. *J. Fluid Mech.* 64, 775–816.

Buschmann M. and Gad-el Hak M. (2005). New mixing-length approach for the mean velocity profile of turbulent boundary layers. *J. Fluids Engg.* 127 (2), 393–396.

Businger J. A., Wyngaard J. C., Izumi Y. and Bradley E. F. (1971). Flux-profile relationships in the atmospheric surface layer. *J.Atmos. Sci.* 28, 190–201.

Camussi R., Felli M. and Pereira F. (2008). Statistical properties of wall pressure fluctuations over a forward-facing step. *Phys Fluids*. 20 (7), 075113.

Cao Z. (1997). Turbulent bursting-based sediment entrainment function. *J. Hydraul. Engg.* 123 (3), 233–238.

Cao Y., Bai. Y., Wang, J., Liao S. and Xu D. (2015). Prediction of scour depth around offshore pipe lines in the south China Sea. *J. Marine Sci. Appl.*14 (01), 83–92.

Cao Z., Pender G. and Meng J. (2006) Explicit formulation of the Shields diagram for incipient motion of sediment. *J. Hydraul. Engg.* 132 (10), 1097–1099.

Cardoso A. H., Graf W. H. and Gust G. (1989). Uniform flow in smooth open channel. *J. Hydraul. Res.* 27 (5), 603–616

Carey W. C. and Keller M. D. (1957) Systematic changes in the beds of alluvial rivers. *J. Hydraul. Div.* 83 (4), 1–24.

Carstens M. R. (1952). Accelerated motion of a spherical particle. *Em. Trans.* 33, 713–721.

Carstens M. R. (1966). *An Analytical and Experimental Study of Bed Ripples Under Water Waves.* Georgia Institute of Technology, Atlanta.

Cataˇno-Lopera Y. A., Demir S. T. and Garcia M.H. (2007). Self-burial of short cylinders under oscillatory flows and combined waves plus currents. *IEEE J. Oceanic Engg.* 32 (1), 191–203.

Cataˇno-Lopera Y. A. and Garcia M. H. (2006). Burial of short cylinders induced by scour under combined waves and currents. *ASCE J. Waterway Engg.* 132 (6), 139–149.

Cataˇno-Lopera Y. A. and Garcia M. H. (2007). Geometry of scour hole around and the influence of the angle of attack on the burial of finite cylinders under combined flows. *Ocean Engg.* 34, 856–869.

Cebeci T. (2004). *Analysis of Turbulent Flows*. Elsevier, London.

Cebeci T. and Chang K. C. (1978). Calculation of incompressible rough-wall boundary layer flows. *AIAA J.* 16, 730.

Cebeci, T. and Smith A. M. O. (1968). Computation of turbulent boundary layers-1968. In S. J. Kline, M. V. Morkovin, G. Sovran and D. S. Cock-rell (Eds.), *AFOSR-IFP-Stanford Conference.* Stanford University Press, Stanford, CA.

Cellino M. (1998). Experimental study of suspension flow in open channel. Doctoral dissertation No. 1824. Ecole Polytechnique Federale de Lausanne.

Cellino M. and Graf W. H. (1997). Measurements of suspension flow in open channels. *Proceedings of IAHR Congress* (pp. 179–184). https://doi.org/10.1080/00221680009498328

Cellino M. and Graf W. H. (2000). Experiments on suspension flow in open channels with bedforms. *J. Hydraul. Res.* 38 (4), 289–298.

Cellino M. and Lemmin U. (2004). Influence of coherent flow structures on the dynamics of suspended sediment transport in open-channel flow. *J. Hydraul. Engg.* 130 (11), 1077–1088.

Chang F. M., Simons D. B. and Richardson E. V. (1967). Total bed-material discharge in alluvial channels. *Proceedings of the twelfth congress of International Association for Hydraulic Research* (pp 132–140). Fort Collins, Colorado.

Chang W.-Y., Constantinescu G., Lien H.-C., Tsai W.-F., Lai J.-S. and Loh C.-H. (2013). Flow structure around bridge piers of varying geometrical complexity. *J. Hydraul. Engg.* 139 (8), 812–826.

Chang Y. (1939) Laboratory investigations of flume traction and transportation. *Trans. ASCE.* 104 (1), 1246–1284.

Chao J. L. and Hennessy P. V. (1972). Local scour under ocean outfall pipelines. *J. Water Poll. Control Fed.* 44 (7), 1443–1447.

Chapman S. and Cowling T. G. (1970). *The Mathematical Theory of Non-Uniform Gases.* Cambridge University Press, Cambridge.

Charru F. (2006). Selection of the ripple length on a granular bed sheared by a liquid flow. *Phys. Fluids.* 18, 121508.

Chatanantavet P., Whipple K. X., Adams M. A. and Lamb M. P. (2013). Experimental study on coarse grain saltation dynamics in bedrock channels. *J. Geophysical Res.* 118, 1161–1176.

Chatterjee D., Ghosh, S., Mazumder B. S. and Debnath K. (2020). Turbulent flow characteristics over forward-facing obstacle. *J. Turbul.* 22(3), 141–179.

Chatterjee D., Ghosh, S., Mazumder B. S. and Sarkar K. (2019). Development of bed forms due to waves blocked by counter-current. *Earth Surf. Process. Landforms.* 44 (6), 1330–1345.

Chatterjee, D., Mazumder B. S. and Ghosh, S. (2018). Turbulence characteristics of wave-blocking phenomena. *Appl. Ocean Res.* 75, 15–36.

Chee R. K.W. (1982). *Live-Bed Scour at Bridge Piers*. University of Auckland, Auckland.

Cheng N. S. (1997). A simplified settling velocity formula for sediment particle. *J. Hydraul. Engg.* 123 (2), 149–152.

Cheng N. S. (2002). Exponential formula for bedloadtransport. *J. Hydraul. Engg.* 128 (10), 942–946.

Cheng N. S. (2003). A diffusive model for evaluating thickness of bedload layer. *Adv. Water Res.* 26 (2003), 875–882.

Cheng N. S. and Chiew Y. M. (1998). Pickup probability for sediment entrainment. *J. Hydraul. Engg.* 124 (2) 232–235.

Cheng N.-S. and Chiew Y.-M. (1999). Analysis of initiation of sediment suspension from bed load. *J. Hydraul. Engg.* 125 (8), 855–861.

Cheng S. J. (1953). On the stability of laminar boundary layer flow. *Quart. Appl. Math.* 11, 346–350.

Chepil W. S. (1945). Dynamics of wind erosion: Initiation of soil movement. *Soil Sci.* **60** (5), 397– 411.

Chepil W. S. (1958). The use of evenly spaced hemispheres to evaluate aerodynamic forces on soil surfaces. *Eos Trans.* 39 (3), 397–404.

Chepil W. S. (1959). Wind erodibility of farm fields. J. *Soil and Water Conserv.* 14 (5), 214–219.

Chepil W. S. (1961). The use of spheres to measure lift and drag on wind-eroded soil grains. *Proc. Soil. Sci. Soc. Am.* 25, 5.

Chien N. (1954). The present status of research on sediment transport. *J. Hydraul. Divn.* 80, 1–33.

Chien N. and Wan Z. (1999). *Mechanics of Sediment Transport*. ASCE Press, Reston, Virginia.

Chiew Y. (1992). Scour protection at bridge piers. *J. Hydraul. Eng.* 118 (9), 1260–1269.

Chiew Y. (1995). Mechanics of riprap failure at bridge piers. *J. Hydraul. Eng.* 121 (9), 635–643.

Chiew Y. M. and Lim F. H. (2000). Failure behaviour of riprap layer at bridge piers under live-bed conditions. *J. Hydraul. Engg.* 126 (1), 43–55.

Chiew Y. M. and Parker G. (1994), Incipient sediment motion on non-horizontal slopes. *J. Hydraul. Res.* 32, 649–660.

Chiew Y. M. (1984). *Local Scour at Bridge Piers*. University of Auckland, Auckland.

Chiew Y. M. (1990). Mechanics of local scour around submarine pipelines. *J. Hydraul. Engg.* 116 (4), 515–529.

Chiew Y. M. (1992). Scour protection around bridge piers. *J. Hydraul. Engg.* 118 (9), 1260–1269.

Chiew Y. M. (1995). Mechanics of riprap failure at bridge piers. *J. Hydraul. Engg.* 121 (9), 635–643.

Choi S. U. and Kwak S. (2001) Theoretical and probabilistic analyses of incipient motion of sediment particles. *Water Engg.* 5 (1), 59–65.

Chow V. T. (1959). *Open Channel Hydraulics*. McGraw-Hill, New York.

Christensen B. A. (1995), Incipient sediment motion on non horizontal slopes. *J. Hydraul. Res.* 33, 725–728.

Christiansen C., Blaesild F. and Dalsgaard K. (1984), Re-interpreting "segmented" grainsize curves. *Geol. Mag.* 121, 47–51.

Christiansen C. and Hartmann D. (1991) The hyperbolic distribution. In J. P. M. Syvitski (Ed.), *Principles, Methods and Application of Particle Size Analysis* (pp. 237–248). Cambridge University Press, Cambridge.

Christiansen J. E. (1935). *Distribution of Silt in Open Channels*. National Research council, Washington, 478–485.

Church M. and Jones D. (1982). Channel bars in gravel-bed rivers. In R. D. Hey, J. C. Bathurst and C. R. Thorne (Eds.), *Gravel-Bed Rivers* (pp. 291–354). John Wiley & Sons, New York.

Church M., Wolcott J. F. and Fletcher W. K. (1991). A test of equal mobility in fluvial sediment transport: behavior of the sand fraction. *Water Resour. Res.* 27, 2941–2951.

Clark J. A. (1968). A study of incompressible turbulent boundary layer in a channel flow. *Trans. Basic Engg.* 90, 455.

Clifford N. J., French J. R. and Hardisty J. (1993). *Turbulence: Perspective on Flow and Sediment Transport*. John Wiley & Sons, Chichester.

Clifford N. J., Hardisty J., French J. R. and Hart, S. (1993). Downstream variation in bed material characteristics: A turbulence-controlled form-process feedback mechanism. In C. Bristow and J. Best (Eds.), *Braided Rivers: Form, Process and Economic Significance*. Geological Society Special Publications, London.

Colby B. R. (1963). *Fluvial Sediments: A Summary of Source, Transportation, Deposition and Measurement of Sediment Discharge*. US Geological Survey, Washington.

Colby B. R. and Hembree C. H. (1955). *Computations of Total Sediment Discharge Niobrara River near Cody, Nebraska*. Geological Survey Water Supply, Washington.

Coleman N. L. (1967). *A Theoretical and Experimental Study of Drag and Lift Forces*. IAHR Congress, Fort Collins.

Coleman N. L. (1981). Velocity profiles with suspended sediment. *J. Hydraul. Res.* 19 (3), 211–229.

Coleman N. L. (1986). Effects of Suspended sediment on the open-channel velocity distribution. *Water Resour. Res.* 22 (10), 1377–1384.

Coleman N. L. and Alanso C.V. (1983). Two-dimensional channel flows over rough surfaces. *J. Hydraul. Engg.* 109 (2), 175–188.

Coleman S. E. (2005). Clearwater local scour at complex piers. *J. Hydraul. Engg.* 131, 330–334.

Coleman S. E. and Melville B. (1996). Initiation of bedforms on a flat sand bed. *J. Hydraul. Engg.* 122 (6), 301–310.

Coleman S. E. and Melville B.W. (1994). Bed-form development. *J. Hydraul. Engg.* 120, 544–560.

Coles D. F. (1956). The law of the wake in turbulent boundary layer. *J. Fluid Mech.* 1, 191–226.

Collinson J. D. and Thompson D. B. (1982). *Sedimentary Structures*. George Allen and Unwin (Publishers) Ltd, Crows Nest.

Colombini M. (2004). Revisiting the linear theory of sand dune formation. *J. Fluid Mech.* 502, 1–16.

Colombini M., Seminara G. and Tubino M. (1987). Finite-amplitude alternate bars. *J. Fluid Mech.* 181, 213.

Cooke R., Warren A. and Goudie A. (1993). *Desert Geomorphology*. UCL Press, London.

Corino E. R. and Brodkey R. S. (1969). A visual investigation of the wall region in turbulent flow. *J. Fluid Mech.* 37 (1), 1–30.

Corrsin S. and Kistler A. L. (1954). The free-stream boundaries of turbulent flows. *NACA Tech.* 31–33.

Cox D. R. and Miller H. D. (1965). *The Theory of Stochastic Processes*. Chapman & Hall, London.

Crimaldi J. P., Koseff J. R. and Monismith S. G. (2006). A mixing-length formulation for the turbulent Prandtl number in wall-bounded flows with bed roughness and elevated scalar sources. *Phys. Fluids.* 18, 095102.

Crowe C. T., Sommerfeld M. and Tsuji Y. (1998). *Multiphase Flows with Droplets and Particles*. CRC Press, Boca Raton, FL.

Dallavalle J. (1943). *Micrometrics*. Pitman, New York.

Dancey C. L., Diplas P., Papanicolaou A. and Bala M. (2002). Probability of individual grain movement and threshold condition. *J. Hydraul. Engg.* 128 (12), 1069–1075.

Dantec Dynamics. (2015). *Particle Image Velocimetry*. http://www.dantecdynamics.com/particle-image-velocimetry

Darghai D. E. (1989). The turbulent flow field around a circular cylinder. *Exper. Fluids.* 8, 1–12.

Das S. and Mazumdar A. (2016). Turbulence flow field around two eccentric circular piers in scour hole. *Int. J. River Basin Manag.* 13 (3), 343–361.

Das V. K., Roy S., Barman K., Debnath K., Chaudhury S. and Mazumder B. S. (2019). Investigations on undercutting processes of cohesive river bank. *Engg. Geol.* 252, 110–124.

Dasgupta R. (2013). Non uniform rates of convergence to normality for two sample *u*-statistics in non ID case with applications. *Adv. Growth Curve*. 46, 60–88.

Davies T. R. H. and Samad M. F. A. (1978) Fluid dynamic lift on a bed particle. *J. Hydraul. Div.* 104 (8), 1171–1182.

Day T. J. (1980a). *A Study of the Transport of Graded Sediments*. Hydraulic Resource Star, Wallingford.

Day, T. J. (1980b). *A Study of Initial Motion Characteristics of Particles in Graded Bed Material*. Geological Survey of Canada, Ottawa.

Debnath K., Manik M. K. and Mazumder B. S. (2012). Turbulence statistics of flow over scoured cohesive sediment bed around circular cylinder. *Adv. Water Resour.* 41, 18–28.

Dey S. (2003). Threshold of sediment motion on combined transverse and longitudinal sloping beds. *J. Hydraul. Res.* 41 (4), 405–415.

Dey S. (2014) *Fluvial Hydrodynamics, Hydrodynamic and Sediment Transport Phenomena*. Springer-Verlag Publications, Berlin, Germany.

Dey S. and Rajkumar V. R. (2007). Clear-water scour at piers in sand beds with an armour layer of gravels. *J. Hydraul. Engg.* 133 (6), 703–711.

Dey S. and Sen D. (2005). *Determination of Scour Depth in Boulder-Bed Rivers under High Stream Velocities*. Indian Institute of Technology, Kharagpur.

Dey S. and Singh N. P. (2008). Clear-water scour below underwater pipelines under steady flow. *J. Hydraul. Eng.* 134 (5), 588–600.

Dey S., Sumer B. M. and Fredsoe J. (2008). Control of scour at vertical circular piles under waves and current. *J. Hydraul. Engg.* 132 (2006), 270–279.

Dhamotharan J. S., Wood A., Parker G. and Stefan H. (1980). *Bed Load Transport in a Model Gravel Stream*. University of Minnesota, Minneapolis.

Dietrich W. E. (1982). Settling velocity of natural particles. *Water Resour. Res.* 18 (6), 1615–1626.

Diplas C. L., Dancey A. O., Celik M., Val yrakis K. and Greer T. A. (2008). The role of impulse on the initiation of particle movement under turbulent flow conditions. *Science*. 322 (5902), 717–720.

Dobbins W. E. (1943). Effect of turbulence on sedimentation. *Trans. Am. Soc.* 109 (1), 1944.

Donate J. (1929). Uber Sohlangriff und Geschiebetrieb, Wasserwirtschaft. *Heft*. 26, 27–36.

Dorrell R. M., Amy L. A., Peakall J. and McCaffrey W. D. (2018). Particle size distribution controls the threshold between net sediment erosion and deposition in suspended load dominated flows. *Geophys. Res. Lett.* 45, 1443–1452.

Dou G. R. (1964). *Bed-Load Transport*. Nanjing Hydraulic Research Institute, Nanjing.

Drake T. G., Shreve R. L., Dietrich, W. E., Whiting, P. J. and Leopold, L. B. (1988). Bedload transport of fine gravel observed by motion-picture photography. *J. Fluid Mech.* 192, 193–217.

Du Boys M. P. (1879). Le Rhone et les Rivieres a Lit affoullable. *Annales de Ponts et Chausses*. 18, 141–195.

Dwivedi A., Melville B. W., Shamseldin A. Y. and Guha T. K. (2011). Analysis of hydrodynamic lift on a bed sediment particle. *J. Geophys. Res.* 116, F02015.

Dyer K. R. (1986). *Coastal and Estuarine Sediment Dynamics*. John Wiley and Sons Ltd., London.

Dzulynski S. and Walton E. K. (1965). *Sedimentary Features of Flysch and Greywackes*. Elsevier Pub. Com, Amsterdam.

Egiazaroff I. V. (1965), Calculation of nonuniform sediment concentrations. *J. Hydraul. Div. Am. Soc. Civ. Engg.* 91, 225–247.

Eiffel G. (1912). Sur la resistance des spheres dans I'airenmouvement. *ComptesRendus*. 155, 1597–1599.

Einstein H. A. (1942). Formulas for the transportation of bed-load. *Trans. ASCE*. 107, 561–597.

Einstein H. A. (1950). *The Bed-Load Function for Sediment Transportation in Open Channel Flows*. Soil Conservation Service, Washington.

Einstein H. A. and Banks R. B. (1950). Fluid resistance of composite roughness. *Trans. Am. Geophys. Union*. 31, 603–610.

Einstein H. A. and Barbarossa N. L. (1952). River channel roughness. *Trans. Am. Soc. Civil Eng.* 117, 1121–1146.

Einstein H. A. and Chien N. (1954). *Second Approximation to the Solution of the Suspended Load Theory*. University of California, Berkely.

Einstein H. A. and Chien N. (1955). *Effect of Heavy Sediment Concentration Near Bed Motion Velocity and Sediment Distribution*. U.S. Army Corps of Engineers, Missouri River Division, Omaha, NE.

Einstein H. A., Anderson A. G. and Johnson J. W. (1940). A distinction between bed-load and suspended load in natural streams. *Trans. Am. Geophys. Union*. 21 (2), 628–633.

Einstein H. A. and El-Samni E. S. A. (1949). Hydrodynamic forces on a rough wall. *Rev. Modern Phys.* 21 (3), 520–524.

Ekman V. W. (1910). On the change from steady to turbulent motion of liquids. *Ark. Mat. Astronoch. Fys.* 6, 12.

Elata C. and Ippen A.T. (1961). *The Dynamics of Open Channel Flow with Suspensions of Neutrally Buoyant Particles*. MIT Hydrodynamic Laboratory, Technical Report n. 45, MIT, Cambridge, Massachusetts, USA.

Elder J. W. (1959) The dispersion of marked fluid in turbulent shear flow. *J. Fluid Mech.* 5, 544–560.

Eldho T. I., Viswanadham B. V. S. and Vijayasree B. A. (2010). Physical model study of scouring effects on pier foundation of bridges. *Proceedings of Indian Geotechnical Conference 2010* (pp. 973–976). IIT, Bombay, India.

Eldho T. I., Viswanadham B. V. S. and Vijayasree B. A. (2012). Physical model studies on extended raft foundation of bridges as a scour protection measure. *Proceedings of the National Conference on Hydraulics, Water Resources, Coastal and Environmental Engineering*, IIT, Bombay, India

Eldho T. I., Viswanadham B. V. S., Vijayasree B. A. and Siva Naga V. N. (2011). Physical model study of scouring effects on rafty foundation of bridge piers. *Proceedings of Indian Geotechnical Conference 2011* (pp. 907–910). Kochi, India.

Engel P. (1981). Length of flow separation over dunes. *J. Hydraul. Divn.* 107, 1133–1143.

Engelund F. (1966). Hydraulic resistance of alluvial streams. *J. Hydraul. Engg.* 92, 2.

Engelund F. (1970). Instability of erodible beds. *J. Fluid Mech.* 42 (2), 225–244.

Engelund F. and Fredsoe J. (1976). A sediment transport model for straight alluvial channel. *Nordic Hydrol.* 7, 293–306.

Engelund F. and Fredsoe J. (1982). Sediment ripples and dunes. *Ann. Rev. Fluid Mech.* 14, 13–37.

Ettema R. (1980). *Scour at Bridge Piers*. University of Auckland, Auckland.

Ettema R., Kirkil G. and Muste M. (2006). Similitude of large-scale turbulence in experiments on local scour at cylinders. *J. Hydraul. Engg.* 32 (1), 33–40.

Euler T. and Herget J. (2011). Obstacle-Reynolds-number based analysis of local scour at submerged cylinders. *J. Hydraul. Res.* 49 (2), 267– 271.

Euler T. and Herget J. (2012). Controls on local scour and deposition induced by obstacles in fluvial environments. *CATENA*. 91, 35– 46.

Exner F. M. (1925). Uber die Wechselwirkungzwischen Wasser und Geschiebe in Flussen. *Sitzber. Akad. Wiss. Wien.* 134(2a), 165–203.

Fael C., Lanca R. and Cardoso A. (2014). Pier shape and alignment effects on local scour. *Proceedings Small Scale Morphological Evolution of Coastal, Estuarine and Rivers Systems Conference*, Nantes. https://core.ac.uk/download/pdf/80534642.pdf

Fage A. and Townend H. G. H. (1934). An examination of turbulent flow with an ultramicroscope. *Proc. Roy. Soc. A.* 135, 656–684.

Fair G. M. and Geyer J. C. (1963). *Water Supply and Waste Water Disposal*. John Wiley & Sons Inc., New York.

Falkner H. (1935). *Studies of River Bed Materials and Their Movement with Special Reference to the Lower Mississippi River.* U.S. Waterways Express Station, Vicksburg.

Fang H., Shang Q., Chen M. and He G. (2014). Changes in the critical erosion velocity for sediment colonized by biofilm. *Sedimentology*. 61 (3), 648–659.

Fehlman H. M. (1985). *Resistance Components and Velocity Distributions of Open Channel Flows Over Bedforms*. Colorado State University, Ft. Collins.

Fer I., McPhee M. G. and Sirevaag A. (2004). Conditional statistics of the Reynolds stress in the under-ice boundary layer. *J. Geophys. Res.* 31, L15311.

Ferguson R. I. (2012). River channel slope, flow resistance, and gravel entrainment thresholds. *Water Resour. Res.* 48, W05517.

Fernandez Luque R. (1974). Erosion and transport of bed-load sediment, Dissertation, Krips Repro B.V., Meppel, The Netherlands.

Fick A. (1855). On liquid diffusion. *Phil. Mag.* 10 (63), 30–39.

Fieller N. R. J. and Flenley E. C. (1992). Statistics of particle size data. *J. Appl. Stat.* 41, 127–146.

Fieller N. R. J., Gillbertson D. D. and Olbricht W. (1984). A new method for environmental analysis of particle size distribution data from shoreline sediments. *Nature*. 311, 648–651.

Finley P. J., Khoo C. P. and Chin J. P., (1966). Velocity measurements in a thin turbulent water layer. *La Houille Blanche.* 21 (6), 713–721.

Flemming B.W. (1988). Zurklassifikationsubaquastistischer, stromungstransversalertransportkoper. *Boch. Geol. U. Geotechn Arb.* 29, 44–47.

Forchheimer P. (1914). *Hydraulik*. Teubner, Leipzig/Berlin.

Fourier J. B. J. (1822). *TheorieAnalytique de la Chaleur*. Didot, Paris.

Francis J. B. (1878). On the cause of the maximum velocity of water flowing in open channels being below the surface. *Trans. Am. Soc. Civil Eng.* 7 (1), 109–113.

Francis J. R. D. (1973). Experiments on the motion of solitary grains along the bed of a water-stream. *Proc. R. Soc. London, Ser. A.* 332, 443–471.

Fredsøe J. (1974). On the development of dunes in erodible channels. *J. Fluid Mech.* 64, 1–16.

Fredsøe J. (1982). Shape and dimensions of stationary dunes in rivers. *J.Hydraul. Divn.* 108 (8), 932–947.

Fredsoe J. (1986). Formation of ripples, dunes, and anti-dunes in river bed. In F. El-Baz (Ed.), *Physics of Desertification*. MartinusNijhoff Publishers, Dordrecht.

Fredsøe J., Andersen K. H. and Sumer B. M. (2000). Wave plus current over a ripple-covered bed. *Coastal Engg.* 38 (4), December 1999, 177–221.

Fredsøe J. and Deigaard R. (1992). *Mechanics of Coastal Sediment Transport*. World Scientific, Singapore.

Frey P., Ducottet C. and Jay J (2003). Fluctuations of bed load solid discharge and grain size distribution on steep slopes with image analysis. *Exp Fluids.* 35, 589–597.

Frijlink H. C. (1952). *Discussion of Bed Load Transport Formulas*. Delft Hydraulics, Delft.

Frisch U. (1985). *Fully Developed Turbulence and Intermittency, Turbulence and Predictability in Geophysical Fluid Dynamics and Climate Dynamics*. Italiana di Fisica, Bologna.

Froude W. (1872). Experiments on the surface friction. *Brit. Ass. Rep.* British Association for the Advancement of Science. The Collected Papers of William Froude, Institution of Naval Architects, 1955, London, Report 42, 118–124.

Fulgosi M., Lakehal D., Banerjee S. and Angelis V. D. (2003). Direct numerical simulation of turbulence in a sheared air–water flow with a deformable interface. *J Fluid Mech.* 482, 319–345.

Gabel S. L. (1993). Geometry and kinematics of dunes during steady and unsteady flows in the Calamus River, Nebraska, USA. *Sedimentology.* 40 (2), 237–269.

Gad-El-Hak M. (2000). *Flow* Q4 *control: Passive, Active, and Reactive Flow Management*. Cambridge University Press, Cambridge.

Gad-el-Hak M. and Bandyopadhyay P. (1994). Reynolds number effects in wall-bounded turbulent flows. *Appl. Mech. Rev.* 47 (8), 307–365.

Galbraith R. A., McSjolander D. and Head M. R. (1977). Mixing length in the wall region in turbulent boundary layers. *Aeronautical Quart.* 27 (2), 97–110.

Garcia M. and Nino Y. (1993). Dynamics of sediment bars in straight and meandering channels: Experiments on the resonance phenomenon. *J. Hydraul. Res.* 31 (6), 739–761.

Garde R. J. (1970). Initiation of motion on a hydrodynamically rough surface – critical velocity approach. *JIP.* 27, 3.

Garde R. J. and Albertson M. L. (1959). Sand waves and regime of flow in alluvial channels. *Proceedings of the IAHR 8th Congress*, Montreal, 4, 7–13.

Garde R. J. and Albertson M. L. (1961). Bed-load transport in alluvial channels. *La Huille Blanche*, 3, 274–286.

Garde R. J. and Isaac N. (1993). *Bed Undulations in Unidirectional Alluvial Streams*. C.W.R.S., Pune.

Garde R. J. and Ranga Raju K. G. (1966). Resistance relationships for alluvial channel flow. *J. Hydraul. Engg.* 92 (4), 77–100.

Garde R. J. and Ranga Raju K. G. (2000). *Mechanics of Sediment Transportation and Alluvial Stream Problems*. New Age International (P) Limited, New Delhi.

Garratt J. R. (1992). *The Atmospheric Boundary Layer.* Cambridge University Press, Cambridge.

Gautam P., Eldho T. I. and Behera M. R. (2021). Effects of pile-cap elevation on scour and turbulence around a complex bridge pier. *Int. J. River Basin Manag.* https://doi.org/10.1080/15715124.2021.1973016

Gautam P., Eldho T. I., Mazumder B. S. and Behera M. R. (2019). Experimental study of flow and turbulence characteristics around simple and complex piers using PIV. *Exper. Thermal Fluid Sci.* 18, 10.

Gautam P., Eldho T. I., Mazumder B. S. and Behera M. R. (2022). Turbulent flow characteristics responsible for current-induced scour around a complex pier. *Can. J. Civil Engg.* 49 (4), 597–606.

Gayan B. and Sarkar S. (2011). Negative turbulent production during flow reversal in a stratified oscillating boundary layer on a sloping bottom. *Phys. Fluids.* 23, 101703.

Gessler J. (1965). *The Beginning of Bed Load Movement of Mixtures Investigated as Natural Armouring in Channels*. California Institute of Technology, Pasadena.

Gessler J. (1967). *The Beginning of Bed Load Movement of Mixtures Investigated as Natural Armoring in Channels*. California Institute of Techology, Pasadena.

Gessler J. (1970) Self-stabilizing tendencies of alluvial channels. *J. Waterways Harbors Div.* 96 (2), 235–249.

Ghosh J. K. (1988). The sorting hypothesis and new mathematical model for changes in size distribution of sand grains. *Ind. J. Geol.* 60 (1), 1–10.

Ghosh J. K. and Mazumder B. S. (1981). Size distribution of suspended particles-unimodality, symmetry and lognormality. *Stat. Dist. Sci. Work.* 6, 21–32.

Ghosh J. K., Mazumder B. S. and Sengupta S. (1981). Methods of computation of suspended load from bed materials and flow parameters, *Sedimentology.* 28, 781–791.

Ghosh J. K., Mazumder B. S., Saha M. and Sengupta S. (1984). Deposition from suspension: experimental and theoretical studies. In B.C. Yen (Ed.), *Stochastic Hydraulics* (55–64). IAHR, Urbana, Illinois.

Ghosh J. K., Mazumder B. S., Saha M. R. and Sengupta S. (1986) Deposition of sand by suspension currents: Experimental and theoretical studies, *J. Sedimentary Petrol.* 56, 57–66.

Ghosh J. K., Mazumder, B. S. and Sengupta S. (1979). *Methods of Computation of Suspended Load from Bed Materials and Flow Parameters*. Indian Statistical Institute, Calcutta.

Ghoshal K. (2004). On velocity and suspension concentration in a sediment-laden flow: Experimental and theoretical studies. PhD. Thesis, Indian Statistical Institute, Kolkata.

Ghoshal K. and Mazumder B. S. (2005) Sediment-induced stratification in turbulent open-channel flow. *Envirometrics.* 16 (7), 673–686.

Ghoshal K. and Mazumder B. S. (2006). Velocity and concentration distributions in sediment-mixed mixed fluid: an approach with mixing length concept. ISH *J. Hydraul. Engg.* 12 (3), 21–29.

Ghoshal K. and Pal D. (2014). An analytical model for bedload layer thickness. *Acta Mech.* 225, 701–714.

Ghoshal K., Mazumder B. S. and Purkait B. (2010). Grain-size distributions of bed load: Inferences from flume experiments using heterogeneous sediment beds. *Sedimentary Geol.* 223 (1–2), 1–14.

Ghoshal K., Mazumder R., Chakraborty C. and Mazumder, B. S. (2013). Turbulence, suspension and downstream fining over sand-gravel beds. *Int. J. Sediment Res.* 12 (2), 194–209.

Ghoshal K., Purkait, B. and Mazumder B. S. (2011). Size distributions in suspension over sand-pebble mixture: An experimental approach. *Sedimentary Geol.* 241, 3–12.

Gilbert G. K. (1914). The transportation of debris by running water. *Geol. Survey* 86, 263.

Gimenez-Curto L. A. and Corniero M. A. (2009). Entrainment threshold of cohesionless sediment grains under steady flow of air and water. *Sedimentology*. 56 (2), 493–509.

Glenn S. M. and Grant W. D. (1987). A suspended sediment stratification correction for combined wave and current flows. *J. Geophys. Res.* 92, 8244–8264.

Gokce T. and Gunbak A. R. (1991). Self-burial and stimulated self-burial of pipelines by waves. *Proceedings of the First International Offshore and Polar Engineering Conference*, Edinburgh, 308–314. https://search.spe.org/i2kweb/SPE/doc/onepetro:5E4BE693

Goldstein S. (1929). The steady flow of viscous fluid past a fixed spherical obstacle at small Reynolds number. *Proc. Roy. Soc.* 123, 225–235.

Goldstein S. (1965). *Modern Developments in Fluid Dynamics.* Dover Pub, New York, vol. I, p. 315.

Gomez B. (1991). Bed load transport. *Earth-Sci. Rev.* 31, 89–132.

Goring D. G. and Nikora V. I. (2002). Despiking acoustic Doppler velocimeter data. *J. Hydraul. Eng.* 128 (1), 117–126.

Grade R. J. and Ranga Raju K. G. (2000) *Mechanics of Sediment Transportation and Alluvial Stream Problems.* New Age International (P) Limited, New Delhi.

Graf W. H. (1971). *Hydraulics of Sediment Transport.* McGraw-Hill, New York.

Graf W. H. (2010). *Hydraulics of Sediment Transport.* Water Resources Publications, Colorado.

Graf W. H. and Acaroglu E. R. (1967). *Remarks on the Rubey Equation for Computing Settling Velocities.* Sedimentological Congress, Great Britain.

Graf W. H. and Istiarto E. (2002). Flow pattern in the scour hole around a cylinder. *J. Hydraul. Res.* 40 (1), 13–20.

Granger R. A. (1985). *Fluid Mechanics*. Holt, Rinehart and Winston, New York.

Granville P. S. (1976). A modified law of the wake for turbulent shear layers. *J. Fluid Engg.* 98, 578–579.

Granville P. S. (1989). A modified van driest formula for the mixing length of turbulent boundary layers in pressure gradients. *J. Fluids Eng. Trans.* 111 (1), 94–97.

Grass A. J. (1970). Initial instability of fine bed sand, *J. Hydraul Div. Am. Soc. Civ. Eng.* 96 (3), 619–632.

Grass A. J. (1971). Structural features of turbulent flow over smooth and rough boundaries. *J. Fluid Mech.* 50, 233–255.

Grass A. J. (1983). The influence of boundary layer turbulence on the mechanism of sediment transport. In B. M. Sumer and A. Muller (Eds.), *Mechanics of Sediment Transport* (pp. 3–17). Balkema, Rotterdam.

Grass A. J., Stuart R. J. and Mansour-Tehrani M. (1991). Vortical structures and coherent motion in turbulent flow over smooth and rough boundaries. *Phil. Trans. R. Soc.* 336, 35.

Griffith O. F. and Grimwood C. (1981). Turbulence measurement study. *J. Hydraulicss Div.* 107 (3), 311–326.

Grifoll J. and Giralt F. (2000). The near wall mixing length formulation revisited. *Inter. J. Heat Mass Transfer.* 43, 3743–3746.

Gris R.B. (2010). Sheath for reducing local scour in bridge piers. *Proceedings of the International Conference on Scour and Erosion (ICSE-5)*, San Francisco, California. https://henry.baw.de/bitstream/handle/20.500.11970/100232/icse-5_Gris_987-996.pdf?sequence=1&isAllowed=y

Guo J. (1998). Turbulent velocity profiles in clear water and sediment-laden flows. PhD Dissertation, Colorado State University, Fort Collins, CO.

Guo J. (2002). Logarithmic matching and its application in computational hydraulics and sediment transport. *J. Hydraul. Res.* 40 (5), 555–566.

Guo J. and Julien P. Y. (2003). Modified Log–Wake Law for Turbulent Flow in Smooth Pipes. *J. Hydraul. Res.* 41 (5), 493–501.

Guo J., Julien P. Y. and Meroney M. N. (2005). Modified log-wake law for zero-pressure gradient turbulent boundary layers. *J. Hydraul. Res.* 43 (4), 421–430.

Gupta A. K. and Gangadharaiah T. (1992). Local scour reduction by a delta wing-like passive device. *Proceeding of the 8th Congress of Asia and Pacific*, CWPRS, Pune, India.

Guy H. P., Simons D. B. and Richardson, E. V. (1966). *Summary of Alluvial Channel Data from Flume Experiments.* Geological Survey, Washington.

Gyr A. and Schmid. (1989). The different ripple formation mechanism. *J. Hydraul. Res.* 27, 61–74.

Haddadchi A., Omid M. H. and Dehghani, A. A. (2013). Bedload equation analysis using bed load-material grain size. *J. Hydrol. Hydromech.*, 61 (3), 241–249.

Hamidifar H., Nasrabadi M. and Omid M. H. (2017). Using a bed sill as a scour countermeasure downstream of an apron. *Ain Shams Engg. J.* 9 (4), 1663–1669.

Hanes D. M. and Bowen A. J. (1985). A granular-fluid model for steady intense bed-load transport. *J. Geophys. Res.* 90 (C5), 9149–9158.

Hansen E. A. (1966). *Bed Load Investigation in Skivkive-Karup River.* Technical University of Denmark, Copenhagen.

Hansen E. A. (1992). Scour below pipelines and cables: A simple model. *Proceedings of the 11th Offshore Mechanics and Arctic Engineering Conference*, ASME, Calgary, Canada.

Hansen, E.A., Fredsae, J., and Mao, Y. (1986). Two dimensional scour below pipelines. *Fifth International Symposium on Offshore Mechanical and Arctic Engineering*, ASME, 670–678.

Happel J. and Brenner H. (1965). *Low ReynoldsNnumber Hydrodynamics*. Prentice-Hall, Englewood Cliffs.

Haque M. I. and Mahmood K. (1985). Geometry of ripples and dunes. *J. Hydraul. Engg.* 111, 48–63.

Harms J. C., Southard J. B. and Walker R.G. (1982). Structures and sequences in clastic rocks. *Soc. Econ. Mineral. Paleontol.* 9, 8–51.

Harris S. A. (1958). Probability curves and the recognition of adjustment to depositional environment. *J. Sedimentary Petrol.* 28, 151.

Harrison A. S. (1963). Computing suspended sand loads from field measurements. *Proceedings of the Federal Inter-Agency Sedimentation Conference.* US Department of Agriculture. https://books.google.co.in/books?id=44MWAAAAYAAJ&printsec=frontcover#v=onepage&q&f=false

Hastie T., Tibshirani R. and Friedman J. (2001). *Elements of Statistical Learning*, Springer Verlag, New York.

Hayashi S., Ohmoto T. and Takikawa K. (2003). Direct numerical simulation of coherent vortex structures in an open-channel flow over dune type wavy bed. *J. Hydrosci. Hydraul. Eng.* 21, 1–10.

Hayashi T. (1970). Formation of dunes and anti-dunes in open channel. *J Hydraul. Div.* 96 (3), 357–366.

Hayashi T., Ozaki S. and Ichibashi T. (1980). Study on bed load transport of sediment mixture. *Proceedings of 24th Japanese Conference on Hydraulics*. 24, 35–43. https://www.jstage.jst.go.jp/article/prohe1975/24/0/24_0_35/_article

Heidarpour M., Afzalimehr H. and Izadinia E. (2010). Reduction of local scour around bridge pier groups using collars. *Int. J. Sedim. Res.* 25 (4), 411–422.

Helley E. J. (1969). Field measurement of the initiation of large bed particle motion in blue Creek. U. S. Geological Survey. https://pubs.er.usgs.gov/publication/pp562G

Herbich J. B. (1981). *Scour Around Pipelines and Other Objects*. Marcell Dekker, Inc., New York.

Hergault P., Frey F., Métivier C., Barat C., Ducottet T. and Böhm, C. (2010). Image processing for the study of bedload transport of two-size spherical particles in a supercritical flow. *Exp. Fluids*. 49, 1095–1107.

Hersen P. (2004). Flow effects on the morphology and dynamics of aeolian and subaqueous barchan dunes. *J. Geophys. Res.* 110, F04S07.

Hersen P. (2005). On the crescentic shape of barchans dunes. *Euro. Phys*. 37, 507–514.

Hersen P., Douady S. and Andreotti B. (2002). Relevant length scale of barchan dunes. *Phys. Rev. Lett.* 89 (26), 264–301.

Hey R. D. (1979). Flow resistance in gravel-bed rivers. *J. Hydraul. Divn.* 105 (4), 365–379.

Hibbeler R. C. (2015). *Fluid Mechanics*. Pearson Publishers, New Delhi.

Hino M. (1963). Turbulent flow with suspended particles. *Am. Soc. Civil Eng*. 92, 2.

Hinze J. O. (1959). *Turbulence*. McGraw-Hill Book Co., New York.

Hinze J. O. (1973). Experimental investigation on secondary currents in the turbulent flow through a straight conduit. *Appl. Sci. Res.* 28, 453–465.

Hinze J. O. (1975). *Turbulence*. McGraw-Hill, NewYork.

Hjulstrom F. (1935) *The Morphological Activity of Rovers as Illustrated by Rivers Fyris*. Geological Institute, Uppsala.

Hjulstrom F. (1939). Transportation of detritus by moving water: Transportation. In P. D. Trask (Ed.), *Recent Marine Sediments. A Symposium* (pp. 5–31). AAPG, Tusla.

Hochstein A. B. and Adams C. E. (1989). Influence of vessel movements on stability of restricted channels. *J. Waterway Engg*. 113 (5), 444–465.

Householder M. K. and Goldschmidt V. W. (1969). Turbulent diffusion and Schmidt number of particles. *J. Waterway Engg.* 1969, 1345–1369.

Houssais M. and Lajeunesse E. (2012). Bedload transport of a bimodal sediment bed. *J. Geophys. Res.* 117, F04015.

Howard A. D., and McLane, C. F. (1988). Erosion of cohesionless sediment by groundwater seepage. *Water Resour. Res.* 24 (10), 1659–1674.

Howarth L. N. (1935). *On the Calculation of the Steady Flow in the Boundary Layer Near the Surface of a Cylinder in Stream*. Aeronautical Research Council, London, UK, Technical Report, pages 68.

Howarth L. N. (1938). On the solution of the laminar boundary layer equations. *Proc. Royal Soc. A*. 164 (919), 547–579.

Hu C. and Hui Y. (1996). Bed-load transport I: Mechanical characteristics. *J. Hydraul. Engg*. 122 (5), 245–254.

Huang S. H., Sun Z. L., Xu D. and Xia S. S. (2008). Vertical distribution of sediment concentration. *J. Zhejiang Univ. Sci*. 9 (11), 1560–1566.

Hui Y. and Hu E. (1991). Saltation characteristics of particle motions in water. *ShuiliXuebao*. 12, 59–64 (in Chinese).

Hunt J. N. (1954). The turbulent transport of suspended sediment in open channel in open channels. *Proc. Soc. London*. 224, 322–335.

Hunt J. N. (1969). On the turbulent transport of a heterogeneous sediment. *Quart. J. Mech. Appl. Math.* 22 (2), 235–246.

Hurst H. E. (1929). The suspension of sand in water. *Proc. Royal Soc. London*. 157, 196–201.

Hussain A. K. M. F. and Reynolds W. C. (1975). Measurements in fully developed turbulent channel flow. *J. Fluids Eng*. 97, 568–580.

Ikeda S. (1982). Incipient motion of sand particles on side slopes. *J. Hydraul. Div. Am. Soc. Civ. Eng*. 108 (1), 95–114.

Ikeda S. (1984). Lateral bed-load transport on side slopes - closure. *J. Hydraul. Engg.* 110 (2), 200–203.
Imamoto H. and Ishigaki T. (1986). Visualization of longitudinal eddies in an open channel flow. *Proceedings of 4th International Symposium on Flow Visualization,* Paris, 323–337.
Ippen A. T. (1973). Transport of suspended sediment. *Proceedings of International Seminar on Hydraulics of Alluvial Streams*, IAHR, New Delhi, India.
IRC 78. (2000). Standard specifications and code of practice for road bridges. The Indian Roads Congress, New Delhi, India. https://archive.org/details/govlawircy201478
IRC 89. (1997). Guidelines for design and construction of river training and control works for road bridges. The Indian Roads Congress, New Delhi, India. https://archive.org/details/gov.in.irc.089.2019
Irmay S. (1949). On steady flow formulae in pipes and channels. *Proceedings 3rd Congress, International Association for Hydraulic Research*, Grenoble, France.
Irvine H. M. (1971). A probabilistic approach to the initiation of movement of non-cohesive sediments in alluvial channels. MS. Thesis, University of Canterbury, Christchurch, New Zealand.
Itakura T. and Kishi T. (1980). Open channel flow with suspended sediments. *J. Hydr. Div.* 8, 1325–1343.
Iversen H. W. and Balent R. (1951). A correlating modulus of fluid resistance in accelerated motion. *J. Appl. Phys*. 22, 3.
Iwagaki Y. (1956). Hydrodynamical study on critical tractive force. *Trans. JSCE.* 41 (in Japanese), 1–21.
Izadinia E. and Heidarpor M. (2012). Simultaneous use of cable and collar to prevent local scouring around bridge pier. *Int. J. Sediment Res.* 27, 394–401.
Jackson R. G. (1975). Hierarchical attributes and a unifying model of bed forms composed of cohesionless material and produced by shearing flow. *Geol. Soc. Am.Bulletin.* 86, 1523–33.
Jackson R. G. (1976). Sedimentological and fluid-dynamic implications of the turbulent bursting phenomenon in geophysical flows. *J. Fluid Mech.* 77 (3), 531–560.
Jaeggi M. N. R. (1984). Formation and effects of alternate bars. *J. Hydraul. Engg.* 110 (2), 142–156.
Jain S. C. (1992). Note on lag in bedload discharge. *J. Hydraul. Engg.* 118 (6), 904–917.
Jain S. C. and Fischer E.E. (1980). Scour around bridge piers at high flow velocities. *J. Hydr. Div.* 106 (11), 1827–1842.
Jain S. C. and Kennedy J. F. (1971). The growth of sand waves. *Proceedings of the International Symposium on Stochastic Hydraulic*, Pittsburgh University Press, Pittsburgh, 449–471.
Jain S. C. and Kennedy, J. F. (1974). The spectral evolution of sedimentary bed forms. *J. Fluid Mech.* 63 (2), 301–314.
Jarocki W. (1963). *A Study of Sediment*. U. S. Department of Science Foundation, Washington.
Jeffreys H. (1929). On the transport of sediment by streams. *Proc. Cambridge Phil. Soc.* 25 (3), 272–276.
Jenkins G. M. and Watts D. G. (1968). *Spectral Analysis and its Applications*. Holden-Day, San Francisco.
Jensen H. R., Jensen B. L., Sumer B. M. and Fredsøe J. (1989). Flow visualization and numerical simulation of the flow around marine pipelines on an erodible bed. *Proceedings of the 8th International Conference on Offshore Mechanics and Arctic Engineering*, ASME, The Hague, The Netherlands.
Jenson V. G. (1959). Viscous flow round a sphere at low Reynolds numbers (<40). *Proc. R.Soc. Lond. Ser. A Math. Phys. Sci*., 249 (1258), 346–366.
Jerolmack D. J. and Mohrig D. (2005), Frozen dynamics of migrating bedforms. *Geology*. 33, 57–60.
Jimenez J. A. and Madsen O. S. (2003), A simple formula to estimate settling velocity of natural sediments. *J. Waterway, Port, Coastal, Ocean Eng*., 129 (2), 70–78.
Jobson H. E. and Sayre W. W. (1970a). Predicting concentration profiles in open channels. *J. Hydraul. Engg.* 96 (10), 1983–1996.
Jobson H. E. and Sayre W. W. (1970b). Vertical transfer in open channel flow., *Proc. Hydraul. Div.* 96 (3), 703–724.
Johansson C. R. (1976). Structural studies of frictional sediments. *Geography Ann.* 58, 201– 300.
Johnson P. A. and McCuen R. H. (1991). *A Temporal Spatial Pier Scour Model*. Transportation Research Board, Washington.
Jones W. P. and Launder B. E. (1993). The calculation of low Reynolds number phenomena with a two equation model of turbulence. *Inter. J. Heat Mass Transfer.* 16, 1119–1130.

Jovanovic J. (2004). *The Statistical Dynamics of Turbulence.* Friedrich-Alexander-University of Erlangen-Nurnber, Germany.

Julien P. Y. (2010). *Erosion and Sedimentation.* Cambridge University Press, Cambridge, UK.

Julien P. Y, and Klaassen G. J. (1995). Sand-dune geometry of large rivers during floods. *J. Hydraul. Engg.* 121 (9), 657–663.

Kadota A. and Nezu I. (1999), Three-dimensional structure of space-time correlation on coherent vortices generated behind dune crest. *J. Hydraul. Res.* 37 (1), 59–80.

Kaftori D., Hetsroni G. and Banerjee S. (1995). Particle behavior in the turbulent boundary layer. II. Velocity and distribution profiles. *Phys. Fluids.* 7 (5), 1107–1121.

Kalinske, A. A. (1947). Movement of sediment as bed load in rivers. *Trans. Am. Geophys. Union.* 28 (4), 615–620.

Kalinske A. A. and Pien C. L. (1943). Experiments on eddy-diffusion and suspended-material transportation in open channels. *Trans. Am. Geophys. Union.* 12, 530–534.

Kamphuis J. W. (1974). Determination of sand roughness for fixed beds. *J. Hydraul. Res.* 12 (2), 193–203.

Kapdasli M.S. and Dyer K.R. (1986). Threshold conditions for sand movement on a rippled bed. *Geo-Mar. Lett.* 6, 161–164.

Karcz I. (1968). Fluviatile obstacle marks from the wadis of the Negev (Southern Israel). *J. Sedimentary Petrol.* 38, 1000–1012.

Karim F. (1995). Bed configuration and flow resistance in alluvial-channel flows. *J. Hydraul. Engg.* 121 (1), 15–25.

Karim F. (1999). Bed-form geometry in sand-bed flows. *J. Hydraul. Engg.* 125, 1253–1261.

Karim M. F. and Kennedy J. F. (1990). Menu of coupled velocity and sediment discharge relations for rivers. *J. Hydr. Eng.* 116 (8), 978–996.

Katul G. G., Angelini C., De Canditiis D., Amato U., Vidakovic B. and Albertson J. D (2003). Are the effects of large scale flow conditions really lost through the turbulent cascade? *Geophys. Res. Lett.* 30 (4), 1164.

Katul G. G., Poggi D., Cava D. and Finnigan J. (2006). The relative importance of ejections and sweeps to momentum transfer in the atmospheric boundary layer. *Boundary-Layer Meteorol.* 120 (3), 367–375.

Kaushal D. K., Seshadri N. and Singh S. N. (2002). Prediction of concentration and particle size distribution in the flow of multi-sized particulate slurry through rectangular duct. *Appl. Math. Model.* 26 (10), 941–952.

Keirsbulck L., Labraga L., Mazouz A. and Tournier, C. (2002). Influence of surface roughness on anisotropy in a turbulent boundary layer flow. *Exp. Fluids.* 33 (3), 497–499.

Kennedy J. F. (1961). *Stationary Waves and Antidunes in Alluvial Channels.* California Institute of Technology, California.

Kennedy J. F. (1963). The mechanics of dunes and anti-dunes in erodible bed channel. *J. Fluid Mech.* 16 (4). 521–544.

Kennedy J. F. (1969). The formation of sediment ripples, dunes and antidunes. *Ann. Rev. Fluid Mech.* 1, 147–168.

Kennedy J. F. and Koh R. C. Y. (1961). The relation between the frequency distributions of sieve diameters and fall velocities of sediment particles. *J. Geophy. Res.* 66, 4233–4246.

Kennedy J. F. and Odgaard A. J. (1991). *Informal Monograph on Riverine sand Dunes.* US Army Waterways Experiment Station, Mississippi.

Keshavarzy A. (1997). Entrainment of sediment particles from a flat mobile bed with the influence of near-wall turbulence. Ph. D. Thesis, The University of New South Wales, Sydney, Australia.

Keshavarzy A. and Ball J. (1997). An analysis of the characteristics of rough bed turbulent shear stresses in an open channel. *Stochastic Hydrol. Hydraul.* 11, 193–210.

Keshavarzi A. and Ball J. (1999). An application of image processing in the study of sediment motion. *J. Hydraul. Res.* 37 (4), 559–576.

Keshavarzi A., Ball J. and Nabavi H. (2012). Frequency pattern of turbulent flow and sediment entrainment over ripples using image processing. *Hydrol. Earth Syst. Sci.* 16, 147–156.

Keshavarzi A., Shrestha C. K., Melville B., Khabbaz H., Ranjbar-Zahedani M. and Ball J. (2018). Estimation of maximum scour depths at upstream of front and rear piers for two in-line circular columns. *Environ. Fluid Mech.* 18 (2), 537–550.

Keshavarzi A., Shrestha C. K., Zahedani M. R., Ball J. and Khabbaz H. (2017). Experimental study of flow structure around two in-line bridge piers. *Proc. Inst.Civil Eng-Water Manag.* 171, 311–327.

Keulegan G. H. (1938). *Laws of Turbulent Flow in Open Channels*. National Bureau of Standards, Washington.

Kirkgoz S. (1989). Turbulent velocity profiles for smooth and rough open-channel flow. *J. Hydr. Engg.* 115 (11), 1543–1561.

Kirkil G. and Constantinescu G. (2015). Effects of cylinder Reynolds number on the turbulent horseshoe vortex system and near wake of a surface-mounted circular cylinder. *Phys. Fluids.* 27, 075102.

Kirkil G., Constantinescu S. G. and Ettema R. (2008). Coherent structures in the flow field around a circular cylinder with scour hole. *J. Hydraul. Eng*. 134, 572–587.

Kironoto B. A. (1993). Turbulence characteristics of uniform and nonuniform, rough open-channel flow, PhD Thesis, Swiss Federal Institute of Technology, Lausanne, Switzerland.

Kironoto B. A. and Graf W. H.(1994). Turbulence characteristics in rough uniform open-channel flow. *Proc. Inst Civ. Engrs Wat. Mari. Energy.* 106 (12), 333–344.

Kjeldsen S. P., GjOrsvtk O., Brtngaker K. G. and Jacobsen J. (1973). Local scour near offshore pipelines. Second International Port and Ocean Engineering under Arctic Conditions, Conference, Iceland. https://trid.trb.org/view/38479

Klebanoff P. S. (1954). Characteristics of turbulence in a boundary layer with zero pressure gradient. NACA Technical note (United States. National Advisory Committee for Aeronautics), 3178. Washington D.C., 56 pages.

Kline S. J., Reynolds W. C., Schraub F. A. and Runstadler P. W. (1967). The structure of turbulent boundary layers. *J. Fluid Mech.* 30 (4), 741–773.

Knaapen M. A. F., Hulscher S. J. M. H., de Vriend H. J. and Stolk A. (2001). A new type of sea bed waves. *Geophys. Res. Lett.* 28 (7), 1323–1326.

Knight D. W., Demetriou J. D. and Mohammed E. H. (1984). Boundary shear in smooth rectangular channels. *J. Hydr. Eng*. 110 (4), 405–422.

Kolmogorov A. N. (1941) Local structure of turbulence in an incompressible viscous fluid at very large Reynolds numbers. *Proc. Royal Soc. Math. Phys. Sci. Engg Sci.* 434, 9–13.

Komar P. D. and Carling P.A. (1991). Grain sorting in gravel bed streams and the choice of particle sizes for flow-competence evaluations. *Sedimentology*. 38, 489–502.

Komar P. D. and Clemens K. E. (1986). The relationship between a grain's settling velocity and threshold of motion under unidirectional currents. *J. Sediment. Petrol.* 56 (2), 258–266.

Kondolf G. M. and Adhikari A. (2000). Weibull vs lognormal distributions for fluvial gravels. *J. Sedimentary Res.* 70 (3), 456–460.

Kostaschuk R. A. (2000). A field study of turbulence and sediment dynamics over subaqueous dunes with flow separation. *Sedimentology.* 47 (3), 519–531.

Kostaschuk R. A., Best J., Villard P., Peakall J. and Franklin M. (2005). Measuring flow velocity and sediment transport with an acoustic Doppler current profiler. *Geomorphology*. 68, 25–37.

Kostaschuk R. A. and Church M. A. (1993). Macroturbulence generated by dunes. *Sedimentary Geol.* 85, 25–37.

Kostaschuk R. A. and Villard P. V. (1996), Flow and sediment transport over large subaqueous dunes. *Sedimentology*. 43, 849–863.

Kothyari U. C. (1995) Frequency distribution of river bed materials. *Sedimentology*. 42, 283–291.

Kothyari U. C., Garde R. J. and Ranga Raju K. G. (1992). Temporal variation of scour around circular bridge piers. *J. Hydraul. Engg.* 118 (8), 1091–1106.

Kovacs A. E. (1998). Prandtl's mixing length concept modified for equilibrium sediment-laden flows. *J. Hydraul. Engg.* 124 (8), 803–812.

Kramer H. (1935). Sand mixtures and sand movement in fluvial models. *Trans. ASCE.* 100, 798–878.

Krey H. (1925). Grenzen der Ubertragbarkeit der Versuchsergebnisse. *Z. angew. Matm Mech.* 5, 6.

Kreyszig E. (1985). *Advanced Engineering Mathematics*. Wiley Eastern Limited, New Delhi.

Kristiansen O. and Torum A. (1989). Interaction between current induced vibrations and scour of pipelines on a sandy bottom. Proceedings of the 8th International Conference on Offshore Mechanics and Arctic Engineering, ASME. The Hague, The Netherlands. https://www.osti.gov/etdeweb/biblio/7271600

Krogstad P. A. and Antonia R. A. (1999). Surface roughness effects in turbulent boundary layer. *Exp. Fluids*. 27, 450–460.

Krumbein W. C. (1934). Size frequency distributions of sediments. *J. Sedim. Petrol*. 4, 65–77.

Krumbein W. C. (1936). Application of logarithmic moments to size frequency distributions of sediments. *J. Sedimentary Petrol.* 6, 35–47.

Krumbein W. C. (1938). Size frequency distributions of sediments and normal Phi curve. *J. Sedimentary Petrol.* 8, 84–90.

Krumbein W. C. (1942). Settling velocities and flume behavior of non-spherical particles. *Trans. Am. Geophys. Union*. 41, 621–633.

Kuenen P. H. and Sengupta S., (1970). Experimental marine suspension currents, competency and capacity. *Geol. Minbouw.* 49 (2), 89–118.

Kuhnle R. A. (1992) Frictional transport rates of bed loea on Goodwin Creek. In P. Bill, R. D. Hey, C. R. Throne and P. Tacconi (Eds.), *Dynamics of Gravel Bed Rivers* (pp. 141–155). John Wiley and sons Ltd., Chichester, UK.

Kuhnle R. A., (1993). Incipient motion of sand–gravel sediment mixtures. *J. Hydraul. Eng*. 119 (12), 1400–1415.

Kumar A. (2016) *Derivation of the Transport Equation of Turbulent Kinetic Energy*. City University, London.

Kumar V., Ranga Raju K. G. and Vittal N. (1999). Reduction of local scour around bridge piers using slot and collar. *J. Hydraul. Engg*. 125 (12), 1302–1305.

Kumbhakar M., Kundu S. and Ghoshal K. (2017) Hindered settling velocity in particle-fluid mixture: a theoretical study using the entropy concept. *J. Hydraul. Eng*. 143 (11), 06017019.

Kumbhakar M., Kundu S. and Ghoshal K. (2018). An explicit analytical expression for bed-load layer thickness based on maximum entropy principle. *Phy. Lett. A* 382, 2297–2304.

Kundu P. K. (1990). *Fluid Mechanics*. Academic Press, NewYork

Kundu S., Kumbhakar M. and Ghoshal K. (2018). Reinvestigation on mixing length in an open channel turbulent flow. *Acta Geophysica*, 66, 93–107.

Kurihara M. (1948). *On the Critical Tractive Force*. Kyushu University, Kyushu, (In Japanese).

Kwan T. F. and Melville B. W. (1994). Local scours and flow measurements at bridge abutments. *J. Hydraul. Res*. 32, 661–673.

Lacey G. (1930). *Stable Channels in Alluvium*. William Clowes & Sons Ltd., London.

Lacey R. W. J. and Rennie C. (2012). Laboratory investigation of turbulent flow structure around a bed-mounted cube at multiple flow stages. *J. Hydraul. Engg*. 138 (1), 71–84.

Lajeunesse E., Malverti L. and Charru E. (2010). Bed load transport in turbulent flow at the given scale: experiments and modeling. *J. Geophys. Res*. 115, 1–16.

Lam K. and Banerjee S. (1992). On the condition of streak formation in a bounded turbulent flow. *Phys. Fluids A* 4(2), 306

Lamb B. H. (1945). *Hydrodynamics*. Dover, New York.

Lamb M. P., Dietrich W. E. and Venditti J. G. (2008). Is the critical Shields stress for incipient sediment motion dependent on channel-bed slope? *J. Geophys. Res*. 113, F02008.

Landahl M. T. and Mollo-Christensen E. (1986). *Turbulence and Random Processes in Fluid Mechanics*. Cambridge University Press, Cambridge.

Lane E. W. (1947). Report of subcommittee on sediment terminology. *Trans. AGU*. 28, 6.

Lane E. W. (1953). Progress report on studies on the design of stable channels of the Bureau of Reclamation. *Proc. ASCE*. 79, 246–261.

Lane E. W. and Kalinske A. A. (1939). The relation of suspended to bed material in rivers. *Trans. Am. Geophys. Union*. 20, 637–641.

Lane E. W. and Kalinske A. A. (1941). Engineering calculation of suspended sediment. *Trans. Am. Geophys. Union*. 20 (3), 603–607.

Lanzoni S. (2000). Experiments on bar formation in a straight flume. *Water Resour. Res*. 36 (11), 3337–3349.

Lau L. and Engel P. (1999). Inception of sediment transport on steep slopes. *J. Hydraul. Engg*. 125, 544–547.

Lauchlan C. S. and Melville B. W. (2001). Riprap protection at bridge piers. *J. Hydraul. Engg*. 127 (5), 412–418.

Laufer J. (1950). Some recent measurements in a two-dimensional turbulent channel flow. *J. Aero. Sci*. 20, 257–287.

Laufer J. (1951). Investigation of turbulent flow in a two-dimensional channel. *NACA Rep.* 1053.

Laufer J. (1954). *The Structure of Turbulence in Fully Developed Pipe Flow*.National Advisory Committee of Aeronautics, Washington, DC.

Laufer J. (1975). New trends in experimental turbulence research. *Ann. Rev. Fluid Mech.* 7, 307.

Launder B. E. and Priddin C. H. (1973). A comparison of some proposals for the mixing length near a wall. *Inter. J. Heat Mass Transfer.* 16, 700–702.

Lavelle J. W. and Thacker W. C. (1978) Effects of hindered settling on sediment concentration profiles. *J. Hydraul. Res.* 16, 507–527.

Le Couturier M. N., Grochowski N. T., Heathershaw A., Oikonomou E. and Collins M. B. (2000) Turbulent and macro-turbulent structures developed in the benthic boundary layer downstream of topographic features. *Est. Coast. Shelf Sci.* 50 (6), 817–833.

Lee D. I. (1969) The viscosity of concentrated suspensions. *Trans Soc Rheol.* 13 (2), 273–288.

Lee H. and Balachandar S. (2012). Critical shear stress for incipient motion of a particle on a rough bed. *J. Geophy. Res.* 117, 1–19.

Lee H. Y., Chen Y. H., You J. Y. and Lin YT. (2000). Investigation of continuous bed load saltating process. *J. Hydraul. Engg.* 126 (9), 691–700.

Lee H. Y. and Hsu I. S. (1994). Investigation of saltating particle motions. *J. Hydraul. Engg.* 120 (7), 831–45.

Leliavsky S. (1955). *An Introduction to Fluvial Hydraulics*. Constable, London.

Leopold L. B., Wolman M. G. and Miller J. P. (1964). *Fluvial Processes in Geomorphology*. Freeman, San Francisco.

Lhermitte R. and Lemmin U. (1994). Open-channel flow and turbulence measurement by high-resolution Doppler sonar. *J. Atmos. Ocean Technol.* 11, 1295–1308.

Li D. and Katul G. G. (2017). On the linkage between the k-5/3 spectral and k-7/3 cospectral scaling in high-Reynolds number turbulent boundary layers. *Phy. Fluids.* 29 (6), 65108.

Li Z. Y. (1983). *Laboratory Investigation on Drag and Lift Forces Acting on Bed Spheres*. Water and Power Press, Beijing.

Liang D. and Cheng L. (2005). "Numerical model for wave induced scour below submarine pipeline. *J. Waterways Port Coastal Ocean Engg.* 131 (5), 193–203.

Lin C. C. (1955). *The Theory of Hydrodynamic Stability*. Cambridge University Press, Cambridge.

Ling, C. H. (1995). Criteria for incipient motion of spherical sediment particles. *J. Hydraul. Engg.* 121, 472–478.

Liu H. K. (1957) Mechanics of sediment ripple formation. *Proc. ASCE.* 83 (2), 1–23.

Liu H. K. (1958) Closer: Mechanics of sediment ripple formation. *Proc. ASCE.* 84 (5), 5–31.

Liu Q. Q., Shu A. P. and Singh V. P. (2007). Analysis of the vertical profile of concentration in sediment-laden flows. *J. Eng. Mech.* 133, 601–607.

Lohrmann A., Cabrera R. and Kraus N. C. (1994). Acoustic-doppler velocimeter (ADV) for laboratory use in fundamental and advancements in hydraulic measurements and experimentation. *Proc. Conf. on Fundamentals and Advancements in Hydraulic Measurements and Experimentation,* Buffalo, NY, American Society of Civil Engineers, 351–365.

Lohse D. and Grossmann S. (1993). Intermittency in turbulence. *Physica A.* 194 (1–4), 519–531.

Long D., Steffler P. M. and Rajaratnam N. (1990). Study of flow structure in submerged hydraulic jump. *J. Hydraul. Res.* 28, 437–460.

Lord R. (1911). On the motion of solid bodies through viscous liquids. *Phil. Mag.* 21, 697–711.

Lorentz H. A. (1907). Abhandlung uber theoretische. *Physik* 1, 43–71.

Lu S. S. and Willmarth W. W. (1973). Measurements of the structure of the Reynolds stress in a turbulent boundary layer. *J. Fluid Mech.* 60, 481–511.

Lu Y., Lu Y. J. and Chiewy M. (2012). Incipient motion of cohesionless sediments on riverbanks with ground water injection. *Int. J. Sed. Res.* 27 (1), 111–119.

Lucassen R. J. (1984). *Scour Underneath Submarine Pipelines*. Delft University of Technology, Delft.

Lumley J. L. and Newman G. R. (1977). The return to isotropy of homogeneous turbulence. *J. Fluid Mech.* 82, 161–178.

Luque R. F. (1974). *Erosion and Transport of Bed Load Sediment*. Delft University of Technology, Delft.

Luque R. F. and van Beek R. F. (1976) Erosion and transport of bedload sediment. *J. Hydraul. Res.* 14 (2), 127–144.

Lyn D. A. (1986). Turbulence and turbulent transport in sediment-laden flows in open channel, PhD. thesis, California Institute of Technology, Pasadena.

Lyn D. A. (1988). A similarity approach to turbulent sediment-laden flows in open channels. *J. Fluid Mech.* 193, 1–26.

Lyn D. A. (1993). Turbulence measurements in open channel flows over artificial bedforms. *J. Hydraul. Eng.* 119 (3), 306–326.

Lyn D. A. (2000). Regression Residuals and mean profioes in uniform open-channels flows. *J. Hydr. Engg.* 126 (1), 24–32.

Maddux T. B., McLean S. R. and Nelson J. M. (2003b) Turbulent flow over three-dimensional dunes: Fluid and bed stresses. *J. Geophys. Res. Earth Surf.* 108, 11.1–11.17.

Maddux T. B., Nelson J. M. and McLean S. R. (2003a). Turbulent flow over three-dimensional dunes: Free surface and flow response. *J. Geophys.* 108 (11), 1–17.

Madsen O. S. (1991). Mechanics of cohesionless sediment transport in coastal waters. In N. C. Kraus, K. J. Gingerich, D. L. Kriedel (Eds.), *Coastal sediment '91* (pp. 15–27), American Society of Civil Engineers, New York.

Maity H., Dasgupta R. and Mazumder B. S. (2013). Evolution of scour and velocity fluctuations due to turbulence around cylinders. In R. Dasgupta (Ed.), *Advances in Growth Curve Models. Proceedings in Mathematics & Statistics* (46, Chapter 7, pp. 131–148). Springer, Berlin.

Maity H. and Mazumder B. S. (2012) Contributions of burst-sweep cycles to Reynolds shear stress over fluvial obstacle marks generated in laboratory flume. *Int. J. Sediment Res.* 27 (3), 378–387.

Maity H. and Mazumder B. S. (2013). Conditional statistics of Reynolds shear stress over obstacle marks. *J. Hydraul. Engg.* 40, 1–11.

Maity H. and Mazumder B. S. (2014). Experimental investigation of the impacts of coherent flow structures upon turbulence properties in regions of crescentic scour. *Earth Surf. Process Landforms.* 39 (8), 995–1013.

Maity H. and Mazumder B. S. (2017). Prediction of plane-wise turbulent events to the Reynolds shear stress in a flow over scour bed. *Environmetrics.* 42, 14.

Mao Y. (1986). *The Interaction Between a Pipeline and an Erodible Bed.* Technology University, Denmark.

Mao Y. (2003) The effects of turbulent bursting on the sediment movement in suspension. *Int. J. Sediment Res.* 18 (2), 148–157.

Markatos N. C. (1986). The mathematical modeling of turbulent flows. *Appl. Math. Modell.* 10, 190–220.

Matthes G. H. (1947). Macroturbulence in natural stream flow. *Trans. Am. Geophys. Union.* **28**, 255–262.

Maude A. D. and Whitemore R. L. (1958). A generalized theory of sedimentation. *British J. Appl. Phys.* 9, 477–481.

Mavis F. T., Liu T. and Soucek E. (1937). *The Transportation of Detritus by Flowing Water-II.* University of Iowa, Iowa.

Maza J. A. (1987). *Introduction to River Engineering. Advanced Course on Water Resources Managemen*t. Universita Italiana per Stranieri, Perugia, Italy.

Mazumder B. S. (1994). Grain-size distribution in suspension from bed materials. *Sedimentology.* 41, 271–277.

Mazumder B. S., Bhattacharya A. and Ojha S. P. (2008) Near-bed particle motion due to turbulent flow using image processing technique. *J. Flow Visual. Image Process.* 15 (1), 1–15.

Mazumder B. S., Bhowmik N. G. and Soong T. W. (1991). Turbulence and Reynolds stress distribution in a natural river. *Proceedings of National Conference HY. DIV/ASCE,* Nashville, Tennesse.

Mazumder B. S., Bhowmik N. G. and Soong T.W. (1993). Turbulence in rivers due to navigation traffics, *J. Hydraul. Engg.* 119, 581–597.

Mazumder B. S. and Dalal D. C. (2003). Saltation layer of particles in water flows related to transport. *Nordic Hydrol.* 34 (4), 343–360.

Mazumder B. S. and Das S. K. (1992). Effect of boundary reaction on solute dispersion in pulsatile flow through a tube. *J. Fluid Mech.* 239, 523–549.

Mazumder B. S., Das S. K. and Das S. N. (2006). Computation of return flows due to navigation traffics in restricted waterways. *Int. J. Sediment Res.* 21 (4), 249–260.

Mazumder B. S. and Ghoshal K. (2002). Velocity and suspension concentration in sediment-mixed fluid. *Int. J. Sediment Res.* 17 (3), 220–232.

Mazumder B. S. and Ghoshal K. (2006). Velocity and concentration profiles in uniform sediment-laden flow. *Appl. Math. Model.* 30, 164–176.

Mazumder B. S., Ghoshal K. and Dalal D. C. (2001). *Influence of Bed Roughness on Sediment Suspension.* ISI Flume Laboratory Data, Kolkota.

Mazumder B. S., Ghoshal K. and Dalal D. C. (2005b). Influence of bed roughness on sediment suspension: Experimental and theoretical studies. *J. Hydraul. Res.* 43 (3), 245–257.

Mazumder B. S., Maity H. and Chadda, T. (2011). Turbulent flow field over fluvial obstacle marks generated in a laboratory flume. *Int. J. Sediment Res.* 26 (1), 62–77.

Mazumder B. S. and Ojha S. P. (2007). Turbulence Statistics of flow due to wave-current interaction. *Flow Measurements Inst.* 18, 129–138.

Mazumder B. S., Pal D. K., Ghoshal K. and Ojha, S. P. (2009) Turbulence statistics of flow over isolated scalene and isosceles triangular-shaped bedforms. *J. Hydraul. Res.* 47 (5), 626–637.

Mazumder B. S., Ray R. N. and Dalal D. C. (2005a) Size distributions of suspended particles in open channel flow over sediment beds. *Environmetrics.* 16 (2), 149–165.

Mazumder B. S. and Sarkar K. (2014). Turbulent flow characteristics and drag over 2-D forward facing dune-shaped structures with two different stoss-side slopes. *Environ. Fluid Mech.* 14 (3), 617–645.

Mazumder R. (2007). Clustering based on geometry and interactions of turbulence bursting rate processes in a trough region. *Environmetrics.* 18 (4) 445–459.

Mazumder R. (2008). Fluid flow pattern analysis in a trough region: a nonparametric approach. *J. Appl. Stat.* 35 (6), 633–645.

Mazumder R. (2000). Turbulence particle interactions and their implications for sediment transport and bed form mechanics under unidirectional current: some recent developments, *Earth - Science Review.* 50, 113–124.

McIlwain S. and Pollard A. (2002). Large eddy simulation of the effects of mild swirl on the near field of a round jet. *Phys. Fluids.* 14 (2), 653–661.

McLaren P. and Bowles D. (1985). The effects of sediment transport on grain-size distributions. *J. Sediment. Petrol.* 55, 457–470.

McLean S. R. (1976). Mechanics of the turbulent boundary layer over sand waves in the Columbia river. PhD. thesis, University of Washington, Seattle.

McLean S. R. (1991). *On the Calculation of Suspended Load for Non-Cohesive Sediments.* University of California, Santa Barbara.

McLean S. R. (1992). On the calculation of suspended load for noncohesive sediments. *J. Geophys. Res.* 97, 5759.

McLean S. R. and Smith J.D. (1986). A model for flow over two-dimensional bed forms. *J. Hydraul. Engg.* 112 (4), 300.

McNown J. S. and Lin P. N. (1952). Sediment concentration and fall velocity. *Proceedings of the 2nd Midwestern Conference in Fluid Mechanics*, Ohio State University. Reprint in Engineering Report No. 109, Iowa State University, 401.

McQuivey R. S. and Richardson E. V. (1969). Some turbulence measurements in open-channel flow. *J. Hydraul. Divn.* 95, 209–223.

Mei R., Adrian R. J. and Hanratty T. J. (1991) Particle dispersion in isotropic turbulence under Stokes drag and Basset force with gravitational settling. *J. Fluid Mech.* 225, 481–495.

Melville B. W. (1975). *Local Scour at Bridge Sites.* University of Auckland, Auckland.

Melville B. W. (1997). Pier and abutment scour—an integrated approach. *J. Hydraul. Engg.* 123 (2), 125–136.

Melville B. W. and Chiew S.C. (1999). Time scale for local scour at bridge piers. *J. Hydraul. Engg.* 125 (1), 59–65.

Melville B. W. and Coleman S. E. (2000), *Bridge Scour.* Water Resources Publications, LLC, Colorodo.

Melville B. W. and Raudkivi A. J. (1977). Flow characteristics in local scour at bridge piers. *J. Hydraul. Res.* 15, 373–380.

Melville B. W. and Raudkivi A. J. (1996). Effects of foundation geometry on bridge pier scour. *J. Hydraul. Engg.* 122 (4), 203–209.

Mendoza C. and Shen H. W. (1985a). Steady two-dimensional flow over dunes. *Proceedings of the 21st Congress of International Association of Hydraulic Research (IAHR)*, Melbourne, Australia.

Mendoza C. and Shen H. W. (1985b). Numerical modelling of turbulent flow over dunes to predict flow resistance. *Proceedings of the International Symposium on Refined Flow Modelling and Turbulence Measurements*. University of Iowa, Iowa City, Iowa.

Mendoza C. and Shen H. W. (1990). Investigation of turbulent flow over dunes. *J. Hydraul. Engg.* 116, 4.

Mendoza C. and Zhou D. (1995). A dynamic approach to sedimentladen turbulent flows. *Water Resour. Res.* 31 (12), 3075–3087.

Meyer-Peter E., Favre H. and Einstein H. A. (1934). NeuereVersuchsresultateüber den Geschiebetrieb. *SchweizerischeBauzeitung*, 103 (13), 147–150. (in German).

Meyer-Peter E. and Mueller R. (1949). *Eine Formel zurBerechnung des Geschiebetriebes*. Schweizer Bauzeitung, Zürich, Switzerland. (In German).

Meyer-Peter E. and Müller R. (1948). *Formulas for Bed Load Transport*. Hydroaulic Environmental Engineering and Research, Madrid.

Middleton G. V. and Southard J. B. (1984). *Mechanics of Sediment Movement*. Society of Economic Paleontologists and Minerals, Tulsa.

Miedema S. A. (2014). Constructing the Shields curve, a new theoretical approach and its applications. In R. H. Curran (Ed.), *9th World Dredging Congress (WODCON XIX)* (pp. 732–750). Eastern Dredging Association, New York.

Milhous R. T. (1973). Sediment transport in a gravel bottomed stream. PhD. thesis. Oregon State University, Corvallis.

Miller M. C., McCave I. N. and Komar P. D. (1977) Threshold of sediment motion under unidirectional currents. *Sedimentology*. 24 (4), 507–527.

Miller R. L. and Byrne R. J. (1966). The angle of repose for a single grain on a fixed rough bed. *Sedimentology*. 6 (4), 303–314.

Milne-Thomson L. M. (1960). *Theoretical Hydrodynamics*. MacMillan, New York.

Mingmin H. and Qiwei H. (1982) Stochastic model of incipient sediment motion. *J. Hydraul. Divn.* 108 (2), 211–224.

Misri R. L. (1981). Partial bed load transport of coarse nonuniform sediment. PhD thesis, Civil Engineering Department, Indian Institute of Technology Roorkee, Roorkee, India.

Misri R. L., Ranga Raju K. G. and Garde R. J. (1984). Bed load transport of coarse nonuniform sediments. *J. Hydraul. Engg.* 110 (3), 312–328.

Misuriya, G., Eldho, T. I., & Mazumder, B. S. (2023a). Experimental investigations of turbulent flow characteristics around different cylindrical objects using PIV measurements. *Euro J Mech-B/Fluids*. 101, 30–41.

Misuriya, G., Eldho, T. I., & Mazumder, B. S. (2023b). Estimation of the local scour around the cylindrical pier over the gravel bed for a low coarseness ratio. *Inter J River Basin Mang*, 1–11.

Misuriya, G., Eldho, T. I., & Mazumder, B. S. (2023c). Turbulent flow field around cylindrical pier on a gravel bed. *J Hydrl Engg* (accepted in May 2023).

Misuriya G., Eldho T. I. Mazumder B. S. (2021). Higher-order turbulence around different circular cylinders using particle image velocimetry. *J. Fluids Engg*. 143, 91–202.

Moncada-M. and Aguirre-Pe A. T. (1999). Scour below pipeline in river crossings. *J. Hydraul. Engg.* 125, 953–958.

Monin A. S. and Yaglom A. M. (1971). *Statistical Fluid Mechanics*. MIT Press, Cambridge.

Monin A. S. and Yaglom A. M. (1973). *Statistical Fluid Mechanics: Mechanics of Turbulence*. MIT Press, Cambridge.

Monin A. S. and Yaglom A. M. (1975). *Statistical Fluid Mechanics*. MIT Press, Cambridge.

More B. S., Dutta S. and Gandhi B. K. (2020). Flow around three side-by-side square cylinders and the effect of the cylinder oscillation. *J. Fluids Engg*. 142, 2.

Moreland C. (1963). Settling velocities of coal particles. *Canad. J. Chem. Eng*. 41, 108–110.

Moreno M., Maia R. and Couto L. (2015). Effects of relative column width and pile-cap elevation on local scour depth around complex piers. *J. Hydraul. Engg.* 142 (2), 04015051.

Mueller E. R., Pitlick J. and Nelson J. M. (2005). Variation in the reference shields stress for bed load transport in gravel-bed streams and rivers. *Water Resour. Res.* 41, W04006.

Müller A. and Gyr A. (1983). Visulization of the mixing layer behind dunes. In B. M. Sumer and A. Müller (Eds.), *Mechanics of Sediment Transport* (pp. 41– 45), CRC Press, New York.

Müller A. and Gyr A. (1986). On the vortex formation in the mixing layer behind dunes. *J. Hydraul. Res.* 24 (5), 359– 375.

Munson B. R., Young D. F., Okiishi T. H. and Huebscs W. W (2006). *Fundamentals of Fluid Mechanics*, John Wiley & Sons. Inc., Singapore.

Muralidhar K. and Biswas G. (2015). *Advanced Engineering Fluid Mechanics.* Narosa Publishers, New Delhi.

Murphy P. J. and Hooshiari H. (1982) Saltation in water dynamics. *J. Hydraul. Divn.* 108 (11), 1251–1267.

Muzzammil M. and Gangadharaiah T. (2003). The mean characteristics of horseshoe vortex at a cylindrical pier. *J. Hydraul. Res.* 41 (3), 285–297.

Nakagawa H. and Nezu I. (1977). Prediction of the contributions to the Reynolds stress from the bursting events in open-channel flows. *J. Fluid Mech.* 80, 99–128.

Nakagawa H., Nezu A. and Ueda H., (1975). Turbulence of open channel flow over smooth and rough beds. *Proc. Japan Soc. Civil Eng.* 241, 155–168.

Nakayama K., Fielding C. R. and Alexander J. (2002). Variations in character an preservation potential of vegetation-induced obstacle marks in the variable discharge burdekin river of North Queensland, Australia. *Sedimentary Geol.* 149, 199–218.

Neill C. R. (1967). *Mean Velocity Criterion for Scour of Coarse Uniform Bed-Material.* IAHR Congress, Fort Collins.

Neill C. R. (1968). Note on initial movement of coarse uniform bed-material. *J. Hydraul. Res.* 6 (2), 173–176.

Neill C. R. and Yalin M. S. 1969. Quantitative definition of beginning of bed movement. *J. Hydraul. Divn.* 95 (1), 585–588.

Nelson J. M. (1990). The initial instability and finite-amplitude stability of alternate bars in straight channels. *Earth Sci. Rev.* 29 (1–4), 97–115.

Nelson J. M., McLean S. R. and Wolfe S. R. (1993). Mean flow and turbulence fields over two-dimensional bed forms. *Water Resour. Res.* 29, 3935–3953.

Nelson J. M., Shreve R. L., McLean S. R. and Drake T. G. (1995). Role of near-bed turbulence structure in bed load transport and bed form mechanics. *Water Resour. Res.* 31, 2071–2086.

Nezu I. (1977). *Turbulent Structure in Open-Channel Flows.* Kyoto University. Kyoto.

Nezu I. (2005). Open-channel flow turbulence and its research prospect in the 21st century. *J. Hydraul. Engg.* 131 (4), 229–246.

Nezu I. and Nakagawa H. (1993). *Turbulence in Open-Channel Flows.* Balkema, Rotterdam.

Nezu I. and Rodi W. (1986). Open-channel flow measurements with laser doppler anemometer. *J. Hyd. Engg.* 112 (5), 335–355.

Ni J. R. and Wang G. Q. (1991). Vertical sediment distribution. *J. Hydraul. Engg.* 117 (9), 1184–1194

Nielson P. (1992). *Coastal Bottom Boundary Layer and Sediment Transport.* World Scientific Publishers, New York.

Nikora V. (2010). Hydrodynamics of aquatic ecosystems: An interface between ecology, biomechanics and environmental fluid mechanics. *River Res. Appl.* 26, 367–384.

Nikora V. I. and Goring D. G. (2000). Flow turbulence over fixed and weekly mobile gravel bed. *J. Hydraul. Engg.* 126 (9), 679–690.

Nikora V. I., Goring D. G. and Biggs B. J. F. (2002). Some observation of the effects of micro-organisms growing on the bed of an open channel on the turbulence properties, *J. Fluid Mech.* 450, 317–341.

Nikora V., Goring D., McEwan I. and Griffiths G. (2001). Spatially averaged open-channel flow over rough bed. *J. Hydraul. Engg.* 127, 123–133.

Nikuradse J. (1930). Untersuchungen uber turbulentestromungen in nichtkreisformigenRohren. *Arch. Appl. Mech.* 1 (3), 306–332.

Nikuradse J. (1933). Strömungsgesetze in rauhenrohren. Verein DeutscherIngenieure. *Forschungsheft Arb. Ing.* 361, 1–22.

Nino Y. and Garcia M. (1998). Using Lagrangian particle saltation observations for bedload sediment transport modeling. *Hydrol. Process.* 12, 1197–1218.

Niño Y. and García M. H. (1994). Gravel saltation II: Modeling. *Water Resour. Res.* 30 (6), 1915–1924.

Niño Y. and García M. H. (1996). Experiments on particle turbulence interactions in the near wall region of an open channel flow: implications for sediment transport. *J. Fluid Mech.* 326, 285–319.

Niño Y., García M. H. and Ayala L. (1994). Gravel saltation I: Experiments. *Water Resour. Res.* 30 (6), 1907–1914.

Noguchi K. and Nezu I. (2009). Particle–turbulence interaction and local particle concentration in sediment-laden open-channel flows. *J. Hydro-Environ. Res.* 3 (2), 54–68.

Nordin C. F. (1963). A preliminary study of sediment transport parameters, Rio Puerco near Bernardo, New Mexico.Geological Survey Professional Paper, 462 C, US Gov. Printing, Washington D.C, 1–21.

Nordin C. F. and Beverage J. P. (1965). *Sediment transport in the Rio Grande, New Mexico.* U.S. Geological Survey, Washington, D.C.

Nordin C. F. and Dempster G. R. (1963). *Vertical Distribution of Velocity and Suspended Sediment, Middle Rie Grandle.* U.S. Geological Survey, Washington DC.

Novak P. and Nalluri C. (1975). Sediment transport in smooth fixed bed channels. *Hydraul. Divn.* 101 (9), 1139–1154.

Novak P. and Nalluri, C. (1984). Incipient motion of sediment particles over fixed beds. *Hydraul. Res.* 22 (3), 181–197.

O'Brien M. P. (1933). Review of the theory of turbulent flow and its relations to sediment transportation. *Trans. Amer. Geophys. Union.* 14, 487–491.

O'Brien M. P. and Rindlaub B. D. (1934). The transportation of bed load by streams, *Trans. Am. Geophys. Union.* 15 (2), 593–603.

Obermeier F. (2006). Prandtl's mixing length model. *Proc. Appl. Math. Mech.* 6, 577–578.

Odgaard A. J. and Wang Y. (1987). Scour prevention at bridge piers. In R. M. Ragan (Ed.), *Hydraulic Engineering* (pp. 523–527). National Conference, Virginia.

Offen G. R. and Kline S. J. (1973). *Experiments on the Velocity Characteristics of "Bursts" and on the Interactions between the Inner and Outer Regions of a Turbulent Boundary Layer.* Stanford University, Stanford.

Offen G. R. and Kline S. J., (1974). Combined dye-streak and hydrogen-bubble visual observations of a turbulent boundary layer. *J. Fluid Mech.* 62, 223–239.

Offen G. R. and Kline S. J. (1975). A proposed model of the bursting process in turbulent boundary layers. *J. Fluid Mech.* 70 (2), 209–228.

Ojha C. S. P., Berndtsson R. and Chandramoulli P. N. (2010). *Fluid Mechanics and Fluid Machinery.* Oxford University Press, New Delhi.

Ojha S. P. and Mazumder B. S. (2008) Turbulence characteristics of flow region over a series of 2D dune shaped structures. *Adv. Water Resour.* 31, 561–576.

Ojha S. P. and Mazumder B. S. (2010). Turbulence characteristics of flow over a series of 2-D bed forms in the presence of surface waves. *J. Geophys. Res. Earth Surf.* 115, F04016.

Ojha S. P., Mazumder B. S., Carstensen S. and Fredsoe J. (2019). Externally generated turbulence by a vertically oscillatiog grid plate and its impact on sediment transport rate. *Coastal Engg. J.* 64 (4), 444–459.

Oliver D. R. (1961). The sedimentation of suspensions of closely-sized spherical particles. *Chem. Engg. Sci.* 15 (3–4), 230–242.

Oliver D. R. and Ward S. G. (1959). Studies of viscosity and sedimentation of suspensions. *Brit. J. Appl. Phys.* 10, 317–321.

Olson R. (1961). *Essentials of Engineering Fluid Mechanics.* International Textbook, Scranton.

Oruç V. (2012). Passive control of flow structures around a circular cylinder by using screen. *J. Fluids Struct.* 33, 229–242.

Oruç V., Akilli H. and Sahin B. (2016). PIV measurements on the passive control of flow past a circular cylinder. *Exp. Therm. Fluid Sci.* 70, 283–291.

Oseen C. W. (1910). Über die Stokes'scheformel, und übereineverwandte Aufgabe in der Hydrodynamik. *Arkivförmatematik, astronomiochfysik*, 6, 29.

Oseen C. W. (1927). *Hydrodynamik*. AkademischeVerlagsgesellschaft, Leipzig.

Owen P. R. (1964). Saltation of uniform grains in air. *J. Fluid Mech.* 20, 225–242.

Pai S. I. (1956). Viscous flow theory I laminar flow. Van Nostrand Co. Inc., New York.

Paintal A. S. (1971a). A stochastic model of bed-load transport. *J. Hydraul. Res.* 9 (4), 527–554.

Paintal A. S. (1971b). Concept of critical shear stress in loose boundary open channels. *J. Hydraul. Res.* 9 (1), 91–113.

Pal D. and Ghoshal K. (2013). Hindered settling with an apparent particle diameter concept. *Adv. Water Resour.* 60, 178–187.

Pani B.S. (2016). *Fluid Mechanics – A Concise Introduction.* PHI India, New Delhi.

Papanicolaou A. N. (1999). Discussion of 'pickup probability for sediment entrainment. *J. Hydraul. Engg.* 125 (7), 788–789.

Papanicolaou A. N., Diplas P., Dancey C. L. and Balakrishnan M. (2001). Surface roughness effects in near-bed turbulence: Implications to sediment entrainment. *J. Eng. Mech.* 127 (3), 211–218.

Papanicolaou A. N., Diplas P., Evaggelopoulos N. and Fotopoulos S. (2002). Stochastic incipient motion criterion for spheres under various bed packing conditions. *J. Hydraul. Engg.* 128 (4), 369–380.

Paphitis D. (2001). Sediment movement under unidirectional flows: An assessment of empirical threshold curves. *Coast. Engg.* 43, 227–245.

Parker G. (1975). Sediment inertia as a cause of river antidunes. *J. Hydraul. Divn.* 101 (2), 211–221.

Parker G., Clifford N. J. and Thorne C. R. (2011). Understanding the influence of slope on the threshold of coarse grain motion: Revisiting critical stream power. *Geomorphology.* 126, 51–65.

Parker G. and Kilingeman P. C. (1982). On why gravel-bed streams are paved? *Water Resour. Res.* 18 (5), 1409–1423.

Parker G., Kilingeman P. C. and McLean D. G. (1982) Bed load and size distribution in paved gravel-bed streams. *J. Hydraul. Divn.* 108 (4), 544–571.

Parola A., Mahavadi S., Brown B. and El Khoury A. (1996). Effects of rectangular foundation geometry on local pier scour. *J. Hydraul. Engg.* 122 (1), 35–40.

Patankar S. V. and Spalding, D. B. (1968). Computation of turbulent boundary layers-1968. In S. J. Kline, M. V. Morkovin, G. Sovran and D. S. Cock-rell (Eds.), *AFOSR-IFP-Stanford Conference*, Stanford University Press, Stanford.

Patel P. L., Porey P. D. and Patel S. B. (2010) Computation of critical tractive stress of scaling sizes in non-uniform sediments. *J. Hydraul. Res.* 48 (4), 531–537.

Patel P. L. and Ranga Raju K. G. (1996). Fraction-wise calculation of bed load transport. *J. Hydraul. Res.* 34 (3), 363–379.

Patel P. L. and Ranga Raju K. G. (1999). Critical tractive stress of non-uniform sediments. *J. Hydraul. Res.* 37(1), 39–58.

Pal D. K. (2011). Investigations on flow and sediment movement effected by waveform structures, PhD. Thesis submitted to the Jadavpur University, Jadavpur, Kolkata, India.

Pettijohn F. J. (1957). *Sedimentary Rocks*. Harper and Row.

Pilotti M., Menduni G. and Castelli E. (1997). Monitoring the inception of sediment transport by image processing techniques. *Exper. Fluids*. 23, 202–208.

Poggi D., Katul G. G. and Albertson J. D. (2004), Momentum transfer and turbulent kinetic energy budgets with in a dense model canopy. *Boundary-Layer Met.* 111, 589–614.

Pope S. (2000). *Turbulent Flows*. Cambridge University Press, Cambridge.

Prandtl L. (1904). *Über flüssigkeitsbewgung bei sehr kleiner reibung*. In A. Krazer (Ed.), *Proc. Third Intern. Math. Congress* (pp. 484–491), Heidelberg.

Prandtl L. (1910). Eine beziehungzwischenwarmeaustauch und stromungswiderstand der flussigkeiten. *Zeitschriftfür Physik.* 11, 1072–1078.

Prandtl L. (1925). Uber die ausgebildeteTurbulenz. *Zeitschriftfür Physik.* 5, 136–139.

Prandtl L. (1926). Ueber die ausgebildeteTurbulenz. *Proceedings of the 2nd International Congress for Applied Mechanics*, Zurich, Switzerland, 12-17 September 1926, page 62.

Prandtl L. (1933). *Attaining a steady air stream in wind tunnels.* National Advisory Committee for Aeronautics Report 726, 4, part 2, 1–35.

Prandtl L. (1935). The mechanics of viscous fluids. In W. F. Durand (Ed.), *AerodynamicTheory III* (pp. 34–208), Julius Springer, Berlin.
Pretsch J. (1941). Die StabilitateinerebenenLaminarstromungbeiDruckgefalle und Druckanstieg. *Luftfahrtforschung*. I, 54–71.
Proffitt G. T. and Sutherland A. J. (1983). Transport of non-uniform sediments. *J. Hydraul. Res.* 21 (1), 33–43.
Puls W., Sunderman J. and Vollmers H. (1977). A numerical approach to solid matter transport computation. *Proceedings of the 17th Congress of International Association of Hydraulic Research (IAHR)*, 1, 129–135, Baden-Baden, West Germany.
Purkait B. (2002). Patterns of grain size distribution in some point bars of the Usri river, India. *J. Sedimentary Res.* 72 (3), 367–375.
Purkait B. (2006). Grain-size distribution patterns of a point bar system in the Usri River, India. *Earth Surf. Process. Landform.* 31, 682–702.
Purkait B. and Mazumder B. S. (2000). Grain size distribution—a probabilistic model for Usri River sediments in India. in Z. Wang and S. Hu (Eds.), *Stochastic Hydraulics* (pp. 291–298). Rotterdam, Balkema.
Purkait B. and Sinha S. (2019). Meander geometry, hydraulics and sedimentary structures—a case study of the Usri River section, Jharkhand, India. *Arab. J. Geosci.* 12, 300.
Qin Y. Y. (1980). Incipient motion of non-uniform sediment. *J. Sediment Res.* 1, 83–91.
Raffel M., Willert C. E., Scarano F., Kähler C. J., Wereley S. T. and Kompenhans J. (2018). *Particle Image Velocimetry: A Practical Guide.* Springer, New York.
Raichlen F. and Kennedy J. F. (1965). The growth of sediment bed forms from an initially flattened bed. *Proceedings of the 11th Congress International Association for Hydraulic Research*, 3, Paper 3.7, Leningrad.
Ramsey A. S. (1920). *A Treatise on Hydromechanics: Hydrodynamics.* Bell and Sons Ltd, London.
Ranga Raju K. G. (1970). Resistance relation for alluvial streams. *La Houille Blanche.* 56, 51–54.
Ranga Raju K. G. and Soni J. P. (1976). Geometry of ripples and dunes in alluvial channels. *J. Hydraul. Res.* 14, 77–100.
Rao K.N., Narasimha R. and Badri Narayanan M. A. (1971), The 'bursting phenomenon' in a turbulent boundary layer. *J. Fluid Mech.* 48, 339–352.
Rashidi M. and Banerjee S. (1988). Turbulence structure in free-surface channel flows. *Phys. Fluids.* 31 (9), 2491–2503.
Raudkivi A. J. (1963), Study of sediment ripple formation. *J. Hydraul. Divn.* 89, 15–33.
Raudkivi A. J. (1967). *Loose Boundary Hydraulics.* Pergamon Press, Oxford.
Raudkivi A. J. (1976). *Loose Boundary Hydraulics.* Pergamon Press, Tarrytown.
Raudkivi A. J. (1997). Ripples on stream bed. *J. Hydraul. Engg.* 116, 58–64.
Raudkivi A. J. and Ettema (1983). Clear-water scour at cylindrical piers. *J. Hydraul. Engg.* 109 (3), 338–350.
Raudkivi A. J. and Witte H. H. (1990). Development of bed features. *J. Hydraul. Engg.* 116 (9), 1063–1079.
Raupach M. R. (1981). Conditional statistics of Reynolds stress in rough-wall and smooth-wall turbulent boundary layers. *J. Fluid Mech.* 108, 363–382.
Rayleigh L. (1887). On the stability of certain fluid motions. Proc. London, Math. Soc. 19, 67.
Rayleigh L. (1895). On the stability of certain fluid motions. *Proc. Math. Soc.*, 4, 203.
Rayleigh L. (1911). On the motion of solid bodies through viscous liquids. *Phil. Mag.* 21, 697–711.
Recking A. (2009). Theoretical development on the effects of changing flow hydraulics on incipient bed load motion. *Water Resour. Res.* 45, W04401.
Reichardt H. (1933, 1938, 1939). MessungenturbulenterSchwankungen. *ZAMM.* 13, 177–180 (1933), Naturwissenschaften 404 (1938), and ZAMM 18, 358–361 (1939).
Reichardt H. (1951). Vollstandigedarstellung der tubulentengeschwindig-keitsverteilung in glattenleitungen. *Z. Angew.Math. Mech.* 31 (7), 208–219, (in German).
Reinaldo M. (2006). A large-scale hydraulic model of riprap-apron performance at a bridge abutment on a floodplain. *International Conference on Civil and Environmental Engineering-2006,* Hiroshima University, Japan. https://www.worldcat.org/title/71558347
Reineck H. E. and Singh I. B. (1973). *Depositional Sedimentary Environments.* Springer, New York.
Reitz W. (1936). Uber Geschiebebewegung. *Wasserwirstsch.* 26, 28–30.

Ren H. and Wu Y. (2011) Turbulent boundary layers over smooth and rough forward facing steps. *Phys. Fluids.* 23 (4), 1–17.

Rennie C. D. and Hay A. (2010). Reynolds stress estimates in a tidal channel from phase-wrapped ADV data. *J. Coastal Res.* 26 (1), 157–166.

Reynolds A. J. (1965). Waves on the bed of an erodible channel. *J. Fluid Mech.* 22 (1), 113–133.

Reynolds A. J. (1976). Decade's investigation of the stability of erodible stream beds. *Nordic Hydrol.* 7, 3.

Reynolds O. (1883). An experimental investigation of the circumstances which determine whether the motion of water shall be direct or sinuous, and of the law of resistance in parallel channels. *Phil. Trans. Roy. Soc.* 174, 935–982.

Reynolds O. (1885). On the dynamical theory of incompressible viscous fluids and the determination of the criterion. *Phil. Trans. Roy. Soc.* 186A, 123–164.

Rice S. P. and Church M. (2010). Grain-size sorting within river bars in relation to downstream fining along a wandering channel. *Sedimentology.* 57, 232–251.

Richards K. J. (1980). The formation of ripples and dunes on an erodible bed. *J. Fluid Mech.* 99, 597–618.

Richards K. J. and Taylor P. A. (1981). A numerical model of flow over sand waves in water of finite depth. *Geophys. J. Ast. Soc.* 1, 103–128.

Richardson E. V. and Davis S. R. (1995). *Evaluating Scour at Bridges.* U.S. Department of Transportation, Washington DC.

Richardson E. V. and Davis S. R. (2001). *Evaluating Scour at Bridges.* U.S. Federal Highway Administration. Washington DC.

Richardson J. E. and Panchang V. G. (1998). Three-dimensional simulation of scour-inducing flow at bridge piers. *J. Hydraul. Engg.* 124, 530–540.

Richardson J. F. and Zaki W. N. (1954) Sedimentation and fuidization. *Trans. Instn. Chemical Engr.* 32, 35–53.

Richardson L. F. (1922). *Weather Prediction by Numerical Process.* Cambridge University Press, Cambridge

Rickenmann D (1991). Bed load transport and hyperconcentrated flow at steep slopes. *Fluvial Hydraul.* 37, 429–441.

Robert A. and Uhlman W. (2001). An experimental study on the ripple dune transition. *Earth Surf. Process. Landforms* 26, 615–29.

Robie Bonilla G. (2010). Sheath for reducing local scour in bridge piers. *Scour Ero.* 987–996.

Rolland T. and Lemmin U. (1997). A two-component acoustic velocity profiler for use in turbulent open-channel flow. *J. Hydraul. Res.* 35 (4), 545–561.

Rotta J. C. (1956). Experimentaller BeitragZurEntstehungturbulenterStromungim. *Rohr. Ing. Arch.* 24, 258–281.

Rotta J. C. (1962). Turbulent boundary layers in incompressible flow. *Prog. Aeronaut. Sci.* 2, 1–219.

Rotta J. C. (1972). *TurbulenteStromungen.* Teubner, Germany.

Rottner, J. (1959). A formula for bed material transport. *Houille Blanche.* 4, 285–307.

Rousar L., ZachovalZ. and Julien P. (2016), Incipient motion of coarse uniform gravel. *J. Hydraul. Res.* 54 (6), 615–630.

Rouse H. (1937). Modern conception of the mechanics of turbulence. *Trans. Am. Soc. Civil Eng.* 102 (1), 461–543.

Rouse H. (1938). Experiments on the mechanics of sediment suspension. *Proc. Int. Cong. Appl. Mech.* 55, 550–554.

Rouse H. (1947). *Elementary Mechanics of Fluids.* Wiley, New York.

Rubey W. W. (1933). Settling velocities of gravels, sand and silt particles. *Am. J. Sci.* 25, 148.

Rubey W. W. (1938). The force required to move particles on a stream bed. *Geol. Survey.* 189, 121–141.

Rubinow S. I. and Keller J. B. (1961). The transverse force on a spinning sphere moving in a viscous fluid. *J. Fluid Mech.* 11, 447–459.

Saffman P. G. (1965). The lift on a small sphere in a slow shear flow. *J. Fluid Mech.* 22, 385–400.

Saffman P. G. (1968). Corrigendum, the lift on a small sphere in a slow shear flow. *J. Fluid Mech.* 31 (3), 624.

Sahu, C., Eldho, T.I. and B. S. Mazumder BS. (2023). Experimental study of flow hydrodynamics around circular cylinder arrangements using Particle Image Velocimetry. *J. Fluid Engg.* 145(1): 011302. https://doi.org/10.1115/1.4055597

Sarkar K., Chakraborty C. and Mazumder B. S. (2015). Space-time dynamics of bed forms due to turbulence around submerged bridge piers. *Stocha. Environ. Res. Risk Assessment.* 29 (3), 995–1017.

Sarkar K., Chakraborty C. and Mazumder B. S. (2016). Variations of bed elevations due to turbulence around submerged cylinders in sand bed. *Environ. Fluid Mech.* 16 (3), 659–693.

Sarkar K. and Mazumder B. S. (2014). Turbulent flow over the trough region formed by a pair of forward-facing bed form shapes. *Eur. J. Mech.* 46, 126–143.

Sarkar K. and Mazumder B. S. (2018a). Space-time evolution of sand bed topography and associated flow turbulence: Experiments with statistical analysis. *Stocha. Environ. Res. Risk Assessment.* 32 (2), 501–525.

Sarkar K. and Mazumder B. S. (2018b). Higher-order moments with turbulent length-scales and anisotropy associated with flow over dune shapes in tidal environment. *Phys. Fluids.* 30, 106602.

Schiller L. (1922). Untersuchungen uber laminare und turbulenteStromung. *Zangew. Math. Mech.* 2, 96–106.

Schiller L. and A. Naumann (1933). Uber die grundlegendenBerechnungenbei der Schwerkraftaufbereitung. *Zangew. Math. Mech.* 77, 03–25.

Schleiss A. J., De Cesare D., Franca M. and Pfister M. (2014). *River Flow*. CRC Press, New York.

Schlichting H. (1966). *Boundary Layer Theory*. McGraw-Hill, New York.

Schlichting H. (1968). *Boundary-Layer Theory.* McGraw-Hill Book Company, New York.

Schlichting H. (1979). *Boundary-Layer Theory.* McGraw Hill, New York.

Schlichting H. and Gersten K. (2000). *Boundary-Layer Theory.* Springer, New York.

Schmeeckle M. W., Nelson J. M. and Shreve R. L. (2007). Forces on stationary particles in near-bed turbulent flows. *J. Geophys. Res.* 112, 1–21.

Schmidt, W. (1925). *Der Massenaustausch in freierLuft und verwandteErscheinungen, Probleme der KosmischenPhysik, 7.* Grand, Hamburg.

Schoklitsch A. (1914) *Uber Schleppkraft und Gschiebebewegung*. Engelman, Leipzig.

Schoklitsch A. (1930). *Handbuch des Wasserbaues*. Springer, Berlin.

Schoklitsch, A. (1962). *Handbuch des Wasserbaus [in German].* Springer, Wien, Germany.

Schulz E. F., Wilde R. H. and Albertson, M. L. (1954*). Influence of Shape on the Fall Velocity of Sedimentary Particles*. SeaTek, Florida.

Sechet P. and Le Guennec B. (1999) Bursting phenomenon and incipient motion of solid particles in bed-load transport. *J. Hydraul. Res.* 37 (5), 683–696.

Sekine M. and Kikkawa H. (1992). Mechanics of saltating grains. *J. Hydraul. Engg.* 118 (4), 536–558.

Sengupta S. (1966). Studies on orientation and imbrication of pebbles with respect to cross-stratification. *J. Sedimentary Petrol.* 36 (2), 362– 369.

Sengupta S. (1967). *Grain-Size Frequency Distribution as Indicator of Depositional Environment in Some Gondwana Rocks.* International Sedimentological Congress, England.

Sengupta S. (1975a) Size-sorting during suspension transportation – lognormality and other characteristics. *Sedimentology*. 22, 257–273.

Sengupta S. (1975b). *Sorting Processes During Transportation of Suspended Sediments – An Experimental-Theoretical Study*. Uppsala University, Uppsala.

Sengupta S. (1979). Grain-size distribution of suspended load in relation to bed materials and flow velocity. *Sedimentology*, 26, 63–82.

Sengupta S. (2007). Introduction to Sedimentology. CBS Publications and Distributors: New Delhi; 674.

Sengupta S., Das S. S. and Gupta A. S. (2005). Current crescent as indicator of flow velocity. In S. N. Bora (Ed.), *Some Aspects of Environmental Fluid Mechanics* (76–77). ICEFM, Guwahati.

Sengupta S., Das S. S. and Maji A. K. (1999) Sediment transportation and sorting processes in streams. *Proc. Ind. Nat. Sc. Acad.* 2, 167–206.

Sengupta S., Ghosh J. K. and Mazumder B. S. (1991). Experimental -theoretical approach to interpretation of grain-size frequency distribution. In J. P. E. Syvitski (Ed.), *Principles, Methods and Applications of Particle Size Analysis* (pp. 264–280). Cambridge Uni. Press, New York.

Sha Y. Q. (1954). Basic principles of sediment transport. *J. Sediment Res.* 1 (2), 1–54. (in Chinese).

Sha Y. Q. (1965) *Introduction to Sediment Dynamics.* Industry Press, Beijing. (in Chinese).

Sheilds A., (1936). *Anwendung der Ahnlichkeitsmechanik und Turbulenzforschung auf Geschiebebewegung. Mitteilungen der Preuss*. Versuchsanst. f. Wasserbau u. Schiffbau, Berlin.

Shen C. and Lemmin U. (1996). Ultrasonic measurements of suspended sediments: a concentration profiling system with attenuation compensation. *Meas. Sci. Technol.* **7**, 1191–1194.

Shen C. and Lemmin U. (1999). Application of an acoustic particle flux profiler in particle-laden open-channel flow. *J. Hydraul. Res.* 37, 407–419.

Shen H. W., Fehlman H. M. and Mendoza C. (1990). Bed form resistance in open channel flows. *J. Hydraul. Engg.* 116, 799–823.

Shen H. W., Schneider V. R. and Karaki S. S. (1966). *Mechanics of Local scour.* U.S. Institute for Applied Technology, Washington, D.C.

Shen H. W., Schneider V. R. and Karaki, S. S. (1969), Local Scour around Bridge Piers. *J. Hyd. Divn.* 95, 1919–1940.

Sheppard D. M. and Glasser T. (2009). Local scour at bridge piers with complex geometries. *Int. Foundation Cong. Equipment Expo*, ASCE, 506–513.

Sheppard D. M., Odeh M. and Glasser T. (2004). Large scale clear-water local pier scour experiments. *J. Hydraul. Engg.* 130 (10), 957–963.

Shields A. (1936). *Application of Similarity Mechanics and Turbulence Research on Bed Load Movement; Mitt. der PreussischenVersuchsanstaltfuer Wasser-, Erd- und Schiffbau Heft.* RWTH Publications, Aachen, Germany. (in German).

Shvidchenko, A. B. and Pender G. (2000a), Flume study of the effect of relative depth on the incipient motion of coarse uniform sediments. *Water Resour. Res.* 36 (2), 619–628.

Shvidchenko, A. B. and Pender G. (2000b), Initial motion of streambeds composed of coarse uniform sediments. *Proc. Inst. Civ. Eng. Water Mar. Eng.* 42, 217–227.

Shvidchenko A. B., Pender G. and Hoey T. B. (2001). Critical shear stress for incipient motion of sand/gravel streambeds. *Water Resour. Res.* 37 (8), 2273–2283.

Simoes F. J. M. (2014). Shear velocity criterion for incipient motion of sediment. *Water Sci. Engg.* 7 (2), 183–193.

Simons D. B. and Albertson M. L. (1963). Uniform water conveyance channels in alluvial material. *Trans. Am. Soc. Civil Eng.* 128, 1.

Simons D. B. and Richardson E. V. (1961). Forms of bed roughness in alluvial channels. *J. Hydr. Divn.* 87 (3), 128.

Simons D. B. and Richardson E. V. (1962). Resistance to flow in alluvial channels. *Am. Soc. Civil Eng. Trans.* 127, 927–1006.

Simons D. B. and Richardson E. V. (1966). *Resistance to Flow in Alluvial Channels.* U.S. Geological Survey, Washington DC.

Simpson R. L., Chew Y. T. and Shivaprasad B. G (1981). The structure of separating turbulent boundary layer. Higher-order turbulence results. *J. Fluid Mech.* 113, 53–73.

Singh K., Sandham N. and Williams J. (2007). Numerical simulation of flow over a rough-bed. *J. Hydraul. Engg.* 133 (4), 386–398.

Singh V. P. (1998). *Entropy-Based Parameter Estimation in Hydrology.* Kluwer, Boston.

Sivashinsky G. I. and Frenkel A. L. (1992). On negative eddy viscosity under conditions of isotropy. *Phys. Fluids* A 4, 1608–1610.

Sivashinsky G. I. and Yakhot V. (1985). Negative viscosity effects in large-scale flows. *Phys. Fluids*. 28, 1040.

Smart G. M. (1984). Sediment transport formula for steep channels. *J.Hydraul. Engg.* 3 (267), 267–276.

Smith J. D. (1977). Modeling of sediment transport on continental shelves. In E. D. Goldberg (Ed.), *The Sea: Ideas and Observations on Progress in the Study of the Seas* (pp. 538–577). John Wiley and Sons, New York.

Smith J. D. and McLean S. R. (1977). Spatially averaged flow over a wavy surface. *J. Geophys. Res.* 82 (12), 1735–1746

Som S. K. and Biswas G. (1998). *Introduction to Fluid Mechanics and Fluid Machines.* Tata McGraw-Hill Publishing Company Limited, New Delhi.

Song C. C. S. (1983). Modified kinematic model: application to bed forms. *J. Hydraul. Engg.* 109 (8), 1133–1151.

SonTek Inc. (2001). *ADV Principles of Operation.* SonTek Inc., San Diego.

Southard J. B. (2006). Threshold of movement. In J. Southard (Ed.), *Special Topics: An Introduction to Fluid Motions, Sediment Transport, and Current-Generated Sedimentary Structures* (pp. 260–284). MIT Open Course Ware, Cambridge, MA.

Southard J. B. (1975). Bed configurations: Depositional environments as interpreted from primary sedimentary structure and stratification sequences. *Soc. Econ. Paleontol. Mineral.* 2, 5–43.

Southard J. B. and Dingler J. R.(1971). Flume study of ripple propagation behinds mounds on flat sand beds. *Sedimentology*. 16, 251–263.

Spalding D. B. (1961). A single formula for the law of wall. *J. Appl. Mech.* 28 (3), 455–458.

Squire H. B. (1933). On the stability of three-dimensional distribution of viscous fluid between parallel walls. *Proc. Roy. Soc. London A*. 142, 621–628.

Stanton T. E. and Pannel J. R. (1914). Similarity of moyion in relation of the surface friction of fluids. *Phil. Trans. Royal Soc. A*. 214, 199.

Stearns F. (1883). On the current meter together with a reason why the maximum velocity of flowing in open channels is below the surface. *Am. Soc. Civil Eng. Trans.* 12, 331–338.

Stevens M. A., Simons D. B. and Lewis G. L. (1976) Safety factors for riprap protection. *J. Hydraul. Divn.* 102 (5), 637–655.

Stokes G. G. (1851) On the effect of the internal friction of fluids on the motion of pendulums. *Trans. Cambridge Philos. S*oc. 9, 80–85.

Stokes G. G. (1856) On the numerical calculation of a class of definite integrals and infinite series. *Trans. Cambridge Philos. S*oc. 9, part I, 166–188.

Straub L. G. (1935). *Missouri River Report.* U.S. Gov. Printing office, Washington DC.

Streeter V. L. and Wylie E. B. (1975). *Fluid Mechanics*. McGraw-Hill Kogakusha Ltd., Tokyo.

Stuart J. T. (1956). On the effects of the Reynolds stress on hydrodynamic stability. *ZAMM, Sonderheft.* 5, 32–38.

Subramanya K. (1997). *Flow in Open Channels*. Tata McGraw-Hill Publishing Company Limited, New Delhi.

Sumer B. M. (2002). *Lecture Note of Turbulence*. Technical University of Denmark, Copenhegan.

Sumer B. M. and Bakioglu M. (1984). On the formation of ripples on an erodible bed. *J. Fluid Mech.* 144, 177–190.

Sumer B. M. and Deigaard R. (1981). Particle motions near the bottom in turbulent flow in an open channel. *J. Fluid Mech.* 109, 311–337.

Sumer B. M. and Fredsoe J. (1990). Scour below pipelines in waves. *J. Waterway Ocean Engg*. 116, 307–323.

Sumer B. M. and Fredsoe J. (1991). Onset of scour below a pipeline exposed to waves. *Int. J. Offshore Polar Engg*. 3, 189–194.

Sumer B. M. and Fredsøe J. (1996). Scour below pipelines in combined waves and current. *Proceeding of the 15th OMAE Conference, Florence, Italy*. https://www.osti.gov/biblio/400959

Sumer B. M. and Fredsøe J. (1997). *Hydrodynamics around Cylindrical Structures*. World Scientific, New Jersey.

Sumer B. M. and Fredsoe J. (2002a). *The Mechanics of Scour in the Marine Environment, Advanced Series in Ocean Engineering*. World Scientific Publishing Company, Hackensack.

Sumer B. M. and Fredsoe J. (2002b). *Hydrodynamics around Cylindrical Structures, Advanced Series in Ocean Engineering*. World Scientific Publishing Company: Hackensack.

Sumer B. M., Jensen B. L. and Fredsoe J. (1991). Effect of a plane boundary on oscillatory flow around a circular cylinder. *J. Fluid Mech.* 225, 271–300.

Sumer B. M., Jensen H. R., Mao Y. and Fredsoe J. (1988). The effect of lee-wake on scour below pipelines in current. *J. Waterway Ocean Engg.* 114 (5), 599–614.

Sumer B. M., Kozakiewicz A., Fredsøe J. and Deigaard R. (1996). Velocity and concentration profiles in sheet-flow layer of movable bed. *J. Hydraul. Engg*. 122 (10), 549–558.

Sumer B. M. and Oguz B. (1978). Particle motions near the bottom in the turbulent flow in an open channel. *J. Fluid Mech.* 86, 109–127.

Sun Z. and Donahue J. (2000). Statistically derived bedload formula for any fraction of nonuniform sediment. *J. Hydraul. Engg*. 126 (2), 105–111.

Sundborg A. (1956). The River Klar€alven: A study of fluvial processes. *GeografiskaAnnaler*. 38 (2–3), 238–316.

Sutherland A. J. (1967) Proposed mechanism for sediment entrainment by turbulent flows. *J. Geophys. Res.* 72 (24), 6183–6194.

Swamee P. K. and Ojha C. S. (1991). Drag coefficient and fall velocity of non-spherical particles. *J. Hydr. Engr. Divn.* 117 (5), 660–667.

Tabarestani M. and Zarrati A. (2012). Effect of collar on time development and extend of scour hole around cylindrical bridge piers. *Int. J. Engg*. 25 (1), 11–16.

Tachie M. F., Balachandar R. and Bergstrom D. J. (2004). Roughness effects on turbulent plane wall jets in an open-channel. *Exper. Fluids*. 37 (2), 281–292.

Tang H. W., Ding B., Chiew Y. M. and Fang S. L. (2009). Protection of bridge piers against scouring with tetrahedral frames. *Int. J. Sedim. Res.* 24 (4), 385–399.

Tani I. (1980). *Progress of Fluid Mechanics: Turbulence.* Maruzen Publisher, Tokyo. (In Japanese).

Tani I. (1984). *Progress of Fluid Mechanics: Boundary Layer.* Maruzen Publisher, Tokyo. (In Japanese).

Tatsumi T. (1986). *Science of Turbulent Phenomena*. Tokyo University Press, Tokyo.

Taylor D. W. (1948). *Fundamentals of Soil Mechanics.* Wiley, New York.

Taylor G. (1935). Statistical theory of turbulence. *Proc. Royal Soc. London.* 151, 421–444.

Taylor P. A. and Dyer K. R. (1977). Theoretical models of flow near the bed and their implications for sediment transport. in E. D. Goldberg, I. N. McCave, J. J. O'Brien and J. H. Steele (Eds.), *The Sea* (pp. 579–601). Interscience, New York.

Techen C. (1947). Mean value and correlation problems connected with the motion of small particles suspended in turbulent fluid. DSc. Dissertation, TechnischeHogeschool, Delft, Holland.

Tennekes H. and Lumley J. L., (1980). *A First Course in Turbulence*. MIT Press, Cambridge, MA.

Tennekes H. and Lumley, J. (1972). *A First Course in Turbulence*. The MIT Press. Cambridge, MA.

Thacker W. C. and Lavelle J. W. (1977). Two-phase flow analysis of hindered settling. *Phys. Fluids*. 20, 1577–1579.

Thomas D. G. (1965). "Transport characteristics of suspension. A note on the viscosity of Newtonian suspensions of uniform spherical particles. *J. Coll. Interface Sci*. 20, 267–277.

Thompson B. G. J. (1965). A new two-parameter family of mean velocity profiles for incompressible turbulent boundary layers on smooth walls. *Aero. Res. Count.* Reports and Memoranda No. 3463, April 1965.

Thompson S. M., and Campbell P. L. (1979). Hydraulics of a large channel paved with boulders. *J. Hydraul. Res.* 17 (4), 341–354.

Thorne C. R., Reed S. and Doornkamp J. C. (1996). *A Procedure for Assessing River Bank Erosion Problems and Solutions*. National Rivers Authority, Almondsbury.

Tiffany J. B. and Bentzel C. B. (1935). Sand mixtures and sand movement in fluvial models: A discussion. *Trans. ASCE*. 61 (1), 101–107.

Tison L. J. (1961). Local scour in rivers. *J. Geoph. Res*. 66, 4227–4232.

Tison L. J. (1953). *Recherches sur la tension limited'Entrainment des materiauxconstitutifs du lit*. International Association Hydraulic Research, Minneapolis, USA.

Tominaga A. and Nezu I. (1992). Velocity profiles in steep open channel flows. *J. Hydraul. Engg.* 118 (1), 73–90

Tominaga A., Nezu I., Ezaki K. and Nakagawa H. (1989). Three-dimensional turbulent structure in open channel flows. *J. Hydraul. Res.* 27 (1), 149–173.

Tomkins M. R., Baldock T. E. and Nielsen P. (2005). Hindered settling of sand grains, *Sedimentology*. 53, 1425–1432.

Torobin L. B. and Gauvin W. H. (1959). Fundamental aspects of solid gas flow. *Canad. J. Chem. Eng*. 37 (4), 129–141.

Townsend A. A. (1976). *The Structure of Turbulent Shear Flow*. Cambridge University Press, Cambridge.

Tsubaki T. and Shinohara K. (1959). *On the Characteristics of Sand Waves Formed upon Beds of Open Channels and Rivers*. Research Institute for Applied Mechanics, Japan.

Tsutsui T. (2008). *Fluid Force Acting on a Cylindrical Pier Standing in a Scour, Bluff Bodies Aerodynamics & Applications*. International Colloquium, Milano, Italy.

Udden J. A. (1914), Mechanical composition of clastic sediments. *Geol. Soc. Am. Bull*. 25, 655–744.

Ueda H., Moller R., Komori S. and Mizushina T. (1977). Eddy diffusivity near the free surface of open channel flow. *Int. J. Heat Mass Transfer.* 20, 1127–1136.

Umeyama M. (1992). Vertical distribution of suspended sediment in uniform open channel flow. *J. Hydraul. Engg*. 118 (6), 936–941.

Umeyama M. (1999). Velocity and concentration fields in uniform flow with coarse sands. *J. Hydraul. Engg*. 125, 6.

Umeyama M. and Gerritsen F. (1992). Velocity distribution in uniform sediment-laden flow. *J. Hydraul. Engg*. 118 (2), 229–245.

Unger J. and Hager W. H. (2006). Down-flow and horseshoe vortex characteristics of sediment embedded bridge piers. *Exp. Fluids*. 42, 1–19.

Unger J. and Hager W. H. (2007). Down-flow and horseshoe vortex characteristics of sediment embedded bridge piers. *Exper. Fluids*. 42, 1–19.

US Interagency Committee. (1957). *Some Fundamentals of Particle Size Analysis: A Study of Methods Used in Measurement and Analysis of Sediment Loads in Streams*. St. Anthony Falls Hydraulic Laboratory, Minneapolis.

Valyrakis M. S. (2011). Initiation of particle movement in turbulent open channel flow. PhD. Thesis, Submitted to Civil and Environmental Engineering, Virginia Polytechnic Institute and State University, USA.

Valyrakis M. S., Diplas P. and Dancey C. L. (2011). Entrainment of coarse grains in turbulent flows: An extreme value theory approach. *Water Resour. Res*. 47, 9.

van Driest E. R. (1956). On turbulent flow near a wall. *J. Aeronaut. Sci*. 23, 1007–1011.

Van Rijn L. C. (1984a). Sediment transport: Bed load transport. *J. Hydraul. Engg.* 110 (10), 1431–1456.

Van Rijn L. C. (1984b). Sediment transport: Suspended load transport. *J. Hydraul. Engg*. 110 (11), 1613–1641.

Van Rijn L. C. (1989). *Handbook: Sediment Transport by Currents and Waves*. Delft Hydraulics, Netherlands.

Vanoni V. A. (1941). Some experiments on the transportation of suspended load. *Trans. AGU*. 4, 608–621.

Vanoni V. A. (1946). Transportation of suspended sediment by water. *Trans. ASCE*. 111, 67–133.

Vanoni V. A. (1953). Some effects of suspended sediment on flow characteristics. Hydraulics Conference, State University of Iowa, Iowa. https://ascelibrary.org/doi/abs/10.1061/JYCEAJ.0000944

Vanoni V. A. (1963). Sediment transportation mechanics: suspension of sediment. *Proc. Am. Soc. Civil Eng.* 89, 5.

Vanoni V. A. (1964). *Measurements of Critical Shear Stress*. California Institute of Technology, California.

Vanoni V. A. and Brooks N. H. (1957). *Laboratory Studies of the Roughness and Suspended Load of Alluvial Streams*. California Institute of Technology, California.

Vanoni V. A. and Hwang L. S. (1967). Relation between bed forms and friction in streams. *J. Hydraul. Divn.* 93 (3), 121–144.

Vanoni V. A. and Nomicos G. N. (1960). Resistance properties of sediment-laden streams. *Trans. Am. Soc. Civil Eng*. 125, 1140–1175.

Vanoni V. A., Norman H. B. and John F. K. (1961). *Lecture Notes on Sediment Transport and Channel Stability*. W.M. Keck Laboratory of Hydraulics and Water Resources, California.

Velikanov M. A. (1936) *Formation of Sand Ripples on the Stream Bottom*. International Association of Scientific Hydrology, Commission de Potamologie, Sec. 3, Report 13, p. 17, Vienna.

Velikanov M. A., (1955). Dynamics of channel flow. In *Sediments and the Channel* (pp. 107–120). State Publishing House for Technical and Theoretical Literature, Moscow.

Venditti J. G. (2003). Initiation and development of sand dunes in river channels, PhD. Thesis, University of British Columbia Library, Vancouver.

Venditti J. G. and Bauer B. O. (2005). Turbulent flow over a dune. Green River Colorado. *Earth Surf. Process Landform*. 30, 289–304.

Venditti J. G. and Bennett S. J. (2000). Spectral analysis of turbulent flow and suspendedsediment transport over fixed dunes. *J. Geophys. Res*. 105 (C9), 22035–22047.

Venditti J. G., Church M. and Bennett S. J. (2005). Morphodynamics of small-scale superimposed sand waves over migrating dune bed forms. *Water Resour. Res*. 41, W10423.

Vijayasree B. A., Eldho T. I. and Mazumder B. S. (2020). Turbulence statistics of flow causing scour around circular and oblong piers. *J. Hydraul. Res*. 58, 673–686.

Vijayasree B. A., Eldho T. I., Mazumder B. S. and Ahmad N. (2019). Influence of bridge pier shape on flow field and scour geometry. *Int. J. River Basin Manag.* 17 (1), 109–129.

Vijayasree B. A., Eldho T. I., Mazumder B. S. and Viswanadham B. V. S. (2018). Effectiveness of combinations of raft foundation with aprons as a protection measure against bridge pier scour. *Sadhana-IAS.* 43, 21.

Vijayasree B. A., Eldho T. I. and Viswanadham B. V. S. (2013). Physical model studies on bridge piers with raft foundation and aprons as a scour protection measure. In V. Sundar et al. (Eds.), *Proceedings of HYDRO 2013 International*, IIT Madras, India

Villaret C. and Trowbridge J. H. (1991). Effects of stratification by suspended sediments on turbulent shear flows. *J. Geophys. Res.* 96 (C6), 10,659–10,680.

Visher G. S. (1965), Fluvial processes as interpreted from ancient and recent fluvial deposits. In G. V. Middleton (Ed.), *Primary sedimentary structures and their hydrodynamic interpretations* (pp. 116–132), *Society of Economic Paleontologists and Mineralogists,* London, Special publications, Technical Report.

Visher G. S. (1969). Grain size distributions and depositional processes. *J. Sed. Pet.* 39, 1074–1106.

Vittal N., Ranga Raju K. G. and Garde R. J. (1977). Resistance of two dimensional triangular roughness. *J. Hydraul. Res.* 15, 1.

von-Kármán T. (1930). Mechanischeähnlichkeit und turbulenz. *Proceedings 3rd International Congress Applied Mechanics*, P. A. Norstedt & Sner, Stockholm Conference.

von-Kármán T. (1934). Turbulence and skin friction. *J. Aeronaut. Sci.* 1 (1), 1–20.

Voropayev S. I., Cense A. W., McEachern G. B., Boyer D. L. and Fernando H. J. S. (1999). Dynamics of sand ripples and burial/scouring of cobbles in oscillatory flow. *Appl. Ocean Res.* 21 (5), 249–261.

Voropayev S. I., Testik F. Y., Fernando H. J. S. and Boyer D. L. (2003). Burial and scour around a short cylinder under progressive shoaling waves. *Ocean Engg.* 30, 1647–1667

Wadell H. (1932). Volume, shape and roundness of rock particles. *J. Geol.* 40 (5), 443–451.

Wadell H. (1933). Sphericity and roundness of rock particles. *J. Geol.* 41, 310–331.

Wang C., Yu J.-Y., Wang P.-F. and Guo P.-C. (2009). Flow structure of partly vegetated open-channel flows with eelgrass. *J. Hydrodyn. Ser. B.* 21 (3), 301–307.

Wang C.-Y., Cheng J.-H., Shih H.-P. and Chang J.-W. (2011). Ring columns as pier scour countermeasures. *Int. J. Sediment Res.* 26 (3), 353–363.

Wang H., Tang H., Liu Q. and Wang Y. (2016a). Local Scouring around twin bridge piers in open-channel flows. *J. Hydraulic. Engg.* 060, 160–168.

Wang H., Tang H., Wu, Xiao J., Wang Y. and Jiang S. (2016b). Clear-water local scouring around three piers in a tandem arrangement. *Science.* 59 (6), 888–896.

Wang S. Y. and Shen H. W. (1985). Incipient sediment motion and riprap design. *J. Hydraul. Engg.* 111(3), 520–538.

Wang X., Yang Q., Lu W. and Wang X. (2011). Effects of bed load movement on mean flow characteristics in mobile gravel beds. *Water Resour Manage.* 25, 2781–2795.

Wang X., Zheng J., Danxun L. and Qu Z. (2008). Modification of the Einstein bed-load formula. *J. Hydraul. Engg.* 134 (9), 1363–1369.

Wang Z. and Plate E. J. (1996). A preliminary study on the turbulence structure of flows of non-Newtonian fluid. *J. Hydrul. Res.* 34 (3), 345–361.

Wang Z. Q. and Cheng N. S. (2005). Secondary flows over artificial bed strips. *Adv. Water Resour.* 28 (5), 441–450.

Ward S. G. (1955). Properties of well-defined suspensions of solids in liquids. *J. Oil Col. Chemists.* 38, 9.

Ward S. G. and Whitmore R. L. (1950). Studies of the viscosity and sedimentation of suspensions: The viscosity of suspension of spherical particles. *British J. Appl. Phys.* 1 (11), 286–290.

Watters G. Z. and Rao M. V. P. (1971). Hydrodynamic effect of seepage on bed particle. *Proc. ASCE.* 97, 3.

Wei T. and Willmarth W. W. (1989). Reynolds number effects on the structure of a turbulent channel flow. *Proc. ASCE.* 204, 57–95.

Weinfurtner S., Tedford E. W. and Penrice M. C. J. (2011). Measurement of stimulated Hawking emission in an analogue system. *Phys. Rev. Lett.* 106 (2), 021302.

Wentworth C. K. (1922). A scale of grade and class terms for clastic sediments. *J. Geol.* 30 (5), 377–392.

Werner P., Unsold G., Koopmann B. and Stefanon A. (1980). Field observations and flume experiments on the nature of comet marks. *Sedimentary Geol.* 26, 233–262.

West J. R., Knight D. W. and Shiono K. (1986). Turbulence measurements in the great ouse estuary. *J. Hydraul. Engg.* 112, 167–181.

White B. R. and Schultz J. C. (1977). Magnus effect in saltation. *J. Fluid Mech.* 81, 497–512.

White C. M. (1940). The equilibrium of grains on the bed of stream. *Proc. RSL. Series A.* 174 (958), 322–338.

Whitehouse R. J. S. and Hardisty J. (1988) Experimental assessment of two theories for the effect of bed slope on the threshold of bedload transport. *Mar. Geol.* 79 (1–2), 135–139.

Whitemore R. L. (1957). The relationship of the viscosity to the settling rate of slurries. *J. Inst. Fuel*, 30 (1957), 238–242.

Wiberg P. L. and Nelson J. M. (1992). Unidirectional flow over asymmetric and symmetric ripples. *J. Geophys. Res.* 97, 12745–12761.

Wiberg P. L. and Rubin D. M. (1989). Bed roughness produced by saltating sediment. *J. Geophys. Res.* 94 (C4), 5011–5016.

Wiberg P. L. and Smith J. D. (1983). A comparison of field data and theoretical models for wave-current interactions at the bed on the continental shelf. *Cont. Shelf Res.* 2, 126–136.

Wiberg P. L. and Smith J. D. (1985). A theoretical model for saltating grains in water. *J. Geophys. Res.* 90 (C4), 7341–7354.

Wiberg P. L. and Smith J. D. (1987). Calculations of critical shear stress for motion of uniform and heterogeneous sediments. Water Resour. Res. 23 (8), 1471–1480.

Wiberg P. L. and Smith J. D. (1989). Model for calculating bedload transport of sediment. *J. Hydraul.Engg.* 115 (1), 101–123

Wieghardt K. (1944). Uber die turbulenteStromunginRohr und langs der platte. *ZAMM.* 24, 294.

Wilcock P. R. (1987). Bed-load transport in mixed-size sediment, PhD. thesis, MIT, Cambridge.

Wilcock P. R. (1993). Critical shear-stress of natural sediments, *J. Hydraul. Eng.*, 119(4), 491–505.

Wilcock P. R. and Crowe J. C. (2003), Surface-based transport model of mixed-size sediment. *J. Hydraul. Engg.* 129 (2), 120–128.

Wilcock P. R. and Suthard J. B. (1988). Experimental study of incipient motion in mixed-size sediment. *Water Resour. Res.* 24 (7), 1137–1151.

Williams G. P. (1967). Flume experiments on the transport of a coarse sand. In T. P. William T.P. (Ed.), *Sediment Transport in Alluvial Channels.* US Government Printing Office, Geological Survey, Washington D.C.

Williams G. P. (1970). *Flume Width and Water Depth Effects in Sediment Transport Experiments.* Geological Survey, Washington, DC.

Williams P. B. and Kemp P. H. (1972). Initiation of ripples by artificial disturbances. *J. Hydraul. Divn.* 98, 1057–1070.

Willis J. C. (1978). Analytical velocity distribution. *J. Hydraul. Divn.* 11, 1541–1549.

Willis J. C. (1979). Suspended load from error-function models. *J. Hydraul. Divn.* 7, 79.

Willis J. C. and Coleman N. L. (1969). Unification of data on sediment transport in flumes by similitude principles. *Water Res.* 5 (6), 1330–1336.

Wilson K. C. (1966). Bedload transport at high shear stresses. *J. Hydraul. Divn.* 92 (6), 49–59.

Wilson K. C. (1987). Analysis of bed-load motion at high shear stress. *J. Hydraul. Engg.* 113 (1), 97–103.

Winant C. D., Broward F. K. (1974). Vortex pairing: the mechanism of turbulent mixing-layer growth at moderate Reynolds number. *J. Fluid Mech.* 63, 237–255.

Woldegiorgis B. T., Griensven A. V. and Bauwens W. (2018). Explicit incipient motion of cohesive and non-cohesive sediments using simple hydraulics. *Depositional Rec.* 4, 78–89.

Wong M, and Parker G (2006). Re-analysis and correction of bedload relation of Meyer-Peter and Muller using their own database. *J. Hydraul. Engg.* 132 (11), 1159–1168.

Woo H. S., Jullien P. Y. and Richardson E. V. (1988). Suspension of large concentrations of sands. *J. Hydraul. Engg.* 114, 888–898.

Worman A. (1992). Incipient motion during static armoring. *J. Hydraul. Engg.* 118 (3), 496–501.

Wu B., Molinas A. and Julien P. Y. (2004). Bed-material load computations for non-uniform sediments. *J. Hydraul. Engg.* 130 (10), 1002–1012.

Wu F.-C. and Chou Y. J. (2003) Rolling and lifting probabilities for sediment entrainment. *J. Hydraul. Engg.* 129 (2), 110–119.

Wu F.-C. and Lin Y.-C. (2002) Pickup probability of sediment under log-normal velocity distribution. *J. Hydraul. Engg.* 128 (4), 438–442.

Wu W. and Wang S. S. Y. (2006) Formulas for sediment porosity and settling velocity. *J. Hydraul. Engg.* 132 (8), 858–862.

Wu W., Wang S. S. Y. and Jia Y. (2000). Non-uniform sediment transport in alluvial rivers. *J. Hydraul. Res.* 38 (6), 427–434.

Wyrwoll K. H. and Smyth G. K. (1985) On using the log-hyperbolic distribution to describe the textural characteristics of eolian sediments. *J. Sedimentary Petrol.* 5, 471–478.

Xie L. Q., Lei H., Yu Y. and Sun, X. (2009) Incipient motion of riverbank sediments with outflow seepage, *J. Hydraul. Engg.* 135 (3), 228–233.

Xie L. Q. and Yu Y. Z. (2009). Incipient motion of riverbank sand subject to seepage. *J. Tsinghua Univ.* 46 (9), 1534–1537 (In Chinese).

Yalin M. S. (1963). An expression for bed-load transportation. *J. Hydraul. Divn.* 89, 221–248.

Yalin M. S. (1964). Geometrical properties of sand waves. *J. Hydraul. Divn.* 90 (5), 105–119.

Yalin M. S. (1972, 1977). *Mechanics of Sediment Transport*, Pergamon, Oxford.

Yalin M. S. (1985). On the determination of ripple geometry. *J. Hydraul. Engg.* 111, 1148–1155.

Yalin M. S. and da Silva A. M. F. (2001). *Fluvial Processes.* IAHR Monograph, IAHR, Delft, The Netherlands.

Yalin M. S. and Karahan E. (1979). Inception of sediment transport. *J. Hydraul. Engg.* 105 (11), 1433–1443.

Yang C. T. (1973). Incipient motion and sediment transport. *J. Hydraul. Divn.* 99 (10), 1679–1705.

Yang C. T. (1996). *Sediment Transport: Theory and Practice.* McGraw-Hill, New York.

Yang S. Q., Lim S. Y. and McCorquodale J. A., (2005). Investigation of near wall velocity in 3-D smooth channel flows. *J. Hydraul. Res.* 43 (2), 149–157.

Yang S. Q., Tan S. K. and Lim S. Y. (2004). Velocity distribution and dip-phenomenon in smooth uniform open channel flows. J. *Hydraul. Engg.* 130 (12), 1179–1186.

Yang Y., Melville B. W., Sheppard D. M. and Shamseldin A. Y. (2018). Clear-water local scour at skewed complex bridge piers. *J. Hydraul. Engg.* 144 (6), 04018019.

Yassin A. (1953). Mean roughness coefficient in open channels with different roughness of bed and sidewalls. PhD. Dissertation, Lemann. Zurich, Switzerland.

Yoon J. Y. and Patel V. C. (1996), Numerical model of turbulent flow over sand dune. *J. Hydraul. Engg.* 122, 10–18.

Yoon T. H. and Kim D. H. (2001). Bridge pier scour protection by sack gabions. *Proceedings of the World Water and Environmental Resources Congress.* https://ascelibrary.org/doi/abs/10.1061/40569%282001%29256

Yu J. and Mei C. C. (2000). Formation of sand bars under surface waves. *J. Fluid Mech.* 416, 315–348.

Yuan S.W. (1970). *Foundation of Fluid Mechanics.* Prentice-Hall of India Pvt. Ltd., New Delhi, India.

Yue W., Lin C.-L. and Patel V. C. (2003). *Numerical Investigations of Turbulent Free Surface Flows Using Level Set Method and Large Eddy Simulation.* Iowa Institute of Hydraulic Research, Iowa City.

Yue W., Lin C.-L. and Patel V. C. (2005a), Large eddy simulation of turbulent open-channel flow with free-surface simulated by level set method. *Phys. Fluids.* 17, 025108, 1–12.

Yue W., Lin C.-L. and Patel V. C. (2005b), Coherent structures in open-channel flows over a fixed dune. *J. Fluids Eng.* 127, 858–864.

Zanke U. C. E. (1977). *Berechung der Sinkgeschwindigkeiten von sedimenten. Mitt. des Franzius-InstitutsfürWasserbau.* Technical University, Hannover, Germany.

Zanke U. C. E. (2003), On the influence of turbulence on the initiation of sediment motion. *Int. J. Sediment Res.* 18 (1), 17–31.

Zarrati A. R., Chamani M. R., Shafaie A. and Latifi M. (2010). Scour countermeasures for cylindrical pier using riprap and combination of collar and riprap. *Int. J. Sediment Res.* 25 (3), 313–321.

Zee C. H. and Zee R. (2017). Formulas for the transportation of bed load. *J. Hydraul. Engg.* 143 (4), 04016101.

Zeller J. (1963). Einfuhrung in den Sediment transport offenerGerinne, Schweiz. *Bauzeitung.* 34, 81.

Zeller M. E. (1969). *Two-Dimensional Hydraulics and Sediment Transport Modeling of the Racetrack Reach of the Mississippi River*. USACE Digital Library, Washington.

Zhang Q., Draper S., Liang C. and An H. (2015). Scour below a subsea pipeline in time varying flow conditions. *J. Appl. Ocean Res.* 55, 151–162.

Zhang R. J. (1989). *Sediment Dynamics in Rivers*. Water Resources Press, Beijing. (in Chinese).

Zhang R. J., Xie J. H., Wang M. F. and Huang J. T. (1989) *Dynamics of River Sedimentation*. Water Power Press, Beijing.

Zhiyao S., Tingting W., Fumin X. and Ruijie L. (2008) A simple formula for predicting settling velocity of sediment particles. *Water Sci. Engg.* 1 (1), 37–43.

Zhong D., Wang G. and Sun Q. (2011). Transport equation for suspended sediment based on two-fluid model of solid/liquid two-phase flows. *J. Hydraul. Engg.* 137, 530–542.

Zhou Y. (1993). Interacting scales and energy transfer in isotropic turbulence. NASA Contractor Report 191477, ICASE Report No. 93–28, pages 1–33, Hampton, Viginia.

Zhu L. J. and Cheng N. S. (1993). *Settlement of Sediment Particles*. Nanjing Hydraulic Research Institute, Beijing. (in Chinese).

Zyserman J. A. and Fredsøe J. (1994). Data analysis of bed concentration of suspended sediment. *J. Hydraul. Engg*, 120 (9), 1021–1042.

Subject Index

For Product Safety Concerns and Information please contact our EU
representative GPSR@taylorandfrancis.com
Taylor & Francis Verlag GmbH, Kaufingerstraße 24, 80331 München, Germany

www.ingramcontent.com/pod-product-compliance
Lightning Source LLC
Chambersburg PA
CBHW080347220326
41598CB00030B/4625